Sterile Drug Products

Recent Titles in Series

Sterile Drug Products: Formulation, Packaging, Manufacturing, and Quality, *Michael J. Akers*

Advanced Aseptic Processing Technology, *James Agalloco and James Akers*

Freeze Drying/Lyophilization of Pharmaceutical and Biological Products, Third Edition, *edited by Louis Rey and Joan C. May*

Active Pharmaceutical Ingredients: Development, Manufacturing, and Regulation, Second Edition, *edited by Stanley H. Nusim*

Generic Drug Product Development: Specialty Dosage Forms, *edited by Leon Shargel and Isadore Kanfer*

Pharmaceutical Statistics: Practical and Clinical Applications, Fifth Edition, *Sanford Bolton and Charles Bon*

Sterile Drug Products

Formulation, Packaging, Manufacturing, and Quality

Michael J. Akers, Ph.D.
Baxter BioPharma Solutions
Bloomington, Indiana, U.S.A.

informa
healthcare

New York London

First published in 2010 by Informa Healthcare, Telephone House, 69-77 Paul Street, London EC2A 4LQ, UK.

Simultaneously published in the USA by Informa Healthcare, 52 Vanderbilt Avenue, 7th Floor, New York, NY 10017, USA.

Informa Healthcare is a trading division of Informa UK Ltd. Registered Office: 37–41 Mortimer Street, London W1T 3JH, UK. Registered in England and Wales number 1072954.

A CIP record for this book is available from the British Library.

Library of Congress Cataloging-in-Publication Data available on application

ISBN-13: 9780849339936

Orders may be sent to: Informa Healthcare, Sheepen Place, Colchester, Essex CO3 3LP, UK
Telephone: +44 (0)20 7017 5540
Email: CSDhealthcarebooks@informa.com
Website: http://informahealthcarebooks.com/

For corporate sales please contact: CorporateBooksIHC@informa.com
For foreign rights please contact: RightsIHC@informa.com
For reprint permissions please contact: PermissionsIHC@informa.com

Typeset by Aptara, Delhi, India
Printed and bound in the United Kingdom.

Preface

This book is based primarily on courses that I taught on the basic principles of sterile dosage formulation, packaging, manufacturing, and quality control and assurance over a span of 35 years. I have basically added written text to the slides that were presented in my courses. So any reader who has participated in one of these courses will likely recognize some of the figures and tables.

This book is written, like the course presented, for the person who either is new to the sterile product field or has some experience, but needs a good refresher tutorial. Although the basics are presented, deeper concepts and principles are given as appropriate. This book is intended to be a helpful resource for individuals working directly and indirectly with sterile dosage forms, be it research, product development (formulation, package, process, analytical), manufacturing, engineering, validation, quality control, quality assurance, regulatory, supply chain, purchasing, scheduling, project management, and any other area that deals with sterile products. This book also is intended to be a reference text for educational courses taught in pharmacy schools or continuing education programs. I have written the book with the intent to remain relevant for the indefinite future even though new technologies and new applications of old technologies will become common.

The advent of biotechnology in the late 1970s increased significantly the stature of the parenteral route of administration as the only way to deliver such large and delicate biomolecules. With continued advances in proteomics, genomics, monoclonal antibodies, and sterile devices, development and manufacture of sterile dosage forms have advanced to new heights with respect to numbers of drug products in clinical study and on the marketplace. All these advances have expanded the need for people to be educated and trained in the field of parenteral science and technology. However, such education and training still does not occur to much extent in university education. Such education and training occur "on the job" via both internal and external courses.

This book is designed to serve as an educational resource for the pharmaceutical and biopharmaceutical industry providing basic knowledge and principles in four main areas of parenteral science and technology:

1. Product development, including formulation, package, and process development (chap. 2–11)
2. Manufacturing, including basic teaching on all the primary unit operations involved in preparing sterile products with emphasis on contamination control (chap. 12–23)
3. Quality and regulatory, with focus on application of good manufacturing practice regulations, sterility assurance, and unique quality control testing methods (chap. 24–30)
4. Clinical aspects, focusing on preparation, use, and administration of sterile products in the clinical setting (chap. 1, 30–33).

Chapters on product development present the basic principles of formulation development of sterile solution, suspension, and freeze-dried (lyophilized) dosage forms. Approaches traditionally used to overcome solubility and stability limitations have been emphasized. Specific formulation components such as vehicles, solubilizers, buffers, antioxidants and chelating agents, cryo- and lyoprotectants, tonicity agents, antimicrobial preservatives, and suspending and emulsifying agents have been covered in good detail. Some coverage of long-acting drug delivery systems, especially the polymers used in commercial formulations, are included. Chapter 11 focuses on overcoming formulation problems, with 14 case studies to help the reader learn how to approach formulation problem solving.

Development of sterile dosage forms not only includes the formulation but also the package and the process. Glass, rubber, and plastic chemistry are covered to some extent, as well as packaging delivery systems and devices, both traditional (e. g., vials, syringes) and more novel (e. g. needleless injectors, dual chambered systems).

The area of manufacturing includes chapters on process development and overview, contamination control, facilities, water, air, personnel practices, preparation of components, sterilization, filtration, filling, stoppering and sealing, lyophilization, aseptic processing, barrier technology, labeling and secondary packaging, and some discussion of manufacturing advances.

The area of quality and regulatory includes chapters on good manufacturing practice, the philosophy of quality as it relates to the sterile dosage form, specific quality control tests unique to sterile products, and some coverage of stability testing.

The final area covered is clinical aspect, general discussion of the use of the injectable dosage form in the clinical setting, advantages and disadvantages of sterile products, hazards of administration, and biopharmaceutical considerations.

I have taken the liberty to use my own published materials, with appropriate approvals, to reproduce in this book. Indeed, several chapters are based on previous book chapter or review article publications, some with coauthors who I have acknowledged and obtained their permission. All in all, this book represents more than 35 years of my teachings, writings, and experience in the sterile product science and technology world. Of course, a singular perspective has its limitations compared with a book that has multiple authors. However, this book does have the advantage of consistency of writing style and the ultimate goal of each chapter being practical to the reader.

Just like I always stated when starting every one of my courses, may you learn as much as possible while at the same time having some fun while reading/studying this book.

Acknowledgments

Since I state that this book represents 35 years of my experience working in the sterile product field, I need to acknowledge those who influenced me the most to remain active in this field all these years. Dr. Gerald Hecht and Dr. Robert Roehrs hired me to join Alcon in 1974 without having any formal training or experience in sterile products so that is where I got my start. Joining the faculty at the University of Tennessee three years later exposed me to the teaching and influence of Dr. Kenneth Avis who for decades was considered the world's leading expert in parenterals. Dr. Joseph Robinson was an influential leader to me primarily through our interactions on the former *Journal of Parenteral Science and Technology* board plus his natural mentoring skills. Dr. Patrick DeLuca kept me involved in teaching sterile products after joining Eli Lilly by asking me to help him teach the Center for Professional Advancement sterile products course that after nearly 30 years I am still teaching. Dr. Steven Nail has been a 30-year colleague and very close friend, plus a coworker these past few years, who has served as a scientific role model for me. Other mentors over these years, scientists whose work I have admired, include Dr. Michael Pikal, Dr. John Carpenter, Dr. Eddie Massey, Dr. Alan Fites, Mr. Bob Robison, and Dr. Lee Kirsch. There are many other scientists, too many to mention, who also have influenced me through their intelligence, creativity, and enthusiasm for the pharmaceutical sciences.

I thank those who helped me write several chapters in this book including Dr. Michael DeFelippis of Eli Lilly and Company (chap. 9), Mr. Mark Kruszynski of Baxter BioPharma Solutions (chap. 19), and Dr. Dana Morton Guazzo who graciously updated chapter 30. I acknowledge many of my Baxter Bloomington R&D colleagues, besides Steve Nail, who helped me to write chapters 4 and 7 (Dr. Gregory Sacha, Ms. Karen Abram, and Ms. Wendy Saffell-Clemmer), or helped me by providing needed figures and photos (Dr. Gregory Sacha, Ms. Lisa Hardwick, and Dr. Wei Kuu).

I greatly appreciate the administrative support I received from Ms. Angie Krusynski who did a lot of the "leg work" helping to obtain reproduction approvals. I thank present and past Baxter executives (Alisa Wright, Lee Karras, Ted Roseman, and Ken Burhop) who have encouraged me to write, even admittedly sometimes on company time. I also appreciate my Baxter Bloomington site head, Mr. Camil Chamoun, for his encouragement and support plus allowing me to use many photos from the Bloomington site.

Finally, of course, the old phrase "behind every good man is even a great woman" is so true in my case as I express my love and respect for my wife and best friend, Mary (Midge) Akers.

Contents

1 Introduction, scope, and history of sterile products

Sterile dosage forms have always been an important class of pharmaceutical products in disease diagnosis, therapy, and nutrition. Certain pharmaceutical agents, particularly peptides, proteins, and many chemotherapeutic agents, can be administered only by injection (with or without a needle), because they are inactivated in the gastrointestinal tract when given by mouth. Administration of drugs by the parenteral (parenteral and injectable will be used interchangeably) route has skyrocketed over the past several years and will continue to do so. A primary explanation for this enormous growth lies with the advent of biotechnology, the products of which are biomolecules that cannot be readily administered by any other route because of bioavailability and stability reasons. Since human insulin became the first biotechnology drug approved by the Food and Drug Administration (FDA) in 1982, over 100 drug products of biotechnological origin have been approved and hundreds more will be approved in the years ahead. Most biotechnology drug products are administered only by the parenteral route. Science is advancing to a time when it is likely that some of these drugs can or will be administered by other routes, primarily pulmonary and perhaps someday even orally, but the mainstay route of administration for these biopharmaceutical drugs will be by injection.

Any statistic given at the time of writing this section will quickly be outdated by the time this book is printed and will continually need to be updated. However, it is safe to state that the number of injectable products being developed, being studied in the clinic, being approved for commercial use, and being administered to humans and animals will significantly increase in the years to come. Perhaps by 2020, the market share of sterile drug products will be approximately the same as that for oral solid dosage forms[1].

This chapter will address some of the basic questions about the sterile dosage form and the parenteral route of administration.

Various definitions and end uses of sterile products will be discussed throughout this book. This book will also address many aspects of formulation development of these dosage forms, how they are manufactured, how they are packaged, how they are tested and what are the acceptable conditions during manufacture, and the uses that assure these unique products maintain their special properties.

There are three terms used interchangeably to describe these products—parenteral, sterile, and injectable. Parenteral and injectable basically have the same meaning and are used interchangeably. Sterile dosage forms encompass parenteral/injectable dosage forms as well as other sterile products such as topical ophthalmic products, irrigating solutions, wound-healing products, and devices. The coverage of devices in this book will be minimal.

Here is a definition of sterile dosage forms:

> A product introduced in a manner that circumvents the body's most protective barriers, the skin and mucous membranes, and, therefore, must be "essentially free" of biological contamination.

Ideally, a sterile dosage form is absolutely free of any form of biological contamination, and, of course, is the ultimate goal of every single unit of sterile product released to the marketplace, either commercial or clinical. Perhaps some day manufacturing procedures and in-process microbiological analysis will guarantee that each and every unit of sterile product will indeed be absolutely free of biological contamination. However, the modifier words "essentially free" are added to this definition because most small-volume ($\leq$100 mL per container)

[1] Among many resources for keeping current with new drug products and trends are Burrill & Company (www.burrillandco.com); Pharmaceutical Research and Manufacturers of America (www.phrma.org); Tufts Center for the Study of Drug Development (http://csdd.tufts.edu); Onesource.com; EvaluatePharma.com; IMS; and Datamonitor, to name a few.

sterile products are produced where the finished product is not terminally sterilized, but rather is aseptically processed. The difference in sterility assurance is far greater (generally at least 3 logs) for terminally sterilized products compared to aseptically processed products. This does not mean that aseptically processed products are frequently contaminated; rather it means that aseptically processed products cannot be validated to the same level of sterility assurance compared to terminally sterilized products. Sterility assurance is covered primarily in chapter 13 while sterilization is covered in chapters 17 and 18 and aseptic processing is covered in chapter 21.

The term "parenteral" comes from two Greek words, "par" meaning "avoid" and "enteral" meaning "alimentary canal." Therefore, the word "parenteral" literally means "beside the intestine." The only way to avoid the alimentary canal and to circumvent the skin and mucous membranes is to inject a pharmaceutical product directly into the body. Parenteral (the author prefers the term "sterile") products must be exceptionally pure and free from physical, chemical, and biological contaminants (microorganisms, endotoxins, particles). These requirements place a heavy responsibility on the pharmaceutical industry to practice current good manufacturing practices (cGMPs) in the manufacture of sterile dosage forms and upon pharmacists and other health care professionals to practice good aseptic practices (GAPs) in dispensing them for administration to patients.

Injections usually are accomplished using needles, but newer technology avoids the use of needles or use of extremely small diameter needles (covered in chap. 4). As stated already, not all sterile dosage forms are administered by injection. Sterile products that are not parenteral or injectable products include the following:

- Topical ophthalmic medications
- Topical wound healing medications
- Solutions for irrigation
- Sterile devices (e.g., syringes, administration sets, and implantable systems)

There are many terms that will be used throughout this book. A glossary of definitions of sterile product terms, not intended to be comprehensive, is given in Table 1-1.

The United States Pharmacopeia (USP)[2] contains several hundred monographs on sterile drugs or diluent preparations. Most products of biotechnology origin are not included because of confidentiality reasons. Some interesting statistics gathered after analyses of these USP monographs are as follows:

- About 22% are solid preparations that require solution constitution prior to use.
- About 9% are diluent preparations, both small and large volume.
- About 10% are radioisotope diagnostic preparations.

Sterile drug products are relatively unstable and are generally highly potent drugs that require strict control of their administration to the patient. Overcoming solubility and stability issues and achieving and maintaining sterility and other purity requirements present great challenges to those developing, manufacturing, and administering sterile drug products.

In this book, the teaching of the principles involved in the product development, product manufacture, and quality control of medicines delivered by the parenteral route will continue to be an important and relevant subject. This book is aimed to provide basic principles and practical applications of the formulation, packaging manufacture, and quality control of injectable dosage forms; in fact, all sterile dosage forms.

HISTORY OF THE STERILE DOSAGE FORM

Avis published probably the most detailed review of the history of the sterile dosage form (1). Turco and King's last book also is a good general resource not only about history but also about clinical applications of sterile dosage forms (2). This chapter will highlight these references plus

[2] In general, referencing the USP also applies to other primary compendia, European Pharmacopeia (EP or PhEur) and Japanese Pharmacopeia (JP).

Table 1-1 Glossary of Terms Related to Sterile Drug Technology

Absolute Rating—The size of the largest spherical particle completely retained on the filter. An absolute filter of 0.2 μ retains all particles ≥0.2 μ.

Action Level—An established microbial or airborne particle level that, when exceeded, should trigger appropriate investigation and corrective action based on the investigation.

Air Lock—A small area with interlocked doors, constructed to maintain air pressure control between adjoining rooms. Used to stage and disinfect large equipment prior to transfer from lesser-controlled room to higher-controlled room.

Alert Level—An established microbial or airborne particle level giving early warning of potential drift from normal operating conditions, and which triggers appropriate scrutiny and follow-up to address the potential problem. Alert levels are always lower than action levels.

Ampule—A final container that is totally glass in which the open end after filling a product is sealed by heat. Also referred as ampul, ampoule, carpule (French).

Antimicrobial Preservative—Solutes such as phenol, meta-cresol, benzyl alcohol, and the parabens that prevent the growth of microorganisms. Must be present in multiple dose parenterals.

Antioxidants—Solutes that minimize or prevent drug oxidation. Examples include sodium bisulfite, ascorbic acid, and butylated hydroxyanisole.

Aseptic—Lack of disease-producing microorganisms. Not the same as sterile.

Aseptic Processing—Manufacturing drug products without terminal sterilization. The drug product is sterile filtered, then aseptically filled into the final package and aseptically sealed.

Autoclave—A system that sterilizes by superheating steam under pressure. The boiling point of water, when pressure is raised 15 psig above atmospheric pressure, is increased to 121°C (250°F). This is the most common means of terminally sterilizing parenteral products.

Barrier—A system having a physical partition between the sterile area (ISO 5) and the nonsterile surrounding area. A barrier is differentiated from an isolator in that the barrier can exchange air from the fill zone to the surrounding sanitized area where personnel are located, whereas an isolator cannot exchange air from the fill zone to the sterilized surrounding area where personnel are located.

Bioburden—Total number of microorganisms detected in or on an article prior to a sterilization treatment. Also called microbial load.

Biological Indicator—A population of microorganisms inoculated onto a suitable medium (e.g., solution, container, closure, paper strip) and placed within an appropriate sterilizer load location to determine the sterilization cycle efficacy of a physical or chemical process. The specific microorganisms are the most resistant to the particular sterilization process.

Bubble Point—Used in filter integrity testing; the pressure where a gas will pass through a wetted membrane filter. Each filter porosity and type has a given bubble point.

Buffers—Solutes used to minimize changes in pH, important for many drugs to maintain stability and/or solubility.

Chelating Agents—Solutes that complex metal ions in solution, preventing such metals from forming insoluble complexes or catalyzing oxidation reactions. Example: ethylenediaminetetraacetic acid (EDTA)

Class X—A Federal Standard for clean room classes. Whatever X is, for example, 100, means that there are no more than X particles per cubic foot ≥ 0.5 μm.

Clean Room—A room designed, maintained, and controlled to prevent particle and microbiological contamination of drug products. Such a room is assigned and reproducibly meets an appropriate air cleanliness classification.

Colony Forming Unit (CFU)—A microbiological term that describes the formation of a single macroscopic colony after the introduction of one or more microorganisms to microbiological growth media.

Coring—The gouging out of a piece of rubber material caused by improper usage of a needle penetrating a rubber closure.

Critical Area—An area designed to maintain sterility of sterile materials.

Critical Surfaces—Surfaces that may come into contact with or directly affect a sterilized product or its containers or closures. Critical surfaces are rendered sterile prior to the start of the manufacturing operation, and sterility is maintained throughout processing.

D-Value—Time in minutes (or dose for radiation sterilization) of exposure at a given temperature that causes a one-log or 90% reduction in the population of specific microorganisms.

Disinfection—Process by which surface bioburden is reduced to a safe level or eliminated. Some disinfection agents are effective only against vegetative microorganisms.

Endotoxin—Extracellular pyrogenic compounds.

HEPA—High Efficiency Particulate Air filters, capable of removing 99.97% of all particles 0.3 μ and higher.

(*continued*)

Table 1-1 Glossary of Terms Related to Parenteral Drug Technology (*Continued*)

Isolator—A decontaminated unit, supplied with Class 100 (ISO 5) or higher air quality that provides uncompromised, continuous isolation of its interior from the external environment. Isolators can be closed or open.

Closed—exclude external contamination from the isolator's interior by accomplishing material transfer via aseptic connection to auxiliary equipment, rather than by use of openings to the surrounding environment.

Open—allow for continuous or semicontinuous ingress and/or egress of materials during operations through one or more openings. Openings are engineered, using continuous overpressure, to exclude the entry of external contamination into the isolator.

Laminar Flow—An airflow moving in a single direction and in parallel layers at constant velocity from the beginning to the end of a straight line vector.

Lyophilization—The removal of water or other solvent from a frozen solution through a process of sublimation (solid conversion to a vapor) caused by combination of temperature and pressure differentials. Also called freeze-drying.

Media Fill—Microbiological evaluation of an aseptic process by the use of growth media processed in a manner similar to the processing of the product and with the same container/closure system being used.

Micron (μ)—One millionth of a meter. Also referred to as micrometer (μm).

Needle Gauge—Either the internal (ID) or external (OD) diameter of a needle. The larger the gauge the smaller the diameters. For example, a 21-G needle has an ID of 510 μ and an OD of 800 μ. A 24-G needle has an ID of 300 μ and an OD of 550 μ. An 18-G needle has an ID of 840 μ and OD of 1,250 μ.

Nominal Rating—The size of particles, which are retained at certain percentages. A 0.2 μ nominal membrane filter indicates that a certain percentage of particles 0.2 μ and higher are retained on the filter.

Overkill Sterilization Process—A process that is sufficient to provide at least a 12-log reduction of a microbial population having a minimum D-value of 1 minute.

Parenteral—Literally, to avoid the gastrointestinal tract. Practically, the administration of a drug product that is not given by mouth, skin, nose, or rectal/vaginal. Parenteral conveys the requirement for freedom from microbiological contamination (sterile), freedom from pyrogens, and freedom from foreign particulate matter.

Pyrogen—Fever producing substances originating from microbial growth and death.

Reverse Osmosis—A process used to produce water for injection whereby pressure is used to force water through a semipermeable membrane where the solute content (ions, microbes, foreign matter) of the solution is retained on the filter while the solvent (pure water) passes through.

Sterile—The complete lack of living (viable) microbial life.

Sterility—An acceptably high level of probability that a product processed in an aseptic system does not contain viable microorganisms.

Sterility Assurance Level—The probability of microbial contamination. A SAL of 10^{-6} means that there is a probability of one in one million that an article is contaminated. Also called probability of nonsterility or sterility confidence level.

Surface Active Agents—Solutes that locate at the surface of water and air, water and oil, and/or water and solid to reduce the interfacial tension at the surface and enable substances to come together in a stable way. Examples include polysorbate 80 and sodium lauryl sulfate.

Terminal Sterilization—A process used to produce sterility in a final product contained in its final packaging system.

Tonicity Agents—Solutes used to render a solution isotonic, meaning similar in osmotic pressure to the osmotic pressure of biological cells. Sodium chloride and mannitol are examples of tonicity agents.

ULPA—Ultra-Low Penetration Air filter with minimum 0.3 μm particle retaining efficiency of 99.999%.

Validation—The scientific study of a process to prove that the process is doing what it is supposed to do and that the process is under control. Establishing documented evidence that provides a high degree of assurance that a specific process will consistently produce a product meeting its predetermined specifications and quality attributes.

Worst Case—A set of conditions encompassing upper and lower processing limits and circumstances, including those within standard operating procedures that pose the greatest chance of process or product failure.

add the author's own research into this area. Table 1-2 summarizes the highlights of the history of the development and application of inventions and advances in sterile drug manufacturing and therapy.

In 1656, the first experimental injection was performed on dogs by Christopher Wren, the architect of St. Paul's cathedral in London. The first primary packaging system was an animal (goose) bladder, and the first type of needle used was the quill of a feather. In 1662, the first recorded injection into man was performed by J. D. Major and Johannes Elsholtz, as depicted

Table 1-2 Summary of the History of Sterile Drug Technology

1656	First experimental injection by C. Wren in dogs (first container was an animal bladder and first needle was a feather quill)
1662	First injection (opium) in man
1796	E. Jenner used intradermal injections of cowpox virus to inoculate children against smallpox
1831	Introduction of IV therapy treatment of cholera with salt, bicarbonate, water
1855	First use of hypodermic syringe for subcutaneous injection
1860s	Pasteur/Lister/Koch all contributed to discovery of germ theory of disease, concerns for sterility and development of sterilization methods (but not accepted for decades)
1884	Use of first autoclave for sterilization
1890s	Crude filters (asbestos) used for filtering drugs
1923	Florence Siebert discovered cause of pyrogenic reactions
1938	Food, Drug, Cosmetic Act passed by Congress (after sulfanilamide disaster). Ethylene oxide sterilization introduced
1940s	Penicillin started being used
1941	Freeze-drying introduced
1961	HEPA filters, laminar airflow introduced in pharmaceutical industry
1963	Clean room standards introduced, FDA first published proposed GMP regulations
1965	Parenteral nutrition introduced
1970s	Emergence of biotechnology, LAL test for endotoxins
1980s	Introduction of controlled IV devices, controlled delivery, home health care First drug product (Humulin®) from recombinant DNA technology approved by FDA
1987	First publication of FDA Aseptic Processing Guidelines and Guidelines for Process Validation
1990s	Barrier isolator technology, aseptic process validation, process validation, pre-approval inspections, biotechnology growth
1992	*The International Conference on Harmonisation (ICH) of Technical Requirements for Registration of Pharmaceuticals for Human Use* Established
1996	European Union published Guidance on Manufacture of the Finished Dosage Form issued
1997	First human monoclonal antibody approved (Rituxan®, rituximab to treat cancer)
2000s	Monoclonal antibodies, impact of genomics and proteomics on new parenteral drug therapy, Quality by Design, disposable technologies
2004	FDA publishes revision to Aseptic Processing Guidelines.
2010	Possibilities include vast new numbers of biosimilar products approved, more advances in aseptic processing to the point that parametric release of products produced by aseptic processing can be done, advances in on-line 100% measurement of quality parameters, oral delivery of proteins, complete automation of filling, stoppering and sealing processes, most product manufacturing outsourced; the possibilities are as many as can be imagined.

in Figure 1-1. The drug injected was opium. While the poor human receiving this injection may have had his pain alleviated, he likely was going to die, eventually from microbial and pyrogenic contamination introduced using this crude means of injection. Other drugs injected into humans during those early days were jalap resins, arsenic, snail water, and purging agents. It is improbable that the initial pioneers of injectable therapy had much appreciation about the needs for cleanliness and purity when injecting these medications. After 1662, injecting drug solutions into humans was not commonly practiced until late in the 18th century.

Intravenous (IV) therapy was first applied around 1831 when cholera was treated by the IV injection of a solution containing sodium chloride and sodium bicarbonate in water. Normal saline was used by Thom Latts to treat diarrhea in cholera patients using intravenous infusions. Intravenous feeding was first tried in 1843, when Claude Bernard used sugar solutions, milk, and egg whites to feed animals. By the end of the 19th century, the intravenous route of administration was a widely accepted practice. Injections of emulsified fat in humans were first accomplished by Yamakawa in 1920 although, not surprisingly, major problems existed in formulating and stabilizing fatty emulsions.

It is conjecture who really was the first person to invent and use a syringe. According to medhelpnet.com, a French surgeon, Charles Gabriel Pravaz (Fig. 1-2), and a Scottish physician, Alexander Wood, independently invented the hypodermic syringe in the mid-1850s. Other references credit G. V. LaFargue for inventing the first syringe used for subcutaneous injections in 1836 with wood, using it to inject morphine. Charles Hunter first used the word "hypodermic"

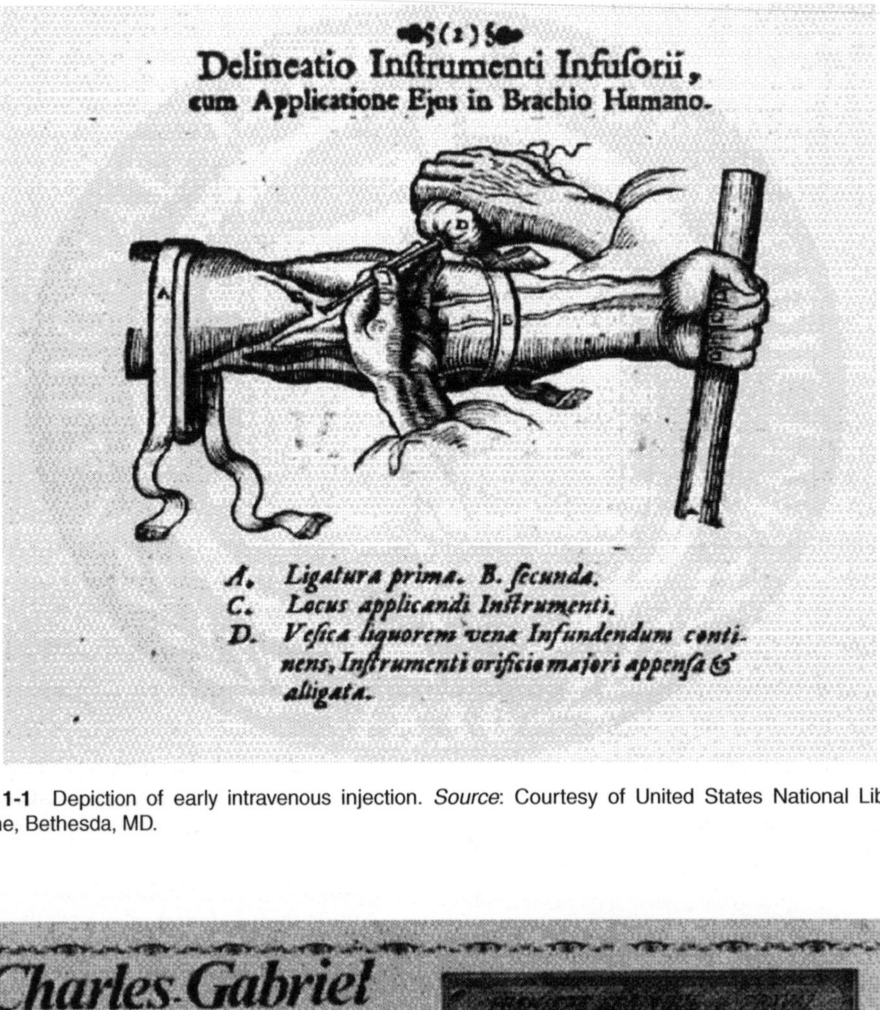

Figure 1-1 Depiction of early intravenous injection. *Source*: Courtesy of United States National Library of Medicine, Bethesda, MD.

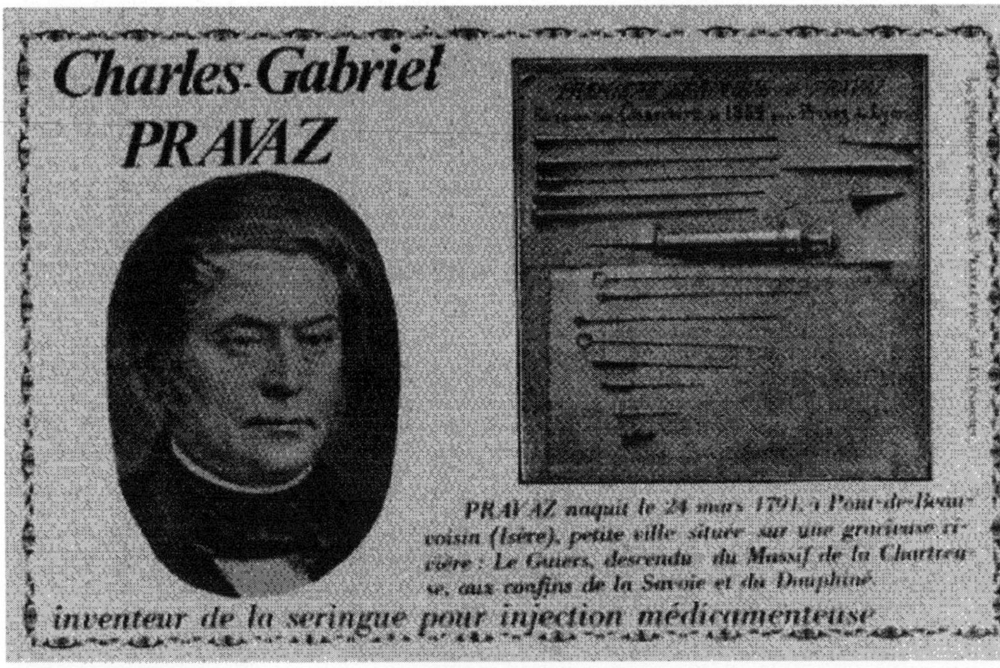

Figure 1-2 Earliest syringes. *Source*: From Ref. 3.

after noting that this route of injection resulted in systemic absorption. Robert Koch in 1888 developed the first syringe that could be sterilized and Karl Schneider built the first all-glass syringe in 1896. Becton, Dickinson and Company created the first mass-produced disposable glass syringe and needle, developed for Dr. Jonas Salk's mass administration of one million American children with the new Salk polio vaccine.

Like many other critical technologies in sterile product manufacturing (e.g., freeze drying, rubber closures, clean rooms), the sterile, prefilled, disposable syringe was developed during World War II. A precursor to the syringe was the Tubex cartridge system developed by Wyeth (4). The injection solution was filled into a glass cartridge having a needle already permanently attached to the cartridge. The prefilled cartridge was then placed in a stainless steel administration device.

Early practice of administering drugs by injection occurred without knowledge of the need for solution sterility plus no one appreciated what caused pain and local irritation while injecting solutions subcutaneously. It was not until around 1880 when a pharmacist named L. Wolff first recognized the role of isotonicity in minimizing pain and irritation when introducing drug solutions to the body. Intramuscular (IM) injections were first performed by Alfred Luton, who believed that this route would be less painful and irritating for acidic, irritating, or slowly absorbed drugs.

Pasteur, Lister, and Koch all contributed to discovery of the germ theory of disease, concerns for sterility, use of aseptic techniques, and development of sterilization methods during the 1860s. However, their concerns for the need to sterilize and maintain sterility of injections were not accepted or implemented for decades. It was not until 1884 that the autoclave was introduced by Charles Chamberland for sterilization purposes. Gaseous sterilization was first discovered using formaldehyde in 1859 and ethylene oxide in 1944. It was also in the early 1940s that radiation, beginning with ultraviolet light, was used as a means of sterilization.

Filtration methods began in the mid-1850s when Fick described "ultrafilter" membranes on ceramic thimbles by dipping them in a solution of nitrocellulose in ether. Crude filters, using asbestos, began to be used in the 1890s. Zsigmondy and Bachmann in 1918 coined the term "membrane filter." Beckhold developed a method to determine the pore size of membrane filters, the method we know now as the "bubble point" method.

Pyrogenic reactions were still commonplace until Florence Siebert in 1923 discovered the cause of these reactions. She was the first person to suggest that fever reactions after injections were microbial in origin. She also proposed that these microbial derivatives were nonliving, nonproteinaceous, and could not be eliminated by sterilization methods. Also, she developed the rabbit pyrogen test, used for decades for the detection of pyrogenic contamination, and still a USP method, although most products today are tested for bacterial endotoxin by the Limulus Amebocyte Lysate (LAL) test discovered by the Johns Hopkins researchers, Levin and Bang in 1964.

Intravenous nutrition using hyperalimentation solutions started in 1937 when W. C. Rose identified amino acids as necessary for the growth and development of rats. This mode of therapy was established first in dogs and then in humans (1967) by S. J. Dudrick who developed a safe method for long-term catheterization of the subclavian vein that permitted these highly concentrated and hyperosmolar solutions to be administered without damaging venous vessels.

Although the first book to be used as a standard for national use, the United States Pharmacopeia, was published in 1820, it was not until the fifth edition of the National Formulary in 1926 that the first parenteral monographs were accepted. In 1938, the Food, Drug, Cosmetic (FD&C) Act was passed by Congress after the sulfanilamide disaster where 107 people including many children died after ingesting a liquid form of this drug dissolved in diethylene glycol. This Act also established the Food and Drug Administration to enforce the Act and required manufacturers to prove to the government that drug products introduced into the marketplace were safe. The legal basis for cGMPs and other FDA regulations are related to the 1938 FD&C Act.

Penicillin started being used in the 1940s, further opening the door for parenteral therapy as a means to save thousands of lives. More companies started to develop parenteral drugs. Because so many injectable drugs were unstable in solution and because of the need to provide blood in a stable form during World War II, freeze-drying was introduced in 1942.

Table 1-3 Injectable Drugs—Therapeutic Classes and Examples

Drug class	Some examples of brand names[a]
Antiemetic agents	Anzemet, Kytril, Zofran
Anti-infective agents	AmBisome, Vancomycin, Zyvox All Cephalosporin Injectables, Nebcin, Garamycin, etc.
Antiparkinsons agents	Apokyn
Antipsoriatic agents	Amevive, Enbrel, Raptiva
Antipsychotic agents	Geodon, Risperdal, Consta
Antiretroviral agents	Fuzeon
Asthma agents	Xolair
Bisphosphonates	Aredia, Zometa
Cardiovascular agents	Dobutamine
Chelating agents	Desferal
Coagulation factors	Advate, Alphanate, AlphaNine-SC, Autoplex T, Bebulin VH, BeneFIX, Feiba VH, Helixate-FS, Hemofil M, Humate-P, Hyate: C, Koate0DVI, Kogenate FS, Monarc-M, Monoclate-P, Mononine, NovoSeven, Profilnine SD, Proplex T, Recombinate, ReFacto
Colony-stimulating factors	Leukine, Neulasta, Neupogen
Contraceptive agents	Depo-Provera
Dystonia agents	Botox, Myobloc
Endocrine and metabolic agents	Humulin, Novolin, Sandostatin, Sandostatin LAR, Thyrogen
Enzyme replacement therapy	Aldurazyme, Aralast, Cerezyme, Fabrazyme, Prolastin, Zemaira
Glucocortoids	Solu-Medrol
Gonadotropin-releasing	Eligard, Lupron, Plenaxis,
Hormone analogues	Trelstar Depot/LA Zoladex
Growth hormone agents	Genotropin, Humatrope, Norditropin, Nutropin, Nutropin AQ/Depot, Saizen, Serostim, Zorbtive,
Growth hormone receptor	Somavert Antagonist:
H2 antagonists	Pepcid, Tagamet, Zantac
Hepatitis C agents	Infergen, Intron-A, Pegasys, Peg-Intron, Rebetron, Roferon-A
Hormone deficiency agents	Deltestryl, Delestrogen, Depo-Estradiol
(androgens and estrogens)	Depo-Testrosterone
Hyaluronic acid derivatives	Hyalgan, Orthovisc, Supartz, Synvisc
Immune globulins	Carimune NF, Flebogamma, Gamimune N S/D, Gammagard S/D, Gammar P.I.V., Gamunex, Iveegam EN, Octagam, Panglobulin NF, Polygam S/D, RhoGAM, Rhophylac, Venoglobulin-S, WinRho SDF
Immunizations	Prevnar
Infertility agents	Antagon, Cetrotide, Chorex, Fertinex, Follistim AQ, Gonal-F, Novarel, Pergonal, Pregnyl, Profasi, Repronex
Interferons	Actimmune, Alferon-N
Migraine agents	D.H.E. 45, Imitrex
Multiple sclerosis	Avonex, Betaseron, Copaxone, Novantrone, Rebif
Ophthalmic agents	Macugen, Visudyne
Osteoporosis agents	Forteo, Miacalcin
Pituitary hormone	DDAVP
Recombinant human erythropoietin	Aranesp, Epogen, Procrit
Respiratory syncytial virus prophylaxis agents	Synagis
Rheumatoid arthritis agents	Enbrel, Humira, Kineret, Methotrexate, Myochrysine, Remicade
Sexual dysfunction agents	Caverject
Thrombocytopenia agents	Neumega
Chemotherapeutic agents	Abraxane, Adriamycin, Adrucil, Alimta, Alkeran, Avastin, BiCNU, Blenoxane, Busulfex, Campath, Camptosar, Cerubidine, Clolar, Cosmegen, Cytosar-U, Cytoxan/Neosar, DaunoXome, DepoCyt, Doxil, DTIC-Dome, Ellence, Eloxatin, Elspar, Erbitux, Fludara, FUDR, Gemzar, Herceptin, Hycamtin, Idamycin, Ifex, Leustatin, Lupron, Methotrexate, Mustargen, Mutamycin, Mylotarg, Navelbine, Nipent, Novantrone, Oncaspar, Ontak, Paraplatin, Platinol-AQ, Plenaxis, Proleukin, Rituxan, Taxol, Taxotere, TheraCys, TICE BCG, Trelstar Depot/LA, Trisenox, Valstar, Vantas, Velcade, VePesid/Toposar, Viadur, Vidaza, Vinblastine, Vincasar, Vumon, Zanosar, Zoladex
Chemotherapeutic adjunctive	Anzemet, Aredia, Ativan, Ethyol,
Agents (not already listed)	Kepivance, Kytril, Osmitrol, Mesnex, Zinecard

[a]All brand name drug products are registered (®).

Clean room technologies, including the use of laminar air flow units, high efficiency particulate air (HEPA) filters, and room classification for particles were not discovered until the early 1950s to the early 1960s. Original clean rooms were used by the United States Biological Laboratories at Fort Detrick, MD, during the 1950s. The HEPA filter was first described in the early 1940s, but not applied to laminar airflow technology until W. J. Whitfield combined HEPA filters and laminar airflow units in 1961. The United States government first proposed clean room classifications in 1962 (Federal Standard 209).

It was also in 1962 that authority was given to the FDA to establish cGMPs, Parts 210 and 211 (21 CFR Parts 210 and 211), issued under section 501(a)(2)(B) of the Federal Food, Drug, and Cosmetic Act (21 U.S.C. 351(a)(2)(B) with the first proposed cGMP regulations published in 1963. In 1976 the FDA proposed to revise and expand these regulations and a final rule by the FDA commissioner was published in the Federal Register on September 29, 1978. Although some changes have occurred since 1978 (e.g., April 2008 changes that included requirement for validation of depyrogenation of sterile containers)[3], and likely minor changes will continue to occur, the great majority of GMP requirements finalized in 1978 remain enforced within the pharmaceutical industry today.

As air classifications became standard for clean rooms, developments in the equipment used in sterile product manufacture also occurred in rapid fashion. Stainless steel and its fabrication into tanks, pipes, and other equipment was refined to provide heliarc welding of joints and fittings as well as the electropolishing of surfaces to reduce potential product reactivity. Clean-in-place and sterilize-in-place technologies were developed in the 1970s that allowed larger equipment to be cleaned and sterilized without dismantling; it also greatly reduced the variability in manual cleaning.

Biotechnology emerged in the 1970s, resulting in significant growth in the development, manufacture, and use of parenteral drugs. Biotechnology, in turn, gave rise to the significant growth of controlled drug delivery systems, convenient delivery systems for home health care, monoclonal antibodies, and the advent of proteomics and genomics. To give one example, the monoclonal antibody market of commercial products is poised to double in number and estimated sales value from 2007 to 2012 (5).

It was also in the 1970s that FDA began to enforce the practice of process validation, starting with validation of sterilization processes. Today, validation of processes, methods, and computers are standard practices because validation practices are continuously being refined and updated.

The 1990s witnessed the advent of barrier isolator technology, preapproval GMP inspections, significant growth of biotechnology processes, and much increased focus and enforcement of aseptic process validation.

Advances will continue in the 21st century in the areas of parenteral drug targeting and controlled release, convenience packaging and delivery systems, aseptic processing, high-speed manufacturing, disposable technologies, rapid methods for chemical and microbiological testing, and GMP regulatory requirements.

Table 1-3 presents a list of therapeutic classes of injectable drugs and some examples of each class. This list will grow not only in number but also in clinical significance and market share. Injectable or parenteral drug science and technology is a wonderful and exciting field of study and endeavor in which to be involved and engaged. It is the author's hope that the readers of this book will readily see the truth of this belief.

REFERENCES

1. Avis KE. The parenteral dosage form and its historical development. In: Avis KE, Lieberman HA, Lachman L, eds. Pharmaceutical Dosage Forms: Parenteral Medications. Vol 1. 2nd ed. New York: Marcel Dekker, 1992:1–16.
2. Turco SJ, King RE. Sterile Dosage Forms: Their Preparation and Clinical Application. 3rd ed. Philadelphia, PA: Lea & Febiger, 1978.
3. http://www.general-anesthesia.com/people/charles-pravaz.html. info@general-anaesthesia.com [David Pearce, BLTC Research, Brighton, United Kingdom, 2004, last updated 2008].

[3] Federal Register /Vol. 73, No. 174 /Monday, September 8, 2008 /Rules and Regulations, starting at page 51919.

4. Turco S, King RE. Sterile Dosage Forms: Their Preparation and Clinical Application. 3rd ed. Philadelphia: Lea & Febiger, 1987:267–269.
5. Monoclonal Therapeutics and Companion Diagnostic Products, Report Code BIO016G, 2008, http://www.bccresearch.com/report/BIO016G.html.

BIBLIOGRAPHY

Allen LV, Popovich NG, Ansel HC, eds. Ansel's Pharmaceutical Dosage Forms and Delivery Systems. 8th ed. Philadelphia, PA: Lippincott Williams & Wilkins, 2005.

Ahern TJ, Manning MC, eds. Stability of Protein Pharmaceuticals, Part A and Part B books. New York: Plenum, 1992.

Akers MJ. Antioxidants in pharmaceutical products. J Parenter Sci Technol 1982; 36:222–228.

Akers MJ. Considerations in selecting antimicrobial preservative agents for parenteral product development. Pharm Tech 1984; 8:36–46.

Akers MJ, Fites AL, Robison RL. Formulation design and development of parenteral suspensions. J Parenter Sci Technol 1987; 41:88–96.

Akers MJ. Parenterals: Small volume. In: Swarbrick J and Boylan J, eds. Encyclopedia of Pharmaceutical Technology. New York: Dekker, 1995.

Akers MJ, DeFelippis MR. Formulation of protein dosage forms: Solutions. In: Hovgaard L, Frokjaer S, eds. Pharmaceutical Formulation Development of Peptides and Proteins. London, UK: Taylor & Francis, 2000.

Akers MJ. Excipient-drug interactions in parenteral formulations. J Pharm Sci 2002; 91:2283–2297.

Akers MJ. Parenterals. In: Remington's Pharmaceutical Sciences. 21st ed. Philadelphia, PA: Lippincott Williams & Wilkins, 2005:802–836.

Bontempo J, ed. Development of Parenteral Biopharmaceutical Dosage Forms. New York: Marcel Dekker, 1997.

Boylan JC, Fites AL, Nail SL. Parenteral Products. In: Banker GS and Rhodes CT, eds. Modern Pharmaceutics. 3rd ed. New York: Dekker, 1995:chap 12.

Carpenter JF, Crowe JH. The mechanism of cryoprotection of proteins by solutes. Cryobiology 1988; 25: 244–250.

Carpenter JF, Pikal MJ, Chang BS, et al. Rational design of stable lyophilized protein formulations: Some practical advice. Pharm Res 1997; 14:969–975.

Carpenter JF, Chang BS, Garzon-Rodriquez W, et al. Rational design of stable lyophilized protein formulations: Theory and Practice. In: Carpenter JF, Manning MC, eds. Rational Design of Stable Protein Formulations. New York: Kluwer Academic, 2002.

DeFelippis MR, Akers MJ. Formulation, manufacture, and control of protein suspension dosage forms. In: Hovgaard L, Frokjaer S, eds. Pharmaceutical Formulation Development of Peptides and Proteins. London, UK: Taylor & Francis, 2000.

ICH: Q8(R2): Pharmaceutical Development, http://www.fda.gov/cder/guidance/index.htm. Accessed August 2009.

Nail SL, Akers MJ, eds. Development and Manufacture of Protein Pharmaceuticals. New York: Kluwers-Plenum, 2002.

Nema S, Ludwig J, eds. Pharmaceutical Dosage Forms: Parenteral Medications. 3rd ed. 3 vols. New York, NY: Informa Healthcare, 2010.

Pearlman R, Wang YJ, eds. Formulation, characterization, and stability of protein drugs, Vol 9. In: Borchardt R, series ed. Pharmaceutical Biotechnology, New York: Plenum, 1995.

Sinko PJ, ed. Martin's Physical Pharmacy and Pharmaceutical Sciences. 5th ed. Philadelphia, PA: Lippincott Williams & Wilkins, 2005.

Tonnesen HH, ed. Photostability of Drugs and Drug Formulations. Boca Raton, FL: CRC Press, 2004.

Wang YJ, Pearlman R, eds. Stability and characterization of protein and peptide drugs. Vol 5. In: Borchardt R, series ed. Pharmaceutical Biotechnology. New York: Plenum, 1993.

2 Characteristics of sterile dosage forms

Sterile dosage forms are unique pharmaceutical dosage forms largely because of their seven primary characteristics that will be featured in this chapter (Table 2-1). Also, specific characteristics of sterile dosage forms that are discussed in the United States Pharmacopeia (USP), primarily general chapter <1> will be featured.

SEVEN PRIMARY CHARACTERISTICS OF STERILE DOSAGE FORMS

Safety

Sterile dosage forms, with some exceptions, are injected directly into the body and, thus, avoid the body's natural barriers for invasion of entities that could harm the body. Therefore, any component of an injectable product must be proven safe at the quantitative level it is injected. Certainly, any substance, if injected in large quantities, can be unsafe.

With respect to safety, formulation of sterile dosage forms can be both easier and more difficult compared to formulation of nonsterile dosage forms. This is because of safety considerations when selecting additives to combine with the active ingredient to overcome one or more problems related to drug solubility, stability, tonicity, and controlled or sustained delivery. If any of these problems exist with a nonsterile dosage form, the formulation scientist has a plethora of choice with respect to additives safe to use for administration other than by injection. However, for overcoming these problems with sterile dosage forms, the requirement for safety prohibits the use of many additives that could be effective.

Under the Kefauver-Harris Amendments to the Federal Food, Drug, and Cosmetic Act, most pharmaceutical preparations are required to be tested for safety in animals. Because it is entirely possible for a parenteral product to pass the routine sterility test, pyrogen and/or endotoxin test, as well as the chemical analyses, and still cause unfavorable reactions when injected, a safety test in animals is essential, particularly for biological products, to provide additional assurance that the product does not have unexpected toxic properties.

The FDA has published guidance for safety evaluation of pharmaceutical ingredients (1) that is periodically updated. Many general chapters of the USP also provide specific instructions for safety evaluation of pharmaceutical excipients. Also, there exists the International Pharmaceutical Excipients Council (IPEC), a federation of three independent regional industry associations headquartered in the United States (IPEC-Americas), Europe (IPEC Europe), and Japan (JPEC). The following is a quote from their Web site:

- Each association focuses its attention on the applicable law, regulations, science, and business practices of its region. The three associations work together on excipient safety and public health issues, in connection with international trade matters, and to achieve harmonization of regulatory standards and pharmacopoeial monographs.
- Over 200 national and multinational excipient makers, producers, and companies, which use excipients in finished drug dosage forms are members of one or more of the three IPEC regional units. Over 50 U.S. companies are IPEC members. (2)

Sterility

Obviously, sterility is what defines/differentiates a sterile product. Achieving and maintaining sterility are among the greatest challenges facing manufacturers of these dosage forms. There are many factors that contribute to achieving and maintaining sterility and these will be covered in more detail in chapters 13, 17, 18, 21, and 23. Suffice to state at this point that the characteristic of sterility is achieved via valid sterilization procedures for all components during manufacturing of the product, valid procedure for sterile (better term is aseptic) filtration, design and maintenance of clean rooms meeting all requirements for preparing sterile products (discussed

Table 2-1 Seven Basic Characteristics of Sterile Product Dosage Forms

1.	Safety (freedom from adverse toxicological concerns)
2.	Sterility (freedom from microbiological contamination)
3.	Nonpyrogenic (freedom from pyrogenic—endotoxin—contamination)
4.	Particle-free (freedom from visible particle contamination)
5.	Stability (chemical, physical, microbiological)
6.	Compatibility (formulation, package, other diluents)
7.	Tonicity (isotonic with biological fluids)

in chap. 14), validation of aseptic processes, training and application of good aseptic practices, use of antimicrobial preservatives for multiple-dose products, and valid testing for sterility of the product and maintenance of container/closure integrity.

Freedom from Pyrogenic Contamination

Pyrogens are discussed extensively in chapters 13 and 28. Pyrogens are fever-producing entities originating from a variety of sources, primarily microbial. In sufficient amounts following injections, pyrogens can cause a variety of complications in the human body. Because of the advent of the in-vitro test, Limulus Amebocyte Lysate (LAL), for the quantitative detection of the most ubiquitous type of pyrogen called bacterial endotoxins, all marketed injectable products must meet requirements for pyrogen (or endotoxin) limits.

To achieve freedom from pyrogenic contamination, like achieving and maintaining product sterility, many factors contribute toward this goal. Depyrogenation methods will be discussed in chapter 13, which include cleaning validation, time limitations, validated depyrogenation cycles for glassware, validation of pyrogen/endotoxin removal from rubber closures and other items that depend on rinsing techniques, validated water systems, and use of endotoxin-free raw materials.

Freedom from Visible Particulate Matter

Most aspects of particulate matter will be discussed in chapters 22 and 29. Visible particulate matter implicates product quality and perhaps safety. It definitely reflects the quality of operations of the product manufacturer. Both ready-to-use solutions and reconstituted solutions are to be free from any evidence of visible particulate matter and must meet compendial specifications for numbers of subvisible particles no greater than certain sizes, those particle sizes being for most compendia no greater or equal to 10 μm and no greater or equal to 25 μm.

Like other product characteristics, several factors contribute to the presence or absence of foreign particulate matter. These include valid cleaning methods of all equipment and packaging materials, valid solution filtration procedures, adequate control of production and testing environments, adequate training of personnel in manufacturing, testing and using sterile product solutions, and employment of required compendial testing procedures for detection of both visible and subvisible particulate matter.

Stability

All dosage forms have stability requirements. All dosage forms are required to be stable under predetermined manufacturing, packaging, storage, and usage conditions. Sterile dosage forms, like all other dosage forms, need to maintain both chemical and physical stability throughout the shelf-life of the product. The achievement of chemical and physical stability is the greatest challenge of scientists responsible for developing sterile dosage forms. With the exception of overcoming solubility challenges, often related to long-term physical stability, addressing and solving stability problems occupies most of the time and effort of scientists in the product development process. With much more complicated chemical structures and vulnerabilities to environmental conditions (temperature, light, pH, shear, metal impurities, oxygen, etc.) stabilization of therapeutic peptides and proteins offer enormous challenges. Achieving and maintaining chemical and physical stability starts with the active ingredient and how it is stored, shipped, and handled. Stability challenges continue with the compounding, mixing, filtration, filling, stoppering, and sealing of the product. So many injectable drugs are so unstable in solution that they must exist in the solid state so lyophilization processes and maintaining stability during lyophilization offer lots of challenges to the development scientist. Maintaining stability in the final container/closure system, while being stored, shipped, and manipulated prior to being administered to people or animals, all present enormous challenges that must be overcome.

Sterile dosage forms also have one extra requirement related to stability and that is maintaining sterility as a function of stability. So, with sterile dosage forms, product stability encompasses not only chemical and physical properties, but also includes microbiological stability (i.e., maintenance of sterility) throughout the shelf-life and usage of the product. Stability aspects of dosage forms are covered in chapters 8 through 11 and stability testing is discussed in chapter 24.

Compatibility

Most pharmaceutical dosage forms are consumed by patients without the patient or health care professional needing to do any manipulation with the dosage form prior to consuming it. While this is also true for many sterile dosage forms, there are also a significant number of sterile dosage forms that must be manipulated prior to injection. For example, freeze-dried products are released by the manufacturer, but must be manipulated by the user and/or health care professional prior to administration. The product must be reconstituted by sterile dilution, withdrawn into a syringe, and, often, then combined with another solution, perhaps a large volume infusion fluid, for administration. What all this means is that the sterile product must be shown to be compatible with diluents for reconstitution and diluents for infusion. Furthermore, many infusions contain more than one drug, so obviously the two or more drugs in the infusion system must be compatible.

Isotonicity

Biological cells maintain a certain "tone"; that is a certain biological concentration of ions, molecules, and aggregated species that give cells specific properties, the most important pharmaceutically of which is its osmotic pressure. Osmotic pressure is a characteristic of semipermeable cell membranes where osmotic pressure is the pressure where no water migrates across the membrane. Osmosis is the phenomenon where solutes will diffuse from regions of high concentration to regions of low concentration. So, if a formulation is injected that has an osmotic pressure less than that of biological cells, that is, the solution is hypotonic, the solvent from the injection will move across the cell membranes and could cause these cells to burst. If the cells are red blood cells, this bursting effect is called hemolysis. Conversely, if the formulation injected has an osmotic pressure greater than that of biological cells, that is, the solution is hypertonic, the solvent or water from the cell interior will move outside the cell membranes and could cause these cells to shrink, for example, crenation.

Ideally, any injected formulation should be isotonic with biological cells to avoid these potential problems of cells bursting or shrinking. Large-volume intravenous injections and small-volume injections by all routes other than the intravenous route must be isotonic to avoid major problems such as pain, tissue irritation, and more serious physiological reactions. Small-volume intravenous injections, while desirable to be isotonic, do not absolutely have to be isotonic because small volumes do not damage an excessive number of red cells that cannot be replaced readily.

It is well known that 0.9% sodium chloride solution and 5% dextrose solution are isotonic with biological cells. Why the difference in isotonic concentrations between these two common large-volume solutions? It has to do with the ability of the solute to dissociate into more than one species. Dextrose is a nonelectrolyte that in solution exists as a single entity; therefore, the osmotic pressure of a nonelectrolyte solution is proportional to the concentration of the solute. Sodium chloride is an electrolyte in solution that dissociates into two ionic species. Thus, the osmotic pressure of a solution containing an electrolyte dissociating into two species would be at least twice that of a solution containing a nonelectrolyte. The fact that the concentration of isotonic dextrose solution is over five times that of isotonic sodium chloride solution may be explained by the fact that ionic species attract solvent molecules, thus holding solvent molecules in solution and reducing their tendency to migrate across the cellular membrane. This, in turn, elevates osmotic pressure of the electrolytic solution such that a lower concentration of electrolyte solute is required to exert that same osmotic pressure as a nonelectrolyte solution. More information about tonicity and formulation is covered in chapters 6 and 8. The United States Pharmacopeia contains general chapter <785> that defines osmotic pressure, osmolality and osmolarity, and measurement of osmolality.

CHARACTERISTICS OF STERILE DOSAGE FORMS FROM THE UNITED STATES PHARMACOPEIA

The first general chapter of the USP is entitled "<1> INJECTIONS." Within this section are the following subcategories with the content under each subcategory summarized. Of course, wording of these characterizations might change over time so the reader must consult the current edition of the USP for current wording.

Introduction

Parenteral products are defined as preparations intended for injection through the skin or other external boundary tissue where the active ingredient is introduced directly into a blood vessel, organ, tissue, or lesion. Parenteral products are to be prepared scrupulously by methods designed to ensure that they meet Pharmacopeial requirements for and, where appropriate, contain inhibitors of the growth of microorganisms.

- Sterility
- Pyrogens
- Particulate Matter
- Other Contaminants

NOMENCLATURE AND DEFINITIONS

There are five general types of parenteral preparations listed in the USP:

- [Drug] Injection: Liquid preparations that are drug substances or solutions thereof.
- [Drug] for Injection: Dry solids that, upon the addition of suitable vehicles, yield solutions conforming in all respects to the requirements of injections.
- [Drug] Injectable Emulsion: Liquid preparations of drug substances dissolved or dispersed in a suitable emulsion medium.
- [Drug] Injectable Suspension: Liquid preparations of solids suspended in a suitable liquid medium.
- [Drug] for Injectable Suspension: Dry solids that, upon the addition of suitable vehicles, yield preparations conforming in all respects to the requirements of Injectable Suspensions.

Definitions included in the USP are as follows:

- Pharmacy Bulk Package: A pharmacy bulk package is a single product containing a sterile drug injection, sterile drug for injection, or sterile drug injectable emulsion (i.e., suspensions cannot be contained in pharmacy bulk packages. A pharmacy bulk package contains many single doses of the active ingredient to be used for the preparation of admixtures for infusion, or, using a sterile transfer device, for filling empty sterile syringes. The closure of the bulk package shall be penetrated only once with a sterile device that will allow measured dispensing of the contents.
- Large- and Small-Volume Injections: The demarcation of volume differentiating a small- from large-volume injection is 100 mL. Any product 100 mL or less is a small-volume injection. The main purpose for differentiating large- from small-volume injections is the method of sterilization. With perhaps a single exception for blood products, all large-volume injections must be terminally sterilized while most small-volume injections are not terminally sterilized.
- Biologics: This definition simply states that pharmacopeial definitions for sterile preparations for parenteral use do not apply to biologics because of their special nature and licensing requirements. Biologic requirements are covered in USP <1041> general chapter.

Ingredients

Three general types of ingredients discussed in this section of the USP are aqueous vehicles, other vehicles, and added substances.

Aqueous vehicles must meet the requirements of the Pyrogen Test <151> or the Bacterial Endotoxins Test <85> whichever is specified. Water for injection is the vehicle unless the individual monograph specifies another aqueous vehicle.

Other vehicles refer to fixed oils that must be of vegetable origin. Oils must meet several compendial requirements, including solid paraffin (under mineral oil); a saponification value between 185 and 200 and an iodine value between 79 and 141 (fats and fixed oils <401>); and tests for unsaponifiable matter and free fatty acids. Saponification is the reaction of an ester with a metallic base and water to produce soap. It is also defined as the alkaline hydrolysis of oil or fat, or the neutralization of a fatty acid to form soap. Unsaponifiable matter is a substance that is incapable of being saponified; that is, it cannot react with a basic substance to form soap.

Added substances are formulated into injectable products to increase stability or usefulness, but must be harmless in the amounts administered and do not interfere with the therapeutic efficacy of the drug or responses to specific assays and tests. USP clearly states that no coloring agent may be added to a parenteral product. Any product with a volume of injection more than 5 mL should not use added substances unless their inclusion is clearly justified and in safe concentrations.

USP provides upper limits (unless higher limits are justified) for several specific additives:

Mercury-containing additives	0.01%
Cationic surfactants	0.01%
Chlorobutanol, cresol, phenol	0.5%
Sulfurous acid salt or equivalent	0.2%

Injections intended for multiple-dose containers must contain an additive to prevent the growth of microorganisms, regardless of the method of sterilization. Three exceptions to this rule are: (*i*) if there are different directions in the individual monograph; (*ii*) if the substance contains a radionuclide with a physical half-life of less than 24 hours; and (*iii*) if the active ingredient itself is antimicrobial. The antimicrobial preservative agent must meet the requirements of Antimicrobial Effectiveness Testing <51> and Antimicrobial Agents—Content <341>.

Labeling

Information that is contained on a product label includes the following:

- For liquid products: Percentage content of the drug or amount of drug in a specified volume.
- For dry products: The amount of active ingredient.
- The route of administration.
- A statement of storage conditions and an expiration date.
- The name and place of business of the manufacturer, packer, or distributor.
- Identifying lot number—The lot number is capable of yielding the complete manufacturing history of the specific package, including all manufacturing, filling, sterilizing, and labeling operations.

If the formulation is not specified in the individual monograph, the product label must contain the specific quantitative amount of each ingredient. For liquid preparations, the percentage content of each ingredient or the amount of each ingredient in a specified volume must be listed. Ingredients that are added to adjust pH or make the solution isotonic do not need to be quantified, but must be declared by name and their effect stated on the label. For dry preparations or those preparations to which a diluent will be added before use, items that must be included on the label include the following:

- The amount of each ingredient
- The composition of the recommended diluent(s)
- The amount to be used to attain a specific concentration of the active ingredient
- The final volume of the solution

- Description of the physical appearance of the constituted solution
- Directions for proper storage of the constituted solution
- An expiration date limiting the storage period during which the constituted solution may be expected to have the required or labeled potency if stored as directed.

Strength and Total Volume for Single- and Multiple-Dose Injectable Drug Products

The primary and prominent expression on the principal display panel of the label needs to be the strength of the active ingredient per total volume, for example:

- Strength per vial: 500 mg/10 mL (or in units per total volume)
- Strength per mL: 50 mg/mL (or in units per mL)

If the container volume is less than 1 mL, then the strength per fraction of 1 mL should be the expression, for example, 12.5 mg/0.0625 mL

Medication errors cannot completely be eliminated by prominent strength labels as insulin is a primary example. However, meeting this requirement for label strength prominence certainly will help to reduce the potential for medication errors.

Aluminum in LVPs, SVPs, and PBPs Used in TPN Therapy

The aluminum content of large-volume parenterals (LVPs) used in total parenteral nutrition (TPN) therapy must not exceed 25 μg per liter. The package insert (see the "Precautions" section) of LVPs used in TPN therapy must state that the drug product contains no more than 25 μg of aluminum per milliliter. For small-volume parenterals (SVPs) and pharmacy bulk packages (PBPs), the immediate container label used in the preparation of TPN parenterals should state, "Contains no more than 25 μg/L of aluminum." The maximum level of aluminum as expiry must be stated on the immediate container label of all SVPs and PBPs used in preparation of TPN parenterals with the statement "Contains no more than ____ μg/L of aluminum" (the USP leaves this blank, to be filled in by the manufacturer). This maximum amount of aluminum must be stated as the highest of either the highest level for the batches produced during the past three years or the highest level for the latest fiveb batches.

The package insert for any and all products used in the preparation of TPN products must contain a warning statement in the "Warning" section of the labeling with the warning statement being word-for-word what is published in this section of the USP <1> Injections.

Packaging

Containers for Injection

The packaging system must not interact physically or chemically with the product stored within the package. The container must be composed of materials that allow inspection of the contents. Individual monographs will state the type of glass (or plastic) preferable for each parenteral preparation.

Containers are closed or sealed to prevent contamination or loss of contents. Container closure integrity testing must be performed to validate the integrity of the packaging system against any kind of microbial contamination or chemical or physical impurities. The packaging system must be able to protect the product when exposed to anticipated extreme conditions of manufacturing, storage, shipment, and distribution.

Closures for multiple-dose containers must permit the withdrawal of the contents without removal or destruction of the closure. The closure must seal itself after the needle is removed to protect the product against contamination. Validation of multiple-dose container integrity must include verification that such a package prevents microbial contamination or loss of product contents under simulated use conditions of multiple entry and use.

Potassium Chloride for Injection Concentrate and Neuromuscular Blocking and Paralyzing Agents

The USP contains two very specific paragraphs for two kinds of injectable products—potassium chloride for injection concentrate and neuromuscular blocking and paralyzing agents. A black closure system on a vial (black flip-off seal and black ferrule to hold the elastomeric closure)

Table 2-2 United States Pharmacopeia General Chapter <1> Recommended Excess Volume in Containers Containing Injectable Solutions

	Recommended Excess Volume	
Label size	**For mobile liquids**	**For viscous liquids**
0.5 mL	0.10 mL	0.12 mL
1.0 mL	0.10 mL	0.15 mL
2.0 mL	0.15 mL	0.25 mL
5.0 mL	0.30 mL	0.50 mL
10.0 mL	0.50 mL	0.70 mL
20.0 mL	0.60 mL	0.90 mL
30.0 mL	0.80 mL	1.20 mL
50.0 mL or more	2%	3%

or a black band at the neck of a glass ampul can only be used for potassium chloride for injection concentrate containers. All injectable preparations of neuromuscular blocking agents and paralyzing agents must be packaged in vials with a cautionary statement printed on the ferrules or cap overseals that warn what product is in the containers.

Containers for Sterile Solids

Containers and closures for sterile dry solids also must not interact physically or chemically with the product. Such containers will permit the addition of a suitable solvent and withdrawal of parts of the resulting solution of suspension without compromising the sterility of the product.

Volume in Container

Each container of an injection must be filled with sufficient excess to allow the labeled amount of volume of product to be withdrawn from the container. General chapter <1151 Pharmaceutical Dosage Forms> is referenced where under "Injections" in that chapter, there is a table that provide the recommended excess volume for injectables labels of various volume sizes to be withdrawn (Table 2-2).

Determination of Volume of Injection in Containers

This section of the USP contains a procedure of how to determine product volume in a container.

1. The labeled volume of the container will determine how many containers are to be used in the test. If the container volume is $\geq$10 mL, three or more containers are used. If the container volume is $\leq$3 mL, five or more containers are used.
2. Individually take up the contents of each container into a dry hypodermic syringe of a rated capacity not exceeding three times the volume to be measured and fitted with a 21-gauge needle not less than 2.5 cm (one inch) in length.
3. Expel any air bubbles from the syringe and needle and then discharge the contents of the syringe, without emptying the needle, into a standardized, dry cylinder (graduated to contain rather than to deliver the designated volumes) of such size that the volume to be measured occupies at least 40% of the cylinder's rated volume.
    a. Alternatively, the contents of the syringe may be discharged into a dry, tared beaker, the volume, in mL, being calculated as the weight, in grams, of injection taken divided by its density.
4. The contents of up to five 1- or 2-mL containers may be pooled for the measurement using a separate dry syringe for each container.
5. The content of containers holding 10 mL or more may be determined by opening them and emptying the contents directly into the graduated cylinder or tared beaker.
6. The volume is not less than the labeled volume in the case of containers examined individually or, in the case of the 1- and 2-mL containers, is not less than the sum of the labeled volumes of the containers taken collectively.

The USP also has specific guidance, not repeated here, for determination of volume of injection for the following special containers: multiple dose, containers with oily contents, cartridges, syringes, and large-volume solutions.

Printing on Ferrules and Cap Overseals

Only cautionary statements are to be printed on these parts of the drug product container. The printing must be of contrasting color and conspicuous under conditions of use. Examples of cautionary statements include "Warning," "Dilute Before Using," "Paralyzing Agent," "IM Use Only," and "Chemotherapy".

Packaging and Storage

This USP segment summarizes the requirements for packaging of different types of injectable products.

1. No more than 1 L of injection volume may be withdrawn and administered at one time.
2. Preparations for intraspinal, intracisternal, or peridural injections may be packaged only in single-dose containers.
3. Unless an individual monograph specifies differently, a multiple-dose container may contain no more than 30 mL volume of injection.
4. Injections packaged for use as irrigation solutions, for hemofiltration or dialysis, or for parenteral nutrition are exempt from the 1-L restriction stated in #1.
5. Containers for injections packaged for use as hemofiltration or irrigation solutions may be designed to empty rapidly (e.g., the closure is a screw-cap rather than a rubber closure) and may contain a volume more than 1 L.
6. Injections labeled for veterinary use are exempt from packaging and storage requirements concerning the limitation to single-dose containers and the limitation on the volume of multiple-dose containers.

Foreign and Particulate Matter

All products intended for parenteral administration shall be prepared in a manner designed to exclude particulate matter as defined in Particulate Matter in Injections <788> and other foreign matter. Versions of the USP through 2005 made the following statement:

> Every care should be exercised in the preparation of all products intended for injection to prevent contamination with microorganisms and foreign material. Good pharmaceutical practice requires also that each final container of injection be subjected individually to a physical inspection, whenever the nature of the container permits, and that every container whose contents show evidence of contamination with visible foreign material be rejected.

The statement was revised in 2006 to read as follows:

> Each final container of all parenteral preparations shall be inspected to the extent possible for the presence of observable foreign and particulate matter ("visible particulates") in its contents. The inspection process shall be designed and qualified to ensure that every lot of all parenteral preparations is essentially free from visible particulates. Qualification of the inspection process shall be performed with reference to particulates in the visible range of a type that might emanate from the manufacturing or filling process. Every container whose contents show evidence of visible particulates shall be rejected. The inspection for visible particulates may take place when inspecting for other critical defects, such as cracked or defective containers or seals, or when characterizing the appearance of a lyophilized product.

Phrases such as "whenever the nature of the container permits," "to the extent possible," "foreign" and "essentially free" are controversial (see chapter 22). Manufacturers who need to use amber or other colored containers to inhibit light from entering the container might use the statement "whenever the nature of the container permits" to justify not performing physical inspections for particles and foreign matter. Regulatory inspectors will greatly frown on this. USP added verbiage requiring supplemental inspections, such as withdrawing contents from containers that limit inspection capabilities. The term "foreign material" has been applied to products of biotechnology where sometimes very small amounts of aggregated protein may be seen visually. While these aggregates are viewed as "particles," they are not viewed as "foreign material." Manufacturers differ on whether such containers with aggregated protein should be rejected. Some protein-containing products do allow for in-line filtration of the product prior to injection into the human body and the FDA permits this as long as the filtration does not filter out protein to the point that it fails potency specifications.

All large- and small-volume injections, unless otherwise specified in individual monographs, are subject to the particulate matter limits set forth under <788> Particulate Matter in Injections. Injections packaged and labeled for use as irrigating solutions are exempt from the requirements for Particulate Matter. Also, at the time of this writing, injections administered by the intramuscular or subcutaneous routes are exempted from the requirements for <788> although this will be changed in future USP editions (see Chap. 29).

Sterility

This section simply states that all preparations for injection must meet the requirements under Sterility Tests <71>.

Constituted Solutions

Dry solids are constituted at the time of use by health care practitioners. Therefore, tests and standards pertaining to the solution as constituted for injection are not included in the individual USP monographs for these products (also true for liquid concentrates). USP states that in the interest of assuring the quality of these preparations as they are actually administered, certain nondestructive tests are to be performed to demonstrate the suitability of constituted solutions prepared.

1. Completeness and Clarity of Solution—The product is reconstituted as directed in the labeling supplied by the manufacturer and observed for:
   a. The solid dissolves completely, leaving no visible residue as undissolved matter or
   b. the constituted solution is not significantly less clear than an equal volume of the diluent or of Purified Water contained in a similar vessel and examined similarly.
2. Particulate Matter—After the sterile dry solid is reconstituted according to the manufacturer's directions, the solution is essentially free from particles of foreign matter that can be observed on visual inspection.

REFERENCES

1. Guidance for Industry—Nonclinical studies for the safety evaluation of pharmaceutical excipients, United States Food and Drug Administration, May 2005. http://www.fda.gov/cber/guidelines.htm.
2. International Pharmaceutical Excipients Council of the Americas, Inc., Arlington, VA, 2010. http://www.ipecamericas.org/index.html.

3 Types of sterile dosage forms

Sterile dosage forms basically can be classified in three broad categories:

1. Conventional small volume injectables
2. Conventional large volume injectables
3. Modified release (depot) injectables

Small volume injectables (SVIs) by definition are products contained to deliver no more than 100 mL from the same container. Large volume injectables (LVIs) are products contained in volumes greater than 100 mL. Modified release or depot injectable drug delivery systems are typically SVI whose formulations are designed to deliver drugs by routes other than intravenous and in regimens less frequent than by conventional therapies.

SMALL VOLUME INJECTABLES

SVIs dosage forms include solutions, suspensions, emulsions, and solids (Fig. 3-1). The number of ready-to-use solution dosage forms far exceeds (perhaps 3–1) the number of the second most frequent type of dosage form, the lyophilized or powder-filled sterile solid dosage form. Suspensions, emulsions, and other dispersed systems are a distant third although with more advancement in sustained release (depot) technologies, suspension dosage forms are increasing in number. While not considered in this chapter, other sterile dosage forms include sterile ophthalmic ointments and gels and implantable depot devices.

Solutions

Solutions are ready-to-use products or can be liquid concentrates (aqueous only) subsequently diluted in a smaller container or within a suitable IV fluid. Solutions can be aqueous or nonaqueous. Aqueous solutions can be completely water based or water combined with a water-miscible organic cosolvent such as ethanol, polyethylene glycol, glycerin, or propylene glycol.

Nonaqueous solutions, also called oleaginous solutions, contain oils as the vehicle. Only oils of vegetable origin are acceptable for injectable products, the most common oils being soybean, sesame, and cottonseed (see chap. 8). Oily solutions must not be administered by the IV route.

Suspensions

Suspensions can be coarse (macro) (Fig. 3-1) or microsized (micro- or nanosuspension) solids dispersed in a suitable vehicle, either water or oil. Insulin, vaccines, and microsphere delivery systems are formulated and delivered as injectable suspensions. In fact, the suspension is the primary dosage form for insulin products (e.g., Neutral Protamine Hagedorn—NPH—and Lente products). Suspensions, unless the dispersed particles are nanoparticles, cannot be administered by the IV route. Chapter 9 is devoted to parenteral suspensions and other dispersed systems.

Emulsions

Emulsions are also dispersed systems combining an oil phase with an aqueous phase. If the oil phase is dispersed in the aqueous phase, the dosage form is called an oil-in-water emulsion. If the aqueous phase is dispersed in the oil phase, the dosage form is called a water-in-oil emulsion. Most, if not all, injectable emulsions are oil-in-water systems. Liposomes are emulsified spherical vesicles composed of a phospholipid bilayer with an aqueous inner phase. Drugs can be incorporated in either the lipid or aqueous phases, depending on solubility. Liposomes structurally are similar to biologic membranes and have high potential as delivery systems for genetic therapeutics. More coverage of liposomes will occur later in this chapter and in chapter 9.

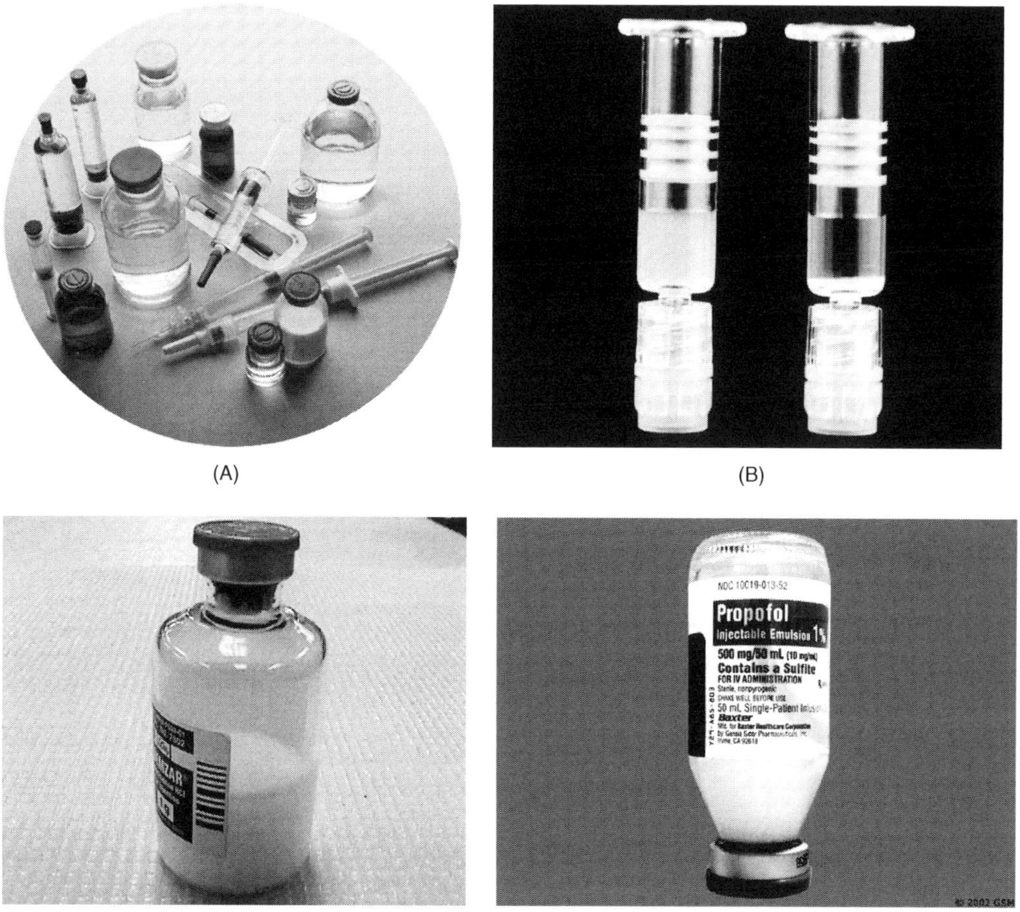

(A) (B)

(C) (D)

Figure 3-1 Examples of injectable dosage forms. (A) Solution. *Source*: Courtesy of Baxter Healthcare Corporation. (B) Suspension. *Source*: Courtesy of Dr. Gregory Sacha, Baxter. (C) Lyophilized powder (Gemzar®). *Source*: Courtesy of Eli Lilly and Company. (D) Emulsion. *Source*: Courtesy of Teva Pharmaceuticals

Parenteral emulsions are milky white in appearance (Fig. 3-2A) and have an average globule size of 1.0 μm to 5 μm. The United States Pharmacopoeia (USP) General Chapter <729> "Globule Size Distribution in Lipid Injectable Emulsions" specifies that "the volume-weighted, large-diameter fat globule limits of the dispersed phase, expressed as the percentage of fat residing in globules larger than 5 μm for a given lipid injectable emulsion, must be less than 0.05% (measured by light-scattering or light obscuration methods)." Emulsions are primarily used for parenteral nutrition and infused intravenously. Parenteral nutrition emulsions indeed are large volume and are terminally sterilized with the sterilization cycle designed to maintain globule size distribution. Small volume injectable emulsions are formulated with an active ingredient, the most common examples being propofol (Fig. 3-2B) and oil soluble vitamins. More coverage of emulsions is found in chapter 9.

Solids

Solids are prepared primarily by lyophilization after liquid filling with secondary preparation by sterile crystallization and powder filling. The reason most sterile solids are prepared by lyophilization is the fact that liquid filling presents less problems than powder filling and for powder filling the product needs to be crystalline in solid state character. Amorphous solids are very difficult to fill accurately because of their relative lack of density (too fluffy).

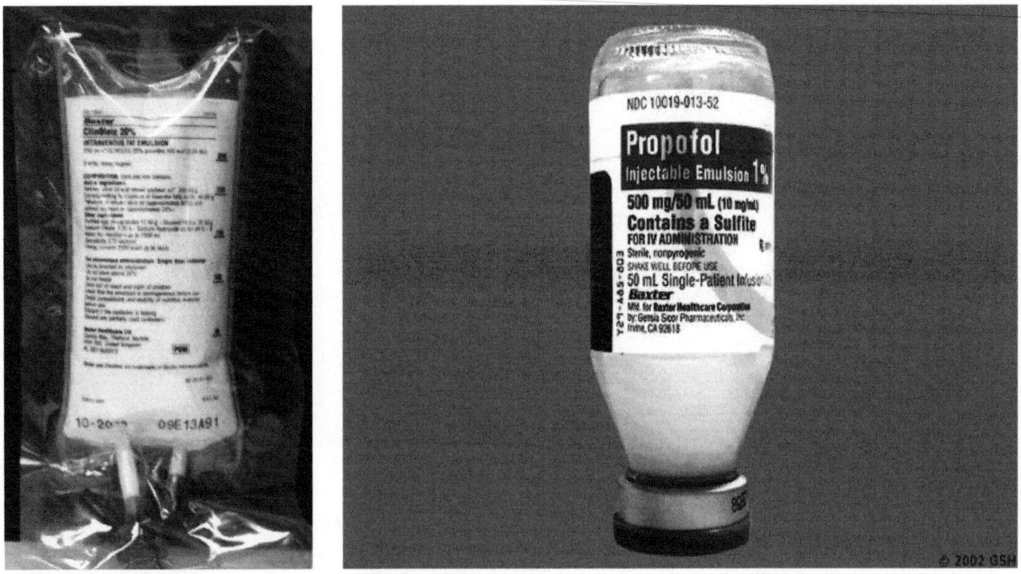

(A) (B)

Figure 3-2 Emulsion formulations. (A) Large volume—typical formulation—soybean oil (10–20%) (linoleic, oleic, palmitic oils; unsaturated fatty acid triglycerides), egg yolk phospholipid (1.2%), glycerin (2.5%), Water for Injection. *Source*: Courtesy of Baxter Healthcare Corporation. (B) Small volume—typical formulation—soybean oil, egg lecithin, glycerin, Water for Injection, pH approximately 8.0—for example, Propofol®, Vitalipid®, Limethason®, Lipfen®, Liple®, Diazemuls®, Fluosol®. *Source*: Courtesy of Teva Pharmaceuticals.

However, if the solid formulation can be crystallized, then powder filling can be a viable alternative to lyophilization. Most injectable cephalosporins, because they can be crystallized, are filled as sterile powders. Some proteins are prepared by spray drying techniques and filled as sterile powders. Sterile solids are reconstituted prior to administration with a suitable diluent. Chapters 10 and 20 cover formulation and processing, respectively, of lyophilized (freeze-dried) solids.

Another category of solids, for a lack of a better place to introduce this type of solid, are the solid implants, surgically inserted within bodily tissue, primarily for prolonged action pharmaceuticals. These are discussed in the following text in section "Polymeric Implants."

LARGE VOLUME INJECTABLES

LVIs include electrolytes, carbohydrates, proteins, fatty emulsions, peritoneal dialysis solutions, and irrigating solutions. Table 3-1 gives a more complete example of commercially available large volume products (courtesy of a Baxter product listing).

Electrolyte Solutions

These solutions are primarily sodium chloride (0.9%) isotonic solutions, other concentrations of sodium chloride (0.45%, 3%), potassium chloride (20–40 mEq/L), Ringer's, lactated Ringer's, sodium lactate, sodium bicarbonate, and various combinations of sodium chloride, potassium chloride, and/or dextrose.

Carbohydrate Solutions

Dextrose 5% in water (D5W) is the most common and popular large volume carbohydrate. Dextran solutions are also included here along with combinations of dextrose and sodium chloride, dextrose and potassium chloride, dextrose and Ringer's or Lactated Ringer's, and other combinations thereof.

Table 3-1 Examples of Commercially Available Large Volume Injections

Dextrose injections
2.5% Dextrose injection, USP in glass container
5% Dextrose injection, USP in glass container
5% Dextrose injection, USP in VIAFLEX plastic container
10% Dextrose injection, USP in VIAFLEX plastic container
Dextrose and electrolyte injections
5% Dextrose and electrolyte no. 48 injection (multiple electrolytes and dextrose injection, Type 1, USP)
Dextrose and sodium chloride injections
5% Dextrose and 0.2% sodium chloride injection, USP
2.5% Dextrose and 0.45% sodium chloride injection, USP
5% Dextrose and 0.9% sodium chloride injection, USP
5% Dextrose and 0.45% sodium chloride injection, USP
5% Dextrose and 0.33% sodium chloride injection, USP
Dextran injections
6% GENTRAN 70 (Dextran 70) in 0.9% sodium chloride injection, USP
10% GENTRAN 40 (Dextran 40) in 0.9% sodium chloride injection, USP
10% GENTRAN 40 (Dextran 40) in 5% dextrose injection, USP
Miscellaneous injections
Ringer's injection, USP
Lactated Ringer's injection, USP
Sterile Water for Injection, USP (for drug diluent use only)
5% Sodium bicarbonate injection, USP
Sodium lactate injection, USP (M/6 sodium lactate)
Ringer's injection, USP
OSMITROL (Mannitol) injections in VIAFLEX plastic container
10% OSMITROL injection (10% Mannitol injection, USP)
15% OSMITROL injection (15% Mannitol injection, USP)
20% OSMITROL injection (20% Mannitol injection, USP)
5% OSMITROL injection (5% Mannitol injection, USP)
PLASMA-LYTE (electrolyte) replenishment solutions in VIAFLEX plastic container
PLASMA-LYTE 148 injection (multiple electrolytes injection, Type 1, USP)
PLASMA-LYTE A injection pH 7.4 (multiple electrolytes injection, Type 1, USP)
PLASMA-LYTE 56 and 5% dextrose injection (multiple electrolytes and dextrose injection, Type 1, USP)
Potassium chloride in 0.45% sodium chloride injections
20 mEq/L potassium chloride in 0.45% sodium chloride injection, USP
Potassium chloride in 0.9% sodium chloride injections
20 mEq/L potassium chloride in 0.9% sodium chloride injection, USP
40 mEq/L potassium chloride in 0.9% sodium chloride injection, USP
Potassium chloride in 5% dextrose injections
20 mEq/L potassium chloride in 5% dextrose injection, USP
Potassium chloride in 5% dextrose and 0.2% sodium chloride injections
10 mEq/L potassium chloride in 5% dextrose and 0.2% sodium chloride injection, USP
20 mEq/L potassium chloride in 5% dextrose and 0.2% sodium chloride injection, USP
40 mEq/L potassium chloride in 5% dextrose and 0.2% sodium chloride injection, USP
Potassium chloride in 5% dextrose and 0.33% sodium chloride injections
20 mEq/L potassium chloride in 5% dextrose and 0.33% sodium chloride injection, USP
Potassium chloride in 5% dextrose and 0.45% sodium chloride injections
10 mEq/L potassium chloride in 5% dextrose and 0.45% sodium chloride injection, USP
20 mEq/L potassium chloride in 5% dextrose and 0.45% sodium chloride injection, USP
30 mEq/L potassium chloride in 5% dextrose and 0.45% sodium chloride injection, USP
40 mEq/L potassium chloride in 5% dextrose and 0.45% sodium chloride injection, USP
Potassium chloride in 5% dextrose and 0.9% sodium chloride injections
40 mEq/L potassium chloride in 5% dextrose and 0.9% sodium chloride injection, USP
20 mEq/L potassium chloride in 5% dextrose and 0.9% sodium chloride injection, USP
Potassium chloride in lactated Ringer's and 5% dextrose injections
20 mEq/L potassium chloride in lactated Ringer's and 5% dextrose injection, USP, VIAFLEX plastic container, 1000 mL
40 mEq/L potassium chloride in lactated Ringer's and 5% dextrose injection, USP

(*continued*)

Table 3-1 Examples of Commercially Available Large Volume Injections (*Continued*)

Ringer's and dextrose injections in VIAFLEX plastic containers
Ringer's and 5% dextrose injection, USP
Lactated Ringer's and 5% dextrose injection, USP
Sodium chloride injections
0.45% Sodium chloride injection, USP in VIAFLEX plastic container
0.9% Sodium chloride injection, USP. VIAFLEX plastic container, 150 mL
0.9% Sodium chloride injection, USP. VIAFLEX plastic container, 250 mL
0.9% Sodium chloride injection, USP. VIAFLEX plastic container, 500 mL
3% Sodium chloride injection, USP in VIAFLEX plastic container
5% Sodium chloride injection, USP in VIAFLEX plastic container
0.9% Sodium chloride injection, USP
0.9% Sodium chloride injection, USP. VIAFLEX plastic container, 1000 mL
Sodium chloride injections in mini-bag plastic containers
0.9% Sodium chloride injection, USP in VIAFLEX plastic container quad pack
0.9% Sodium chloride injection, USP in VIAFLEX plastic container single pack
0.9% Sodium chloride injection, USP in VIAFLEX plastic container multi pack

Source: From Ref. 1.

Nutritional Proteins

These are synthetic amino acids, ranging from 2.5% to 10% concentrations of a mixture of L-amino acids, nearly all of the 20 main types of amino acids. A wide variety of products are available and usage depends on patient situation (starvation, renal and/or hepatic failure) and level of stress (e.g., trauma, infection, degree of illness, and burns). Computers are used to calculate final formulation requirements.

Fatty (Lipid) Emulsions

Large volume emulsions serve as a source of nutrient fat for patients under parenteral nutritional therapy. Emulsions are composed of soybean oil (usually 10–20%), water (pH usually around 8), egg yolk phosopholipid (1.2%) that serves as the emulsifying agent/stabilizer, and glycerin (2.5%) for isotonicity adjustment.

Peritoneal Dialysis

Dialysis solutions require large volumes of glucose (dextrose) (0.5–4.25%) to remove waste such as urea and potassium from the blood, as well as excess fluid, when the kidneys are incapable of this (i.e., in renal failure). Peritoneal dialysis works on the principle that the peritoneal membrane that surrounds the intestine can act as a natural semipermeable membrane, and that if a specially formulated dialysis fluid is instilled around the membrane then dialysis can occur, by diffusion. Excess fluid can also be removed by osmosis, by altering the concentration of glucose in the fluid.

Irrigating Solutions

There are a variety of irrigating solution formulations, containing various components such as electrolytes and some organics (e.g., glutathione in BSS Plus ophthalmic irrigating solution). Irrigating solutions differ from injectable solutions with respect to the package closure. Injectable solutions are sealed with a rubber closure where the only entry point is through the rubber closure via a needle or injection spike. Irrigating solutions are closed with a screw cap that is twisted open just like a soda screw cap. Irrigating solutions, like injectable solutions, must be sterile, pyrogen, and particulate free.

INJECTION CATEGORIES

There are six main categories of injectable products:

1. Solutions ready for injection
2. Dry, soluble products ready to be combined with a solvent prior to use
3. Suspensions ready for injection
4. Dry, insoluble products ready to be combined with a vehicle prior to use

5. Emulsions
6. Liquid concentrates ready for dilution prior to administration.

SUSTAINED RELEASE INJECTABLE DELIVERY SYSTEMS

An explosion of advances and commercial successes in controlling and/or sustaining the delivery of injectable drugs has occurred in the past few years (2–10). Major technologies developed for injectable controlled release include primarily microspheres, implants, or hydrogels. For pharmaceutical protein controlled or sustained release, microsphere or hydrogel technologies are the most likely choices. These systems include classical microcrystalline suspensions (e.g., NPH or Lente insulin formulations), biodegradable microspheres, nondegradable implants, gel systems, pegylated protein formulations, and hyperglycosylated protein formulations. Sustained- or controlled-release injectable delivery systems are desirable for three main reasons:

1. Increased duration of release, reduced number of injections, and increased compliance
2. Localized delivery in the case of cancer therapy and vaccinations
3. Protection against in vivo degradation of the active ingredient.

Polymeric Implants

Polymeric implants are sterile, solid drug products manufactured by compression, melting, or sintering processes. The implant consists of the drug and a biodegradable or replaceable polymeric system, with the polymeric system generally being the rate-controlling key to sustained and prolonged drug delivery. Commercial examples of polymeric implants include

1. Norplant®—Levonorgestrel in silastic capsules deposited subdermally into the upper part of the arm within one week of the onset of menses. Drug delivery can last up to five years.
2. Duros®—A titanium cylindrical osmotic pump implanted in the upper arm that delivers drug for weeks to months. Viadur® is an example.
3. Gliadel® wafer—Polifeprosan plus carmustine are formulated with a biodegradable polyanhydride copolymer with the wafer being 1.45 cm in diameter and 1 mm in thickness. This wafer is implanted into the cavity created by a brain tumor resection with up to eight wafers (61.6 mg carmustine) implanted that provides up to three weeks of antineoplastic therapy.
4. Compudose®—composed of silicone rubber for subcutaneous estradiol implantation behind the ear of cattle.

Polymeric implants are difficult to manufacture, drug stability sometimes is questionable, and surgical procedures are required to implant and remove the device.

Microspheres

Microspheres are injectable suspensions containing particles of diameters of 1 to 100 μm and are supplied as dried powders. Prior to injection, the particles are mixed with an appropriate vehicle, dispersed, and administered. Release kinetics are controlled by polymer degradation and diffusion of the drug, and the duration can be adjusted from days to months.

Microsphere encapsulation involves rather harsh conditions that may involve high shear, organic solvents, or high temperatures. In addition, the encapsulated molecules will be exposed to high body temperature over extended periods of time. As a result of these processing requirements and potential stability issues, the technology was not thought to be appropriate for peptides and proteins, but indeed there are several commercial examples of long-acting microspheres containing peptides and proteins. An example of a peptide that has been encapsulated is leuprolide acetate, a synthetic nonapeptide analog of LHRH (leutenizing hormone-releasing hormone). The microencapsulated peptide is marketed as Lupron® Depot and is used for the treatment of advanced prostatic cancer. Reconstitution of the dried particles with vehicle results in a suspension that is administered intramuscularly at monthly intervals.

Another example is microencapsulated human growth hormone. By exploiting the stabilizing effect of zinc ion complexation and using a low temperature method for incorporation during encapsulation, degradable microspheres are prepared containing structurally intact human growth hormone. Various formulations and manufacturing processes have been published although a primary preparation technique is the double emulsion solvent evaporation method.

The polymeric systems used to fabricate drug-containing microspheres operate under at least five different mechanisms for sustained or controlled drug release.

1. Bioerodible release—The microsphere erodes layer-by-layer like an onion with equal amounts of drug localized within each layer. Bioerodible polymers include hydrophobic materials such as poly(ortho esters) with acid-labile linkages.
2. Biodegradable release—The microsphere erodes gradually as a whole (bulk erosion) with equal amounts of drug released per unit time. The most widely used biodegradable polymer is poly(glycolic acid-co-DL-lactic acid) copolymer. This polymer is most often and widely used because it is very safe (it is the component of surgical suture material), not phagocytosed by macrophages, and the ratios of polylactic acid and polyglycolic acid can be easily altered to change the rate of polymer degradation. Polylactic acid degrades over several years while polyglycolic acid degrades over several weeks. Other biodegradable polymers include poly(hydroxybutyrate), poly(hydroxyvalerate), polyanhydrides, collagen gels, dextran, albumin, and gelatin. Lupron® Depot and Atrigel® formulations use biodegradable technologies. Table 3-2 contains a partial listing of commercial formulations that use the lactide/glycolide biodegradable copolymer microsphere system.
3. Swelling-controlled release—The microsphere hydrates and swells with drug diffusing out of the polymer due to internal pressure produced by the swelling. There are dozens of swelling-controlled polymers including natural materials such as alginates, chitosans, collagen, dextrans, and gelatin and synthetic polymers such as cross-linked hydrophilic polymers like poly(2-hydroxyethylmethacrylate) and poly(*N*-isopropylacrylamide).
4. Osmotically controlled release—The microsphere consists of semipermeable membranes that swell, but do not burst. The drug is propelled out of the polymer through an orifice in the polymer produced by a laser.
5. Diffusion-controlled release—The microsphere permits constant diffusion of the incorporated drug through the polymeric membrane. Hydrophilic polymers such as hydroxypropyl cellulose or hyaluronic acid are examples of diffusion-controlling polymers. The SABER™ system from Southern BioSystems uses high viscosity polymers such as sucrose acetate isobutyrate to control drug diffusion from the microsphere.

Injectable gel formulations, such as Atrigel®, and other formulations containing natural materials such as alginates, chitosans, or collagens, rely on environmental changes, primarily temperature, to convert a subcutaneously injected liquid to a semisolid or solid depot. The

Table 3-2 Lactide/Glycolide Injectable Microsphere Extended Release Products

Product	Dosage form	Distributor	Active	Duration (mo)
Decapeptyl®	Microparticle	Ferring	Triptorelin acetate	1
Decapeptyl® SR	Microparticle	Ipsen-Beaufour	Triptorelin acetate	1, 3
Zoladex®	Implant	AstraZeneca	Goserelin acetate	1, 3
LupronDepot®	Microparticle	Takeda Pharma NA	Leuprolide acetate	1, 3, 4
Sandostatin LAR® Depot	Microparticle	Novartis	Octreotide acetate	1
Profact® Depot	Implant	Sanofi-Aventis	Buserelin acetate	2, 3
Suprecur® MP	Microparticle	Sanofi-Aventis	Buserelin acetate	1
Eligard®	Liquid	Sanofi-Aventis	Leuprolide acetate	1, 3
Luprogel®	Liquid	MediGene AG	Leuprolide acetate	1
Trelstar™ Depot	Microparticle	Watson	Triptorelin acetate	1
Trelstar™ LA	Microparticle	Watson	Triptorelin acetate	3
Arestin®	Microparticle	OraPharma	Minocycline HCl	0.5
Atridox®	Liquid	CollaGenex	Doxycycline hyclate	0.25
Risperdal® Consta™	Microparticle	J&J	Risperidone	0.5
SMARTShot B12	Microparticle	Stockguard	Vitamin B12	4, 8
Vivitrol®	Microparticle	Alkermes	Naltrexone	1
Revalon®-XS	Implant	Intervet	Trenbolone acetate/estradiol	6
Ozurdex™	Implant	Allergen	Dexamethasone	1.5–2

Source: Courtesy of Dr. Tom Tice, Surmodics, March 2010.

active pharmaceutical ingredient subsequently is slowly released as the polymer degrades. For Atrigel® formulations, a biodegradable polymer is dissolved in a biocompatible carrier. Biodegradable polymers include primarily poly(DL-lactide), lactide/glycolide copolymers, or lactide/caprolactone copolymers. Solvents used to dissolve these polymers include *N*-methyl-2-pyrrolidone (primary), polyethylene glycol (PEG), tetraglycol, glycofurol, triacetin, ethyl acetate, and benzyl benzoate. Indeed, any organic solvent used must be safe, biocompatible, water miscible, and easily used in a manufacturing environment.

Dextran-based microspheres encapsulate liposomes and proteins using an aqueous-based emulsion technique tailored for solvent-unsuitable drugs. ProMaxx® (Baxter-Epic) is based on completely aqueous systems to form well-controlled, uniform microspheres allowing high drug loading. Microspheres, containing the active and excipients such as dextran sulfate, hydroxyethyl starch, and albumin, are formed through patented adjustments of ionic strength, pH, active and polymer concentrations, and temperature. Promaxx® microsphere technology is unique because microspheres are manufactured without the need for organic solvents.

Other microsphere formulations meeting clinical or commercial success include Chroniject™, ProLease®, Medisorb®, and SABER™.

Some additional coverage of microspheres is found in chapter 9.

Liposomes

In recent years more liposomal formulations have been commercially available. Table 3-3 shows examples of marketed liposome products where the application of liposome technology has moved beyond formulations containing either doxorubicine or amphotericin. In 1995, Sequus marketed the first stealth liposome (Doxil). Stealth liposomes are nanoparticles with special polyethylene derivatives that allow the liposome to avoid detection by the reticuloendothelial system that normally would update these injected particles and minimize their circulation to the appropriate receptor sites. Earlier problems with economic and reproducible large-scale production of liposomes have been largely solved.

Liposomal-based technologies have been used to deliver genetically engineered, nonviral plasmids across cellular barriers that target brain cancer. This is also called RNAi (RNA interference) technology that inhibits a growth factor responsible for keeping cancer cells alive.

Other examples of liposome technology—Pacira's multivesicular liposome formulation (DepoFoam™), Neopharm's NeoLipid™, and Genzyme's Lipobridge™. DepoFoam™

Table 3-3 Examples of Commercial Injectable Liposome Products

Drug product	Drug substance	Delivery matrix[a]	Other excipients	Delivery technology
Abelcet® (Enzon)	Amphotericin B	DMPC, DMPG	Sodium chloride	Lipid complex
AmBisome® (Astellas)	Amphotericin B	HSPC, cholesterol, DSPG	Vitamin E, disodium succinate hexahydrate	Liposome
Amphotec®	Amphotericin B	Cholesterol sulfate	Tromethane, disodium EDTA, lactose (lyophilized powder)	Colloidal dispersion
DepoCyte® (Pacira/Enzon)	Cytarabine	DOPC, DPPG, cholesterol, triolein	Triolein	Liposome
DepoDur (Pacira/EKR)	Morphine	DOPC, DPPG, cholesterol	Tricaprylin, triolein	Liposome
Doxil® (Ortho Biotech)	Doxorubicin	MPEG-DSPE, HSPC, cholesterol	Ammonium sulfate, histidine	Stealth liposomes
Visudyne® (Novartis)	Verteporfin	Ascorbyl palmitate, BHT, DMPC	Lactose	Liposome

[a]DMPC, 1,2-dimyristoyl-sn-glycero-3-phosphocholine, DMPG, 1,2-myristoyl-sn-glycero-3[phospho-rac-(1-glycerol...)]; HSPC, fully hydrogenated soy phosphatidylcholine; DSPG, 1,2-distearoyl-sn-glycero-3[phospho-rac-(1-glycerol...)]; DOPC, 1,2-dioleoyl-sn-glycero-3-phosphocholine; DPPG, 1,2-dipalmitoyl-sn-glycero-3[phospho-rac-(1-glycerol...)]; MPEG-DSPE, *N*-(carbonyl-methoxy PEG 2000)-1,2-distearoyl-sn-glycero-3-phosphoethanolamine: liposome coating that keeps immune system from recognizing liposome as a foreign body.

technology includes at least two marketed products—DepoDur for controlled release of morphine and DepoCyt, an intrathecally injected sustained release anticancer product.

Some further coverage of liposomes as a dispersed pharmaceutical system is found in chapter 9.

REFERENCES

1. http://www.ecomm.baxter.com/ecatalog/browseCatalog.do?lid=10001&hid=10001&cid=10016&key=a17fb6d9dd83be5836e1adffd2d249f.
2. Brown L, Qin Y, Hogeland K, et al. Water-soluble formation of monodispersed insulin microspheres. In: Svenson S, ed. Polymeric Drug Delivery II, Polymeric Matrices and Drug Particle Engineering, ACS Symposium Series 924. : Washington, D.C.: American Chemical Society, 2006; chap 22.
3. Dunn R. Application of ATRIGEL® implant drug delivery technology for patient-friendly, cost-effective product development. Drug Deliv Technol 2003; 3:38–43.
4. Harrison RC. Development and applications of long-acting injection formulations. Drug Deliv Technol 2006; 6:36–40.
5. Johns G, Corbo G, Thanoo BC, et al. Broad applicability of a continuous formulation process for manufacture of sustained-release injectable microspheres. Drug Deliv Technol 2004; 4:60–63.
6. Martini A, Lauria S. Sustained release injectable products. Am Pharm Rev 2003; 6(Fall):16–20.
7. Morar AS, Schrimsher JL, Chavez MD. PEGylation of proteins: A structural approach. BioPharm Int 2006; 19(4):34–49.
8. Reimer D, Eastman S, Flowers C, et al. Liposome formulations of sparingly soluble compounds: Liposome technology offers many advantages for formulation of sparingly soluble compounds. Pharmaceutical Formulation and Quality. August/September 2005:42–44.
9. Tice T. Delivering with depot formulations. Drug Deliv Technol 2004; 4:44–47.
10. Verrijk R, Gos B, Crommelin BJA, et al. Controlled release of pharmaceutical proteins from hydrogels. Pharm Manuf Packing Sourcer 2003 (Summer), 27–33.

4 Sterile product packaging systems

This chapter deals with sterile product container systems, both conventional and more advanced systems. In chapter 7, more attention is devoted to the specific chemical and physical properties of glass, rubber, and plastic and issues surrounding extractables and leachables. Also packaging systems with respect to container/closure integrity testing are discussed in chapter 30.

STERILE PRODUCT CONTAINER SYSTEMS

There are six basic primary packaging or container systems:

1. Ampoules—glass
2. Vials—glass and plastic
3. Prefilled syringes—glass and plastic
4. Cartridges—glass
5. Bottles—glass and plastic
6. Bags—plastic

Generally, vials comprise about 50% of small volume injectable packaging, syringes 30% and ampoules 10%, and cartridges and bottles/bags filling the rest.[1] Usage of all packaging types, except ampoules, are trending upward, especially prefilled syringes. Each of these packaging systems for parenteral drug delivery has significant advantages and disadvantages. Generally, advantages involve user convenience, marketing strategy, handling during production and distribution, volume considerations, and compatibility with the product. The primary disadvantage with all these packaging systems is the potential reactivity between the drug and other ingredients in the formulation (e.g., antimicrobial preservatives) and the packaging components. The reactivity is typically manifested through the appearance of particulate matter, detection of extractables, evidence of protein aggregation, and other physical and chemical incompatibilities.

This chapter covers each of these primary packaging systems, advances in primary packaging for special delivery systems, and needle technology.

Selection of the packaging system not only depends on compatibility with the product formulation and the convenience to the consumer, but also on the integrity of the container/closure interface to ensure maintenance of sterility throughout the shelf-life of the product. Container/closure integrity testing has received significant attention and usually is an integral part of the regulatory submission and subsequent regulatory good manufacturing practice (GMP) inspections. While it is beyond the scope of this chapter to discuss the various container/closure integrity testing methods (these methods are covered in chap. 30), it is emphasized that formulation scientists developing the final product including the final package must appreciate the need to develop appropriate methods to ensure that the selected packaging system possesses the proper seal integrity to protect the product during its shelf-life from any ingress of microbiological contamination.

Ampoules

For decades, glass-sealed ampoules (Fig. 4-1) were the most popular primary packaging system for small volume injectable products. To the formulator, ampoules offer only one type of material (glass) to worry about for potential interactions with the drug product compared with other packaging systems that contain both glass or plastic and rubber.

Two disadvantages of glass ampoules are the assurance of the integrity of the seal when the glass tip is closed by flame and the problem of glass particles entering the solution when the ampoule is broken to remove the drug product. There exist "easy-opening ampoules,"

[1] Based solely on author's experience and perception.

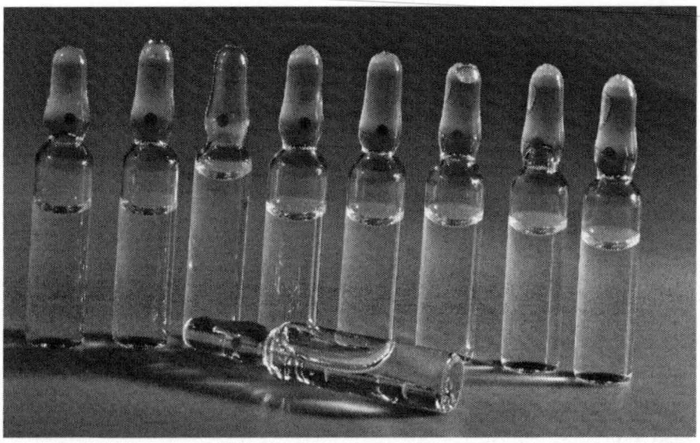

Figure 4-1 Glass-sealed ampoules.

weakened at the neck by scoring or applying a ceramic paint with a different coefficient of thermal expansion, that permit the user to break off the tip at the neck constriction without the use of a file. Nevertheless, it is the glass particle problem and the need for use of a filter to withdraw product from the ampoule that caused the American Society of Hospital Pharmacists (now called American Society of Health-System Pharmacists) in the late 1980s to appeal to the pharmaceutical industry not to use ampoules for any new sterile product. Glass-sealed ampoules still exist, but they are not the choice for new products in America. Elsewhere in the world, ampoule products are still widely used and still a popular package of choice for new sterile product solutions.

Glass ampoules are Type I tubing glass (see chap. 7 for further elaboration of Type I and tubing glass) in sizes ranging from 1 to 50 mL. After solution is filled into the top opening of the ampoule, the glass is heat sealed by one of two techniques—tip sealing or pull sealing. Tip sealing has the open flame directed toward the top of the ampoule that melts and seals itself while the ampoule is rotating on the sealing machine. Pull sealing has the open flame directed at the middle of the portion of the ampoule above the neck where the glass is melted while rotating and the top portion is physically removed during rotation. Thus the tip-sealed ampoule has a longer section above the neck while the pull-sealed ampoule has a more blunt, 'fatter" top.

Modifications of ampoules are available, for example, wide-mouth ampoules with flat or rounded bottoms to facilitate filling with dry materials or suspensions.

Vials

The most common packaging for liquid and freeze-dried injectables is the glass vial (Fig. 4-2). Plastic vials have made some ingress as marketed packages for cancer drugs, but may require

Figure 4-2 Glass vials with rubber closures and aluminum seal.

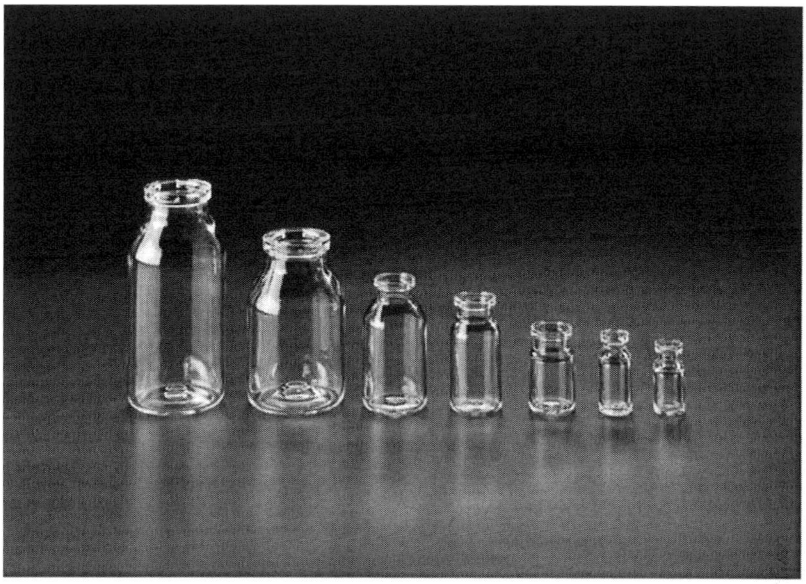

Figure 4-3 Plastic Vials—Daikyo Crystal Zenith Cyclic Polyolefin (Daikyo Crystal Zenith® is a registered trademark of Daikyo Seiko, Ltd). *Source*: © 2010 by West Pharmaceutical Services.

some time before being commonplace in the injectable market. Plastic vials are made of cyclic olefin polymer (COP) or cyclic olefin copolymer (COC). The appearance of a plastic vial looks identical to a glass vial (Fig. 4-3).

The main reason why plastic vials have not become as commonplace as glass vials is associated with the ease of introducing the container into a classified (ISO 5) aseptic environment. Glass vials are sterilized and depyrogenated in dry heat tunnels that convey the vials directly into the aseptic environment without the need for manual transfer. Plastic vials are presterilized (typically irradiation) at the vial manufacturer and the finished product manufacturer needs to figure out how to aseptically transfer plastic vials into the aseptic environment. This is not an easy solution, especially compared to the convenient way glass vials are introduced via the dry heat tunnels.

Two other potential disadvantages of plastic vials are: (1) challenges in handling and movement of much lighter weight containers compared with glass along conveyer systems on high-speed filling lines, with smaller vials (1–5 mL) especially difficult to process; and (2) concerns about potential interactions with the drug product (absorption, adsorption, migration, leachables) especially over a two to three year shelf-life.

Manufacture of glass vials, either tubing vials or blow-molded vials, is covered in chapter 7. Vial openings are 13, 20, or 28 mm.

Syringes

Syringes are very popular delivery systems and growing in market share more than any other injectable primary packaging system (1–5). They are used either as empty sterile container systems where solutions are withdrawn from vials into the empty syringe prior to injection or as prefilled syringes (Fig. 4-4). Prefilled syringes can be presterilized by the empty syringe manufacturer or can be cleaned and sterilized by the finished product manufacture. Other options regarding syringe size, components, formats, treatment of rubber materials, and manufacturing methods are summarized in Table 4-1. Most of the world's vaccines are packaged and delivered in syringes. The growth rate for products filled and packaged in prefilled syringes increases about 13% per year (6). This growth is related to the top factors that influence a physician's choice of a drug delivery type, including the ease of use by patients, convenience, and comfort.

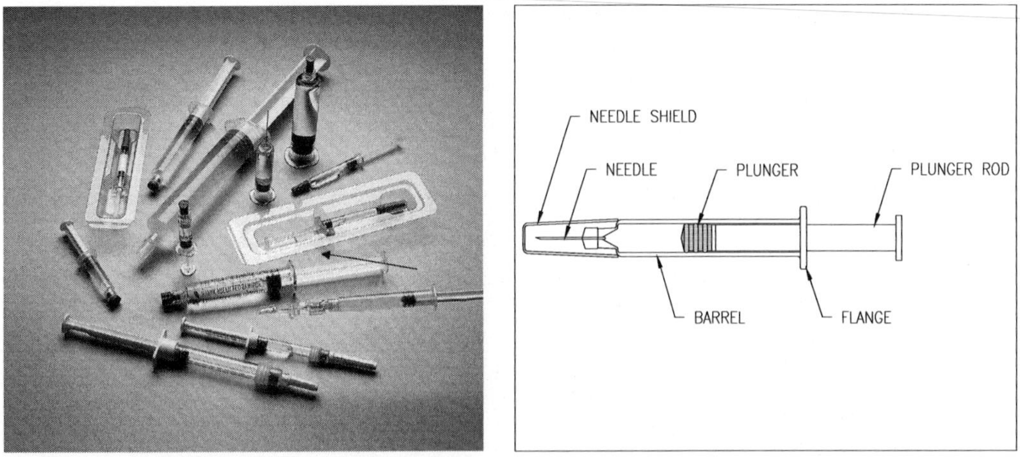

Figure 4-4 Syringe examples. *Source*: Courtesy of Baxter Healthcare Corporation.

Primary reasons for syringe popularity include the following:

- The emergence of biotechnology and the need to eliminate overfill (reduced waste) of expensive biomolecules compared with vials and other containers. Vaccines, antithrombotics, and various home health care products such as growth hormone and treatments for rheumatoid arthritis and multiple sclerosis are much more conveniently used and administered using prefilled syringes

Table 4-1 Prefilled Syringe Options

Sterilization	Presterilized by empty syringe manufacturer and ready-to-fill Supplied nonsterile, washed and sterilized by product manufacturer
Barrel size	0.5–100 mL; typically 0.5–10 mL
Needle format	Luer tip, use needle of choice Staked needle affixed to syringe Hub, not used often
Needle gauge	21–32
Needle length	$^1/_2$ to $^5/_8$ in.
Needle shield	Natural or synthetic rubber
Silicone application	Silicone oil or silicone emulsion Applied at syringe manufacturer Applied at finished product manufacturer
Silicone level	Varies, 0.6–1.0 mg per 1 mL syringe
Type of rubber plunger	Synthetic rubber (halobutyl)
Type of rubber septum (tip)	Natural or synthetic rubber Plastic covers
Coating of rubber	Absent or use of fluoropolymer
Filling machine	Rotary piston Peristaltic Time pressure Rolling diaphragm Single head up to 10 heads Up to 600 syringes filled per minute
Rubber plunger insertion	Insertion tube system Vacuum

- Availability of enormous (millions) quantities of presterilized ready-to-fill syringes such as BD Hypak® SCF and BunderGlas RTF
- The advent of contract manufacturers specializing in syringe processing with lower costs and high-speed filling equipment
- Elimination of dosage errors because unlike vials, syringes contain the exact amount of deliverable dose needed
- Ease of administration because of elimination of several steps required before injection of a drug contained in a vial. Because fewer manipulations are required, sterility assurance is increased
- More convenient for health care professionals and end users; easier for home use; easier in emergency situations
- Reduction of medication errors, misidentification; better dose accuracy
- Better use of controlled drugs such as narcotics
- Lower injection costs—less preparation, fewer materials, easy storage, and disposal
- Elimination of vial overfill for products transferred to syringes for direct injection or addition to primary diluents.

Syringe barrels can either be glass or plastic while syringe plunger rods are usually plastic. Plastic polymers for the syringe barrel include polypropylene, polyethylene, and polycarbonate. However, newer technologies are being developed in the area of "glass-like" composite materials.

Syringes with needles may also have needle protectors (Fig. 4-5) to avoid potential dangers of accidental needle sticks postadministration. Such protectors either can be part of the assembly or can be assembled during the finishing process. The use of these protection devices is increasing due to the 2001 United States Federal Needle Stick Safety and Prevention Act (7). Needle stick prevention can be manual (shield activated manually by the user although there can be risk of accidental sticking), active (automated needle shielding activated by user), or passive (automated needle shielding without action by the user).

Issues that must be addressed in selecting and qualifying components of a syringe include

- Container/closure integrity testing
- Plastic component extractables
- Sterilizability, especially if needle is part of the package to be sterilized
- Siliconization of barrel and plunger (although silicone-free syringes now exist that provide both lubricity and inert drug-contact surfaces)
- Compatibility of product with syringe contact parts, especially the rubber plunger

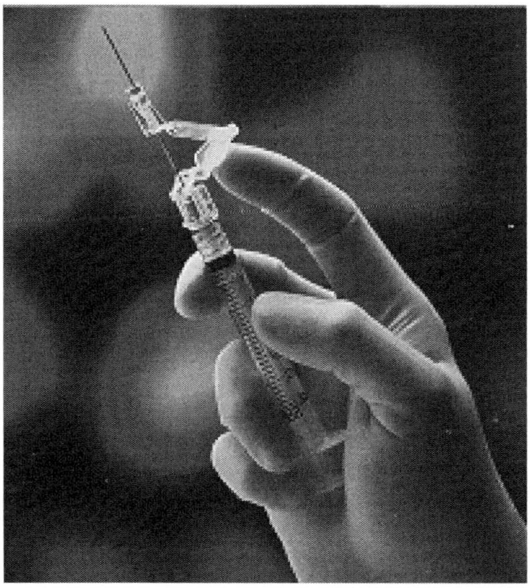

Figure 4-5 Syringe with needle guard. *Source*: Courtesy of © Becton, Dickinson and Company.

- Appropriate gauge size of needle for product and its indication within the syringe. Syringe needle gauges range from 21 to 32 gauge (G). It is important to note that some suspensions may not syringe properly if the needle gauge is not carefully considered.

One of the most challenging aspects of syringe quality control is the assurance of container/closure integrity during and after filling and terminal sterilization (chap. 30).

Siliconization Issues with Syringes

Like rubber closures, syringes require a "slippery surface." Rubber requires such a surface for facile movement of closures along the stainless steel tracks of a rubber closure hopper or feeding machine to deposit the rubber on top of a container at a rate of hundreds per minute. Without the slippery surface, rubber closures would move haltingly, if at all, and filling at any speed could not be accomplished. For syringes, the rubber plunger must move easily within the syringe barrel with the "glide force" being the same throughout the barrel (from distal to proximal end).

There are several concerns related to siliconization of syringes—functionality, potential for protein aggregation, and increased potential for particulate matter. Syringe functionality involves forces both to initiate movement of the plunger rod within the syringe barrel and to maintain movement of the plunger rod throughout the barrel to the end of the syringe. Siliconization significantly facilitates both forces. However, excess silicone is a problem from a physical stability standpoint both with respect to visible appearance of silicone droplets in the product and greater potential for protein interaction with these hydrophobic droplets. Therefore, great effort is made by syringe manufacturers to minimize the amount of silicone applied within the inner surface area of the syringe. However, sometimes not all the inner surface of the barrel is coated with silicone. This will potentially lead to an effect called "chattering" where the syringe barrel will "stick" and require greater force to make it move again. This may not be a problem with manual injections where the health care professional or the patient giving self-injections will simply apply more pressure with the fingers to overcome the lack of siliconization. However, if autoinjectors are used, sometimes the spring or compressed gas force will be insufficient and incomplete delivery of medication will occur.

The FDA added a requirement for functionality testing as part of long-term stability testing of drug products contained in syringes and cartridges because of the possibility of inadequate/incomplete siliconization of syringes resulting in potential inadequate/incomplete drug delivery (8). Articles are being published about technologies that apply optical techniques such as confocal Raman spectroscopy, Schlieren optics, and thin film interference reflectometry to visualize and characterize (in situ morphology, thickness, and distribution) of silicone oil in prefilled syringes (9). The articles demonstrate that these techniques show that uneven distribution of silicone oil within syringe glass barrels as potential sources of chattering and stalling of the syringe plunger during injection using autoinjectors.

Syringe siliconization raises the potential for protein aggregation. This is a primary driver for plastic syringes perhaps becoming more popular for use with biopharmaceutical products because the plastic surface does not require silicone for facile movement of the rubber plunger and plunger rod through the plastic barrel. Manufacturers of plastic syringes have developed alternatives to silicone to provide lubricity within the plastic composition of the syringe to achieve acceptable functional performance. Studies have been published that implicate silicone as the cause of turbidity and particle formation in insulin products (10) and other protein products (11). Until plastic syringes without the presence of silicone become more common, continuous improvements in the consistent application and distribution of silicone in syringe barrels must be pursued.

Siliconization also increases the potential for increased particulate matter, either real or the fact that electronic particle counters detect a silicone droplet as a particle. Thus, products in syringes could experience higher levels of particles as measured by light obscuration compared with the same product in a vial. Typically, the levels of particulate matter for syringes still fall way below the required limits for subvisible particles as defined by the United States Pharmacopoeia (USP) General Chapter <788>. However, if the USP ever decides to require

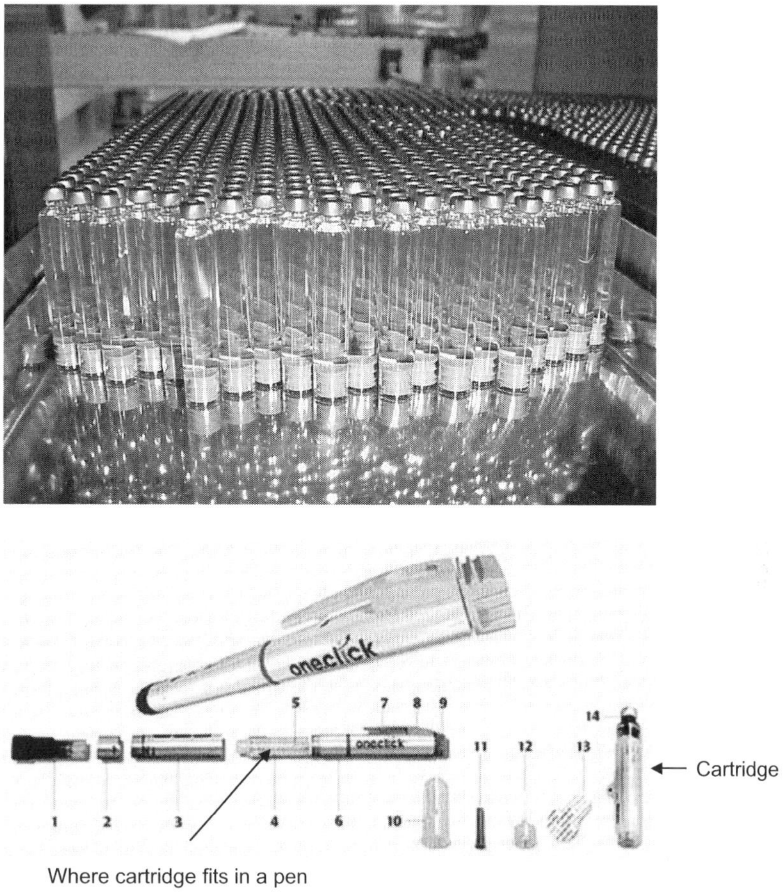

Figure 4-6 Cartridges and Saizen® pen example. *Source*: Saizen® courtesy of Merck Serono.

measurement of particles less than the current lowest level of 10 μm, then particle levels might be much higher for syringe products due to the presence of silicone droplets in the range of <10 μm.

Cartridges

Cartridges are similar to syringes with respect to having a product filled into a glass tube closed on either side by a rubber plunger and a rubber disk seal and inserted into a delivery pen device (Fig. 4-6).

Cartridge/pen delivery systems are used primarily for multiple dose proteins such as insulin and growth hormone. Advantages of cartridge/pen delivery systems include dose accuracy and patient convenience while its disadvantages are slight increased costs unless the pen system is subsidized.

Cartridges were used for years in the dental field, but did not grow markedly until insulin was manufactured in a cartridge and delivered in a specialized pen. Pens are the predominant insulin delivery system in most countries of the world, except in the United States, where syringes and insulin vials still dominate (12). Some pens use replaceable insulin cartridges while some pens use a nonreplaceable cartridge and are disposed of after use. All pens use replaceable needles. Most pens use special pen needles that can be extremely short and thin (28–33 G). Cartridges in delivery pens offer repeatability in dosing accuracy compared with syringes. Also, because dosing with a pen involves dialing a mechanical device and not looking at the side of a syringe, insulin users with reduced visual acuity can be assured of accurate dosing with a pen.

Bottles

Bottles typically refer to containers larger than 100 mL; thus, large volume injectable solutions or emulsions are contained in bottles (or bags) rather than vials. Bottles are manufactured by the blow-molded process. Bottles can be glass or plastic, both are commonly used in hospital pharmacy practice.

Bags

Bags used for IV fluids include prefilled or empty containers that range in size from 25 mL to greater than 1 L. Sizes that are 1 L or greater are often used in hospital settings for delivery of total parenteral nutrition. Bags of all sizes are often used for ease of delivery and ease of transport. However, maintaining identification of the bags can be a problem. Printing on plastic bags is a challenge because of the flexibility of the bag material and labels adhered to the bags can become difficult to read. This was mostly resolved by the introduction of bar coding that allows traceability of bags from filling to patient use. Compatibility issues between the bag polymer and the drug solution have plagued the industry over the years. Polyvinyl chloride (PVC) was the polymer material of choice for many years because of the important collapsibility characteristic of PVC. However, PVC was notorious for leaching a plasticizer used to add flexibility, that material being di(2-ethylhexyl) phthalate (DEHP). Since the Environmental Protection Agency classified DEHP as a probable human carcinogen (13,14), governments and industry have labored to provide a similar type of bag material that is non-PVC, typically mixtures of polyalkenes (polyethylene and polypropylene).

Plastic bags are manufactured by form-fill-finish processes where strips of plastic polymer are sealed on three sides, solution filled into the "pouch," then the bag is sealed with the fourth side that contains the spike and needle outlets.

NEEDLES

Historically, stainless steel needles have been used to penetrate the skin and introduce a parenteral product inside the body. The advent of needleless injection systems has obviated the need for the use of needles for some injections (e.g., vaccines) and is gaining in popularity over the conventional syringe and needle system. However, needleless injections are generally more expensive, can still produce pain on injection, are potentially a greater source of contamination (and cross-contamination from incessant use), and may not be as efficient in dose delivery.

Needles are hollow devices composed of stainless steel or plastic. Needles are available in a wide variety of lengths, sizes, and shapes. *Needle lengths* range from $^1/_4$ in to 6 in. *Needle size* is measured both in length (usually inches in the United States; centimeters in the rest of the world) and gauge. Needle gauge includes both internal or inner diameter (ID) and external or outer diameter (OD) of the needle. The larger the gauge, the smaller the diameter. For example, a 21 G needle has an ID of 510 μm and an OD of 800 μm. Table 4-2 provides a listing of ID and OD lengths as a function of gauge.

The ID is important especially for dispersed system formulations containing insoluble particles suspended in a vehicle and for highly viscous formulations (usually viscosity greater than 4 centipoise). Obviously the smaller the ID, the potential greater difficulties encountered in needles clogging due to bridging of particles or insufficient force per unit area to eject viscous solutions. The term "syringeability" is an important property to consider in determining what needle ID gauge can be used for suspensions and viscous solutions. Syringeability is covered in chapter 9.

The OD is important for the obvious reason of the potential degree of discomfort, pain, and tissue irritation when the needle penetrates the skin. The smallest possible gauge needle is always used as long as the product can be easily ejected from the syringe or other delivery device into the appropriate bodily location. For deep intramuscular (IM) injections, typically (and unfortunately for the patient) a long (1.5–2 in) needle of a typical gauge of 18 to 20 must be used. For subcutaneous (SC) injections requiring injection of very small volumes of drug product, a short ($^1/_4$ to $^1/_2$ in) high gauge (27–33) needle can be used, causing a minimal amount of pain or discomfort.

Table 4-2 Inner and Outer Diameters of Various Needle Gauge Sizes

Gauge	Outer diameter (μ)	Inner diameter (μ)
11	3048	2388
12	2769	2159
13	2413	1803
14	2108	1600
15	1829	1372
16	1651	1194
17	1473	1067
18	1270	838
19	1067	686
20	908	603
21	819	514
22	717	413
23	641	337
24	565	311
25	514	260
26	463	260
27	412	210
28	362	184
29	336	184
30	311	159
31	260	133
32	235	108

Gauge ranges are 11 to 32 G in practice (there are smaller and larger gauge sizes, but uncertain of their application for injectables) with the largest gauge for injection usually being no greater than 16 G. Sixteen gauge needles have an OD of 0.065 in (1.65 mm) while 32 G have an OD. of 0.009 in (0.20 mm). *Needle shape* includes regular, short bevel, intradermal, and winged. Needle shape typically is defined by one end of a needle enlarged to form a hub with a delivery device such as a syringe or other administration device. The other end of the needle is beveled, meaning that it forms a sharp tip to maximize ease of insertion. *Bevels* can be standard or short (Fig. 4-7).

The route of administration, type of therapy, and whether the patient is a child or adult dictate the length and size of needle used. Intravenous injections typically use 1 to 2 in 15 to 25 G needles. IM injections use 1 to 2 inch 19–22 G needles. SC injections use $^1/_4$ to $^5/_8$ in 24 to 25 G needles. Needle gauge for children rarely is larger than 22 G, usually 25 to 27 G. Winged needles are used for intermittent heparin therapy. Many different types of therapies (e.g., radiology, anesthesia, biopsy, cardiovascular, ophthalmic, transfusions, and tracheotomy) have their own peculiar types of needle preferences. Needles are purchased either alone (e.g., Luer-Lok) to be attached to syringes, cartridge, and other delivery systems, or, for syringes, can be part of the syringe set (stake needle).

Microneedles have been designed, called "proboscis-mimicking" microneedles, to mimic a mosquito's proboscis in dimensions where the OD is only 60 μ. These needles are composed of

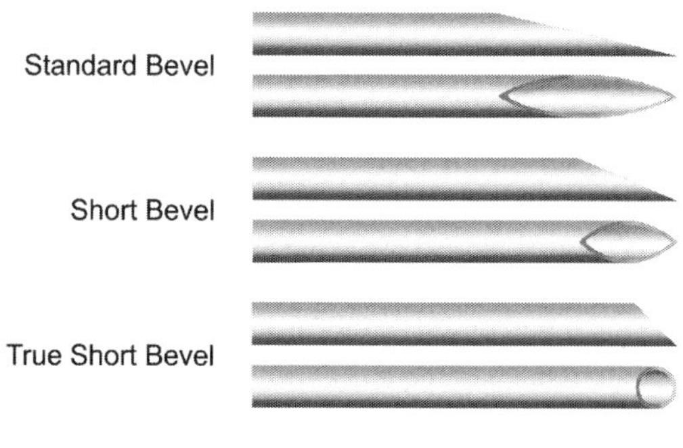

Figure 4-7 Needles with different bevels. *Source*: Courtesy of Richard Wheeler (http://en.wikipedia.org/wiki/File:NeedleBevels.svg).

titanium and related alloys, rendering them sufficiently strong to administer medicines without the risk of breaking. These needles are used for intradermal and SC injections, penetrating up to 3 mm beneath the skin.

NEEDLELESS INJECTORS

For SC injections, needleless injection systems have become popular. Because of the great fear and frequent discomfort of using needles to inject parenteral products, needleless injection systems offer an alternative delivery mode for SC parenteral therapy. Besides eliminating the fear factor of needle injectors, needle-free administration adds inherent compliance to Sharps laws.

IM and SC routes are possible with needle-free injectors depending on the pressure and orifice size. The effect of pressure on protein stability must be determined and controlled.

Needleless injection systems can be reusable or disposable. Manufacturers of reusable needle-free devices include Bioject, Antares, The Medical House, and others. Manufacturers of disposable needle-free devices include Bioject, Valeritas, and Zogenix. Pressurized gas (e.g., nitrogen) provides sufficient power to drive the drug product through the skin into the SC tissue. Needleless injections can also be powered by spring-loaded devices. Typically, injection volumes with needleless injectors are less than 1.0 mL. In some cases, needleless injections can be more painful than using needles, but the fear factor of "seeing" the needle penetrate the skin is eliminated. Vendors claim that pain experienced with needle-free injectors is no more than that experienced with a 27 to 30 G needle. Other possible side effects of needle-free injections include bruising and bleeding, depending on the injection site and the individual receiving the injection.

While the concept behind needle-free injection has been around for decades, it has only been recently, with the convergence of synthetic materials and computerized design software that reliable and cost-effective devices have begun to appear. Many vendors manufacture devices that depend on compressed carbon dioxide or nitrogen gas or on a mechanical spring to eject drug product from device to tissue subcutaneously. The pressure required supposedly does not damage proteins. There is some pain, bruising, and bleeding, depending on person and the injection site. Needle-free injection systems do not present an advantage with respect to patient discomfort accompanying injection, because the needle-free system causes greater dispersion of the formulation within tissue at the injection site than injection with a needle and syringe. Also, costs are higher than needle/syringe delivery.

PACKAGING ADVANCES

Advances related to sterile dosage forms have been concentrated in these areas (4,15–17):

1. Prefilled syringes
2. Use of plastics
3. Reducing or eliminating the use of silicone
4. More user-friendly packaging systems for home health care and self-administration of injectable drugs—next section
5. Reducing the level of leachable substances.

As previously emphasized, *prefillable syringes* have and will experience the greatest growth in the marketplace based on infection prevention and response time advantages in the delivery of critical and emergency care medication. Vials, ampoules, and intravenous containers will generate below average demand gains, with competition from prefillable syringes holding back growth for vials and ampoules. Trends toward less invasive surgical procedures and advances in alternative drug delivery systems will soften market growth for IV containers.

Use of *plastic packaging* for vials and syringes is increasing. Plastics have an obvious advantage of eliminating the risk and consequences of broken glass, especially if containing cytotoxic drugs. Plastics also eliminate concerns regarding glass delamination and alkali leachates. Furthermore, plastic syringes may offer advantages for proteins because of less surface adsorption and for hydrophobic proteins because of little to no silicone coating compared with glass syringes. Disadvantages of plastic containers include aseptic transfer issues of introducing presterilized containers into classified production environments and the potential for product–package interactions. There is less compendial standardization for plastics used in parenteral product packages compared with glass. Thus, subtle changes in plastic composition, vendors for plastic components, and/or manufacture of those components may cause new problems, requiring extra time and effort to study these changes.

Silicone historically has been required to provide surface lubricity for rubber closures and cylindrical containers such as syringes and cartridges. Silicone facilitates ease of movement of closures in stoppering equipment and plungers to glide smoothly through syringes and cartridges. While silicone on glass typically is "baked" (chemically bonded) onto glass surfaces during dry heat sterilization/depyrogenation, there still exist trace amounts of "free" silicone that can interact with hydrophobic protein domains, causing insoluble aggregate formation. Silicone on rubber is not chemically bonded to the rubber surface and can be more of a problem interacting with formulation components.

Advances in rubber closure technologies have introduced closures that do not require siliconization because of a special polymer coating applied to the outer surface of the closure. Examples are the Daikyo/West closures (FluroTec®) and the Helvoet (Omniflex®) closures. The Daikyo/West FluroTec® is a laminated stopper containing a coating of copolymer film of tetrafluoroethylene and ethylene (ETFE). The Omniflex® stopper is coated with a mixture of polyethylene and tetrafluoroethylene (PTFE) film. These coated stoppers offer the following advantages compared with stoppers that must be siliconized

- Eliminates the need for adding silicone oil
- Provides lubricity for machinability
- Reduces rubber stopper clumping problems
- Decreases particulate matter levels
- May reduce potential for formulation adsorption and absorption
- Reduces chemical extractable levels.

Plastic containers, such as RESIN CZ® resin (Daikyo Seiko), can be combined with FluoroTec®-coated stoppers to produce a silicone-free syringe.

Most traditional plastic materials such as PVC and polyethylene have the disadvantage that they are not as transparent as glass and, therefore, inspection of the contents is impeded. However, recent technologies have overcome this limitation, evidenced by plastic resins such as CZ® (polycyclopentane, Daikyo Seiko) and Topas® COC (Ticona). In addition, many of these materials will soften or melt under the conditions of thermal sterilization. However, careful selection of the plastic used and control of the autoclave cycle have made thermal sterilization of some products possible, large volume parenterals in particular. Ethylene oxide or radiation sterilization may be employed for the empty container with subsequent aseptic filling. However, careful evaluation of the residues from ethylene oxide or its degradation products and their potential toxic effect must be undertaken. Radiation sterilization also carries the risk of discoloration of the plastic. Investigation is required concerning potential interactions between the formulation and the plastic surface and other problems that may be encountered when a parenteral product is packaged in plastic.

PACKAGING ADVANCES INVOLVING CONVENIENT INJECTABLE DRUG DELIVERY

The injectable drug product market has significantly grown, largely because of biotechnology that can produce protein medicines to treat both acute and chronic diseases such as diabetes, growth hormone deficiencies, multiple sclerosis, osteoporosis, rheumatoid arthritis, anemia, hemophilia, cancer, stroke, and many other disease states. Such growth has promulgated the need for improvements and advances in the ease of use and convenient delivery of injectable drug products, especially those self-administered. Because of market demands for more "user-friendly" injectable delivery systems and the consistent need to reduce costs and wastes, several advances in convenient injectable drug delivery packaging devices have been marketed (see Table 4-3 for a listing of many examples of injectable drug packaging advances with some of these systems discussed in the following text).

Table 4-3 Examples (Not Exhaustive) of Novel Parenteral Packaging Systems

- ADD-Vantage® vial/flexible container admixture system (Abbott Laboratories)
- Heparin dextrose double bag (Baxter)
- Inter-Vial® (Duoject)
- Vari-Vial® (Duoject)
- Clip'n'Ject (West)
- Bio-Set™ transfer system (Baxter)
- Viringe™ vascular access flush device (Avitro)
- Pen delivery systems
  - PEG-Intron® Redipen (disposable)
  - Roferon®-A (reusable)
  - Puregon® (follicle stimulating hormone, reusable)
  - Gonal-f® (follicle stimulating hormone, disposable)
  - Human growth hormone systems (Table 4-4)
  - Human insulin systems, e.g.
    - Optipen® (Insulin glargine, Aventis) (reusable)
    - Optiset® (Insulin glargine, Aventis) (prefill)
    - Humalog® (Insulin analog, Lilly) (disposable)
    - Humulin® (Lilly) (prefill)
    - Novopen® (Novo-Nordisk) (metal reusable)
    - Novomix® (Novo-Nordisk) (prefill)
    - Innolet® (Novo-Nordisk) (plastic reusable)
    - Innovo® (Novo-Nordisk) (plastic prefill)
  - Byetta®
  - Forteo™
- Auto injector systems
  - AutoJect 2® (Copaxone, Glatiramer acetate) (reusable)
  - Betaject Light® (Betaseron, Interferon Beta-1b) (reusable)
  - Rebiject II™ (Rebif, Interferon Beta-1a) (reusable)
  - SimpleJect™ (Kineret™, Anakinra) (reusable)
  - EpiPen®
  - See disposable pen manufacturer websites—e.g., Ypsomed, The Medical House, Scandinavian Health Limited, Owen-Mumford

Prefilled syringes and cartridges have already been discussed. Indeed, the market trend continues to be prefilled syringes for liquid stable products, outpacing new products in any other packaging system. For liquid-unstable products, combination systems, for facilitating transfer of lyophilized drug powders in vials to syringes or bags, are systems that easily connect a diluent syringe to a lyophilized drug vial or connect one vial to another (discussed further in next section). For multidose liquid or suspension products, pen and/or autoinjector delivery systems, either single use or reuseable, have gained significantly in popularity (Table 4-3). In fact, many recent new product launches have occurred using only these type of delivery systems (e.g., Forteo® and Byetta®).

Delivery systems for human growth hormone provide best market example of the use of packaging (device) injection delivery advances to meet changing patient/customer needs and gain some advantage over competition. At the time of this writing, there were five manufacturers of human growth hormone products—Genentech, Lilly, Novo-Nordisk, Pfizer, and Merck-Serono—all marketing various device systems to deliver this important biopharmaceutical (Table 4-4).

Many other pens and autoinjectors are now available (Table 4-3). Historically, autoinjectors were used for emergency antidote requirements, but today most are used to deliver biopharmaceuticals. Pens and autoinjectors can be reusable or disposable. Human insulin and human growth hormone pens have been used since the 1980s. Pens have become popular for anemia, hepatitis, infertility, and osteoporosis markets. Pens require the user to manually attach a needle, insert the needle into the skin, and press a button to inject the drug product. Some automatic systems exist, but not as popular and perhaps not as dependable as the manual systems. Pens require cartridges with Figure 4-6 showing where a cartridge fits in a typical pen delivery system. Pens are typically used for drugs self-administered in the home. Cartridge–pen delivery systems can be single or multiple dose and the dose can be varied because pens

Table 4-4 Novel Injectable Packaging Delivery Systems for Human Growth Hormone

- Eli Lilly (Humatrope®)
  - HumatroPen™
    - Reusable pen with lyophilized powder in a cartridge with diluent connector and prefilled diluent syringe
- Genentech (Nutropin®)
  - Nutropin AQ® Pen (reusable)
- Novo-Nordisk (Norditropen®)
  - NordiPen®
    - Reusable pen with cartridges
  - NordiPenmate®
    - Slides over pen for automated insertion and injection
  - Nordiflex®
    - First and only disposable pen for human growth hormone
- Pfizer (Genotropin®)
  - MiniQuick
    - Lyophilized powder in dual chamber syringe
  - Reusable pen with clip-on color panels
  - Reusable mixer for dual chamber cartridge
  - Intra-Mix®
    - Reconstitution device prefilled with dual-chambered cartridge
- Serono (Saizen®)
  - One.click™ autoinjector
  - Click.easy™ reconstitution aid
  - Cool.click™ resusable needle-free device

have a "dial-a-dose" feature. Of course, the drug manufacturer is responsible for validating the dependability of the device to deliver the right dose every time, all the time. The device manufacturer is not responsible for submitting these data to regulatory authorities. Single-dose disposable pens and fixed-dose pens also are commercially available. Major pen manufacturers included Becton-Dickinson (Fig. 4-8), Owen-Mumford (Fig. 4-9), West Pharma (Fig. 4-10), and Ypsomed (Fig. 4-11).

Reusable autoinjector, like reusable pens, requires several steps for preparation and injection. Autoinjectors can inject a fixed dose of 1 mL or less. Autoinjectors have been used primarily for treatment of multiple sclerosis and osteoarthritis, as they are quite suitable for home health care. Disposable autoinjector systems have been used historically for emergency uses, for example, the EpiPen®, but now are used for rheumatoid arthritis, anemia, and oncology purposes. Disposable injectors are single use and simple to operate. They are relatively expensive ($1–$4 per injection) and better serve less frequent administered drug products.

RECONSTITUTION PACKAGING SYSTEMS

Historically, lyophilized drugs were available in vials where the diluent, used to reconstitute the freeze-dried powder, was either provided with the vial package (combination package of vial and syringe) or the pharmacist used a common diluent in a vial (usually Sterile Water for Injection) to withdraw the appropriate volume of diluent from a vial using an empty sterile syringe and then reconstituting the drug-containing vial product.

Currently, while the classic way of reconstituting freeze-dried powders still is routinely practiced, two advances have gained popularity. One is the use of vial adapters and preassembled systems (Figs. 4-12 and 4-13) that facilitate the combination and transfer of diluents into the freeze-dried product vial. The other is the design of dual-chambered vials or syringes where the freeze-dried powder and the diluent are contained in the same packaging system separated by a rubber septum, where at the time of reconstitution the rubber septum is moved toward the powder compartment and the diluent combines with the powder via a bypass design in the syringe. The Vetter Lyo-Ject® (Fig. 4-14) has been the major player in this market although by the time of the publication of this book, other systems might be available. Dual-chambered vials such as Solu-Medrol® Mix-O-Vial™ are marketed, but other dual-chambered vials like Redi-Vial™ were removed from the market due to excessive costs.

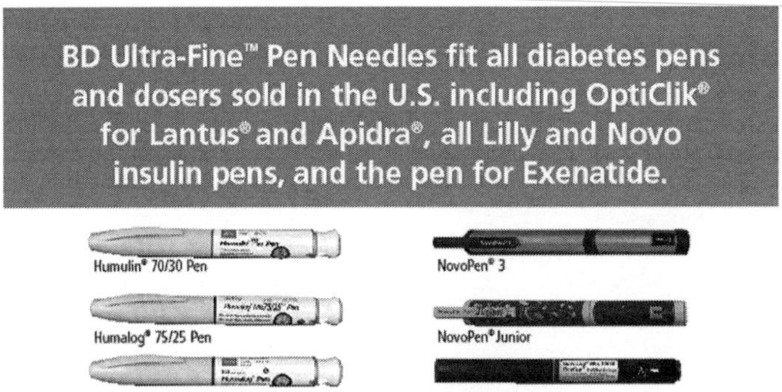

Figure 4-8 Delivery pen and needle examples. *Source*: Courtesy of © Becton, Dickinson and Company.

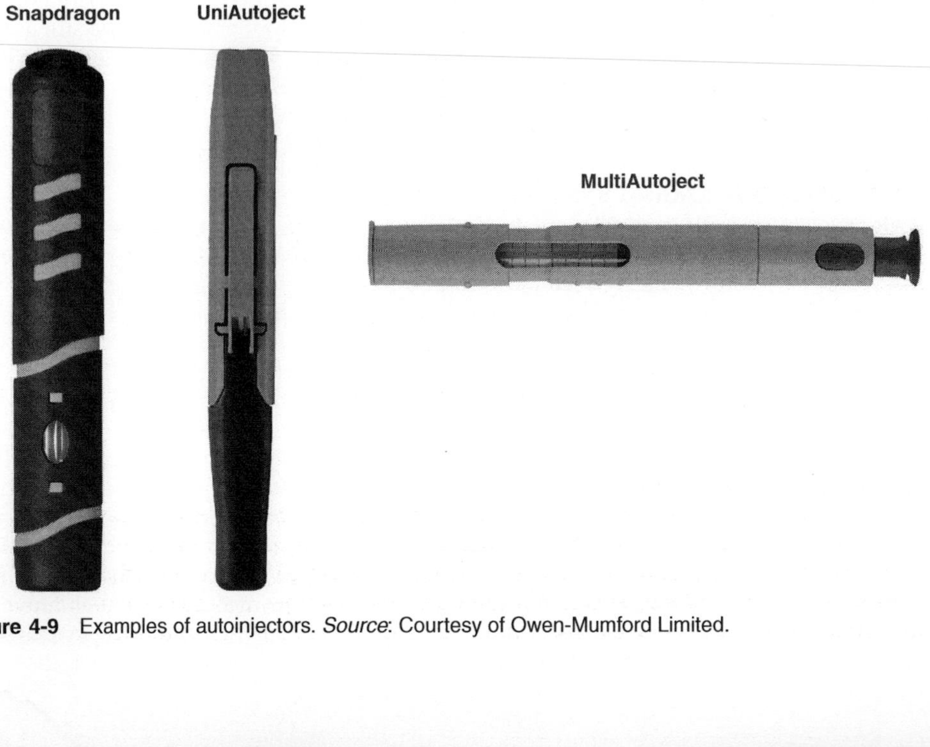

Figure 4-9 Examples of autoinjectors. *Source*: Courtesy of Owen-Mumford Limited.

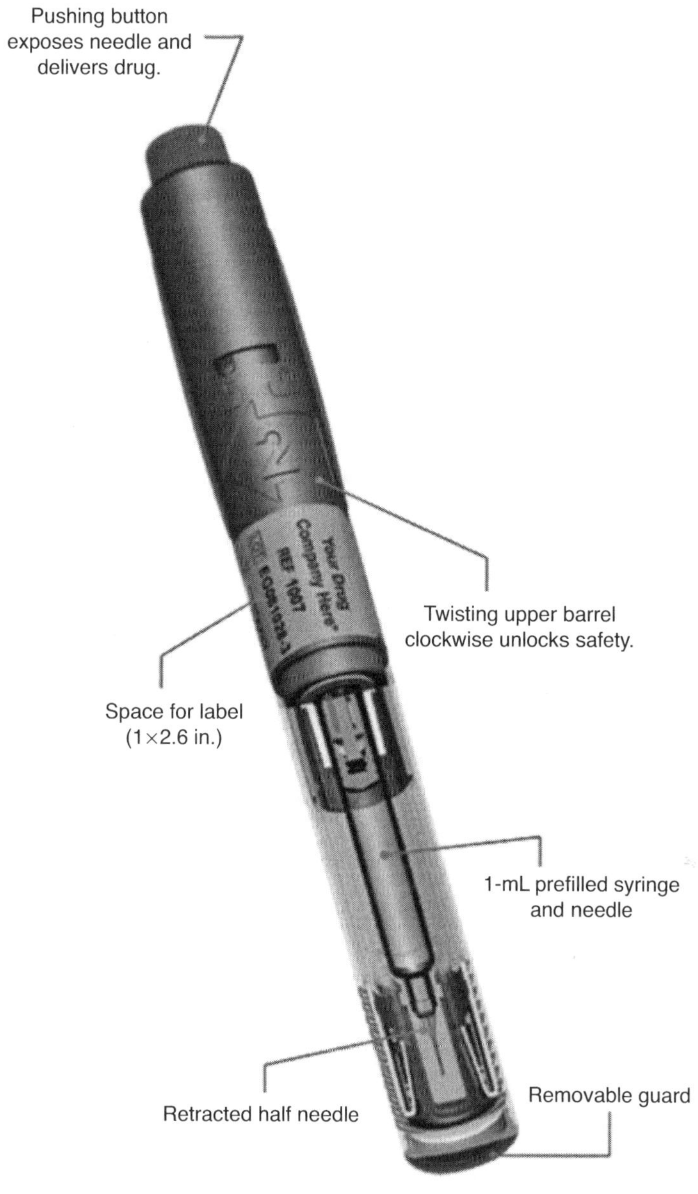

Figure 4-10 Example of autoinjector using glass or plastic syringe. ConfiDose® system—uses 1-mL long prefilled syringe, either glass or Daikyo Crystal Zenith® plastic prefilled syringe. *Source*: © 2010 by West Pharmaceutical Services Inc.

While all of these reconstitution systems add convenience to the user and may enhance sterility assurance (although no studies have been published), their main disadvantage is added cost for such convenience.

PARENTERAL COMBINATIONS

Most dosage forms, when released to the marketplace by the manufacturer, are consumed by the patient without any significant manipulation of the product. For example, tablets and capsules are ingested in the same form as they were when released by the manufacturer. For many parenteral drug products, this is not the case. For example, products in vials must be withdrawn into a syringe prior to injection and often combined with other products in infusion

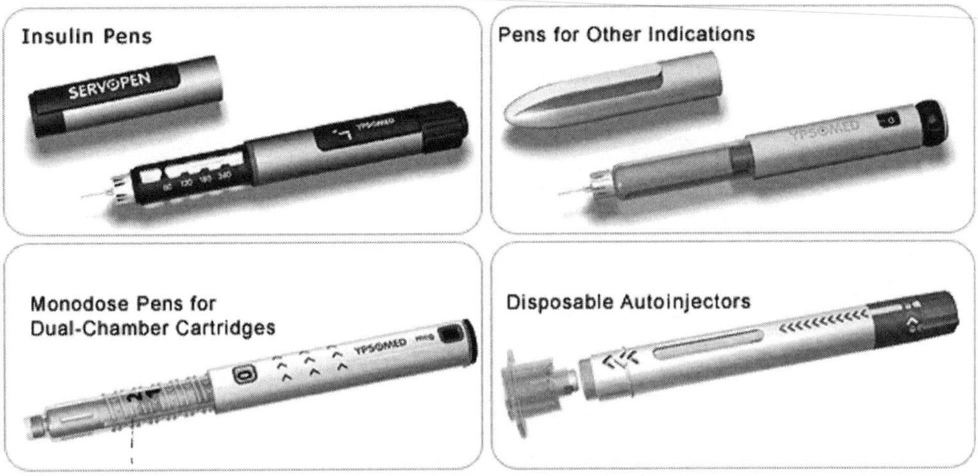

Figure 4-11 More examples of pens and autoinjectors. *Source*: Courtesy of Ypsomed AG.

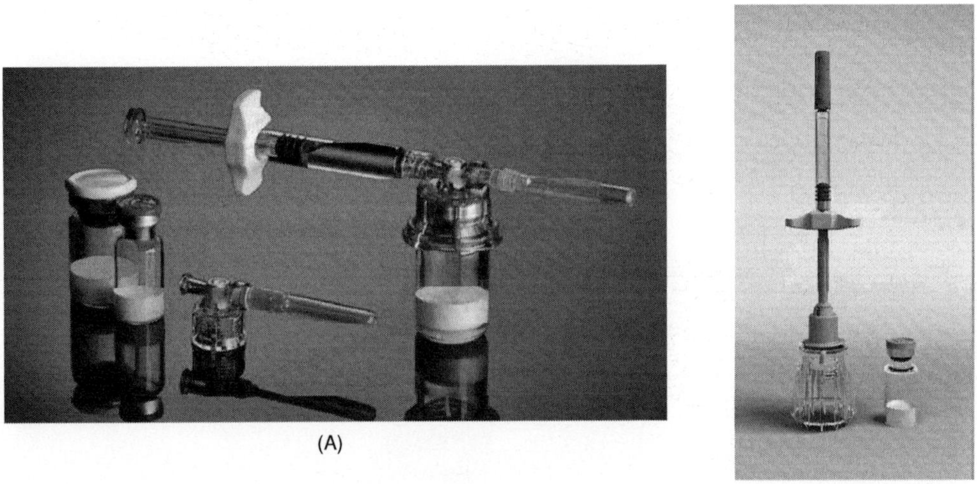

(A)

(B)

Figure 4-12 Examples of reconstitution and transfer sets. (A) Mix-Ject®. *Source*: Courtesy of West Pharmaceutical Services. (B) Duoject Smart-Rod XR. *Source*: Courtesy of Duoject Medical Systems Inc.

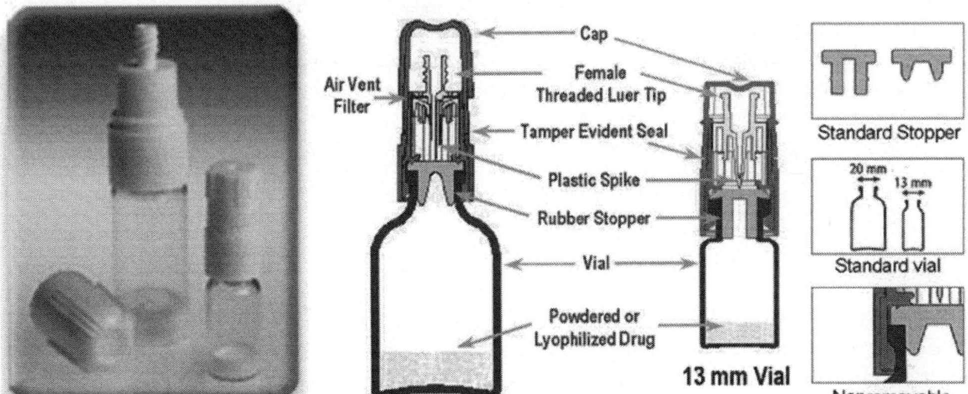

Figure 4-13 More examples of reconstitution and transfer sets—BIO-SET Luer admixture system. *Source*: Courtesy of Baxter Healthcare Corporation.

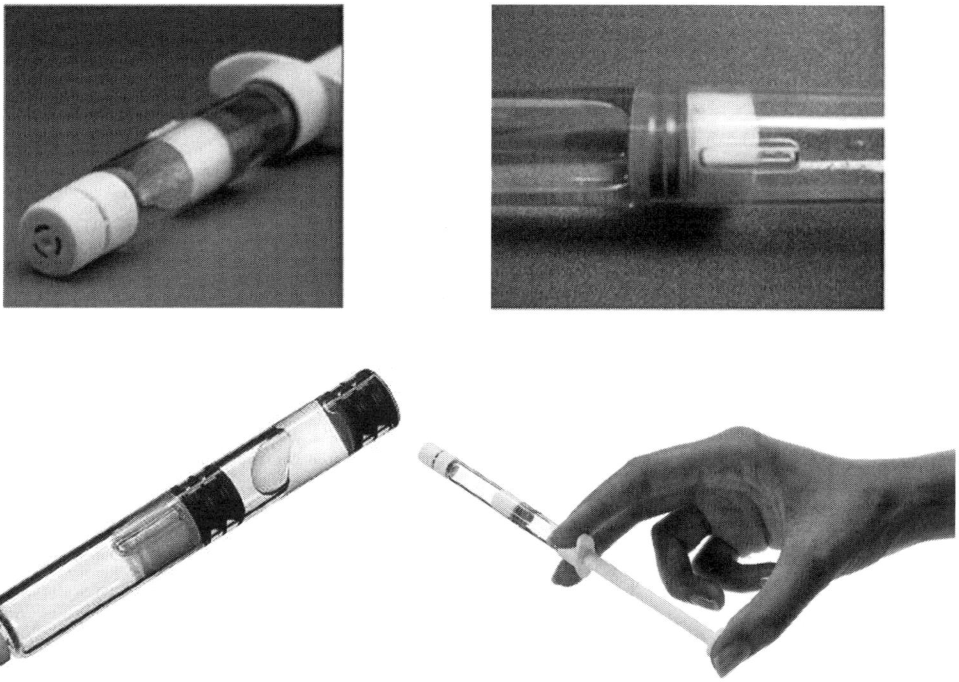

Figure 4-14 Dual-chambered syringes (e.g., Lyo-Ject®). *Source*: Courtesy of Vetter Pharma Int'l, Inc.

solutions prior to administration. Freeze-dried products first have to be reconstituted with a specific or nonspecific diluent prior to being withdrawn from the vial (see previous discussion). Other combination packages include

- Two or more vials and/or syringes containing two or more different drug products
- Two or more vials and/or syringes containing different additives of a formulation that are combined right before use
- Large volume bottle of diluent and small volume vial of active ingredient
- Cartridge and pen delivery combinations
- Dual-chambered systems such as the Vetter syringe (Fig. 4-14) and the B. Braun Duplex® double bag (Fig. 4-15).

It is common practice for a physician to order the addition of a small volume therapeutic injection (SVI), such as an antibiotic, to large volume injections (LVIs), such as 1000 mL of 0.9% sodium chloride solution, to avoid the discomfort for the patient of a separate injection. Certain aqueous vehicles are recognized officially because of their valid use in parenterals. Often they are used as isotonic vehicles to which a drug may be added at the time of administration. The additional osmotic effect of the drug may not be enough to produce any discomfort when administered. These vehicles include sodium chloride injection, Ringer's injection, dextrose injection, dextrose and sodium chloride injection, and lactated Ringer's injection.

While the pharmacist is the most qualified health professional to be responsible for preparing such combinations, interactions among the combined products can be troublesome even for the pharmacist. In fact, incompatibilities can occur and cause inactivation of one or more ingredients or other undesired reactions. Patient deaths have been reported from the precipitate formed by two incompatible ingredients. In some instances, incompatibilities are visible as precipitation or color change, but in other instances there may be no visible effect.

The many potential combinations present a complex situation even for the pharmacist. To aid in making decisions concerning potential problems, a valuable compilation of relevant data has been assembled by Trissel (17) and is updated regularly. Further, the advent of computerized

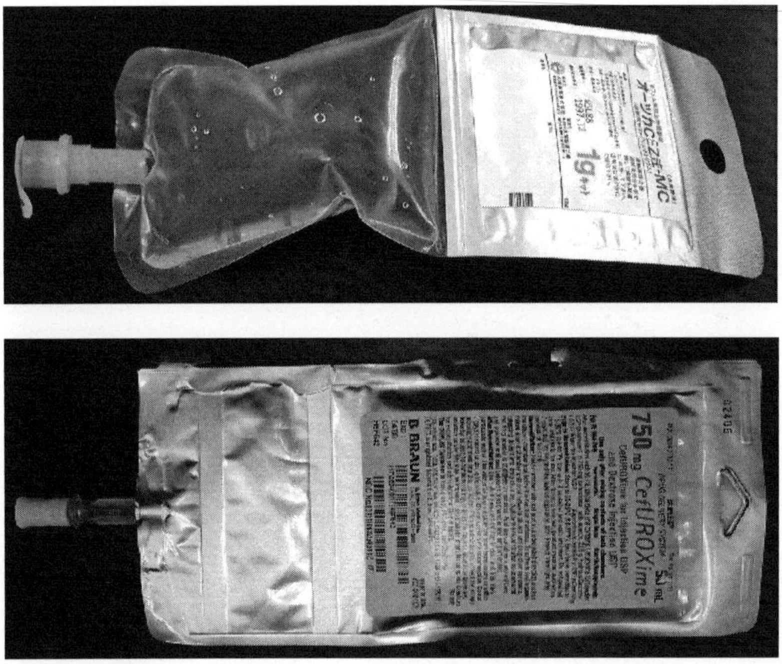

Figure 4-15 Duplex® (B. Braun). *Source*: Photos courtesy of Dr. Wei Kuu.

data storage and retrieval systems has provided a means to organize and gain rapid access to such information.

Ideally, no parenteral combination should be administered unless it has been studied thoroughly to determine its effect on the therapeutic value and the safety of the combination. However, such an ideal situation may not exist. Nevertheless, it is the responsibility of the pharmacist to be as familiar as possible with the physical, chemical, and therapeutic aspects of parenteral combinations and to exercise the best possible judgment as to whether or not the specific combination extemporaneously prescribed is suitable for use in a patient.

ACKNOWLEDGMENT

Gregory Sacha, Karen Abram, and Wendy Saffell-Clemmer of Baxter BioPharma Solutions are acknowledged for their help in writing this chapter.

REFERENCES

1. Harrison B, Rios M. Developments in prefilled syringes. Pharm Tech 2007; 31(March):11.
2. Overcashier DE, Chan EK, Hsu CC. Technical considerations in the development of pre-filled syringes for protein products. Am Pharm Rev 2007; 9(7):77–82.
3. Polin JB. The ins and outs of prefilled syringes. Pharmaceutical and Medical Packaging News. May 2003.
4. Polin J. The evolution of parenteral drug packaging. Pharmaceutical and Medical Packaging News. March 2004.
5. Karras L, Wright L, Cox T, et al. Current issues in manufacturing and control of sterile prefilled syringes. Pharm Tech 2000; 24:188–196.
6. Soikes R. Moving from vial to prefilled syringe: A project manager's perspective. Pharm Tech 2009; (September):S12–S17.
7. Needlestick Safety and Prevention Act. 2000. http://frwebgate.access.gpo.gov/cgi-bin/getdoc.cgi?dbname=106_cong_public_laws&docid=f:publ430.106. Accessed June 3, 2010.
8. ICH. Stability testing of new drug substances and products, Q1A (R2). 2003. http://www.ich.org/LOB/media/MEDIA419.pdf, 13. Accessed June 3, 2010.

9. Wen Z-Q, Vance A, Vega F, et al. Uneven distribution of silicone oil in pre-filled glass syringes and its implications to delivery of protein therapeutics. PDA J Pharm Sci Tech 2009; 63(March):149–158.
10. Chantelau E. Silicone oil contamination of insulin. Diabet Med 1989; 6:278.
11. Jones L, Kaufmann A, Middaugh C. Silicone oil induced aggregation of proteins. J Pharm Sci 2004; 94:918–927.
12. Wood L. Research and markets: Insulin delivery devices market analysis and future forecast. Reuters 2009. http://www.reuters.com/article/pressRelease/idUS70251+09-Jun-2009+BW20090609. Accessed June 3, 2010.
13. EPA. Bis (2-ethylhexyl) phthalate (DEHP). 1992. http://www.mindfully.org/Pesticide/DEHP-PVC-Toxicity.htm. Accessed June 3, 2010.
14. FDA. Safety assessment of di(2-ethylhexyl) phthalate (DEHP) released from PVC medical devices. Rockville, MD: Center for Devices and Radiological Health, 2002. http://www.fda.gov/downloads/MedicalDevices/DeviceRegulationandGuidance/GuidanceDocuments/UCM080457.pdf. Accessed June 3, 2010.
15. French D. Advances in parenteral drug delivery devices. West Point, NY: Arden House Conference presentation, 2007.
16. Akers MJ, Nail SL, Saffell-Clemmer W. Top ten hot topics in parenteral science and technology. PDA J Pharm Sci Tech 2007; 61:337–361.
17. Trissel L.A. Handbook on Injectable Drugs. 15th ed. Bethesda, MD: American Society of Health-Care Pharmacists, 2009; book updated every two years.

5 Overview of product development

The purpose of this chapter is to provide guidance for the pharmaceutical product development of sterile dosage forms. This chapter introduces specific formulation chapters 6 through 10. Chapter 11 provides some formulation case studies to help the reader review learning points.

THE PRODUCT DEVELOPMENT PROCESS

Drug product development is a lengthy, expensive, and risky process. It is estimated that on average it takes $1.1 billion and 12.5 years to bring a single drug from concept to commercialization (1–3). Only 1 in 5000 new chemical compounds makes it to market and 80% of all investigational new drugs fail. Figure 5-1 summarizes the pathways and Table 5-1 summarizes the various specific tasks involved in taking a drug from its discovery through the product development and clinical testing stages until it is approved as a commercial product. This is an oversimplification of a very complex procedure for moving an active pharmaceutical ingredient from conception to commercialization. The key points to realize from this figure are the following:

1. Developing the analytical method(s) precedes being able to develop the final dosage form. Analytical methods should be developed as early as possible. Some validation effort should be expended prior to investigational new drug (IND) filing, but final assay validation cannot take place until the final formulation has been determined.
2. The figure is not drawn to scale. Clinical phases may require many years, with Phase III being the longest. Phase I studies can occur within a relatively short period of time as these studies are designed to determine the safety profile of the new compound in a relatively small number of patients. Phase II studies begin the clinical determination of drug efficacy plus learning more about safety. Phase III involves a very large number of subjects spread over several clinical sites to provide statistical verification of the safety and efficacy of the new drug.
3. The final formulation, package, and process should be locked in prior to or during early Phase III clinical studies. This allows clinical data to be obtained using the exact product and process that will be marketed, if approved.
4. Final product specifications are finalized during Phase III clinical batch production. This allows definitive stability batches and subsequent long-term studies to be initiated prior to new drug application (NDA) submission so that at least one-year real-time stability data on the final product and process will be available.
5. Process validation batch production also should be initiated prior to the submission of the NDA. The validation report does not necessarily have to be completed at the time of the NDA submission, but protocols and initial data on validation batches will be available. Often manufacturers plan on validation batches to be available for market sale assuming NDA approval.

Figure 5-2 compares the general schemes of obtaining NDA and abbreviated new drug application (ANDA) approval before and after the generic drug scandal in the late 1980s that gave rise to the need for preapproval good manufacturing practice (GMP) compliance inspections. Prior to 1990, a drug product manufacturer would submit an NDA, wait for FDA approval, and then, once approved, begin the investment of time and money to build facilities, buy production equipment, scale-up the process, and collect data on stability and validation batches. Moreover, prior to 1990, a company did not have to have proof that it could manufacture a product at full scale repetitively to meet NDA commitments with respect to processes, procedures, and specifications. It would need to do so during a biannual FDA current good manufacturing practice (cGMP) inspection audit, but that audit might occur more than two years after an NDA was approved.

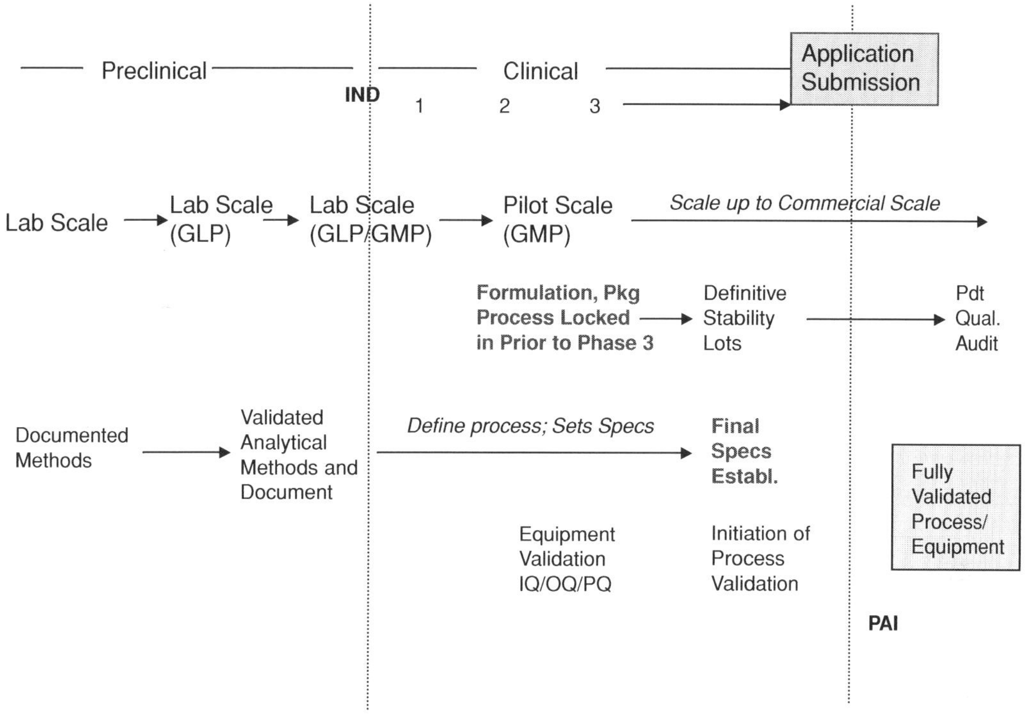

Figure 5-1 The product development process. *Abbreviations:* IND, investigation drug application; GLP, good laboratory practices; GMP, good manufacturing practices; IQ, installation qualification; OQ, operational qualification; PQ, performance qualification; PAI, pre-approval inspection; Pkg, Package; Pdt, Product.

However, the generic drug scandal of the late 1980s changed all this. Essentially certain manufacturers were obtaining ANDA approval based on data on relatively small batches; then when commercial batches were prepared, certain specifications were not met, but that did not stop these manufacturers from releasing the batches. The FDA eventually caught this improper practice, plus some firms were also hiding or covering up information and data that had an impact on the safety, identity, strength, purity, and quality of their drug products. This scandal affected the entire pharmaceutical industry since FDA could no longer trust the industry that what was stated on documentation submitted to the agency was exactly what was being adhered to. This gave rise to the preapproval inspection program where approval or withholding approval of all NDAs and ANDAs was dependent not only on FDA review of the documentation in the application but also on the results of a GMP inspection at the plant site of the firm filing the application. As one high-ranking FDA official was quoted regarding preapproval NDA/ANDA inspections: "We have moved from a system of trust to a system of trust that must be verified." Preapproval inspection gave local GMP compliance field offices and inspections sudden new and significant power in determining the approvability of NDAs and ANDAs.

Preapproval inspections also led to the requirement for a document available to the GMP inspection called the Development History Report. The purpose of this report primarily was to show due diligence on the part of the manufacturer that batches of drug product used in early clinical studies could be correlated as essentially having the same quality parameters as batches prepared and used in late Phase III clinical studies and batches to be commercialized.

FORMULATION PRINCIPLES

Parenteral drugs are formulated as solutions, suspensions, emulsions, liposomes, microspheres, nanosystems, and powders to be reconstituted as solutions. Each dosage form is formulated differently, although formulation components besides the active ingredient are added only when absolutely necessary. In other words, ideally, a formulation will contain active ingredient and water with no added substances. While this is true for high-dose products such as cephalosporin

Table 5-1 Example of Flow of Tasks in New Drug Product Development (Several Unpublished Sources Used to Create This Table)

Prior to first human dose	Phases I and II	Prior to Phase III	Prior to registration stability initiation review	NDA submission readiness review	PAI readiness review
API characterization proof of structure	Characterization of several API lots	Proof of structure update	API definitive stability protocol	Update on API validation/batches	NDA filing update
Analytical methods development	Reference standard characterization	Final API lots characterized and purity profile	Final API methods/ validation specifications	API available stability data	API and product stability update
API purity/impurity profile; acceptance limits	API stress testing	API stress data	API methods transfer	Proposed NDA specifications for API	PAI readiness plan for API
API properties critical to formulation and process	API final methods, acceptance limits	Related subs ID	Product demonstration lot	Available product stability data	Update from plant validation batches
Proposed sourcing strategy (CT, commercial)	Long-term stability studies for API	Phase III formulation/ package process/control strategy	Formulation/process/ package readiness for stability	Product methods transfer status	PAI readiness for product
Preformulation data	Finish preformulation	Unit operations studies/ lyophilization cycle	Final control strategy	Production experience	Strategy to address submission deficiencies
Toxicology supplies	CT formulation/package process/commercial dose	Final equipment/facility needs	Product/methods transfer status/issues	Full-scale batches/ validation plans	Monograph plans
Early phase dosage form, package, process	Technology transfer plans	Final cleaning methods	Product stability protocols	Affiliate review	
Dosage form strategy for subsequent CTs	Methods/acceptance inactive ingredients	Product methods/validation	Final product methods/ validation and specs	NDA package—total analysis	
Anticipated special equipment process, facility needs	Marketing/manufacturing agreement	Final marketing/ manufacturing agreement		Review of NDA modules status	
Product test methods acceptance limits	Product methods and acceptance limits	Package extractable C/C integrity		DHR and other reports	
Cleaning methods	CT lot stability studies	Product stability and specs		Control strategy in place	
Preliminary stability studies	Micro requirements	Product stress in use			
Regulatory strategy	Legal/patent	Product-related subs and qualification for toxicology			
Documentation strategy	Documentation system established				

Abbreviations: PAI, Pre-approval inspection; API, active pharmaceutical ingredient; CT, clinical trial; C/C, container/closure; ID, identification; DHR, development history report.

Prior to 1990s	Today
Submit application	Processes finalized
↓	↓
FDA reviews at headquarters	Submit application
↓	↓
Approval	FDA reviews at headquarters
↓	↓
Company finalizes processes	GMP preapproval inspection
↓	↓
General GMP inspection	Recommendation from local office to FDA headquarters
	↓
	Approval

Figure 5-2 Comparison of new drug application/abbreviated new drug application (NDA/ANDA) approval process.

antibiotics, the large majority of parenteral drug products do contain added substances. Nevertheless, it is always the goal of a formulator of a sterile drug product to keep that formulation as simple as possible with a minimum of added substances (excipients).

GENERAL GUIDANCE FOR DEVELOPING FORMULATIONS OF PARENTERAL DRUGS

The final formulation of a parenteral drug product depends on understanding the following factors that dictate the choice of formulation and dosage form.

Route of Administration

Injections may be administered by such routes as intravenous, subcutaneous, intradermal, intramuscular, intra-articular, and intrathecal (chap. 31). The type of dosage form (solution, suspension, etc.) will determine the particular route of administration that may be employed. Conversely, the desired route of administration will place requirements on the formulation. For example, suspensions would not be administered directly into the bloodstream because of the danger of insoluble particles blocking capillaries. Solutions to be administered subcutaneously require strict attention to tonicity adjustment; otherwise irritation of the plentiful supply of nerve endings in this anatomical area would give rise to pronounced pain. Injections intended for intraocular, intraspinal, intracisternal, and intrathecal administration require stricter standards of such properties as formulation tonicity, component purity, and limit of endotoxins because of the sensitivity of tissues encountered to irritant and toxic substances.

If the route of administration must be intravenous, then only solutions or microemulsions can be the dosage form. If the route of administration is to be subcutaneous or intramuscular, then the likely type of dosage form is a suspension or other microparticulate delivery system.

Pharmacokinetics of the Drug

Rates of absorption (for routes of administration other than intravenous or intra-arterial), distribution, metabolism, and excretion for a drug will have some effect on the selected route of administration and, accordingly, the type of formulation. For example, if the pharmacokinetic profile of a drug is very rapid, modified release dosage formulations may need to be developed. The dose of drug and the dosage regimen are affected by pharmacokinetics, so the size (i.e., concentration) of the dose will also influence the type of formulation and amounts of other ingredients in the formulation. If the dosage regimen requires frequent injections, then a multiple dose formulation must be developed, if feasible. If the drug is distributed quickly from the injection site, complexing agents or viscosity-inducing agents may be added to the formulation to retard drug dissolution and transport.

Drug Solubility

If the drug is insufficiently soluble in water at the required dosage, then the formulation must contain a cosolvent or a solute that sufficiently increases and maintains the drug in solution. If relatively simple formulation additives do not result in a solution, then a dispersed system dosage form must be developed. Solubility also dictates the concentration of the drug in the dosage form.

Figure 5-3 presents a schematic decision tree that formulation scientists typically follow in overcoming drug-solubility problems for drugs intended for IV administration, and this will be further elaborated in chapters 6 and 8. Basically, the approaches used to overcome solubility problems start first with the drug itself (salt formation), then simple approaches with the formulation (pH adjustment, addition of cosolvent, complexing agent, and/or surface-active agent), and, if none of these produce the desired result, the last approach is to change the dosage form to a dispersed system or other more complicated formulation.

Drug Stability

If the drug has significant degradation problems in solution, then a freeze dried or other sterile solid dosage form must be developed. Stability is sometimes affected by drug concentration, which, in turn, might affect the size and type of packaging system used. For example, if concentration must be low due to stability and/or solubility limitations, then the size of primary container must be larger and this might preclude the use of syringes, cartridges, and/or smaller vial sizes. Obviously, stability dictates the expiration date of the product, which, in turn, will determine the storage conditions. Storage conditions might dictate the choice of container size, formulation components, and type of container. If a product must be refrigerated, then the container cannot be too large and formulation components must be soluble and stable at colder conditions. Achieving drug stability is overall the number one reason for adding solutes to an injectable formulation and subsequent chapters on formulation of solutions, dispersed systems, and freeze-dried products (chaps. 8–10) will show this emphasis.

Compatibility of Drug with Potential Formulation Additives

It is well known that drug-excipient incompatibilities frequently exist and these will be pointed out in subsequent chapters. Initial preformulation screening studies are essential to assure that formulation additives, while possibly solving one problem, will not create another. Stabilizers, such as buffers and antioxidants, while chemically stabilizing the drug in one way, may also catalyze other chemical degradation reactions. Excipients and certain drugs can form insoluble complexes. Impurities in excipients can cause drug degradation reactions. Peroxide impurities in polymers may catalyze oxidative degradation reactions with drugs, including proteins, that are oxygen sensitive.

Desired Type of Packaging

Selection of packaging (type, size, shape, color of rubber closure, label, and aluminum cap) often is based on marketing preferences and competition. Knowing the type of final package early in the development process aids the formulation scientist in being sure that the product formulation will be compatible and elegant in that packaging system.

Tables 5-1 and 5-2 provide two different views of the steps and tasks involved in the formulation of a new parenteral drug product. This information can also be viewed as a list of questions, the answers of which will facilitate decisions on the final formulation that should be developed.

Basic Guidelines to Consider in the Development of Parenteral Solutions of Proteins and Peptides

Table 5-3 lists some important basic guidelines or principles specific for development of protein and peptide formulations that should be combined with what is presented in Tables 5-1 and 5-2. The first principle to realize in approaching formulation development of a product containing a protein or other biopharmaceutical is to obtain a thorough understanding of the physical and chemical properties of the protein or peptide bulk drug substance. There are now well-documented analytical techniques available for studying these properties in solution. Effects of temperature, pH, shear, oxygen, buffer type and concentration, ionic strength, and

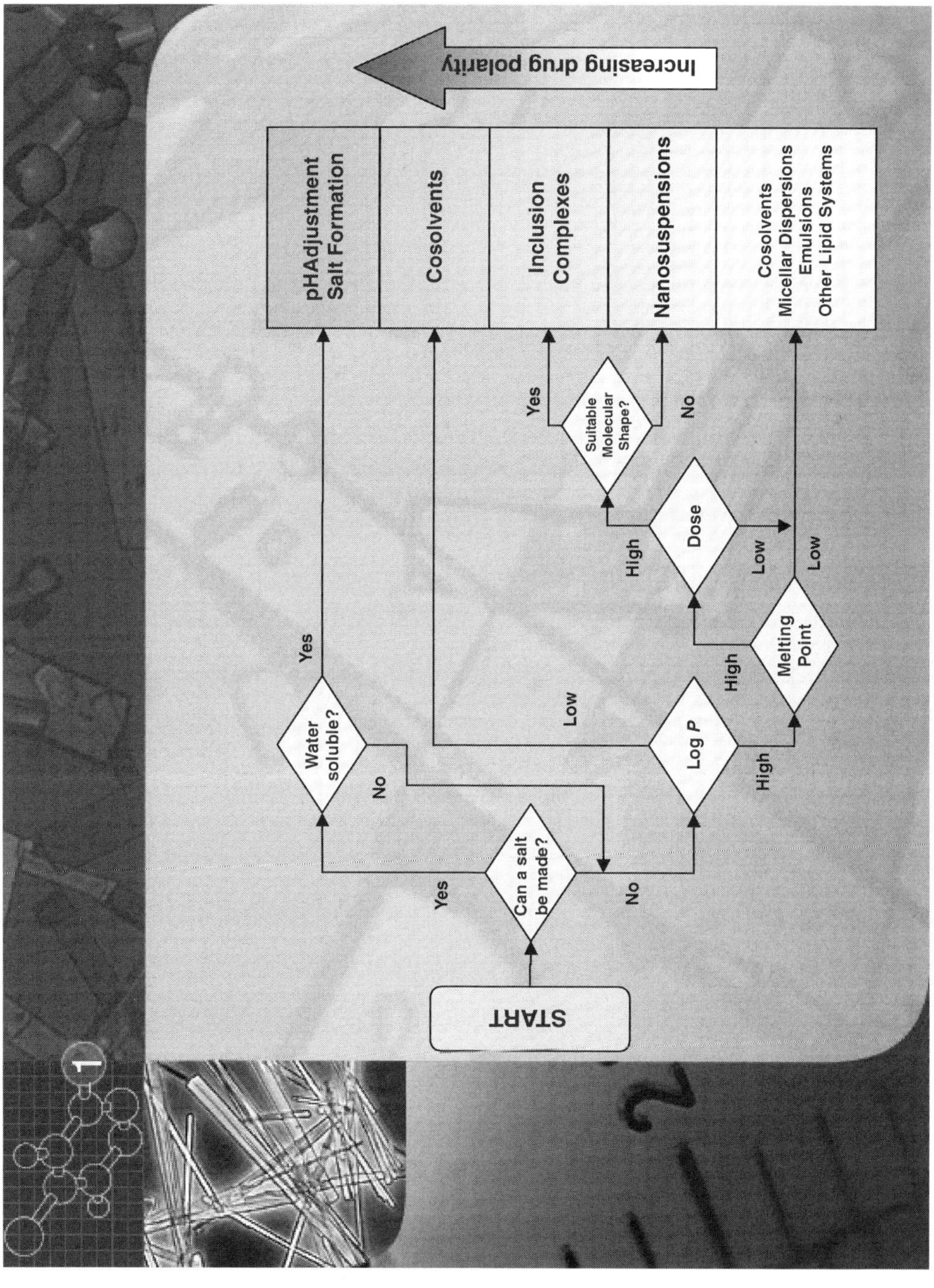

Figure 5-3 Solubilization strategies. Source: Courtesy of Dr. James Kipp, Baxter Healthcare Corporation.

Table 5-2 Main Steps Involved in the Formulation of a New Sterile Drug Product

1. Obtain physical properties of active drug substance
    a. Structure, molecular weight
    b. "Practical" solubility in water at room temperature
    c. Effect of pH on solubility
    d. Solubility in certain other solvents
    e. Unusual solubility properties
    f. Isoelectric point for a protein or peptide
    g. Hygroscopicity
    h. Potential for water or other solvent loss
    i. Aggregation potential for protein or peptide
2. Obtain chemical properties of active drug substance
    a. Must have a "validatable" analytical method for potency and purity
    b. Time for 10% degradation at room temperature in aqueous solution in the pH range of anticipated use
    c. Time for 10% degradation at 5°C
    d. pH stability profile
    e. Sensitivity to oxygen
    f. Sensitivity to light
    g. Major routes of degradation and degradation products
3. Initial formulation approaches
    a. Know timeline(s) for drug product
    b. Know how drug product will be used in the clinic
        i. Single dose vs. multiple dose
        ii. If multiple dose, will preservative agent be part of drug solution/powder or part of diluent?
        iii. Shelf-life goals
        iv. Combination with other products, diluents
    c. From knowledge of solubility and stability properties and information from anticipated clinical use formulate drug with components and solution properties that are known to be successful at dealing with these issues. Then perform accelerated stability studies
        i. High-temperature storage
        ii. Temperature cycling
        iii. Light and/or oxygen exposure
        iv. For powders, expose to high humidities
    d. May need to perform several short-term stability studies, as excipient types and combinations are eliminated
    e. Understand need for any special container and closure requirements
    f. Design and implement an initial manufacturing method of the product
    g. Finalize formulation
        i. Need for tonicity adjusting agent
        ii. Need for antimicrobial preservative
    h. Approach to obtain sterile product
        i. Terminal sterilization
        ii. Sterile filtration and aseptic processing

Table 5-3 Basic Guidelines to Consider in the Development of Parenteral Solutions of Proteins and Peptides (in Addition to Considerations Presented in Tables 5-1 and 5-2)

1. Learn and understand the basic physical and chemical properties of the biopharmaceutical active ingredient
2. Know the intended route(s) of administration and formulation requirements unique to each route (e.g., pH, osmolality, freedom from particles, viscosity, and volume)
3. Rationale and selection of formulation components
4. Effects of manufacturing process on stability of the active ingredient
5. Selection of final container and closure system
6. Effects of storage and distribution on product stability

protein/peptide concentration must be understood. From preformulation studies, protein/peptide chemical and physical degradation pathways will be better understood so that the final formulation, manufacturing process, and packaging system will be rationally developed.

The second principle is that the route of administration must be known in order to select the final dosage form, vehicle, volume, and tonicity requirements for the product. For example, if the primary route of administration is intravenous, the vehicle has to be water although some water-miscible cosolvents can be used. The volume can be limitless (unless an antimicrobial preservative is part of the formulation in which case the volume is limited to 15 mL), and the tonicity does not necessarily have to be isotonic because the injected solution will be rapidly diluted. However, if the route of administration will be subcutaneous or intramuscular, then the vehicle can be aqueous or nonaqueous, the volumes are limited (usually no more than 2 mL for subcutaneous, 3 mL for intramuscular), and the tonicity of the product needs to be more tightly controlled since the product is not quickly nor readily diluted. The rate of injection also is a factor to be considered in the selection of final formulation ingredients in that some ingredients, including the protein/peptide itself, can be irritating and even cause local inflammatory reactions if injected too quickly and/or at too high a concentration.

The third principle involves careful screening for selection of solutes for solubilization, stabililization, and preservation, and tonicity adjustment must take place. This will be covered in detail in chapter 8.

The fourth principle asks what are the potential effects of the manufacturing process on the stability of the protein/peptide in the final formulation? Proteins/peptides cannot withstand terminal sterilization techniques (heat, gas, radiation) and, thus, must be sterilized by aseptic filtration. The filter used must be qualified so that it does not bind the protein/peptide. The effect of flow rate during filtration and filling on solution stability must be studied. Also, the effect of shear (mechanical stress) that is encountered during manufacturing must be known. Time limitations must be established from the time the protein solution is compounded until it is sterile filtered in order to avoid any increase in endotoxin levels from whatever the bioburden, however small, may be in the nonsterile solution.

The fifth principle concerns the importance of the selection of the most compatible container/closure system. Formulation scientists must appreciate that the container and closure system is just as important as the final solution formulation in assuring long-term stability and maintenance of sterility and other quality parameters of the product. Proteins and peptides are well known to adsorb to glass, so experiments must be designed to study this possibility and, if adsorption occurs significantly, additives such as albumin must be considered to reduce the adsorption. Glass leachates and particulates are possible and the formulator must be aware of this. Experiments must be conducted to ensure elimination of this potential problem. The choice of rubber closure is particularly important because of known potential for the closure to leach some of its own ingredients into a solution, to adsorb components of the protein/peptide formulation, to core (rubber particulates) when penetrated by a needle, to generate particulates, and to leak because of problems with the fitment on the glass vial, or resealability of the elastomer after needle penetration. Studies on adsorption of the protein to plastic surfaces will be necessary if the final product will be a plastic container. Even if plastic is not part of the primary container, protein–plastic compatibility studies should be done since plastic tubing, such as silicone or polyvinyl chloride, will be used in pharmaceutical process equipment (e.g., filling machines) and the final dosage form might be added to large volume parenteral solutions contained in plastic bags.

The final principle requires studies to be conducted to understand the effects of distribution and storage on the stability of the final product. Temperature excursions during shipping, mechanical stress, exposure to light, and other simulated shipping and storage conditions must be studied. From these studies, appropriate procedures for distribution and long-term storage of these relatively unstable dosage forms can be developed.

SPECIAL PROBLEM

Assume that you are a new formulation scientist, recently hired. You obviously want to make a favorable impression on your management and peers. You are given an assignment to develop a parenteral formulation of a brand new molecule. Let us assume that there is very little known about the molecule and it comes to you as a bulk freeze-dried powder. Let us also assume that quantities are limited and the drug is expensive, so you have to be very smart in how you study the drug and develop its formulation. What questions would you ask and what studies would you propose to do? Do this before checking the below mentioned examples that provide some of the questions and learning points that should be part of the development of a new parenteral drug dosage form:

- Target dose? Of course, need to know the target dose and/or range to design formulation experiments.
- Target patient population—This relates to knowing the therapeutic activity of the active pharmaceutical ingredient (API) and who will be receiving the product. Will there be a pediatric indication that could affect formulation component choices? Will there be an elderly indication that could affect ease of use of the medication?
- Route of administration and mode of therapy—How will the drug product be administered ... bolus dose, intermittent dosing, infusion? This impacts toxicity, safety, acute versus long-term usage of the drug product.
- Type of delivery system (e.g., vial, syringe, and infusion).
- Safety concerns ... classification of the active ingredient, material safety data sheet (MSDS), personnel precautions.
- Analytical method development
  - Potency method
  - Purity method
  - Stability-indicating method
  - In-process assay
  - Identity method
- Basic chemistry of active ingredient
  - Structure, hydrophilicity, stability questions, etc.
  - Physical and chemical properties ... salt form, pK_a, partition coefficient, solubility, etc.
  - Solubility studies ... structure dictates solvents to study
  - Solution stability—function of temperature, pH
- Compatibility with other materials (excipients, packaging).
- Initial formulations—depends on solubility, stability, intended clinical usage. Let us assume that stability limitations require the drug to be freeze dried:
  - Freeze dry the drug alone, determine what happens
  - If excipients needed, start with commonly known excipients that
    - Produce acceptable cakes with rapid reconstitution rates
    - Have minimal collapse temperatures
    - Provide the desired finished product with respect to the nature of the solid (crystalline vs. amorphous).
- Formulations should have solids content between 5% and 30% with a target of 10% to 15%.
- Determine the maximum allowable temperature (chap. 10) permitted during freezing and primary drying
  - ($T_e/T_{g'}/T_c$) of tentative formulation
- Select the appropriate size of vial and product fill volume.
- Select the appropriate rubber closure
  - Low water vapor transmission, no oil vapor absorption, top design minimizes shelf sticking
- Conduct initial stress tests (e.g., freeze-thaw cycling, agitation studies) to screen initial formulations.
- Small molecules
  - May not need an excipient (depends on nature and amount of active ingredient)
  - Might need one or more of the following: bulking agent, buffer, salt
  - Most stability problems with small molecules are moisture related, not the effects of freezing and/or drying as is the case with large molecules
  - Generally, the drier, the better

- Large molecules typically need
  - Bulking agents, stabilizers (protectants, surfactants, buffers)
  - Salts should be avoided (low T_e and $T_{g'}$) plus concentration effects on proteins
  - Tonicity modifiers (mannitol and/or sucrose best)
- Once formulation and package tentatively selected, determine appropriate freeze dry process parameters, for example,
  - Rate of freezing
  - Need for annealing
  - Temperature and pressure during primary drying
  - Temperature and pressure during secondary drying
  - Sealing under vacuum or nitrogen
- Optimize formulation and process based on stability information both during and after lyophilization process and after storage in dry state.

REFERENCES

1. Patterson J. Can big pharma produce the next generation of medicines? Pharm Tech 2008; 32:114.
2. Blaisdell P. Twenty-First Century Pharmaceutical Development. Englewood, CO: Interpharm Press, 2001.
3. European Medicines Agency. ICH Topic Q8, Note for guidance on pharmaceutical development, EMEA/CHMP/167068/2004, May 2006.

6 Formulation components (solvents and solutes)

Sterile formulations, by necessity, must be as simple as possible. Safety considerations limit the number and choices of additives to use in formulations besides the active and, if stability is sufficient, a vehicle. The ideal parenteral formulation would contain the active ingredient and water and nothing else. In reality, the author is unaware of any sterile formulation that contains only active ingredient in a ready-to-use 2-year stable aqueous solution. Formulations that contain only the active ingredient typically are freeze-dried with the therapeutic dose of the active sufficient to produce an elegant cake that is constituted with a diluent vehicle prior to administration. Most sterile formulations contain at least one additive besides the active ingredient and a majority of formulations contain two or more additives. This chapter will describe the types and purposes of additives (solutes) and vehicles used in sterile formulations. References 1 through 8 provide reviews and listings of approved additives in marketed sterile product formulations including a valuable Food and Drug Administration Web site.

Care must be taken in selecting active pharmaceutical ingredients and excipients to ensure that their quality is suitable for parenteral administration. A low microbial level will enhance the effectiveness of either the aseptic or terminal sterilization process used for the drug product. Likewise, nonpyrogenic ingredients enhance the nonpyrogenicity of the finished injectable product. It is now a common GMP procedure to establish microbial and endotoxin limits on active pharmaceutical ingredients and most excipients. Chemical impurities should be virtually nonexistent in active pharmaceutical ingredients for parenterals, because impurities are not likely to be removed by the processing of the product. Depending on the chemical involved, even trace residues may be harmful to the patient or cause stability problems in the product. Therefore, manufacturers should use the best grade of chemicals obtainable and use its analytical profile to determine that each lot of chemical used in the formulation meets the required specifications.

Reputable chemical manufacturers accept the stringent quality requirements for sterile products and, accordingly, apply good manufacturing practices to their chemical manufacturing. Examples of critical bulk manufacturing precautions include the following:

- Using dedicated equipment or properly validated cleaning to prevent cross-contamination and transfer of impurities
- Using WFI for rinsing equipment
- Using closed systems wherever possible for bulk manufacturing steps not followed by further purification
- Adhering to specified endotoxin and bioburden testing limits for the substance

VEHICLES (SOLVENTS)

The solvent in injectable formulations typically is the largest component. Of course, the preferred solvent or vehicle is water for injection (WFI). For drugs that are not sufficiently soluble in water, water-miscible organic co-solvents may be used with limitations on the acceptable amounts from a safety viewpoint. For drugs completely insoluble in water and not required to be injected intravenously, oily (oleaginous) solvent systems of vegetable origin may be used.

Water

Since most liquid injections are quite dilute, the component present in the highest proportion is the vehicle. The vehicle of greatest importance for sterile products is water. Water of suitable quality for compounding and rinsing product contact surfaces may be prepared either by distillation or by reverse osmosis, to meet United States Pharmacopeia (USP) specifications for WFI. Preparation and quality standards of WFI and description of other types of compendial water are covered in chapter 15.

Table 6-1 Most Commonly-Used Water-Miscible Co-Solvents in Injectable Products (Percent Range Approved by FDA[a])

Co-solvent	Per cent range	Product examples (all are trademarks or registered)
Ethanol (alcohol, dehydrated alcohol, ethyl alcohol)	0.6–80	Prograf, BiCNU, Nitro-Bid, Alkeran for Injection, Septra, Valium, VePesid, Triostat, Lanoxin, D.H.E 45, Nembutal Sodium, Dilantin, Toradol, Vumon, Taxol, Sandimmune
Propylene glycol	0.1–75	Terramycin, Loxitane, Septra, Lanoxin, Nembutal, Dalgan, Dilantin, Valium, Nitro-Bid, Alkeran,
Polyethylene glycol 300 or 400	0.15–100	VePesid, Robaxin, Bioclate, Ativan
Glycerin	0.04–70	Multitest CMI, D.H.E. 45, Sus-Phrine
Polyethylene glycol 3350	0.3–3.0	Depo-Medrol
Cremophor® EL	50–65	Taxol, Vumon, Sandimmune
Dimethylsulfoxide	$<$0.06	Eminase
Dimethylacetamide	2.0–6.0	Vumon
Sorbitol	0.2–50	Cardene, Aristospan, several vaccine products

[a]Must know the dose of injection to determine actual amount (mg/mL or %) of co-solvent injected per dose of active in product.

Water-Miscible Co-Solvents

A number of solvents that are miscible with water have been used as a portion of the vehicle in the formulation of parenterals. These solvents are used primarily to solubilize certain drugs in an aqueous vehicle and to reduce hydrolysis. The most important solvents in this group are ethyl alcohol, liquid polyethylene glycol, and propylene glycol. Ethyl alcohol is used particularly in the preparation of solutions of cardiac glycosides and the glycols in solutions of barbiturates, certain alkaloids, and certain antibiotics. Such preparations usually are given intramuscularly. There are limitations with the amount of these co-solvents that can be administered because of cellular toxicity concerns, greater potential for hemolysis, and potential for drug precipitation at the site of injection (9). Formulation scientists needing to use one or more of these solvents must consult the literature and toxicologists to ascertain the maximum amount of co-solvents allowed for their particular product (10). Several references provide information on concentrations of co-solvents used in approved commercial parenteral products. An alphabetical listing of acceptable co-solvents, based on their presence in one or more FDA-approved commercial products, is given in Table 6-1 along with some commercial examples.

Nonaqueous Vehicles

Oily vehicles cannot be administered by the intravenous route. The most important group of nonaqueous vehicles is the fixed oils. The USP provides specifications for such vehicles, indicating that the fixed oils must be of vegetable origin so that they will be metabolized, will be liquid at room temperature, and will not become rancid readily. The USP also specifies limits for the free fatty acid content, iodine value, and saponification value (oil heated with alkali to produce soap, i.e., alcohol plus acid salt). The oils most commonly used are corn oil, cottonseed oil, peanut oil, and sesame oil. Fixed oils are used particularly as vehicles for certain hormones (e.g., progesterone, testosterone, deoxycorticicosterone) and vitamin (e.g., Vitamin K, Vitamin E) preparations. The label must state the name of the vehicle so that the user may beware in case of known sensitivity or other reactions to it.

ADDED SUBSTANCES

The USP includes in this category all substances added to a preparation to improve or safeguard its quality. An added substance may

- Increase and maintain drug solubility. Examples include complexing agents and surface-active agents. The most commonly used complexing agents are the cyclodextrins, including Captisol®. The most commonly used surface-active agents are polyoxyethylene sorbitan monolaurate (Tween 20) and polyoxyethylene sorbitans monooleate (Tween 80).

- Provide patient comfort by reducing pain and tissue irritation, as do substances added to make a solution isotonic or near physiological pH. Common tonicity adjusters are sodium chloride, dextrose, and glycerin.
- Enhance the chemical stability of a solution, as do antioxidants, inert gases, chelating agents, and buffers.
- Enhance the chemical and physical stability of a freeze-dried product, as do cryoprotectants and lyoprotectants.
- Enhance the physical stability of proteins by minimizing self-aggregation or interfacial induced aggregation. Surface-active agents serve nicely in this capacity.
- Minimize protein interaction with inert surfaces such as glass and rubber and plastic. Competitive binders such as albumin and surface-active agents are the best examples.
- Protect a preparation against the growth of microorganisms. The term *preservative* sometimes is applied only to those substances that prevent the growth of microorganisms in a preparation. However, such limited use is inappropriate, being better used for all substances that act to retard or prevent the chemical, physical, or biological degradation of a preparation.
- While not covered in this chapter, other reasons for adding solutes to parenteral formulations include sustaining and/or controlling drug release (polymers), maintaining the drug in a suspension dosage form (suspending agents, usually polymers and surface-active agents), establishing emulsified dosage forms (emulsifying agents, usually amphiphilic polymers and surface-active agents), and preparation of liposomes (hydrated phospholipids).

While added substances may prevent a certain reaction from taking place, they may induce others. Not only may visible incompatibilities occur, but hydrolysis, complexation, oxidation, and other reactions may decompose or otherwise inactivate the therapeutic agent or other added substances. Therefore, added substances must be selected with due consideration and investigation of their effect on the total formulation and the container/closure system.

Table 6-2 presents some examples of problems that excipients may cause in formulation development. Formulation scientists must be aware of these potential problems and work around them. In general, any solute added to a parenteral formulation has the potential to cause a problem. In formulation, it is true that something added to solve one problem likely would give rise to another problem. So, formulation scientists must be careful and observant in developing formulations and not be ignorant or blind to potential problems that added substances may create.

Introduction of various categories of added substances used in sterile dosage forms takes place in this chapter with additional coverage found in chapter 8 specific to formulation of solution dosage forms and chapter 10 for freeze-dried dosage forms. Added substances for dispersed systems are covered exclusively in chapter 9.

Table 6-2 Examples of Potential Problems Caused by Use of Certain Excipients in Sterile Drug Product Formulations

Potential problem	Some examples
Source of impurities	Peroxides in polymers, e.g., Polysorbate 80
	Heavy metals (all excipients contain some heavy metal impurity content)
React with and cause drug degradation	Buffers
	Bisulfite
React with packaging components	Antimicrobial preservatives
	Surface-active agents
Crystallize during freezing and potential shift pH	Dibasic sodium phosphate
Crystallize during long-term storage and change rate of dissolution	Mannitol
	Glycine
Change pH in solution via carbon dioxide evolution	Sodium carbonate
Concerns about animal sourced materials	Glycerin
	Polysorbate 80

Source: From Ref. 11.

Table 6-3 Dielectric Constants for Various Solvents

Solvent	Dielectric constant (ε) at 25°C
Water	78.5
Glycerol	42.5
Propylene glycol	32.0
Polyethylene glycol 400	13.6
Dimethyl sulfoxide	46.7
Dimethylacetamide	37.8
Ethanol	24.3
N-Octanol	10.3
Cottonseed oil	3.0

Solubilizing Agents

Solubilizing agents are either co-solvents (strictly speaking, part of the solvent system, not solutes, but still considered as solubilizing agents in this discussion) or amphiphilic compounds classified as either complexing agents or surface-active agents.

Co-solvents already have been covered along with examples given earlier in this chapter Table 6-1). A survey of injectable formulations containing co-solvents finds that ethanol and propylene glycol are the most commonly used co-solvents. Both have high solvent power for organic molecules because of a dielectric constant (measure of electric current conductance) in the 24 to 32 range (water is 78, cottonseed oil is 3). Both are relatively nontoxic in the ranges used in parenteral products (Table 6-1).

Table 6-3 gives dielectric constant (ε) values for several solvents. Dielectric constant is a measure of the electric current conductivity property of solvents. The higher the dielectric constant, the better electric current will travel through the solvent. Thus, water has the highest ε while oil has the lowest. Poorly soluble drugs will have greater solubility in solvents whose ε is not as high as water. Thus, mixtures of water and one or more water-miscible co-solvents will solubilize slightly polar drugs.

An example of the power of a co-solvent to increase solubility of a poorly water-soluble drug is given in Figure 6-1.

The primary problem in using co-solvents is the toxicity of these solvents. Table 6-4 shows the LD_{50} of the four major co-solvents and advantages and disadvantages of using them. In general, small amounts of co-solvents are acceptable, but if the drug dosage is large (i.e., greater than 5 mL), then the usage of co-solvents is limited.

Another disadvantage of using co-solvents is the concern for precipitation at the site of injection if the solution is administered too quickly and the blood stream does not have adequate

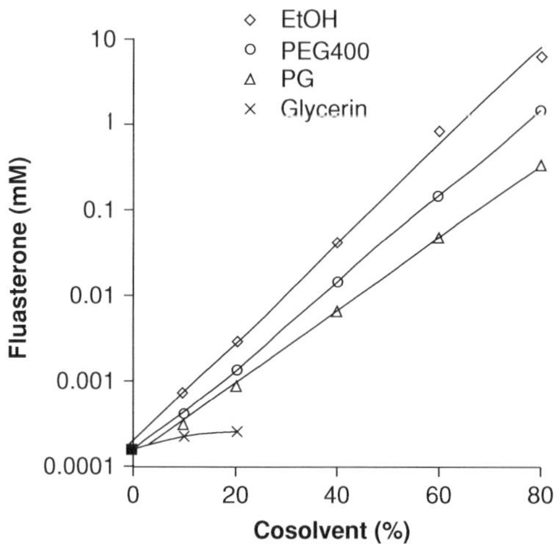

Figure 6-1 Example of co-solvent effect on drug solubility. *Source*: From Ref. 12.

Table 6-4 LD_{50} and Advantages and Disadvantages of the Major Injectable Co-Solvents Used in Commercial Sterile Product Formulations

Co-Solvent	LD_{50} (IV dose in mice)	Advantages and uses	Disadvantages
Ethanol	1.97 g/kg	High solvent power	Can be painful
Glycerin	4/25 g/kg	Popular with insulin and other protein products	Low solvent power compared to others
Polyethylene glycol 400	8.6 g/kg	Stable, low irritation	Viscous, can contain peroxide impurities
Propylene glycol	6.63 g/kg	Stable, wide usage	Moderate solvent power compared to ethanol

Source: From Ref. 6.

time to dilute the drug. Also, all co-solvents have hemolytic effects on red blood cells that can be minimized simply by minimizing the amount of co-solvent administered.

If the co-solvent approach either is unsuccessful or not preferable, then the next formulation approach to increase drug solubility is the use of solubilizing solute additives such as complexing agents or surface-active agents. Some drugs will interact with certain additives to form more soluble complexes. Such additives typically are polymeric amphiphilic molecules, a listing of which is given in Table 6-5. The most commonly used surface-active agents from a safety standpoint for injectables are the nonionic polyoxyethylene fatty acids (Polysorbates or Tweens). The most commonly used complexing agent in recent years has been Captisol®. The chemical structures of Captisol® and Polysorbate 80 are shown in Figures 6-2 and 6-3, respectively, where the amphiphilic nature of both solubilizing agents can be seen.

Complexing agents increase the solubility of drugs anywhere from 2- to 10-fold, but they are not as powerful as co-solvents. In Figure 6-4 it can be seen that nonionic hydroxypropyl beta cyclodextrin (α form contains six glucopyranose rings, β form contains seven rings, and the γ form contains eight rings) and anionic sulfobutylether beta cyclodextrin (Captisol) increases the solubility of the same steroid as was shown in Figure 6-1 for co-solvent solubilization. However, comparison of the ordinates of the two figures shows that the solubilization effect of the cyclodextrins on the drug is linear while the solubilization effect of the co-solvents was logarithmic.

Examples of marketed injectable products containing cyclodextrins as solubilizing agents include Sporanox® (hydroxypropyl-beta-cyclodextrin) and Vfend®, Geodon®, and Zeldox® (sulfobutylether-beta-cyclodextrin). Hydroxypropyl-beta-cyclodextrins suffer from potential renal toxicity problems while sulfobutylether beta cyclodextrin does not accumulate and is more easily eliminated by the renal system.

Surface-active agents will solubilize drugs via micellar solubilization where the drug molecule is "encapsulated" with the hydrophobic core of the agent. A primary use of surface-active agents in the biopharmaceutical product development arena is to help stabilize large molecules from aggregating due to hydrophobic interactions at liquid–air and liquid–solid

Table 6-5 Examples of Added Solute Substances Used In Commercial Sterile Dosage Forms To Increase Injectable Drug Solubility

Hydroxypropyl-beta-cyclodextrin
Sulfobutylether-beta-cyclodextrin (Captisol®)
Polyvinylpyrrolidone (PVP)
Polyethylene glycol 3350
Ethyl lactate
Niacinamide
Desoxycholate sodium
Gelatin
Sodium lauryl sulfate

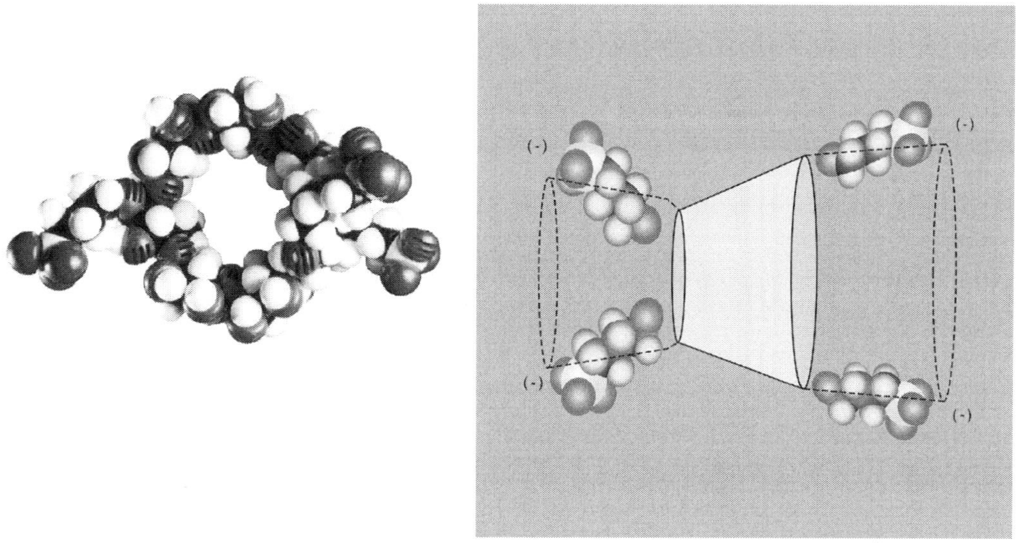

Figure 6-2 Chemical structures of Captisol®. *Source*: Courtesy of CyDex Pharmaceuticals, Inc.

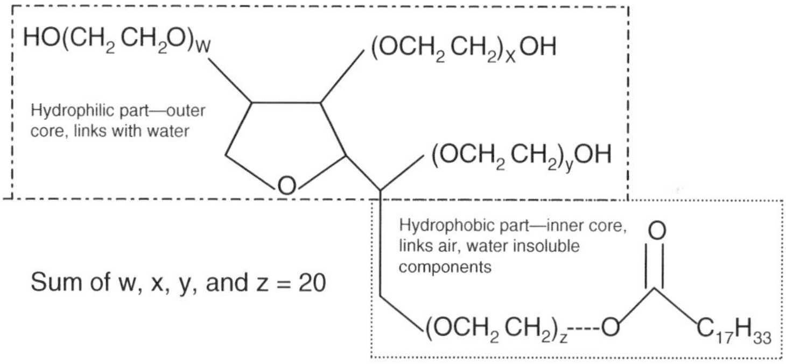

Figure 6-3 Chemical structure of polysorbate 80 (Polyoxyethylene sorbitan monooleate).

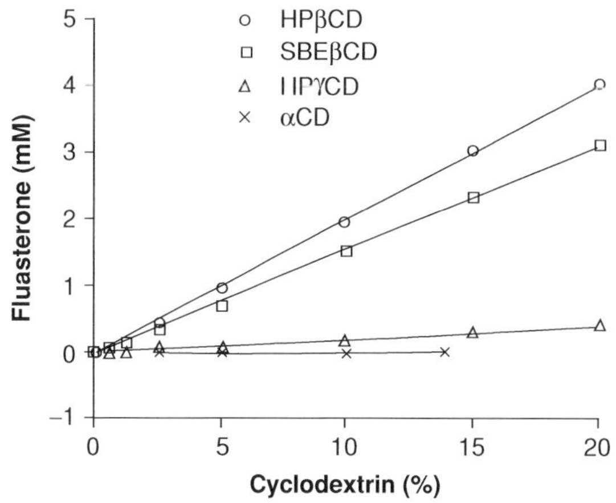

Figure 6-4 Effect of cyclodextrin on fluasterone solubility. *Source*: From Ref. 12.

Table 6-6 Examples of Polysorbates Contained in Commercial Protein Formulations

Product	Active ingredient amount or concentration	Dosage form	Polysorbate concentration
Aranesp®	25–500 μg	Liquid	0.05 mg/mL
ReoPro®	2 mg/mL	Liquid	0.001%
Humira®	40 mg/mL	Liquid	0.8 mg/mL
Avastin®	25 mg/mL	Liquid	1.6 mg/mL
Remicade®	10 mg/mL	Lyophilized	0.05 mg/mL
Aralast®	600 mg/mL	Lyophilized	0.05 mg/mL
Activase®	1 mg/mL	Lyophilized	0.09 mg/mL
Koate®	1.5 mg/mL	Lyophilized	0.025 mg/mL
Advate®	250–1500 IU	Lyophilized	0.17 mg/mL
Kogenate®	1000 IU	Lyophilized	600 μg
NovoSeven®	0.6 mg/mL	Lyophilized	0.1 mg/mL
BeneFix®	250–1000 IU	Lyophilized	0.01%
Tisseel® VH	45 mg	Lyophilized	60 mg
WinRho® SDF	600–5000 IU	Lyophilized	0.01%
PEG-Intron®	0.106–0.307 mg/mL	Lyophilized	0.106 mg/mL
Retavase®	1.81 mg/mL	Lyophilized	0.52 mg/mL
TNKase®	5.25 mg/mL	Lyophilized	0.43 mg/mL
Herceptin®	22 mg/mL	Lyophilized	0.09 mg/mL

interactions. Polysorbates 20 (polyoxyethylene sorbitan monolaurate) and 80 (polyoxyethylene sorbitan monooleate) are amphiphilic, nonionic, and the most widely used surface-active agents used in parenteral formulations. Concentration of polysorbate 20 or polysorbate 80 in commercial protein formulations range from 0.001% to 0.5%. Examples of commercial protein formulations containing polysorbates are listed in Table 6-6.

Surface-active agents may cause pseudoallergic reactions due to release of histamine in mast cells caused by oleic acid, the free fatty acid component of polysorbate 80. This is especially true in the dog model, such that dogs cannot be used to perform toxicity studies of products containing polysorbate 80.

Formulation scientists must be aware of the autooxidation possibility with polysorbates (13). Autooxidation results in cleavage at the ethylene oxide subunits and hydrolysis of the fatty acid ester bond. Hydroperoxides form along with cleavage of side chains and formation of short chain acids such as formic acid, all of which could adversely affect the stability of sensitive biopharmaceutical products.

Examples of surfactants maintaining solubility (and, hence, physical stability) of proteins are found in Figures 6-5 and 6-6 for human growth hormone and factor VIII, respectively.

Antimicrobial Agents

The USP states that antimicrobial agents in bacteriostatic or fungistatic concentrations must be added to preparations contained in multiple-dose containers[1]. They must be present in adequate concentration at the time of use to prevent the multiplication of microorganisms inadvertently introduced into the preparation while withdrawing a portion of the contents with a hypodermic needle and syringe. The USP provides a test for antimicrobial preservative effectiveness to determine that an antimicrobial substance or combination adequately inhibits the growth of microorganisms in a parenteral product. Because antimicrobials may have inherent toxicity for the patient, the USP prescribes maximum volume and concentration limits for those that are used commonly in parenteral products (Table 6-7), for example, phenylmercuric nitrate and thimerosal 0.01%, benzethonium chloride and benzalkonium chloride 0.01%, phenol or cresol 0.5%, and chlorobutanol 0.5%.

[1] The European Pharmacopeia requires multiple-dose products to be bacteriocidal and fungicidal.

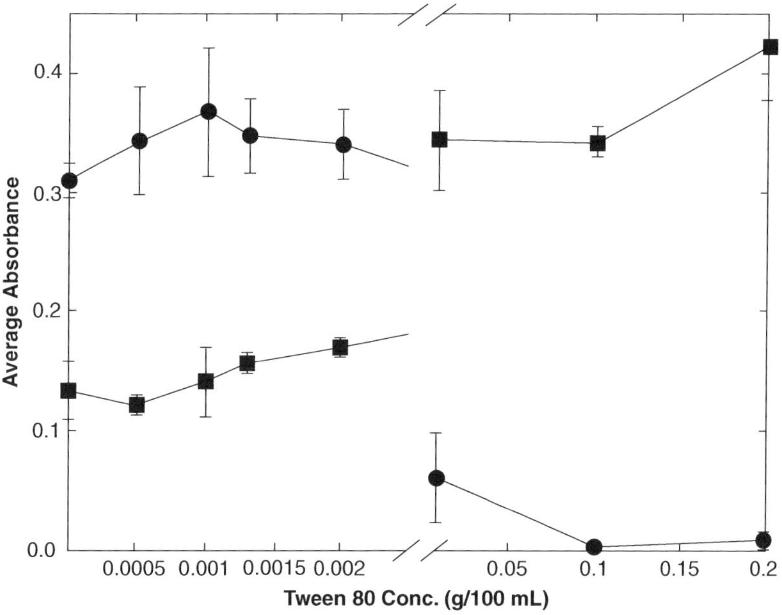

Figure 6-5 Effect of Tween 80 on physical stability of human growth hormone. *Source*: From Ref. 14.

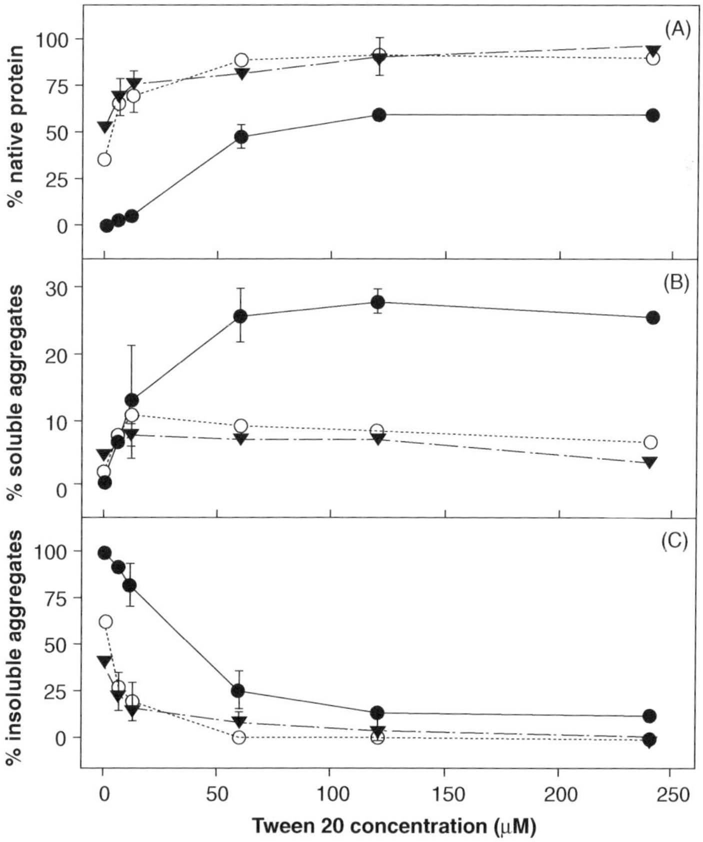

Figure 6-6 Effect of Tween 20 on physical stability of factor VIII. (**A**) Recovery of rXIII Open circle: 1 mg/mL; Closed circle: 5 mg/mL; Triangle: 10 mg/mL. (**B**) Formation of Soluble Aggregates. (**C**) Formation of Insoluble Aggregates. After 10 freeze-thaw cycles as function of concentration of Tween 20. *Source*: From Ref. 15.

Table 6-7 Examples of Commercial Sterile Dosage Forms Containing Antimicrobial Preservative Agents and Their Concentrations

AP agent	Concentration	Examples (all®)
Benzyl alcohol	0.9%–3.0%	VePesid and Vumon
Phenol	0.002%–0.5%	Hydeltrasol and Sus-Phrine
Meta-Cresol	0.25%–0.3%	Humatrope and Genotropin
Phenoxyethanol	0.5%–1.0%	Poliovax and Ipol™
Thimerosal	0.0002%–0.012%	Recombivax and Hyperab
Chlorobutanol	0.25%–0.5%	Aquasol and Oxytocin
Methylparaben	0.02%–0.2%	Intron A and Gentamicin
Propylparaben	0.002%–0.02%	Bicillin L-A and Tobramycin
Phenylmercuric acetate, borate, nitrate	0.001%–0.002%	Several topical ophthalmic medications
Benzalkonium/Benzethonium chloride	0.01%–0.02%	Benadryl and many topical ophthalmic medications

Source: From Ref. 2.

The above limit is rarely used for phenylmercuric nitrate, most frequently employed in a concentration of 0.002%. Methyl *p*-hydroxybenzoate 0.18% and propyl *p*-hydroxybenzoate 0.02% in combination, and benzyl alcohol 2% also are used frequently. Benzyl alcohol, phenol, and the parabens are the most widely used antimicrobial preservative agents used in injectable products. While the mercurials are still allowed to be used in older products, they are not used for new products because of concerns regarding mercury toxicity. In multiple-dose oleaginous preparations, such as Progesterone Injection USP in sesame oil, antimicrobial preservatives such as benzyl alcohol, are part of the formulation. The preservative must be able to partition from the oil phase to the aqueous-based microbial challenge inoculum in order to pass the required compendial preservative effectiveness test.

Antimicrobial agents must be studied with respect to compatibility with all other components of the formula. In addition, their activity must be evaluated in the total formula. It is not uncommon to find that a particular agent will be effective in one formulation but ineffective in another. This may be due to the effect of various components of the formula on the biological activity or availability of the compound; for example, the binding and inactivation of esters of *p*-hydroxybenzoic acid by macromolecules such as Polysorbate 80 or the reduction of phenylmercuric nitrate by sulfide residues in rubber closures. A physical reaction encountered in which bacteriostatic agents sometimes are removed from solution by rubber closures.

Protein pharmaceuticals, because of their cost and/or frequency of use, are preferred to be available as multiple-dose formulations (e.g., human insulin, human growth hormone, interferons, vaccines). However, several proteins are reactive with antimicrobial preservative agents (e.g., tissue plasminogen activator, sargramostim, interleukins) and, therefore, are only available as single dosage form units (see chap. 8). Phenol and benzyl alcohol are the two most common antimicrobial preservatives used in peptide and protein products. Phenoxyethanol is the most frequently used preservative in vaccine products. Table 6-7 lists some examples of protein and small-molecule injectables that are multiple-dose products with antimicrobial preservative type and concentration in the formulation.

Single-dose containers and pharmacy bulk packs that do not contain antimicrobial agents are expected to be used promptly after opening or to be discarded. The ICH/CPMP guidelines[2] require that products without preservatives must be used immediately (within 3 hours after entering the primary package) or a longer usage period must be justified. Large-volume, single-dose containers may not contain an added antimicrobial preservative. Therefore, special care must be exercised in storing such products after the containers have been opened to prepare an admixture, particularly those that can support the growth of microorganisms, such as total parenteral nutrition (TPN) solutions and emulsions. It should be noted that while refrigeration slows the growth of most microorganisms, it does not prevent their growth.

[2] www.eudra.org/emea/pdfs/CPMP_QWP_159_96.pdf.

Buffers

Buffering agents are used primarily to stabilize a solution against chemical degradation or, especially for proteins, physical degradation, i.e., aggregation and precipitation that might occur if the pH changes appreciably. Buffer systems employed should normally have as low a buffering capacity as feasible so as not to disturb significantly the body's buffering systems when injected. In addition, the buffer type and concentration on the activity of the active ingredient must be evaluated carefully. Buffer components are known to catalyze degradation of drugs. The acid salts most frequently employed as buffers are citrates, acetates, and phosphates.

Antioxidants

Substances called antioxidants or reducing agents are required frequently to preserve products because of the ease with which many drugs are oxidized. Sodium bisulfite and other sulfurous acid salts are used most frequently. Ascorbic acid and its salts also are good antioxidants. The sodium salt of ethylenediaminetetraacetic acid (EDTA) has been found to enhance the activity of antioxidants in some cases, apparently by chelating metallic ions that would otherwise catalyze the oxidation reaction.

The oxidation reaction of an oxygen-sensitive drug in general follows the reaction scheme in Figure 6-7. There are three main learning points from this simple schematic that are as follows:

1. The oxidation process is initiated by the formation of a free radical due to the loss of a hydrogen atom that is catalyzed by one or more of the following environmental or product factors:
   a. High temperature—Ambient temperature can be problematic for some oxygen-sensitive drugs. Manufacturing environments for processing oxygen-sensitive products should be in the temperature range of 15° to 21°C.
   b. High pH—The Nernst equation:

   $$E = E_0 + \frac{0.06}{2\log[\mathrm{H}^+]} \rightarrow \frac{[\mathrm{Ox}]}{[\mathrm{Rd}]}$$

   shows the relationship between pH and oxidation potential. The higher the pH the lower the potential for oxidation, meaning that drugs will more readily oxidize at higher pH. The example of morphine oxidative degradation as a function of pH is shown in Figure 6-8. Epinephrine borate, formulated at pH 7.4 requires the synergistic effects of two antioxidants (ascorbic acid and acetylcysteine) for stabilization whereas epinephrine hydrochloride, formulated at pH 3.5, requires no antioxidant addition for stabilization.

Drug (R) Oxidation Process

$$\mathrm{RH} \xrightarrow{\text{Catalysts}^a} \mathrm{R}\bullet + \mathrm{H}\bullet$$

$$\mathrm{R}\bullet + \mathrm{O_2} \longrightarrow \mathrm{RO_2}\bullet$$

$$\mathrm{RO_2}\bullet + \mathrm{RH} \longrightarrow \mathrm{RO_2H} + \mathrm{R}\bullet$$

[a]Catalysts include
high temperature
high pH
heavy metals
peroxides
light

Therefore, oxidation requires two main factors:
-- free radical formation caused by catalysts
-- molecular oxygen

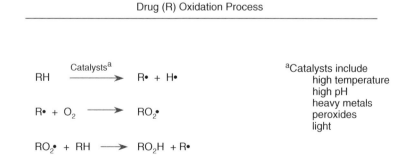

Figure 6-7 Basic schematic of an oxidation reaction.

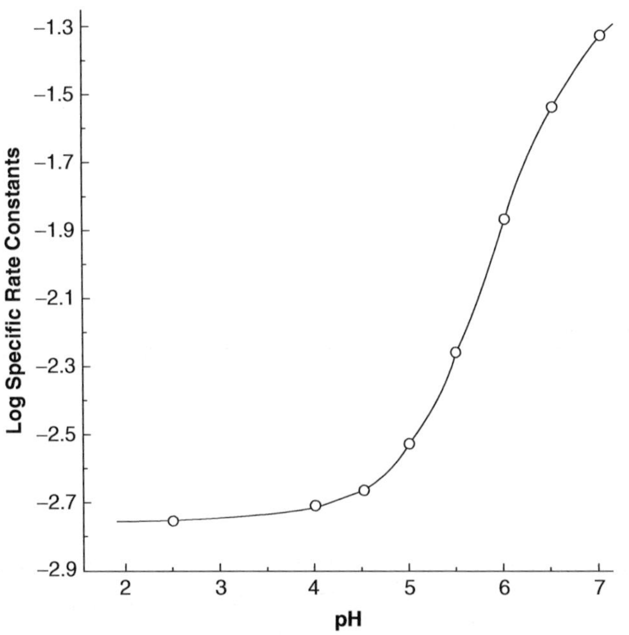

Figure 6-8 Example of pH effect on morphine oxidation. *Source*: From Ref. 16.

c. Light exposure—Light itself can photolytically degrade many drugs (examples include many cephalosporins, aminophylline, diazepam, gentamicin, atropine sulfate, dexamethasone, haloperidol, nitroprusside, many others) and/or can catalyze the formation of free radicals leading to oxidative degradation. Minimizing light exposure is accomplished by such measures as minimizing light intensities in manufacturing areas, minimizing exposure of solution to the environment, and using primary and/or secondary packaging that blocks light transmission into the container interior.
d. Heavy metals—Heavy metal contamination is a common problem because trace heavy metal extractables originate from glass, rubber, plastic, water, and raw materials. Because heavy metals cannot be completely eliminated, heavy metal chelating agents such as disodium ethylenediaminetetraacetic acid (Fig. 6-9) or citrate solutes are added to the formulation.
e. Peroxides—Peroxides primarily originate from polymers used for plastic tubing, filters, packaging, or polymeric solutes such as polyoxyethylene surfactants. Obviously, the best way to minimize the potential for peroxide contamination is to avoid exposure or presence of any of these substances known to contain peroxides. Solutes such as polysorbate 80 that can contain peroxides can be purchased in grades where peroxides have been removed.

Therefore, all measures and precautions taken to remove these catalysts during drug product manufacture will minimize or eliminate the formation of free radicals.

2. The free radical that is formed then reacts with available molecular oxygen to propagate the oxidation reaction. Thus, efforts to reduce or eliminate the presence of molecular oxygen will minimize the oxidation reaction from proceeding extensively.
3. The presence of an antioxidant will either prevent the free radical formation of the drug (oil-soluble free radical inhibitor antioxidants) or preferentially form the free radical and react with oxygen (water-soluble antioxidants).

Displacing the air (oxygen) in and above the solution by purging with an inert gas, such as nitrogen, also can be used as a means to control oxidation of a sensitive drug. Process control is required for assurance that every container is deaerated adequately and uniformly. However, conventional processes for removing oxygen from liquids and containers

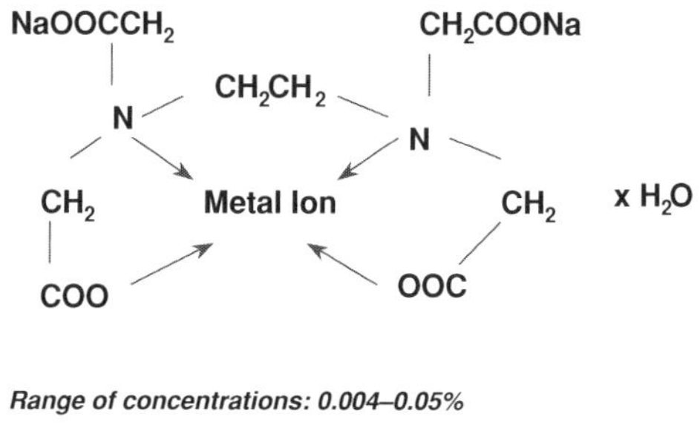

Figure 6-9 Structure of ethylenediaminetetraacetic acid (EDTA) disodium salt and how it binds metal ions.

do not absolutely remove all oxygen. The only approach for completely removing oxygen is to employ isolator technology where the entire atmosphere can be recirculating nitrogen or another nonoxygen gas.

Elaboration of the use of antioxidants and other approaches employed to stabilize oxygen-sensitive protein drugs in solution may be found in chapter 8.

Tonicity Agents

While it is the goal for every injectable product to be isotonic with physiologic fluids, this is not an essential requirement for small-volume injectables that are administered intravenously. However, products administered by all other routes, especially into the eye or spinal fluid must be isotonic. Injections into the subcutaneous tissue and muscles also should be isotonic to minimize pain and tissue irritation. Tonicity-adjusting agents most commonly used are electrolytes (sodium chloride most common), glycerin, and mono- or disaccharides.

Cryoprotectants and Lyoprotectants

These substances serve to protect biopharmaceuticals from adverse effects due to freezing and/or drying of the product during freeze-dry processing. Sugars (nonreducing) such as sucrose or trehalose, amino acids such as glycine or lysine, polymers such as liquid polyethylene glycol or dextran, and polyols such as mannitol or sorbitol all are possible cryo- or lyoprotectants. Several theories exist to explain why these additives work to protect proteins against freezing and/or drying effects. Excipients that are preferentially excluded from the surface of the protein are the best cryoprotectants, and excipients that remain amorphous during and after freeze-drying serve best as lyoprotectants. These concepts of additive stabilization of biopharmaceuticals during freezing, drying, and/or in the dry state are covered in chapter 10.

Competitive Binders

These additives are used if the active ingredient is known to bind excessively to container and manufacturing equipment surfaces. Such additives compete with the active ingredient for the surface-binding sites and keep the active ingredient from losing potency or activity in the dosage form. Historically, the best or most commonly used competitive binder has been human serum albumin (HSA) at concentrations ranging from 0.1% to 1.0%.

Concerns used to exist over potential viral contamination of natural substances such as HSA. Attempts to identify other potential competitive binding agents as effective as HSA have generally been unsuccessful, although it has been reported that Polysorbate 80, albeit at fairly high concentrations, inhibited recombinant Factor VIII adsorption at solid–water surfaces (17). Recombinant HSA removed the viral contamination fears and is now used in commercial products.

Other Additives

Other purposes for solute additives in sterile product formulations include bulking agents for freeze-dried products, suspending agents and wetting agents for suspensions, emulsifying agents for emulsions, viscosity-inducing agents for topical ophthalmic products, and the specialized polymers used to formulate advanced sustained-, prolonged-, extended-, delayed-, or

Table 6-8 Examples of Additives Used in Specialized Sterile Dosage Forms

Dosage form	Purpose of additive	Primary examples
Freeze-dried products	Bulking agents	Mannitol Glycine Sodium phosphate
Suspensions	Wetting agents	Surfactants (e.g., Polysorbate 80) Lecithin Sorbitol trioleate
	Suspending agents	Sodium methylcellulose Sodium carboxymethylcellulose Polyethylene glycol Propylene glycol Polyvinylpyrrolidone Sodium alginate
Emulsions	Emulsifying agents	Egg yolk phospholipid Lecithin Surfactants
Topical ophthalmic solutions	Viscosity-inducing agents	Hydroxypropyl methylcellulose Polyvinyl alcohol
Liposomes	Incorporate active ingredient for targeting or other distribution mechanism	See Table 3-3
Extended-release products	Affect release of drug from injected formulation	Polylactic-polyglycolic polymers Polyanhydrides Poly(orthoesters) Poly(2-hydroxyethyl methacrylate) Hyaluronic acid Polyethylene glycol (pegylation)

controlled-release dosage forms (microspheres, liposomes, gels, and other specialized injectable delivery systems). While most of the dosage forms and formulation additives are covered in other chapters, a summary of the examples of additives used in specialized sterile dosage forms is given in Table 6-8.

REFERENCES

1. Nema S, Washkuhn R, Brendel RJ. Excipients and their use in injectable products. PDA J Pharm Sci Technol 1997; 51:166–171.
2. Powell MF, Nguyen T, Baloian L. Compendium of excipients for parenteral formulations. PDA J Parenter Sci Technol 1998; 52:238–311.
3. Strickley RG. Parenteral formulations of small molecules therapeutics marketed in the United States, Part I. PDA J Parenter Sci Technol 1999; 53:324–349.
4. Strickley RG. Parenteral formulations of small molecules therapeutics marketed in the United States, Part II. PDA J Parenter Sci Technol 2000; 54:69–96.
5. Strickley RG. Parenteral formulations of small molecules therapeutics marketed in the United States, Part III. PDA J Parenter Sci Technol 2000; 54:152–169.
6. Rowe RC, Sheskey PJ, Quinn ME. Handbook of Pharmaceutical Excipients. 6th ed. London/Chicago: Pharmaceutical Press, 2009.
7. Katdare A, Chaubal MV, eds. Excipient Development for Pharmaceutical, Biotechnology, and Drug Delivery Systems. New York/London: Informa Healthcare, 2006.
8. FDA/Center for Drug Evaluation and Research, Office of Generic Drugs, Division of Labeling and Program Support. http://www.accessdata.fda.gov/scripts/cder/iig/index.cfm. Last Updated April 22, 2010.
9. Yalkowsky SH, Krzyzaniak JF, Ward GH. Formulation-related problems associated with intravenous drug delivery. J Pharm Sci 1998; 87:787–796.
10. Mottu F, Laurent A, Rufenacht DA, et al. Organic solvents for pharmaceutical parenterals and embolic liquids: A review of toxicity data. PDA J Pharm Sci Technol 2000; 54:456–469.
11. Akers MJ. Excipient-drug interactions in parenteral formulations. J Pharm Sci 2002; 91:2283–2300.
12. Zhao L, Li P, Yalkowsky SH. Solubilization of fluasterone. J Pharm Sci 1999; 88:967.

13. Kerwin BA. Polysorbates 20 and 80 used in the formulation of protein biotherapeutics: Structure and degradation pathways. J Pharm Sci 2008; 97:2924–2935.
14. Katakam M, Bell LN, Banga AK. Effect of surfactants on the physical stability of recombinant human growth hormone. J Pharm Sci 1995; 84:713.
15. Krielgaard L, Jones LS, Randolph T, et al. Effect of tween 20 on freeze-thawing and agitation-induced aggregation of recombinant human factor XIII. J Pharm Sci 1998; 87:1597–1603.
16. Yeh S, Lach JL. Stability of morphine in aqueous solution III. Kinetics of morphine degradation in aqueous solution. J Pharm Sci 1961; 50:35–42.
17. Joshi O, McGuire J, Wang DQ. Adsorption and function of recombinant factor VIII at solid-water interfaces in the presence of Tween-80. J Pharm Sci 2008; 97:4741–4755.

7 Sterile products packaging chemistry

Parenteral products are filled into primary packaging that is either glass or plastic. Many primary packaging systems, including vials, all bottles except for solutions for irrigation, syringes, and cartridges, are closed with some kind of rubber stopper, be it the closure on the vial or bottle or the septum and plunger for the syringe and cartridge. Irrigating solutions are packaged in glass bottles with screw caps rather than rubber closures. Products for topical application to the eye are packaged into plastic droptainers with plastic screw caps or, for ophthalmic ointments, into aluminum tubes and capped with plastic screw caps. Of course, all primary packaging is sterilized either prior to filling for aseptic processed products or terminally sterilized. This chapter focuses on some of the basic chemistry principles of glass (1–4), rubber (5–9), and plastic materials (10–12) and will highlight concerns about extractables and leachables from these surfaces (13–19). A review paper on which this chapter was based can also be a good source of information with additional references and coverage of convenient packaging delivery systems (20).

GLASS[1]

Glass is primarily composed of the element silicon. Silicon is a chemical element, one of the 109 known substances that constitute the universe's matter. Second only to carbon in its presence on earth, one-quarter of the earth's crust is silicon. Carbon is also the only element capable of producing more compounds than silicon.

However, one does not find silicon alone in nature. It always exists as silica or silicates. Silica is silicon dioxide (SiO_2), commonly found in sand and quartz. A silicate is a compound made of silicon, oxygen, and at least one metal, sometimes with hydrogen, sometimes without it. The most widely recognized synthetic form is sodium silicate, or water glass, a combination of silica with sodium and hydrogen. Materials lacking the molecular lattice structure of a solid state are amorphous, for example, all liquids. Thus, an amorphous form of a material possesses the same atomic makeup as the crystalline version, but without a highly ordered geometry.

The Assyrian King Ashurbanipal (669–626 BC) described glass as "Take 60 parts sand, 180 parts ashes of sea plants, 5 parts chalk–and you have glass" (21). Glass is an inorganic product of melting, which when cooled without crystallization, assumes a solid state. Glass is structurally similar to a liquid but has a viscosity so great at normal ambient temperatures that it is considered a solid.

Glass is employed as the container material of choice for most small-volume injectables. It is composed principally of silicon dioxide, with varying amounts of other oxides such as sodium, potassium, calcium, magnesium, aluminum, boron, and iron. The basic structural network of glass is formed by the silicon oxide tetrahedron. Boric oxide will enter into this structure, but most of the other oxides do not. The latter are only loosely bound, present in the network interstices, and are relatively free to migrate. These migratory oxides may be leached into a solution in contact with the glass, particularly during the increased reactivity of thermal sterilization. The leaching process is a diffusion controlled ion-exchange process involving exchange of hydrogen ions for the alkali ions present in the glass. The result is an increase in solution pH. This is especially problematic for packaged water products (e.g., Sterile Water for Injection) or dilute drug products that have little to no buffer capacity. Additionally, some glass compounds will be attacked by solutions and, in time, dislodge glass flakes into the solution. Such occurrences can be minimized by the proper selection of the glass composition and appropriate control of the container manufacturing process (discussed later).

[1] The technical information provided by Schott and Alcan are greatly appreciated.

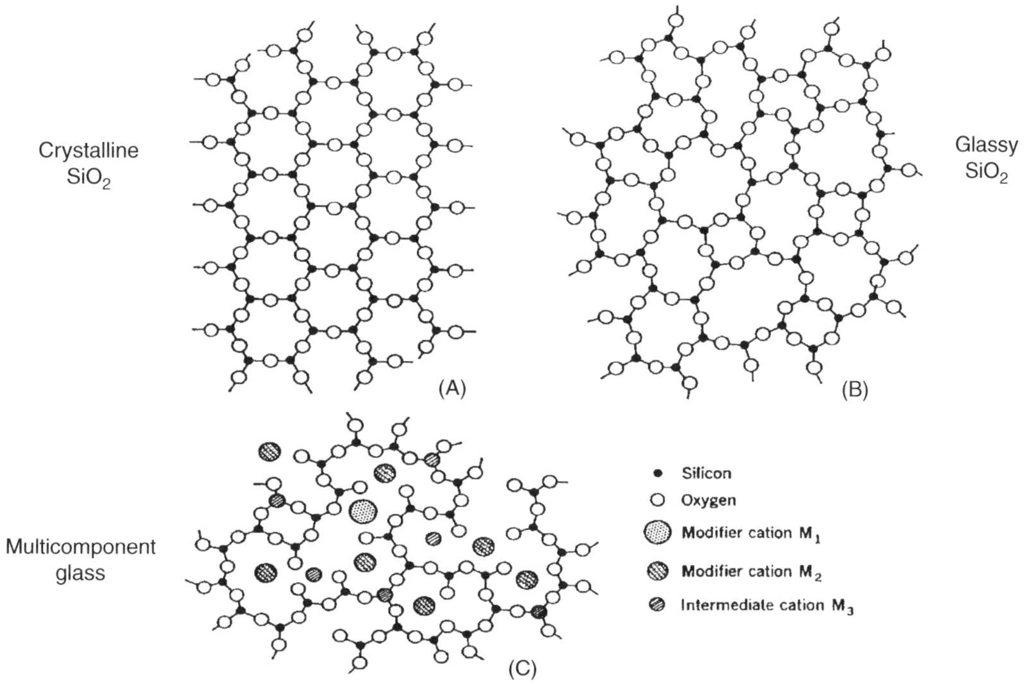

Figure 7-1 Molecular structures of glass. *Source*: Courtesy of Schott Glass.

Types of glass used in parenteral packaging are mixtures of crystalline oxides and carbonates (Fig. 7-1). Glass is melted by heating into a viscous liquid state becoming increasingly resistant to flow as it is cooled. Glass is considered a solid below ~500°C. Glass is composed of the network former—SiO_2 tetrahedron plus network modifiers (disodium oxide, boron oxide, and lead oxide) that lower the melting point. Stabilizers such as calcium oxide, aluminum oxide, and more disodium oxide are added to improve durability. Some glass contains colorants such as iron or titanium oxides.

Types

The United States Pharmacopeia <661> provides four classifications of glass based on chemical resistance:

- Type I, a borosilicate glass
- Type II, a soda-lime treated glass
- Type III, a soda-lime glass
- NP, a soda-lime glass not suitable for containers for parenterals.

Type I glass is composed principally of silicon dioxide (~81%) and boric oxide (~13%), with low levels of the non-network-forming oxides (such as sodium and aluminum oxides) (Fig. 7-2). It is a chemically resistant glass (low leachability) also having a low thermal coefficient of expansion (33×10^{-7} cm/cm/°C or $49–54 \times 10^{-7}$ cm/cm/°C). The former is called "Type I 33 expansion glass" and the latter is called "Type × (typically 51) expansion glass."

Types II and III glass types (both are soda-lime glass with Type II being chemically treated to reduce alkali leachables) are composed of relatively high proportions of sodium oxide (~14%) and calcium oxide (~8%) (Fig. 7-3). This makes the glass chemically less resistant. Both types melt at a lower temperature, are easier to mold into various shapes, and have a higher thermal coefficient of expansion than Type I (e.g., 90×10^{-7} cm/cm/°C for Type III). While there is no one standard formulation for glass among manufacturers of these United States Pharmacopeia (USP) type categories, Type II glass usually has a lower concentration of the migratory oxides than Type III. In addition, Type II has been treated under controlled temperature and humidity

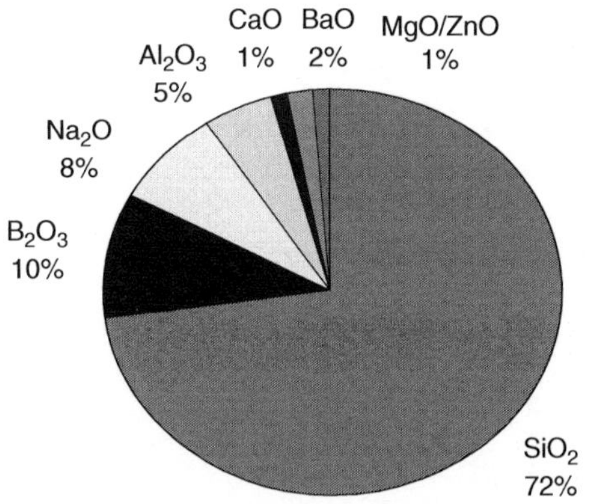

Figure 7-2 Typical composition of Type I borosilicate glass.

conditions with sulfur dioxide or other dealkalizers to neutralize the interior surface of the container. This surface will substantially increase the chemical resistance of the glass while the surface remains intact. However, repeated exposures to sterilization and alkaline detergents will break down this dealkalized surface and expose the underlying soda-lime compound.

The glass types are determined from the results of two USP tests: the Powdered Glass Test and the Water Attack Test (USP <660>). The Water Attack Test is used only for Type II glass and is performed on the whole container, because of the dealkalized surface; the former is performed on powdered glass, which exposes internal surfaces of the glass compound. The results are based on the amount of alkali titrated by 0.02 N sulfuric acid after an autoclaving cycle with the glass sample in contact with a high-purity distilled water. Thus, the *Powdered Glass Test* challenges the leaching potential of the interior structure of the glass while the *Water Attack Test* challenges only the intact surface of the container. Compendial references include USP <661>, European Pharmacopeia (EP) 3.2.1, and Japanese Pharmacopeia (JP) <57>. It is important to note that although the glass powder test challenges the leaching potential of the glass structure, it does not provide any information on the resistance of the inner surface of the container (22).

Selecting the appropriate glass composition and reaching agreement with the supplier on the final requirements are critical facets of determining the overall specifications for each parenteral formulation.

Physical Properties

Glass, as already described, is extremely viscous and deforms very slowly under external forces. Viscosity is temperature dependent with glass formation occurring at 10^3 to 10^8 poise. The annealing viscosity of glass is approximately 10^{13} poise.

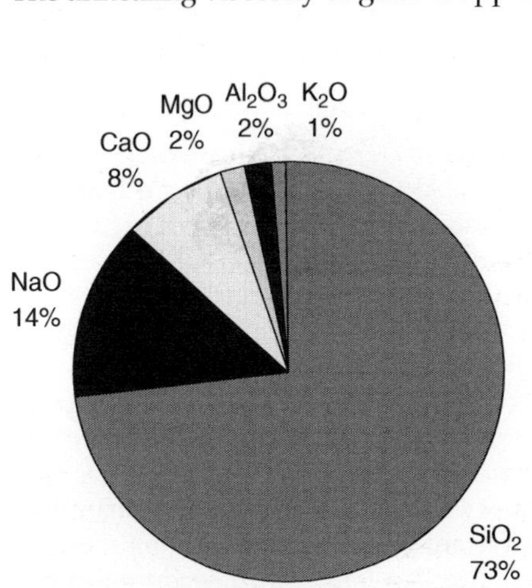

Figure 7-3 Typical composition of Type II and III soda-lime glass.

Figure 7-4 Example of tubing (left two) and molded (right two) glass vials. *Source*: Photo courtesy of Dr. Gregory Sacha, Baxter BioPharma Solutions.

Other important glass physical properties include chemical durability (as determined by compendial alkalinity tests), thermal expansion, color, and density. Glass color is produced using different metal oxides. Amber colored glass is created using ferric oxide and is the most frequently used colored glass for parenteral products. Glass density depends somewhat on type of glass, but the range of density is 2.44 to 2.50 g/cc.

Commercially available containers vary in size from 0.5 to 1000 mL. Sizes up to 100 mL may be obtained as ampoules and vials, and larger sizes as bottles. The latter are used mostly for intravenous and irrigating solutions. Smaller sizes also are available as syringes and cartridges. Ampoules, syringes, and cartridges are formed from glass tubing. The smaller vials may be made by molding or from tubing. Larger vials and bottles are made only by molding. Containers produced from glass tubing are generally optically clearer and have a thinner wall than molded containers (Fig. 7-4). Compared with molded glass, tubing glass also has better wall and finish dimensional consistency, no seams, easier to label, weighs less, facilitates inspection, and has lower tooling costs. Tubing glass is preferable to molded glass for freeze-dried products because of more efficient heat transfer from the shelf into the product. Molded containers can be more uniform in external dimensions, stronger, and heavier.

Glass containers must be sufficiently strong to withstand the physical shocks of handling and shipping and the pressure differentials that develop, particularly during the autoclave sterilization cycle. They must be able to withstand the thermal shock resulting from large temperature changes during processing, for example, when the hot bottle and contents are exposed to room air at the end of the sterilization cycle. Therefore, a glass with a low coefficient of thermal expansion is necessary. The container also must be transparent to permit inspection of the contents.

Preparations that are light sensitive must be protected by placing them in amber glass containers or by enclosing flint glass containers in opaque cartons labeled to remain on the container during the period of use. It should be noted that the amber color of the glass is imparted by the incorporation of potentially leachable heavy metals, mostly iron and manganese, which may act as catalysts for oxidative degradation reactions. Silicone coatings, typically silicone emulsions, are sometimes applied ("baked") to the inner surfaces of vials to produce a hydrophobic surface. One example for using the application is to facilitate the drainage of injectable suspension products.

The size of single-dose containers is limited to 1000 mL by the USP and multiple-dose containers to 30 mL (23) unless stated otherwise in a particular monograph. Multiple-dose vials are limited in size to reduce the number of punctures for withdrawing doses and the accompanying risk of contamination of the contents. As the name implies, single-dose containers are opened or penetrated with aseptic care, and the contents used at one time. These may range in size from 1000-mL bottles to 1-mL or less ampoules, vials, or syringes. The integrity of both single and multiple-dose vials is compromised whenever the rubber plunger is punctured with a needle, but the presence of an antimicrobial preservative agent allows the multiple-dose container to be punctured multiple times.

A multiple-dose container is designed so that more than one dose can be withdrawn at different times while maintaining a seal between uses. It should be evident that with full aseptic precautions, including sterile syringe and needle for withdrawing the dose and disinfection of the exposed surface of the closure, there is still a substantial risk of introducing contaminating microorganisms and viruses into the contents of the vial. Because of this risk, the USP requires that all multiple-dose vials must contain an antimicrobial agent or be inherently antimicrobial, as determined by the USP *Antimicrobial Preservatives Effectiveness* tests (24). There are no comparable antiviral effectiveness tests, nor are antiviral agents available for such use. In spite of the advantageous flexibility of the dosage provided by multiple-dose vials, single-dose, disposable container units provide the clear advantage of greater sterility assurance and patient safety.

Manufacturing

Tubing glass is manufactured by starting with a tube of glass of the appropriate diameter formed by either the Danner or Vello processes (22,25–26). In both processes, glass flows vertically from the bottom of the furnace. Liquid glass is drawn, horizontally, away from a mandrel in the Danner process and is drawn vertically in the Vello process. The Danner process is typically used for glass formulations containing 10% or less B_2O_3 and the Vello process is not suitable for diameters of approximately 45 to 50 mm.

In the forming part of the process (Fig. 7-5), the tubing is preheated to form the shoulder of the container and form the finish of the glass opening. The tube is then cut to form the bottom, heated to smooth the bottom, treated if desired, and cooled. The heat used during smoothing vaporizes the glass and the vapors condense on the inside, producing a rough surface (27). These rough spots are chemically different from the rest of the surface and can be more reactive and less durable. A sulfur treatment using ammonium sulfate can be used to make the sodium borate deposited at the surface of the vial more soluble. The deposits are then washed away during the cleaning process. The treatment can help to reduce pH shifts in solutions resulting from the sodium ion but has no effect on the smoothness or durability of the vial surface.

The heat used when forming the vials directly affects the level of extractables at the surface of the vials. Therefore, some manufacturers offer vials produced at lower temperatures, referred to as a cold forming process. This improves the resistance of the glass to reduce the level of extractables.

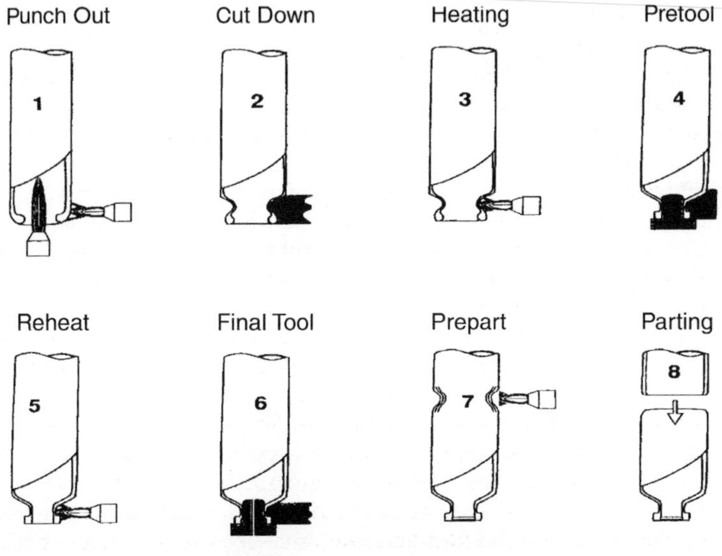

Figure 7-5 Formation of tubing glass. *Source*: Courtesy of Schott Glass.

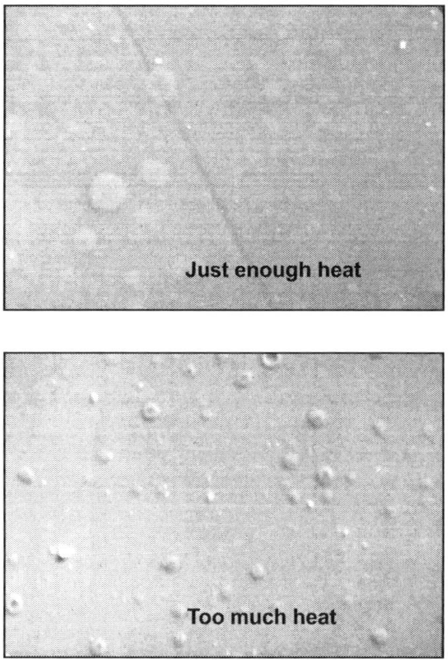

Figure 7-6 Comparison of quality of glass interior surface (1500× magnification). Just enough heat—minimal vaporization, little ion exchange, durable surface; too much heat—extensive vaporization, much ion exchange, vulnerable surface. *Source*: Courtesy of Schott Glass.

Molded glass is prepared from extruded molten glass. The mouth and initial shape are formed, air is blown to shape the mold, then annealed and cooled.

Tubing glass has fewer cosmetic defects and is more aesthetically pleasing. Tubing glass is utilized more for lyophilized and/or costly drug products. Molded glass tends to be heavier, more durable, less costly, and better for larger fill volumes. Molded glass is easier to process on production lines since it is heavier and does not tip or wobble so easily. If the tubing glass process is poorly controlled, for example, too much heat applied in glass formation and shaping, the surface characteristics will be adversely affected (Fig. 7-6).

Cleaning and Sterilization

With the exception of ready-to-fill glass syringes, glass containers are cleaned using Water for Injection, then sterilized and depyrogenated using dry heat, usually with tunnel sterilizers where the temperatures reach at least 300°C, necessary for depyrogenation. Depyrogenation does not follow the same time–temperature model used for sterilization (28). A minimum processing temperature must be reached to destroy pyrogens. Therefore, longer exposure times at lower temperatures are not acceptable. Glass syringes and cartridges need to be siliconized with the siliconization occurring before sterilization procedures.

The following is a typical procedure: Vials are received from the warehouse and part numbers verified as required on the master batch record. Vials are wrapped in shrinkwrap to minimize particulate matter. Vials are washed (actually rinsed, there is typically no detergent used) using washing equipment such as Calumatic or Metromatic machines. After rinsing, wet vials are placed either in a dry heat oven (e.g., Despatch) or on a conveyor line (e.g., Strunck, Bosch + Strobel) for sterilization and depyrogenation. Depending on the size, vials may proceed through a tunnel set at 340°C in approximately eight minutes, continue through a cooling area within Class 100/Grade A/ISO 5 conditions, and then flow into the vial filling equipment.

Glass syringes are prewashed, depyrogenated by the washing and rinsing process thus capitalizing on the high temperature annealing process previously described. The syringes are presterilized using ethylene oxide by the manufacturer and come in "tubs" (Fig. 7-7) (29) that are ready for filling in the Class 100 clean area. For glass cartridges, cartridges are loaded onto a conveyor system where they are rinsed, siliconized, sterilized, and depyrogenated, then filled with product all on the same preparation and filling equipment.

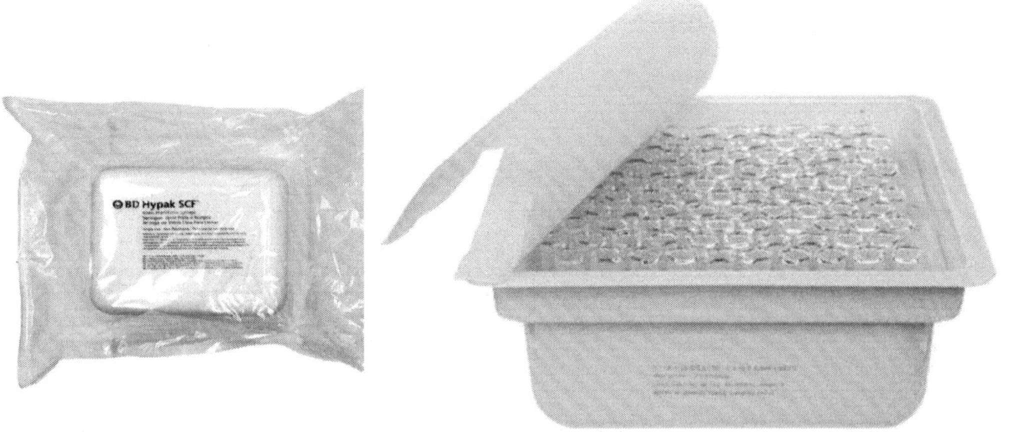

Figure 7-7 Hypak™ syringe "tub" containing SCF™ (sterile, clean, ready-to-fill) syringes. *Source*: Courtesy of © Becton, Dickinson and Company.

Glass Defects and Particulates

Glass defects such as microcracks and strains (Fig. 7-8), if not detected early, will eventually result in glass breakage later. Close attention to the process controls during the forming of glass vials and careful handling prior to depyrogenation and filling are necessary to reduce the potential for glass breakage. Rough handling or significant glass-to-glass contact can result in damage to the vials. This results in significant economic loss, especially for expensive biopharmaceutical products, and, worse, if not detected prior to release, could result in product complaints and, more seriously, resulting in microbial contamination and patient harm. While it is impossible to prevent glass defects completely, additional controls can be in place to minimize or eliminate any defective glass package from being released. First, several companies have resorted to 100% inspection of all glass units during incoming QC inspection prior to release to manufacturing. Second, every effort must be expended to minimize glass-to-glass contact during processing. This is especially the case when glass is loaded onto tunnel conveyor systems prior to dry heat sterilization and depyrogenation as well as glass units moving on accumulation tables post depyrogenation prior to filling.

Glass flaking is another kind of defect that must be controlled. Flaking is due to embedded glass on the inner surface that is delaminated over time and handling. Such flakes are alkali

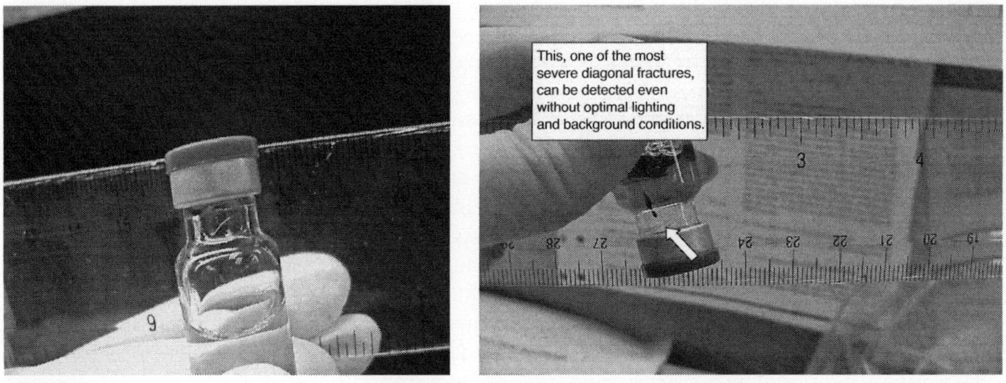

(A) (B)

Figure 7-8 Examples of glass cracks. (A) Characteristic fracture. (B) This, one of the most severe diagonal fractures, can be detected even without optimal lighting and background conditions. *Source*: Courtesy of Kristy Fraizer, Baxter BioPharma Solutions.

borates that migrated to the inner surface, evaporated, then recondensed during final glass preparation at the manufacturer. Flaking can be minimized by the glass manufacturer using proper times and temperatures during preparation and annealing (30). The risk of delamination increases with solutions formulated at high pH ($\geq$ pH 8) and with certain buffers formulated at pH 7 or greater. Common buffers that can cause delamination include citrate, tartrate, and phosphate (30).

When developing solutions containing these buffers and/or when the solution pH is alkaline, using chemically treated glass may help, but this does not always solve the problem of glass delamination. Forming the container, especially the bottom, with as little heat as possible, may be a better approach. Although this requires close interactions and relationships with the glass container manufacturer. Rinsing and depyrogenation also need to be optimized by first removing as much excess water as possible prior to heat depyrogenation and applying an optimal depyrogenation time/temperature exposure cycle.

Treated Glass

When glass is heated, metal ions, primarily sodium, and potassium will increase at the surface of the glass and become potential serious leachates. Trace amounts of calcium and potassium sulfates may also form at the glass surface. One common treatment of glass to reduce these potential leachates is ammonium sulfate. Acidic ammonium sulfate in the vapor state (>490°C) will react with these cationic metals forming soluble salts with sodium and potassium and displace calcium with hydrogen from decomposing acid ammonium sulfate. This pretreatment of glass is relatively inexpensive and effective in reducing potential metallic leachates, but may not reduce delamination.

Schott developed a technology called Plasma Impulse Chemical Vapor Deposition (PICVD) that coats the inner surface of Type I glass vials with an ultra-thin film of silicon dioxide (31). This film forms a highly efficient diffusion barrier that practically eliminates glass leachables. This kind of glass is especially useful for drug products having high pH values, formulations with complexing agents, or products showing high sensitivity to pH shifts.

Glass Leachables (Glass–Drug Product Interactions) and Extractables

All types of glass have the potential to leach alkali-based substituents into the product. In general, the following guidelines apply with respect to glass leachables:

- Relatively low levels of leachables at pH 4–8
- Relatively high levels of leachables at pH > 9
- Major extractables are silicon, sodium, and boron
- Minor extractables may include potassium, barium, calcium, and aluminum, depending on the specific glass formulation
- Trace extractables include iron, magnesium, and zinc
- Treated glass gives less extractables if pH < 8, although there is always the possibility of having sulfate leachables.

The presence of phosphate (e.g., phosphate buffer) anions make formulations especially vulnerable because of the distinct possibility of phosphate forming insoluble salts with divalent metal cations (e.g., calcium, iron, zinc, and magnesium) potentially present at the surface of the glass interior. The amount of potential extractable ions at the inner glass surface is a function of how the glass was manufactured, what temperatures were used, and exposure to high temperatures (e.g., glass sterilization).

Type I glass will be suitable for all products, although sulfur dioxide treatment sometimes is used for even greater resistance to glass leachables. Because cost must be considered, one or the other, less-expensive types may be acceptable. Type II glass may be suitable, for example, for a solution that is buffered, has a pH below 7, or is not reactive with the glass. Type III glass usually will be suitable principally for anhydrous liquids or dry substances. However, some manufacturer-to-manufacturer variation in glass composition should be anticipated within each glass type. Therefore, for highly chemically sensitive parenteral formulations it may be necessary to specify both USP Type and a specific manufacturer.

Incompatibilities between glass and product may include the following:

1. Ion exchange of metal ions if the product contains sodium, magnesium, calcium, aluminum, or lithium.
2. Dissolution of glass and resultant particles if the product contains phosphate or citrate.
3. Pitting of glass resulting in particles if the product contains a metal chelating agent such as disodium ethylenediaminetetraacetic acid (EDTA).
4. Adsorption of the active ingredient at the glass surface, a major problem for many biomolecules requiring the use of competitive binding excipients in the formulation.

If any of these problems are found to occur during product development, then treated glass must be used or the formulation modified to remove or reduce the amount of ingredient reacting with the glass surface.

Testing Methods

Glass extractables are always the primary concern and this is reflected in the required compendial test requirements. Test requirements vary depending on the compendia (USP vs. EP vs. JP). All require light transmission, arsenic, and the alkalinity tests (powdered glass or water attack). Other tests include hydrofluoric acid testing (EP), soluble iron (JP), and appearance (JP). The USP and EP require either a crushed-glass test that determines the bulk composition of the glass or a surface test to examine the composition and durability of the glass as a result of the forming process.

Glass syringes present an interesting case where an additional extractable did not directly originate from the glass. The inner needle channel in glass syringes is often formed using a tungsten pin (19,32). Residual tungsten can remain on the glass depending on the processing conditions. The residual tungsten can interact with proteins and lead to aggregation (see later discussion).

RUBBER

In the injectable drug product business, rubber is used for many applications—closures for vials and bottles, seals and plungers for syringes and cartridges, gaskets in manufacturing equipment, and ports on plastic bags and intravenous administration sets (3,33–35).

Basic Chemistry and Composition

The physical properties to be considered in the selection of a particular rubber formulation include elasticity, hardness, tendency to fragment, and permeability to vapor transfer. The elasticity is critical in establishing a seal with the lip and neck of a vial or other opening and in resealing after withdrawal of a hypodermic needle from a vial closure. The hardness should provide firmness but not excessive resistance to the insertion of a needle through the closure, while minimal fragmentation of pieces of rubber should occur as the hollow shaft of the needle is pushed through the closure. While vapor transfer occurs to some degree with all rubber formulations, appropriate selection of ingredients makes it possible to control the degree of permeability.

Depending on how the rubber material will be used (e.g., as a closure, septum, plunger) and properties of the drug product closed by the rubber, other important physical and chemical properties will dictate the best choice of rubber formulation for the product. Examples of specific physical and chemical properties of the rubber closure include oxygen transmission, water vapor transmission, durometer (hardness), pressure to puncture, coring, resealability, breakforce, vacuum retention, and specific leachables/extractables. Typically the rubber manufacturer generates these data although certain functionality tests, for example, breakforce required to begin the movement of a rubber plunger in a syringe or cartridge, is performed by the product manufacturer.

Elastomers

Rubber formulations contain a variety of components. The elastomer determines most of the physical and chemical characteristics of the rubber formulation. The base material for the rubber is the elastomer that is composed of either natural or synthetic rubber. The earliest source of the

Table 7-1 Examples of Rubber Closure Components

Component type	General purpose
Elastomer	Base material—natural, butyl or halobutyl, silicone
Curing (vulcanizing) agent	Forms cross-links to shape the rubber. Common agents are sulfur, zinc oxide, and peroxide
Accelerator	Increases curing rate. One example used to be 2-mercaptobenzothiozole
Activator	Increases efficiency of accelerator. Common agents are zinc oxide and stearic acid
Antioxidant	Resists aging (e.g., phenol)
Plasticizer	Aids in the shaping process
Filler	Modifies hardness (e.g., carbon black)
Pigments	Provides color

elastomer material was the natural latex rubber liquid obtained from the *Hevea brasiliensis* tree. An increase in demand for rubber during World War II as well as the threat of strategic blockade of rubber shipments led to developments in the production of synthetic rubbers. Synthetic rubbers include styrene-butadiene rubber (SBR), neoprene [poly-(2-chloro-1,3-butadiene)], nitrile rubber, and butyl rubber. Butyl rubber is the most commonly used elastomer for pharmaceutical applications today because of its superior oxygen/moisture barrier (36). Neoprene is a halogenated synthetic rubber, which is available for more oil-based products.

Chemistry

Rubber stoppers are composed of the elastomer, a curing agent, an activator, a filler, as well as additional compounds to control cure rate, color, and resistance properties (Table 7-1). The curing agent, or vulcanizing agent, forms the cross-links in the rubber that provides the shape, the elasticity, and the resiliency to the rubber. Sulfur is the most common curing agent for elastomers with chemically unsaturated backbones (37). An activator, typically a metal oxide with a fatty acid, is used to accelerate the rate at which the sulfur reacts with the unsaturated polymer. The most common activator is zinc oxide combined with stearic acid (38). An accelerator, typically a sulfenimide, and a retarder, often benzoic acid, salicylic acid, and phthalic anhydride, can be added to further control the rate of vulcanization. Fillers are added to reduce tack, adjust color, and often to increase hardness and durability. Common fillers include carbon black, clay, calcium carbonate, and precipitated silica. Additional additives can include antioxidants and antiozonates as well as colorants.

Manufacturing

The basic steps in the manufacture of rubber closures involve the following (3):

1. Raw materials are tested usually at a minimum for identity and purity.
2. Batch ingredients are weighed usually within $\pm$ 1% tolerances.
3. Batch ingredients are mixed.
4. The mix is tested to insure cure characteristics.
5. The mix is placed on an extruder to create pellets, strips, or sheets.
6. The rubber is molded by injection, compression, or transfer.
7. The molded sheets are trimmed.
8. The stoppers are washed to remove trim and mold lubricants.
9. During post extraction, the stoppers are baked and autoclaved, if applicable.
10. The stoppers are tested for conformance by chemical and physical testing.
11. The stoppers are packed and shipped.

The two major steps in the production of rubber products are the compounding of the components followed by curing. In the compounding process, the rubber is masticated, or broken down by heat and shearing with a mixer. Mastication, breaks down the polymer, increasing its viscoelasticity, and enables the incorporation of additives such as fillers. Following mastication, the remaining additives, with the exception of the curing agents are added and mixed during a process known as masterbatching. Remilling may follow masterbatching if required to improve dispersion of additives or to modify viscosity. The curing system is added during the finish mixing step and the hot mixture is then extruded through a die to form pellets or through a pair of rollers to form a sheet. The rubber is further formed by injection molding or more commonly for stoppers, compression molding, and is then cured or vulcanized. Vulcanization consists of three stages: induction, curing, and reversion (or overcure). The process occurs through a variety of methods involving the application of heat and pressure depending on the desired final product form. The process chosen depends on the necessary dimensional tolerances and the cost. Injection molding provides the best dimensional tolerances, but is the most expensive.

Cleaning and Sterilization

Rubber closures are cleaned and depyrogenated by rinsing with copious amounts of Water for Injection and, if necessary, a cleaning agent such as sodium hydroxide, Liquid Safe-Kleen (LSK-9), or trisodium phosphate (TSP). Many rubber formulations contain polymer surfaces that do not require siliconization and process without difficulty. However, if siliconization is required, like with glass, it is done prior to sterilization, but after the depyrogenation procedure, and usually in the stopper washer. A predetermined amount of silicone is added to the stopper washer during a specified period of the washing cycle.

Sterilization of rubber closures occurs by steam sterilization in an autoclave using a validated cycle. Rubber plungers used in presterilized, ready-to-fill syringes are sterilized by gamma radiation.

Examples of manufacturers of stopper preparation equipment are DCI[2], Getinge[3], and Icos[4]. The DCI machines clean, siliconize, and depyrogenate stoppers within the same unit and the stoppers are batched and sterilized in an autoclave. The Getinge machines clean, siliconize, depyrogenate, and sterilize the stoppers within the same unit.

Alternatively, stoppers may be purchased directly from the stopper manufacturer already washed, siliconized, depyrogenated, and/or sterilized. Stoppers may be purchased from the stopper manufacturers as:

1. Raw stoppers—have not been processed and must be washed, siliconized (if applicable), and sterilized.
2. Ready-to-sterilize (RTS) stoppers—have been washed and siliconized (if applicable) in bags but have not been sterilized.
3. Ready-to-use (RTU) stoppers—have been washed, siliconized (if applicable), and sterilized.

Qualification

Physicochemical and toxicological tests for evaluating rubber closures are described in compendia such as the USP. Biological tests are both in vitro (USP <87>) and in vivo (USP <88>) tests. In vitro biological reactivity tests for rubber include the agar diffusion test, the direct contact test, and the elution test. In vivo biological reactivity tests include the systemic injection test, the intracutaneous test, and the implantation test. Physicochemical tests (USP <381> and EP 3.2.9) involve extractable studies using water. Approximately 100 cm^2 of rubber material in these solvents is autoclaved and then the following tests are done, each with specifications: turbidity, color, reducing substances, heavy metals, acidity or alkalinity, absorbance, extractable zinc, ammonium, and volatile sulfides. Functionality tests, such as penetrability, fragmentation, and self-sealing capacity, are also performed per the USP and the EP. Refer to the appropriate sections of the USP and EP for complete descriptions of these tests.

[2] http://www.dciinc.com/products/chemical.php.

[3] http://www.getinge.com/productPage.aspx?m1=115028548064&m2=115884183296&productGroupID=115035475063&divisionID=6&languageID=1.

[4] http://www.icosusa.com/stopperwashersterilizer.html.

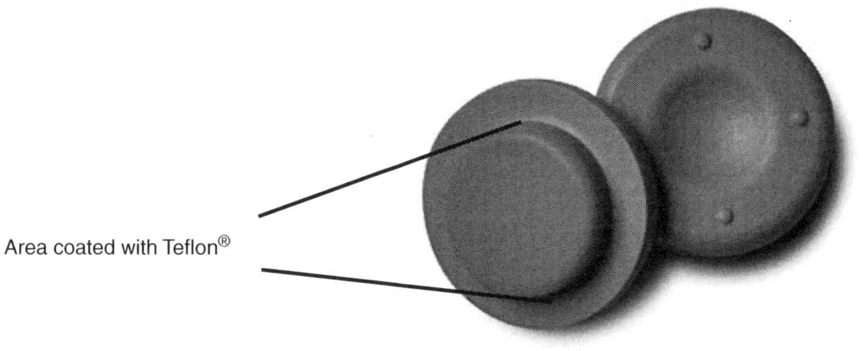

Figure 7-9 Examples of Teflon® (Teflon® is a registered trademark of E.I. du Pont de Nemours and Company) coated rubber closure. *Source*: © 2010 by West Pharmaceutical Services, Inc.

Extractables and Interactions with Drugs and Excipients Including Testing Methods

Leaching of the ingredients dispersed throughout the rubber formulation may occur when the product contacts the rubber closure. These ingredients pose potential compatibility interactions with product ingredients if leached into the product solution, and these effects must be evaluated. Further, some ingredients must be evaluated for potential toxicity. To reduce the problem of leachables, laminates have been applied to the product contact surfaces of closures, with various polymers, the most successful being Teflon® [DuPont polytetrafluoroethylene (PTFE)] (Fig. 7-9) and FluroTec® (West/Daikyo fluorinated ethylene propylene film) (Fig. 7-10). Recently, polymeric coatings have been developed that are claimed to have more integral binding with the rubber matrix, but details of their function are trade secrets.

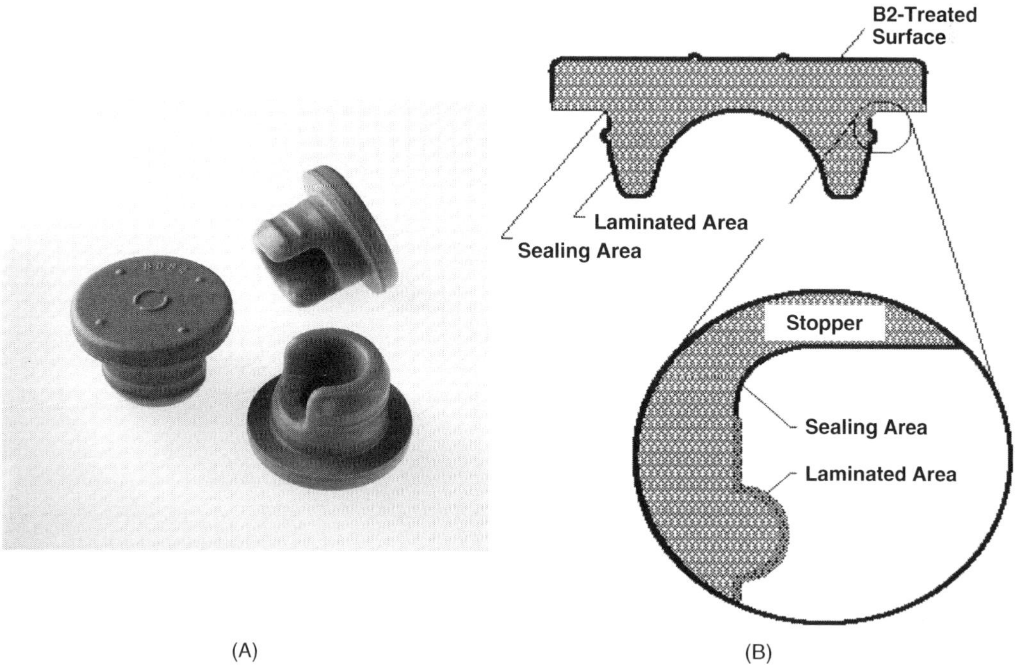

Figure 7-10 Daikyo FluroTec® closures. (A) Single vent lyophilization stopper. (B) FluroTec® (FluroTec® is a registered trademark of Daikyo Seiko Ltd) laminated plug with B2-treated top surface. *Source*: © 2010 by Daikyo/West Pharmaceutical Services, Inc.

There are four general types of rubber interactions with the drug product:

1. *Adsorption* of the active ingredient at the surface of the rubber. Proteins are well known to adsorb to rubber surfaces.
2. *Absorption* of one or more formulation components into the rubber. Components with high partition coefficients are prone to absorb into rubber.
3. *Permeation* of a formulation component through the rubber. Phenolic preservatives are a well-known example.
4. *Leaching* of rubber components into the drug product. The well-known example is 2-mercaptobenzothiazole; also aluminum, nitrosamines, and zinc are common rubber leachates.

Siliconization

Rubber closures must be "slippery" in order to move easily through a rubber closure hopper and other stainless steel passages until they are fitted onto the filled vials. Traditionally, rubber materials are "siliconized" (silicone oil or emulsion applied onto the rubber) in order to provide lubrication. The traditional practice of applying silicone to the rubber closures is acceptable as long as the silicone application process is effective (i.e., not too much nor too little silicone applied to each rubber closure) and the product does not have any interactions with silicone. However, as implied, siliconization of rubber closures presents many potential problems:

1. Oversiliconization provides excess silicone that may react adversely with a product component that is sensitive to hydrophobic interactions, as is the case with many biomolecules, causing precipitation and/or haze.
2. Undersiliconization may result in problems with high speed filling equipment and a greater tendency for rubber closures to stick together. Undersiliconization may result in a problem called "chattering." This occurs when the plunger rod and rubber resists facile movement throughout the syringe or cartridge barrel.
3. Excess silicone on the rubber will migrate into the product, causing potential increases in particulate matter counts since electronic particle counters detect silicone droplets as particles and a potential increase in air bubble formation that is not easily dissipated.
4. Silicone is difficult to remove in the manufacturing area, so cleaning becomes more of an issue where siliconization processes exist.

Coating

Coatings are utilized for one or two main purposes.

1. As a barrier between the stopper and the drug product to reduce leachables and extractables
2. To eliminate the requirement of silicone for processing.

A coated rubber closure consists of monomers applied directly to the rubber, then polymerized and bonded during processing. A laminated rubber closure consists of a polymeric coating applied to part or all of the closure as a laminated film. However, the two terms—coating and laminate—are used interchangeably in describing rubber closures that do not require siliconization.

As already discussed on page 39, but repeated for convenience here, the Daikyo/West FluroTec® laminated stopper contains a coating of copolymer film of tetrafluoroethylene and ethylene (ETFE). FluroTec® is a barrier laminate that protects the drug product from interacting with the rubber formulation. The FluroTec® coating does not cover the rubber/glass seal portion of the closure. The Omniflex® stopper is a true coated stopper containing a mixture of polyethylene and tetrafluoroethylene (PTFE) film.

Another advancement with the FluroTec® coating is a stopper called LyoTec™ (Fig. 7-10). The top surface of the LyoTec™ stopper is treated with FluroTec®. This prevents the stopper from sticking to the top pressure plates of the freeze-dryer shelves and either coming completely off of the vial or having the stopper plus the vial stick to the shelf above.

Other coatings are available that result in a barrier with the product and the stopper formulation but still require siliconization or may contain silicone. For example, the West B2 coating is a polymerized silicone coating for silicone oil replacement, which aids in the stoppering process. B2 stoppers do not require additional siliconization for processing and reduce particle and extractable silicone in the finished product. The B2 coating is most commonly used as a surface coat to the top of stoppers and is usually used in conjunction with an inner FluroTec® coating (39).

Another laminated film available only to stoppers with a flat inner surface is Teflon®. Teflon® coated stoppers require additional siliconization for processing. Recently, Teflon® has come under scrutiny as a possible carcinogen. The Environmental Protection Agency (EPA) has initiated a major investigation to determine if perfluorooctanoic acid, which is a chemical used to make Teflon®, is a possible carcinogen (40).

Coatings are a barrier to prevent leachables and extractables from the drug product but coatings may create processing issues. The coatings may cause the stoppers to become more rigid that can result in more equipment adjustments to be able to stopper the product. The coating does not level out imperfections of the containers. Upon compression, the coatings may cause wrinkles to form between the stopper and the components, which creates a cosmetic defect resulting in the rejection of the component or may affect container/closure integrity. Some coatings have caused the container/closure dye ingress test to fail when compared with the same uncoated stopper. On average, the dye ingress test failed 38% more when a coated versus uncoated stopper was evaluated. The preference is to have a coated stopper but not in the flange/neck area of the stopper (41).

Major Formulations Used in the Parenteral Industry

Table 7-2 describes a few stopper formulations available. The list is not comprehensive to all formulations and it is recommended to contact the manufacturer for further information. As noted, some formulations are preferred for different applications. For example, the Daikyo D777 and the Stelmi 6720 are recommended for lyophilized formulations due to lower moisture content and low extractables.

Table 7-2 Common Pharmaceutical Rubber Materials

Rubber formulation	Manufacturer	Coatings available	Comments
4432/50	West	B2, Teflon®, FluroTec®	Preferred in United States. Low levels of metal extractables
4023/50	West	B2, Teflon®, FluroTec®	Preferred in EU. Low levels of metal extractables
D777–1 lyophilization	Daikyo	B2, FluroTec®	Preferred for lyophilization and protein formulations, hydrophobic, very low moisture, extremely low organic, and metal ion extractables
D21	Daikyo	B2, FluroTec®	Low metal ion extractables and virtually no free sulfur content. Low gas and moisture permeability
D713	Daikyo	B2, FluroTec®	Good for oil based and hygroscopic materials. Extremely low free sulfur content, organic, and metal ion extractables
6720	Stelmi	None	Low residual moisture content, good for lyophilization. Excellent functional properties (self-sealing and resistance to coring)
6950 or 6955	Stelmi	None	Chlorobutyl-based, zinc-free high-purity formulation, extremely low extractables
FM257	Helvoet	OmniFlex®	Standard bromobutyl compound, latex free
FM460	Helvoet	OmniFlex®	Low moisture bromobutyl, low extractables
FM457	Helvoet	OmniFlex®	For syringes, ultra-low extractables bromobutyl stopper

Table 7-3 Most Common Polymers Used in Sterile Product Packaging

- Polyethylene (PE)
- Polyvinylchloride (PVC)
- Polypropylene (PP)
- Polyamide (Nylon)
- Polycarbonate (PC)
- Ethylene vinyl acetate (EVA)
- Polyolefin (mixtures of low-density PE, high-density PE, polypropylene, and EVA)

PLASTIC

Basic Chemistry and Composition

Plastics are widely used for parenteral drug containers and administration devices. Plastics are polymers, synthetic or natural, that can be shaped when softened, then hardened to the desired final appearance (42). The major types of polymers used in parenteral plastic packaging are given in Table 7-3. A polymer is a large organic molecule built from the repetitious joining of smaller, simpler, molecules (monomers), linked together by carbon-to-carbon bonds with a variety of complex organic groups attached. The polymerization process uses heat, pressure, and catalysts. Many common polymers used in pharmaceutical packaging, such as polyvinylchloride and polystyrene, are formed by addition polymerization. In addition polymerization, the double bond of the monomer unit is opened and the resulting free valences participate in bond formation with an additional monomer unit. No side products are formed from this reaction (43). Basic chemical structures of some of these polymers are shown in Figure 7-11.

Plastic Additives

The bulk properties of a polymer can be significantly altered by the additives incorporated into the plastic material. For example, the addition of plasticizers such as di-iso-octyl phthalate (DOP) to polyvinylchloride can change its physical characteristics from a hard, rigid solid to a rubber-like material (42). Plasticizers are typically nonvolatile solvents with a molecular weight of at least 300 Da. Other additives include those in the category of fillers, lubricants, anti-aging additives, flame retarders, colorants, blowing agents, cross-linking agents, and UV-degradable additives (43). Typical additives used in plastics are listed in Table 7-4.

Most plastics contain impurities, although in very small amounts (<1%), the result of unpolymerized monomers or residues of materials used in the manufacturing of the plastic product.

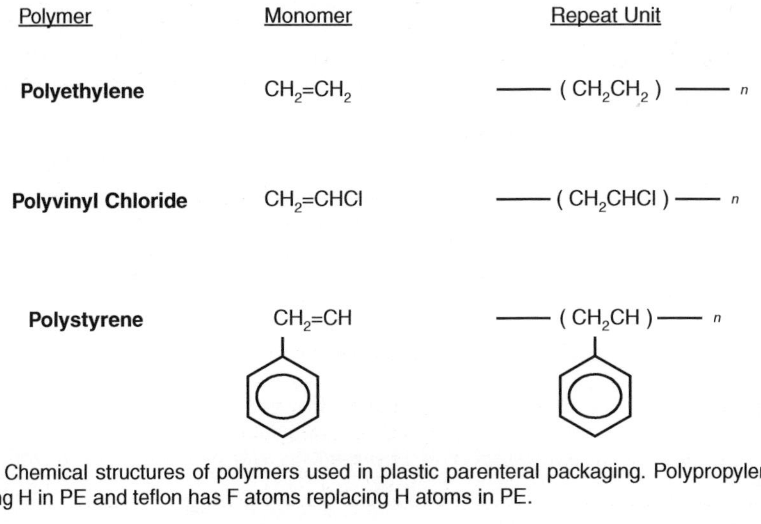

Polymer
Monomer
Repeat Unit
Polyethylene
$CH_2=CH_2$
—— (CH_2CH_2) —— n
Polyvinyl Chloride
$CH_2=CHCl$
—— (CH_2CHCl) —— n
Polystyrene
$CH_2=CH$
—— (CH_2CH) —— n

Figure 7-11 Chemical structures of polymers used in plastic parenteral packaging. Polypropylene has a CH_3 group replacing H in PE and teflon has F atoms replacing H atoms in PE.

Table 7-4 Examples of Plastic Additives

Additive type	Examples
Antioxidants	BHT, thioesters, phosphates
Heat stabilizers	Metallic stearates, epoxidized soybean oil, barium benzoate
Lubricants	Fatty acid amides, polyethylene waxes, silicones, fluorocarbon, zinc stearate
Plasticizers	Phthalates (30–40% added to polyvinyl chloride)
Colorants	Dyes, ultramarine blue, other pigments

Abbreviation: BHT, butylated hydroxytoluene.

Basic Types of Plastics

Plastics have been divided into two classes, thermoplastics and thermosets, based on their behavior when heated and cooled. Thermoplastics are polymers that soften upon heating and solidify upon cooling with these processes being reversible. Most parenteral packaging are thermoplastics and have been established as packaging materials for sterile preparations such as large-volume parenterals, ophthalmic solutions, and, increasingly, small-volume parenterals. Thermosets are chemically reactive polymers in the fluid state, are hardened irreversibly by cross-linking, and form three-dimensional networks that break when exposed to subsequent heating and cooling cycles. Examples of thermosets are epoxies, melamine resin, cross-linked polyesters, and phenolics. Table 7-5 provides a listing of the typical polymers used in different types of containers and devices.

Large-Volume Flexible Containers

Recent innovations in plastic parenteral packaging are in the areas of flexible container systems for the administration of intravenous (IV) fluids and premixed IV administered medications. The principle advantages of using plastic packaging materials are their durability and the substantial weight reduction of the material. The flexible bags currently in use for large-volume IV fluids have the added advantage that no air interchange is required; the flexible wall simply collapses as the solution flows out of the bag, preventing air from being infused. Comparative properties of the major plastic polymers used in flexible containers are listed in Table 7-6.

Three common problems exist when using these materials:

1. Permeation of vapors and other molecules in either direction through the wall of the plastic container.
2. Leaching of constituents from the plastic into the product.
3. Sorption (absorption and/or adsorption) of drug molecules or ions on the plastic material.

One of the more extensive problems is permeation. Permeation results in the loss of container contents by permitting volatile constituents, water, or specific drug molecules to migrate through the wall of the container. This problem has been resolved, for example, by the use of an overwrap in the packaging of IV solutions in polyvinylchloride (PVC) bags to prevent the loss of water during storage. Reverse permeation also may occur in which oxygen or other

Table 7-5 Plastic Polymer Applications in Injectable Drug Delivery

Type of plastic device	Typical type of polymer used
Plastic vials	Polycyclopentane, cyclic olefin copolymer
Containers for blood products	Polyvinyl chloride, polyolefin, others
Disposable syringes	Polycarbonate, polyethylene, polypropylene
Irrigating solution container	Polyethylene, polypropylene, polyolefin
Intravenous infusion container	Polyvinyl chloride, polyester, polyolefin
Administration set	Acrylonitrile butadiene
Administration set spike	Nylon
Administration tubing	Polyvinylchloride, other
Needle adapter	Polymethylmethacrylate
Clamp	Polypropylene
Catheter	Teflon, polypropylene

Table 7-6 Comparative Properties of Major Plastic Polymers

Property	PVC	LDPE	HDPE	PP	EVA
Compatibility with contained drug products	Poor	Good	Good	Good	Fair
Moisture permeation	Very poor	Good	Excellent	Good	Very poor
Heat sterilization	Fair	Poor	Good	Excellent	Very poor
Transparency characteristics	Good	Fair	Poor	Fair	Fair
Collapsibility characteristics	Excellent	Poor	Poor	Poor	Good
Disposability capability	Poor	Good	Good	Good	Fair

Abbreviations: PVC, polyvinyl chloride; LDPE, low-density polyethylene; HDPE, high-density polyethylene; PP, polypropylene; EVA, ethylene vinyl acetate.

molecules may penetrate to the inside of the container and cause oxidative or other degradation of susceptible constituents. *Leaching* may be a problem when certain constituents in the plastic formulation, such as plasticizers or antioxidants, migrate into the product. Thus, plastic polymer formulations should have as few additives as possible, an objective characteristically achievable for most plastics being used for parenteral packaging. *Sorption* is a problem on a selective basis, that is, sorption of a few drug molecules occurs on specific polymers. For example, sorption of insulin and other proteins, diazepam, methohexital sodium, procainamide, vitamin A acetate, warfarin sodium, and other drugs, has been shown to occur on PVC bags and tubing when these drugs were present as additives in IV admixtures (44).

Flexible Container Film Types

PVC was the first polymer used in the manufacture of collapsible containers for IV administration. However, PVC performs poorly in tests of all properties, with the exception of collapsibility and transparency. In addition, the presence of plasticizers such as DEHP [di(2-ethylhexyl) phthalate] limits the type of fluids suitable for storage in PVC because of concerns that the plasticizers can leach into the contained product. Concerns over the long-term safety of exposure to DEHP have led several nations to ban the use of PVC materials (e.g., Germany, Sweden, France, Canada, Spain, South Korea, and the Czech Republic, among others). Lastly, environmentalists have condemned PVC because it produces dioxin when incinerated (45). To improve characteristics such as compatibility and moisture permeation, ethylene vinyl acetate (EVA) films were developed. While EVA does not contain plasticizers, it, like PVC, has poor resistance to moisture permeation and both types of single-layer bags required secondary overwraps. The problem of moisture permeation was solved with the development of multilayer films. In one case, ethylene (vinyl) alcohol (EVOH), which has a high gas barrier, is used as a core and is physically bonded between two layers of EVA film (46). Additional films are available that combine a polyethylene (PE) inner product contact film, with other barrier films such as EVOH. The PE product contact layer provides excellent chemical compatibility and is rated as having excellent resistance to acids, alcohols, aldehydes, alkalis, amines, esters, glycols, vegetables oils, and salts. Poor resistivity was observed for hydrocarbons, and resistance to essential oils and ketones was fair and good, respectively (47).

Innovations in Flexible Containers

A number of suppliers now offer IV solutions, premixed medications, and custom manufacturing of IV solutions in flexible containers free of PVC and DEHP. These include Baxter's AVIVA and Galaxy lines, B. Braun's Excel and PAB containers, and Hospira's VisIV and CR3 container (48–50). While most large-volume flexible containers require terminal sterilization, Baxter's Galaxy lines offer proprietary aseptic filling processes for unstable and heat-sensitive drugs such as therapeutic proteins and monoclonal antibodies (51).

Flexible Container Sterilization

The impact of sterilization methods on container properties must be carefully considered during package and process development. Many plastic materials will soften or melt under the conditions of thermal sterilization. However, careful selection of the plastic used and control of the autoclave cycle have made thermal sterilization of large-volume parenterals possible. Ethylene oxide sterilization may be employed for the empty container with subsequent aseptic filling. However, careful evaluation of the residues from ethylene oxide or its degradation products and their potential toxic effect must be undertaken. Additionally, gamma radiation of plastic containers can have negative effects on the appearance of the plastic as well as the stability of the product. Depending on the film composition, irradiation of PVC can result in a change in color from clear to yellow, pH shifts, and an increase in extractables. In addition, some polypropylene formulations experienced postradiation degradation upon storage (47).

Small-Volume Plastic Containers

Many large-volume injectable products are now available in flexible plastic containers, but the movement of small-volume injectables into plastic vials and syringes has been slow. Until several years ago, plastic materials had the disadvantage that they are not as clear as glass and, therefore, inspection of the contents is impeded. The major suppliers of pharmaceutical glass have developed containers of cyclic olefins, such as COC (cyclic olefin copolymer) and COP (cyclic olefin polymer), relatively inert plastics with glass-like transparency (52).

Currently Marketed Small-Volume Containers

The polyolefins used in these resins are polymers of ethylene, propylene, and up to 25% of larger hydrocarbons (C4 to C10) or carboxylic acids or esters. Polyolefins may also contain up to three antioxidants and a lubricant or unblocking agent (53). West Pharmaceutical Services offers the Diakyo Crystal Zenith (CZ) ready-to-use prefillable syringe (Fig. 7-12) and will soon offer a staked needle system made from the same CZ brand COP. Daikyo manufactures vials from the same CZ COP (54). Schott provides vials made from COC in sizes from 2 to 10 mL and ready-to-use prefillable TopPac® syringes also made of COC. Becton Dickenson (BD) offers BD Sterifill SCF™ a ready-to-fill made of a proprietary "Crystal Clear Polymer (CCP)." BD also produces a line of prefilled syringes for flush applications made of polypropylene (55).

Potential Advantages of Plastic Small-Volume Containers

The obvious benefit of plastic containers for small-volume injectables over glass is their resistance to breakage. However, there are many other advantages. First, COC and COP are both relatively inert materials that minimize leaching of container components. This is particularly important for nonbuffered diluents such as water or saline that may be affected by pH when stored in glass as a result of alkali leachables (53). Second, some protein therapeutics,

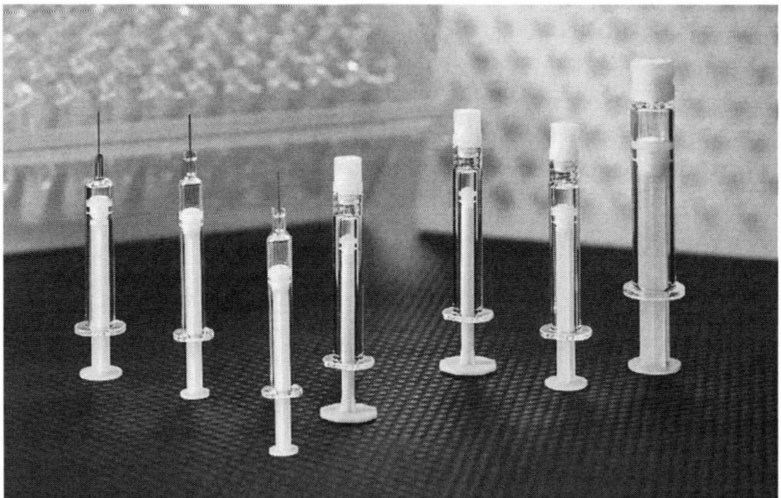

Figure 7-12 Daikyo Crystal Zenith® (Daikyo Crystal Zenith® is a registered trademark of Daikyo Seiko, Ltd) syringe systems. *Source*: © 2010 by Daikyo/West Pharmaceutical Services, Inc.

particularly those formulated at low concentrations may be less likely to adsorb to plastic vial or syringe surfaces than a glass container. Third, most glass syringes use a silicone lubricant to facilitate movement of the plunger. The West Daikyo CZ syringe is silicone free, and the cap and plunger are coated with Daikyo's FluroTec® brand coating to reduce extractables and provide lubrication (54). Silicone may be incompatible with certain drug formulations and can cause protein aggregation and possibly increase immunogenicity risk (56). Baxter compared the stability of three model protein therapeutics packaged in BD Hypak® and a silicone-free copolymer syringe by measuring absorption, aggregation, silicone levels, and tungsten levels. Absorption was low for both formats and aggregation was equivalent or lower for some proteins when stored in the copolymer syringe format. Silicone levels were equivalent or lower in the formulations stored in the copolymer syringe, while tungsten levels were significantly lower in the products stored in the copolymer versus plastic syringe (57).

Sterilization of Small-Volume Plastic Containers

As with flexible containers, care must be taken during thermal sterilization of plastic syringes and vials to prevent melting. The West Daikyo CZ resin can withstand autoclave temperatures of 121°C (56). BD also claims that their Sterifill SCF™ can be autoclaved after filling (58). While many glass syringes require EtOH sterilization, the Schott ready-to-fill TopPac COC® syringes are gamma sterilized and can be guaranteed stable for up to two years (52). In a review of polyolefin composite materials, it was concluded that a polyolefin's leachables profile is not dramatically impacted by irradiation (53).

Marketed Products Filled in Small-Volume Plastic Containers

Plastic syringes have gained wide use in nondrug applications such as contrast media and viscoelastics (52). BD offers the PosiFlush™ line of polypropylene syringes prefilled for saline flush and heparin lock flush applications. APP Pharmaceuticals produces Acyclovir in 10 and 20 mL plastic vials for IV administration (59). Hospira offers 0.9% sodium chloride injection, USP, Sterile Water for Injection, USP, as well as sodium lactate injection, USP in plastic flip-top vials (60). However, plastic syringes and vials have been slow to enter the marketplace. Companies are hesitant to move existing drugs currently packaged in glass into plastic containers because of regulatory filing barriers. Table 7-7 is a listing of injectable products packaged in plastic syringes, vials, and other containers according to Daikyo/West (61).

Environmental Impact of Plastic Containers

Beyond benefits of less weight and breakage, the use of plastic containers for parenterals also has environmental benefits and a potential cost savings to hospitals in the area of waste reduction. The increase in availability and use of disposable products has increased the quantity

Table 7-7 Products Packaged in Daikyo CZ® Vials and Syringes

Location	Container	Product
Japan	Syringes	Contrast media MRI Hyaluronic acid Calcitonin
Japan	Vials	Fluconazole Oncology products Anticoagulant products
United States	Vials	Acyclovir Hyaluronic acid
Europe	Syringes	Contrast media MRI
Europe	Vials	Oncology products Blood plasma

Source: Courtesy of Daikyo/West Pharmaceutical Services, Inc.

of waste generated by hospitals. The environmentally preferred means of waste disposal is recycling. While most plastic vials are recyclable, the high melting temperature of borosilicate glass makes it unsuitable for recycling in municipal recycling programs (62). In addition, plastic parenterals can be incinerated while glass parenterals cannot. Collapsible containers have lower waste disposal costs than glass, simply because the container is lighter and occupies less space when empty. In a comparison of Flexbumin, packaged in a collapsible plastic container, to the equivalent amount of Albumin packaged in glass, disposal costs of the empty containers are estimated to be $130 for the Flexbumin versus $220 for the Albumin packaged in glass (63).

Testing Requirements

Physicochemical and biological test requirements for plastic containers for parenteral use vary according to different compendia (the USP reference is chap. <661>). These data would be contained in a master file that a company references when registering a product packaged in plastic.

EXTRACTABLES AND LEACHABLES FROM CONTAINERS

Extractables and leachables (or leachates) have been covered to some extent in previous sections on glass, rubber, and plastic. However, the topic is so important and under so much regulatory scrutiny that additional coverage is justified. According to Kauffman (17), extractables are compounds that can be extracted from a packaging material usually requiring the presence of harsh solvents or elevated temperatures. Leachables are compounds that leach into the drug product formulation from the packaging material because of an interaction between the material and the product formulation under normal conditions. Leachables are a subset of extractables. Sources of these compounds include plastic components, elastomers, coatings, accelerants, antioxidants, inks, and vulcanizing agents.

Compounds migrating from the container–closure or any product contact surface into the product itself have always been a concern of dosage form development scientists and packaging engineers. Good manufacturing practice (GMP) regulations require that "Drug product containers and closures shall not be reactive, additive, or absorptive so as to alter the safety, identity, strength, quality or purity of the drug beyond the official or established requirements" (21CFR 211.94).

FDA's concerns about the potential for extractables and leachables beyond GMP regulation requirements have led to the publication of a guidance document (64). Section 3.3(f) on Specific Tests and Criteria in this document states:

> Extractables: Generally, where development and stability data show evidence that extractables from the container/closure systems are consistently below levels that are demonstrated to be acceptable and safe, elimination of this test can normally be accepted. This should be reinvestigated if the container/closure system or formulation changes. Where data demonstrate the need, tests and acceptance criteria for extractables from the container/closure system components (e.g., rubber stopper, cap liner, plastic bottle, etc.) are considered appropriate for oral solutions packaged in non-glass systems, or in glass containers with non-glass closures. The container/closure components should be listed, and data collected for these components as early in the development process as possible.

The FDA published a guidance document on container–closure systems in 1999. There are a couple of interesting statements in this document stressing the FDA's concerns on extractables (12):

> The potential effects of packaging component/dosage form interactions are numerous. Hemolytic effects may result from a decrease in tonicity and pyrogenic effects may result from the presence of impurities. The potency of the drug product or concentration of the antimicrobial preservatives may decrease due to adsorption or absorption. A cosolvent system essential to the solubilization of a poorly soluble drug can also serve as a potent extractant of plastic additives. A disposable syringe may be made of plastic, glass, rubber, and metal components, and such multicomponent construction provides a potential for interaction that is greater than when a container consists of a single material.
>
> For elastomeric and plastic components, data showing that a component meets the requirements of the USP Biological Reactivity Tests will typically be considered sufficient

evidence of safety. If the extraction properties of the drug product vehicle may reasonably be expected to differ from that of water (e.g., due to high or low pH or due to a non-aqueous solvent or a solubilizing excipient), then drug product should be used as the extracting medium. If the drug substance significantly affects extraction characteristics, it may be necessary to perform the extractions using the drug product vehicle. If the total extracts significantly exceed the amount obtained from water extraction, then an extraction profile should be obtained. It may be advisable to obtain a quantitative extraction profile of an elastomeric or plastic packaging component and to compare this periodically to the profile from a new batch of the packaging component. Extractables should be identified whenever possible. For a glass packaging component, data from USP Containers: Chemical Resistance—Glass Containers will typically be considered sufficient evidence of safety and compatibility. In some cases (e.g., for some chelating agents), a glass packaging component may need to meet additional criteria to ensure the absence of significant interactions between the packaging component and the dosage form.

Examples of leachables from container–closure materials that are found in sterile products include:

1. Phthalates, such as diethylhexylphthalate (DEHP), a plasticizer in polyvinyl chloride plastic (PVC) bags.
2. Zinc, nitrosamines, stearates, and polynuclear aromatic hydrocarbons (PAHs) are known potential extractables from rubber materials.
3. Orthophenylphenol from silicone tubing was found to be extracted by formulations containing sulfobutylether cyclodextrin (13).
4. Tungsten leachate from staked needles in syringes caused protein aggregation (19,65).
5. Additional examples provided by the FDA (14) (Table 7-8).

Leachates can originate from anything that comes into contact with the drug product. Some examples are interactions between the drug product and container–closure system as well as with materials used during manufacturing.

A well-known example of a leachate originating from an uncoated rubber stopper is the case of the Eprex® formulation (15,66,67). Polysorbate 80 in the formulation, an effective surface-active agent used to minimize protein aggregation, was believed to be the cause for a product–rubber interaction. The interaction resulted in a leachate that was only found in drug product sealed with an uncoated rubber stopper. The leachate was not found in any product manufactured with coated stoppers or with product containing human serum albumin instead of polysorbate 80. Furthermore, these leachates were believed to act as an adjuvant and stimulate the formation of anti-erythropoietin antibodies that lead to pure red blood cell aplasia.

Flexible containers may also be a source of contaminants. The contaminants may originate from the container itself or from secondary sources. For example, cyclohexanone is a bonding agent used for connecting tubing, ports, and other materials (68). Cyclohexanone can be found in products that come into contact with areas of the container that use the bonding agent.

Potential glass leachates can be reduced using either "treated" glass or using special prepared coated glass (Schott Type I Plus®). Rubber leachates can be reduced using special

Table 7-8 Examples of Leachable Materials

Leachable	Source	Analytical methods used to detect the leachable
Phosphorus, zinc in a buffer diluent vial	Rubber closure	Electron microscopy X-ray fluorescence
Butylated hydroxytoluene in a reconstituted lyophilized vial	Rubber closure	HPLC, UV, mass spectroscopy
2-(2-butoxyethoxy) ethyl acetate and 2-(2-butoethoxy)-ethanol	Silicone tubing	HPLC-MS Evaporative light scattering
Unknown particles	Sterilizing filter	USP <788> subvisible particulate matter test

Abbreviations: HPLC, High Pressure Liquid Chromatography; UV, Ultraviolet spectroscopy; HPLC-MS, High Pressure Liquid Chromatography-Mass Spectroscopy.
Source: From Ref. 14.

coatings such as copolymer films of tetrafluoroethylene (EFTE) and ethylene (FluroTec®) or a mixture of polyethylene and tetrafluoroethylene (PTFE) film (Omniflex®). Plastic leachates can be reduced or removed by using plastic materials that are known to contain low levels of potential extractable materials. A primary example is replacing polyvinylchloride requiring relatively high levels of the extractable plasticizer, DEHP with combinations of PE, polypropylene, or other polyolefinic materials that have low levels of extractable material.

Even with strict control of the packaging materials, some contaminants may still be present. These contaminants originate from the materials used during manufacturing. One example is the extraction of polyvinyl alcohol from the autoclave paper used to cover equipment and the open ends of tubing prior to sterilization. Another example was discovered during the technology transfer of a new formulation when new impurities were observed after the transfer (14). The new impurities originated from platinum-cured tubing that contained 2-(2-butoxyethoxy) ethyl acetate that is used as a carrying agent for the platinum in the tubing.

Scientists at Genentech have developed a thorough and holistic extractable and leachable program that contains elements of risk assessment, literature review, and consolidation of the best industry practices (69). Their extractable-leachable program includes six stages:

1. Selection of components
2. Determination of analytical procedures including extraction conditions and analytical methods to be used on components identified based on a risk-assessment approach
3. Selection of target leachables
4. Performing the leachable study
5. Health-based risk assessment of leachables
6. Life-cycle management.

ACKNOWLEDGMENT

Gregory Sacha, Karen Abram, and Wendy Saffell-Clemmer of Baxter BioPharma Solutions are acknowledged for their help in writing this chapter.

REFERENCES

1. Borchert SJ, Ryan MM, Davidson RL, et al. Accelerated extractable studies of borosilicate containers. J Parenter Sci Technol 1989; 43:67–69.
2. Abendroth RP, Clark RN. Glass containers for parenterals. In: Avis KE, Lieberman HA, Lachman L, eds. Pharmaceutical Dosage Forms: Parenteral Medications. Vol 1. New York: Marcel Dekker, 1992:361–385.
3. Smith EJ, Nash RJ. Elastomeric closures for parenterals. In: Avis KE, Lieberman HA, Lachman L, eds. Pharmaceutical Dosage Forms: Parenteral Medications. Vol 1. New York: Marcel Dekker, 1992:445–512.
4. Wang WJ, Chien YW. Sterile Pharmaceutical Packaging: Compatibility and Stability. Technical Bulletin No. 5. Bethesda, MD: Parenteral Drug Association, 1984.
5. Anon. PDA Task Force on lubrication of packaging components. Technical Bulletin No. 12. J Parenter Sci Technol 1988; 42.
6. Danielson JW, Oxborrow GS, Placencia AM. Quantitative determination of chemicals leached from rubber stoppers into parenteral solutions. J Parenter Sci Technol 1988; 42:90–93.
7. Bontempo JA. Considerations for elastomeric closures for parenteral biopharmaceutical drugs. In: Bontempo JA, ed. Development of Biopharmaceutical Parenteral Dosage Forms. New York: Marcel Dekker (now Informa), 1997:223–240.
8. Anes JM, Nase RS, White CH. Use of plastics for parenteral packaging. In: Avis KE, Lieberman HA, Lachman L, eds. Pharmaceutical Dosage Forms: Parenteral Medications. Vol 1. New York: Marcel Dekker, 1992:387–444.
9. Lambert P. Packaging of intravenous solutions. Med Device Technol 1991; 4:48–53.
10. Bhowmick AK, Stephens HL, eds. Handbook of Elastomers, New Developments and Technology. New York: Marcel Dekker, 1988.
11. International Conference on Harmonisation of Technical Requirements for Registration of Pharmaceuticals for Human Use. Guidance for Industry, Specifications: Test Procedures and Acceptance Criteria for New Drug Substances and New Drug Products, ICH Q6A. Silver Spring, MD: FDA, 1999.
12. U.S. Department of Health and Human Services Food and Drug Administration Center for Drug Evaluation and Research (CDER) Center for Biologics Evaluation and Research (CBER). Container Closure Systems for Packaging Human Drugs and Biologics: Chemistry, Manufacturing, and Controls Documentation. Silver Spring, MD: FDA, 1999:23.

13. Zimmerman JA, Ballard JM, Wang H, et al. Extraction of 0-phenylphenol from silicone tubing by a sulfobutylether cyclodextrin formulation. Int J Pharm 2003; 267:113–120.
14. Castner J, Williams N, Bresnick M. Leachables found in parenteral drug products. Am Pharm Rev 2004; 7(March/April):70–75.
15. Sharma B, Bader F, Templeman T, et al. Technical investigations into the cause of the increased incidence of antibody-mediated pure red cell aplasia associated with Eprex®. Eur J Hosp Pharm 2004; 12:88–91.
16. Markovic I. Challenges associated with extractables and/or leachable substances in therapeutic biologic protein products. Am Pharm Rev 2006; 9:20–27.
17. Kauffman JS. Identification and risk-assessment of extractables and leachables. Pharm Technol 2006; 30(Analytical Methods Supplement):s16–s22.
18. Thirumangalathu R, Krishnan S, Ricci MS, et al. Silicone oil- and agitation-induced aggregation of a monoclonal antibody in aqueous solution. J Pharm Sci 2009; 98:3167–3181.
19. Bee JS, Nelson SA, Freund E, et al. Precipitation of a monoclonal antibody by soluble tungsten. J Pharm Sci 2009; 98:3290–3301.
20. Sacha GA, Abram K, Saffell-Clemmer W, et al. Practical fundamentals of glass, rubber, and plastic sterile packaging systems. Pharm Dev Technol 2010; 15:6–34.
21. Pfaender HG. Schott Guide to Glass. 2nd ed. London: Chapman and Hall, 1996.
22. Woerder H. Pharmaceutical primary packaging materials made of tubular glass from the aspect of drug safety and product applications. Eur J Parenter Pharm Sci 2004; 9(4):123–128.
23. The United States Pharmacopeial Convention. United States Pharmacopeia, Pharmacopeia National Formulary. 31 ed. Rockville, MD: The United States Pharmacopeial Convention, 2008:33–36, 67.
24. The United States Pharmacopeial Convention. U.S. Pharmacopeia National Formulary. 32 ed. Rockville, MD: The United States Pharmacopeial Convention, General Chapter <51>, 2009.
25. Avis K, Lieberman H, Lachman L. Pharmaceutical Dosage Forms: Parenteral Medications. 2nd ed. New York: Marcel Dekker, Inc., 1992.
26. Bauer E, ed. Pharmaceutical Packaging Handbook. New York: Informa Healthcare, 2009.
27. Forcinio H. Not all Type I glass is created equal. Pharm Technol 1998; 22:30–34.
28. Anderson N, Kildsig D. Alternative analysis of depyrogenation processes. Bull Parenter Drug Assoc 1983; 37:75–78.
29. Kettelhoit S. Industrial production of glass syringes. Pharm Indus 2008; 70:1261–1269.
30. Ennis RD, Pritchard R, Nakamura C, et al., Glass vials for small volume parenterals: influence of drug and manufacturing processes on glass delamination, Pharm Dev 001, 6: 393–405.
31. Schott Pharmaceutical Packaging Coating Technology. Schott, Schott Type 1 Plus Product Brochure. Elmsford, NY: Schott Pharmaceutical Packaging Coating Technology, 2005.
32. Jenke D. Suitability-for-use considerations for pre-filled syringes. Pharm Technol 2008; 32: s30–s33.
33. DeGrazio F. Closure and container considerations in lyophilization. In: Rey L, May J, ed. Freeze-Drying/Lyophilization of Pharmaceutical and Biological Products. 2nd ed. New York: Marcel Dekker, 2004:277–297.
34. Hora M, Wolfe S. Critical steps in the preparation of elastomeric closures for biopharmaceutical freeze-dried products. In: Rey L, May J, eds. Freeze-Drying/Lyophilization of Pharmaceutical and Biological Products. 2nd ed. New York: Marcel Dekker, 2004:309–323.
35. Smith E. Elastomeric components for the pharmaceutical industry. In: Swarbrick J, ed. Encyclopedia of Pharmaceutical Technology, 3rd ed. New York: Informa, 2007:1466–1481.
36. Garzella L, ed. Presentation at the University of Tennessee Health Science Center. Memphis, TN: West Pharmaceuticals, 2008.
37. Ciullo P, Hewitt N, eds. Rubber Formulary. Norwich, NY: Willam Andrew Publishing, 1999:743.
38. Rabinow B, Roseman T. Plastic packaging materials. In: Remington: The Science and Practice of Pharmacy. 21st ed. Philadelphia: Lippincott, Williams & Wilkins, 2005:1047–1057.
39. West Pharmaceutical Services. A critical assessment of B2, FluroTec, and Omniflex coatings. West Pharmaceutical Services Technical Report 1999/040, 1999.
40. Eilperin J. Compound in teflon a likely carcinogen. Washington Post, 2005:4.
41. Sigg J. An oral presentation at Arden House Europe. London, 2007, American Association of Pharmaceutical Scientists and Royal Pharmaceutical Society of Great Britain (Co-sponsors).
42. Byrdson J. Plastic Materials. 7th ed. Oxford: Butterworth-Heinemann, 1999.
43. Cooper J. Plastic containers for drugs. WHO Chronicle 1974; 28:395–401.
44. Turco S, King R. Sterile Dosage Forms, 3 ed. Philadelphia: Lea & Febiger, 1987:275–277.
45. PVC Waste and Recycling. http://archive.greenpeace.org/toxics/html/content/pvc3.html. Accessed June 4, 2010.
46. Thermo-Scientific. HyClone BioProcess Containers. Waltham, MA: Thermo-Scientific, 2009. http://www.thermo.com/eThermo/CMA/PDFs/Various/File_51362.pdf. Accessed June 4, 2010.

47. Portnoy R. Medical Plastics—Degradation Resistance and Failure Analysis. Norwich, NY: Plastics Design Library, 1998.
48. Hospira. http://www.one2one.hospira.com/default.aspx. Accessed June 4, 2010.
49. Braun B. http://www.bbraunusa.com. Accessed June 4, 2010.
50. Solutions BB. www.baxterbiopharmasolutions.com. Accessed June 4, 2010.
51. Solutions BB. Frozen Premix. http://www.baxterbiopharmasolutions.com/enhanced_packaging/infusion_delivery/frozen_premix.html. Accessed June 4, 2010.
52. Vaczek D. Promoting dosing accuracy with prefilled syringes. Pharmaceutical & Medical Packaging News 2007; 15(4):42–47.
53. Jenke D. Extractable/leachable substances from plastic materials used as pharmaceutical product containers/devices. PDA J Pharm Sci Technol 2002; 56:332–371.
54. Crain Communications Inc. http://plasticsnews.com/. Accessed June 4, 2010.
55. Systems BM. PosiFlush Syringes. http://www.bd.com/ca/pdfs/safety/products/injection/prefilled_syringes/posiflush_sell_sheet_rev.pdf. Accessed June 4, 2010.
56. Services WP. Solutions for Prefillable Syringes. http://www.westpharma.com/na/en/products/Pages/CrystalZenithRU.aspx. Accessed June 4, 2010.
57. Shah D, Moore E, Leesch V, et al. Enhanced Compatibility of Various Model Protein Formulations in Glass and CLEARSHOT Syringe Formats. Round Lake, IL: Baxter Healthcare Corporation, 2007.
58. BD. Plastic Prefillable Syringe Systems. http://www.bd.com/pharmaceuticals/products/plastic-prefillable.asp. Accessed June 4, 2010.
59. APP Pharmaceuticals L. Acyclovir Sodium Injection. 2008. http://www.apppharma.com/ProductDetail.aspx?Id = 2. Accessed June 4, 2010.
60. Hospira. Product Catalog. Lake Forest, IL: Hospira.
61. Brucker B. brian_brucker@westpharma.com. West Pharmaceutical Services. Oral Presentation, 2009.
62. Baetz B. Parenteral packaging waste reduction. Can J Hosp Pharm 1990; 43(4):179–181.
63. Corporation BH. Albumin Therapy. http://www.albumintherapy.com/product_information/flexbumin/flexbumin_vs_albumin_glass.html. Accessed June 4, 2010.
64. International Conference on Harmonization of Technical Requirements for Registration of Pharmaceuticals for Human Use. ICH Q6A Guideline, Specifications: Test Procedures and Acceptance Criteria for New Drug Substances and New Drug Products. 2005:1–122.
65. Fries A. Drug delivery of sensitive biopharmaceuticals with pre-filled syringes. J Drug Deliv Sci Technol 2009; 9(5):22–27.
66. Bennett C, Luminari S, Nissenson A. Pure red-cell aplasia and Epoetin therapy. New Engl J Med 2004; 14:1403–1408.
67. Boven K, Stryker S, Knight J, et al. The increased incidence of pure red cell aplasia with an Eprex formulation in uncoated rubber stopper syringes. Kidney Int 2005; 67:2346–2353.
68. Jenke D. Linking extractables and leachables in container/closure applications. PDA J Pharm Sci Technol 2005; 59:265–281.
69. Wakankar A, Wang Y, Canova-Davis E. On developing a process for conducting extractable-leachable assessment of components used for storage of biopharmaceuticals. J Pharm Sci. 2010; 99(5):2209–2218.

8 Formulation and stability of solutions

Ready-to-use solution dosage forms comprise the largest percentage of sterile dosage forms in the marketplace. The solution formulation must be resistant both to physical and chemical degradation. Drugs in solution are subject to several major mechanisms of degradation–hydrolytic, oxidative, photolytic, and, for proteins, covalent and noncovalent aggregations, deamidation, cleavages, oxidation, and surface denaturation reactions.

Optimal formulations can minimize or prevent these degradation reactions. Typical additives that help to stabilize injectable drugs in solution include surface-active agents, buffers, sugars, salts, antioxidants, chelating agents, competitive binders, and amino acids. Also, storing solutions at colder temperatures (i.e., refrigerated or even frozen) can help to minimize drug degradation. Drugs in solution also may have a tendency to form insoluble forms, therefore, physical stabilization is vitally important. This chapter focuses on formulation and stabilization of sterile drugs in solution, particularly biopharmaceutical drugs with more complex structures that present greater or a wider variety of challenges (1,2). There are many other primary literature resources for sterile solution drug formulation including an exhaustive updated review article on protein stability by Manning et al. (3). Proteins and other biopharmaceutical molecules not only readily degrade chemically, but also, and perhaps more readily, are prone to physical instabilities such as aggregation and precipitation.

OPTIMIZING HYDROLYTIC STABILITY

Hydrolysis is the reaction between water and the drug molecule resulting in the loss of potency and stability. One of the first major studies to be conducted in early drug dosage form development is to determine the solubility and stability of the drug in solution as a function of pH. Therapeutic proteins, being structurally more complex with secondary and tertiary structures and amino acids of differing properties being potentially exposed to an aqueous environment, experience a variety of potential degradation pathways over a broad pH range. While small molecules do not have this range of potential degradation pathways, many follow pH-stability profiles like the one depicted for penicillin in Figure 8-1. It is common for weak electrolytes to have "V-shaped" degradation versus pH profile where the objective with such molecules is to identify the pH range where drug stability is greatest. However, typically, the pH range where stability is greatest also is where drug solubility is lowest, again clearly shown in Figure 8-1. Solution pH and type of solvent used also significantly matters for minimizing protein aggregation, an example of which is shown in Figure 8-2 for recombinant human granulocyte colony-stimulating factor (rhGCSF).

Proteins and some small molecules may degrade in solution by more than one mechanism and each degradation mechanism has a different pH-stability profile. Tissue plasminogen activator undergoes dimer formation, loss of clot lysis or peptidolytic activity, each of which have slightly different pH-stability profiles. Glucagon in solution will degrade by hydrolysis, oxidation, and aggregation; the same is true for growth hormone. Insulin degrades by hydrolysis (deamidation) and formation of higher molecular weight forms as do many other protein molecules.

Hydrolysis or deamidation occurs with peptides and proteins containing susceptible asparagines (Asn) and glutamine (Gln) amino acids, the only two amino acids that are primary amines. The side chain amide linkage in a Gln and Asn residue may undergo deamidation to form free carboxylic acid. Deamidation can be promoted by a variety of factors including high pH, temperature and ionic strength (1).

Minimizing hydrolytic stability of drugs, particularly peptides and proteins, can be accomplished through one or more of the following approaches:

1. Optimization of amino acid sequence; that is, engineering protein structures to remove unstable amino acids or insert amino acid that sterically hinder Asn or Gln

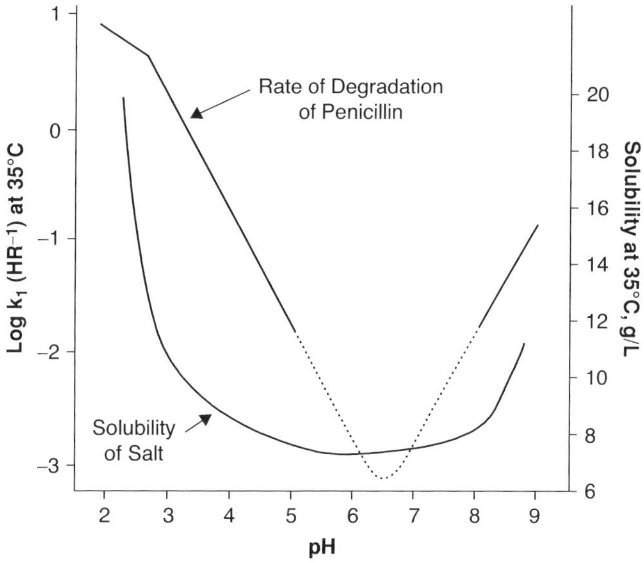

Figure 8-1 Effect of pH on solubility and stability of procaine penicillin G. *Source*: From Ref. 4.

deamidation, as long as this does not affect protein activity, potency, toxicity, or any other quality attribute.

2. Formulate at optimal solution pH. For example, human epidermal growth factor 1–48 demonstrates some interesting pH-dependent stability in that at pH <6, succinimide formation at Asp^{11} is favored while at pH >6, deamidation of Asn^1 is favored (6). The optimal pH, therefore, is right at pH of 6. Generally, deamidation occurs above pH 5.0 with the optimal pH range to minimize deamidation being between 3.0 and 5.0.
3. Store at low temperatures although this will always create difficulties in complying with the requirement during distribution and long-term storage of the product. However, with the advent of cold storage distribution businesses, this is less of a problem than in previous years.
4. Optimize the effects of ionic strength using empirical approaches to determine the effects of added electrolytes.

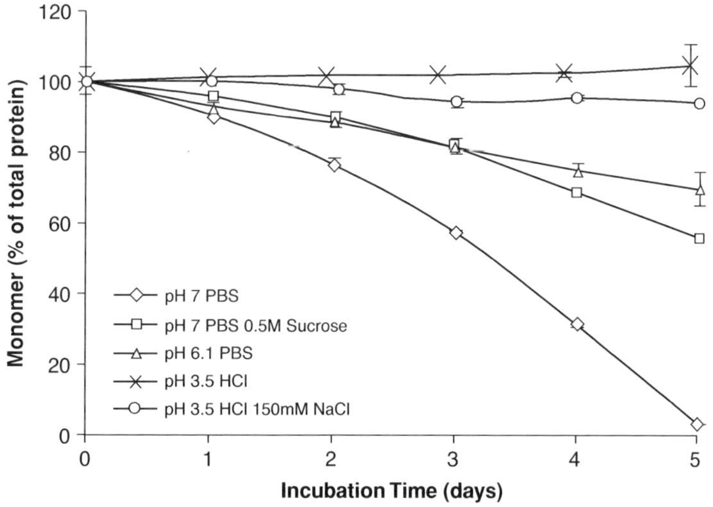

Figure 8-2 Aggregation profiles of recombinant human granulocyte colony-stimulating factor (rhGCSF) as a function of pH and type of solution. *Source*: From Ref. 5.

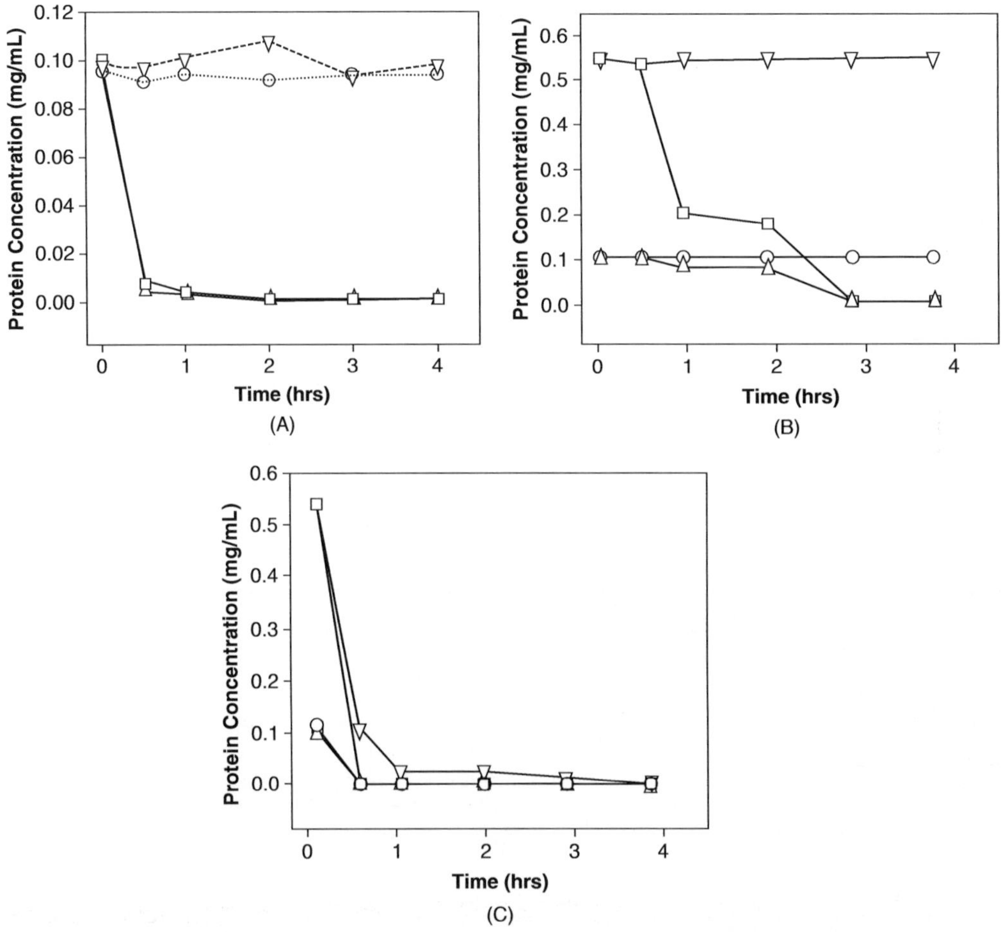

Figure 8-3 Protein concentration in the supernatant after isothermal incubation of rBoNTA(Hc) at 50°C at pH 5, 6, and 8. (**A**) rBoNTA(Hc) in 20 mM Na-succinate, pH 5 at 0.1 mg/mL protein concentration in buffer alone (circles) and with 150 mM NaCl (triangles), at 0.55 mg/mL protein concentration in buffer alone (downward triangles), and with 150 mM NaCl (squares); (**B**) rBoNTA(Hc) in 20 mM histidine, pH 6 at 0.1 mg/mL protein concentration in buffer alone (circles), with 150 mM NaCl (triangles), at 0.55 mg/mL protein concentration in buffer alone (downward triangles), and with 150 mM NaCl (square).(**C**) rBoNTA(Hc) in 20 mM K-phosphate, pH 8 at 0.1 mg/mL protein concentration in buffer alone (circles), and with 75 mM NaCl (triangles), at 0.55 mg/mL protein concentration in buffer alone (downward triangles), and with 75 mM NaCl (square). *Source*: From Ref. 7.

Buffers and Hydrolytic Stability

Buffers prevent small changes in solution pH. Even pH changes of 0.1 can affect drug solubility and stability. Buffers are composed of salts of ionic compounds. The most common buffers used in sterile product formulations are acetate, citrate, and phosphate. Recent buffer systems, especially effective for monoclonal antibody formulations, are amino acid-based buffers. Buffer systems acceptable for use in sterile solutions are listed in Table 8-1, both ionic compounds and amino acids.

The proper selection of buffer type and concentration is determined by performing solubility and stability studies as a function of pH and buffer species, an example of which is shown in Figure 8-3 for pH stability of recombinant botulinum serotype A vaccine in three different buffers with varying sodium chloride levels at three different pHs (7). However, a pH range that is a good compromise between solubility and stability can be selected and that pH range maintained with the proper selection of the appropriate buffer component.

Generally, deamidation is much slower at acidic pH than at neutral or alkaline pH. For example, ACTH deamidation in the pH range of 7 to 11 was catalyzed by increasing buffer concentrations, whereas no buffer catalysis occurred at pH 5 to 6.5 (8).

Table 8-1A Most Common Buffers Used in Sterile Drug Solutions

Buffer system	pK_a	Typical buffer pH range
Lactic acid/lactate	3.1	2.0–4.0
Tartaric acid/tartrate	3.0, 4.2	2.0–5.3
Glutamic acid/glutamate	2.1, 4.3, 9.7	2.0–5.3
Malic acid/malate	3.4, 5.1	2.5–5.0
Citric acid/citrate	3.1, 4.8, 5.2	2.5–6.0
Gluconic acid/gluconate	3.6	2.6–4.6
Benzoic acid/benzoate	4.2	3.2–5.2
Succinic acid/succinate	4.2, 5.6	3.2–6.6
Acetic acid/acetate	4.8	3.5–5.7
Histidine	1.8, 6.1, 9.2	5.5–7.4
Phosphoric acid/phosphate	2.1, 7.2, 12.7	6.0–8.2
Glycine/glycinate	2.4, 9.8	6.5–7.5
Tromethamine (TRIS, THAM)	8.1	7.1–9.1
Diethanolamine	8.0	8.0–10.0
Carbonic acid/carbonate	6.4, 10.3	5.0–11.0

Table 8-1B Dissociation Constants of Amino Acids Used as Buffers in Sterile Drug Solutions, Especially Monoclonal Antibody Products

Amino acid	α-Carboxylic acid	α-Amino group	Side chain
Alanine	2.35	9.87	–
Arginine	2.01	9.04	12.48
Aspartic acid	2.10	9.82	3.86
Cysteine	2.05	10.25	8.00
Glycine	2.35	9.78	–
Histidine	1.77	9.18	6.10
Lysine	2.18	8.95	10.53

One of the great challenges in scale-up and technology transfer from laboratory scale to production scale batch sizes is the adjustment of pH. Despite the presence of a buffer, target pH often is not met following addition of all components. Buffers are typically not used to adjust pH of production batches; rather dilute solutions of strong acids (e.g., hydrochloric, acetic or phosphoric acids) and strong bases (e.g., sodium hydroxide) are used. Careful pH adjustment with these dilute acids and/or bases is very important, because if target pH is missed, additional use of these strong acids and bases may alter buffer capacity and ionic strength of the final formulation.

General acid and/or general base buffer catalysis can accelerate the hydrolytic degradation. An example is given in Figure 8-4 where the inactivation rate of an experimental drug was affected by both type and concentration of buffer component (9). The deamidation rate of a small peptide using different buffers found that the peptide was most unstable in a phosphate buffer and most stable in Tris buffer (10). Buffer type and concentration will affect aggregation of basic fibroblast growth factor depending on pH (11). At pH 5, aggregation increased as citrate buffer concentration increased. Citrate buffer at pH 3.7 caused aggregation, whereas acetate buffer at pH 3.8 did not. At pH 5.5 to 5.7, phosphate, acetate, and citrate buffers all showed similar aggregation rates.

Histidine has been found to be an excellent buffer component for monoclonal antibodies (e.g., Synagis®, Herceptin®, Xolai®, Raptiva®) maximally stable in the pH 6 range. The pK_a of histidine is 6.0 that makes it an ideal buffer at pH of 6.0. Histidine is the only amino acid with pH 7.4 within its buffering range, therefore, it has found importance in parenteral formulations requiring buffering in the physiological pH range (12).

High concentrations of monoclonal antibodies ($\geq$50 mg/mL) have the ability to self-buffer (13). IgG_2 was found to be more stable at pH 5 after accelerated stability studies as a self-buffered formulation than in formulations containing conventional buffers such as acetate, glutamate, and succinate.

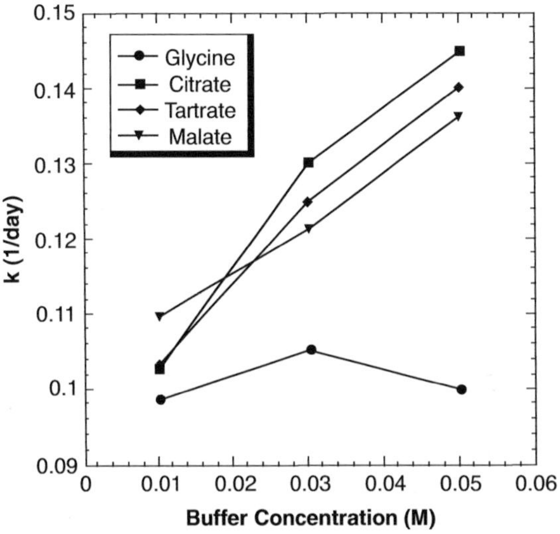

Figure 8-4 Rate of hydrolysis of GW280430 as function of buffer type and concentration. *Source*: From Ref. 9.

Ionic Strength

Ionic strength is a measure of the intensity of the electrical field in a solution. Ionic strength depends on the total concentration of ions in solution and the valence of each ion. The ionic strength of a 0.1 M solution of sodium chloride is 0.1. The ionic strength of a 0.1 M solution of sodium sulfate is 0.3, because sulfate ions have a valence of 2 added to the valence of 1 for the sodium ions. Ionic strength may have an effect on drug stability in solution. The Debye–Hückel theory predicts that increased ionic strength would be expected to decrease the rate of degradation of oppositely charged reactants and increase the rate of degradation of similarly charged reactants. For example, increasing ionic strength will increase the stability of recombinant alpha-1-antitrypsin (14). Conversely, increasing ionic strength will increase the rate of deamidation of human growth hormone (hGH) (15), bovine somatotropin (BST) (16) and lead to opalescence and higher viscosity of a monoclonal antibody (17).

OPTIMIZING OXIDATIVE STABILITY

Drugs containing such functional groups as phenols, catechols, and thioethers will be subject to oxidative degradation. Epinephrine, phenylephrine, dobutamine, dopamine, morphine, Terramycin, ascorbic acid, and many others are examples of small molecule drugs that will oxidize in solution. Proteins containing amino acids methionine, cysteine, cystine, histidine, tryptophan, and tyrosine are susceptible to oxidative and/or photolytic degradation depending on the conformation of the protein and resultant exposure of these sensitive amino acids to the solvent and environmental conditions. Environment conditions that catalyze oxidative degradation include the presence of dissolved oxygen in solution, light exposure, high temperature, low solution pH, metal ions (ppm, even ppb levels), and impurities such as peroxide. Oxidation of sulfhydryl-containing amino acids (e.g., methionine and cysteine) will lead to disulfide bond formation and loss of biological activity. The free-thiol group that is present in a cysteine residue of any native biologically active protein may oxidize not only to produce an incorrect disulfide bridge, but also can result in other degradation reactions such as alkylation, addition to double bonds, and complexation with heavy metals.

Human growth hormone, chymotrypsin, lysozyme, parathyroid hormone, human granulocyte colony-stimulating factor, insulin-like growth factor I, acidic and basic fibroblast growth factors, relaxin, the monoclonal antibody OKT3, interleukin 1β, and glucagon are a few of the examples of proteins that may undergo oxidative degradation.

For protection against oxidation, choice of an effective antioxidant is one of the several precautions that must be practiced in formulation development and final product manufacture. Indeed, minimizing drug oxidative degradation requires a combination of several approaches,

Table 8-2 Various Approaches Used To Minimize Oxidative Drug Degradation

- Preparation and storage at low temperatures
- Use of chelating agents to eliminate metal catalysis
- Increasing ionic strength
- Elimination of peroxide and metallic contaminants in formulation additives
- Protection from light
- Awareness of possible interaction of light exposure and phosphate buffer in forming free radicals
- Replacing oxygen with nitrogen or argon during manufacturing
- Removing oxygen from the headspace of the final container
- Formulation established at the lowest pH possible while still maintaining adequate solubility and overall stability
- Use of a container/closure system that allows no oxygen transmission through the package during distribution and storage
- Assuring that phenolic or other oxidizing cleaning agent residues are minimal in the production environment

not only formulation, but also hermetic packaging, oxygen-free processing, and all other precautions listed in Table 8-2.

Formulators should be aware of the potential for polysorbate 80 to adversely affect the oxidative stability of proteins. Polysorbate 80 is a commonly used surface-active agent in protein formulations to minimize surface aggregation problems. However, it has the tendency to produce peroxides that can oxidize methionine and cysteine residues. This phenomenon was reported in studies involving formulation development of Neupogen® (18) and recombinant human ciliary neurotrophic factor (19).

Antioxidants

There are several choices of antioxidants that can be used in sterile formulations. Those used most frequently are ascorbic acid, salts of sulfurous acid (sodium bisulfite, sodium metabisulfite or sodium thiosulfate), and thiols such as thioglycerol and thioglycolic acid. Dithiothreitol, reduced glutathione, acetylcysteine, mercaptoethanol, and thioethanolamine are thiols which usually oxidize too readily to be of practical use in pharmaceutical formulations requiring long-term storage.

Precautions must be applied when considering certain antioxidants in certain drug formulations. Here is one example. Ascorbate in the presence of Fe^{3+} and oxygen actually induces the oxidation of methionine in small-model peptides (20). Ascorbate is a powerful electron donor in that it is readily oxidized to dehydroascorbate. It also generates highly reactive oxygen species such as hydrogen peroxide and peroxyl radicals. These, in turn, will accelerate the oxidation of methionine. Phosphate buffer accelerated the degradation of methionine in the presence of ascorbic acid. The addition of EDTA did not enhance stability even though ferric ion and other transition metals were components in the formulation, either purposely added or as trace components of the buffer and peptide. This pro-oxidant effect of ascorbate methionine oxidation was concentration dependent and occurred most readily at pH 6 to 7.

Chelating Agents

Chelating agents are used in formulations to aid in inhibiting free radical formation and resultant oxidation of active ingredients caused by trace metal ions such as copper, iron, calcium, manganese, and zinc. There are several examples of commercial formulations (Nebcin®, Decadron-LA®, Versed®, Cleocin®, and others) where a chelating is all that is needed, that is, no antioxidant in the formulation, to protect the active ingredient against metal-catalyzed oxidation. The most common chelating agent used is disodium ethylenediaminetetraacetic acid (DSEDTA), typically at very low concentrations, for example, $\leq$0.04%. DSEDTA tends to dissolve slowly and is usually among the first of formulation ingredients to be dissolved during compounding before adding other ingredients, including the active. Citrate buffer can also serve as a chelating agent although not as effective as DSEDTA.

EDTA should not be used in formulations of metalloproteins such as insulin or hemoglobin or fibrolase as the chelating agent will attack the metal that is part of the stable conformation of the protein. EDTA will accelerate the oxidative degradation of methionine in human insulin-like growth factor I solutions (21). Oxidation of methionine 59 was catalyzed by light and ferric ions in combination with EDTA. It was suggested that EDTA actually enables ferric ions to be active by stabilizing the transfer of electrons from ferric ions to ferrous ions. Methionine in this protein is radicalized by light and then oxidized to methionine sulfoxide. Light may also trigger the generation of hydroxyl radicals by decomposition of water that may oxidize the methionine. Thus, the formulator must not indiscriminately include EDTA in protein formulations without carefully determining that its presence aids in oxidative stabilization of the protein.

To illustrate how several multiple approaches can be applied to minimize oxidative degradation, parathyroid hormone was used as a model protein to investigate stabilization of methionine, tryptophan, and histidine amino acids from oxidative degradation (22). Successful approaches included using polysorbate free from peroxide contamination; mannitol also helped protect against peroxide-induced oxidation, EDTA to complex heavy metals originating from stainless steel surfaces, and free-radical scavenger stabilizers such as Trolox (6-hydroxy-2,5,7,8-tetramethylchroman-2-carboxylic acid) and pyridoxine.

Inert Gases

Inert gases are frequently used in production of sterile dosage forms. The most commonly used inert gas is nitrogen. Other inert gases used, although not often primarily because of expense, include argon and helium. Argon, however, has been shown to be more efficient in displacing oxygen because it is heavier than air and will more readily stay in the vial compared to nitrogen. The normal use of inert gases in sterile product manufacturing involves solution and headspace saturation. Addition to water and compounding solutions prior to aseptic filtration saturates the solution and minimizes the level of dissolved oxygen. However, oxygen is never completely displaced with an inert gas when the solution is sparged. Many manufacturers use a dual needle that permits simultaneous filling of a liquid and purging of gas at the same time. Inert gas introduced into the headspace of a filled vial right before the vial is stoppered with a rubber closure theoretically displaces oxygen in the headspace. Again, a dual needle can be used to fill solution and purge gas into the final container at the same time.

The inert gas must be high quality grade and must be sterilized, usually with a 0.22-μm hydrophobic membrane filter. The integrity of the gas filter is tested before and after use by diffusion flow methods.

Packaging and Oxidation

All the appropriate formulation and processing procedures can be in place for stabilizing protein solutions against oxidation, but if the packaging system is inadequate from an integrity standpoint, the product will readily degrade. Most injectable products are packaged in glass vials or syringes with rubber closures or plungers. The rubber-glass interface and the oxygen transmission coefficient of the rubber closure will dictate the quality of the container/closure system (chap. 30).

Oxygen transmission coefficients are determined for a particular rubber closure formulation by the rubber closure manufacturer. Rubber formulations having the lowest oxygen transmission coefficients are the synthetic butyl and halobutyl types. The formulator should determine from the rubber manufacturer how the halobutyl rubber is cured (shaped, molded) since common curing agents are zinc oxide, aluminum, and peroxide, which potentially can leach out of the rubber formulation with time and catalyze oxidative degradation.

Many drugs (catecholamines, cephalosporins, aminoglycosides, some steroids, iron-containing molecules, and many others) are sensitive to light. Effective packaging is the primary (in most cases only) way to protect drugs from light degradation. Good light protective secondary packaging, use of amber-colored primary packaging (although more expensive and difficult to inspect for particulate matter), and maintaining product storage in the dark are the ways that sensitive drugs are protected from light degradation. There is no practical formulation approach to stabilize light-sensitive drugs; good packaging is the key to protect against light degradation.

OTHER STABILIZERS TO MINIMIZE DRUG DEGRADATION

The literature contains many examples of excipient stabilization phenomena with injectable drugs. The following are only a few examples. Sugars and polyols, such as ethylene glycol, glycerol, glucose, and dextran, at high concentrations, can inhibit the metal-catalyzed oxidation of human relaxin (23). All but dextran act as chelating agents in complexing transition metal ions, whereas dextran, which has a higher binding affinity to metal ions and undergoes depolymerization in a metal-catalyzed oxidation, protects relaxin by a radical scavenging mechanism.

Mannitol has been shown to inhibit the iron-catalyzed oxidation of Met-containing peptides (22,24). Mannitol is the most commonly used excipient in freeze-dried formulations often serving a dual role as a bulking agent and a stabilizer.

Fibroblast growth factors, both acidic and basic, possess nearly identical three-dimensional structures of 12 antiparallel β-strands arranged with approximate threefold-internal symmetry (25). Acidic fibroblast growth factor was found that its tendency to aggregate in solution was inhibited by a variety of polyanionic additives such as inositol hexasulfate or sulfate β-cyclodextrin and by a number of commonly used excipients such as sucrose, dextrose, trehalose, glycerol, and glycine. In all cases, these interactions between acidic fibroblast growth factor and various excipients resulted in an increase in the protein's Tm, the midpoint of the temperature of the transition from the folded to unfolded protein. Basic fibroblast growth factor has a major degradation pathway that involves not only aggregation and precipitation, but also a succinimide replacement of aspartate at position 15 of the protein sequence (26). Adjusting solution pH from 5 to 6.5 and storage at low temperatures will help to avoid this reaction.

A variety of co-solvents can stabilize proteins in solution because the co-solvent is preferentially excluded from surface interaction with the protein (27). Co-solvents behaving this way include glycerol and sorbitol. Polyethylene glycol (PEG) also is preferentially excluded from the protein, yet will still denature or destabilize proteins in solution.

OPTIMIZING PHYSICAL STABILITY

Physical stability problems are rare with small molecules except with sparingly soluble molecules that are borderline soluble in the formulation vehicle. However, proteins, because of their unique ability to adopt higher order secondary and tertiary three-dimension structures, tend to undergo a number of physical changes, independent of chemical modifications. Physical instability of proteins is sometimes a greater cause for concern and is more difficult to control compared to chemical instability. Many proteins, particularly when exposed to stressful conditions, for example, extremes in temperature, will unfold such that the hydrophobic portions become exposed to the aqueous environment. Such exposure will promote aggregation or self-association, possibly leading to physical instability and loss of biological activity since the interaction with the receptor site requires folded structures with correct conformation. The relationship of the different pathways of physical destabilization of proteins is shown in Figure 8-5.

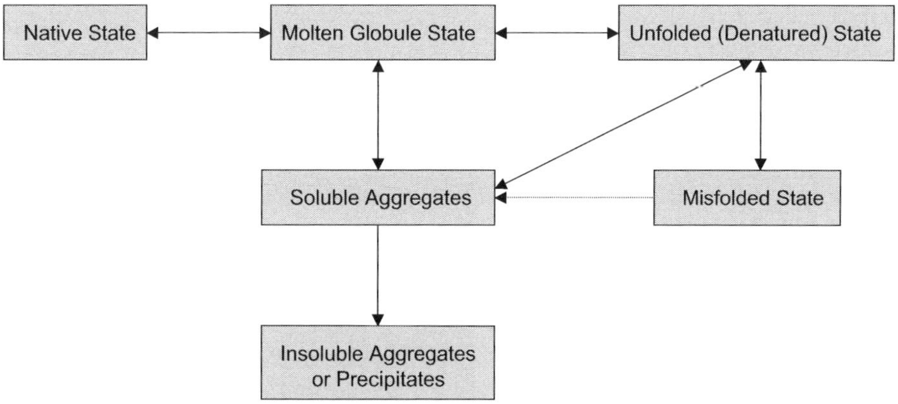

Figure 8-5 Schematic pathway of physical degradation of a protein. *Source*: From Ref. 28.

Denaturation

Protein denaturation occurs when native higher order structure is disrupted. Denaturation can lead to unfolding and the unfolded polypeptide chain may undergo further reactions. Such inactivation could be association with surfaces and/or interaction with other protein molecules leading to aggregation and precipitation. Denaturation may be reversible or irreversible. Reversible denaturation results from high temperature exposure, or in purposeful experimental conditions when the protein is exposed to chaotropic agents (e.g., urea, guanidine hydrochloride). When the denaturing condition is removed, the protein will regain its native state and maintain its activity. Reversible denaturation can be decreased by the use of additives such as salts that bind to nonspecific binding sites on the proteins (29–31). Preferential hydration of proteins in the presence of a glycerol-water mixed solvent system is a prerequisite for stabilizing the native structure of several globular proteins (32).

Irreversible denaturation means that the protein, once unfolded, will not regain its native form and activity. Aggregation phenomena lead to irreversible denaturation.

Protein Aggregation

Aggregation of peptides and proteins is caused mainly by hydrophobic interactions that eventually lead to denaturation. Sources of hydrophobic conditions include exposures to air–liquid and solid–liquid interfaces, light, temperature fluctuations, impurities, and foreign particles. When the hydrophobic region of a partially or fully unfolded protein is exposed to water, a thermodynamically unfavorable situation is created. The normally "buried" hydrophobic interior is now exposed to a hydrophilic aqueous environment. Consequently, the decrease in entropy from structuring water molecules around the hydrophobic region forces the denatured protein to aggregate, mainly through the exposed hydrophobic regions. Thus, solubility of the protein may also be compromised. In some cases self-association of protein subunits, either native or misfolded, may occur under certain conditions and this may lead to precipitation and loss in activity (Fig. 8-6). Irreversible aggregation can be minimized, even prevented, through expert formulation approaches involving stabilizers such as surfactants, polyols, or sugars.

Factors that affect protein aggregation in solution generally include protein concentration, pH, temperature, other excipients, and mechanical stress. Some factors (e.g., temperature) can be easily controlled during compounding, manufacturing, storage, and use. Other factors, (e.g., mechanical stress, temperature excursions during shipping and distribution, inherent instability of the active ingredient) cannot be so easily controlled. Formulation studies will dictate appropriate choice(s) of pH and excipients that will not induce aggregation and/or, in fact, will aid in the prevention of aggregation. A new class of alkyl saccharide excipients originally intended to dramatically enhance transmucosal absorption of peptide and protein drugs was found to be highly effective in preventing protein aggregation (33). These alkyl saccharide excipients stabilize and reduce aggregation of peptides or proteins in therapeutically useful formulations, and they may provide solutions for aggregation-related manufacturing or formulation problems and/or unwanted immunogenicity. Examples of proteins stabilized by these excipients (0.125% concentrations) include insulin and growth hormone.

Figure 8-6 Example of aggregated protein.

The desire to identify stable solution preparations of insulin for use in novel delivery systems such as continuous infusion pumps, led to the development of test methodology for assessing the impact of various additives on physical stability. Insulin (and many other proteins) physical stability typically is evaluated using thermomechanical procedures involving agitation or rotation of protein solutions at elevated temperature. Turbidity resulting from aggregation is usually determined as a function of time by visual inspection or light scattering analysis. Alternatively, reductions in the soluble protein content due to precipitation can be quantified by HPLC assay as a function of time. Relative stability is defined by the length of time a preparation remains on the test without showing a change in either parameter. It should be noted that the greatest difficulty in applying such testing strategies is interpreting the experimental data and correlating it in a practical way to "real life" conditions that the formulation may actually experience. Nevertheless, regulatory agencies may request data from such testing to support dating periods or other product claims. Physical stress testing, however, is more appropriately used as a development screening tool to identify the capability of various additives to prevent aggregation.

Analytical methods used for determining protein aggregation are listed on pages 177–178 (chapter 11 under "Answers for Case Study 10").

Foreign Particles, Protein Aggregation, and Immunogenicity

The reality of protein aggregation has raised the concerns about such aggregates, even at subvisible levels, leading to an immune response resulting from antibody-mediated neutralization of the protein's activity or alterations in bioavailability (34,35). Among many causes for protein aggregation are protein particles resulting either from the protein alone or resulting from heterogenous nucleation on foreign micro- or nanoparticles originating from the manufacturing process (mixing tanks, process tubing, filter systems, filling machines) and from the container/closure system (36). Silicone oil, used as a lubricant for rubber closures on vials and rubber plungers in prefilled syringes also can induce protein aggregation (37).

Large protein aggregates are subvisible particles (smaller than 10 micrometers) that are not currently monitored and quantified by compendial subvisible particulate matter measurement systems. Carpenter, et al. (34) have questioned this current practice and have proposed that (*i*) scientists from industry and academia work together to define the quantitative capabilities of particle counting instruments for particles as small as 0.1 μm, (*ii*) develop new particle counting instruments for more reliable measurement of particles at sizes approaching 0.1 μm, and (*iii*) more studies be conducted and published on the impact of protein aggregation on immunogenicity including the role of protein class, amount of aggregate, size of aggregates, and protein conformation in aggregates.

Also, the reader is referred to the end of chapter 29 where there is some discussion about the huge variety of biopharmaceutical commercial product package insert language regarding acceptability of visible particulate matter and use of different types of transfer and in-line filters.

ADSORPTION

Proteins exhibit a certain degree of surface activity; that is, they adsorb to surfaces due to their innate nature of being amphiphilic polyelectrolytes. Consequently biological activity may be either reduced or totally lost if such adsorption occurs during manufacturing, storage, or use of the final product. Insulin has been the most studied protein with respect to surface adsorption. Potential problems may be encountered while delivering insulin because of its ability to adsorb onto the surfaces of delivery pumps, glass containers, and to the inside of the intravenous bags. Insulin adsorption usually is finite once binding sites are covered and such adsorption is usually not clinically significant.

Adsorption to surfaces depends on protein–protein interactions, time, temperature, pH, and ionic strength of the medium and the nature of the surface (38). Interactions that determine the overall adsorption process between a protein and a surface include redistribution of charged groups in the interfacial layer, changes in the hydration of the sorbent and the protein surface, and structural rearrangements in the protein molecule. Surface denaturation which commonly takes place at the liquid–solid and liquid–air interface to involve conformational changes such as loss of α-helices to β-sheets and certain random structures (39). These structural changes,

Table 8-3 Possible Strategies to Overcome Protein Adsorption

- Increase protein concentration during filtration and/or use extra volume to saturate the filter with protein solution
- Modify (e.g., siliconize) the surface of the glass containers, providing a resistant barrier to protein-surface interaction
- Decrease the rate of mixing when it is known that shear will affect protein adsorption
- Add excipients such as surfactants that have higher surface activity
- Add macromolecules such as albumin and gelatin (must be synthetic) to complete for binding sites on the surface.

determined by the nature of the interfaces, are similar to those observed with aggregation caused by heat, high pressure, or chemical denaturants. In the case of proteins, sources such as the polymer of the membrane filter, the administration set, agitation that occurs during the purification process as well as the method of manufacture are known or at least suspected to cause surface denaturation. Strategies often used to overcome protein denaturation due to adsorption are presented in Table 8-3.

SURFACTANTS

Surface-active agents (surfactants) exert their effect at surfaces of solid–solid, solid–liquid, liquid–liquid, and liquid–air because of their chemical composition containing both hydrophilic and hydrophobic groups (see chap. 6). Surfactants effectively compete against proteins for these interfacial hydrophobic locations, thus helping to minimize protein adsorption and potential aggregation.

Generally, ionic surfactants can denature proteins. However, nonionic surfactants usually do not denature proteins even at relatively high concentrations (1% w/v) (40). Most parenterally acceptable nonionic surfactants come from either the polysorbate (sorbitol-polyethylene oxide polymers) or polyether (polyethylene oxide-polypropylene oxide block co-polymers) groups. Polysorbate 20 and 80 and sodium dodecyl sulfate are effective and acceptable surfactant stabilizers in marketed protein formulations (Table 6-6). The chemical structure of polysorbates, factors affecting micelle formation and degradation pathways of polysorbates 20 and 80 are the subject of a review article by Kerwin (41). Effectiveness of polysorbate stabilization is dependent on the structure of polysorbate (monomer or micelle) and polysorbate–protein ratio (42). Other surfactants that have been used in protein formulations for clinical studies and/or found in the patent literature include Pluronic F68, and other polyoxyethylene ethers (e.g., the "Brij" class).

The choice of surfactant and the final concentration optimal for stabilization is quite dependent on a variety of factors including other formulation ingredients, for example, sugars, protein concentration, headspace in the container, the type of container, and test methodology.

Recombinant hGH will aggregate readily under mechanical and thermal stress. Aggregation from mechanical stress can be substantially reduced in the presence of surfactants (43). Mechanical stress may cause proteins to be more exposed to air–water interfaces where denaturation is more likely to occur than in the bulk phase of water. Surfactants will preferentially compete with proteins for accumulation at the air–water interface and keep the protein from undergoing interfacial denaturation resulting from mechanical stress. Pluronic F68 and Brij 35 will stabilize hGH at their critical micelle concentrations (0.1% and 0.013%, respectively), whereas stabilization with polysorbate 80 requires a concentration of 0.1%, higher than the critical micelle concentration value for polysorbate 80 of 0.0013%. The reasons for these differences in stabilizing concentrations are not clear, but simply reflect differences in interactions between different surfactants and proteins. It is interesting to note that these surfactants do not stabilize hGH from aggregation due to high temperature stress.

Surface-active agents, particularly polysorbate 80, protect proteins against surface-induced denaturation during freezing (44). A strong correlation exists between freeze denaturation (quick freezing of the protein) and surface denaturation (shaking the protein in solution). Proteins that tend to denature under these conditions are protected by the addition of polysorbate 80 (0.1%). Other surfactants—Brij 35, Lubrol-px, Triton X-10, and even the ionic surfactant,

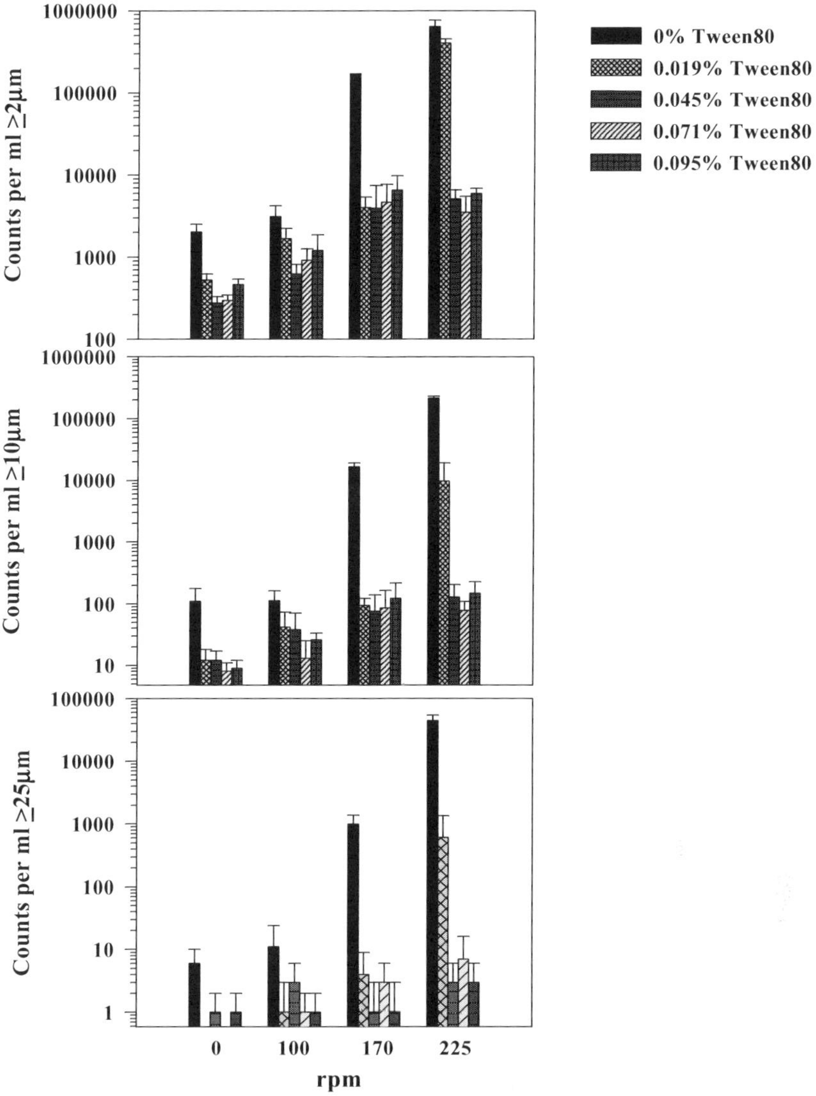

Figure 8-7 Effect of Tween 80 concentration on particle formation in solutions of recombinant human hemoglobin as a function of shear stress. *Source*: From Ref. 46.

sodium dodecyl sulfate—also protected the protein from denaturation although these surfactants have not yet been approved for use in injectable formulations. The authors pointed out that surfactants may be needed to protect proteins from denaturation during the freezing step only, and that other stabilizers, for example, sucrose, may be needed to further protect the protein during freeze drying.

Surfactants were ineffective in preventing BST aggregation and precipitation in solution at elevated temperature[1], whereas other stabilizers such as sucrose were more effective (45). Tween 80 was more effective in reducing the amount of measurable particles due to aggregation of recombinant human hemoglobin (Fig. 8-7).

[1] While polysorbate 80 was not effective in stabilizing BST at elevated temperature, it was effective when the applied stress was agitation. Also, the authors noted that polysorbate 80 destabilization of BST was not observed at ambient or refrigerated temperatures as other decomposition pathways, for example, deamidation, became more predominant at lower temperatures.

Peroxides are known contaminants of nonionic surfactants (19). Peroxide levels from different sources of polysorbate 80 ranged from less than 1 mEq/kg to more than 27 mEq/kg. Peroxide levels increased upon storage at ambient temperatures probably due to headspace oxygen and/or the container/closure interface allowing ingress of air. Peroxides in polysorbate can result in oxidative degradation of proteins. Improvements have been made in the manufacturing of polysorbate, for example, certified peroxide-free polysorbates are now readily available.

Electron paramagnetic resonance (EPR) spectroscopy has been used to determine the binding stoichiometry of the surfactant to the protein and, thus, what potentially is the optimal amount of surfactant to use to stabilize the protein against surface denaturation and other physical instability reactions (47).

CYCLODEXTRINS

Cyclodextrins are cyclic (α-1,4)-linked oligosaccharides of α-D-glucopyranose containing a relatively hydrophobic central cavity and hydrophilic outer surface. Cyclodextrins come in a wide variety of structural derivatives, the most common being α-, β-, and γ-cyclodextrins, which consist of six, seven, and eight glucopyranose units, respectively. Two parenteral cyclodextrins are Encapsin™, a hydroxylpropyl-β-cyclodextrin, and Captisol™, a sulfobutylether-β-cyclodextrin. They have been used widely for increasing the solubility stability, and bioavailability of small drug molecules (see chap. 6). Peptides and proteins can also be stabilized in cyclodextrin complexes. β-cyclodextrins at a 25-fold excess stabilized leucine enkephalin against enzymatic degradation in sheep nasal mucosa (48). Hydroxypropyl-β-cyclodextrin at a 1% concentration was shown to enhance the reconstituted solution stability of keratinocyte growth factor (49) and several other proteins in solution (50). Glucagon will form inclusion complexes with γ-cyclodextrin in acidic solution that results in enhancement of glucagon's physical and chemical stability (51).

ALBUMIN

Serum albumin is a widely used stabilizer in protein formulations for minimizing protein adsorption to glass and other surfaces (Table 8-4). Albumin preferentially competes with other proteins for binding sites on surfaces, but why this is so is not clear.

Because albumin is a natural protein, concerns have been raised about potential contamination of albumin with human prion protein that is thought to be the infectious agent in bovine spongiform encephalopathy (BSE). Indeed, the use of animal-source excipients (and this includes not only albumin, but also glycerol and polysorbate 80) is no longer practiced. The development of synthetic (e.g., recombinant HSA) versions of these materials has eliminated concerns over potential disease transmission.

OTHER PHYSICAL COMPLEXING/STABILIZING AGENTS

PEG is a common co-solvent for solubilizing small nonproteinaceous molecules and may minimize the aggregation of several peptides and proteins (52). PEG modification of proteins for sustained-release purposes has seen wide application. The concentration of PEG needs to be

Table 8-4 Some Examples of Commercial Protein Dosage Forms Containing Human Serum Albumin

Generic	Brand®	% HSA in product
Alglucerase	Ceredase	1.0
Erythropoietin	Epogen	0.25
Interferon Alpha-2a	Roferon-A	0.5
Interferon Alpha-2b	Intron-A	0.1 (after reconstitution)
Urokinase	Abbokinase	5.0 (after reconstitution)
Alpha-1-Proteinase	Aralast	0.5 (after reconstitution)
Antihemophilic factor	Recombinate	1.0 (after reconstitution)
Botulism toxin	Myobloc	0.05
Streptokinase	Streptase	2.0 (after reconstitution)
Hyaluronidase	Halozyme	0.1

fairly low (< 1%, w/v) to serve as a stabilizer; otherwise at higher concentrations (>10% w/v) it can cause precipitation (53).

Poly(vinylpyrrolidone) (PVP) also is like PEG in that at low concentrations it can stabilize proteins, whereas at high concentrations it may help lead to protein aggregation and precipitation. PVP at low concentrations (≤2.0%) effectively stabilizes human IgM monoclonal antibody against heat-induced aggregation, whereas PVP concentrations ≥5.0% will cause aggregation (54).

Fibroblast growth factors, acidic and basic, are prone to acid and thermal inactivation and can be stabilized by a number of heparin and heparin-like molecules (25). Human keratinocyte growth factor, also prone to aggregation at high temperature, is stabilized by heparin, sulfated polysaccharides, anionic polymers, and citrate ion (55).

OPTIMIZING MICROBIOLOGICAL ACTIVITY

Antimicrobial Preservatives

Many products (perhaps around 25%) are commercially available as multiple-dose formulations. If a sterile product is intended for multiple dosing, then it must contain an effective antimicrobial preservative (AP) agent. AP agents are formulated with the active pharmaceutical ingredient if the product is a ready-to-use solution or is part of the diluent used to reconstitute freeze-dried products intended for subsequent multiple dosing. While rare, there are examples of AP agents formulated within the freeze-dried product and not part of the diluent.

Of 145 peptide and protein drug products listed in 2006 Physicians' Desk Reference, 36 contained preservatives (56). Most vaccine products used to contain AP agents, especially thimerosal, but by 2006, only 8 vaccine products still were formulated as multidose products.

The most common APs used in multiple-dose formulations are phenol, meta-cresol, and benzyl alcohol. Less common, especially for new formulations, but still used APs include methyl and propylparaben. Some, although very few, vaccines still contain APs with phenoxyethanol being the most common. Thimerosal used to be commonly used for vaccine products, but not today. Examples of use of these preservatives are listed in chapter 6 (Table 6-7).

Use of antimicrobial agents requires passing a preservative efficacy test (PET) (USP chap. <51> provides the directions for conducting this test). Unfortunately, the United States Pharmacopeia (USP) and the British and/or European Pharmacopeial (BP/EP) tests for PET are different in their requirements. Table 8-5 summarizes the differences between the tests. The USP basically requires a bacteriostatic preservative system while the BP/EP requires a bacteriocidal system. For example, the USP requires a 3-log reduction in the bacterial challenge by the 14th day after inoculation, while criteria A of the BP/EP test requires the same 3-log reduction within 24 hours. This great difference in compendial requirements for preservative efficacy has caused many problems in the formulation of protein dosage forms for various markets. One unpublished example involved a new protein product where the scientist developing the formulation was unaware of the different compendial requirements. The focus was minimizing instability of the new protein in the presence of the AP and used a minimal amount of AP in the formulation. The phase 1 clinical study was scheduled for a European clinic so the EP PET was performed. The formulation failed miserably and the product had to be reformulated with start of the clinical study delayed by almost a year.

Passing the BP/EP PET requires the use of relatively high amounts of phenol or cresol or other AP that may have an impact on the stability of the formulation and could result in sorption of the preservative into the rubber closure. The formulator must keep in mind that

Table 8-5 Comparison of USP and EP Preservative Efficacy Tests

Test	USP <51>	EP <Chapter 5.1.3>
Bacterial	1-log reduction within 7 days	2-log reduction within 6 hours
Challenge	3-log reduction with 14 days	3-log reduction within 24 hours
Fungal challenge	No increase after 28 days	2-log reduction with 7 days
Overall requirement	Bacteriostatic	Bacteriocidal

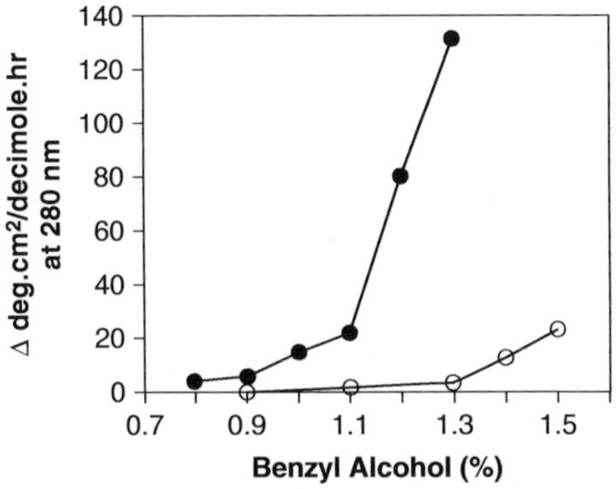

Loss in ellipticity at 280 nm 280 of rhIFN-γ as function of benzyl alcohol concentration in 16 mM acetate buffer at pH 5.0 (○) and 16 mM succinate buffer at pH 5.0 (●).

Figure 8-8 Effect of benzyl alcohol on recombinant human interferon gamma aggregation. *Note*: Tobler, et al. J Pharm Sci, June, 2004 used hydrogen-deuterium isotope exchange detected by MS to detect tertiary structure changes that involve only a limited part of this protein still causing irreversible loss of activity. Benzyl alcohol causes protein to unfold forming very large aggregates. *Source*: From Refs. 62 & 64.

increasing the concentration of APs may have a negative impact on protein physical stability (precipitation, aggregation, etc). Increasing AP levels will increase the hydrophobicity of the formulation and could affect the aqueous solubility of the protein. Increasing AP concentrations also increases the potential for toxicological hazards.

It is well known that APs not only protect insulin formulations against inadvertent contamination, but also may have a significant effect on protein stability. For example, phenolic preservatives have a profound effect on the conformation of insulin in solution (57) and the assembly of the specific type of LysPro insulin hexamer (58). Furthermore, phenol and/or *m*-cresol in insulin solutions will have a tendency to be adsorbed by and permeate rubber closures (59). Therefore, rubber formulations must be designed to minimize these potential problems.

APs are known to interact with proteins and can cause stability problems such as aggregation. For example, phenolic compounds will cause aggregation of hGH (60). Phenol will produce a significant decrease in the α-helix content of insulinotropin resulting in aggregation of β-sheet structures (61). Benzyl alcohol, above certain concentrations and depending on other formulation factors, will interact with recombinant human interferon-γ causing aggregation of the protein (Fig. 8-8) (62). Other examples are granulocyte-stimulating factor and recombinant interleukin-1R (56). These examples point out the need for the formulation scientist to understand the importance of potential effects of preservative type, concentration, and other formulation additives on the interaction with proteins in solution while balancing the needs for antimicrobial efficacy.

In determining the appropriate AP agent or agents, insulin was studied as the protein to be preserved and combining insulin with different types of AP agents either alone or in combination (63). These formulations were challenged with the five USP PET organisms and D values[2] determined. The D-value determination allows a single-quantitative estimate of the AP effectiveness of a certain agent or combination of agents in a specific formulation against a specific microorganism. The preservative combination of 0.2% phenol and 0.3% *m*-cresol gave the lowest D-value (fastest time required for a 1-log reduction in the initial inoculum of *S. aureus* and, thus, was the most effective AP system in this particular insulin formulation.

There are instances where a manufacturer, because of concerns regarding aseptic processing and sterility assurance of the product throughout its shelf-life, will add an AP agent in

[2] D value = Time required for a 1-log reduction in the microbial population due to the effect of the antimicrobial preservative system. The smaller the D value, the greater the effect of the preservative on the microorganism in question. Covered in chapter 18.

the protein formulation even though it is intended only for a single-dose injection. This is a very controversial practice. Regulatory agencies worldwide object to this approach if, in their opinion, the use of APs in a single-dose injectable product is practiced in order to "cover up" for inadequate aseptic manufacturing practices and controls.

Many countries require PET be performed for routine stability protocols and for special stability studies. Also there may be requests from agencies to do PET on containers that have been used (i.e., penetrated; partial volume withdrawn) to demonstrate that the product can still kill microorganisms. In mid-1995, the Australian Drug Evaluation Committee (ADEC) passed resolutions that in light of safety concerns with contamination and cross-contamination, the use of injectable products in multi-dose packages is discouraged. In order to support the use of a multidose product and the shelf-life once a package has been reconstituted or opened for use, AP efficacy data are required for approval.

OSMOLALITY (TONICITY) AGENTS

Salts or nonelectrolytes (e.g., glycerin) are added to protein formulations in order to achieve an isotonic solution. Nonelectrolytes often are preferred over salts as tonicity adjusters because of the potential problems salts cause in precipitating proteins. Generally, solutions containing proteins administered IV, IM, or SC do not have to be precisely isotonic because of immediate effects from dilution by the blood. Intrathecal and epidural injections into the cerebrospinal fluid require very precise specifications for the product to be isotonic and at physiological pH. This is because extremes in osmolality and/or pH can damage or destroy cells and cerebrospinal cells cannot be reproduced or replaced.

SIMPLE EXERCISE

For each of these commercial sterile solution formulations, name the purpose of each excipient.

Nebcin® (Lilly)
Tobramycin 80 mg
Sodium bisulfite 5 mg
Disodium EDTA 0.1 mg
Phenol 5 mg

Valium Injection (Roche)
Diazepam 5 mg
Propylene glycol 40%
Ethanol 10%
Benzoic acid/Sodium benzoate 5%
Benzyl alcohol 1.5%

Nutropin AQ® (Genentech)
Somatropin 10 mg
Sodium chloride 17.4 mg
Phenol 5 mg
Polysorbate 20.4 mg
Sodium citrate 10 mM

Rebif® (Serono)
Interferon beta-1 a 44 mcg
Human albumin 4 mg
Mannitol 27.3 mg
Sodium acetate 0.4 mg

REFERENCES

1. Akers MJ, Vasudevan V, Stickelmeyer M. Protein dosage form development. In: Nail SL, Akers MJ, eds. Borchardt RT, series editor. Development and Manufacture of Protein Pharmaceuticals. Volume in series on Pharmaceutical Biotechnology. New York, NY: Plenum, 2002:47–127.

2. Akers MJ, DeFelippis MR. Formulation of protein dosage forms: Solutions. In: Hovgaard L, Frokjaer S, eds. Pharmaceutical Formulation Development of Peptides and Proteins. London, UK: Taylor and Francis, 2000:145–177.
3. Manning M, Chou D, Murphy B, et al. Stability of protein pharmaceuticals: An update. Pharm Res 2010. doi:10.1007/s11095–009-0045–6.
4. Schwartz MA, Buckwalter FH. Pharmaceutics of penicillin. J Pharm Sci 1962; 51:1119–1128.
5. Chi EY, Kirshnan S, Randolph TW, et al. Physical stability of proteins in aqueous solution: Mechanism and driving forces in nonnative protein aggregation. Pharm Res 2003; 20:1331.
6. Senderoff RI, Wootton SC, Boctor AM, et al. Aqueous stability of human epidermal growth factor 1–48. Pharm Res 1994; 11:1712–1720.
7. Roy S, Henderson I, Nayar R, et al. Effect of pH on stability of recombinant botulinum serotype A vaccine in aqueous solution and during storage of freeze-dried formulations. J Pharm Sci 2008; 97:5132–5146.
8. Patel K. Stability of adrenocorticotropic hormone (ACTH) and pathways of deamidation of asparaginyl residue in hexapeptide segments. In: Wang YJ, Pearlman R, eds. Stability and Characterization of Protein and Peptide Drugs: Case Histories. New York, NY: Plenum Press, 1993:207–212.
9. Zhu H, Meserve K, Floyd A. Preformulation studies for an ultrashort-acting neuromuscular blocking agent GW280430A. I. Buffer and cosolvent effects on the solution stability. Drug Dev Ind Pharm 2002; 28:135–142.
10. Capasso S, Mazzarella L, Zagari A. Deamidation via cyclic imide of asparaginyl peptides: Dependence on salts, buffers, and organic solvents. Pept Res 1991; 4:234.
11. Wang YJ, Shahrokh Z, Vemuri S, et al. Characterization, stability, and formulations of basic fibroblast growth factor. In: Pearlman R, Wang YJ, eds. Formulation, Characterization, and Stability of Protein Drugs: Case Histories. New York, NY: Plenum Press, 1996:164–165.
12. Daugherty AL, Mrsny RJ. Formulation and delivery issues for monoclonal antibody therapeutics. Adv Drug Del Rev 2006; 58:686–706.
13. Gokam YR, Kras E, Nodgaard C, et al. Self-buffering antibody formulations, 2008. J Pharm Sci 2008; 97:3051–3066.
14. Vemuri S, Yu CT, Roosdorp N. Formulation and stability of recombinant a1-antitrypsin. In: Wang YJ, Pearlman R, eds. Stability and Characterization of Protein and Peptide Drugs: Case Histories. New York, NY: Plenum Press, 1993:269–270.
15. Pearlman R, Bewley TA. Stability and characterization of human growth hormone. In: Wang YJ, Pearlman R, eds. Stability and Characterization of Protein and Peptide Drugs: Case Histories. New York, NY: Plenum Press, 1993:44.
16. Davio SR, Hageman MJ. Characterization and formulation considerations for recombinant derived bovine somatotropin. In: Wang YJ, Pearlman R, eds. Stability and Characterization of Protein and Peptide Drugs: Case Histories. New York, NY: Plenum Press, 1993:76–80.
17. Wang N, Hu B, Ionescu R, et al. Opalescence of an IgG1 monoclonal antibody formulation is mediated by ionic strength and excipients. Biopharm Int 2009; 22:36–47.
18. Herman AC, Boone TC, Lu HS. Characterization, formulation, and stability of Neupogen® (Filgrastim), a recombinant human granulocyte-colony stimulating factor, characterization, stability, and formulations of basic fibroblast growth factor. In: Pearlman R, Wang YJ, eds. Formulation, Characterization, and Stability of Protein Drugs: Case Histories. New York, NY: Plenum Press, 1996:324–325.
19. Knepp VM, Whatley JL, Muchnik A, et al. Identification of antioxidants for prevention of peroxide-mediated oxidation of recombinant human ciliary neurotrophic factor and recombinant human nerve growth factor. PDA J Pharm Sci Tech 1996; 50:163.
20. Li S, Schoneich C, Wilson GS, et al. Chemical pathways of peptide degradation. V. Ascorbic acid promotes rather than inhibits the oxidation of methionine to methionine sulfoxide in small model peptides. Pharm Res 1993; 10:1572.
21. Fransson JR. Methionine oxidation and covalent aggregation in aqueous solution. J Pharm Sci 1997; 86:1046–1050.
22. Junyan JA, Zhang B, Cheng W, et al. Methionine, tryptophan, and histidine oxidation in a model protein. PTH J Pharm Sci 2009; 98:4485–4500.
23. Li W, Patapoff TW, Nguyen TH, et al. Inhibitory effect of sugars and polyols on the metal-catalyzed oxidation of human relaxin. J Pharm Sci 1996; 85:868–872.
24. Li S, Schoneich C, Borchardt RT. Chemical pathways of peptide degradation. VIII. J Pharm Sci 1995; 64:356–357.
25. Tsai PK, Volkin DB, Dabora JM, et al. Formulation design of acidic fibroblast growth factor. Pharm Res 1993; 10:649–659.
26. Shahrokh Z, Eberlein G, Buckley D, et al. Major degradation products of basic fibroblast growth factor: Detection of succinimide and iso-aspartate in place of aspartate. Pharm Res 1994; 11:936–944.

27. Arakawa T, Kita Y, Carpenter JF. Protein-solvent interactions in pharmaceutical formulations. Pharm Res 1991; 8:285–291.
28. Brange J. Physical stability of proteins. In: Frokjaer S, Hovgaard L, eds. Pharmaceutical Formulation Development of Peptides and Proteins. London: Taylor & Francis, 2000:95.
29. Arakawa T, Timasheff SN. Preferential interaction of protein with salts in concentrated solution. Biochemistry 1982; 21:6545.
30. Arakawa T, Timasheff SN. Mechanism of protein salting in and salting out by divalent cation salts: Balance between hydration and salt binding. Biochemistry 1984; 23:5913.
31. Arakawa T, Prestrelski SJ, Kenney WC, et al. Factors affecting short-term and long-term stabilities of proteins. Adv Drug Del Rev 1993; 10:1–29.
32. Gekko K, Timasheff SN. Mechanism of protein stabilization by glycerol: Preferred hydration in glycerol-water mixtures. Biochemistry 1981; 20:466.
33. Maggio ET. Novel excipients prevent aggregation in manufacturing and formulation of protein and peptide therapeutics. Bioprocess Int 2008; 2–5.
34. Carpenter JF, Randolph TW, Jiskoot W, et al. Overlooking subvisible particles in therapeutic protein products: Gaps that may compromise product quality. J Pharm Sci 2009; 98:1201–1205.
35. Rosenberg AS. Effects of protein aggregates: An immunologic perspective. AAPS J 2006; 8:E501–E507.
36. Tyagli AK, Randolph TW, Dong A, et al. IgG particle formation during filling pump operation: A case study of heterogeneous nucleation on stainless steel nanoparticles. J Pharm Sci 2008; 98:94–104.
37. Sharma DK, Oma P, Krishnan S. Silicone microdroplets in protein formulations—Detection and enumeration. Pharm Tech 2009; 33(4):74–79.
38. Norde W. Adsorption of proteins at solid-liquid interfaces. Cells Materials 1995; 5:97–112.
39. Lenk JR, Ratner BD, Gendreau RM, et al. IR spectral changes of bovine serum albumin upon surface adsorption. J Biomed Mater Res 1989; 23:549–569.
40. Cleland JL, Powell MF, Shire SJ. The development of stable protein formulations: A close look at protein aggregation, deamination and oxidation. Crit Rev Ther Drug Carrier Sys 1993; 10:307–377.
41. Kerwin BA. Polysorbates 20 and 80 used in the formulation of protein biotherapeutics: Structure and degradation pathways. J Pharm Sci 2009; 97:2924–2935.
42. Deechongkit S, Wen J, Narhi LO, et al. Physical and biophysical effects of polysorbate 20 and 80 on darbepoetin alfa. J Pharm Sci 2009; 98:3200–3217.
43. Katakam M, Bell LN, Banga AK. Effect of surfactants on the physical stability of recombinant human growth hormone. J Pharm Sci 1995; 84:713–716.
44. Chang BS, Kendrick BS, Carpenter JF. Surface-induced denaturation of proteins during freezing and its inhibition by surfactants. J Pharm Sci 1996; 85:1325–1330.
45. Hageman MJ, Tinwalla AY, Bauer JM. Kinetics of temperature-induced irreversible aggregation/precipitation of bovine somatropin as studied by initial rate methods. Pharm Res 1993; 10:S-85.
46. Kerwin BA, Akers MJ, Apostol I, et al. Acute and long-term stability studies of deoxy hemoglobin and characterization of ascorbate-induced modifications. J Pharm Sci 1999; 88:79–99.
47. Bam NB, Randolph TW, Cleland JL. Stability of protein formulations: Investigation of surfactant effects by a novel EPR spectroscopic technique. Pharm Res 1995; 12:2–11.
48. Irwin WJ, Dwivedi AK, Holbrook PA, et al. The effect of cyclodextrins on the stability of peptides in nasal enzymatic systems. Pharm Res 1994; 11:1698–1703.
49. Zhang MZ, Wen J, Arakawa T, et al. A new strategy for enhancing the stability of lyophilized protein: The effect of the reconstitution medium on keratinocyte growth factor. Pharm Res 1995; 12:1447–1452.
50. Brewster ME, Hora MS, Simpkins JW, et al. Use of 2-hydroxypropyl-β-cyclodextrin as a solubilizing and stabilizing excipient for protein drugs. Pharm Res 1991; 8:792–795.
51. Matilainen L, Larsen KL, Wimmer R, et al. The effect of cyclodextrins on chemical and physical stability of glucagon and characterization of glucagon/γ-CD inclusion complexes. J Pharm Sci 2007; 97:2720–2729.
52. Bhat R, Timasheff SN. Steric exclusion is the principle source of the preferential hydration of proteins in the presence of polyethylene glycols. Protein Sci 1992; 1:1133–1143.
53. Cleland JL, Randolph TW. Mechanism of polyethylene glycol interaction with the molten globule folding intermediate of bovine carbonic anhydrase B. J Biol Chem 1992; 267:3147–3153.
54. Gombotz WR, Pankey SC, Phan D. The stabilization of a human IgM monoclonal antibody with poly(vinylpyrrolidone). Pharm Sci 1994; 11:624–632.
55. Chen B, Arakawa T, Morris CF, et al. Aggregation pathway of recombinant human keratinocyte growth factor and its stabilization. Pharm Res 1994; 11:1581–1587.
56. Meyer BK, Ni A, Hu B, et al. Antimicrobial Preservative Use in Parenteral Products: Past and Present. J Pharm Sci 2007; 96:3155–3167.
57. Wollmer A, Rannefeld B, Johnasen BR. Phenol-promoted structural transformation of insulin in solution. Biol Chem Hoppe-Seyler 1987; 368:903–911.

58. Birnbaum DT, Kilcomons MA, DeFelippis MR, et al. Assembly and dissociation of human insulin and Lys Pro- insulin hexamers: A comparison study. Pharm Res 1997; 14:25–36.
59. Brange J. Galenics of Insulin. New York, NY: Springer-Verlag, 1997:41.
60. Maa YF, Hsu CC. Aggregation of recombinant human growth hormone induced by phenolic compounds. Int J Pharm 1996; 140:155–168.
61. Kim Y, Rose CA, Liu Y, et al. RT-IR and near-infared FT-Raman studies of the secondary structure of insulinotropin in the solid state: α-helix to β-sheet conversion induced by phenol and/or high shear force. J Pharm Sci 1994; 83:1175–1180.
62. Lam XM, Patapoff TW, Nguyen TH. The effect of benzyl alcohol on recombinant human interferon-g. Pharm Res 1997; 14:725–729.
63. Akers MJ, Boand AV, Binkley D. Preformulation method for parenteral preservative efficacy evaluation. J Pharm Sci 1984; 73:903–907.
64. Tobler SA, Holmes BW, Cromwell ME, Fernandez EJ. Benzyl alcohol-induced destabilization of interferon-gamma: A study by hydrogen-deuterium isotope exchange. J Pharm Sci 2004; 93:1605–1617.

9 Dispersed systems

*Michael DeFelippis**

Dispersed systems are heterogeneous formulations containing insoluble drug particles in aqueous or oil vehicles. Broadly, dispersed systems include suspensions (coarse or colloidal), emulsions, liposomes, and micro- or nanoparticulate systems. Dispersed systems also can be aqueous, nonaqueous, polymeric, or insoluble salt forms/complexes. Reasons why dispersed systems are formulated and marketed are given in Table 9-1. Dispersed systems compose a relatively small segment of the injectable drug product market compared with solutions and freeze-dried products. However, the most widely used insulin dosage forms (Table 9-2) and several other therapeutic proteins (Table 9-3) are formulated as injectable suspensions. Several important small molecule products and most vaccine products are formulated as dispersed systems (Table 9-4). There is significant growth in the commercialization of dispersed systems used as depot sustained-release injectables (Tables 9-5 and 3-2). Liposomal dispersed system examples are given in chapter 3 (Table 3-3).

The route of administration of injectable dispersed systems depends on the particle size range of the drug particles. If the range is in the micron range (1 μm and above), the product cannot be administered intravenously (IV) and must be injected by intramuscular (IM) or subcutaneous (SC) routes. However, microemulsions and nanoparticulate systems can be injected IV because their globule or particle size range is below 1 μm.

For the IM route, injections occur in the gluteal, deltoid, and vastus lateralis (thigh) muscles. The typical volume of injection range is 1 to 3 mL. The typical needles used are 1 to $1\frac{1}{2}$ in., 19 to 22 G needles. Additional information about IM injections is found in chapters 31 and 33.

SUSPENSIONS

The majority of injectable suspensions are aqueous-based (drug dispersed in an aqueous vehicle) (1). Suspensions formulated with an oily vehicle include bovine somatotropin (bST) in sesame oil (with the oil thickened by incorporating a wax such as beeswax for sustained-release purposes) and a long-acting parenteral suspension of adrenocorticotropic hormone (ACTH) that is also formulated in sesame oil gelled with aluminum monostearate.

If the drug can be crystallized, then a crystalline suspension can be prepared that offers several advantages over an amorphous suspension. A main advantage of preparing crystals for pharmaceutical suspensions is the fact that a suspension composed of these crystals in the size range of about 1 to 40 μm will likely have desirable pharmaceutical properties such as resuspendability, syringeability, and injectability. Successful, reproducible drug crystallization is significantly dependent on drug substance purity since impurities will likely influence the crystallization outcome.

Two practical methods for preparing crystals can be described using the classic insulin suspensions as the models—neutral protamine Hagedorn (2) and Ultralente (3). The NPH insulin method involves crystals being grown from a solution containing all of the ingredients at the proper concentrations that comprise the final formulation. The Ultralente insulin method involves preparing a concentrated crystal suspension that is then diluted with a suitable, aqueous suspension vehicle to produce the final formulation.

A schematic representation of the NPH insulin crystallization process is shown in Figure 9-1. Insulin is cocrystallized with the basic peptide protamine. Two solutions are used for the crystallization. One solution contains insulin and protamine dissolved in water at acidic pH and the other contains dibasic sodium phosphate adjusted to slightly basic conditions. Both solutions also contain the additional ingredients necessary to complete the crystallization and

* This chapter updated by Dr. Michael DeFelippis of Eli Lilly and Company.

Table 9-1 Reasons for Sterile Dispersed System Products

- The drug is not sufficiently soluble in aqueous or, sometimes, oily solution
- There is a marketing advantage to formulating a product that requires a fewer number of injections for therapy
- Delayed, sustained, or controlled-release injectables must be formulated as dispersed systems
- More localized treatment can be accomplished
- Adverse systemic events/effects can be minimized
- Dosing compliance is improved because of less frequent administration

Table 9-2 Examples of Insulin Suspensions

Peptide or protein	Suspension characteristics
Insulin Ultralente[a]	Crystalline
Insulin NPH	Crystalline
Lys^{B28}, Pro^{B29} human insulin protamine suspension	Crystalline
Asp^{B28} insulin protamine suspension	Crystalline
Insulin Semilente[a]	Amorphous
Insulin Lente[a]	Amorphous/crystalline mixture
Regular/NPH insulin mixtures	Soluble/crystalline mixture
Lys^{B28}, Pro^{B29} human insulin analog mixtures	Soluble/crystalline mixture
Asp^{B28} insulin analog mixtures	Soluble/crystalline mixture

[a]Production discontinued by major manufacturers of insulin products.

Table 9-3 Other Examples of Protein Injectable Suspensions

Peptide or protein	Suspension characteristics
Insulinotropin	Crystalline
Interleukin-4	Crystalline
Zinc-Interferon Alpha-2B	Crystalline
Infliximab	Crystalline
Rituximab	Crystalline
Trastuzumab	Crystalline
Alpha interferon	Protamine complex
Insulinotropin	Protamine complex
Somatostatin	Protamine complex
Glucagon	Protamine complex
Gonadotropins	Protamine complex
Octanoyl-*N*ε-LysB29-human insulin; human insulin; protamine cocrystals	Protamine complex
Bovine somatotropin	Oleaginous suspension
Porcine somatotropin	Oleaginous suspension
Growth hormone-releasing hormones	Oleaginous suspension
Adrenocorticotropic hormone	Oleaginous suspension
Leutenizing hormone-releasing hormone	Suspension of degradable microspheres
Human somatotropin	Suspension of degradable microspheres
Human OB protein	Crystalline or amorphous

Table 9-4 Examples of Injectable Nonprotein Dispersed Systems

Product	API	Indication	Route of administration
Depo-Medrol®	Methylprednisolone acetate	Steroidal anti-inflammatory agent	IM
Depo-Provera®	Medroxyprogesterone acetate	Contraceptive	IM
Prevnar®[a]	Pneumococcal 7-valent conjugate	Vaccine	IM
Gardisil®[a]	Human papillomavirus quadrivalent	Vaccine, recombinant	IM
Diprivan®	Propofol emulsion	Sedative-hypnotic	IV
Amphotec®	Amphotericin B	Invasive aspergillosis	IV
Haldol®	Haloperidol decanoate	Antipsychotic	IM
Primaxin® IM	Imipenem and cilastatin	Antibacterial	IM

[a]These are but two of many commercial vaccines, most of which are suspension dosage forms.

Table 9-5 Examples of Commercial Sustained-Release Injectable Suspensions

Product name	Active ingredient
Lupron Depot®	Luprorelin
Trelstar™ Depot	Triptroelin
Sandostatin LAR®	Octreotide
Arestin®	Minocycline
Somatuline® LA	Lanreotide
Risperidal® Consta™	Risperidone

All available in biodegradable DL-lactic and glycolic acids copolymer (PLGA) microspheres.

final formulation. Precipitation is initiated by combination of the solutions in a 1:1 ratio causing a rapid change to neutral pH conditions. Amorphous material forms immediately which then transforms over time (approximately 24 hours) to form rod-shaped crystals about 3 to 6 μm long and 1 to 1.5 μm wide.

Ultralente suspensions are prepared without using protamine sulfate (Fig. 9-2). Precipitation is initiated by adjusting conditions to the isoelectric point of insulin in the presence of zinc ions, sodium chloride, and sodium acetate accomplished by mixing an acidic solution of insulin with buffer such that the appropriate pH is achieved. Most of the ingredients required for the final preparation are present during crystallization except for preservative. Concentrations for insulin and other ingredients are 10-fold higher during crystal growth and a diluent containing preservative is used to dilute the concentrated suspension to produce the final preparation. Because a monodisperse particle size distribution is desired for the final preparation, predetermined amounts of seed crystals are added during the crystallization phase. Seeding also effectively eliminates self-nucleation. Commercial Ultralente preparations contain rhombohedral crystals in the approximate size range of 20 to 30 μm.

Crystalline suspensions have been put forth as potential approaches for overcoming certain challenges associated with delivery of relatively large amounts of monoclonal antibodies (4). Solubility constraints may necessitate administering higher volumes of the preparation that can only be delivered by intravenous infusion. For SC administration, higher concentrations of antibodies are required resulting in unmanageable viscosity. Forming crystals of these antibodies

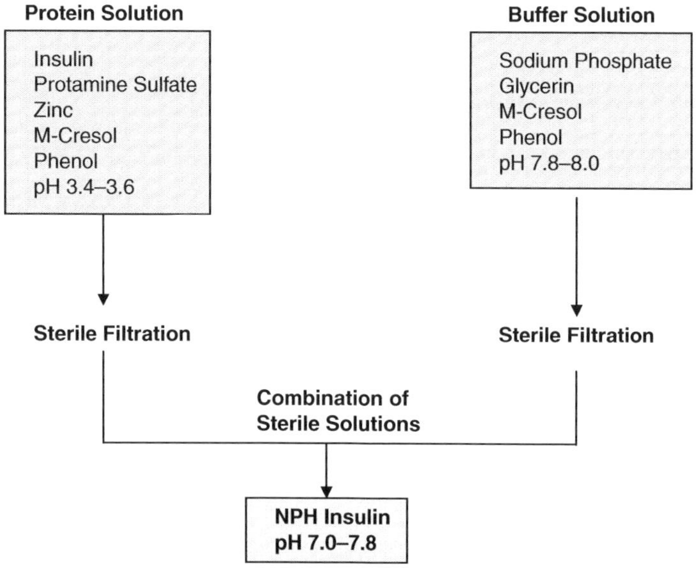

Figure 9-1 Insulin NPH Pharmaceutical Process. The procedure involves cocrystallizing insulin with the basic peptide protamine in the presence of all necessary excipients used in the final suspension preparation. Additional details can be found in the text.

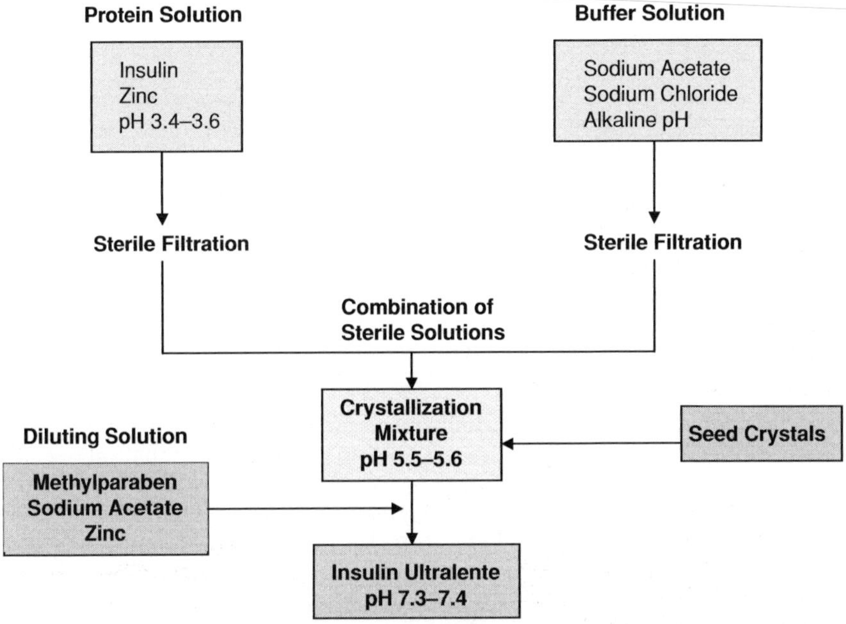

Figure 9-2 Insulin Ultralente Pharmaceutical Process. The procedure involves preparing a concentrated suspension of insulin crystals that is then diluted with vehicle containing additional excipients (e.g., preservative) required for the final suspension preparation.

and injecting them as suspensions subcutaneously in animal models, similar pharmacokinetic results were obtained compared with solutions administered by SC or IV routes.

Semilente insulin is an example of a suspension composed of flocculated amorphous particles. The suspension is prepared by performing the Ultralente crystallization under less than optimum pH conditions. Physical stability has been correlated to the degree of flocculation (5) suggesting that optimization of a final preparation will be necessary even if amorphous precipitation is easily accomplished. While it is difficult to predict if amorphous suspensions of a given peptide or protein will have the desired pharmacological properties and necessary stability, Semilente insulin preparations highlight the fact that development of a completely crystalline suspension may not be an absolute requirement. For example, an amorphous suspension formulation of the human OB protein has been described in the patent literature (6). Particles were formed by precipitating the protein in the presence of zinc chloride at neural pH. This preparation was claimed to have improved stability at higher concentrations and physiological pH compared with a solution of the protein. The suspension further demonstrated a sustained-release effect in mice.

The Lente insulin preparation is an example of a suspension containing a mixture of particles having different morphology derived by combination of Ultralente and Semilente to produce a 3:7 mixture of amorphous to crystalline material. This insulin preparation was specifically designed to produce a midrange insulin release profile to minimize the number of injections.

PREPARATION OF PARTICLES

Suspension particles not only can be prepared by crystallization techniques but also by lyophilization (see chap. 20), spray drying, atmospheric spray freeze-drying (7), and super critical fluid processes. Many other techniques are available to produce drug particles, but these are the main ones for producing sterile suspension particles. Spray drying involves the active ingredient dissolved in solvent and sprayed into a drying chamber. Rapid solvent evaporation is accomplished using a hot stream of sterile gas resulting in the formation of uniform spherical particles. Atmospheric spray freeze-drying first spray freezes the drug solution against a fluidized bed of pulverized ice, the frozen powder conveyed to the exit filter, and then freeze-drying performed by continuously flowing a dry gas stream through the packed powder bed at a temperature below the eutectic point of water. Super critical fluid particle formation requires a supercritical gas-like carbon dioxide that is used as either solvent or antisolvent to achieve supersaturation and subsequent generation of particles.

Crystallization and lyophilization are more generally applicable because of less stressful conditions and greater ability to maintain the sterility of the dried material. The spray drying process might result in degradation either at the liquid–air interface or resulting from the high temperature required to evaporate solvent. However, the technique may be appropriate for small peptides that lack higher order structure or selected proteins. Increasing attention is focusing on the potential of supercritical fluid for pharmaceutical processing in particular for its suitability to peptide and protein particle formation.

Regardless of the method employed to produce particles, some additional characterization studies should be performed to confirm that particle processing procedures do not adversely affect the properties of the molecule. Besides particle-preparing procedures, particle size reduction procedures (e.g., milling and sieving) also may be required, depending on the particle size and particle distribution of the solid material by milling and sieving. Particle-producing procedures and particle size reduction procedures can impact other properties of the active ingredient, especially if the active ingredient is a biomolecule. Examples of characterization methods might include electrophoretic and spectroscopic procedures, peptide mapping, amino acid analysis, protein sequencing, circular dichroism, mass spectrometry, or other activity or functional assays depending on the specific properties of the active ingredient.

PREPARATION OF VEHICLE AND COMBINATION

Aqueous or nonaqueous vehicles are prepared separately from particle formation. Examples of nonaqueous vehicles include many highly purified natural or synthetic oils such as sesame, peanut, or other vegetable oils. Depending on the solubility of the constituents and overall viscosity of the vehicle, sterilization can be accomplished by either filtration or autoclaving. The sterile combination approach offers more flexibility in the choice of vehicle (aqueous or nonaqueous) since particle growth is accomplished independently.

Once processing of each section is completed, the dry particles and vehicle are aseptically combined. Some form of agitation is required to achieve a homogenous dispersion of particles. In the case of peptides or proteins, appropriate controls should be in place to ensure that the dispersion process does not result in denaturation or other physical changes.

EXCIPIENT SELECTION

Choice of excipient additives is a major consideration in the design of suspension products. Some excipients are integral to the production of particles and their presence is essential for maintaining specific properties. This is especially true for the insulin preparations described earlier. Other excipients, such as preservatives, are included to meet various regulatory requirements for parenteral pharmaceutical products. Not only should these ingredients perform intended functions, but optimization studies along with thorough evaluations for compatibility with the other constituents of the preparation and the container/closure system are essential. The quality of each excipient and consistency batch-to-batch is another important consideration. Any trace levels of metal ions, salts, organic agents, or other impurities could compromise particle generation and/or impact stability. Therefore, defined specifications and an incoming raw material testing plan for excipients should be part of the overall control strategy for the suspension manufacturing process. These requirements impose limitations on excipient choices in addition to the prerequisite that they be acceptable for use in pharmaceutical products for injection.

The various excipients that are used in parenteral suspensions are categorized as adjuvants, buffering agents, isotonicity modifiers, preservatives, stabilizers, complexing agents, or other auxiliary agents. Discussion of these additives will supplement coverage in chapter 8.

Adjuvants

Adjuvants are combined with vaccine conjugates to form suspensions that will allow the vaccine, after SC or IM injection, to be slowly released from the injection site. Historically, vaccine adjuvants used in commercial U.S. human products have been one of the aluminum salts, either hydroxide, phosphate, or potassium aluminum sulfate (alum) (8). Other adjuvants used in Europe have included calcium phosphate, MF-59 (a variation of biodegradable oil squalene), and other mineral oil emulsions or liposomes.

Buffering Agents

Physiologically tolerated buffers are added to maintain pH in a desired range and some examples include sodium phosphate, sodium bicarbonate, sodium citrate, and sodium acetate. The addition of a buffer is not absolutely necessary if it can be demonstrated that the formulation maintains the desired target pH range. In certain cases, these agents are present as a result of the process for achieving particle formation, yet have no significant buffering capacity at the pH of the final preparation. Ultralente insulin provides an example of such a situation. The sodium acetate is present during crystal growth at pH 5.5, but the final suspension is adjusted to neutral pH conditions where the buffering capacity is minimal. Potential interactions between buffers and metal ions must be considered as reaction products can lead to compromised stability.

Isotonicity Modifiers

These agents are added to minimize pain that can result from cell damage due to osmotic pressure differences at the injection depot. Glycerin and sodium chloride are examples used in insulin suspensions. Effective concentrations can be determined by osmometry using an assumed osmolality of 285 mOsmol/kg. Typical concentrations of 7 mg/mL and 16 mg/mL are used for sodium chloride and glycerin, respectively. Which agent is chosen may be dictated by the need to have a particular ingredient present during particle formation, as is the case for sodium chloride in the Lente insulin preparations. The two examples of isotonicity modifiers differ in ionic strength (sodium chloride: high ionic strength; glycerin: low ionic strength) and these properties might influence the choice of one over the other depending upon compatibility and stability considerations.

Preservatives (Antimicrobial Agents)

Multidose parenteral preparations require the addition of preservatives at sufficient concentration to minimize risk of patients becoming infected upon injection. Regulatory requirements for antimicrobial effectiveness have been established that take into account whether the formulation has inherent bacterial growth inhibition properties. Typical preservatives for parenteral suspensions include: m-cresol, phenol, methylparaben, ethylparaben, propylparaben, butylparaben, chlorobutanol, benzyl alcohol, phenylmercuric nitrate, thimerosol, sorbic acid, potassium sorbate, benzoic acid, chlorocresol, and benzalkonium chloride. Use of mercury-containing preservatives, especially for vaccine preparations, has been curtailed because of safety concerns. Indeed, toxicological issues will impose limitations on the use of other chemicals especially for chronic use applications.

It is well known that most vaccine suspensions contained organomercurial (thimerosal was most common) preservative agents for many years. However, there became an increasing awareness of the theoretical potential for neurotoxicity of even low levels of these organomercurials, especially in infants receiving multiple immunizations. The Food and Drug Administration worked with vaccine manufacturers to reduce or eliminate the inclusion of thimerosal from vaccine preparations. FDA websites (9) should be consulted for the latest discussions and decisions concerning preservative use in vaccine suspensions.

The type of preservative and concentration chosen may also be influenced by factors related to crystal growth, maintaining acceptable suspension stability, or compatibility with already grown crystals in addition to achieving necessary antimicrobial effectiveness. For example, insulin Ultralente cannot be formulated with phenol as the crystal morphology is destroyed over time (5), but methylparaben does not exhibit this effect. In contrast, insulin NPH crystals require phenolic preservative for crystal growth (10), and a mixture of m-cresol and phenol in a defined ratio is present in commercial preparations.

Stabilizers

Stabilizers include a variety of agents that impart stability to particles themselves or the entire suspension. General categories include: metal ions (zinc, calcium, etc.), salts used to produce crystals, or organic molecules. Divalent metal ions play a pivotal role in insulin self-assembly and bringing about crystallization.

The various salts necessary to achieve crystal growth may also serve as stabilizers in the final suspension. Insulin Ultralente suspensions cleverly exploit the requirement for sodium chloride for crystal growth by also using the ingredient as a tonicity modifier. Organic ligands such as phenolic preservatives, in addition to serving a role as antimicrobial agents, may additionally function as stabilizers.

Complexing Agents

Protamine sulfate is an example of a complexing agent used to prepare suspensions. As excess protamine is undesirable from an immunogenicity standpoint and may impact the stability of biphasic (solution:suspension) mixtures by complexing some of the soluble component, the exact ratio required to completely complex all of the available peptide or protein needs to be determined. Under appropriate conditions, no detectable free protamine or peptide/protein remains in the supernatant.

Aluminum salts are complexing agents that were covered in the preceding text as adjuvants.

Wetting and Suspending Agents

Drugs in immediately acting injectable suspension formulations have limited aqueous solubility and do not easily "wet" because of their hydrophobicity. To enable these particles to be suspended in an aqueous vehicle, wetting agents such as surface-active agents, lecithin, or sorbitan trioleate are used to form an initial "slurry" of the insoluble particles prior to adding suspending agents. Basically wetting agents serve as dispersants to separate particles that otherwise would clump together and not easily separate to enable suspendability and dose homogeneity.

Suspending agents serve to maintain the insoluble particles in a suspended state for a period of time to allow for uniformity of dosage filled into each primary container and uniformity of dosage after each withdrawal of dose from the container. Suspending agents typically are polymers such as sodium carboxymethylcellulose (sodium CMC), polyvinylpyrrolidine (PVP), polyethylene glycol (PEG), or propylene glycol. Suspending agents may also be simple electrolytes such as sodium chloride. Suspending agents usually have an effect on increasing the viscosity of the vehicle although with injectable suspensions, viscosity cannot be too high or there will be problems with syringeability and injectability.

Other additives to injectable suspensions might include acids and bases such as hydrochloric acid and sodium hydroxide, necessary for pH adjustment during particle formation and of the final suspension.

It should be apparent to the pharmaceutical scientist involved in development of suspensions that the formulation and process are integrally related especially in the case of in situ particle growth and dose uniformity. Therefore, excipient selection must be considered with both formulation factors (stability, homogeneity, particle size, viscosity, tonicity, preservation, etc.) and processing factors (flowability, viscosity, dose uniformity, scale-up capability, etc.) in mind.

GENERAL REQUIREMENTS FOR SUSPENSION PRODUCTS

In addition to demonstrating appropriate chemical, physical, and microbiological stability over shelf life and during its intended in-use period, a well-formulated suspension should have the following characteristics:

1. Resuspension of particles is accomplished with reasonable agitation.
2. Rapid settling of dispersed particles does not occur.
3. Particles can be homogeneously dispersed such that consistent doses are obtained repeatedly.
4. The particles do not cake or pack at the bottom of the container over the shelf life period making it difficult to redisperse the product.

5. The suspension manufacturing process reproducibly produces particles with properties that are maintained batch to batch and during the preparation intended shelf life and in-use periods.
6. The suspension product must be easily drawn up into a syringe through a 20 to 25 gauge needle and readily expelled.

Optimization of these characteristics is an essential part of the development process for suspension products.

Characteristics 1 to 5 are concerned with special requirements for suspensions relating to elegance, physical attributes, and stability. Requirement 5 is especially important for suspensions intended to have specific delayed or controlled-release profiles, as the properties of the particles will govern pharmacology.

Characteristic 6 relates to suspension properties essential for administration of proper doses. Needles for parenteral injections have become increasingly smaller in diameter in an attempt to minimize pain and increase patient compliance. While 20 to 25 gauge needles are acceptable, certain insulin injector devices are currently utilizing 28 to 30 gauge needles, perhaps even 32 gauge needles. Suspensions composed of large particles could clog these narrow gauge needles affecting both the ability to withdraw a proper dose from a container into a syringe (syringeability) or inject the dose into the patient (injectability). In addition to these particle size considerations, suspensions that are composed of particles that tend to agglomerate or employ highly viscous vehicles can also affect syringing and injection operations or patient acceptability. Exceptionally viscous suspensions are not necessarily out of the question as the bST product demonstrates; however, the preparation requires 14 to 16 gauge needles for SC administration. Clearly, such a suspension is of limited practicality for human health products.

There are unique considerations that also must be taken into account for delivery systems intended for suspensions. Multiuse injectors resembling a writing pen have become very popular for delivering insulin products including the NPH and mixtures preparations. These devices use a special cartridge container–closure system that is filled with the formulation such that there is essentially no headspace between the seal and plunger. Because suspensions typically require manipulation to disperse settled particles, the injector must be designed to allow various hand manipulations by the patient that are not too difficult or burdensome to perform. Unlike a suspension in a vial that will have a headspace above the aqueous vehicle and requires minimal effort to disperse particles, cartridges generally need more agitation to achieve a homogenous suspension. To assist with the resuspension operation, a single glass bead may be added to the suspension preparations contained in cartridges. Because it is essential to ensure complete suspension homogeneity prior to dosing, the device must provide the health care practitioner and patient a clear viewing window to enable inspection of the suspension.

TESTING AND OPTIMIZATION OF CHEMICAL, PHYSICAL, AND MICROBIOLOGICAL PROPERTIES

As with any pharmaceutical product, there is a need to both optimize and test the chemical, physical, and microbiological properties of the preparation and demonstrate appropriate stability over shelf life and during the intended in-use period. Suspensions are somewhat more complicated in this regard because optimization of physical properties is extremely challenging and additional testing is usually required. However, many of the concepts relating to other dosage forms also apply to suspensions so that in some cases only a general overview is provided.

Chemical Properties

Suspended drugs, particularly peptides and proteins, are subject to a variety of chemical modifications resulting in the formation of specific degradation products just like drugs in solution. Chemical degradation reactions and stabilization approaches are covered in chapters 8 and 11.

Physical Properties

Suspensions are thermodynamically unstable systems and the goal is to design a preparation that is kinetically stable for a sufficient period of time (i.e., shelf life) so that product performance is not compromised by gross changes in physical properties. Two common physical instability

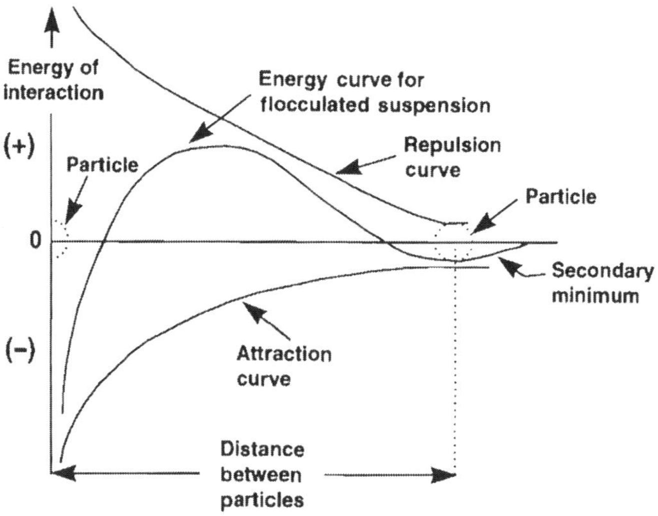

Figure 9-3 Potential energy curves for particle interactions in coarse suspensions. Particles in suspension are subject to van der Waals attractive and electrostatic repulsive forces. For coarse dispersions that are flocculated, weaker attractive forces occur at significant interparticle distances at the region referred to as the secondary minimum resulting in the formation of loosely aggregated particles. *Source*: From Ref. 11.

problems are caking and crystal growth. In order to understand the behavior of suspensions and methods for optimization, a basic understanding of theoretical concepts explaining these physical transformations is required. However, a detailed discussion on theory is beyond the scope of this review, and the reader is referred to appropriate texts on this subject for additional details.

The Derjaguin, Verwey, Landau, and Overbeek (DVLO) theory was originally devised to explain the stability of colloidal systems, but the principles have also been invoked to explain particle interactions in coarse dispersions such as suspensions. According to DVLO theory, the forces on particles in a dispersion are due to electrostatic repulsion and van der Waals attraction, although other forces are usually included to adequately explain interactions in dispersed systems. Potential energy curves for particle interactions are shown in Figure 9-3. The forces at particle surfaces will affect the degree of flocculation and agglomeration observed for suspensions. Thus, DVLO theory provides a framework for understanding the interactions of particles controlling physical properties of suspensions.

Referring to the composite curve in Figure 9-3, the collision of particles will be opposed if the repulsion energy is high (e.g., low electrolyte concentration in aqueous suspensions). Such a system is referred to as deflocculated. When the particles settle, the energy barrier is overcome and strong attractive forces in the potential well cause a densely packed sediment to form. Eventually, a hard cake results that is difficult to disperse using normal agitation procedures for resuspension. Such a condition is highly undesirable, as a nonuniform dispersion of particles can impact dosing reliability.

For coarse dispersions that are flocculated, the potential energy barrier is still too large to be surmounted by approaching particles. However, weaker attractive forces occur at significant interparticle distances at the region referred to as the secondary minimum in Figure 9-3. Particle interactions in this case result in the formation of loose aggregates (floccules). Flocculation can be induced in a suspension by the addition of a flocculating agent such as an electrolyte. Suspensions that are flocculated are considered pharmaceutically stable because sedimented material is readily redispersed upon normal agitation procedures.

The properties of flocculated and deflocculated suspension are compared in Table 9-6 (12,13). For flocculated suspensions, the sedimentation properties may result in a preparation that appears to contain a majority of clear vehicle upon settling. This condition is not a serious problem provided caking does not occur making the particles difficult to disperse with minor

Table 9-6 Relative Properties of Deflocculated and Flocculated Particles in Suspension

State	Particle characteristics	Sedimentation rate	Appearance	Cake	Resuspension
Deflocculated	Exist as separate entities	Slow	Initially suspended, but settles to a small volume	Yes	Difficult to redisperse
Flocculated	Exist as weak aggregates (floccules)	Fast	Settling results in the presence of an obvious, clear vehicle region. Final volume may be large or small	No	Easy to redisperse

agitation. Ultralente insulin is an example of a suspension that displays a very small sedimentation volume, but the particles are easily resuspended to homogeneity with gentle shaking of the vial. Because Ultralente insulin crystal growth conditions are defined and the suspension has appropriate stability, there is little value in making adjustments to improve its physical appearance upon settling. Thus, suspension formulation design may require compromises between aesthetic aspects and other desirable physical attributes of the preparation.

Sedimentation volume and zeta potential measurements are useful for optimizing the physical properties of suspensions by providing information on the degree of flocculation. Sedimentation volume is determined by measuring the equilibrium volume of settled particles relative to the total suspension volume after resuspension. The quantity is expressed as a ratio:

$$F = V_u / V_o \qquad \text{(Equation 1)}$$

where V_u is the equilibrium volume of sediment and V_o the total suspension volume.

Zeta potentials are determined to estimate surface charges. The relationship between sedimentation volume and zeta potential is illustrated in Figure 9-4. The addition of a flocculating agent causes a progressive reduction in zeta potential and changes in sedimentation volume. There exists a region where the sedimentation volume is maximized (flocculated) and no caking is observed. Note that if too much flocculating agent is added overflocculation and caking can occur. Exposing suspensions to extremes in temperature or mechanical stress can also produce this effect. This example indicates that the zeta potential must be controlled in order to produce a suspension with desirable physical properties.

Another method besides controlled flocculation for achieving optimal physical stability of suspensions is termed the structured vehicle approach. In this case, viscosity-imparting suspending agents, such as sodium alginate, glycerin, or sodium CMC, are added to the vehicle to reduce sedimentation and maintain the particles in suspended state. The vehicle is described as having pseudoplastic or plastic flow, is preferably thixotropic, entraps suspended drug particles to prevent or slow settling properties, and its shear thinning properties facilitates particle resuspension. In most situations, structured vehicles are not appropriate for parenteral preparations because vehicle viscosity is too high, adversely affecting syringeability.

Crystal Growth, Caking, and Syringeability

Three primary problems occur with dispersed systems—crystal growth, caking, and syringeability. Crystal growth occurs when drug particles stick together to form larger particles such that they cannot be redispersed easily and the uniformity of drug particles per unit volume is unequal. Caking occurs when the drug particles settle to the bottom of the container and pack so tightly that no amount of agitation (shear force) can cause the particles to resuspend. Syringeability refers to the ease or difficulty in withdrawing the suspension from the container through a narrow needle into the syringe.

To overcome crystal growth, selection of appropriate suspending agents and viscosity-inducing agents to coat drug particles and reduce the rate of particle settling according to

$$K_{cg} = Ae^{-\alpha\eta+\beta} \qquad \text{(Equation 2)}$$

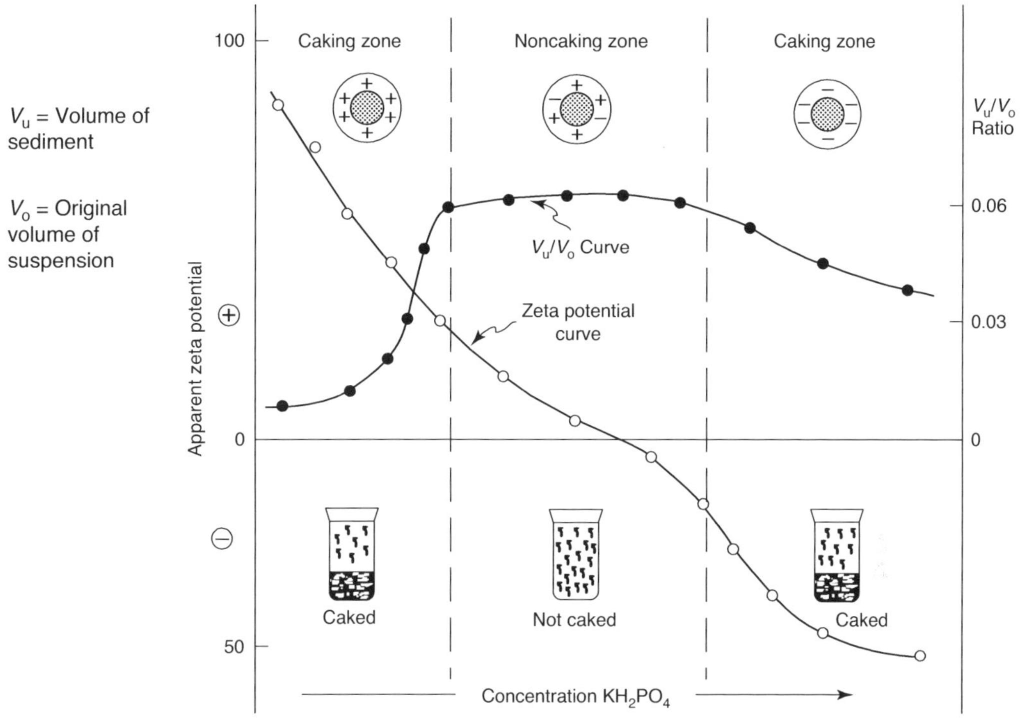

Figure 9-4 Relationship between zeta potential and sedimentation volume. The diagram depicts the effect of adding an anionic flocculating agent to a positively charged system. Maximum flocculation, as determined by sedimentation volume measurements, occurs within a narrow range of zeta potential values. The region designated as "noncaking zone" defines formulation conditions where caking of the suspension is less likely to occur. *Source*: From Ref. 14.

where crystal growth (K_{cg}) is inversely proportion to the log of viscosity (η). Particle size distribution, dissolution and recrystallization, pH and temperature changes, and polymorphism and solvate formation are factors that can affect crystal growth. The close contact of particles in settled deflocculated suspensions will favor crystal growth by a process referred to as Ostwald ripening (15).

Retarding crystal growth by the addition of viscosity-imparting agents usually is not appropriate for parenteral suspensions. Increasing viscosity of the vehicle (continuous phase) will help to a point, beyond which syringeability issues may be prominent. The best approaches for minimizing crystal growth in suspensions are to control the particle size distribution, select the correct polymorph (if applicable), or use the controlled flocculation approach. Appropriate testing should also be conducted to determine the impact of pH and temperature changes on physical stability. Milling methods that produce a narrow particle size distribution range also help to minimize packing, caking, and crystal growth of particles. Using amorphous additives like polymers that do not crystallize will help.

Caking of suspensions can be minimized by following the principles of Stokes Law

$$Y = \frac{2r^2(\Delta\rho)g}{9\eta} \quad \text{(Equation 3)}$$

where Y is the "terminal velocity" or the rate of settling (m/sec) leading to caking, r^2 is the radius squared (m^2), $\Delta\rho$ is the difference in density between the dispersed phase (drug particle) and the continuous phase (vehicle) (kg/m^3) g is gravitational acceleration (m/s^2), and η is the vehicle viscosity ($kg\ m^{-1}\ sec^{-1}$). Therefore, the rate of setting can be reduced if the particle size is reduced, the density differences between suspended particles and vehicle are minimal, and the vehicle viscosity is increased.

Wetting agents are useful to coat all drug particles to minimize forces of attraction. Electrolytes are used to increase zeta potential to a point where controlled flocculation will take place.

Testing of Physical Properties

As highlighted in the preceding text, the physical properties of suspensions are rather unique compared with other pharmaceutical preparations and much of the testing will be focused in this area. One of the simplest attributes to evaluate is physical appearance. This qualitative assessment is performed in stability programs and after exposing the suspension to various stresses of temperature or mechanical agitation. The preparation must remain "elegant" (no clumping of particles, uniform dispersion of particles upon agitation, particle characteristics remain unchanged, and material does not adhere to container/closure surfaces) after exposure to reasonable conditions. In instances where product elegance is compromised, the information can be used to draft instructions for proper storage and use of the product.

Other more quantitative evaluations of physical stability involve measurement of sedimentation volume and rate. The procedure for determining sedimentation volume has been previously described, with Figure 9-5 providing a basic illustration of measuring sedimentation rate and volume. Sedimentation rate is used to estimate the rate of particle settling, and can be done in conjunction with sedimentation volume determinations by measuring the top boundary of particles progressing downward as a function of time. In addition to determining sedimentation rate for samples exposed to extreme conditions, this important parameter must be evaluated for flocculated suspensions where particles sediment rapidly. After proper resuspension manipulations, the particles must remain dispersed in vehicle for sufficient time to allow injections of proper doses.

To simulate the various stresses a suspension might encounter during shipping or patient handling, thermomechanical testing is routinely performed to evaluate product performance under these conditions. Methods of this type include some agitation of the suspension in its container induced by either shaking or rotation. Temperatures are generally elevated during mechanical stress testing as well. Such treatment can result in a rather unsightly appearance of the suspension due to clumping or result in damage to the particles, changes in sedimentation properties, adherence of solids to the container/closure system, and loss or change in

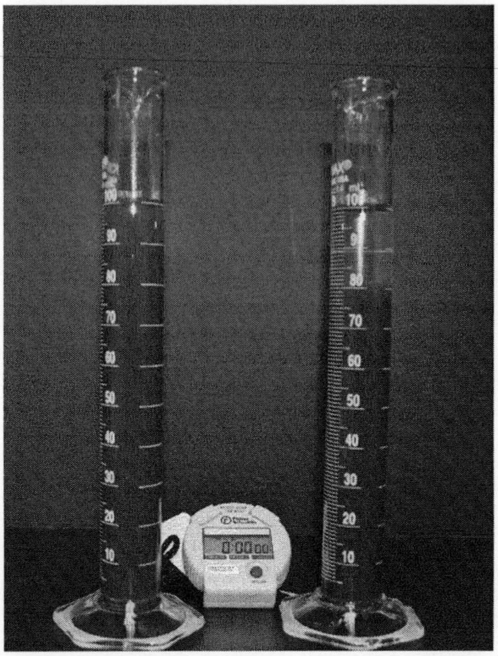

Figure 9-5 Example of measurement of sedimentation volume. At time zero, record the volume of suspended material and then remeasure that volume over time. Also check for changes in sedimentation rate over time. *Source*: Courtesy of Dr. Gregory Sacha, Baxter BioPharma Solutions.

activity depending on the duration of exposure. These conditions can result from extended exposure to high temperatures alone as shown for insulin suspensions (5). All of these aspects should be evaluated for potential significance to the final product. Shnek et al. (16) reviewed physical stress testing methods designed to study insulin suspensions and solutions in cartridge containers intended for portable pen-based injector devices. Protocols generally involve exposing product samples to high temperature (e.g., 25 and/or 37°C) combined with mechanical agitation at periodic intervals, and may include simulated dosing following resuspension over the course of several days. The extreme conditions employed with these procedures far exceed those encountered during typical patient use. Nevertheless, regulatory authorities may request data from such experiments to support recommended in-use periods for the product.

In the same manner that aberrant temperature studies are performed to evaluate chemical stability, the impact of extreme conditions on physical attributes must be determined. The effect of excessively high temperatures has been mentioned, but physical changes can also occur at extreme cold temperatures. Studies exploring extreme temperature excursions (low and high) are also useful for understanding the impact of potential aberrant shipping conditions since refrigeration is typically required for peptide and protein products to maintain appropriate stability.

Syringeability or syringing studies are generally required for suspension preparations stored in vials to ensure that the preparation can be used reliably. Syringing can also be performed to evaluate physical stability. Procedures involve performing daily resuspension manipulations and withdrawal of doses from product containers under conditions that mimic the intended use profile. Evaluations are routinely conducted at room temperature and continue for a length of time corresponding to the proposed in-use dating period. The physical appearance of the suspension is examined throughout as well as conducting a thorough chemical analysis of the material remaining at the end of test. An apparatus was devised and used in the industry that allows quantitative measurements of syringeability (17). The apparatus was shown to be appropriate for parenteral solutions and suspensions, especially those that are nonaqueous. A common occurrence for suspensions evaluated in this manner is the adherence of solids to the wall of the container primarily in the region closest to the stopper. This material forms as the suspension flows downward upon settling. Vehicle drains, leaving solid particles on the interior glass surface.

Dosing studies with suspensions contained in cartridge containers are also necessary. The approaches are similar to those performed for vials except that doses are expelled following resuspension using a delivery device. In addition to evaluating the physicochemical properties of material remaining within the cartridge, concentration determinations are made on expelled samples to ensure suspension homogeneity of the dose.

Microbiological Properties

In designing a suspension formulation and manufacturing process for producing it, steps must be taken to ensure that the microbiological properties are optimized and maintained in the final product. The process must include appropriate procedures for sterilization taking into consideration the way in which particles are produced. In certain cases, sterile filtration of solutions prior to initiating particle growth is appropriate while other processes involve aseptic combination of preformed sterile particles with a sterilized vehicle. Because of the sensitivity of peptides and proteins to extreme conditions, careful thought must be given to the manner in which sterilization is accomplished. For example, terminal sterilization is unlikely to be tolerated by peptides and proteins.

Phenolic preservative concentrations can be decreased due to absorption into tubing during mixing and recirculation operations. Permeation through the closure and chemical degradation of these agents is also possible. To account for possible degradation over shelf life, formulations varying in preservative concentration should be subjected to antimicrobial effectiveness testing to define the effective range wherein efficacy is maintained and to establish specifications. In addition, preservative excesses may be necessary to account for loss during manufacturing and to achieve final target concentrations.

TECHNIQUES FOR CHARACTERIZING AND OPTIMIZING SUSPENSIONS

A number of techniques can be applied throughout development of parenteral suspensions to achieve the optimal properties. Characterization and optimization efforts can be directed toward the particle formation process, the particles themselves, or the suspension properties of the preparation.

Microscopy

Perhaps one of the most basic of techniques, the importance of microscopy cannot be overstated. White light microscopy can be used to visualize particles in the size range of 0.4 to 100 μm, and the technique is particularly useful for characterizing nonspherical or amorphous particles. Microscopy can also be used to monitor particle growth over time.

Particle Sizing

Information on particle size is essential to have throughout all stages of suspension development. In addition, particle size can be an important property for evaluating product stability and ensuring process control. A number of techniques are available to determine particle size and size distributions, each having associated pros and cons for a given application (18). Regardless of the methodology employed in the various commercial instruments, measured parameters are usually reported in terms of the diameter of spherical particles. In many cases, particles may not be spherical or could be irregularly shaped so that sizes are approximate and will vary amongst techniques. Particle size distributions, usually included as part of the analysis, may be more useful as they provide information on the population of species present in the sample. The choice of technique will depend on the nature of the sample to be measured and appropriateness of the methodology.

Electrophoretic Light Scattering

Electrophoretic light scattering (ELS) provides a direct measure of the velocity of particles moving in an electric field. Velocity is obtained by measuring the Doppler shift of laser light scattered from a moving particle electrophoresing in liquid. The electrophoretic mobility (U), which is proportional to the surface charge density of the particle, is determined by using the following equation:

$$U = V/E \qquad \text{(Equation 4)}$$

where V is the electrophoretic velocity and E the applied electric field.

The zeta potential (ξ) is derived from U by using the relationship:

$$U = \xi\varepsilon/\eta \qquad \text{(Equation 5)}$$

where ξ is the zeta potential, expressed in millivolts, η the viscosity, $\varepsilon = \varepsilon_0 D$, ε_0 the permittivity of free space, and D the dielectric constant.

As surface charge governs particle interactions, zeta potentials are useful to determine during development of pharmaceutical suspensions as the quantities can be correlated with physical attributes and stability of coarse dispersions.

Dynamic Light Scattering

Although primarily applicable to solution studies, the technique can be employed during suspension development as a tool to study the potential for systems to form certain particles. It is widely recognized that protein crystallization begins with aggregation of individual molecules in solution. Aggregates formed during the early prenucleation phase will determine the potential of the system to form crystals. The goal of dynamic light scattering experiments is to determine whether precipitation or crystallization will occur based on aggregation behavior in solution. Such information can assist with screening activities for appropriate crystallization conditions.

Calorimetry

Differential scanning calorimetry (DSC) has been applied to study the crystal growth mechanism of a protein suspension (19). In another example of the application of calorimetry, isothermal titrating calorimetry (ITC) has been used to study adsorption of soluble insulin onto NPH crystals and obtain estimates for the thermodynamic parameters associated with this process (20).

Scanning Probe Microscopy

The scanning probe microscopy (SPM) techniques of scanning tunneling microscopy (STM) and atomic force microscopy (AFM) provide added advantages with specific applications to pharmaceutical systems. The AFM technique is appropriate for characterizing protein crystal packing and growth mechanisms. For example, tapping-mode AFM (TMAFM) was used to identify polymorphs of bovine insulin (21), study crystal growth characteristics of $Lys^{B28}Pro^{B29}$ insulin (22), characterize Ultralente crystals prepared from human, porcine, and bovine insulin (23), and assess interfacial structure, morphology, and growth characteristics associated with $Lys^{B28}Pro^{B29}$ protamine crystals (24). The in situ imaging capabilities of the technique allows the direct visualization of the effects of additives and other parameters as crystal growth occurs.

In Vitro Dissolution

In the development of sustained or controlled-release suspensions, it is useful to have an in vitro assay available for quickly approximating dissolution properties. Analogous to dissolution testing for solid dosage forms, the procedure requires some detection method to continuously monitor release of drug. As an example, a continuous-flow spectrophotometric method was developed that can categorize insulin suspension preparations based on clinical time-action classifications (25). Prabhu et al. (26) describe the use of a spin-filter device to study the factors controlling dissolution of zinc-complexed insulin suspensions.

SUSPENSION MANUFACTURE

Developing and validating parenteral commercially viable suspension manufacturing processes present significant challenges. A schematic example of suspension manufacture is seen in Figure 9-6. As pointed out earlier, crystallization of drugs at small scale is not simple, but the difficulty of the problem is magnified by virtue of the large volumes needed and the strict controls required for the preparation of pharmaceutical products. Generally, incremental increases in scale are attempted starting from the bench process and progressing upward in volume to the required batch size. Changes in container composition (e.g., glass vs. stainless steel) and geometry will occur during the transition and this could impact the crystallization. In addition, one must consider how certain operations performed with ease in the laboratory such as additions, mixing, transfers, and temperature control will be conducted under aseptic conditions of a manufacturing facility.

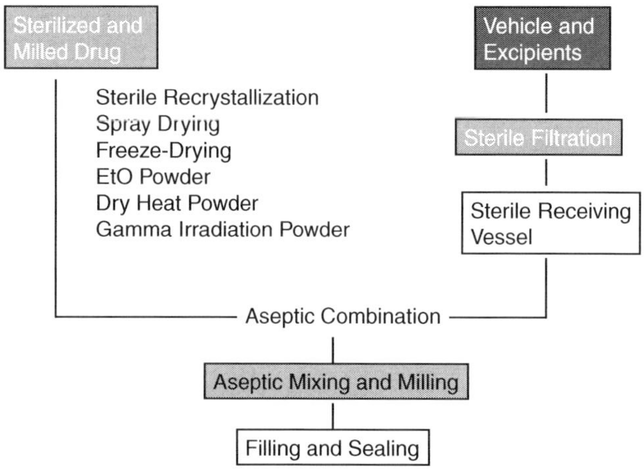

Figure 9-6 Flow diagram schematic of sterile suspension manufacture.

The other methods of particle generation are not any easier to scale-up (27), especially when peptides or proteins are the target molecule. Seemingly sound lab or pilot scale procedures can produce undesirable outcomes at larger scale. Therefore, steps must be taken to ensure that particle generation and size reduction operations are accomplished without affecting the properties of the molecule or reproducibility of the process. Milling operations must be conducted aseptically because no practical means of resterilization is feasible if sterility is compromised. Nonaqueous (oil) vehicles do require special consideration if they are to be sterilized by filtration as filter composition, pore size, and flow rates can impact capture efficiency. Finally, an appropriate strategy needs to be devised for aseptically combining the particles and vehicle.

Following each successive scale-up, it is important to consider comparability[1] of the product properties relative to known characteristics at the previous scale. Expanded physicochemical evaluation beyond routine testing will need to be employed to ensure, for example, that molecular integrity, particle morphology, suspension/sedimentation aspects, dissolution profile, and stability performance (including accelerated and stress conditions) are comparable between scales. Depending upon the stage of clinical development or whether the product is already licensed, it may be necessary to also include nonclinical and/or clinical studies to evaluate in vivo pharmacological performance. Comparability assessments should be considered for other types of changes to the process, such as introducing new raw materials, parameter modifications, or transfer of the process to a different manufacturing facility due to the potential to influence properties of the suspension.

Filling

Suspension homogeneity must be maintained throughout the filling operation to ensure content uniformity in the finished units. Continuous mixing and recirculation are typically conducted to keep particles homogeneously dispersed. The specific type of agitation required is highly dependent on the sedimentation properties of the particles and nature of the vehicle. Careful examination of parameters, such as mixer configuration, mixing/recirculation speed and duration, is necessary to determine optimal conditions. Computational modeling approaches may be useful for defining agitation parameters necessary to achieve optimal particle dispersion. The issues associated with mixing peptide and protein suspensions have already by elaborated. While the concerns are similar for recirculation, there are additional considerations. The recirculation operation involves pumping the suspension through tubing and the impact of this agitation on molecular structure and/or particle integrity needs to be assessed. Product interactions with contact surfaces of equipment used for recirculation should be additionally explored since the duration of filling may last several hours. The potential for leachables from recirculation line tubing also exists raising the same concerns described earlier for the container–closure. One final consideration for suspension filling involves line stoppages. If this situation does occur, stopping the agitation may be advisable in order to minimize exposure of the product to these physical stress conditions. Sufficient time must be allowed upon restart to ensure homogeneity and some population of the filled units will likely be discarded once filling commences to ensure uniformity has been reestablished.

Since some form of agitation is necessary to properly fill a suspension product, a balance must be achieved so that suspension homogeneity is accomplished without impacting the molecule or the particles. One approach to overcome the filling issues associated with suspensions involves particle formation in individual product containers. In this case, fixed volumes of two solutions may be combined together in the vial initiating particle formation. This filling strategy is limited to suspension products where particle formation in aqueous vehicle is feasible. Furthermore, since a commercial batch size could conceivably yield in excess of ten thousands individual units, a thorough understanding of the particle formation process and the influence of associated parameters is essential. Validation of the process must demonstrate that consistency of suspension properties is achieved for each individual unit.

[1] For further details concerning the concept of comparability, consult the following reference: Comparability of biotechnological/biological products subject to changes in their manufacturing process. International Conference on Harmonisation, Q5E, June 2005. This guidance document is available at www.ich.org.

The control strategy for the filling operation involves confirmation of uniformity, and for peptide or protein-based preparations measurement can be achieved by nitrogen determination, colorimetric test, or an HPLC assay. A statistically defined set of samples across the entire filling operation of the batch is typically evaluated due to the destructive nature of testing. Continuous on-line measurement of optical density is also a possibility, and offers the advantage of nondestructively examining every container for appropriate uniformity. However, process analytical technology (PAT) approaches require development, validation and maintenance of measurement equipment, and associated computer models.

Control Strategy

Final batch release testing to confirm quality consists of a set of attributes, test methods, and acceptance criteria that comprise the product specification. However, it is important to appreciate that the batch specification only represents one part of the overall control strategy. There are a number of required tests for sterile injectable products including assays for identity, content, purity, extractable volume, sterility, and endotoxin that must be included before product can be released to the market. Any additional testing that is included at this point depends on the design of the overall control strategy. Many options for implementing various control elements throughout a suspension process have been highlighted in the text of this section so there may not be a need to repeat certain tests at batch release. For suspension products, it might be appropriate to include in the specification measurement of particle morphology, particle size and distribution, or rheological properties depending on the nature of the final suspension. Multiuse suspension products containing a preservative may include a content determination for this excipient to ensure that the concentration remains in the range effective for antimicrobial effectiveness. Finally, this discussion only considered a subset of the unit operations involved in suspension manufacture and does not represent a complete description of a suitable control strategy for a pharmaceutical product.

Additional coverage of suspension manufacture is found in chapter 12.

EMULSIONS

Emulsions for injection exist both as large-volume and small-volume products. Injectable emulsions are oil in water systems with the oil phase as the internal or dispersed phase and the water phase as the external or continuous phase. Globule size for emulsions range from 0.1 to 50 μm with emulsions administered by intravenous injection or infusion needing to be of globule size less than 1 μm. Large-volume emulsions are used for parenteral nutrition purposes while small-volume emulsions are considered alternative dosage forms for poorly water-soluble drugs.

Large-volume fatty lipid emulsions are used in parenteral nutrition therapy. Formulations are typically as follows:

Soybean oil	10–20%
Egg yolk phospholipid	1.2%
Glycerin	2.5%
Water	QS

Large-volume emulsions have a pH around 8 and are terminally sterilized using patented steam sterilization cycles that maintain the globule size distribution of the product so that it can be safely administered IV.

Small-volume emulsions for injection have the general formulation of the drug, soybean oil, egg lecithin, glycerin, and water. Drugs that are formulated into emulsion dosage forms include propofol, vitamins, dexamethasone, flurbiprofen, prostaglandin E1, diazepam, and perfluorocarbon.

It is very difficult to extemporaneously incorporate a water-insoluble drug into an existing emulsion formulation (e.g., large-volume emulsion) and have a stable product. The solubilization of the drug is marginal and the drug usually causes the emulsion to destabilize. Drug-containing emulsions should be prepared where the emulsion is formed after the drug is dissolved in the oil phase.

Table 9-7 Physical Stability of Emulsions

- Flocculation or creaming
  - Upward creaming if the density of the dispersed (oil) phase is less than the density of the continuous (aqueous) phase
  - Downward creaming if the opposite is true
- Coalescence or breaking
  - Depends on the phase–volume ration
  - If the oil phase is greater than 74% of the emulsion composition, the emulsion is susceptible to breaking
- Phase inversion
- Other physical or chemical changes

Like all dispersed systems, emulsions are thermodynamically unstable and must be carefully formulated, processed, and packaged so instability problems are minimized. Table 9-7 summarizes potential physical stability problems of emulsions.

Emulsion manufacture consists of dispersing the oil phase and all dissolved components into the aqueous phase with its dissolved components. Methods of dispersion range from simple agitation/mixing to the use of high shear equipment such as colloid mills, ultrasonifiers, and homogenizers.

LIPOSOMES

Liposomes are hydrated phospholipid vesicles where the active ingredient is incorporated into the inner hydrophobic core of the liposome (Fig. 9-7).

They are spontaneously formed by dispersion of lipid films in an aqueous environment. They can be unilamellar or multilamellar. Liposomal formulations can deliver low-molecular-weight compounds (e.g., amphotericin), proteins, and peptides (e.g., Epaxal®—Hepatitis A), and are extremely important in the formulation and delivery of DNA-based therapeutics. The two most successful and long-term liposome products have contained amphotercin B and doxorubicin, with exemplary formulations presented in Table 9-8.

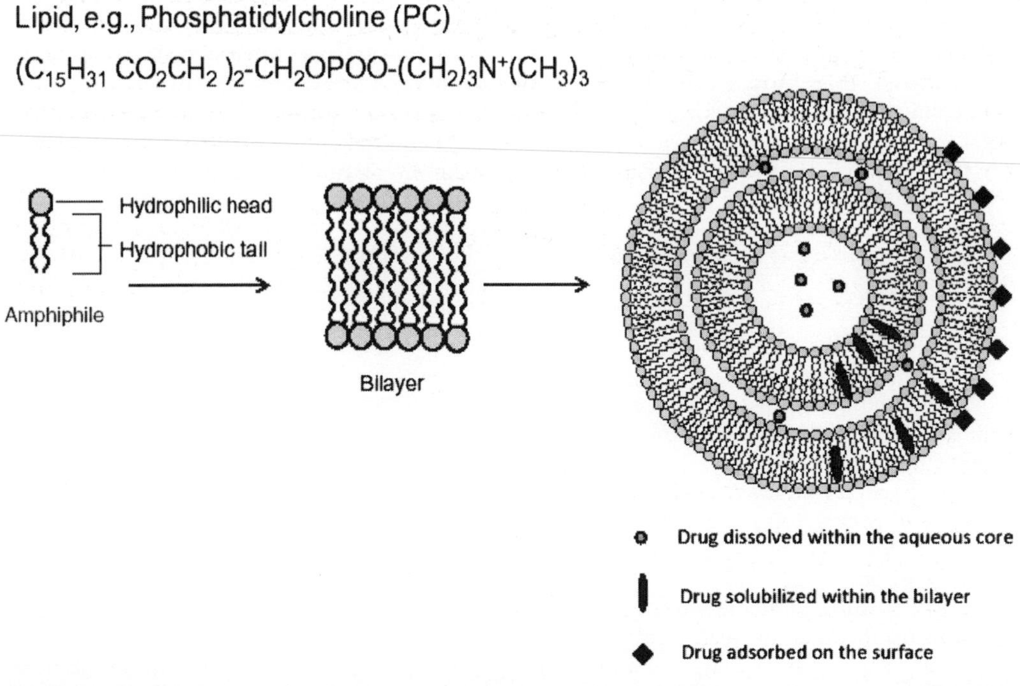

Figure 9-7 Schematic of a multilamellar liposome. *Source:* Courtesy of Professor Yvonne Perrie, Aston Pharmacy School, United Kingdom.

Table 9-8 Examples of Liposome Formulations

Amphotec® (Sequus) lyophilized	Mg/vial
Amphotericin B	50
Sodium cholesteryl sulfate	26.4
Tromethamine	5.64
DSEDTA	0.372
Lactose	950
Doxil (Sequus) dispersion 10 mL vials (Stealth® liposome)	Mg/mL
Doxirubicin hydrochloride	2
MPEG-DSPE[a]	3.19
HSPC[b]	9.58
Cholesterol	3.19
Ammonium sulfate	2
Histidine (buffer)	pH 6.5
Sucrose (isotonicity)	

[a] *N*-(carbonyl-methoxyPEG 2000)-1,2-distearoyl-sn-glycero-phosphoethanolamine.
[b] Fully hydrogenated soy phosphatidylcholine.

Conventional liposomes are either neutral formulations containing lecithins and/or phosphatidylcholine anionic formulations containing phosphatidylglycerol or the phosphatidylinositols. Liposomes typically are delivered to liver and spleen after phagocytosis. Liposome composition affects phase transition temperature (T_m).

The T_m for lipids with phosphatidylcholine polar head groups vary from −15°C for dioleoyl chains to 55°C for distearoyl chains. Cholesterol in liposome formulations orders the fluid phase and improves stability by decreasing leakage.

Gene delivery (DNA-based therapeutics) by liposomal systems combine cationic and zwitterionic lipids. Cationic lipids facilitate DNA complexation while zwitterionic lipids help in membrane perturbation and fusion. Cationic lipids have cytotoxic effects on cells.

Liposomal formulations have been pursued and studied for decades because of the advantages of prolonged circulation and targeting potential being good carriers and stabilizers for peptides, proteins, monoclonal antibodies, and genetic material. However, there have been significant problems with reproducible drug loading efficiencies, chemical and physical stability issues, and liver uptake of the drug after IV injection or lymphatic update after SC injection. Indeed, liposomes are sterically stabilized systems that extend in vivo circulation times. Uptake issues have been circumvented by formulations called "stealth liposomes," where certain liposomal carriers containing polyethylene components (pegylation) (Table 9-8) are not "recognized" by liver enzymes and the product is not as readily inactivated.

The main challenge in liposome formulation is encapsulation of the active ingredient into the liposome matrix. Encapsulation methods are summarized in Table 9-9. Another huge challenge is liposome stability. Chemical stability problems include liposome hydrolysis to fatty acids and lysolipids that affect membrane permeability and oxidative degradation. Physical

Table 9-9 Liposome Encapsulation Methods

Passive
- Low retention
- Charged drugs released rapidly
- Leakage is common, inversely proportional to acyl chain length
- Encapsulated during formation of liposome

Active
- pH gradient between center of liposome and external environment
- pH of external phase adjusted to ensure drug is nonionized
- Transported to internal phase and ionized
- High loading efficiency and less leakage

Complexation
- Drug–lipid complexes

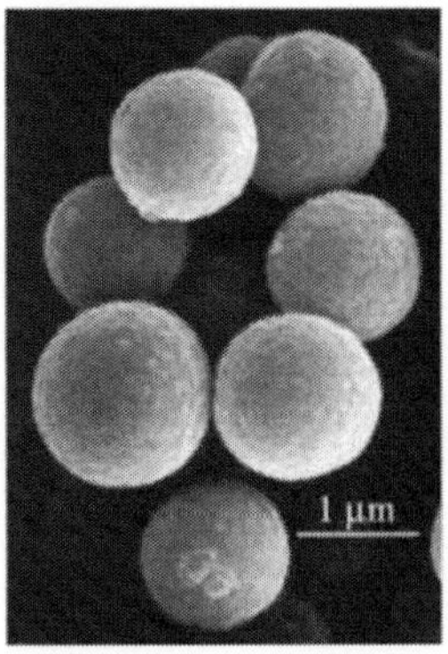

Figure 9-8 Example of physical appearance of microspheres. *Source*: Courtesy of Dr. Larry Brown, Baxter Healthcare Corporation.

degradation reactions include aggregation that often occurs with neutral liposomes of large size, promoted by trace elements and sedimentation and fusion where there occurs irreversible formation of larger vesicles promoted by high-stress curvature of small vesicles.

Liposomes are manufactured like other dispersed systems where compatible systems are compounded and sterile filtered, then combined and appropriately mixed by high shear techniques.

The Food and Drug Administration has published a guidance document worth reviewing on the chemistry, manufacturing, and controls for liposome drug products submitted in new drug applications (28).

MICROSPHERES

Microspheres have already been described in chapter 3, but since they are dispersed systems, some redundancy will occur here. Microspheres are solid, spherical particles, 1 to 1000 μm, usually polymeric formulations (Fig. 9-8). They are formulated primarily to sustain the release of drugs at the site of injection and have become popular controlled-release drug delivery systems. They also can be formulated to deliver a drug to a specific target and improve the safety profile of the drug. Microspheres serve as adjuvants for vaccines.

The most common polymer choice for microsphere formulations are the copolymers of lactic acid and glycolic acid (poly-lactic-co-glycolic acid; PLGA). Different levels and viscosities of the two polymers affect the rate of degradation and nature of erosion of the microsphere. Polyanhydride is another example of a microsphere polymer that gradually erodes. Other examples and discussion of microsphere formulations were covered in chapter 3.

Microspheres can be manufactured by a variety of techniques, summarized in Table 9-10.

Table 9-10 Microsphere Manufacturing Methods

1. Coacervation
   - Mix two immiscible liquids (concentrated polymer phase and a dilute liquid phase)
   - Alter conditions to favor polymer–polymer interactions, dehydration of polymer, and cross-linking
   - Polymer spheres will form around any material present
2. Divalent ion gelling
   - Form gel by mixing polymer with a divalent ion to cause cross-linking
   - An example is mixing of alginate with calcium chloride solution
3. Spray drying
4. Solvent evaporation
5. Precipitation
6. Freeze-drying
   - Organic, continuous phase sublimated at low temperature
   - Followed by sublimation of dispersed phase solvent
7. Supercritical fluid precipitation

NANOSUSPENSIONS

As the name implies, nanosuspensions are dispersed systems where the insoluble drug particles are in the nanometer size range and can be administered by IV injection. Several patented technologies exist for formulating nanosuspensions (29). Basically, the drug is dissolved in a water-miscible cosolvent (e.g., ethanol, *N*-methyl-2-pyrrolidinone) containing a surface-active agent and sterile filtered, then combined with a sterile filtered aqueous-based formulation containing whether water-soluble additives (e.g., buffers) are required for the finished product. This combination produces an unstable, friable microprecipitate (called "presuspension") that, in turn, is homogenized. The homogenization step is the key step that produces the stable nanosuspension. The role of the surface-active agent also can be important in stabilizing the finished product by coating the drug particles during the homogenization process. A schematic example of the preparation of a nanosuspension is shown in Figure 9-9.

SUMMARY

Dispersed systems include the following:

- Coarse and colloidal suspensions
- Emulsions
- Liposomes
- Microspheres
- Nanosuspensions.

Dispersed systems are complicated delivery systems that require significant knowledge of the physical/chemical behavior of the constituents and the dosage form. Such knowledge includes particle size and distribution, factors affecting sedimentation and aggregation, effect of administration site on release and absorption, and effect of physical/chemical characteristics on the in vivo release profile. Batch-to-batch variation can significantly change the in vivo release profile.

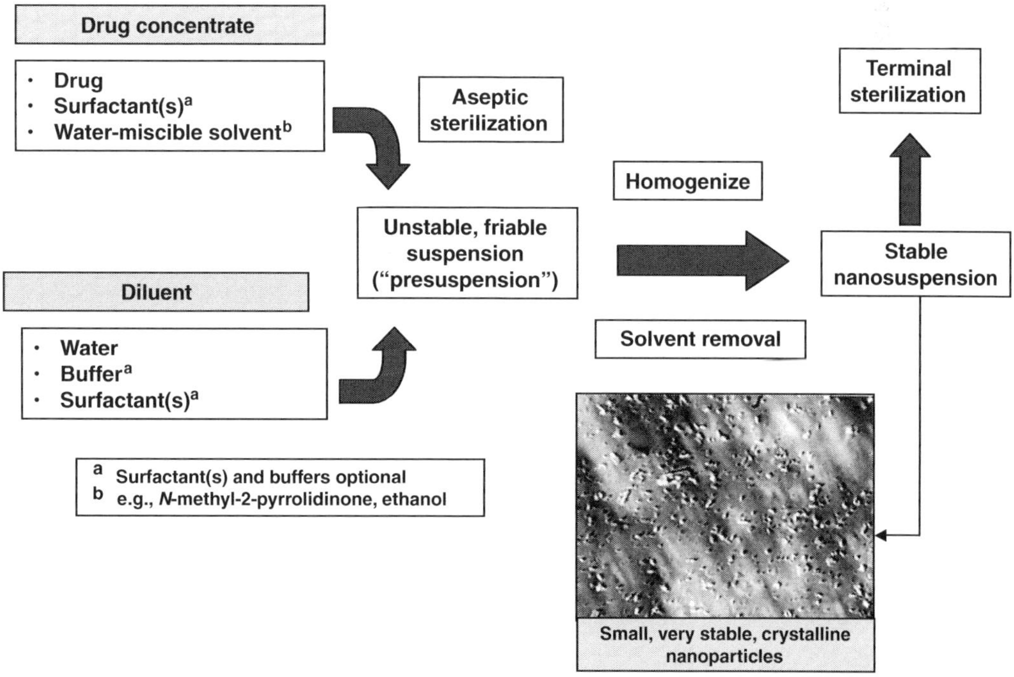

Figure 9-9 Example of nanosuspension manufacture. *Source*: Courtesy of Dr. James Kipp, Baxter Healthcare Corporation.

Dispersed systems offer advantages in drug delivery when drugs are too insoluble or unstable to be formulated as solutions or freeze-dried powders. Dispersed systems also offer advantages of controlling drug release, decreasing toxicity, and targeting drug delivery. A thorough understanding of the factors affecting the stability and delivery of drugs in these types of dosage form is required to ensure safety, efficacy, and reproducibility.

REFERENCES

1. DeFelippis MR, Akers MJ. Formulation, manufacturing and control of protein suspension dosage forms. In: Hovgaard L, Frokjaer S, eds. Pharmaceutical Formulation Development of Peptides and Proteins. London: Taylor and Francis, 2000:113–144.
2. Hagedorn HC, Jensen BN, Krarup NB, et al. Protamine insulinate. J Am Med Assoc 1936; 106:177–180.
3. Hallas-Møller K, Petersen K, Schlichtkrull J. Crystalline and amorphous insulin-zinc compounds with prolonged action. Science 1952; 116:394–399.
4. Yang MX, Shenoy B, Disttler M, et al. Crystalline monoclonal antibodies for subcutaneous delivery. Proc Natl Acad Sci U S A 2003; 100:6934–6939.
5. Brange J. Galenics of Insulin: Physico-chemical and Pharmaceutical Aspects of Insulin and Insulin Preparations. Berlin Heidelberg: Springer-Verlag, 1987.
6. Brems DN, French DL, Speed MA. Stable, active, human OB protein compositions and methods. United States Patent Application. US 2002/0019352 A1. 2002.
7. Wang ZL, Finlay WH, Peppler MS, et al. Powder formation by atmospheric spray-freeze-drying. Powder Technol 2006; 170:45–52.
8. Wilson-Welder JH, Torres MP, Kipper MJ, et al. Vaccine adjuvants: Current challenges and future approaches. J Pharm Sci 2009; 98:1278–1316.
9. Thimerosal in Vaccines, U.S. Food and Drug Administration, Vaccines, Blood and Biologics, http://www.fda.gov/biologicsbloodvaccines/safetyavailability/vaccinesafety/ucm096228.htm. Accessed June 6, 2010.
10. Krayenbühl C, Rosenberg T. Crystalline protamine insulin. Rep Steno Mem Hosp Nord Insulinlab 1946; 1:60–73.
11. Martin AN. Physical chemical approach to the formulation of pharmaceutical suspensions. J Pharm Sci 1961; 50:513–517.
12. Nash RA. Pharmaceutical suspensions. In: Lieberman HA, Rieger MM, Banker GS, eds. Pharmaceutical Dosage Forms Disperse Systems. Vol. 1. New York: Marcel Dekker, Inc., 1988:151–198.
13. Swarbrick J, Rubino JT, Rubino OP. Coarse dispersions. In: Gennaro AR, ed. Remington—The Science and Practice of Pharmacy. 21st ed. Philadelphia: Lippincott Williams & Wilkins, 2005:319–337.
14. Martin A, Swarbrick J. Pharmaceutical Suspensions: A Review. In: Beal HM, Sprowls JB, eds. Sprowl's American Pharmacy. 6th ed. Philadelphia: Lippincott Williams & Wilkins, 1966:205.
15. Nash RA. Pharmaceutical Suspensions. In: Lieberman HA, Rieger MM, Banker GS, eds., Pharmaceutical Dosage Forms: Dispersed Systems, Vol 2, 2nd ed. Marcel Dekker, New York, 1996; 6–7.
16. Shnek DR, Hostettler DL, Bell MA, et al. Physical stress testing of insulin suspensions and solutions. J Pharm Sci 1998; 87:1459–1465.
17. Chien YW, Przybyszewski P, Shami EG. Syringeability of nonaqueous parenteral formulations-development and evaluation of a testing apparatus. J Parenter Sci Technol 1981; 35:281–284.
18. Kelly RN, Lerke SA. Particle size measurement technique selection within method development in the pharmaceutical industry. Am Pharm Rev 2005; 8:72–81.
19. Ooshima H, Urabe S, Igarashi K, et al. Mechanism of crystal growth of protein: Differential scanning calorimetry of thermolysin crystal suspension. In: Botsaris GD, Toyokura K, eds. Separation and Purification by Crystallization. Washington, DC: American Chemical Society, 1997:18–27.
20. Dodd SW, Havel HA, Kovach PM, et al. Reversible adsorption of soluble hexameric insulin onto the surface of insulin crystals cocrystallized with protamine: An electrostatic interaction. Pharm Res 1995; 12:60–68.
21. Yip CM, Ward MD. Atomic force microscopy of insulin single crystals: Direct visualization of molecules and crystal growth. Biophys J 1996; 71:1071–1078.
22. Yip CM, Brader ML, DeFelippis MR, et al. Atomic force microscopy of crystalline insulins: The influence of sequence variation on crystallization and interfacial structure. Biophys J 1998; 74:2199–2209.
23. Yip CM, DeFelippis MR, Frank BH, et al. Structural and morphological characterization of Ultralente insulin crystals by atomic force microscopy: Evidence for hydrophobically driven assembly. Biophys J 1998; 75:1172–1179.

24. Yip CM, Brader ML, Frank BH, et al. Structural studies of a crystalline insulin analog complex with protamine by atomic force microscopy. Biophys J 2000; 78:466–473.
25. Graham DT, Pomeroy AR. An in-vitro test for the duration of action of insulin suspensions. J Pharm Pharmacol 1984; 36:427–430.
26. Prabhu S, Jacknowitz AI, Stout PJ. A study of factors controlling dissolution kinetics of zinc complexed protein suspensions in various ionic species. Int J Pharm 2001; 217:71–78.
27. Akers MJ, Fites AL, Robison RL. Formulation design and development of parenteral suspensions. J Parenter Sci Technol 1987; 41:88–96.
28. Food and Drug Administration. Guidance for Industry. Liposome Drug Products. Rockville, MD: FDA, 2002.
29. Kipp J. The role of solid nanoparticle technology in the parenteral delivery of poorly water-soluble drugs. Int J Pharm 2004; 284:109–122.

BIBLOGRAPHY

Burgess DJ, ed. Injectable Dispersed Systems: Formulation, Processing, and Performance. New York: Informa Healthcare USA, Inc., 2007.

10 Formulation of freeze-dried powders

With the advent of biotechnology medicines, freeze-drying formulation and process development have embarked on new heights of importance in the parenteral industry. Roughly 40% of commercial biopharmaceutical products are freeze-dried; this percentage likely will keep increasing with time. Freeze-drying and lyophilization mean the same thing. Freeze-drying perhaps is more accurate because the process involves both freezing of a solution and then removing the solvent from that solution that involves drying procedures. Lyophilization means to "love the dry state," but the title does not emphasize the cooling/freezing segment. Freeze-drying involves:

1. Compounding, filtering, and filling drug formulations as solutions into vials historically although now more syringes are being used as primary container for lyophilized products. Most of the discussion in this chapter will focus on the vial being the primary package.
2. Inserting a partially slotted rubber closure on the neck of the vial (Fig. 10-1) and transferring the containers into a freeze-drying chamber. If the vial, as well as syringes or cartridges, is to be part of a dual-chambered device (lyophilized powder in one compartment, diluent solution in the other compartment, separated by a rubber plunger), then no rubber closure is inserted prior to lyophilization.
3. Cooling the product to a predetermined temperature that assures that the solution in all containers in the freeze-dryer become frozen.
4. Adjusting the temperature of the shelf/shelves of the freeze-dryer that is as high as possible without causing the temperature of any product container to be above its "critical temperature" (eutectic temperature, glass transition temperature, collapse temperature).
5. Applying a predetermined vacuum that establishes the required pressure differential between the vapor pressure of the sublimation front of the product and the partial pressure of gas in the freeze-drying chamber that allows the removal of frozen ice from all product containers—the process of sublimation.
6. Increasing the shelf temperature once all the ice is sublimed in order to remove whatever remaining water is part of the solute composition to a residual moisture level predetermined to confer long-term stability of the drug product.
7. Completely inserting the rubber closure into the container via hydraulic-powered lowering of the dryer shelves.
8. Removing all freeze-dried containers, completing the sealing (or for syringes/cartridges adding the rubber septum), and carefully inspecting each product unit (inspection criteria for lyophilized products covered in chap. 22, Table 22-4).

This chapter will focus on the formulation of freeze-dried products, whereas chapter 20 will focus on the process of freeze-drying.

ADVANTAGES AND DISADVANTAGES OF FREEZE-DRYING

Freeze-drying is required for active pharmaceutical ingredients that are insufficiently stable in the solution state. Insufficiently stable means that the drug will excessively degrade in solution within a period of time not amendable to marketing the product as a ready-to-use solution. Many small and large molecules are labile in the presence of water and within several days to several weeks will degrade to a point that is unacceptable, usually more than 10% loss of activity or potency compared to the label claim amount of active ingredient. Were it not for freeze-drying technology, many important therapeutic agents would not be commercially available.

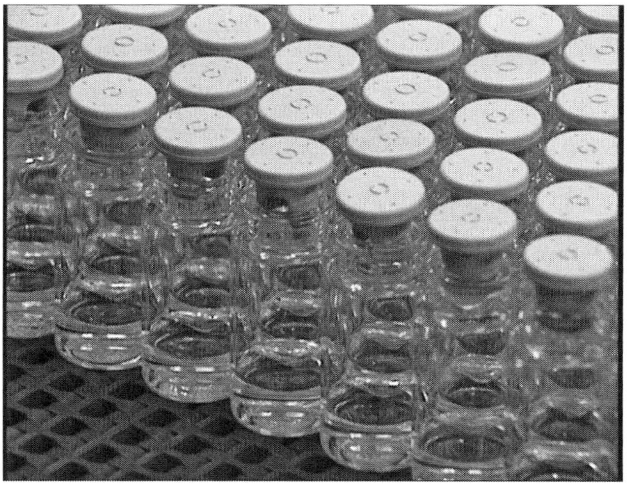

Figure 10-1 Partially slotted stoppers in solution vials prior to loading into freeze-dryer.

Tables 10-1 and 10-2 present two lists of commercial freeze-dried products. Table 10-1 presents general information about these products, whereas Table 10-2 focuses more on the specific quantitative formulations for each product. They are not exhaustive and will not be up to date at the time of this publication, but provide excellent representative information about freeze-dried formulated products being successfully used to save and affect lives.

Besides overcoming stability problems by converting a solution to a dry powder, freeze-drying also offers the advantages of processing the product in the liquid form. Sterile powders can also be produced by other processes (not covered in this book) such as spray-drying, spray-freeze drying, or sterile crystallization followed by powder filling. However, freeze-drying offers certain advantages over other powder production processes including the fact that the product can be dried without the need for elevated temperatures, product sterility is more easily achieved and maintained, the contents of the dried material remain homogeneously dispersed, and the reconstitution times generally are faster. Also, for drugs that are oxygen sensitive, freeze-drying is a better powder-producing alternative, because the environment during the freeze-drying process can be an oxygen-free condition and an inert gas can fill the headspace of the container prior and during closing of the container.

Freeze-drying also has certain limitations, perhaps the foremost being cost compared to other powder-producing processes and certainly more expensive than liquid filling and stoppering. Volatile compounds in the formulation could be removed if high vacuum levels are required and high vacuum has been known to increase the extractable levels from the rubber closure. The freezing and drying steps are known to cause stability problems with some proteins that usually can be overcome using stabilizers called cryo- or lyoprotectants. Because the product has been previously sterilized prior to loading into the freeze-drying chamber, sterility must be maintained during the loading and unloading process and also during the freeze-drying process itself. The ability to maintain aseptic conditions during these processes as well as validating the sterilization of the freeze-dryer chamber and all connections and gases leading into the chamber must be demonstrated.

ATTRIBUTES AND REQUIREMENTS OF A FREEZE-DRIED PRODUCT

The ideal freeze-dried product has a very pleasing aesthetic appearance (i.e., intact cake, uniform color, and appearance) (Fig. 10-2), sufficient strength of active ingredient, chemical and physical stability, sufficient dryness and other specifications that are maintained throughout the product shelf-life, sufficient porosity that permits rapid reconstitution times, and freedom from microorganisms (sterility), pyrogens, and particulate matter after reconstitution. Also, after the drug is in solution, it must remain within certain predetermined specifications (e.g., potency,

(Text continues on page 154.)

Table 10-1 Commercial Lyophilized Products

Drug (Brand) (all are registered®)	Common drug name	Company	Route of administration	Container	Indications	Lyo/ Liquid	Light sensitive	Refrigerated?	Diluent first/ final mL	Albumin?	Special additive (not intended to becomplete)	pH[a]	Launch
Abbokinase	Urokinase	Abbott	IV Infusion	Vial	Pulmonary emboli	Lyo	No	Yes	5/	Yes			March 1982
Activase	Alteplase	Genentech	IV Infusion	Vial	Acute myocardial infarction	Lyo	Yes	Yes	/50, 100	No		7.3	December 1987
Advate	Octocog Alfa	Baxter	IV Infusion	Vial	Hemophilia	Lyo	No	Yes	5/	No	Polysorbate 80		
Amevive	Alefacept	Astellas	IV Bolus IM	Vial	Psoriasis	Lyo	Yes	Yes	10/	No		6.9	February 2003
Aralast	α-1 Proteinase Inhibitor	Baxter	IV Infusion	Vial	Emphysema	Both	No	Yes	/25, 50	Yes	PEG <112 μg/mL	7.4	November 2004
Avonex	Interferon β-1a	Biogen Idec	IM	Vial PFS	Multiple sclerosis	Both	Yes	Yes	10	No		7.3	May 1996
BeneFIX	Coagulation factor IX	Wyeth	IV Infusion	Vial	Hemophilia B	Lyo	Yes	Yes		No		6.8	June 1997
Betaferon/ Betaseron	Interferon β-1b	Schering AG	SC	Kit Vial PFS	Multiple sclerosis	Lyo	No	No	1.2	Yes		7.4	October 1993
Botox	Clostridium botulinum toxin type A	Schering AG/ Allergan	IM	Vial	Cervical dystonia	Lyo	No	Yes	1, 2, 4, 8	Yes			February 1990
Bravelle	Urofollitropin	Ferring	SC IM	Vial	Fertility	Lyo	Yes	Yes	1	No	Polysorbate 20		
Cerezyme	Imiglucerase	Genzyme	IV Infusion	Vial	Gaucher disease	Lyo	No	Yes	100–200	No		6.1	May 1997
CroFab	Crotalidae Polyvalent Immune Fab (Ovine)	Protherics	IV Infusion	Vial	Gaucher disease	Lyo	No	Yes	5/250	No			December 2000
DigiFab	Digoxin Immune Fab (Ovine)	Protherics	IM	Vial	Digoxin toxicity	Lyo	No	Yes	4/	No			February 2002
Eligard	Leuprolide acetate	Wyeth	SC	PFS	Prostrate cancer	Both	No	Yes		No			May 2002
Elitek	Rasburicase	Roche	IV Infusion	Vial	Hyperuricemia	Lyo	Yes	Yes	1, 5/50	No	Poloxamer 188		September 2002
Elspar	Asparaginase	Merck	IM and IV Bolus	Vial	Acute lymphocytic leukemia	Lyo	No	Yes	5	No			
Enbrel	Etanercept	Amgen	SC	Vial PFS	RA (Rheumatoid arthritis)	Both	Yes	Yes	1/	No		7.4	November 1998
Exubera	Insulin human	Pfizer	Inhalation	cartridge	Diabetes	Lyo	No	No	n/a	No			December 2006

Fabrazyme	Agalsidase beta	Genzyme	IV Infusion	Vial	Fabry's disease	Lyo	No	Yes	1.1, 72/500	No			July 2003
Fuzeon (t-20)	Enfuvirtide	Chugai	SC	Vial	HIV/AIDS	Lyo	No	Yes	1.2	No		7.5	March 2003
Genotropin	Somatropin	Pfizer	SC	cartridge	Human growth factor	Lyo	Yes	Yes	n/a	No		6.7	January 1997
GlucaGen	GlucaGen	Novo Nordisk	SC, IM, IV Bolus	Kit Vial PFS	Hypoglycemia	Lyo	Yes	Yes	n/a	No			
Helixate	Factor VIII	CSL Behring	IV Infusion	Vial	Hemophilia A	Lyo	Yes	Yes	2.5, 5/	No		6.9	November 1994
Hemofil M	Antihemo Factor	Baxter	IV Bolus	Vial	Hemophilia A	Lyo	No	No		Yes	PEG 0.07 mg/IU, Octonol 9		
Herceptin	Trastuzumab	Takeda	IV Infusion	Vial	Breast cancer	Lyo	No	Yes	20/250	No		6	October 1998
Humate-P	Humate-P; Factor VIII multimers, Centeon; Humate-P; von Willebrand's Factor multimers	CSL Behring	IV Infusion	Blood Product	Hemophilia	Lyo	No	Yes	5, 10, 15/	Yes			
Humatrope	Somatropin	Eli Lilly	SC IM	Vial Cartridge	Human growth factor	Lyo	No	Yes	n/a	No		7.5	July 1987
Humira	Adalimumab	Abbott	SC	Syringe, Pen	Rheumatoid arthritis	Liquid	Yes	Yes	n/a	No			
Integrilin	Eptifibatide	Millennium/ Schering Plough	IV-infusion	Vial	Acute coronary syndrome (ACS), percutaneous coronary intervention (PCI)	Liquid	Yes	Yes	100 mL	No		5.35	January 1998
Intron A	Interferon alfa-2b	Schering Plough	IV Bolus SC IM	Vial Pen	Hepatitis B & C	Both	No	Yes	10/100	Yes			June 1986
Kepivance	Palifermin	Amgen	IV Bolus	Vial	Severe oral mucositis	Lyo	Yes	Yes	1.2	No		6.5	January 2005
Kogenate	Antihemophilic Factor	Bayer	IV Infusion	Vial	Hemophilia A	Lyo	Yes	Yes	2.5	No	PEG (Koate) 1.5 mg/mL	6.4	February 1994
Leukine	Sargramostim	Shionogi	SC IV infusion	Vial	Chron disease	Both	No	Yes	1	Yes			March 1991

(*continued*)

Table 10-1 Commercial Lyophilized Products (*Continued*)

Drug (Brand) (all are registered®)	Common drug name	Company	Route of administration	Container	Indications	Lyo/ Liquid	Light sensitive	Refrigerated?	Diluent first/ final mL	Albumin?	Special additive (not intended to becomplete)	pH[a]	Launch
Lupron Depot	Leuprolide acetate	Takeda	IM	Dual syringe	Uterine disorders, prostate cancer	Lyo	No	Yes	1	No	PLGA microspheres		
Luveris	Lutropin alfa	Serono	SC	Vial	Fertility	Lyo	Yes	Yes	1	No		8	November 2004
Menopur	Menotropin	Ferring	SC IM	Vial	Fertility	Lyo	Yes	Yes	1	No			April 2005
Mylotarg	Gemtuzumab ozogamicin	GSK	IV Infusion	Vial	Leukemia	Lyo	Yes	Yes	5/100	No	Dextran 40		May 2000
Myozyme	Alglucosidase alfa	Daiichi Sankyo Group	IV Infusion	Vial	Pompe disease	Lyo	No	Yes	50–600	No	Polysorbate 80	6.2	April 2006
Natrecor	Nesiritide	Serono	IV Infusion	Vial	Acutely decompensated congestive heart failure	Lyo	No	Yes	5/250	No	Complex recon procedure		August 2001
Neumega	Oprelvekin (IL-11)	GSK	SC	Vial	Thrombocytopenia	Lyo	Yes	Yes	1	No		7	January 1998
NovoSeven	Coagulation Factor VII a (Eptacog alfa)	Novo Nordisk	IV Bolus	Vial	Hemophilia A	Lyo	Yes	Yes	2.2, 4.3, 8.5/	No		5.5	April 1999
Novarel	Human chorionic gonadotropin	Ferring	IM	Vial	Obesity, Infertility	Lyo	No	No	n/a	No			
Nutropin	Somatropin	Genentech	SC	Vial	Growth hormone	Lyo	No	Yes	n/a	No		7.4	September 1994
Orencia	Abatacept	BMS	IV Infusion	Vial	RA (Rheumatoid arthritis), Lymphoma	Lyo	Yes	Yes	10/100	No		7.5	December 2005
Peg-Intron	Peginterferon alfa-2b	Schering Plough	SC	Vial Pen	Hepatitis C	Lyo	No	Yes	0.7, 1.25	No	PEG 12000 covalent	7	February 2001
Pregnyl	Human chorionic gonadotropin	Organon	IM	Vial	Infertility	Lyo	Yes	Yes	1	No			
Prolastin	α-1 Proteinase Inhibitor	Talecris	IV Infusion	Vial	α-1 antitrypsin deficiency	Lyo	No	Yes	20, 40	No		7	February 1988
Proleukin	Aldesleukin	Chiron	IV Infusion	Vial	Renal cell carcinoma	Lyo	Yes	Yes	50	No	Sodium dodecyl sulfate	7.5	June 1992

Raptiva	Efalizumab	Eisai	SC	Vial	Psoriasis	Lyo	Yes	Yes	1.2	No		6.2	November 2003
Recombinate	Antihemophilic Factor	Baxter	IV Infusion	Vial	Hemophilia	Lyo	No	Yes	10/	Yes	PEG 3350 1.5 mg/mL		May 1993
ReFacto	Antihemophilic Factor (Recombinant)	MedImmune	IV Bolus	Kit Vial PFS	Hemophilia A	Lyo	Yes	Yes	4	No	Polysorbate 80		February 2001
Refludan	Lepirudin	Bayer	IV Bolus or Infusion	Vial	Heparin-induced thrombocytopenia	Lyo	No	Yes	1/250	No		7	June 1998
ReoPro	Abciximab	J&J	IV Infusion	Vial	Percutaneous coronary intervention (PCI)	Liquid	No	Yes	n/a	No	Polysorbate 80	7.2	February 1995
Repronex	Menotropins	Ferring	SC IM	Vial	Fertility	Lyo	Yes	Yes	1, 2	No			July 1997
Retavase	Reteplase	J&J	IV Bolus	Vial	Acute myocardial infarction	Lyo	Yes	Yes	10/	No		6	January 1997
Remicade	Infliximab	J&J	IV Infusion	Vial	Chron Disease and RA	Lyo	No	Yes	10/250	No		7.2	September 1998
Saizen	Somatropin	Serono	SC	Vial Cartridge	Human growth factor	Both	No	Yes	n/a	No		7.5	August 1997
Serostim	Somatropin	Wyeth	SC	Vial	AIDS wasting or cachexia	Both	No	No	n/a	No		8	March 1997
Simulect	Basiliximab	Novartis	IV Bolus or Infusion	Vial	Transplantation	Lyo	No	Yes	2.5, 5/25, 50 mL	No		6.5	June 1998
Somavert	Pegvisomant	Pfizer	SC	Vial	Acromegaly	Lyo	No	Yes	1	No	PEG-5000 covalent		April 2003
Stemgen	Ancestim	Ipsen	SC	Vial	Stem cell mobilization	Lyo	No	Yes	1.2	No			
Streptase	Streptokinase	CSL Behring	IV	Vial	Acute evolving transmural myocardial infarction, pulmonary embolism, deep vein thrombosis	Lyo	No	No	50	No			
Synagis	Palivizumab	MedImmune (USA)	IM	Vial	RSV diseases	Both	No	Yes	n/a	No		6	April 1998

(continued)

Table 10-1 Commercial Lyophilized Products (*Continued*)

Drug (Brand) (all are registered®)	Common drug name	Company	Route of administration	Container	Indications	Lyo/ Liquid	Light sensitive	Refrigerated?	Diluent first/ final mL	Albumin?	Special additive (not intended to becomplete)	pH[a]	Launch
Targocid	Teicoplanin	Sanofi-Aventis/ Wyeth	IV/IM	Vial	Sepsis	Lyo	No	No	3/20, 50	No	Cannot store in syringe		July 1998
TEV-TROPIN	Somatropin	Savient	SC	Vial	Human growth factor	Lyo	No	Yes	1, 5	No		8	February 2005
Thymoglobulin	Antithymocyte globulin	SangStat	IV Infusion	Vial	Acute renal graft rejection	Lyo	Yes	Yes	5/50, 500	No		7	February 1999
Thyrogen	Thyrotropin alfa	Genzyme Corporation	IM	Vial	Thyroid cancer	Lyo	Yes	Yes	1.2	No		7	February 1999
Tisseel	Fibrin Sealant	Baxter	Topical	Syringe	Adjunct to hemostasis	Lyo	No	Yes	2, 4, 10	Yes			
TNKase	Tenecteplase	Genentech	IV Bolus	Kit Vial PFS	Myocardial infarction	Lyo	No	Yes	10/	No		7.3	June 2000
WinRho SDF	Immune Globulin	Baxter	IV Bolus IM	Vial	Immune thrombocytopenic purpura (ITP)	Lyo	No	Yes	1.25, 2.5, 8.5/	No			September 1995
Xolair	Omalizumab	Abbott/ Shionogi	SC	Vial	Asthma	Lyo	Yes	Yes	1.4	No		6	July 2003
Xigris	Drotrecogin alfa	Eli Lilly	IV Infusion	Vial	High-risk severe sepsis	Lyo	Yes	Yes	2.5, 5/	No			November 2001
Zemaira	α-Antitrypsin	CSL Behring	IV Infusion	Vial	α-1 antitrypsin deficiency	Lyo	No	Yes	20/	No			

[a]Blank spaces indicate that data not given in package insert.

Table 10-2 Formulations of Some Commercial Biopharmaceutical Products

Generic name	Trade Name	Manufacturer	Physical form	Route of administration	Component	Concentration
Aldesleukin (Metastatic Renal Cell Carcinoma)	Proleukin	Chiron Corp	Lyophilized 22 106 IU/vial (use within 48 h after reconst)	IV/IM	Proleukin	1.1 mg/mL
					Mannitol	50 mg/mL
					Sodium dodecyl sulfate	0.18 mg/mL
					Sodium phosphate monobasic	0.17 mg/mL
					Sodium phosphate dibasic	0.89 mg/mL
α-1 Proteinase Inhibitor (α-1 Proteinase inhibitor deficiency with emphysema)	Zemaira	Aventis Behring	Lyophilized (Recons with sterile wfl)	IV Infusion	A1-PI	1000 mg/vial
					Sodium	81 mM
					Chloride	38 mM
					Phosphate	17 mM
					Mannitol	144 mM
α-1 Proteinase Inhibitor (α-1 Proteinase inhibitor deficiency with emphysema)	Aralast	Alpha Therapeutics	Lyophilized	IV	Elastase Inhibitory Activity	400–800 mg Active a1-PI
					Albumin	NMT 5 mg/mL
					Polyethylene Glycol	NMT 112 μg/mL
					Polysorbate 80	NMT 50 μg/mL
					Sodium	NMT 230 mEq/L
					Tri-n-butyl Phosphate	NMT1.0 μg/mL
					Zinc	NMT 3 ppm
Alteplase (Acute MI; pulmonary embolism)	Activase®	Genentech	Lyophilized 50, 100 mg (Recons. with sterile wfl)	IV	Alteplase	100 mg = 58 million IU
					L-Arginine	3.5 g
					Polysorbate 80	≤ 11 mg
					Phosphoric Acid	1 g
Antihemophilic Factor	Koate®-DVI	Bayer	Lyophilized 2505001000 U (Recons with Sterile Wfl)	IV	PEG	1.5 mg/mL
					Glycine	0.05 M
					Polysorbate 80	25 μg/mL
					Tri-n-Butyl Phosphate	5 μg/g
					Calcium	3 mM
					Albumin	10 mg/mL
					Aluminum	1 μg/mL
					Histidine	0.06 M

(continued)

Table 10-2 Formulations of Some Commercial Biopharmaceutical Products (*Continued*)

Generic name	Trade Name	Manufacturer	Physical form	Route of administration	Component	Concentration
Antihemophilic Factor	Recombinate®	Baxter Genetics Inst	Lyophilized 2505001000 U (Recons with Sterile Wfl)	IV	PEG	1.5 mg/mL
					Polysorbate 80	1.5 µg/IU
					Calcium	0.2 mg/mL
					Albumin	12.5 mg/mL
					Sodium	180 mEq/L
					Histidine	55 mM
Antihemophilic Factor	ReFacto	Wyeth	Lyophilized 250, 500, 1000 IU/vial (Recons with NaCl diluent)	IV	Sodium chloride	(Not available)
					Sucrose	
					L-histidine	
					Calcium chloride	
					Polysorbate 80	
Antihemophilic Factor (monoclonal antibody)	Hemofil M	Baxter	Lyophilized (Recons with Sterile Wfl)	IV	Albumin	12.5 mg/mL
					PEG	0.07 mg/IU
					Histidine	0.39 mg/IU
					Glycine	0.1 mg/IU
					Mouse protein	0.1 ng/IU
					Tri-n-butyl phosphate	18 ng/IU
					Octoxynol 9 (detergent)	50 ng/IU
Antihemophilic Factor	Advate	Baxter	Lyophilized (Recons with Sterile Wfl)	IV	rAHF-PFM	250–1500 IU/vial
					Mannitol	38 mg/mL
					Trehalose	10 mg/mL
					Sodium	108 mEq/L
					Histidine	12 mM
					Tris	12 mM
					Calcium	1.9 mM
					Polysorbate 80	0.17 mg/mL
					glutathione	0.10 mg/mL
Antihemophilic Factor	Kogenate	Bayer	Lyophilized (Recons with Sterile Wfl)	IV	FVIII	250–1000 IU/vial
					Glycine	10–30 mg/mL
					Calcium chloride	2–5 mM
					Sodium	100–130 mEq/L
					Chloride	100–130 mEq/L
					Polysorbate 80	600 µg/1000 IU
					Imidazole	500 µg/1000 IU
					Albumin	4–10 mg/mL

Antihemophilic Factor	Kogenate FS	Bayer	Lyophilized (Recons with Sterile Wfl)	IV	FVIII	250–1000 IU/vial
					Sucrose	0.9–1.3%
					Glycine	21–25 mg/mL
					Histidine	18–23 mM
					Calcium chloride	2–3 mM
					Sodium	27–36 mEq/L
					Chloride	32–40 mEq/L
					Polysorbate 80	NMT 35 μg/mL
					Imidazole	NMT 20 μg/1000 IU
					Tri-n-butyl phosphate	NMT 5 μg/1000 IU
					Copper	NMT 0.6 μg/1000 IU
Antihemophilic Factor	Humate-P	Centeon	Lyophilized (Recons with Sterile Wfl)	IV	Factor VIII	20–40 IU/mL
					Glycine	15–33 mg/mL
					Sodium citrate	3.5–9.3 mg/mL
					Sodium chloride	2–5.3 mg/mL
					Other proteins	1–7 mg/mL
Anti-inhibitor Coagulant Complex Vitamin K-dependent clotting factors (Heparin)	Autoplex® T	Nabi	Lyophilized (Recons with Sterile Wfl)	IV	Heparin	2 units/mL
					Polyethylene glycol	2 mg/mL
					Sodium Citrate	0.02 M
Antithymocyte Globulin (transplant rejection)	Thymoglobulin	SangStat	Lyophilized (Recons with Sterile Wfl)	IV	Antithymocyte Globulin	25 mg
					Glycine	50 mg
					Mannitol	50 mg
					NaCl	10 mg
Asparaginase (acute lymphocytic leukemia)	Elspar	Merck	Lyophilized (Recons with Wfl, NaCl, or D5W)	IV/IM	Asparaginase	10000 IU
					Mannitol	80 mg
Basiliximab (organ rejection)	Simulect	Novartis	Lyophilized (Recons with Sterile Wfl)	IV	Basiliximab	10 mg
					K phosphate monobasic	3.61 mg
					Disodium hydrogen phosphate	0.5 mg
					NaCl	0.8 mg
					Sucrose	10 mg
					Mannitol	40 mg
					Glycine	20 mg

(continued)

Table 10-2 Formulations of Some Commercial Biopharmaceutical Products (*Continued*)

Generic name	Trade Name	Manufacturer	Physical form	Route of administration	Component	Concentration
BCG, live (urinary bladder cancer)	Pacis	BioChem Pharma	Lyophilized (recons with 0.9% NaCl Inj USP)	Intravesicular	Mycobacterium bovis	2.4–12 × 108 CFU/ampule
					Lactose	15%
Botulism Immune globulin (infant botulism)	BabyBIG	CA Dept. of Health	Lyophilized	IV infusion	IgG antibodies, type A	15 IU/mL
					IgG antibodies, type B	4 IU/mL
					Human albumin	1%
					Sucrose	5%
Coagulation Factor VIIa (hemophilia treatment)	NovoSeven	Novo Nordisk	Lyophilized 60, 240 IU (Recons with sterile Wfl)	IV	NovoSeven	0.6 mg/mL
					NaCl	3 mg/mL
					Calcium chloride dihydrate	1.5 mg/mL
					Glycylglycine	1.3 mg/mL
					Polysorbate 80	0.1 mg/mL
					Mannitol	30 mg/mL
Coagulation Factor IX (hemophilia treatment)	BeneFix®	Wyeth	Lyophilized 2505001000 U (Recons with sterile Wfl)	IV	Histidine	10 mM
					Sucrose	1%
					Glycine	260 mM
					Polysorbate 80	0.01%
Crotalidae Polyvalent Immune Fab (rattlesnake antivenom)	CroFab	Protherics	Lyophilized (Recons with sterile Wfl)	IV	Protein	1 g/vial
					Dibasic Na phosphate	buffer
					NaCl	buffer
					Thimerosal	mercury NGT 0.11 mg/vial
Digoxin Immune Fab (Digoxin toxicity)	DigiFab	Protherics	Lyophilized (Recons with Sterile Wfl)		Digoxin immune Fab	40 mg
					Mannitol	75 mg/vial
					Sodium acetate	2 mg/vial
Drotrecogin Alfa (sepsis)	Xigris™	Eli Lilly	Lyophilized 5 & 20 mg (Recons with sterile Wfl)	IV Infusion	Drotrecogin alfa	5.3 or 20.8 mg
					Sodium Chloride	40.3 or 158.1 mg
					Sodium Citrate	10.9 or 42.9 mg
					Sucrose	31.8 or 124.9 mg
Etanercept (arthritis)	Enbrel®	Immunex	Lyophilized (Recons with Bact. Wfl)	SC	Etanercept	25 mg
					Mannitol	40 mg
					Sucrose	10 mg
					Tromethamine	1.2 mg

Fibrin Sealant	TISSEEL VH Kit	Baxter	Lyophilized Sealer protein and Thrombin (Recons with special diluents)	Topical	*Sealer protein (0.5 mL vial)*	
					Fibrinogen	37.5–57.5 mg
					Polysorbate 80	50–65 mg
					Sodium Chloride	1–2 mg
					Trisodium citrate	2–4 mg
					Glycine	7.5–17.5 mg
					Thrombin (0.5 mL vial)	
					Thrombin	250 IU
					Sodium Chloride	4–6 mg
					Glycine	1.2–1.8 mg
Glucagon (emergency use diabetic coma; gastric relaxant for endoscopy)	GlucaGen	Bedford	Lyophilized 1 mg/vial (Recons with sterile Wfl)	IV/IM/SC	Glucagon hydrochloride	1 mg = 1 IU
					Lactose monohydrate	107 mg
	Glucagon	Eli Lilly	Lyophilized (Special diluent with glycerine, WFI, HCl)	IV/IM/SC	Glucagon	1 mg = 1 IU
					Lactose	49 mg
Human chorionic gonadotropin	Pregnyl®	Organon	Lyophilized 10000 U/vial (Recons with special diluent)	IM	Sodium Phosphate monobasic	5 mg
					Sodium Phosphate dibasic	4.4 mg
					Sodium Chloride (diluent)	0.56%
					Benzyl Alcohol (diluent)	0.90%
Human chorionic gonadotropin	Profasi®	Serono	Lyophilized 10000 U/vial (Recon with Bact Wfl)	IM	Sodium Phosphate monobasic	4 mg
					Sodium Phosphate dibasic	16 mg
					Mannitol	100 mg
					Benzyl Alcohol (diluent)	0.90%
Human chorionic gonadotropin	Pergonal	Serono	Lyophilized 75 U/2 mL ampule (Recon with NaCl for inj.)	IM	Lactose	10 mg
Human chorionic gonadotropin	Novarel	Ferring	Lyophilized 10000 units (Recon with bact. Wfl)	IM	Mannitol	100 mg
					Benzyl Alcohol (diluent)	0.90%
Human chorionic gonadotropin	Repronex	Ferring	Lyophilized 75 U/2 mL ampule (Recon with NaCl for inj.)	IM/SC	Lactose monohydrate	20 mg

(*continued*)

Table 10-2 Formulations of Some Commercial Biopharmaceutical Products (*Continued*)

Generic name	Trade Name	Manufacturer	Physical form	Route of administration	Component	Concentration
Human growth hormone	Humatrope®	Lilly	Lyophilized 2,5,10000 U/vial (Recons with special diluent)	SC/IM	Somatropin	5 mg = 15 IU
					Mannitol	25 mg
					Glycine	5 mg = 15 IU
					Sodium Phosphate dibasic	1.13 mg
					Meta Cresol (diluent)	0.30%
					Glycerin (diluent)	1.70%
Human growth hormone	Nutropin®	Genentech	Lyophilized (Recons with Bact Wfl)	SC	Somatropin	5 mg = 15 IU
					Mannitol	45 mg
					Sodium Phosphate monobasic	0.4 mg
					Sodium Phosphate dibasic	1.3 mg
					Glycine	1.7 mg
					Benzyl Alcohol (diluent)	0.90%
Human growth hormone	Nutropin Depot™	Genentech	Microspheres Powder (Recons with special diluent)	SC (1–2 × monthly)	Somatropin	13.5 mg
					Zinc acetate	1.2 mg
					Zinc carbonate	0.8 mg
					Polylactic/polyglycolic co-polymers	68.9 mg
Human growth hormone	Saizen® for growth def.	Serono	Lyophilized (Recons with Bact Wfl)	IM/SC	Somatropin	5.0 mg = 15 IU
					Sucrose	34.2 mg
					o-phosphoric acid	1.165 mg
Human growth hormone	Serostim® for AIDS	Serono	Lyophilized (Recons with SWFI or Bact Wfl)	SC	Somatropin	4.0 mg = 12 IU
					Sucrose	27.3 mg
					Phosphoric acid	0.9 mg
Human growth hormone	Genotropin	Pharmacia Upjohn	Lyophilized 1.5 mg (Recons with 1.13 mL SWFI)	SC	Somatropin	1.5 mg = 4.5 IU
					Glycine	27.6 mg
					Sodium dihydrogen phosphate anhydrous	0.3 mg
					disodium phosphate anhydrous	0.3 mg
Immune globulin (Gamma globulin IgG)	WinRho SDF	Nabi	Lyophilized 600,1500,5000 IU (Recons with 0.9% NaCl	IV/IM	Glycine	0.1 M
					Sodium Chloride	0.04 M
					Polysorbate 80	0.01%

Infliximab (rheumatoid arthritis and Chron disease)	Remicade	Centocor, Inc	Lyophilized (Recons with sterile Wfl)	IV infusion	Infliximab	100 mg
					Sucrose	500 mg
					Polysorbate 80	0.5 mg
					Sodium phosphate monobasic	2.2 mg
					Sodium phosphate dibasic	6.1 mg
Interferon Alfa-2b (hairy cell leukemia)	Intron A®	Schering	Lyophilized 10,18,25,50 MIU (Recons with Bact. Wfl)	IM/SC/IV Intralesion	Glycine	20 mg/mL
					Sodium Phosphate monobasic	0.55 mg/mL
					Sodium Phosphate dibasic	2.3 mg/mL
					Human Serum Albumin	1.0 mg/mL
					Benzyl Alcohol (diluent)	0.90%
Peginterferon alfa-2b (hepatitis C)	PEG-Intron	Schering	Lyophilized (Recons with sterile Wfl)	SC	PEG-Intron	74–222 μg
					Sodium phosphate monobasic	1.11 mg
					Sodium phosphate dibasic	1.11 mg
					Sucrose	59.2 mg
					Polysorbate 80	0.074 mg
Interferon β-1a (multiple sclerosis)	Avonex	Biogen	Lyophilized (Recons with sterile Wfl)	IM	Interferon β-1a	33 μg = 6.6 million IU
					Human albumin	16.5 mg
					Sodium chloride	6.4 mg
					Sodium phosphate monobasic	1.3 mg
					Sodium phosphate dibasic	6.3 mg
Interferon β-1B (multiple sclerosis)	Betaseron®	Berlex	Lyophilized (Recons with NaCl 0.54%)	SC	Interferon β-1B	0.3 mg = 9.6 million IU
		Chiron			Human Serum Albumin	15 mg
					Mannitol	15 mg
Interleukin-2 (metastatic renal cell carcinoma)	Proleukin®	Chiron	Lyophilized (Recons with Sterile Wfl)	IV	Proleukin	1.1 mg/mL = 18 million IU
				Infusion	Mannitol	50 mg/mL
					Na dodecyl sulfate	0.18 mg/mL
					Na phosphate dibasic	0.89 mg/mL
					Na phosphate monobasic	0.17 mg/mL

(continued)

Table 10-2 Formulations of Some Commercial Biopharmaceutical Products (*Continued*)

Generic name	Trade Name	Manufacturer	Physical form	Route of administration	Component	Concentration
Interleukin-11 (prevention of severe thrombocytopenia; reduce need for platelet transfusions)	Neumega®	Genetics Institute	Lyophilized 5 mg (Recons with 1 mL Sterile Wfl)	SC	Oprelvekin	5 mg = 4 × 107 units
					Glycine	23 mg
					Na phosphate dibasic	1.6 mg
					Na phosphate monobasic	0.55 mg
Leuprolide (prostate cancer)	Lupron Depot® 3.75 mg, 1 mo to 30 mg, 4 mo	TAP	Lyophilized (Recons with special diluent)	IM	Leuprolide acetate	3.75 mg
					Purified gelatin	0.65 mg
					DL-lactic/glycolic acid copolymer	33.1 mg
					Mannitol	6.6 mg
					CMC sodium (diluent)	5 mg
					Polysorbate 80 (diluent)	1 mg
					Mannitol (diluent)	50 mg
Leuprolide (prostate cancer)	Eligard™ 7.5 mg 1 mo	Atrix/Sanofi	Two syringes (one lyophilized, one solution)	SC	Leuprolide acetate	7.5 mg
					DL-lactide-co-glycolide poly	82.5 mg
					N-methyl-2-pyrrolidone (solvent)	160 mg
Palivizumab (respiratory syncytial virus treatment)	Synagis	MedImmune	Lyophilized 50, 100 mg (Recons with sterile Wfl)	IM	Palivizumab	100 mg/mL
					Histidine	47 mM
					Glycine	3.0 mM
					Mannitol	5.6%
Rasburicase (Mgmt plasma uric acid levels in ped. chemo.)	Elitek™	Atrix/Sanofi	Lyophilized (Recons with special diluent)	IV	Rasburicase	1.5 mg
					Mannitol	10.6 mg
					L-alanine	15.9 mg
					Dibasic sodium phosphate	12.6–14.3 mg
					Poloxamer 188 (diluent)	1.0 mg/mL
Reteplase (ventricular function following myocardial infarction)	Retavase®	Centocor, Inc	Lyophilized (Recons with sterile Wfl)	IV	Reteplase	18.1 mg/vial
					Tranexamic acid	8.32 mg/vial
					Dipotassium hydrogen phosphate	136.24 mg/vial
					Phosphoric acid	51.27 mg/vial
					Sucrose	364.0 mg/vial
					Polysorbate 80	5.20 mg/vial

Sargramostim (granulocyte macrophage CSF for bone marrow transplant)	Leukine®	Berlex	Lyophilized	IV Infusion	Sargramostim	250 µg = 1.4 x 106 IU/vial
			250, 500 µg (Recons with sterile or bact. Wfl)	SC	Mannitol	40 mg/mL
					Sucrose	10 mg/mL
					Tromethamine	1.2 mg/mL
Sermorelin acetate (Stimulates pituitary gland to release growth hormone)	Geref® Ampuls	Serono	Lyophilized (Recons with accomp. 0.9% sodium chloride	SC	Sermorelin acetate	50 µg
					Mannitol	5 mg
					Sodium Phos monobasic	0.66 mg
					Sodium Phos dibasic	0.04 mg
Streptokinase (Ventricular function following myocardial infarction)	Streptase	Aventis Behring GmbH	Lyophilized	IV	Streptokinase	2.5–15 × 105 IU
			0.25, 0.75, 1.5 MU (Recons with NaCl or 5% dextrose)		Gelatin polypeptide	25 mg
					Sodium L-glutamate	25 mg
					Human albumin	100 mg
					NaOH	pH adjust
Tenecteplase (following myocardial infarction)	TNKase	Genentech	Lyophilized (Recons with sterile Wfl)	IV	Tenecteplase	52.5 mg
					L-arginine	0.55 g
					Phosphoric acid	0.17 g
					Polysorbate 20	4.3 mg
Thyrotropin alfa (Thyroid releasing hormone)	Thyrogen®	Genzyme	Lyophilized (Recons with sterile Wfl)	IM	Thyrotropin alfa	1.1 mg = 4 IU
					Mannitol	36 mg
					Sodium Phosphate	5.1 mg
					Sodium Chloride	2.4 mg
Trastuzumab (Binds to HER2 protein; treats breast cancer)	Herceptin®	Genentech	Lyophilized	IV	Trastuzumab	440 mg
			440 mg/vial (Recons with bact. Wfl)		L-histidine HCl	9.9 mg
					L-histidine	6.4 mg
					Trehalose	400 mg
					Polysorbate 20	1.8 mg
Urokinase	Abbokinase®	Abbott	Lyophilized (Recons with sterile Wfl)	IV, Intra-Coronary Infusion	Urokinase	250000 IU/vial
					Mannitol	25 mg
					Sodium Chloride	50 mg
					Human Serum Albumin	250 mg

Abbreviations: Bact., bacteriostatic; Chemo, chemotherapy; IM, intramuscular; Inj., injection; IV, intravenous; Mgmt, management; Ped., pediatric; Recons., reconstitution; SC, subcutaneous; SWFI, sterile water for injection; Wfl, water for injection; Lisa Hardwick of Baxter BioPharma Solutions, Bloomington, is acknowledged for preparing this table.

(A) (B)

Figure 10-2 Examples of a pharmaceutically elegant freeze-dry cakes. *Source:* Courtesies of Eli Lilly and Company (A) and Dr. Gregory Sacha, Baxter BioPharma Solutions (B).

pH, freedom from particulate matter) for a certain period of time prior to administration. The desired minimum time for solution stability after reconstitution is 24 hours at ambient temperature although many products, especially biopharmaceuticals, are insufficiently stable at ambient temperature and must be refrigerated even for these short periods of time. Also, European requirements that generally have been applied throughout the world require products without antimicrobial preservatives to be used (administered) "immediately," generally meaning within three hours after reconstitution. Freeze-dried products reconstituted with diluents containing antimicrobial preservatives can be stored for much longer times depending more on drug stability in solution than on potential microbial contamination concerns.

Freeze-dried formulation requirements usually are different depending on whether the active ingredient is a small molecule or large molecule. Formulation of a freeze-dried product containing a small molecule often does not need any additives, depending on the amount of active ingredient per container. For example, many freeze-dried antibiotic products contain only the antibiotic. If the active constituent of the freeze-dried products is present in a small quantity (usually less than 100 mg) where, if freeze-dried alone, its presence would be hard to detect visually, then additives are used. This is true for many small-molecule freeze-dried products, for example, those containing anticancer agents, and practically always true for large-molecule freeze-dried products. The solid content of the original product ideally should be between 5% and 30%. Therefore, excipients often are added to increase the amount of solids. Such excipients are called "bulking agents"; the most commonly used bulking agent in freeze-dried formulations is mannitol. However, most freeze-dried formulations must contain other excipients because of the need to buffer the product and/or to protect the active ingredient from the adverse effects of freezing and/or drying. Thus, buffering agents such as sodium or potassium phosphate, sodium acetate, and sodium citrate are commonly used in freeze-dried formulations. Sucrose, trehalose, dextran, and amino acids such as glycine are commonly used lyoprotectants. Other types of stabilizing excipients often required in freeze-dried formulations are surface-active agents or competitive binding agents. Other reasons for adding excipients freeze-dried compositions, although typically these are part of the diluent formulation rather than the freeze-dried formulation, are tonicity-adjusting agents and antimicrobial preservatives for multiple-dose applications.

Each of these substances contribute to the appearance characteristic of the finished dry product (plug), such as whether the appearance of the finished product is dull and spongy or sparkling and crystalline, firm or friable, expanded or shrunken, or uniform or striated. Therefore, the formulation of a product to be freeze-dried must include consideration not only of the nature and stability characteristics required during the liquid state, both freshly prepared and when reconstituted before use, but also the characteristics desired in the dried product as it is released for commercial use and distributed to the ultimate user.

A "rule-of-thumb" for freeze-dried products containing small molecules is "the drier, the better" because most stability problems with small molecules are moisture-related. However, for freeze-dried products containing large molecules, "drier is not necessarily better." Each molecule is different, but in general for large molecules, the effects of freezing and drying may be as much or more deleterious to the active constituent as the potential for hydrolytic degradation.

FORMULATION COMPONENTS IN FREEZE-DRIED PRODUCTS

Freeze-dried drug molecules, evidenced by the requirement to be freeze-dried, are relatively unstable molecules. Even in the dry state, freeze-dried formulations typically require additives for maintaining pH, isotonicity, or protection against adverse effects of the freezing and/or drying process. Additives may also be required, not for dry-state purposes, but to maintain stability and, in some cases, solubility of the drug in solution after adding a reconstitution diluent. Such additives to enhance solution stability and solubility include buffers, surface-active agents, and complexing agents. For drugs reconstituted to serve as multiple-dose products, an antimicrobial preservative system must be part of the freeze-dried formulation or part of the reconstitution diluent. Table 10-3 lists examples of formulation additives in freeze-dried formulations.

Freeze-dried formulations containing small molecules either do not require any additive excipients because of the large quantity of drug to be freeze-dried, (typically more than 100 mg per container), or additives required are for relatively simple purposes such as adding bulk to the powder, buffering the formulation, providing isotonicity, or perhaps helping to maintain solubility of the drug. Formulation challenges for small molecule formulations are relatively simple at least for the experienced formulation scientist.

Stabilizing large molecules during freeze-drying requires much more formulation expertise and challenge. Freeze-dried formulations of large molecules typically contain one or more of the following additives: bulking agents, lyoprotectants, surfactants, and buffers. Some large-molecule freeze-dried formulations, typically when the protein content is so dilute (low mg to ng/mL levels), contain human serum albumin or some other component to serve as competitive binders to minimize loss of protein due to adsorption to manufacturing surfaces (filters, tubing, disposable mixing bags, stainless steel) and primary container surfaces (glass and rubber). Certain additives such as mannitol and sucrose also may serve as tonicity modifiers. Salts usually are avoided because they decrease the critical temperature of the formulation (lower eutectic or glass transition temperature) and are known to cause concentration-dependent destabilization effects on proteins. Table 10-2 presents a listing of freeze-dried protein formulations, not to be exhaustive but to give the reader an idea of the qualitative composition of these formulations.

Some protein molecules can be adversely affected by the freeze-drying process, that is, the process of freezing and/or drying can cause the protein to denature and aggregate and lose potency. Certain excipient stabilizers have been found to minimize or prevent the problems caused by freezing and/or drying. Excipients that stabilize the protein against the effects of freezing are called cryoprotectants. The primary theory, although not completely accepted,

Table 10-3 Additive Categories and Examples for Freeze-Dried Formulations

Category	Example(s)
Bulking agents	Mannitol, lactose, glycine
Stabilizers	
"Ridigizers" (prevent collapse)	Mannitol, glycine
Minimize aggregation	Polysorbate 20 or 80; poloxamer 188
Cryoprotection	Polyethylene glycol, some sugars
Lyoprotection	Sucrose, trehalose
Minimize surface adsorption	Human serum albumin, polysorbates
Buffers	Acetate, citrate, phosphate, Tris, amino acids
Collapse temperature modifiers	Dextran, polyethylene glycol, disaccharide sugars
Tonicity modifiers	Mannitol, sodium chloride, glycerin

for explaining the cryoprotective effects of certain additives, is called the "excluded solute" or "preferential exclusion" theory (1–3). Some scientists have suggested that solutes that help protect the protein from dissociating during freezing do so because they are excluded from the surface of the protein, as can be demonstrated by dialysis experiments (where the protein and the excipient are not found together in the dialysate). When solutes are excluded from the protein surface, the chemical potential of both the protein and the solute increase. This presents a thermodynamically unfavorable environment for the denatured form of the protein as the denatured form is an unfolded form and yields a greater surface area to the solvent. The native form, with less surface area, is therefore thermodynamically favored.

Another way of explaining the effects of cryoprotectants is the fact that they induce preferential hydration of the surface of the protein because by not binding at the protein surface, this favors water molecules to bind preferentially and this helps to stabilize the native protein state.

Sugars (sucrose, lactose, glucose, trehalose), polyols (glycerol, mannitol, sorbitol), amino acids (glycine, alanine, lysine), and polymers (polyethylene glycol, dextran, polyvinylpyrrolidone) all serve as potential cryoprotectants. The best or, at least, most preferred cryoprotectants appear to be polyethylene glycol (PEG) (molecular weight 3350 Daltons), sucrose, and trehalose.

For proteins requiring both cryo- and lyoprotection, it may be judicious to employ both an agent such as PEG along with a sugar. An example of a marketed therapeutic with this combination is Venoglobulin-S, which contains PEG and sorbitol. A potential caveat to using PEG in lyophilized formulations is the possibility of a liquid–liquid phase separation induced by freeze-concentration, an event implicated in protein unfolding (4).

Proteins may not denature or experience any loss of potency during freezing or in the frozen state, but may experience adverse effects when the sublimation process occurs and when stored in the dry state. Such proteins need stabilizers called lyoprotectants. Lyoprotectants appear to stabilize proteins from the effects of drying and the dry state by what is referred to as the "water-substitute" hypothesis or the "vitrification" hypothesis. Sugars are excellent lyoprotectants. They provide a glassy matrix that retards molecular motions and reduces the rates of deleterious reactions (5,6). They also decrease protein–protein contacts and inhibit deleterious reactions depending on such contacts (e.g., aggregation) (7–9). Sugars serve as water-replacement substrates that form hydrogen bonds to proteins in the dried state (4). The water-replacement or substitute hypothesis is supported by solid-state studies exploring techniques such as Fourier-transform infrared (10), water sorption (11,12), and dissolution calorimetry (13). It is likely sugars have all these possible mechanistic roles in their ability to stabilize proteins.

Often the same excipient can provide both cryo- and/or lyoprotection. An example of cryoprotection is the stabilization effect of sucrose, trehalose, sorbitol, and gelatin on a recombinant adenoviral preparation (14). An example of lyoprotection is the stabilizing effect of lactose and other sugars on recombinant human growth hormone (rhGH) (15). However, lactose, a reducing sugar, is not preferred because of its potential adduct formation.

In both dry state theories, it is important that the excipient stabilizer, the lyoprotectant, exist in the amorphous state, hence the name "vitrification" (glass formation). Protein stability in the dry state results from the protein existing with an amorphous solute in an inert, rigid amorphic matrix where the water content in the matrix also helps to stabilize the protein. Obviously, too much excess water and the protein will degrade by chemical processes (e.g., deamidation), but proteins need a certain amount of water to maintain secondary and higher structure. Thus, excipients that remain amorphous during the freeze-dry process molecularly interact with the amorphous protein and together the matrix confers stability on the protein for long-term stability in the dry state. It has been shown that, for optimal stabilization, the sugar excipient should remain in the same amorphous phase containing the peptide or protein (all of the above mechanisms are consistent with this observation). For example, crystallization of mannitol has been implicated to explain incomplete stabilization of lyophilized rhGH (16) and the structure of bovine serum albumin, ovalbumin, β-lactoglobulin, and lactate dehydrogenase (LDH) upon freeze-drying (17). In addition to crystallization, separation of amorphous phases can also occur, particularly in the frozen state.

Once excipients crystallize, they no longer molecularly interact with the protein and cannot protect it. Amorphous excipients, combined with the protein, have a unique glass transition temperature (T_g) in the dry state. If storage temperature exceeds the glass transition temperature, the physical state of the dried matrix changes from a glassy solid to a rubbery solid that can result in collapse or partial collapse of the freeze-dried cake. Product collapse is not necessarily detrimental to the stability of some proteins, although pharmaceutical elegance still remains an important quality parameter of freeze-dried products.

Gradual conversion of excipients from the amorphous to the crystalline state occurs when there is adequate molecular mobility for nucleation and crystal growth (12). Molecular mobility is a term used to describe the movement of molecules in a formulation. Water will act as a plasticizer to lower the glass transition temperature of amorphous solids and increase the molecular mobility of the amorphous system (18). Molecular mobility typically occurs when the amorphous solid is stored at a temperature greater than its glass transition temperature, but can also occur at temperatures below the glass transition temperature of certain amorphous solids (19,20). Molecular mobility of protein molecules can be measured by solid-state ^1H nuclear magnetic resonance (21), nuclear magnetic resonance relaxation based critical mobility temperature (22), and ^{13}C solid-state nuclear magnetic resonance (23). All these techniques measure water mobility in lyophilized formulations and this can be correlated to protein stability.

Additives in a formulation can prevent crystallization of carbohydrates. Examples include high molecular weight polymers (e.g., dextran and polyvinylpyrrolidone, (24) and proteins (e.g., recombinant bovine somatotropin (BST) (12). Polymers can increase the glass transition temperature, thereby decreasing the mobility of the amorphous solute, whereas proteins such as BST are thought to interfere with either nucleation rates or number of nuclei formed to support a crystal.

Alpha$_1$-antitrypsin (rAAT) is an example of a recombinant protein that must be freeze-dried, yet does not need cryo- or lyoprotection (25). It is interesting that some proteins use cryo-and/or lyoprotectants while others do not (Table 10-2). Formulations without cryo-and/or lyoprotectants either truly are sufficiently stable (e.g., alteplase, α-1 proteinase inhibitor, glucagon, human chorionic gonadotropin) without the need for these stabilizers or may not be all that stable and must be refrigerated in the solid state (e.g., aldesleukin, asparaginase).

CONCENTRATIONS OF STABILIZERS

If cryoprotection is needed, the relevant concentration of the stabilizer in solution prior to freezing should be on the order of 0.3 M or above. If lyoprotection is required, the relevant concentration depends on the level of protein present. The sugar stabilizer needs to be in the proper ratio to the protein (either mole ratio to satisfy water "binding" sites or volume ratio if the relevant degradation mechanism(s) relate directly to glass dynamics and dilution of the protein in an inert solid matrix). In practice, a mass ratio of about 1:1 (sugar:protein) is usually needed for proper stabilization, regardless of the mechanism(s) that might be operating.

There are some data to suggest that the amount of sugar for optimum protection should be enough to satisfy sites on the dried protein that have a strong affinity for water (e.g., charged or polar residues). Studies of water vapor sorption on solid proteins describe this level as a "water monolayer" (11,26). Satisfying these sites (by providing enough amorphous sugars to interact with them in the dried state) stabilizes proteins. For instance, rhGH contains approximately sixty-six strongly water binding sites per molecule protein; about this amount of various sugars was required for maximum stabilization of the lyophilizate upon accelerated storage (16). For recombinant humanized monoclonal antibody (rhuMAb), a much larger molecule containing approximately 500 such strongly water-binding sites, a ratio of about 360 to 500 moles of sugar per mole protein afforded the best storage stability for freeze-dried protein (27). Formulations with combinations of sucrose (20 mM) or trehalose (20 mM) and mannitol (40 mM) had comparable stability to those with sucrose or trehalose alone at 60 mM. Formulations with mannitol alone were less stable.

In order to provide protection in the dried state, a stabilizing sugar should generally remain amorphous in the same phase as the protein. Even so, crystallizing sugars (e.g., mannitol) are widely used, either as a bulking agent or to promote stability. In this regard, a combination of amorphous and crystallizing stabilizing excipients may be employed. In this case, the presence of the amorphous agent could serve to retard crystallization of the other. For example, it was reported for lyophilized rhuMAb that a combination of sucrose and mannitol provided stabilization, provided that the total amount of sugar satisfied the level cited above (28). ENBREL provides an example of a marketed biopharmaceutical product in a freeze-dried form and containing a combination of mannitol and sucrose.

In addition to sugars, proteins such as gelatin and albumin are also employed to provide general stabilization in lyophilized biopharmaceutical products. In particular, human albumin (purified from plasma) is widely found in biopharmaceuticals (and, by itself, is also considered a biopharmaceutical product).

CRYSTALLINE AND AMORPHOUS EXCIPIENTS

In general, a freeze-dried formulation that is predominately crystalline will "look" better, that is, look more pharmaceutically elegant than a formulation that is predominately amorphous. Crystalline solutes are also much easier to dry because the water is adsorbed on the surface of the solute rather than within the molecular structure of the solute, as is the case with amorphous solutes. However, amorphous formulations offer stability advantages if the active ingredient is a protein, minimize the potential for overdrying of the product, and can adsorb moisture that over time may be released from the rubber closure.

Moisture content of a freeze-dried product is obviously an important property to monitor. In general, too much moisture means that the product will eventually collapse and the active ingredient will degrade chemically. Each freeze-dried product must be studied for the moisture specifications required for long-term stability in the dry state. Residual moisture specifications for most products fall within the range of 0.5% to 3.0%. The amount of residual moisture is not as important as where the water resides in the freeze-dry matrix and what kind of solid morphology exists. If the matrix is crystalline, water exists as surface water and is not likely problematic. However, if the matrix is amorphous, or if the active is amorphous with the rest of the matrix crystalline, then excess water may interact molecularly with the drug and cause unacceptable degradation.

The judicious use of excipients can greatly influence product stability. Methylprednisolone was freeze-dried in the presence of mannitol or lactose (29). Although moisture content in the two cakes was identical, the rate of hydrolysis was higher when mannitol was the bulking agent. Mannitol crystallized during freeze-drying and had little to no interaction with water in the microenvironment of the drug. In fact, crystallized mannitol is essentially anhydrous, and any residual water will localize in the amorphous drug phase only. Lactose, however, did not crystallize and served to interact with residual water, thus preventing it from interacting with and hydrolyzing the drug. The degree of crystallinity of the bulking agent can have significant effect on the distribution of water in the freeze-dried matrix.

Distribution of residual moisture in the finished dried product is as important as the overall water content. Moisture content was measured immediately after freeze-drying in three sections of a lyophilized product (top, middle, and bottom of the plug) as well as moisture along the vial wall (30). They found that moisture content in the top section was less than moisture content in the bottom section and that the lowest moisture content of the entire plug existed along the walls in the vial (Fig. 10-3). Thus, drying along the vial walls occurs faster than drying in the plug core. They proposed that faster drying along the vial walls is a result of observed product shrinkage during drying, providing a low resistance pathway for vapor escape along the vial wall.

Many freeze-dried formulations contain three solid components—the active, a crystalline bulking agent, and an amorphous stabilizer. A good example is the formulation for human growth hormone where both mannitol and glycine are additives with mannitol crystallizing during freeze-drying and glycine remaining amorphous. A review of the formulations listed in Table 10-2 shows that several contain more than one bulking agent/stabilizer that helps to maintain an amorphous component and form a protective amorphic matrix with the protein.

MANNITOL

Mannitol, being a major excipient used in lyophilized formulations, is the subject of many papers. Mannitol crystallization is highly influenced by freezing rate (31–33), concentration (32), and other excipients present in the formulation such as sucrose, trehalose, citric acid, hydroxypropyl-β-cyclodextrin, polysorbate 80 (34) and phosphate buffers and polymers (33).

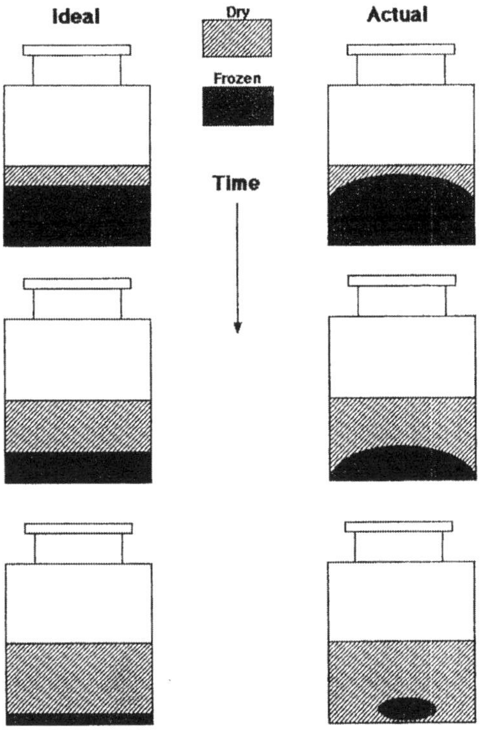

Figure 10-3 Schematic of suggested geometry of the ice–vapor interface during primary drying: A comparison of the "ideal" planar geometry with the curved interface geometry proposed. *Source*: From Ref. 30.

Sucrose especially will affect degree of mannitol crystallization and cause higher levels of mannitol hydrate and resultant residual moisture (35).

Mannitol crystallizes into different polymorphic forms as a function of concentration (relative to other components in the formulation) and freezing rate. Low concentrations of mannitol formed the δ polymorph while higher concentrations favored the formation of the β polymorph (36). At least three polymorphs of mannitol are present at different ratios in the lyophilized product depending on the freezing rate. Rapid freezing produces the α polymorph predominantly while slower rates (0.5°C/min) favor the formation of the δ polymorph. Annealing the frozen product will result in the β polymorph being most prominent. One-year storage will cause the δ polymorph to convert to a combination of α and β polymorphs. There is no evidence that the formation of the α, β, or γ polymorphic forms of mannitol, alone or in various combinations, has any effect on drying/processing characteristics, cake appearance, and or product stability.

A hydrate of mannitol can form during freeze-drying, particularly in conditions not conducive for producing well-developed crystalline mannitol, for example, low temperatures and concentrated solutions (37). This seems impossible, but the authors show thermal and crystallographic evidence for a hydrated form of mannitol that survives a freeze-dry cycle. This hydrate is metastable, able to convert to anhydrous polymorphs of mannitol upon heating. While not specifically investigated, these authors theorized that mannitol hydrates can potentially reduce the drying rates of mannitol-containing formulations and can redistribute residual water to the drug substance upon mannitol crystallization during storage at accelerated conditions. Because mannitol hydrate formation varied greatly from vial to vial, even in the same batch, this also potentially could lead to problems with vial-to-vial variation in moisture levels. Annealing during the freezing stage is the best approach to promote crystallization of the anhydrous form and reduce or eliminate the mannitol hydrate.

The formation of mannitol hydrate formed during freezing may be desolvated and converted to the anhydrous form by conducting secondary drying at 40°C or higher (38). In this paper it was also emphasized that mixtures of mannitol and sucrose in a 4 to 1 ratio successfully produced stable lyophilized formulations of four proteins (daniplestim, leridistim, promegapoietin, and progenipoietin) using a primary drying product temperature of −10°C. The crystalline mannitol allows primary drying to be performed at temperatures above the T_g' of amorphous sucrose in the formulation.

Maintaining mannitol in the amorphous state during freeze-drying for optimal stabilization can be accomplished (39). All the enzymes studied were protected when mannitol remained amorphous, but become unstable with an increase in mannitol crystallinity. Mannitol in freeze-dried cakes containing enzyme and sodium phosphate buffer remained amorphous at lower concentrations (< 200 mM), although annealing the frozen solution resulted in mannitol crystallization. However, mannitol at higher concentrations (> 250 mM) in this enzyme-phosphate formulation crystallized and had no protective effects on preserving enzyme activity after freeze-drying.

MORE ON STABILIZING EXCIPIENTS IN LYOPHILIZED FORMULATIONS

Plasticizers (examples include glycerol, propylene glycol, ethylene glycol, or DMSO) will modify disaccharide and polymeric lyoprotective glasses (40). The proposed mechanism of protein stabilization was attributed to the following: the plasticizers fill small volumes left open by the larger (or stiffer) host glass-former, restricting motion, and thereby slowing the fast dynamics of the glass and subsequent protein degradation. This approach has narrow application because of the relatively large amounts of plasticizer required, and also because of the suggestion that the lower molecular weight oligomers like ethylene glycol are better stabilizers.

Raffinose will not lyoprotect an unstable drug as well as sucrose or trehalose (41). This observation was in contrast to other studies (42–44) showing raffinose to be as effective as trehalose and superior to lactose, maltose, and sucrose in stabilizing several enzymes. A possible explanation for these differences may involve differences in dehydration conditions and storage temperatures.

Mannitol and glycine in frozen solutions will influence the crystallization of each other (45). Glycine was shown to have a stronger initial tendency to crystallize, while it was easier to influence the crystallization of mannitol. Buffer salts, such as sodium phosphate, inhibited crystallization of both mannitol and glycine. The activity of LDH correlated with the extent of crystallization of these excipients (Fig. 10-4).

LDH formulations containing maltodextrin protected LDH against inactivation during freeze-drying because of the amorphous nature of these partially hydrolyzed starches (46). Maltodextrins were also reported to be better lyoprotectants for LDH than sucrose or maltose, although the mechanism is unknown.

The stabilization effects of amorphous additives on freeze-dried proteins are well accepted on the basis of several publications. β-Galactosidase was stabilized with inositol as long as this excipient stayed amorphous, but if inositol crystallized during storage, the enzyme activity declined (47). Inositol crystallization was prevented by addition of polymers such as dextran, Ficoll, and sodium carboxymethylcellulose.

Stabilization of lyophilized proteins not only depends on the formulation and dehydration process parameters, but also depends sometimes on the formulation of the reconstitution medium (48). Keratinocyte growth factor aggregation upon reconstitution with water can occur readily, but several additives such as sulfated polysaccharides, surfactants, polyphosphates, and amino acids in the reconstitution medium will significantly reduce this aggregation.

$$\text{The Gordon–Taylor equation: } T_g = \frac{[W_a T_{ga} + k(1 - W_a)T_{gb}]}{[W_a + k(1 - W_a)]}$$

was applied to predict the glass transition temperature of a model tripeptide in the presence of different sugars (sucrose, lactose, trehalose, and maltose) in both frozen solutions and in lyophilized products (49). Correlation was excellent between the predicted and actual T_g for various ratios of tripeptide to sugars. The authors also showed a significant effect of sodium chloride on the T_g' of the tripeptide in frozen solution, but no effect after lyophilization. Figure 10-5 from this paper demonstrates the plasticizing effect of water on decreasing the glass transition temperature.

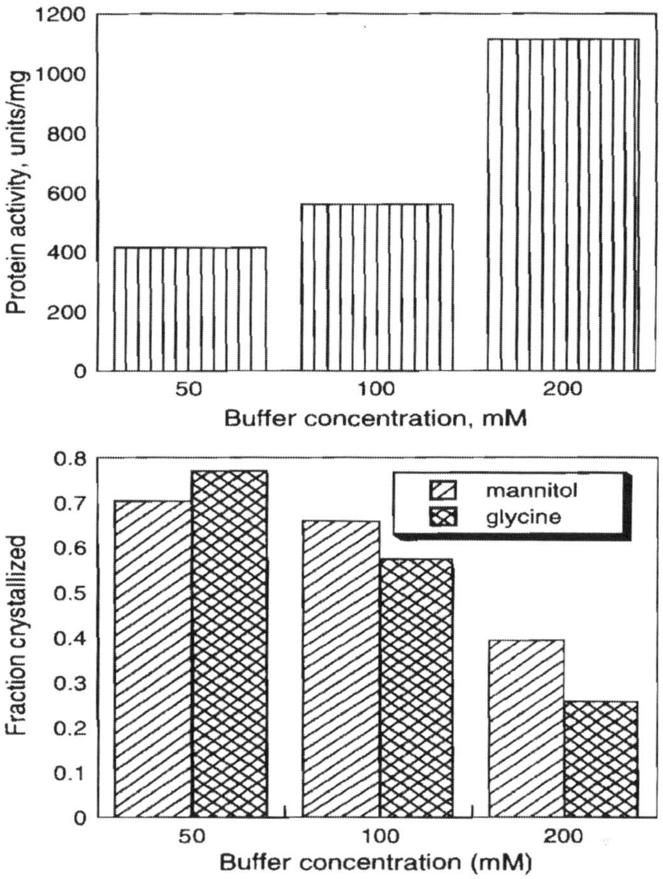

Figure 10-4 Effect of the solid state of excipients on the activity of LDH after freeze-drying. (**A**) Protein activity as a function of buffer concentration in formulations containing 5% w/w each of mannitol and glycine. (**B**) Crystalline mannitol and glycine fractions in the final product. *Source*: From Ref. 45.

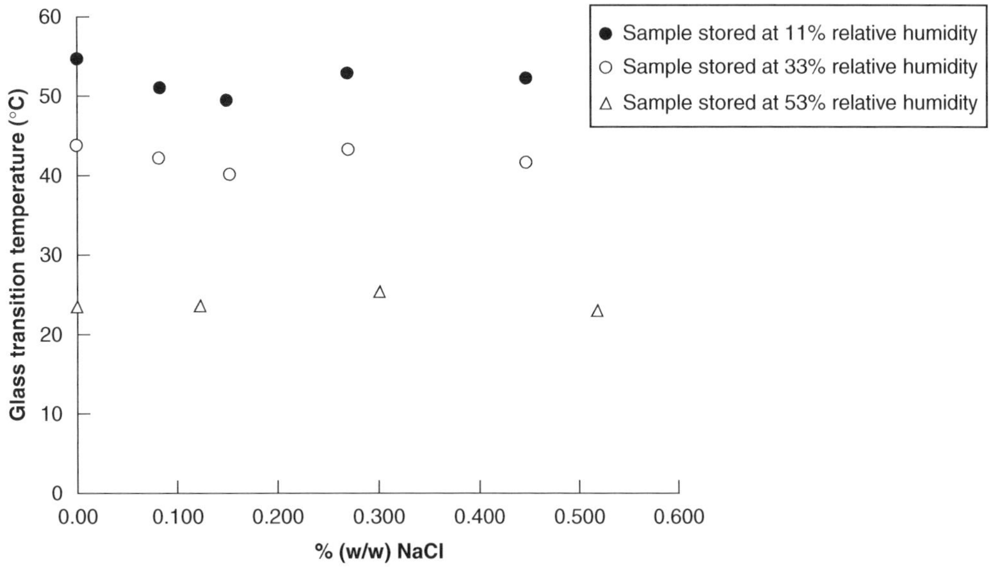

Figure 10-5 Effect of the solid state of excipients on the activity of LDH after freeze-drying. *Source*: From Ref. 49.

All solutes are concentrated during the freezing process. If solutes crystallize, they are separated from solutes that do not crystallize. Therefore, only solutes that do not crystallize have any molecular contact with a protein in the frozen state. Concentration during freezing can have several possible adverse effects. These include the denaturing effects of salts on proteins, crystallization of buffer components that change the pH, and freeze denaturation of some proteins if not protected by cryoprotecting excipients.

Sugar protection of freeze-dried protein (bovine serum albumin) was proposed by applying a Langmuir-type equation in analyzing stability data to suggest a dependency of the α-helix content of the protein in the dry state on the ability of the sugar to form an amorphous matrix with protein molecules (50). The hypothesis was an assumption that interaction of a sugar and protein is a saturable adsorption-like interaction that a Langmuir equation can model. The ability of the amorphous sugar matrix to preserve the α-helix content of the protein was determined by fourier transform infrared spectroscopy (FTIT) that measured the magnitude of the change in the secondary structure contents of the protein relative to that for the native protein. The dependence of the α-helix content ($C_{\alpha\text{-helix}}$) on sugar content (C_{sugar}) could be represented by the equation:

$$C_{\alpha\text{-helix}} = K \times \left(\frac{\left(C^{\max}_{\alpha\text{-helix}} - C^{0}_{\alpha\text{-helix}}\right) \times C_{\text{sugar}}}{(1 + K \times C_{\text{sugar}}) + C^{0}_{\alpha\text{-helix}}} \right)$$

where K indicated the ability of the amorphous sugar matrix to associate to an equilibrium point with the protein, and $C^{0}_{\alpha\text{-helix}}$ and $C^{\max}_{\alpha\text{-helix}}$ indicated the α-helix content in the absence of sugar and saturated levels of sugar, respectively. Preservation effects of sugars could be characterized by K and $C^{\max}_{\alpha\text{-helix}}$, showing that both these values tended to be higher with decreasing T_g values for the amorphous sugar. An amorphous sugar matrix with lower T_g values is structurally more flexible. In this study, sucrose (T_g 65°C) showed a greater stabilizing effect compared to trehalose (T_g 80°C). However, it cannot be concluded from this one study that sucrose is a superior stabilizer.

Not all sugars are wise choices as stabilizers for lyophilized proteins and peptides. Reducing sugars, such as glucose, lactose, and other monosaccharides may form adducts with the protein in the first step of the Maillard reaction. Glucose caused rapid covalent modifications of human relaxin, where covalent adducts of glucose formed with various amino acids on the side chains of the protein (51). Glucose also caused a significant amount of serine cleavage from the C-terminal of the B-chain of relaxin. These degradation reactions did not occur if mannitol or trehalose were used instead of glucose as bulking agents and stabilizers.

Hemoglobin is an example of a protein that is not protected by amorphous excipients during freeze-drying (52). Sucrose, trehalose, Tween 80, or Triton X-100 had little protective effect against phase separation–induced damage during freeze-drying of hemoglobin solutions. However, crystallizing solutes such as mannitol were able to stabilize hemoglobin by segregating the freeze concentrated solution into microscopic domains that blocked propagation and nucleation of phase separating events.

STEPS IN THE DEVELOPMENT OF A FREEZE-DRIED FORMULATION AND PROCESS

Development of freeze-dried products involves the combination of formulation science and process development. One cannot develop the formulation without coordination with the development of the freeze-dried cycle. And, of course, the cycle cannot be finalized without a final formulation being selected. General steps involved in freeze-dry product development are as follows:

1. Determine the critical temperature of the drug alone using thermal analytical methods like differential scanning calorimetry, thermoelectric resistance, or freeze-dry microscopy. This temperature will be the initial basis for determining the initial freezing and primary drying temperature set points. Figure 10-6 shows a thermogram where the glass transition temperature for amorphous mannitol and eutectic temperature for crystalline mannitol are shown. Figure 10-7 shows two photos of a product being freeze-dried with freeze-drying microscopy where collapse of the product in the second photo can be seen easily. Thus,

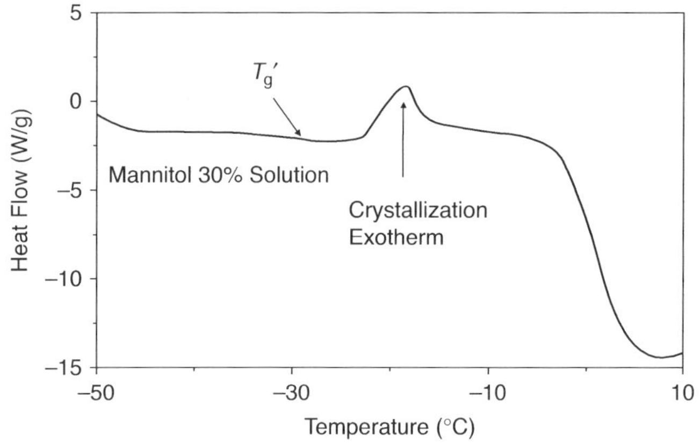

Figure 10-6 Thermogram showing glass transition temperature, T_g', and crystallization exotherm of mannitol.

the product must be maintained at temperatures below the temperature where collapse is observed until all of the ice has been sublimed during primary drying.

2. Freeze-dry the drug alone and determine what happens.
3. If excipients are needed, start with commonly known excipients that produce acceptable cakes with rapid reconstitution rates, have minimal collapse temperatures, and provide the desired finished product with respect to the nature of the solid (crystalline vs. amorphous).
4. Conduct initial stress tests (e.g., freeze-thaw cycling, agitation studies) to screen initial formulations to eliminate worse ones. Freeze-thawing was used to determine effects of solution conditions (pH, salt), protein concentration, cooling and warming rates, and container materials on aggregation potential of an IgG_2 antibody (53). Samples stressed in plastic or glass containers contained low amounts of aggregate. Storage in Teflon or commercial freezing containers, however, led to significantly higher levels of aggregate formation.
5. Formulations should have solid content between 5% and 30% with a target of 10% to 15%.
6. Determine the maximum allowable temperature permitted during freezing and primary drying ($T_e/T_g'/T_c$) of tentative formulation. Again, refer to Figures 10-6 and 10-7 that show how these critical temperatures are obtained.
7. Select the appropriate size vial and fill volume.
8. Select the appropriate rubber closure, one that has low water vapor transmission, no absorption of oil vapor, and a top design that minimizes sticking to shelf.

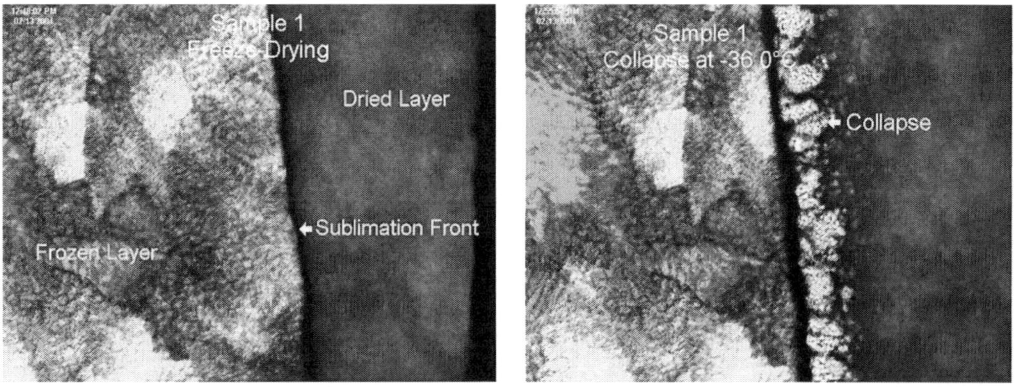

Figure 10-7 Freeze-dry photomicrographs showing uncollapsed sample (*left*) and collapsed sample (*right*).

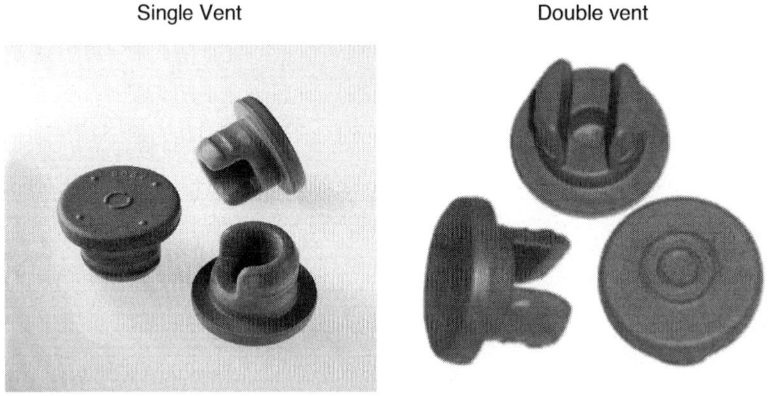

Figure 10-8 Lyophilization rubber closure configurations. *Source*: Courtesy of West Pharmaceutical Services.

9. Once formulation and package are tentatively selected, determine appropriate freeze-dry process parameters, for example, rate of freezing, need for annealing, pressure during primary drying, pressure during secondary drying, and sealing under vacuum or nitrogen.
10. Optimize formulation and process based on stability information both during and after lyo process and after storage in dry state.

Experts (54) in freeze-dry formulation development of biopharmaceuticals provide some basic formulation rules to follow:

1. Use a nonreducing disaccharide (sucrose or trehalose) that forms an amorphous matrix with the protein in the solid state. Do not use reducing sugars (lactose, glucose, maltodextrins) as these may degrade proteins via the Maillard reaction.
2. Use a nonionic surface-active agent such as polysorbate 20 or 80 (55).
3. Use a crystallizing bulking agent such as mannitol or glycine, but do not use crystallizing bulking agents alone with the protein.
4. Do not use bulky additives like dextran even though they have high collapse temperatures. They are too bulky to interact with the protein to form an amorphous matrix.

PACKAGING CONSIDERATIONS

The primary container for lyophilized products is the glass vial. Plastic vials and syringes (dual-chambered) may also be primary package systems for lyophilized products, but the overwhelming majority of lyophilized dosage forms are contained in glass vials.

Tubing vials are preferred to blow-molded vials because the vial bottom configuration is flatter and thinner, allowing for better contact with the shelf surface and more efficient heat transfer from the shelf into the product.

The most common rubber closures for lyophilization are single-vent (igloo style) or double-vent (two-legged) configurations (Fig. 10-8), although there are triple-vent and even up to 12-vent closures. According to DeGrazio (56), even though the total vent area is the same for these configurations, the 1-, 2-, or 3-vent configurations are more efficient in conductance of water vapor out of the vial compared to the 12-vent closure.

A common problem with vented (slotted) stoppers (closure) is the tendency to fall off the opening of the vial as the stopper may not be sufficiently secured as it is partially inserted. This problem exists right after the stopper is partially inserted on the filling line, then will continue as the product is loaded into the freeze-dryer. There is a "groove" within the neck of the closure that is supposed to aid in the "gripping" of the partially inserted closure with the neck of the vial. Such "grooves" are less prominent with multiple-vent closures compared to the single-vent closure that simply has more groove surface area.

Table 10-4 Advantages and Limitations in Using Co-Solvents in Lyophilization (59)

Advantages	Limitations[a]
Increased drug solubility	Safe handling of flammable solvents
Increased sublimation rates	Special facilities and/or equipment
Increased predried solution stability	Determination and control of residual solvent levels
Increase dry product stability	Toxicity of remaining solvent
Decreased reconstitution time	Qualification of appropriate GMP purity
Possible enhancement of sterility assurance	Qualification of supplier
	Overall cost benefit
	Possible adverse environmental impact
	Potential increased regulatory scrutiny

[a]Another possible limitation not mentioned in Ref. 59 is the loss of product during sublimation of solvent.

Two other common problems with rubber closures and lyophilized products during manufacturing are (*i*) the closure not being completely inserted into the vial opening after the lyophilization cycle is completed, and (*ii*) the closure sticking to the shelf after the shelves have been lowered to enable full insertion of the closure into the vial.

The West Company has developed a special stopper for lyophilized products (LyoTec™) where the top surface of the stopper is treated with Flurotec® that prevents the stopper from sticking to the top pressure plates of the freeze-dryer shelves, that otherwise would require manual intervention to remove vials from the freeze-dryer. The stopper segment that makes contact with the opening of the glass vial is uncoated to allow for maximum container-closure integrity (see Fig. 7-10).

CO-SOLVENTS

Tertiary butyl alcohol (TBA) offers advantages when used as a co-solvent with water (57). TBA was used to enable the freeze-dried cake of the antibiotic, tobramycin, to be readily loosened as a free-flowing powder so that the contents could be poured from the vial in orthopedic surgery (58). Applying a slow rate of freezing allowed the crystallization of the TBA solvent prior to drying, with annealing after freezing, significantly reducing the level of residual TBA in the final cake.

Teagarden published an excellent review paper on the use of nonaqueous co-solvents in freeze-drying (59). While co-solvent usage in freeze-dried preparations has been used in at least 20 products (Table 1 in Teagarden's paper), few are commercially available. One product that is currently marketed, Caverject® Sterile Powder, is prepared using a co-solvent solution (20% v/v tert-butanol/water) that is freeze-dried.

Co-solvent use in freeze-dry formulations offers some advantages, but significant limitations must be considered as reviewed by Teagarden and summarized here and in Table 10-4. Residual solvent levels in the finished product can be a problem, but steps can be taken to control the amounts. While residual tert-butanol levels in a crystalline matrix are generally in the 0.01% to 0.03% range, amorphous systems can contain 3% residuals or more. These levels can be decreased in amorphous formulations by (*i*) humidifying the dried solid to decrease the glass transition temperature, allowing crystallization of the matrix and subsequent rapid release of the residual tert-butanol; (*ii*) adding an annealing step to enable any remaining unfrozen tert-butanol hydrate to crystallize and produce a more uniform product; or (*iii*) actually increasing the amount of tert-butanol in the formulation to a level above the threshold concentration required for eutectic crystallization of the solvent. These steps have been shown to decrease the amount of residual tert-butanol to levels of 1% or lower.

ACKNOWLEDGMENT

The author thanks Dr. Wei Kuu of Baxter BioPharma Solutions for some of the references on lyo- and cryoprotection.

REFERENCES

1. Arakawa T, Prestrelski SJ, Kinney W, et al. Factors affecting short-term and long-term stability of proteins. Adv Drug Deliv Rev 1993; 10:1–28.
2. Carpenter JF, Crowe JH. The mechanism of cryoprotection of proteins by solutes. Cryobiology 1998; 25:244–255.
3. Timasheff SN. Protein hydration, thermodynamic binding, and preferential hydration. Biochemistry 2002; 41:13473–13482.
4. Heller MC, Carpenter JF, Randolph TW. Manipulation of lyophilization-induced phase separation: Implications for pharmaceutical formulations. Biotechnol Prog 1997; 13:590–596.
5. Franks F. Long-term stabilization of biologicals. Biotechnology 1994; 12:253–256.
6. Chang LL, Pikal MJ. Mechanisms of protein stabilization in the solid state. J Pharm Sci 2009; 98:2886–2908.
7. Liu WR, Langer R, Klibanov AM. Moisture-induced aggregation of lyophilized proteins in the solid state. Biotechnol Bioeng 1991; 37:177–184.
8. Jordan GM, Yoshioka S, Terao T. The aggregation of bovine serum albumin in solution and in the solid state. J Pharm Pharmacol 1993; 46:182–185.
9. Costantino HR, Langer R, Klibanov AM. Aggregation of a lyophilized pharmaceutical protein, recombinant human albumin: Effect of moisture and stabilization by excipients, Biotechnology 1995; 13:493–496.
10. Costantino HR, Griebenow K, Mishra P, et al. Fourier-transform infrared spectroscopic investigation of protein stability in the lyophilized form. Biochim Biophys Acta 1995; 1253:69–74.
11. Costantino HR, Curley JG, Hsu CC. Determining the water sorption monolayer of lyophilized pharmaceutical proteins. J Pharm Sci 1997; 86:1390–1393.
12. Sarciaux J-ME, Hageman MJ. Effects of bovine somatotropin (rbSt) concentration at different moisture levels on the physical stability of sucrose in freeze-dried rbSt/sucrose mixtures. J Pharm Sci 1997; 86:365–371.
13. Souillac PO, Costantino HR, Middaugh CR, et al. Investigation of protein/carbohydrate interactions in the dried state. 1. Calorimetric studies. J Pharm Sci 2002; 91:206–216.
14. Croyle MA, Roessler BJ, Davidson BL, et al. Factors that influence stability of recombinant adenoviral preparations for human gene therapy. Pharm Dev Technol 1998; 3:373–383.
15. Costantino HR, JCurley JG, Wu S, et al. Water sorption behavior of lyophilized protein-sugar systems and implications for solid-state interactions. Int J Pharm 1998; 166:211–221.
16. Costantino HR, Carrasquillo KG, Cordero RA, et al. Effect of excipients on the stability and structure of lyophilized recombinant human growth hormone. J Pharm Sci 1998; 87:1412–1420.
17. Izutsu K, Kojima S. Excipient crystallinity and its protein-structure stabilizing effect during freeze drying. J Pharm Pharmacol 2002; 54:1033–1039.
18. Tong P, Zografi G. Effects of water vapor absorption on the physical and chemical stability of amorphous sodium indomethacin. AAPS PharmSciTech 2004; 5(2):e26.
19. Dudd SP, Zhang G, Dal Monte PR. The relationship between protein aggregation and molecular mobility below the glass transition temperature of lyophilized formulations containing a monoclonal antibody. Pharm Res 1997; 14:596–600.
20. Hancock BC, Shamblin SL, Zografi G. Molecular mobility of amorphous pharmaceutical solids below their glass transition temperatures. Pharm Res 1995; 12:799–806.
21. Yoshioka S, Aso Y, Kojima S. Determination of molecular mobility of lyophilized bovine serum albumin and γ-globulin by solid state ^{1}H NMR and relation to aggregation-susceptibility. Pharm Res 1996; 13:926–930.
22. Yoshioka S, Aso Y, Kojima S. The effect of excipients on the molecular mobility of lyophilized formulations, as measured by glass transition temperature and NMR relaxation-based critical mobility temperature. Pharm Res 1999a; 16:135–140.
23. Yoshioka S, Aso Y, Kojima S, et al. Molecular mobility of protein in lyophilized formulations linked to the molecular mobility of polymer excipients, as determined by high resolution 13C solid state NMR. Pharm Res 1999b; 16:1621–1625.
24. Suleki-Gerhardt A, Zografi G. Non-isothermal and isothermal crystallization of sucrose from the amorphous state. Pharm Res 1994; 11:1166–1173.
25. Vemuri S, Yu C-D, Roosdorp N. Effect of cryoprotectants on freezing, lyophilization, and storage of lyophilized recombinant alpha1-antitrypsin formulations. PDA J Pharm Sci Technol 1994; 48:241–246.
26. Rowe AJ. Probing hydration and the stability of protein solutions – a colloid science approach. Biophys Chem 2001; 93:93–101.
27. Cleland JL, Lam X, Kendrick B, et al. A specific molar ratio of stabilizer to protein is required for storage stability of a lyophilized monoclonal antibody. J Pharm Sci 2001; 90:310–321.

28. Breen ED, Curlery JG, Overcashier DE, et al. Effect of moisture on the stability of a lyophilized humanized monoclonal antibody formulation. Pharm Res 2001; 18:1345–1353.
29. Herman BD, Sinclair BD, Milton N, et al. The effect of bulking agent on the solid state stability of freeze dried methylprednisolone sodium succinate. Pharm Res 1994; 11:1467–1473.
30. Pikal MJ, Shah S. Intravial distribution of moisture during the secondary drying state of freeze drying. PDA J PharmSciTechnol 1997; 51:17–24.
31. Williams NA, Guglielmo J. Thermal mechanical analysis of frozen solutions of mannitol and some related stereoisomers: Evidence of expansion during warming and correlation with vial breakage during lyophilization. PDA J Pharm Sci Technol 1993; 47:119–123.
32. Kim AI, Akers MJ, Nail SL. The physical state of mannitol after freeze-drying: Effects of mannitol concentration, freezing rate and a non-crystallizing co-solute. J Pharm Sci 1998; 87:931–935.
33. Cavatur RK, Vemuri NM, Pyne A, et al. Crystallization behavior of mannitol in frozen aqueous solutions. Pharm Res 2002; 19:894–900.
34. Martini A, Kume S, Crivellente M, et al. Use of subambient differential scanning calorimetry to monitor the frozen-state behavior of blends of excipients for freeze-drying. PDA J Pharm Sci Technol 1997; 51:62–67.
35. Schneid S, Riegger X, Gieseler H. Influence of common excipients on the crystalline modifications of freeze-dried mannitol. Pharm Technol 2008; 32:178–184.
36. Cannon AJ, Trappler EH. The influence of lyophilization on the polymorphic behavior of mannitol. PDA J Pharm Sci Technol 2000; 54:13–22.
37. Yu L, Milton N, Groleau EG, et al. Existence of a mannitol hydrate during freeze-drying and practical implications. J Pharm Sci 1999; 88:196–198.
38. Johnson RE, Kirchhoff CF, Gaud HT. Mannitol-sucrose mixtures—versatile formulations for protein lyophilization. J Pharm Sci 2002; 91:914–922.
39. Izutsu K, Yoshioka S, Terao T. Effect of mannitol crystallinity on the stabilization of enzymes during freeze-drying. Chem Pharm Bull 1994; 42:5–8.
40. Cicerone MT, Tellington A, Trost L, et al. Substantially improved stability of biological agents in the dried form. The role of glassy dynamics in preservation of biopharmaceuticals. Bioproc Int 2003; 1:36–47.
41. Davidson P, Sun WQ. Effect of sucrose/raffinose mass ratios on the stability of co-lyophilized protein during storage above the *Tg*. Pharm Res 2001; 18:474–479.
42. Kajiwara K, Franks F, Echlin P, et al. Structural and dynamic properties of crystalline and amorphous phases in raffinose-water mixtures. Pharm Res 1999; 16:1441–1448.
43. Rossi S, Buera P, Moreno S, et al. Stabilization of the restriction enzyme, *Eco*RI dried with trehalose and other selected glass-forming solutes. Biotechnol Prog 1997; 13:609–616.
44. Suzuki T, Imamura K, Yamamoto K, et al. Thermal stabilization of freeze-dried enzymes by sugars. J Chem Eng Japan 1997; 30:609–613.
45. Pyne A, Chatterjee K, Suryanarayanan R. Solute crystallization in mannitol-glycine systems—implications on protein stabilization in freeze-dried formulations. J Pharm Sci 2003; 92:2272–2283.
46. Corveleyn S, Remon J-P. Maltodextrins as lyoprotectants in the lyophilization of a model protein, LDH. Pharm Res 1996; 13:146–150.
47. Izutsu K, Yoshioka S, Kojima S. Physical stability and protein stability of freeze-dried cakes during storage at elevated temperature. Pharm Res 1994; 11:995–999.
48. Zhang MZ, Wen J, Arakawa T, et al. A new strategy for enhancing the stability of lyophilized protein: The effect of the reconstitution medium on keratinocyte growth factor. Pharm Res 1995; 12:1447–1452.
49. Jang J-W, Kitamura S, Guillory JK. The effect of excipients on glass transition temperatures for FK906 in the frozen and lyophilized states. J Parenter Sci Technol 1995; 49:166–174.
50. Imamura K, Ogawa T, Sakiyama T, et al. Effects of types of sugar on the stabilization of protein in the dried state. J Pharm Sci 2003; 92:266–274.
51. Li S, Patapoff TW, Overcashier D, et al. Effects of reducing sugars on the chemical stability of human relaxin in the lyophilized state. J Pharm Sci 1996; 85:873–877.
52. Heller MC, Carpenter JC, Randolph TW. Protein formulation and lyophilization cycle design: Prevention of damage due to freeze-concentration induced phase separation. Biotechnol Bioeng 2000; 63:166–174.
53. Kueltzo LA, Wang W, Randolph TW, et al. Effects of solution conditions, processing parameters, and container materials on aggregation of a monoclonal antibody during freeze-thawing. J Pharm Sci 2008; 97(5):1801–1812.
54. Carpenter JF, Pikal MJ, Chang BS, et al. Rational design of stable lyophilized protein formulations: Some practical advice. Pharm Res 1997; 14:969–975.

55. Jones LS, Randolph TW, Kohnert U, et al. The effects of tween 20 and sucrose on the stability of anti-l-selectin during lyophilization and reconstitution. J Pharm Sci 2001; 90:1466–1477.
56. DeGrazio FL. Closure and container considerations in lyophilization. In: Rey L, May JC, eds. Freeze-Drying/Lyophilization of Pharmaceutical and Biological Products, 2nd ed. New York: Marcel Dekker (now Informa), 2004:277–297.
57. Van Drooge DJ, Hinrichs WLJ, Frijlink HW. Incorporation of lipophilic drugs in sugar glasses by lyophilization using a mixture of water and tertiary butyl alcohol as solvent. J Pharm Sci 2004; 93:713–725.
58. Wittaya-Areekul S, Needham GF, Milton N, et al. Freeze-drying of tert-butanol/water cosolvent systems: A case report on formation of a friable freeze-dried powder of tobramycin sulfate. J Pharm Sci 2002; 91(4):1147–1155.
59. Teagarden DL. Practical aspects of freeze-drying of pharmaceutical and biological products using non-aqueous co-solvent systems. Am Pharm Rev 2003:32–44.

BIBLIOGRAPHY

Carpenter JF, Chang BS. Lyophilization of protein pharmaceuticals. In: Avis KE, Wu VL, eds. Biotechnology and Biopharmaceutical Manufacturing, Processing, and Preservation. Boca Raton, FL: Interpharm Press (now CRC Press), 1996:199–264.

Costantino HR, Pikal MJ. Lyophilization of Biopharmaceuticals. Arlington, VA: AAPS Press, 2005.

Gatlin LA, Auffret T, Shalaev E, et al. Freeze-drying concepts: The basics. In: McNally EJ, Hastedt JE, eds. Protein Formulation and Delivery. London: Informa Healthcare, 2007:177–196.

Jennings T. Lyophilization: Introduction and Basic Principles. Boca Raton, FL: CRC Press, 1999.

Nail, S.L., Jiang, S., Chongprasert, S., and Knopp, S.A., Fundamentals of freeze-drying. In Nail, S.L. and Akers, M.J., eds. Development and Manufacture of Protein Pharmaceuticals. New York: Kluwers Academic/Plenum, 2002:281–360.

Pikal MJ. Freeze-drying of proteins. In: Cleland JL, Langer R, eds. Formulation and Delivery of Proteins and Peptides. 1994. Washington, DC: ACS Symposium Series, 567, 120–133.

Rey L, May JC, eds. Freeze-Drying/Lyophilization of Pharmaceutical and Biological Products, 2nd ed. New York: Marcel Dekker, 2004; revised and expanded.

Schwegman JJ, Hardwick LM, Akers MJ. Practical formulation and process development of freeze-dried products. Pharm Dev Technol 2004; 10:151–173.

Shalaev EY, Wang W, Gatlin LA. Rational choice of excipients for use in lyophilized formulations. In: McNally EJ, Hastedt JE, eds. Protein Formulation and Delivery. London: Informa Healthcare, 2007:197–218.

11 Overcoming formulation problems and some case studies

Challenges and problems to solve when developing sterile dosage forms are innumerable. However, while not claiming to be totally comprehensive, Table 11-1 summarizes the formulation problems to overcome in sterile product development. This chapter is intended to provide some practical guidance on approaching and solving common formulation problems in sterile product development on the basis of actual experiences of the author.

OVERCOMING SOLUBILITY PROBLEMS

There are seven general approaches to overcome solubility problems (1).

The first approach is to determine if the optimal salt form of the drug can be used, provided the drug is an electrolyte (can be ionized at different pH values) and depending on properties of the salt form, especially drug stability. Often the most soluble form is too unstable for further development so a compromise must be made between acceptable solubility and stability. It is beyond the scope of this book to discuss optimal drug molecular structure for optimal solubility, but before a formulation scientist begins to develop the final product formulation, he/she should be using the optimal salt form of the drug (Table 11-2).

The next approach that can be tried to render drugs more soluble in aqueous solution is to adjust solution pH. This is a relatively simple approach that is widely used. Either the drug salt form is isolated and used, which when dissolved in solution, solubilizes the drug without adjustment of pH, or the drug is added to water to form a slurry (drug in suspension). The pH is then adjusted by adding the appropriate acid or base form to reach the pH at which the drug will dissolve and remain in solution.

The relationship of drug solubility and solution pH is described by the following equation:

$$\text{pH} = \text{p}K_a + [\log(S\text{-So})/\text{So}] \quad \text{Henderson–Hasselbalch Equation.}$$

Where S is the total solubility in water, So is the solubility of the drug in its natural (undissociated) form, and pK_a is the dissociation constant of the drug. A simple example will demonstrate the use of this equation. A drug has a natural solubility of 0.61 mg/mL and a pK_a of 7.12. A total solubility of 50 mg/mL is required for the final dosage form. What must the solution pH be adjusted to in order to obtain this concentration?

$$\text{pH} = \text{p}K_a + [\log(S\text{-So})/\text{So}] = 7.12 + [\log(50 - 0.61)/0.61] = 9.03.$$

Typically, increasing drug solubility will also increase its instability in water. Figure 8-2 presented the classic example of the solubility and stability of penicillin as a function of pH.

Adjustment of pH in situ can render insoluble drugs soluble in water. Vancomycin-free base is relatively insoluble in water. A slurry of vancomycin-free base in water is prepared and then the slurry is slowly titrated with hydrochloric acid. As the hydrochloride salt of vancomycin is formed in situ, vancomycin hydrochloride dissolves completely in solution.

If a salt form of the drug cannot be used (or is still insufficiently soluble) or pH adjustment does not increase drug solubility sufficiently, then slightly more complicated, albeit still relatively simple, formulation approaches are possible. These approaches, using co-solvents or complexing agents or surface-active agents, are presented in chapter 6.

If the addition of co-solvents or solubilizing solutes do not solve the solubility problem for a particular drug, or there are safety/toxicologic issues using these additives for a particular product and/or route of administration and/or clinical indication, then an alternative dosage form must be developed. In such dosage forms the drug is not in a true solution but can be formulated and processed where the mean particle (micro- or nanosuspension) or globule (microemulsion or liposome) size is 1 μm or less, allowing for intravenous (IV) administration. If the route of administration does not include IV, then the dosage form can be a conventional (macro) suspension (size ranges > 5 μm) or an emulsion.

Table 11-1 Summary of Sterile Product Formulation Challenges

1. Overcoming solubility problems, especially for aqueous insoluble drugs intended for IV administration.
2. Overcoming stability problems, the greatest area of challenges and opportunities for development scientists. The major stability issues include:
   a. Instability in water, influenced by pH and ionic strength.
   b. Oxidation stability problems
   c. Photolytic stability problems
   d. Stability problems during processing, e.g.,
      i. During freeze-drying, especially for biomolecules
      ii. During processing of time/temperature sensitive molecules
      iii. During processing of dispersed systems where significant physical instability problems can arise
3. Overcoming compatibility problems between drug and added substance, between one or more added substances, between drug and/or added substance and packaging. This is especially significant when formulating biomolecules because of their high reactivity with almost any other component in the formulation.
4. Overcoming problems that formulations may cause, with pain and tissue irritation upon injection.
5. Overcoming homogeneity problems with dispersed systems, especially suspensions.
6. Overcoming drug delivery problems when targeting a certain dose to be delivered at or over a certain time and depending on the formulation matrix to achieve this goal consistently.

For first human dose studies for cancer drugs typically, "heroic" approaches are taken to render drugs soluble for IV administration. These approaches may be suitable for very early clinical phase studies of life-saving drugs to patients who otherwise may die, but are not suitable for commercial injectable formulations. Examples of heroic approaches for drug solubilization include:

- Using relatively toxic solvents such as dimethyl sulfoxide (DMSO)
- The vehicle being 100% ethanol
- High concentrations of surface-active agents or other solubilizing agents whose concentrations exceed what is known to be safe for injectable administration

Table 11-2 Examples of Salt Forms Used To Improve Drug Solubility (in order of frequency of appearances in the United States Pharmacopeia)

Cationic salts	Anionic salts
Sodium	Hydrochloride
Potassium	Sulfate
Calcium	Acetate
Aluminum	Phosphate/diphosphate
Zinc	Chloride
Meglumine	Maleate
Benzathine	Citrate
Tromethamine	Tartrate/Bitartrate
Magnesium	Mesylate
Lysine	Nitrate
Procaine	Succinate
Ethylenediamine	Gluconate
	Bromide
	Fumarate
	Hydrobromide
	Carbonate
	Iodide
	Benzoate
	Pamoate
	Valerate

Source: From Ref. (2).

OVERCOMING STABILITY PROBLEMS

Because pharmaceutical dosage form stability is perhaps the most important property and one requiring the most effort from formulation development scientists, overcoming stability problems and formulating stable products, especially for proteins, has been covered in chapters 8, 9, and 10 for the specific type of dosage form.

OVERCOMING COMPATIBILITY PROBLEMS

Compatibility problems with injectables cover a wide range of possible problems.

- Drug and additive incompatibilities that result in inactivation of the drug or precipitation of an insoluble complex.
- Drug or additive interacting with packaging components such as
  - binding to glass
  - adsorption to rubber
  - adsorption to plastic
  - binding to silicon coatings from glass syringes or cartridges or rubber closures or plungers
  - reactions with needle components
- Leachates from the packaging system caused by certain properties of the formulation.
- Drug binding to the surfaces of devices used to deliver the product, such as syringes, IV sets, and catheters

OVERCOMING PAIN AND TISSUE IRRITATION PROBLEMS

Injections are painful, either for real or imaginary. Pain is associated with the needle primarily, but besides the needle, the properties of the drug product can cause pain and irritation. Obviously the pharmaceutical product formulation scientist cannot help to avoid needle pain, but can do some things to minimize the pain caused by the product.

Both in vitro and in vivo methods exist (Table 11-3) to study pain and irritation potential of drug products (3). These methods can be used to pinpoint one or more of the component of the product, be it drug, additive, or solvent and properties therein, that might be causing or contributing to the pain sensation. Data from these studies can help to determine whether a component can be reduced or eliminated or a property such as pH, tonicity, volume percent of co-solvent, ionic strength, or particle size, whatever the case, be changed.

There is precedence for altering the amino acid sequence of polypeptide drugs in order to minimize the self-aggregation of the protein at the site of injection, the aggregation of which caused patients to complain of injection pain.

Once everything has been done to optimize the formulation to minimize or eliminate the pain, yet pain response still results, the main approach that can be used is to minimize the rate of drug product injection or infusion. Other possible aids, although not as effective as injection/infusion rate, include dilution of the drug product prior to injection and use of an

Table 11-3 In Vitro and In Vivo Methods for Determination of Formulation Potential for Pain, Irritation, and/or Other Problems Following Intravenous Administration of the Dosage Form

Method type	Specific description of method	What potential problem is determined?
In Vitro	Mixing formulation and blood (1:10 ratio)	Hemolysis
	Slow addition and mixing of formulation with blood or isotonic Sorensen's phosphate buffer	Precipitation in blood
	Depletion of creatine phosphokinase when formulation mixed with muscle cells	Pain
	Precipitation studies	Phlebitis
In Vivo	Rabbit ear vein model	Phlebitis, precipitation
	Blood level of hemoglobin	Hemolysis
	Rat or rabbit paw lick	Pain
	Serum levels of creatine phosphokinase	Pain, cell damage

Source: From Ref. (3).

in-line filter if there's a possibility that particulate load, however small, might be contributing to the problem.

OVERCOMING HOMOGENEITY PROBLEMS

Dispersed system homogeneity problems, manifested typically through nonuniformity of active ingredient dose from container to container, are almost always the result of poor formulation design and/or nonuniform particle size distribution of the suspended drug particles.

Well-designed parenteral suspensions are formulation according to the principles described in chapter 9. Perhaps the overriding key factor to good homogeneous suspension formulations is adequate wetting of all drug particles and particle size reduction of particles within a very narrow size distribution range.

OVERCOMING DRUG DELIVERY PROBLEMS

Drug release from microspheres or other polymer delivery devices must be constant as a function of time. If zero-order rates of release and availability of drugs from depot injections does not occur, serious therapeutic problems will occur.

There are three approaches used to help assure that drug delivery problems from depot injections do not occur:

1. Understand polymer chemistry. This includes understanding the following factors:
   a. How the active ingredient is entrapped within the polymer matrix
   b. Understanding how the polymer degrades in vivo
   c. Knowing the safety profile of the polymer
   d. Understanding how the polymer used is itself manufactured
   e. Knowing the polymer physical properties
2. Understanding the engineering of microsphere or other delivery formulation
   a. What is the process used in fabricating the microspheres and how is each step controlled and characterized.
3. Understanding the stability of the active ingredient in the polymer matrix:
   a. During encapsulation
   b. Shelf life in microsphere
   c. After administration
   d. Verification of protein conformation in microsphere
   e. Protein–polymer interactions

CASE STUDIES IN FORMULATION DEVELOPMENT OF STERILE DOSAGE FORMS

These case studies are based on actual problems encountered by the author or problems that have been published elsewhere. The case study problems will be presented first with the idea that the reader will think about how to approach solving the problem. The answers are given starting on page 174.

Case Study 1

Stability samples of a ready-to-use sterile solution show visible evidence of extremely small particulates after storage for 1 month at ambient temperature. What are possible causes or sources of these particulates? For each cause or source, what can be done to eliminate the problem?

Case Study 2

A new small molecule you are to formulate contains a phenolic functional group that you know will be sensitive to oxidation. What should be done to minimize its oxidation potential in solution? What do you do if your new product shows discoloration during storage?

Case Study 3

During initial dosing of a phase I clinical study, patients experience a high incidence of thrombophlebitis after IV injection of your formulation. What can you do or suggest that will minimize or eliminate this problem?

Case Study 4

A new drug has been assigned to you that has significant water insolubility yet the clinic desires an IV dosage form. What can be done to provide a soluble IV dosage form?

Case Study 5

During initial freeze-drying experiments, you find that your protein undergoes some degradation including aggregation. What causes this during freeze-drying? What can you do to solve this problem?

Case Study 6

Your newly developed drug suspension tends to clog within a 24-gauge needle. Manufacturing also has reported some problems with filling needles not dosing properly. The product can be difficult to resuspend. How do you solve these problems?

Case Study 7

A new protein product is being developed. Marketing has requested that the product be a multidose product. You will be attending a project team meeting to discuss the issues with multidose formulations containing proteins. What would you share? What antimicrobial preservative choices do you have to develop a multidose parenteral solution or suspension that can be used globally?

On a separate note, discuss potential of using antimicrobial preservatives in single use injections that are aseptically processed.

Case Study 8

You are developing a new product containing a drug that must be freeze-dried. Freeze-drying will produce the amorphous form of the drug initially, but the drug will tend to crystallize over time after freeze-drying. What formulation and process approaches can you try to minimize the crystallization of the drug? What are some of the potential problems if the freeze-dried drug crystallizes over time? When is it preferable to have the drug in the crystalline form?

Case Study 9

Your product is a protein formulated with phosphate buffer (10 mM), sodium chloride (0.9%), EDTA (0.01%), and pH = 6.5. Normal storage is at refrigeration. At 3 months, visual inspection showed small (1–2 mm), gray-whitish particles at 25°C storage while no particles seen with the refrigerated vials. However, at 6 months, the refrigerated vials showed these same types of particles. What could have happened to this product? What methods would you use to support your conclusion(s) with data? What recommendations would you make to your management?

Case Study 10

You are the formulation development scientist of a protein that significantly self-associates in solution, adversely affecting in vivo bioavailability and efficacy. This protein does not have any significant chemical stability problems and does not bind to glass. What analytical techniques might be useful? What formulation approaches might inhibit aggregation?

Case Study 11

A drug product intended for IV bolus injection is formulated in aqueous solution containing a buffer, tonicity adjuster, and nonionic surface-active agent. The drug is stable to hydrolysis, but degrades by an oxidative mechanism at ambient temperature. Because of rapid loss of solubility at pH > 3 and its sensitivity to light and oxygen, the pH is adjusted to 2.5. It also loses some activity after 1 month when stored in Type I tubing glass containers. What kinds of problems would you anticipate is this drug product experiencing and what can be done to solve these problems?

Case Study 12

The manufacturer of a commercial product filled in a 1 mL glass syringe begins to receive complaints from the marketplace that the product contains particles or haze or "something floating." The product is a polypeptide formulated in water containing mannitol with pH adjusted in the range of 5.0 to 6.5 (no buffer). What could be the causes and possible solutions for this problem?

Case Study 13

How do you determine what kind and how much buffer component to incorporate into your final formulated product? What choices are available if preformulation API stability indicates that pH 5.5 to 6.3 is desirable? What about pH 6.8 to 7.6?

Case Study 14

An anesthetic drug is to be administered by IV bolus injection. It is very soluble at very low pH, but solubility rapidly declines at pH 3 and virtually insoluble at pH > 5. It is stable in solution to hydrolysis, but susceptible to oxidation at room temperature at pH > 3.5. The formulation consists of active, buffer, tonicity adjuster, nonionic surfactant, and WfI.

a. Phase I clinical trial was conducted with a dose of 10 mg/mL active. No problems reported. A Phase Ia clinical trial was conducted using a dose of 20 mg/mL, no changes in the formulation. There was a problem. What might this have been and what was the solution?
b. The filter integrity test (bubble point) after filtration was lower than the prefiltration integrity test. Why?
c. For long-term stability a freeze-dried product was developed. However, reports from the market complained of a haze appearing within the walls of the vials. Production personnel also had observed and reported a film or haze upon the interior door of the lyophilizer at the completion of the cycle. The shelf temperature of the secondary drying cycle is +58°C for 11 hours at a pressure of 20 μm. Review of the thermogram from DSC shows an exotherm at +20°C. What caused the haze and what was the solution?

ANSWERS AND SOME DISCUSSION OF CASE STUDIES

Case Study 1

What is the first thing to do? Isolate and attempt to identify the composition of the particulates. This is much easier stated than done. However, if the particle can be identified, corrective and preventative actions can much more easily be enacted.

Table 11-4 lists possible causes and potential elimination or, at least, minimization of the source of particles found in injectable solutions:

Case Study 2

It has been emphasized in this text that overcoming oxidation problems involves more than developing an optimal formulation. This problem must be attacked simultaneously:

1. Evaluate packaging choices. Does the rubber closure used have an acceptable air transmission coefficient (to discuss with the rubber closure vendor)? If plastic packaging is used, what are its oxygen transmission properties? How tight is the container closure integrity? Is there a need to use better secondary packaging? Would the use of oxygen scavengers in the secondary package help?
2. Eliminate all sources of oxygen exposure during manufacturing:
   a. Inert gassing of water
   b. Inert gas flushing of final solution and container prior to filling
   c. Inert gas overlay in container prior to sealing
3. Other manufacturing precautions
   a. Use lowest temperatures possible in room and solution preparation
   b. Protect from light
4. Use as low a pH as possible
5. Use of antioxidants and chelating agents
6. Use excipients free from peroxide contamination, metal impurities

Table 11-4 Possible Causes and Solutions of Particles in Solution

Cause/source of particles in solution	Minimization/elimination of source of particles
Degradation product	Understand mechanism of drug degradation and use approaches to minimize such degradation
Change in pH causing particle formation	Increase buffer capacity although beware of salting out effect or potential for injection pain if pH not physiologic. Also could try a different buffer system
Low solubility of salt form	Try different salt Increase dilatation Use a co-solvent, surfactant, or a cyclodextrin
Exposure to lower temperature caused drug to irreversibly fall out of solution	Use better temperature controls and monitors
Interaction with container/closure system	Use coated rubber closure Use treated Type I glass
Extractables and leachables from any kind of product contact surface (manufacturing equipment, packaging components)	Cleaning methods Impeccable handling and proper functioning of equipment, especially filling machines
Personnel	Proper hygiene, gowning, and aseptic practices and techniques
Complexation with formulation component	Need to do empirical studies to determine source of interaction
Impurity	Identify impurity, modify API manufacturing process
If a protein product, aggregation	Optimize formulation using surfactants Reduce sources of particles coming from contact surfaces by improving cleaning methods and adequate rinsing Minimize shear effects during processing and distribution

Case Study 3

Factors that can affect potential for phlebitis during/after injection include:

1. pH
2. Tonicity
3. Infusion rate
4. Concentration of infused drug
5. Inherent irritation properties of the drug molecule
6. Impurities in bulk drug substance
7. Precipitation at injection site due to loss of solubility
8. Extractable material from rubber closures
9. Particulate matter

The review article by Yalkowsky et al. (3) can be helpful in understanding how problems with pain and irritation upon injection can be tackled and solved.

Case Study 4

The approaches discussed in this chapter can be applied in solving this solubility problem

1. Use the optimal salt form of the API
2. Solution pH adjustment *in situ* or via modest reformulation
3. Use of co-solvent (PEG, PG, glycerin, alcohol)
4. Use of surface-active agents (polysorbate 80, Pluronics)
5. Use of complexing agents (Beta cyclodextrins)
6. Going to an alternative dosage form (emulsion, liposome, nanosuspension)

The article by Sweetana and Akers (1) can be a helpful reminder in dealing with solubility problems for drugs intended for IV administration.

Case Study 5

Freezing and drying will denature some proteins. Freezing will concentrate solutes such as salts, which will unfold some proteins. Drying may cause excessive loss of water within the structure of the protein causing distortion of the protein.

1. Use cryoprotectants (amino acids, polyols, sugars, PEG) that stabilize the native conformation of the protein by being excluded from the surface of the protein, thereby inducing "preferential hydration" of the protein.
2. Freeze as rapidly as possible in order to keep solute stabilizer in the amorphous state, for example, mannitol, glycine.
3. Use more than one cryoprotectant in order to inhibit one cryoprotectant from crystallizing

Case Study 6

1. Mill or micronize the drug. This will require sterile milling and the issue of validation of sterility assurance of the particle size reduction operation(s).
2. Modify the method(s) of drug crystallization in order to produce a smaller particle size distribution of the drug
3. Use surface-active agents to help disperse the drug
4. Develop a syringe/needle handling technique that will minimize bridging of the suspension particles in the needle.
5. Use a larger gauge needle, for example, 21-G (although impossible if infants are included in indication for use of drug product)
6. Since the suspension is likely to have some elegance problems if it is clogging needles, use silicon to coat the glass walls and a slotted stopper to minimize drug deposits.

The article by Akers et al. (4) can be helpful in dealing with suspension formulation problems.

Case Study 7

Discussion points to share with the team would include:

1. Examples of incompatibilities of proteins and preservatives
2. Lack of ideal preservative; many problems including odor, stability, solubility, purity, effectiveness, interaction with closures, and tubing
3. Problems in passing EP preservative challenge test
4. Limitations in choice of preservatives and maximum dose

1. Phenol	0.5%
2. Meta-cresol	0.3%
3. Benzyl alcohol	1.0%
4. Methylparaben	0.2%
5. Propylparaben	0.02%
6. Combinations of the above	
7. Chlorobutanol	0.5%
8. Thimerosal	0.02%
9. Benzalkonium chloride	0.01%
(for topical ophthalmics)	

In general, preservatives are not to be used for single-dose products. Only a couple of marketed examples and these are older products where a preservative is in the formulation of a single dose product.

See Ref. (5) for further review.

Case Study 8

1. Use high ratios of excipient to drug to cause preferential crystallization of excipient while keeping drug in amorphous state
2. Primary drying step conducting in very cold ($< -30°C$) temperature conditions
3. Storage of finished product in refrigerator

Crystallization over time can create problems both in solubilization of reconstituted freeze-dried product (rate and extent) and/or in solid state stability.

Crystalline form is usually preferable as long as crystalline form is in a sufficient soluble and stable. Crystalline form is usually more stable than amorphous form. Changes over storage time are less likely to occur. Crystalline mannitol preferred because amorphous mannitol will want to crystallize over time anyway and crystallization process can cause vial breakage.

Crystalline powder is usually more elegant and the collapse temperature is usually higher, thus shortening freeze-dry cycles.

Case Study 9 (6)

Potential causes include:

- protein denaturation, precipitation
- bacterial contamination
- leachates from glass vial
- silicon interactions
- impurities in packaging or raw materials

Refer to preformulation studies:

- particulates should be known via elemental analysis using SEM and energy–dispersive X-ray technology
- known routes of degradation

Use SDS–PAGE to investigate potential unfolding.

Use ICP–MS and Electron Microscopy to determine possible inorganic leachates.

Case Study 10

Summary of Analytical Techniques (7–9)

1. Measuring changes in secondary structure: Far UV CD, FTIR, Raman
2. Measuring changes in tertiary structure: Intrinsic fluorescence spectroscopy, near UV CD, near UV absorbance, Raman
3. Measuring changes in quaternary structure: SEC, light scattering, sedimentation

Measuring Aggregation:

For Aggregate Size up to 0.05 μm

1. Size-exclusion chromatography—Can be a stability-indicating assay for soluble aggregation, not for insoluble aggregation.
2. Analytical ultracentrifugation—Can measure large changes in molecular weight and can distinguish between reversible and irreversible aggregation.
3. Field flow fractionation—Characterizes both soluble and insoluble aggregation. Can be used with MALLS detector to determine molecular weight.
4. Electrophoresis (CE, gel, CE-SDS)—More useful for protein purity and quantity, especially at extremely low concentrations.
5. Light scattering (DLS, MALLS, SEC-MALLS)– Measures diameter and molecular weight of protein in solution, but cannot quantify insoluble aggregates.
6. Spectroscopy:
    a. Absorption—If protein contains aromatic amino acids (Phe, Try, Trp). Can detect unfolding since aromatic amino acids are folded into the protein interior.
    b. Fluorescence—Aromatic amino acids (tryptophan) also can fluoresce (excited at one wavelength; fluoresce at another). Reveals information about microenvironment generated by folding of protein.
    c. Circular Dichroism—Measures alterations in side chain absorption due to alpha helices and β sheets. β sheets indicative of aggregation. Changes in secondary structure changes CD shape and signal intensity.

   d. IR Spectroscopy—Normal IR; also FTIR for evaluation of solid state protein structure (e.g., bulk lots)
7. Mass spectroscopy
8. Calorimetry

For Aggregate Size Greater Than 0.5 μm

1. Light obscuration
2. Turbidity
3. Microscopy
4. Visual testing

Formulation and Other Approaches

1. Surface-active agents: Polysorbates 20 and 80; Pluronics
2. Reduce ionic strength: Reduce buffer concentration
3. Co-solvents: Glycerol, ethanol (usually too high concentrations are required)
4. Reduce temperature
5. Optimize the protein structure: LysPro Insulin, an example

Case Study 11

Potential problems:	Potential solutions:
1. Peroxide impurities in surfactant	1. Work with vendor to use peroxide-free surfactant raw material
2. Leachates from rubber	2. Use coated rubber closures
3. Delamination of glass	3. Consider coated glass (e.g., Schott Plus)
4. Activity loss may be due to heavy metal contaminants	4. See if adding chelating agent helps to stabilize the drug against metal-catalyzed oxidation
5. Too much oxygen in headspace solution and/or water	5. Overlay with inert gas, purge in with inert gas; add antioxidant (Sodium metabisulfite best at low pH)

Case Study 12

The particles or haze or whatever is floating needs to be identified first. If it is nonproteinaceous, then it is likely a particle coming from the glass or rubber components of the syringe. It could also be some excess silicon that looks like a foreign particle.

If the particles are proteinaceous, then the protein is self-aggregating due to one or more of the following:

1. Air-liquid interfacial interactions: Need a surface-active agent
2. Mechanical shear during manufacturing and/or distribution and use—Need a surface-active agent, gentler handling
3. Interactions with siliconon glass—reduce silicon amount, improve the bonding and siliconization process
4. pH shift might cause protein to be close to pI—need a buffer

Case Study 13

1. Must know pH-solubility and pH-stability profile and choose a target pH range that provides overall the best solubility and stability properties for the drug product.
2. Use the literature to be reminded of what buffer systems have dissociation constants (pK_a) that fall well within the desired pH range, preferably as close to the mid-point as possible. Also search the literature for what buffer systems have been approved in marketed products within the same pH region.

3. The amount of buffer component can be determined by asking the following questions:
   a. How much does the solution pH drift in solution without a buffer?
   b. What is the route of product administration?
   c. What is generally known about concentration levels of preferred buffer?
   d. What are concentrations of similar buffer components in other products?
4. Start with the lowest amount of buffer that you believe will control pH, then perform short-term accelerated stability studies to determine if concentration needs to be increased.

Case Study 14

a. Increasing concentration without modifying ratios with stabilizing excipients could lead to stability issues, both chemical and physical. Control of pH can be affected by changing the ratio of drug to buffer and hydrophobic surface interactions can be affected by changing the ratio of drug to surfactant.
b. Typically, the presence of a surfactant or any amphiphilic compound (e.g., proteins) will affect the surface tension property of filter surfaces.
c. Some (perhaps many) drugs and solutes can exist both as amorphous or crystalline solids. Conversion of amorphous to crystalline solid state structure can lead to changes in rate and extent of drug solubility. Also the combination of relatively high temperature and very low pressure during secondary drying could lead to rapid crystallization and even degradation.

REFERENCES

1. Sweetana S and Akers MJ. Solubility principles and practices for parenteral drug dosage form development. PDA J Pharm Sci Technol 1996; 50:330–342.
2. Kumar L, Amin A, Bansal AK. Salt selection in drug development. Pharm Tech 2008; 32:128–146.
3. Yalkowsky SH, Krzyzaniak JF, Ward GH. Formulation-related problems associated with intravenous drug delivery. J Pharm Sci 1998; 87:787–796.
4. Akers MJ, Fites AL, Robison RL. Formulation Design and Development of Parenteral Suspensions," J Parenter Sci Technol 1987; 41:88–96.
5. Akers MJ, Vasudevan V, Stickelmeyer M. Protein dosage form development. In Nail SL and Akers MJ, eds. Development and Manufacture of Protein Pharmaceuticals. New York: Kluwer Academic/Plenum, 2002:80–84.
6. Scatt C. Formulation case studies: a tutorial. BioPharm 2001; 14:34–35.
7. Nguyen LT, Wiencek JM, Kirsch LE. Characterization methods for the physical stability of biopharmaceuticals. PDA J Pharm Sci Technol 2003; 57:429–445.
8. Chan CP. Biochemical and biophysical methods currently used for therapeutic protein development. Amer Pharma Rev 2008; 11:48–54.
9. Cordoba-Rodriquez RV. Aggregates in MAbs and recombinant therapeutic proteins: A regulatory perspective. BioPharm Int 2008; 21:44–53.

12 Overview of sterile product manufacturing

All pharmaceutical manufacturing operations are complicated, requiring utmost organization and control to ensure that every dosage form produced meets all quality attributes and specifications. Sterile pharmaceutical manufacturing has the added complication of assuring that all dosage forms produced are free from microbial contamination, endotoxin contamination, and visible particulate matter. Not only do all sterile manufacturing operations require mechanical excellence, but also require absolute cleanliness, sanitization, and sterilization of all product-contact components. Unit processes involved in the manufacturing of sterile dosage forms include compounding and mixing, filtration, filling, terminal sterilization (when possible), lyophilization (freeze-drying), closing and sealing, sorting and inspection, labeling, and final packaging for distribution (Table 12-1). Each of these unit processes will be the subject of subsequent chapters with this chapter presenting a general overview.

MANUFACTURING PROCEDURES

The processes required for preparing sterile products constitute a series of events initiated with the procurement of approved raw materials (drugs, excipients, vehicles, etc.) and primary packaging components (containers, closures, etc.), and ending with the sterile product sealed in its dispensing package (Fig. 12-1 for solution and lyophilized products). Each step in the process must be controlled very carefully so that the product will have its required quality. To ensure the latter, each process should be validated to be sure that it is accomplishing what it is intended to do. For example, any sterilization process must be validated by producing data showing that it effectively kills resistant forms of microorganisms (or removes them with filtration processes); or, a cleaning process for rubber closures should provide evidence that it is cleaning closures to the required level of cleanliness; or a filling process that repeatedly delivers the correct fill volume per container. The validation of processes requires extensive and intensive effort to be successful and is an integral part of current good manufacturing practices (cGMP) requirements.

TYPES OF PROCESSES

The preparation of sterile products may be categorized as small-scale dispensing, usually one unit at a time, or large-scale manufacturing, in which hundreds of thousands of units may constitute one lot of product. Small-scale processing involves early phase clinical batches, although there are some commercial products whose batch sizes are in the hundreds rather than tens of thousands. Small-scale processing also occurs in some hospital pharmacies who need to follow the requirements of the United States Pharmacopeia (USP) general chapters <797> and <1206>. Often small-scale processes use presterilized components and equipment to simplify all steps and increase the assurance of sterility.

Large-scale processing will be the focus of this chapter and other chapters related to manufacturing. Such processing begins with nonsterile components in large (thousands of square feet) facilities with state-of-the-art equipment, instrumentation, lines, and all other required technology to produce sterile products at massive numbers of unit. Underlying all sterile product production, regardless of batch size, is strict adherence to cGMP principles. The following are examples of cGMP compliance and include (Code of Federal Regulation reference(s) in parenthesis):

- Ensuring that the personnel responsible for assigned duties are capable and qualified to perform them. GMP emphasizes the need for personnel to have a combination of education, experience, and training to do their jobs (CFR 211.25).
- Ensuring that ingredients used in compounding the product have the required identity, quality, and purity (CFR 211.80, 211.84, 211.86).

Table 12-1 Major Sterile Manufacturing Topics

- GMP basics
- Overview of processing flow
- Facilities
- Environmental monitoring
- Equipment
- Packaging/siliconization
- Cleaning and sanitization
- Sterilization
- Depyrogenation
- Raw materials
- Water for Injection
- Documentation
- Personnel
- Gowning
- Compounding/mixing
- Filtration
- Filling/stoppering
- Barrier isolator technology
- Lyophilization
- Process validation
- Closing/sealing
- Finishing/inspection
- QC tests and release
- New/advancing technologies

- Validating critical processes to be sure that the equipment used and the processes followed will ensure that the finished product will have the qualities expected (CFR 211.63, 211.65, 211.67, 211.68, 211.182).
- Maintaining a production environment suitable for performing the critical processes required, addressing such matters as orderliness, cleanliness, asepsis, and avoidance of cross contamination (CFR 211.42, 211.46, 211.56, 211.113).
- Confirming through adequate quality-control procedures that the finished products have the required potency, purity, and quality (CFR 211.160, 211.165, 211.167).
- Establishing through appropriate stability evaluation that the drug products will retain their intended potency, purity, and quality until the established expiration date (CFR 211.137, 211.166).
- Ensuring that processes always are carried out in accordance with established written procedures (CFR 211.100; many segments of CFR 211 make the statement, "There shall be written procedures. . . ").
- Providing adequate conditions and procedures for the prevention of mix-ups or cross-contamination (CFR 211.80, 211.105, 211.113, 211.176).
- Establishing adequate procedures, with supporting documentation, for investigating, correcting, and preventing failures or problems in production or quality control (CFR 211.100, 211.192, 211.194).
- Providing adequate separation of quality control responsibilities from those of production to ensure independent decision making (CFR 211.22).
- Establishment of time limits for each phase of production (CFR 211.111).
- Requirements for the ability to reprocess unit operations (CFR 211.115).
- Requirements for label control (CFR 211.122, 211.125, 211.130).
- Documentation requirements for master production records and batch records (CFR 211.186).

The pursuit of cGMP is an ongoing effort that must blend with new technological developments and new understanding of existing principles. Because of the extreme importance of quality in health care of the public, the United States Congress has given the responsibility of regulatory scrutiny over the manufacture and distribution of drug products to the Food and Drug Administration (FDA). Other nations have done similar actions with their own FDA-like

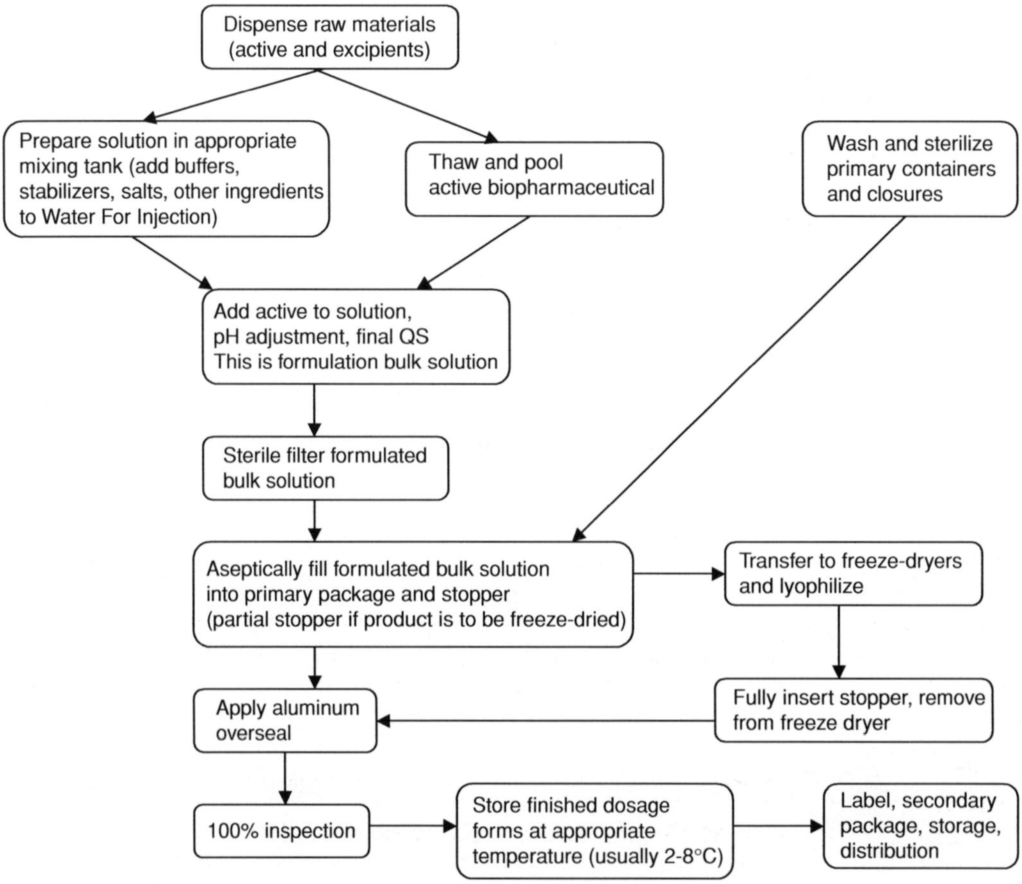

Figure 12-1 Schematic overview of processing solution and freeze-dried biopharmaceutical dosage forms.

overseers of pharmaceutical manufacturing. Therefore, the operations of the pharmaceutical industry are subject to the oversight of the FDA and, with respect to manufacturing practices, to the application of the cGMPs. These regulations are discussed more fully in chapters 25 and 26.

In concert with the pursuit of cGMPs, the pharmaceutical industry has shown initiative and innovation in the extensive technological development and improvement in quality, safety, and effectiveness of parenteral dosage forms in recent years. Examples include developments in the following:

- modular facility design and construction
- container and closure cleaning, siliconization (if applicable) and sterilization
- disposable technologies
- sterilization technologies
- filling technologies
- aseptic processing technology including barrier isolator technology
- aseptic connections and sampling
- freeze-drying technologies including automated loading and unloading
- control of particulate matter
- automation in weight checking, inspection technologies, and labeling and finishing operations.

SCHEDULING

Scheduling functions are the key deliverables of the planning role. Scheduling is the function of coordinating all of the logistical issues required to organize an efficient period of time to prepare and produce sterile products. Scheduling involves proper coordination of facilities, lines, equipment maintenance and repairs, raw materials, and human resources. Scheduling must be flexible to work quickly and efficiently around unplanned events such as testing failures, facility problems, change in forecasts, and production delays.

BATCH RECORD AND OTHER DOCUMENTATION

The batch record is the most important document in sterile product manufacturing, although many other documents and records are part of the batch record. The batch record is the complete record of the manufacture and control, distribution of a single batch of a product.

It is critical and essential that the documentation of a batch record is accurate, complete, readily followed, and reviewed. Table 12-2 lists the essential information that must be contained in a batch record.

Other key pieces of documentation include process and material records, deviation reports, validation records, complaint records, and all standard operation procedures (SOPs) associated with the batch production.

Good documentation practices (GDP) are activities that require that all raw data, written entries, and records to be accurate, legible, traceable (defensible), reproducible, complete, and verified. All entries must be signed and dated by the individual who made the entry on the date that the entry was made. All records must contain proper identification on each page. When mistakes are made, there are proper practices to follow, such as never erasing a mistake, single-line cross-outs, adding the correct entry with one's initials, date, and reason for the correction. All original records must be archived. There is the classic FDA position that if something is not written down or written correctly, it was not done. Documentation failures are perhaps the most frequently cited observations during GMP regulatory inspections by FDA and other government inspections. Both good training practices and good employee attitudes are key to following GDP.

Electronic batch records have replaced paper records at many pharmaceutical companies. GMP regulations require that electronic batch records must substantiate that every significant step (e.g., batch dates, identity of equipment/facilities used, components, in-process and laboratory control results, all labeling records, sampling, personnel and supervision, verification of all steps, etc.) in the production, packaging, and hold of each batch of drug product was accomplished according to GMP. The FDA began to accept electronic batch records after the 1997 publication of 21CFR Part 11—Electronic Records and Electronic Signatures—that define criteria under which electronic records and electronic signatures are considered to be trustworthy and equivalent to paper documentation. Updates and/or revisions of the original 1997

Table 12-2 Example of Batch Record Information

- Product identification
- Document identification
- Company name
- Dates of manufacturing
- Step-by-step list of unit operations and tests
- Specifications/limits of each step
- Data for blanks to be filled in during the process
- Formulation information—names, quantities, ID numbers
- Control numbers for each component with quality approval
- Start and completion times for each operation
- Chemical weight checks; QA counter checks
- Identity of all processing equipment
- Process details
- Deviation investigations/reports
- Labeling requirements
- In process sampling procedures, test requirements
- Final test results
- Material accountability
- Signatures

Part 11 requirements have been expected, but at the time of this publication, still have not been published.

FACILITIES AND EQUIPMENT PREPARATION

Sterile product facilities need to be constructed of materials that are smooth, cleanable, and resistant to physical and chemical deterioration. Elaboration of facility construction, requirements, and controls are presented in chapter 14.

There are a significant number of equipment items that are involved in sterile product manufacturing. Examples of typical equipment items used in sterile product manufacturing are listed in Table 12-3. For each of these types of equipment, accurate documentation much be maintained that includes maintenance, cleaning, sterilization, and usage. Advances in manufacturing equipment must be monitored by appropriate personnel so that state-of-the-art equipment is used, particularly to enhance quality of the resulting product.

GENERAL MANUFACTURING PROCESS

The preparation of a parenteral product includes procurement of all components, applying all appropriate manufacturing process to compound, mix, fill, and close the unit dosage form, packing and labeling of the dosage forms, and controlling the quality of the product at all steps of the process. The general flow of the manufacturing process along with the environmental quality of each step of the process is given in Figure 12-2.

Procurement encompasses selecting and testing according to specifications of the raw-material ingredients and the containers and closures for the primary and secondary packages. Microbiological purity, in the form of bioburden and endotoxin levels, have become standard requirements for raw materials. Raw material providers need to be audited by sterile product manufacturers using their products. It is acceptable for sterile product manufacturers to accept raw material certificates of analysis without the need to repeat any specification tests except for an identity test upon receipt of the material.

Processing includes cleaning containers and equipment to validated specifications, compounding the solution (or other dosage form), filtering the solution, sanitizing or sterilizing the containers and equipment, filling measured quantities of product into the sterile containers, stoppering (either completely or partially for products to be freeze-dried), freeze-drying, terminal sterilization if possible, and final sealing of the final primary container.

Table 12-3 List of Production Equipment for Production of Sterile Products

- Washing equipment for packaging components
    - Glass container washer
    - Rubber closure washer (plus siliconizer, depyrogenator, sterilizer)
    - Plastic tubing washer
- Mixing tanks
- Laminar air flow units with HEPA filters
- Dry heat ovens and tunnel sterilizers
- Steam sterilizers
- Clean-in-place and steam-in-place systems for large equipment
- Filter equipment—liquid and gas
- Storage tanks
- Filling equipment—ampoules, vials, syringes, cartridges, bags, bottles
- Stoppering equipment
- Sealing equipment
- Freeze-dryers
- Barrier isolators
- Packaging and labeling equipment
- Particle detectors
- Environmental monitoring systems
- Homogenizers/mills/micronizers

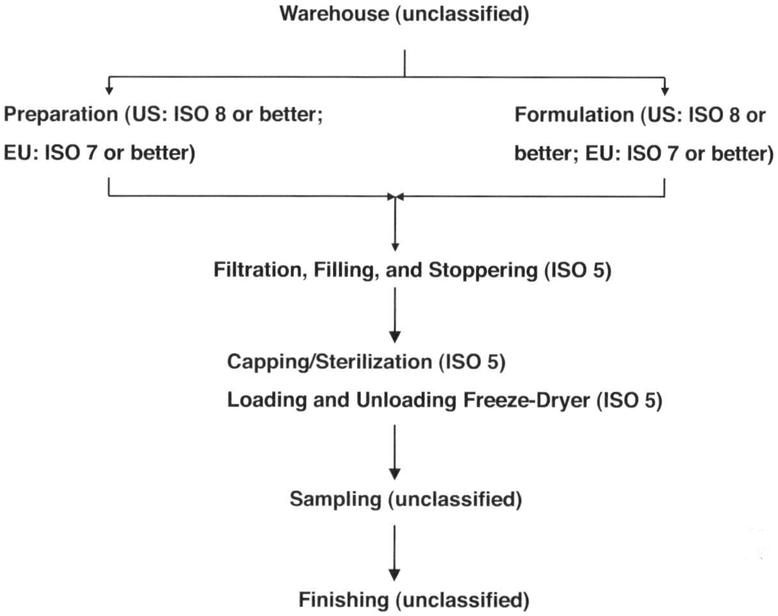

Figure 12-2 Flow of sterile manufacturing (Air classification of each area).

Packaging normally consists of the labeled, cartoned, filled, and sealed primary containers. The control of quality begins with the incoming supplies, being sure that specifications are met. Careful control of labels is vitally important as errors in labeling can be dangerous for the consumer. Each step of the process involves checks and tests to be sure that the required specifications at the respective step are being met. Labeling and final packaging operations are becoming more automated.

The quality control unit is responsible for reviewing the batch history and performing the release testing required to clear the product for shipment to users. A common FDA citation for potential violation of cGMP is the lack of oversight by the quality control unit in batch testing and review and approval of results.

PRODUCT PREPARATION

Preparation of Components

Components of sterile products include the active ingredient, formulation additives, vehicle(s), and the primary container and closure. Establishing specifications to ensure the quality of each of these components is essential. For sterile preparations, two of the most critical specifications are microbial and endotoxin levels in each raw material. All cleaning operations for all components must be validated that the cleaning processes remove all extraneous chemical material and particulate matter. Cleaning validation of equipment is subject of many publications, seminars, and regulatory documents and had not been covered in this book other than to indicate it.

The most stringent chemical-purity requirements normally will be encountered with aqueous solutions, particularly if the product is to be sterilized at an elevated temperature where reaction rates will be accelerated greatly. Dry preparations pose relatively few reaction problems but may require definitive physical specifications for ingredients that must have certain solution or dispersion characteristics when a vehicle is added.

Compounding (or Formulation)

Compounding is the preparation of the product to be filled into the primary container. Facility classification for compounding typically is Grade D (ISO 8) or better. The development of the compounding procedure occurs in the formulation development laboratory initially. The development scientist determines the optimal order of addition of ingredients to ensure complete solubilization (or dispersion) of components with minimal loss (degradation) of active ingredient. For example, buffer components might be dissolved first in order to obtain the pH range optimal for drug solubilization and/or stability.

A master formula would have been developed and be on file. Each batch formula sheet should be prepared from the master and confirmed for accuracy. All measurements of quantities should be made as accurately as possible and checked by a second qualified person. Frequently, formula documents are computer-generated, and the measurements of quantities of ingredients are computer controlled. Although most liquid preparations are dispensed by volume, they are prepared by weight, since weighings can be performed more accurately than volume measurements and no consideration needs be given to the temperature.

Care must be taken that equipment is not wet enough to dilute the product significantly or, in the case of anhydrous products, to cause a physical incompatibility. The order of mixing of ingredients may affect the product significantly, particularly those of large volume, where attaining homogeneity requires considerable mixing time. For example, the adjustment of pH by the addition of an acid, even though diluted, may cause excessive local reduction in the pH of the product so that adverse effects are produced before the acid can be dispersed throughout the entire volume of product.

Parenteral dispersions, including colloids, emulsions, and suspensions, provide particular problems. In addition to the problems of achieving and maintaining proper reduction in particle size under aseptic conditions, the dispersion must be kept in a uniform state of suspension throughout the preparative, transfer, and subdividing operations.

Proteinaceous solutions are especially reactive when preparing these products. Proteins are usually extremely sensitive to many environmental and processing conditions that they are exposed to during production such as temperature, mixing time and speed, order of addition of formulation components, pH adjustment and control, and contact time with various surfaces such as filters and tubing. Development studies must include evaluation of manufacturing conditions in order to minimize adverse effects of the process on the activity of the protein.

Cold temperature control can be accomplished to aid in the minimization of protein instability during manufacturing. Compounding and mixing can be performed under cold room and cold tank conditions. Temperature sensors can be associated with filling needles that enable the filling machine to reject vials that do not meet temperature acceptance criteria. For lyophilized proteins, cold temperature control can be maintained on the accumulation table or transfer bed prior to freeze-dryer loading and after unloading from the dryer prior to capping.

The formulation of a stable product is of paramount importance. Certain aspects of this are mentioned in the discussion of components of the product. It should be mentioned here, however, that the thermal sterilization of parenteral products increases the possibility of chemical reactions. Such reactions may progress to completion during the period of elevated temperature in the autoclave or be initiated at this time but continue during subsequent storage. The assurance of attaining product stability requires a high order of pharmaceutical knowledge and responsibility.

Typical compounding and mixing problems (Table 12-4) may include incomplete dissolution of sparingly soluble components, excess foaming during the mixing procedure, problems in pH adjustment (over or under shooting pH target, having to go back and forth), adding components too quickly before the water for the injection solvent has cooled, and mistakes in quantities added or the order of addition.

Preparation of Containers and Closures

Containers and closures are in prolonged, intimate contact with the product and may release substances into, or remove ingredients from, the product. Rubber closures are especially problematic (sorption, leachables, air and moisture transmission properties) if not properly evaluated for its compatibility with the final product. Assessment and selection of containers and closures are essential for final product formulation, to ensure that the product retains its purity, potency, and quality during the intimate contact with the container throughout its shelf-life. Administration devices (syringes, tubing, transfer sets) that come in contact with the product should be

Table 12-4 Common Compounding Problems in Sterile Product Manufacturing

- Mistakes in calculations of active ingredient
  - E.g., calculating actual amount of free acid or free base of an electrolytic drug where label claim based acid or base, not salt
  - E.g., calculating correct anhydrous amount of active from hydrated or solvated forms
- Incorrect order of addition of components
- Rate of addition of components too fast, incomplete dissolution
- Not allowing sufficient time or mixing force for complete dissolution of all components
- Problems with pH adjustment
  - Over-shooting or under-shooting pH
  - Increasing ionic strength with excessive addition of strong acid and/or base
- Errors in final QS (Quantity Sufficient) step, making final solution too dilute or too concentrated
- Excessive sampling having effect on final volume
- Errors in sampling or actual measurement of in-process samples
- Components compounded separately, then combined with final product where mistakes are made in calculations, volumes combined, and improper mixing conditions
- Introduction of contamination during aseptic addition of formulation components
- Excessive foaming due to excessive shear force or sloppy technique in component addition
- Errors in weighing of ingredients
- "Down times" excessive, time limits may be exceeded
- Maintaining dose homogeneity during and after compounding for dispersed systems

assessed and selected with the same care as are containers and closures, even though the contact period is usually brief.

All cleaning processes of all containers and closures (and equipment) must be validated for removal of all extraneous chemical materials and all particulate matter. Foreign particulates in final products in majority of the instances originate from the containers and closures. Cleaning validation is not covered in this book because there are many resources for this subject. Suffice to indicate, however, that validation of cleaning processes continues to be a major focus in regulatory inspections in the pharmaceutical industry and many manufacturing companies continue to receive 483 observations and Warning Letters for their problems and failures to prove adequate validation of cleaning processes. More coverage of cleaning of containers and closures can be found in chapter 13.

Glass

Glass containers are cleaned using Water for Injection. Glass sterilization and depyrogenation are accomplished using dry heat, usually with tunnel sterilizers where the temperatures reach 300°C, which is necessary for depyrogenation (see chap. 13). Glass syringes and cartridges need to be siliconized with the siliconization occurring before sterilization procedures.

The following is a typical procedure: Vials are received from the warehouse and part numbers verified as required on the master batch record. Vials are wrapped in shrink-wrap to minimize particulate matter. Vials are washed (actually rinsed, there is typically no detergent used) using washing equipment discussed in chapter 13. After rinsing, the wet vials are placed either in a dry heat oven (e.g., Despatch) or on a conveyor line (e.g., Strunck) for sterilization and depyrogenation.

Generally, glass syringes are prewashed and presterilized and depyrogenated by the manufacturer and come in "tubs," (e.g., Hypak™, see Fig. 7-7) that are ready for filling in the Class 100 clean area. For glass cartridges, the cartridges are loaded onto a conveyor system where they are rinsed, siliconized, sterilized, and depyrogenated, then filled with the product all on the same preparation and filling equipment.

Rubber

Rubber closures are cleaned and depyrogenated by rinsing with copious amounts of Water for Injection. Sterilization of rubber occurs by steam sterilization. Rubber must be "slippery" to move easily during production. Modern rubber formulations contain polymer surfaces that do

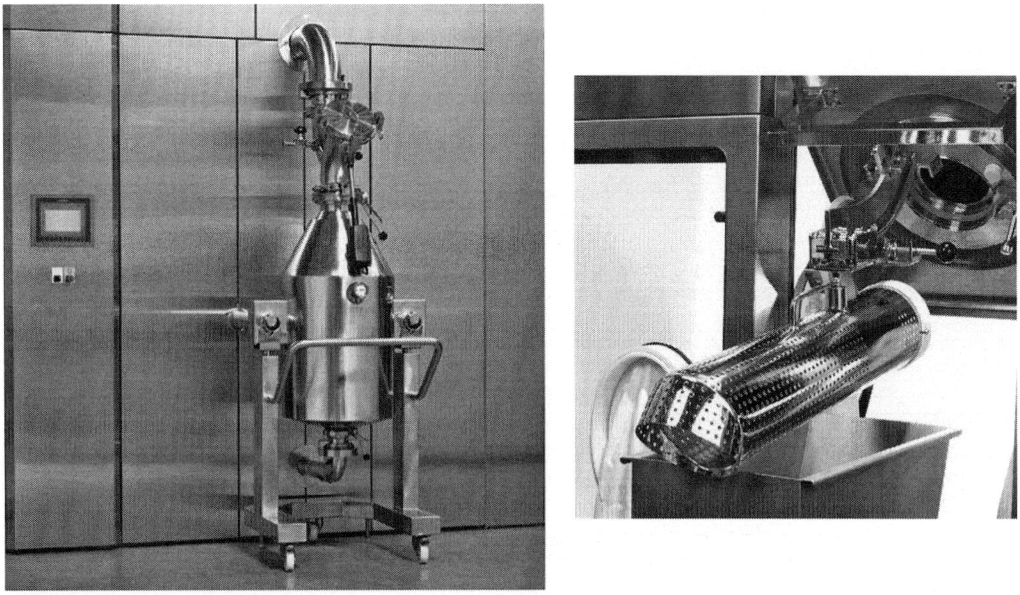

Figure 12-3 Examples of rubber stopper preparation systems that clean, siliconize, and depyrogenate stoppers. *Source*: Courtesy of Getinge, Inc.

not require siliconization. However, if siliconization is required, like with glass, it is done prior to sterilization.

Examples of stopper preparation equipment are DCI, Getinge (Fig. 12-3), and Icos. These machines clean, siliconize, and depyrogenate stoppers within the same unit and then the stoppers are batched and sterilized in an autoclave.

Plastic

Plastic packaging (as well as other plastic components such as filters and tubing) are cleaned and depyrogenated by rinsing with Water for Injection. Plastic containers usually are sterilized

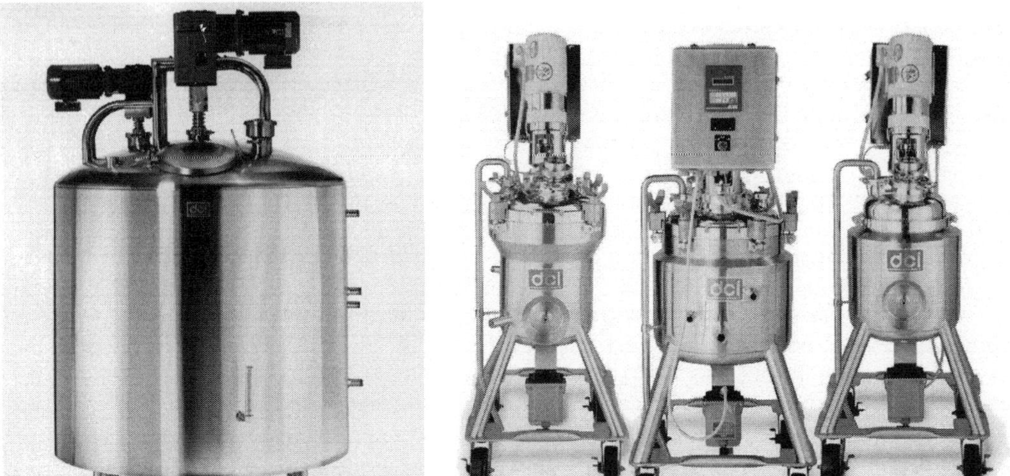

Figure 12-4 Examples of electropolished mixing tanks. *Source*: Courtesy of DCI, Inc.

by gas or radiation sterilization. Plastic components used in manufacturing are sterilized by steam sterilization with depyrogenation accomplished by prerinsing.

Siliconization

Silicone historically has been used to coat rubber closures in order to provide sufficient lubrication for the closures to feed and flow readily in high-speed filling equipment. Silicone application to syringes and cartridges also has been necessary for facile movement of a plunger rod with a rubber tip through the barrel of the container.

The amount of silicone applied per rubber stopper depends on the formulation size, weight, and design of stopper. Typically, the formulas used are established by trial and error. One example is that for 60000 13-mm lyophilization stoppers being processed in a Getinge stopper processor, with an average stopper weight of 630 mg, 31 mL of silicone oil is used[1]. Typically, the larger the stopper, the smaller the amount of silicone applied per stopper. For example, 13-mm liquid stoppers will contain a target of 0.85% silicone, 20-mm liquid stoppers will contain a target of 0.1% silicone, and 28-mm liquid stoppers will contain a target of 0.05% silicone. However, these are examples and not standards so the amount of silicone applied per stopper load will vary according to manufacturer, stopper processor, and filling speed of product fill.

Despite the ubiquitous usage of silicone in parenteral packaging preparation, it is surprising that very little has been published on the subject. While silicone coating on closures and syringe/cartridge barrels certainly offers significant advantages, there are many disadvantages with the use of silicone. Among these are the following:

1. Cleanability
2. Balance between too much and too little silicone applied
3. Potential incompatibilities with biomolecules
4. Viewed as a particle by electronic particle measuring devices

Mixing

Effective mixing must assure that the entire solution is thoroughly mixed and that there are no areas of stagnation where mixing is minimal or none. Mixing procedures are relatively straight forward for readily soluble components, but much more of a challenge for poorly soluble or slow-to-dissolve components and for biopharmaceutical active ingredients sensitive to the effects of mixing shear. Excessive foaming or entrapped air should be avoided, as denaturation at the air–liquid interface is possible. Precautions in mixing must be taken to prevent foaming. For suspension mixing and re-circulation during filling, a balance must be established between adequate mixing to achieve suspension homogeneity without impacting particle size distribution.

Primary mixing parameters to be controlled are shear rates (rpms), time, and temperature. Electropolished mixing tanks are available in different sizes and shapes with volumes ranging from roughly 100 L to 2000 L (Fig. 12-4). Mixing equipment must be designed to be cleaned-in-place (CIP) and sterilized-in-place (SIP). There should be no retention of liquid when the mixing tank is emptied and must have no "dead" areas or crevices. Materials of construction must be product compatible and corrosion resistant.

Mixing mechanisms include shear (propellers, blades, even magnetic stirring bars), diffusive, and vibratory. High shear mixers (e.g., Ross) are used for dissolving "hard-to-wet" components (e.g., polymers like carboxymethylcellulose). Bottom mount tank mixers (e.g., NovAseptic®, Fig. 12-5), are good general mixing systems where the blade type and configuration can dictate whether mixing includes vortexing. Suspension formulations where compounding and mixing must be aseptic use Rütten magnetic mixers. Rütten also produces the Vibromixer®, an intensive but gentle mixer applicable for shear-sensitive formulations. Mixing systems have now become disposable (e.g., ATMI LifeSciences) based on a single-use mixing bag containing a bottom mounted disposable magnetic impeller on a disposable bearing.

[1] These and other examples used in this book do not necessarily reflect what is true at Baxter BioPharma Solutions, but represent a general view of what is true in the entire sterile products manufacturing industry.

Figure 12-5 NovAseptic® GMP mixer. *Source*: Courtesy of Millipore Corporation.

In-process Testing

In-process controls are those operations within a manufacturing process where a critical parameter is controlled to a proven acceptable range or an analytical assay is conducted to determine proper interim conditions prior to progressing onto the next step. Examples of possible in-process controls are drug potency, pH, clarity, appearance, bioburden, and filter integrity. For suspension processes additional process controls may include order and rate of addition of components, location of addition, temperature, heat gain/loss (rate and overall time), agitation (type, rate, intensity, and duration), particle size control, suspension homogeneity, and morphology.

All in-process controls need to be supported by studies designed to establish appropriate specifications that will ensure batch-to-batch reproducibility. The order, rate, and location of additions and temperature control can be extremely important, especially for producing the desired crystalline form of a peptide or protein. Heat gain or loss can result in denaturation or influence particle generation if strict controls are not in place. Agitation is another critical parameter requiring precise control when dealing with peptide or protein suspensions as just described.

As with any finished dosage form, the product specifications must define key attributes of the preparation that ensure safety, identity, strength, purity, and quality. Besides potency, purity, and stability, suspension preparations will also require, at a minimum, specifications for content uniformity, particle morphology, and physical appearance. Specifications might also be established for other parameters such as particle size and distribution, sedimentation rate, and sedimentation volume, or rheological properties depending on the nature of the final suspension.

Filtration

Chapter 18 is devoted to filtration, but lets discuss a few introductory remarks here. The primary purposes of filtration are to clarify (remove all visible particulate matter) and to sterilize. Even if the final product is to be terminally sterilized, the solution is filtered through a sterilizing filter.

Filters are either flat disks or cartridge-type filters of different polymer compositions, sizes (diameters or surface areas), and porosities. Filters are selected based on three primary validations—microbial retention, product compatibility, and low extractables. Filter integrity during processing is determining via nondestructive integrity tests such as the bubble point test or the diffusion test.

Filling/Closing/Stoppering/Sealing

Chapter 19 is devoted to the processes of filling and stoppering. These operations must occur under Grade A/B (ISO 5) clean room conditions. Sterile products are filled into final containers as liquids, dispersed systems, or powders. Liquid filling machines operate under peristaltic, piston, or time-pressure mechanisms. Dose control is imperative, with factors such as filling speed, product viscosity, and product potential to foam thereby affecting dose accuracy. Filling rates, depending on type of product, filling volume, and container type and size, can be as high at 600 units filled per minute. Filling accuracy should be within the range of $\pm$ 0.25%. Filling machine vendors include Bosch/TL, Chase-Logeman, Cozzoli, Mateer-Burt, Inova, National Instrument, and Perry.

The main issue with dispersed system filling is maintaining dose homogeneity—a huge challenge. Actually, at the point of filling the product into the container, the potential for clogging of the filling needle or nozzle is a concern. Dose homogeneity is a function of the ability of recirculation system supporting the filling system to prevent suspension particle or emulsion globule interaction and growth.

Powder filling also must control dose uniformity and accuracy that is a function both of the engineering of the powder filling machine and the particle size characteristics dictated by methods used to produce the solid product. Control of relative humidity during filling and minimizing foreign particle contamination also are challenges with powder filling. Primary vendors of powder filling machines are Perry and Chase-Logeman.

With respect to closing/stoppering of product-filled containers, ampoules, of course, do not require rubber closures and are sealed with a flame. Vials are closed with rubber stoppers (or, for vials containing solution to be freeze-dried, the stopper is partially inserted into the vial opening), and syringes and cartridges are closed with rubber plungers at the distal end (with rubber septa sealing the proximal end except for staked-needle syringes). Rubber stoppers and plungers need to be lubricated either with applied silicone oil or emulsion or with special coatings (see chap. 7) that permit and facilitate rubber units to move easily from the closure feeder (hopper) along stainless steel tracks or rails to the openings of the primary containers (Fig. 12-6). For vial openings, the closure must fit snugly, not "pop out." Often, filling efficiencies are dependent more on the stoppering process than on the actual filling process as there are tendencies for rubber closures to slip or pop off the openings of vials. For syringes and cartridges, the placement of the rubber plunger is dictated by the desired position of the plunger within the barrel of the syringe or cartridge to deliver the claimed volume of product.

Figure 12-6 Examples showing rubber closures moving along stainless steel railings from the feeder to the container opening. *Source*: Courtesy of Baxter Healthcare Corporation.

The closing of primary containers will affect the final integrity of the container/closure interface. Container/closure integrity testing and validation is covered in chapter 30.

For syringes and cartridges, no further sealing is done, although units are either placed in secondary packaging for unit dosing or part of a tray system, for example, Hypak™ (Becton-Dickinson). For vials and bottles, aluminum seals are crimped around the rubber closure and top of the container. Seal force integrity can be measured by a torque-testing device.

Terminal Sterilization

The desired scenario for filled containers after closing and sealing is to transfer them to a steam sterilizer or other sterilization system (e.g., radiation) and have the entire batch terminally sterilized. Terminal sterilization offers the greatest assurance of product sterility. Unfortunately, a large majority (perhaps >80%) of all small-volume injectable products contain active pharmaceutical ingredients that are heat- or radiation-labile and cannot be terminally sterilized.

It should always be the goal to develop sterile products that can be terminally sterilized. Discussion of possibilities for sterilization cycle modifications that will allow heat-sensitive products to be terminally sterilized are discussed in chapter 17.

Freeze-Drying

Freeze-drying or lyophilization is a major sterile process operation with approximately 40% of all biopharmaceutical products requiring freeze-drying for product stability. Products to be freeze-dried are processed as solutions up to the stoppering step. Following partial stoppering, vials are accumulated in a tray either for manual or automated transfer to a shelf of a freeze-dryer. Following the lyophilization process, the freeze-dryer shelves are hydraulically lowered to fully insert stoppers into the vials. Vials are manually or automatically removed from the dryer, loaded onto a capping line for sealing and inspection. Freeze-dry processing is discussed in chapter 20.

Finishing and Inspection

Finishing includes all the operations following the closure of the primary package that are as follows:

- Sealing or capping (chap. 19)
- Attachment of a plunger rod for syringes
- Labeling (chap. 22)
- Secondary packaging (chap. 22)
- Storage and distribution (chap. 24)

Inspection of finished units of sterile products requires every single unit to be evaluated for visible foreign particulate matter and any other defect. This subject is covered in detail in chapters 22 and 29.

Quality Control Testing

An appropriate number of finished product samples are removed using a statistically valid sampling plan for final product testing before the batch of product can be released for clinical or commercial use. Testing includes a variety of chemical and physical tests and assays, whatever is required to assure product safety, purity, strength, and quality.

Three special quality control tests are unique for sterile products—test for sterility, freedom from pyrogens or endotoxins, and freedom from visible particulate matter, and excessive subvisible particulate matter. These tests are discussed in detail in chapters 27, 28, and 29, respectively.

The pharmaceutical scientist must be aware of the various issues involved in the manufacturing arena that can impact the stability and quality of the pharmaceutical product, especially protein or peptide formulation. Among the more relevant areas of concern include shear rate and stress during compounding, filtration, and filling adsorption onto process tubing and filter surfaces, and the effects of time and temperature during each step of the manufacturing process. Formulation scientists and process engineers should work together to design and implement experiments to determine processing effects on protein stability and establish an appropriate control strategy. In most cases, for example, protein adsorption onto filter surfaces, the potential problems can be avoided or minimized once understood through experimentation by alternative choices of filter material or predicting the amount of solution to be passed through the filter to saturate the binding sites.

The surge of potential heat-labile products from biotechnology and the inability to terminally sterilize these molecules has accelerated the development of barrier/isolator technology (chap. 23). This technology, when perfected, will enable the processing of protein and peptide solutions to occur under a much higher degree of sterility assurance than what is now achievable with conventional aseptic processing. The main features of barrier/isolator technology are the ability to sterilize, not just sanitize, the environment under which sterile solution is exposed during filling and stoppering, and the removal of humans from direct contact with the exposed sterile solution.

BIBLIOGRAPHY

Akers MJ. Parenteral products. In: Remington: The Science and Practice of Pharmacy. 21st ed. Philadelphia, PA: Lippincott, Williams & Wilkins, 2005:802–836.

Carleton FJ, Agalloco JP, eds. Validation of Aseptic Pharmaceutical Processes. New York: Informa USA, Inc., 2007.

Harwood RJ, Portnoff JB, Sunbery EW. The processing of small volume parenterals and related sterile products. In: Avis KE, Lieberman HA, Lachman L, eds. Pharmaceutical Dosage Forms: Parenteral Medications. Vol 2. 2nd ed. New York: Marcel Dekker, 1992:1–92.

Groves MJ, Murty R, eds. Aseptic Manufacturing II. Buffalo Grove, IL: Interpharm Press, 1995.

Groves MJ, Olson WP, Anisfeld MH, eds. Sterile Pharmaceutical Manufacturing. Vols 1 & 2. Buffalo Grove, IL: Interpharm Press, 1991.

Nail SL, Akers MJ, eds. Development and Manufacture of Protein Pharmaceuticals. New York: Wolters-Kluwers, 2002.

Nema S, Ludwig J, eds. Pharmaceutical Dosage Forms: Parenteral Medications. 3rd ed. New York: Informa USA, 2010.

13 Contamination control

After understanding what is involved in the manufacturing of sterile dosage forms, it is rather daunting to imagine how facilities, equipment, utilities, personnel, and the product itself can be controlled such that no microbiological contamination can occur. Yet, this is exactly what must be achieved for sterile product manufacturers to produce safe and high-quality sterile dosage forms to serve and protect the public as well as to stay in business and not be in trouble with regulatory authorities. Contamination control is not only required in the production activities for sterile products but also in the testing for contamination and in the manipulation of sterile products in hospital and home environments.

With the advent of biotechnology and all the current and future therapeutic protein products being manufactured, contamination control takes on even greater prominence and concern. Why? Because proteins are natural nutrient sources for microorganisms plus typically formulated at neutral pH and isotonic, all ideal conditions for microbial growth. The only possible advantage of biotechnology products over small molecule products with respect to microbial growth resistance is the fact that biotechnology products either are freeze-dried and exist as solids rather than solutions or, if they are sufficiently stable as solution dosage forms, typically they are refrigerated.

There are obvious serious consequences if contamination control is not enforced and contaminated products are released to the market (1–4). In 1970, nonsterile intravenous fluids resulted in 9 deaths and over 400 cases of septicemia. Between 1965 and 1975, Food and Drug Administration (FDA) had 608 product recalls involving over 43 million containers suspected of microbial contamination resulting in at least 54 deaths and hundreds of injuries. Between 1981 and 1991, the FDA recalled over 50 products involving millions of containers. Since every batch recalled had been aseptically processed and not terminally sterilized, the FDA issued a proposal that current good manufacturing practices (cGMPs) be revised to require all sterile products to be terminally sterilized unless sufficient data exists justifying why they cannot (5). While the revision was not enacted as a new regulation, the industry must prove to FDA why new sterile products being registered cannot be terminally sterilized. In any given year, product recalls due to evidence of contamination or suspicion of potential contamination (GMP inspections that result in the auditors seriously questioning the microbiological safety of products released by the company being inspected) are amongst the most common of all types of product recalls.

Contamination control basically involves at least 14 entities to control or that help to determine the extent (quality) of control. Whenever there is a failure of a microbiologically related test on a batch purported to be sterile (e.g., sterility test failure, media fill failure, consistent failures to meet microbial limits with environmental monitoring samples), investigations to determine the cause of failure must include recovery and evaluation of all data related to each of these entities:

1. Facilities
2. Utilities (water, steam, compressed gases)
3. Air handling
4. Cleaning and sanitization
5. Equipment
6. Personnel
7. Environmental monitoring
8. Quality of raw materials
9. Prefiltration bioburden
10. Sterilization of all items used in sterile product manufacture
11. Filtration
12. Sterility testing

13. Endotoxin testing
14. Maintenance of facilities and equipment

This chapter focuses on cleaning, sanitization, environmental monitoring, and control of microbial and pyrogenic contamination. Other subjects are covered in subsequent chapters.

SOURCES OF MICROBIAL CONTAMINATION

Personnel must have appreciation and knowledge regarding where contamination can originate (6) and participate in making sure that these sources are minimized or eliminated. One basic truth about microbial growth that personnel must appreciate is that microorganisms must have food, moisture, and the right temperature to grow. Food sources are typically organic or proteinaceous material that can originate anywhere. Moisture depends on the relative humidity conditions in the environment plus any water source that is not controlled. The temperature requirement for most organisms is room temperature, although it is well known that microorganism may survive over a wide range of temperatures.

The Atmosphere

Air is not a natural environment for microbial growth because usually within the confines of a production facility the air is too dry and too clean (absence of nutrients). Microorganisms that can tolerate a dry environment include *Bacillus, Clostridium, Staphylococcus, Streptococcus, Penicillin, Aspergillus, and Rhodotorula.* The degree of contamination in the atmosphere depends on the level of dust and particles and humidity.

Water

Microorganisms indigenous to fresh water include *Pseudomonas sp., Bacillus sp.* (especially from soil erosion during/after heavy rains), and *Escheridia coli* (from sewage). Other than personnel, water is the main source of microorganisms that must be controlled to minimize or eliminate their presence in the clean room.

Raw Materials

Many raw materials originate from plants or animals and, thus, can be contaminated with pathogens such as *E. coli* and *Salmonella.* Fortunately, great efforts have been made to avoid the use of plant or animal-sourced raw materials in parenteral formulations, with only synthetic components used.

Packaging

When stored in cardboard containers and with any moisture present, glass supports the growth of mold spores (*Penicillin, Aspergillus*) and bacteria (*primarily Bacillus*). Rubber closures and paper also serve as good supports for mold. Plastic materials usually have relatively low-surface microbial contamination.

Buildings

Walls and ceilings can contain mold contamination (*Cladosporium sp, Aspergillus sp.*). Nutrients for microbial growth originate from the plaster on which paint is applied. Contamination of floors will occur if water is present and any cracks develop. Inadequate sealing at edges and joints will be a potential source of microbial life and growth.

Equipment

Hard-to-clean locations can be harbors for microbial contamination. Such locations include screw threads, agitator blades, valves, and pipe joints. Brooms and mops that are used to clean facilities themselves can be sources of contamination if not cleaned and used properly.

People

People, not surprisingly, are the largest single source of contamination. Indeed, contamination risk is almost completely related to human activity. Skin, hair, moisture from breathing, coughing, sneezing, cosmetics, and clothing are all sources of microbial contamination. Ordinary movements such as simple walking can emit more than 10,000 particles per minute. The greater the particle contamination, the more likely microbial contamination may be present.

MAINTENANCE OF CLEAN ROOMS

Maintaining the clean and sanitized conditions of clean rooms, particularly the aseptic areas, requires diligence and dedication of expertly trained custodians. Assuming the design of the facilities to be cleanable and easily sanitized, a carefully planned schedule of cleaning should be developed, ranging from daily to monthly, depending on the location and its relation to the most critical Class 100 areas. Tools used should be nonlinting, designed for clean room use, held captive to the area and, preferably, sterilizable.

Sanitization/Disinfection Agents

Liquid solutions (or foams) that serve as antimicrobial agents may be classified either as a sanitizing solution or a disinfectant solution. The FDA defines a sanitizer as a chemical or physical agent used to reduce microbial contamination levels present on inanimate environmental surfaces. A disinfectant is used to kill potential infectious agents. Thus, the term sanitizing agent (or solution) more properly defines the role of these agents since the primary aim is to reduce the microbial bioburden on pharmaceutical surfaces and equipment. Sanitizing agents are NOT sterilizing agents.

Sanitizing agents should be selected carefully because of data showing their reliable activity against inherent environmental microorganisms. Examples of bacteriocidal sanitizing agents (Table 13-1A) (7,8) include quarternary ammonium compounds (e.g., cetrimide or cetylpyridinium), phenolic mixtures, alcohols, biguanides (e.g., chlorhexidine), formaldehyde, chlorine, peroxide, glutaraldehyde, and combinations of these and other agents. Examples of sporicidal agents (Table 13-1B) include aldehydes and halogen-releasing agents (e.g., sodium hypochlorite and iodophors). A very common sanitizing agent is called LpH® (Steris Corporation), an acid phenolic disinfectant made specifically for use on hard nonporous surfaces. Sanitizing agents should be recognized as supplements to good housekeeping, never as substitutes. Sanitizing agents normally are not sporicides.

Table 13-1A Common Sanitizing Agents Used for Contamination Control in Parenteral Manufacturing Facilities

	Advantages	Disadvantages	Bacterial efficiency	Fungicidal efficiency	Sporicidal efficiency
Isopropyl alcohol (70%)	Inexpensive Safe No residue Best for skin	High conc. required Inactivated by organic matter	Good	Not reliable	Ineffective
Phenolics (1–3%)	Not affected by organic matter	Pungent odor	Excellent	Excellent	Ineffective
Hypochlorite (1–5%)	Best for floor cleaning	Corrosive, odor Inactivated by organic matter	Good	Good	Fair
Quaternary ammonium compounds (1:750)	Safe Odorless	Inactivated by anionic compounds	Fair	Not reliable	Ineffective
Hydrogen peroxide[a] (3–6%)	Safe Odorless		Excellent	Excellent	Excellent at 3–10% concs.
Peracetic acid[a] (1–3%)	Low concs Neutral pH	Corrosive Irritating	Excellent	Excellent	Excellent

[a]A widely used commercially available sanitizing agent, also effective against spores, is Spor-Klenz® (Steris Corp), a stabilized blend of peracetic acid, hydrogen peroxide, and acetic acid. *Source*: From Refs. 7 and 8.

Table 13-1B Common Sporocidal Agents Used for Contamination Control in Parenteral Manufacturing Facilities

	Advantages	Disadvantages
Sodium hypochlorite (Bleach) (typically 5% solution)	Very rapid activity, especially freshly prepared and buffered to pH 7.6	Organic matter reduces effectiveness
Vaporized hydrogen peroxide	Excellent activity	Prone to decomposition
Glutaraldehyde	Excellent activity Superior to formaldehyde	Less effective at acid pH
Iodophors	Depends on availability and concentration of iodine	Less potent than glutaraldehydes Activity less at alkaline pH

Source: From Ref. 9.

The United States Pharmacopeia (USP) contains a chapter <1072> entitled "Disinfectants and Antiseptics" that addresses subjects such as:

- Selecting chemical disinfectants and antiseptics
- Demonstrating the effectiveness of disinfectants as bactericidal, fungicidal, or sporicidal agents
- Applying disinfectants in manufacturing areas.

Choices of sanitizing agents must be validated for their effectiveness. Validation primarily involves reproducible bactericidal (or sporicidal) activity of the sanitizing agent. Manufacturers differ with respect to frequency of revalidation after initial validation, with some of them doing no revalidation and others revalidating on an annual basis. Microorganisms used in validation are typically environmental isolates. A 2 to 3-log reduction in inoculum challenge is required. Validation is performed using membrane filters or surface testing. Some manufacturers use polysorbate 80 solutions to neutralize surfaces during validation.

Organic materials inactivate most sanitizing agents. Sanitizing effectiveness depends on such factors as time exposure, concentration of agent, pH, hydration, and temperature. The physical removal when using sanitizing agents on hard surfaces is as much or possibly more important as chemical destruction.

The largest use of sanitizing agents is for decontaminating floors where most floors are sanitized (mopped) at least daily, perhaps multiple times a day. Walls are sanitized less frequently than floors. Equipment surfaces, workstations, chairs, communication systems, and other surfaces that cannot be sterilized are sanitized regularly. Sanitization can be classified as

- "Deep-cleaning"—generally done after an area shutdown
- Routine—daily or some other frequency during normal operations
- Continuous—for example, frequent sanitization of gloved hands and utensils used during manufacturing.

It has become standard practice that disinfectants should be rotated with sufficient frequency to avoid the development of resistant strains of microorganisms (10). The European Commission's Good Manufacturing Practice Guidelines on the Manufacture of Sterile Medicinal Products advocates disinfectant rotation while the FDA's Aseptic Processing Guidelines do not. Most sterile product manufacturers rotate sanitizing agents either on a weekly or monthly basis. However, in reality, there is no need to rotate disinfectants unless there are data from environmental monitoring samples that suggest rotation must be implemented. Data that suggest rotation is needed would include

- A trend in breaches in alert/action levels in the EM program
- Recovery of repetitive isolates subsequently shown not to be inactivated by the disinfectant.

The material used to apply sanitizing solutions to surfaces must not only hold a certain amount of solution, but, as importantly, must also deposit the solution readily and evenly on the surface. A variety of fabric wipers are available, but the most common type are wipers

Figure 13-1 Three-bucket sanitizing system. *Source*: Courtesy of Contec, Inc, www.contecinc.com.

composed of hydrophilic polyurethane foam. Other wiper materials include knitted polyester, woven cotton, and polyvinyl alcohol foam.

An example of the "three-bucket" system used to sanitize facilities is shown in Figure 13-1.

- The first bucket contains the sanitizing solution where the mop or sponge system is dipped and then the floor or other surfaced is mopped.
- The second bucket contains water for injection or the same sanitizing solution as the first bucket. After mopping the floor or other surface, the mop head is rinsed in this bucket.
- The third bucket is empty with a wringer where the rinsed mop/sponge is squeezed "dry" so that it can be effectively soaked with sanitizing solution from bucket one.

The sanitizing solution should be rendered sterile prior to use although, of course, once in use, it will no longer be sterile.

It should be noted that ultraviolet (UV) light rays of 237.5 nm wavelength, as radiated by germicidal lamps, are an effective surface disinfectant. But, it must also be noted that they are only effective if they contact the target microorganisms at a sufficient intensity for a sufficient time. The limitations of their use must be recognized, including no effect in shadow areas, reduction of intensity by the square of the distance from the source, reduction by particulates in the ray path, and the toxic effect on epithelium of human eyes. It is generally stated that an irradiation intensity of 20 $\mu w/cm^2$ is required for effective antibacterial activity.

CLEANING CONTAINERS AND EQUIPMENT

Containers and equipment coming in contact with parenteral preparations must be cleaned meticulously. New, unused containers and equipment will be contaminated with such debris as dust, fibers, chemical films, and other materials arising from such sources as the atmosphere, cartons, the manufacturing process, and human hands. Residues from previous use must be removed from used equipment before it will be suitable for reuse. Equipment should be reserved exclusively for use only with sterile products and, where conditions dictate, only for one product in order to reduce the risk of contamination. For many operations, particularly with biologic and biotechnology products, equipment is dedicated for only one product.

Calumatic

Bosch + Strobel

Metromatic

Figure 13-2 Examples of glass container washers. *Source*: Courtesy of Baxter Healthcare Corporation.

Cleaning of Containers

A variety of container washers of various ranges of sizes and automation are available for cleaning sterile product container (Fig. 13-2). The selection of the particular type will be determined largely by the physical type of containers, the type of contamination, and the number to be processed in a given period of time.

Validation of cleaning procedures for equipment is another "hot topic" with respect to cGMP regulatory inspections. Inadequate cleaning processes have been a frequent citing by FDA and other regulatory inspectors when inspecting both active ingredient and final product manufacturing facilities. It is incumbent upon parenteral manufacturers to establish scientifically justified acceptance criteria for cleaning validation. If specific analytical limits for target residues are arbitrarily set, this will cause concern for quality auditors. Validation of cleaning procedures can be relatively complicated because of issues with sample methods (e.g., swab, final rinse, testing of subsequent batch), sample locations, sensitivity of analytical methods, and calculations used to establish cleaning limits.

Cleaning containers requires adherence to some very basic principles:

1. The liquid or air treatment must be introduced in such a manner that it will strike the bottom of the inside of the inverted container, spread in all directions, and smoothly flow down the walls and out the opening with a sweeping action. The pressure of the jet stream should be such that there is minimal splashing that might affect adequate cleaning and turbulence that might redeposit debris loosened during the process.
2. The container must receive a concurrent outside rinse.
3. Cleaning cycle should alternate hot and cool treatments with the final rinse treatment using water for injection (WFI).

4. All metal components of the washing/cleaning equipment coming in contact with the containers need to be constructed of stainless steel or some other noncorroding and noncontaminating material.

Detergents rarely are used for new containers because of the risk of leaving detergent residues. Thermal-shock sequences in the washing cycle usually is employed to aid in the loosening of debris that may be adhering to the container wall. Sometimes only an air rinse is used for new containers, if only loose debris is present. In all instances, the final rinse, whether air or WFI, must be ultraclean so that no particle residues are left by the rinsing agent.

Only new containers are used for parenterals with such container wrapped in tight, low-particle (nonshedding) blister packaging that minimizes any build up of dirt and debris during shipment and storage.

Containers may be manually loaded, cleaned, and then removed. The wet, clean containers must be handled in such a way that contamination will not be reintroduced. They must be protected by storing in a laminar flow of clean air until covered, within a stainless steel box, or within a sterilizing tunnel. Modern cleaning machines move wet and clean containers immediately into a sterilization and depyrogenation dry-heat tunnel. Wet, clean containers should be dry-heat sterilized as soon as possible after washing. Doubling the heating period generally is adequate also to destroy pyrogens; for example, increasing the dwell time at 250°C from one to two hours, but the actual time-temperature conditions required must be validated (see chap. 17).

Increases in process rates have necessitated the development of continuous, automated line processing with a minimum of individual handling, still maintaining adequate control of the cleaning and handling of the containers. The clean, wet containers are protected by filtered, laminar flow air from the preparation area through the tunnel sterilizer and until they are delivered to the filling line.

Cleaning of Closures

Rubber closures are more difficult to clean than glass containers because of their shape with several convoluted surfaces. The normal procedure calls for gentle agitation in a hot solution of a mild water softener or detergent. The closures are removed from the solution and rinsed several times, or continuously for a prolonged period, with filtered WFI. The rinsing is to be done in a manner that will flush away loosened debris. The wet closures are carefully protected from environmental contamination, sterilized, usually by steam sterilization (autoclaving), and stored in closed containers until ready for use. This cleaning and sterilizing process also must be validated with respect to rendering the closures free from pyrogens. Depyrogenation of rubber closures relies on the cleaning and rinsing process since steam sterilization does not depyrogenate. If the closures were immersed during autoclaving, the solution is drained off before storage to reduce hydration of the rubber compound. Closures must be dry for use, they may be subjected to vacuum drying at a temperature in the vicinity of 100°C. Longer times and even higher drying temperatures are used for closures that are used with freeze-dried products.

The equipment used for washing large numbers of closures is usually an agitator or horizontal basket-type automatic washing machine. Because of the risk of particulate generation from the abrading action of these machines, some procedures simply call for heating the closures in kettles in detergent solution, followed by prolonged flush rinsing. The final rinse always should be with low-particulate WFI. Modern closure processors will simultaneously wash, siliconize, sterilize, and transport closures directly to the filling line (Fig. 12-3). It is also possible to purchase rubber closures already cleaned and lubricated in sterilizable bags supplied by the rubber closure manufacturer (e.g., Westar®, West Pharma).

Cleaning of Sterile Processing Equipment

The main concern with cleaning equipment is accessing all the internal parts of the equipment that are most difficult to clean. This requires particular attention to joints, crevices, screw threads, and other structures where debris is apt to collect. Exposure to a stream of clean steam will aid in dislodging residues from the walls of stationary tanks, spigots, pipes, and similar structures. Thorough rinsing with distilled water should follow the cleaning steps.

Computer-controlled cleaning systems (usually automated), called Clean-in-Place (CIP) systems, are now a standard approach to clean equipment. Stainless-steel equipment must be designed with smooth, rounded internal surfaces and without crevices. The cleaning is accomplished with the scrubbing action of high-pressure spray balls or nozzles delivering hot detergent solution from tanks captive to the system, followed by thorough rinsing with WFI. The system often is extended to allow sterilizing-in-place (SIP) to accomplish sanitizing or sterilizing as well.

Rubber tubing, rubber gaskets, and other rubber parts may be washed in a manner such as described for rubber closures. Thorough rinsing of tubing must be done by passing WFI through the tubing lumen. However, because of the relatively porous nature of rubber compounds and the difficulty in removing all traces of chemicals from previous use, it is considered by some inadvisable to reuse rubber or polymeric tubing. Rubber tubing must be left wet when preparing for sterilization by autoclaving.

ENVIRONMENTAL CONTROL EVALUATION

Environmental monitoring programs include obtaining microbiological and particulate matter information on the following:

- Room/facility surfaces and air
- People
- Utilities, for example, water, compressed gases, clean steam
- High Efficiency Particulate Air (HEPA) filters
- Filling nozzles after a microbiological culture media fill
- Performance qualification studies.

Manufacturers of sterile products use extensive means to control the environment so that these critical products can be prepared free from contamination. Nevertheless, tests should be performed to determine the level of control actually achieved. Normally, the tests consist of counting viable and nonviable particles suspended in the air or settled on surfaces in the workspace. A baseline count, determined by averaging multiple counts when the facility is operating under controlled conditions, is used to establish the optimal test results expected. During the subsequent monitoring program, the test results are followed carefully for high individual counts, a rising trend, or other abnormalities. If they exceed selected alert or action levels, a plan of action must be put into operation to determine if or what corrective and follow-up measures are required.

The tests used generally measure either the particles and microorganisms existing in a volume of sampled air or the particles and microorganisms that are settling or are present on surfaces. Table 13-2 summarizes the standard methods used to monitor classified clean room environments for viable and nonviable particles in the air and on surfaces including personnel monitoring. Tables 13-3A, 13-3B, and 13-3C provide comparisons of standards (limits) for particles and microorganisms (colony-forming units, CFUs) for the three standard classifications of clean rooms (Class 100, Class 10,000, and Class 100,000) (11).

PARTICLE COUNTERS

In order to measure the total particle content in an air sample, electronic particle counters are available (an example shown in Fig. 13-3), operating on the principle of the measurement of light

Table 13-2 Environmental Monitoring Systems

Type of particle	Air	Surfaces
Viable	Nutrient agar plates (settle plates, fallout plates) Slit samplers Rotary centrifugal samplers	RODAC (replicate organism detection and counting) plates
Nonviable	Electronic particle counters Air membrane filters	Garment sampler

Table 13-3A Class 100 Monitoring

Country document	U.S. standard	USP <1116>	EU (at rest)	EU (dynamic)	EU (dynamic)	ISO 14644-1
Classification	M3.5 (100)	M3.5	A and B	A	B	5
Frequency of monitoring	Not stated	Each operating shift	Not stated	Frequent, using a variety of methods	Frequent, using a variety of methods	Not stated
Total particle count	$3500/m^3$ or $100/ft^3 \geq 0.5\ \mu m$	$100/ft^3 \geq 0.5\ \mu m$	$3500/m^3 \geq 0.5\ \mu m$	$3500/m^3 \geq 0.5\ \mu m$	$350,000/m^3 \geq 0.5\ \mu m$	$3520/m^3 \geq 0.5\ \mu m$
			$0/m^3 \geq 5\ \mu m$	$0/m^3 \geq 5\ \mu m$	$2000/m^3 \geq 5\ \mu m$	$29/m^3 \geq 5\ \mu m$
Airborne viable units	Not stated	$0.1\ CFU/ft^3$	Not stated	$<1\ CFU/m^3$ Settle plate (90 mm) <1 CFU/4 hr	$<10\ CFU/m^3$ Settle plate (90 mm) 5 CFU/4 hr	Not stated
Surface viable units (except floors)	Not stated	3 CFU/plate	Not stated	<1 CFU/ plate	5 CFU/plate	Not stated
Surface viable units—floors	Not stated	3 CFU/plate	Not stated	<1 CFU/ plate	5 CFU/plate	Not stated
Personnel gown	Not stated	3 CFU/plate	Not stated	Not stated	Not stated	Not stated
Personnel gloves	Not stated	3 CFU/pl	Not stated	Glove-5 fingers <1 CFU/glove	Glove-5 fingers 5 CFU/glove	Not stated
Air velocity unidirectional	Not stated	Not stated	0.45 m/sec ± 20%	0.45 m/sec ± 20%	Not appropriate	Not stated
Frequency of ΔP monitoring	Not stated	Each shift	Not stated	Continuous	Continuous	Not stated

Grade A—Terminally sterilized products—filling of these products.
Grade B—Aseptically prepared products—aseptic preparation and filling; handling of sterile starting materials and components; transfer of partially closed containers in open trays.
Source: From Ref. 11.

scattered from particles as they pass through the cell of the optical system (e.g., of suppliers: *Climet, HIAC Royco, Met One, Particle Measuring Systems*).

Principle of Light Scattering

When a beam of light strikes a solid object, three events occur: some of the light is absorbed, some of the light is transmitted, and the rest of the light is scattered. Scattered light is a composite of diffracted, refracted, and reflected light. Particle counters that operate on the basis of light scattering are designed to measure the intensity of light scattered at fixed angles to the direction of the light beam (see chap. 29 for more details).

As liquid flows into a light-sensing zone, particles in the fluid scatter light in all directions. The scattered light is directed onto a system of elliptical mirrors which then focus the light onto a photodetector. The light trap is designed to absorb most of the main light beam photons.

Met One and Climet particle counters represent examples of counters operating under this principle. Met One particle counters are laser-based particle counters that have become very popular instruments in the pharmaceutical industry both for airborne and liquid-borne particles. These instruments not only count particles but also provide a size distribution based on the magnitude of the light scattered from the particle. While a volume of air measured by an electronic particle counter will detect all particles instantly, these instruments cannot differentiate between viable (e.g., bacterial and fungal) and nonviable ones. However, because of the need to control the level of microorganisms in the environment in which sterile products are processed, it is also necessary to detect viable particles.

Table 13-3B Class 10,000 Monitoring

Country document	U.S. standard	USP <1116>	EU (at rest)	EU (Dynamic)	ISO 14644–1
Classification	M5.5 (10,000)	M5.5	C	C	7
Frequency of monitoring	Not stated	Each operating shift	Not stated	Not stated	Not stated
Total particle count	353,000/m^3 or 1000/ft^3 ≥ 0.5 μm	10,000/ft^3 ≥ 0.5 μm	350,000/m^3 ≥ 0.5 μm	3,500,000/m^3 ≥ 0.5 μm	352,000/m^3 ≥ 0.5 μm
			2000/m^3 ≥ 5 μm	20,000/m^3 ≥ 5 μm	930/m^3 ≥ 5 μm
Airborne viable units	Not stated	0.5 CFU/ ft^3	Not stated	$<$100 CFU/ m^3 Settle plate (90 mm) 50 CFU/4 hrs	Not stated
Surface viable units (except floors)	Not stated	5 CFU/plate	Not stated	25 CFU/plate	Not stated
Surface viable units—floors	Not stated	10 CFU/plate	Not stated	Not stated	Not stated
Personnel gown	Not stated	20 CFU/plate	Not stated	Not stated	Not stated
Personnel gloves	Not stated	10 CFU/pl	Not stated	Not stated	Not stated
Frequency of ΔP monitoring	Not stated	Each shift twice/week	Not stated	Not stated	Not stated

Grade C—Terminally sterilized products—preparation of solutions and filling of these products.
Aseptically prepared products—preparation of solutions to be sterile filtered.
Adjacent to Class 100.
Source: From Ref. 11.

Table 13-3C Class 100,000 Monitoring

Country document	U.S. standard	USP <1116>	EU (at rest)	EU (dynamic)	ISO 14644-1
Classification	M6.5 (100)	M6.5	D	D	8
Frequency of monitoring	Not stated	Twice a week	Not stated	Not stated	Not stated
Total particle count	3,533,000/m^3 or 100,000/ ft^3 ≥ 0.5 μm	100,000/ft^3 ≥ 0.5 μm	3,500,000/m^3 ≥ 0.5 μm	Not stated	3,520,000/m^3 ≥ 0.5 μm
			20,000/m^3 ≥ 5 μm		29,300/m^3 ≥ 5 μm
Airborne viable units	Not stated	2.5 CFU/ft^3	Not stated	200 CFU/m^3 Settle plate (90 mm) 100 CFU/4 hr	Not stated
Surface viable units (except floors)	Not stated	Not stated	Not stated	50 CFU/plate	Not stated
Surface viable units—floors	Not stated	Not stated	Not stated	Not stated	Not stated
Frequency of ΔP monitoring	Not stated	Weekly	Not stated	Continuous	Not stated

Grade D—Terminally sterilized products—preparation of solutions and components for subsequent filling.
Aseptically prepared products—handling of components after washing.
Source: From Ref. 11.

VIABLE PARTICLES

Viable particles usually are fewer in number than nonviable ones and are detectable as CFUs after a suitable incubation period at, for example, 30°C to 35°C for up to 48 hours. Thus, test results will not be known for 48 hours after the samples are taken, although eventually rapid-acting microbiological test methods will obviate the need for incubation time beyond an hour or two.

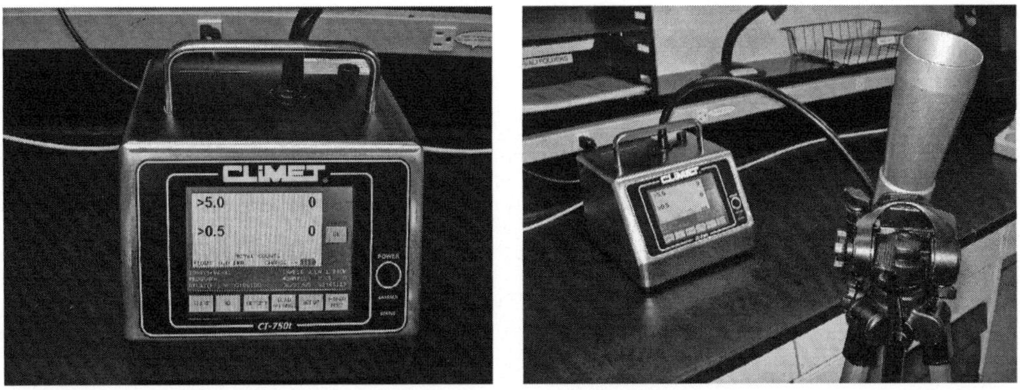

Figure 13-3 Air particle counter. *Source*: Courtesy of Climet Instruments.

Locations for sampling should be planned to reveal potential contamination levels that may be critical in the control of the environment. For example, the most critical process step is usually the filling of containers, a site obviously requiring monitoring. Other examples include the gowning room, high-traffic sites in and out of the filling area, the penetration of conveyor lines through walls, and sites near the inlet and exit of the air system.

The sample should be large enough to obtain a meaningful particle count. At sites where the count is expected to be low, the size of the sample may need to be increased. For example, in Class 100 areas, the sample should be at least 30 ft^3 and, probably, much more (12). Many firms employ continuous particle monitoring in Class 100 areas to study trends and/or to identify equipment malfunction.

Several air-sampling devices are used to obtain a count of microorganisms in a measured volume of air. A slit-to-agar (STA) sampler (suppliers: *Mattson-Garvin, New Brunswick, Vai*) draws by vacuum a measured volume of air through an engineered slit, causing the air to impact on the surface of a slowly rotating nutrient agar plate (Fig. 13-4). Microorganisms adhere to the surface of the agar and grow into visible colonies that are counted as CFUs, since it is not known whether the colonies arise from a single microorganism or a cluster. A centrifugal sampler (supplier: *Biotest*) pulls air into the sampler by means of a rotating propeller and slings the air by centrifugal action against a peripheral nutrient agar strip. The advantages of this unit are

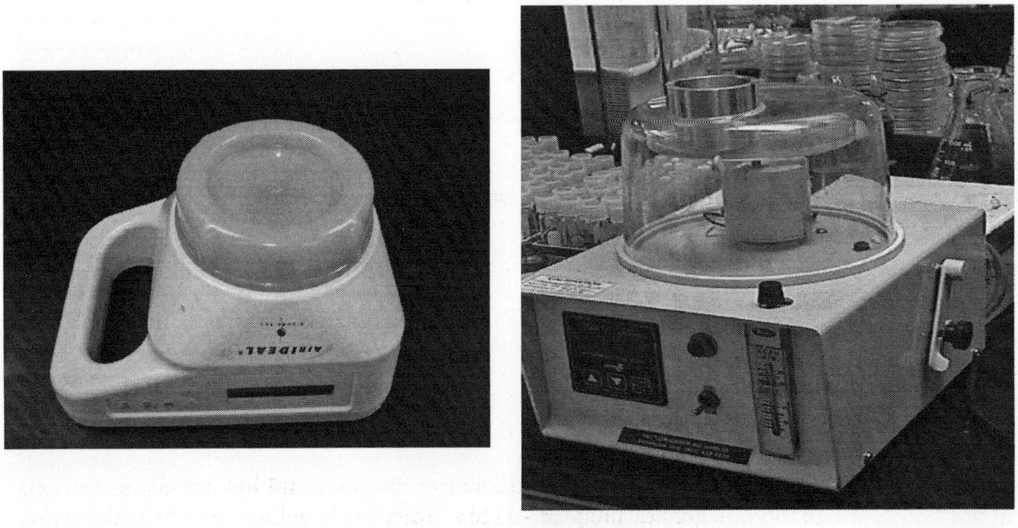

Figure 13-4 Examples of a slit-to-agar (STA) quantitative air sampler.

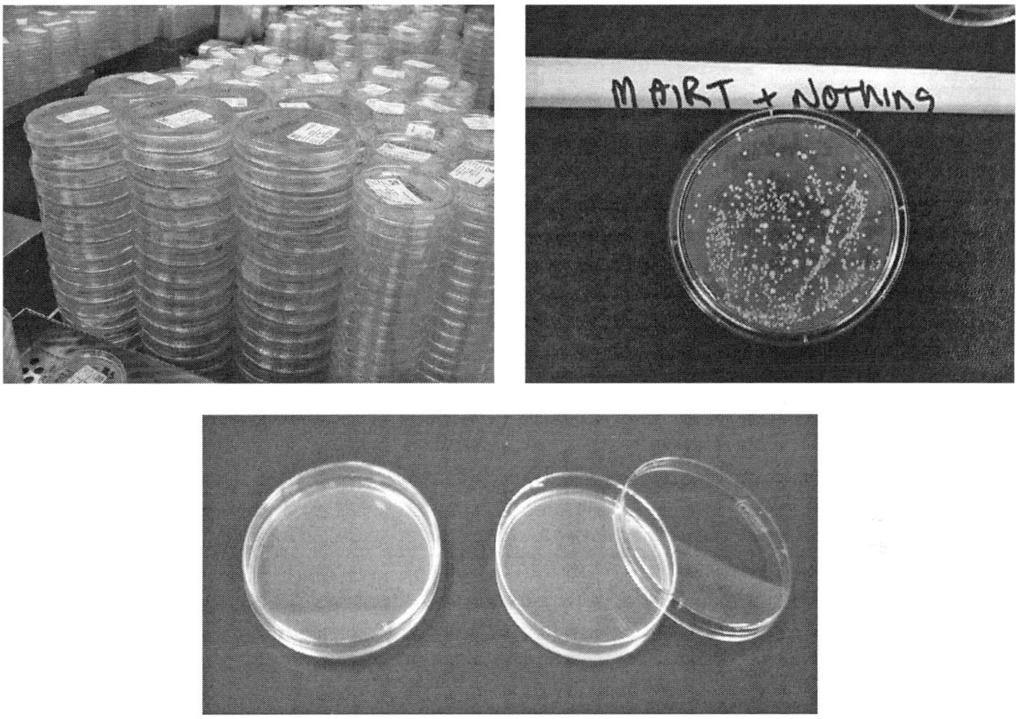

Figure 13-5 Examples of settle plates (fallout plates).

that it can be disinfected easily and is portable, so that it can be hand-carried wherever needed. These two methods are used quite widely.

A widely used method for microbiological sampling consists of the exposure of nutrient agar culture plates to the settling of microorganisms from the air (Fig. 13-5). This method is very simple and inexpensive to perform but will detect only those organisms that have settled on the plate; therefore, it does not measure the number of microorganisms in a measured volume of air (a nonquantitative test). Some companies have questioned the value of this passive air monitoring method, but European regulatory agencies have consistently supported their use (13). If the conditions of exposure are repeated consistently, a comparison of CFUs at one sampling site from one time to another can be meaningful. However, the use of settle plates, especially if there is agency pressure to increase the number of plate samples, may increase the risk for contamination. There certainly is no advantage of performing both passive air settle plate monitoring and active air-sampling techniques to detect clean room contamination (13).

The European Union GMP guidelines for sterile manufacture of medicinal products suggest an exposure period of not more than four hours for settling plates. Any longer periods, while perhaps desirable to monitor the environment during more or all of the filling cycle, run the risk of media dehydration and obtaining invalid microbial results.

The number of microorganisms on surfaces can be determined with nutrient agar plates having a convex surface (Fig. 13-6). With these it is possible to roll the raised agar surface over flat or irregular surfaces to be tested. Organisms will be picked up on the agar and will grow during subsequent incubation. This method also can be used to assess the number of microorganisms present on the surface of the uniforms of operators, either as an evaluation of gowning technique immediately after gowning or as a measure of the accumulation of microorganisms during processing. Whenever used, care must be taken to remove any agar residue left on the surface tested.

Further discussion of proposed viable particle test methods and the acceptable particle limits will be found in the USP general chapter <1116> "Microbial Evaluation and Classification of Clean Rooms and Other Controlled Environments."

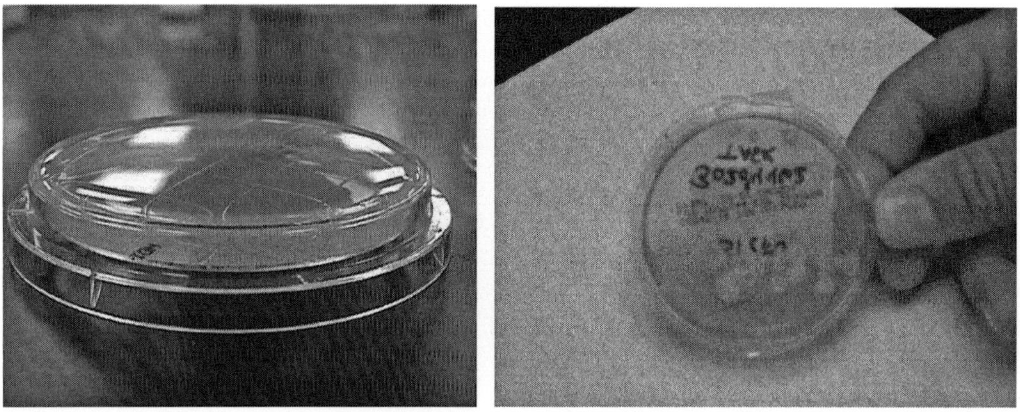

Figure 13-6 Example of a rodac plate (touch plate).

Results from the tests mentioned earlier, although not available until two days after sampling, are valuable to keep cleaning, production, and quality-control personnel apprised of the level contamination in a given area and, by comparison with baseline counts, will indicate when more extensive cleaning and sanitizing is needed. The results also may serve to detect environmental control defects such as failure in air-cleaning equipment or the presence of personnel who may be disseminating large numbers of bacteria without apparent physical ill effects.

Issues regarding environmental monitoring remain among the most controversial aspects of cGMP regulatory inspections of parenteral manufacturing and testing environments. Regulatory trends include requiring an increase in the number and frequency of locations monitored in the clean room and on clean room personnel, enforcing numerical alert and action limits, and linking environmental monitoring data to the decision to release or reject the batch. It has been pointed out that fully gowned personnel will still release a finite number of microorganisms (typically 10–100 CFR per hour) so that it is unreasonable to impose the requirement of zero microbial contamination limits at any location in the clean room (14).

Table 13-4 lists some practical or sensible realities about environmental monitoring (15,16). A sampling of FDA observations on environmental monitoring violations culled from warning letters available to the public on the FDA website is provided in Table 13-5 (17).

PYROGENS/ENDOTOXINS

While pyrogen and endotoxin testing are covered in detail in chapter 28, some coverage needs to occur in a chapter devoted to contamination control. Control of pyrogens and endotoxins means that contamination control is functional and successful.

Pyrogens are products of metabolism of microorganisms. The most potent pyrogenic substances are called endotoxins and are constituents (lipopolysaccharides, LPS) of the cell wall of gram-negative bacteria (e.g., *Pseudomonas sp, Salmonella sp, E. coli*). Endotoxins are LPS that typically exist in high molecular weight aggregate forms. However, the monomer unit of LPS is less than 10,000 Da, enabling endotoxin to easily pass through sterilizing 0.2 μ filters. Pyrogens or endotoxins, when present in parenteral drug products and injected into patients, can cause fever, chills, pain in the back and legs, and malaise. The Limulus Amebocyte Lysate (LAL) test, discussed in the following text, can detect the presence of LPS.

Control of Pyrogens

In general, it is impractical, if not impossible, to remove pyrogens once present without adversely affecting the drug product. Therefore, the emphasis should be on preventing the introduction or development of pyrogens in all aspects of the compounding and processing of the product.

Pyrogens may enter a preparation through any means that will introduce living or dead microorganisms. However, current technology generally permits the control of such

Table 13-4 Practical Realities of Environmental Monitoring

- Controlling the environment is not the same thing as monitoring the environment. Control exists at all times; monitoring is only a snapshot
- Can monitor too much where monitoring can be a source of contamination (the units themselves, more people in the clean room, disruption of laminar flow)
- There are many sources of error in EM sampling operations—media, recovery efficiency, incubation conditions, technicians contaminate plates or strips in their handling
- The enormous potential for variability in sampling, incubation, and other laboratory operations means that any CFU value could have a huge range associated with it. Air samplers, especially, are not quantitatively accurate
- Ordinary statistical methods cannot be applied to microbiological data with values typically 0 or 1 plus with all the variability in sampling and testing
- The source of microbial contamination is not from the air or facility, if properly maintained, but from people, of whom design engineers of facilities cannot control
- Microorganisms cannot survive in a clean room very long and cannot easily proliferate. Therefore, monitoring the environment at the end of a fill or shift does not mean that it will pick up the accumulation of microorganisms generated during that fill or shift
- Should not over-react should one plate or sample have a usually high number of CFUs. Looking more for trends, not single point data
- Anaerobes, molds, and yeast are common contaminants in aseptic processing areas and monitoring for their presence is essential

Source: From Refs. 15,16.

contamination, and the presence of pyrogens in a finished product indicates processing under inadequately controlled conditions. It also should be noted that time for microbial growth to occur increases the risk for elevated levels of pyrogens. Therefore, compounding and manufacturing processes should be carried out as expeditiously as possible, preferably planning completion of the process, including sterilization, within the maximum allowed time according to process validation studies. Aseptic processing guidelines require establishment of time

Table 13-5 FDA Warning Letter Statements Related to Environmental Monitoring (17).

- Monitoring is not conducted routinely nor concurrently with manufacturing. Sampling should be done daily during both shifts, both inside and outside of the LAF (laminar air flow) areas, and sample times should be varied to cover all parts of the production period. Sampling frequencies and locations must be defined
- Microbial air samples under laminar flow modules are collected only under static conditions
- Less than 10% of the microbial air samples were collected after noon, although production routinely continues until 3:00 p.m.
- Room air microbial samples are collected with the RCS (rotary centrifugal sampler) on a tripod at a height of 5 ft, which does not represent working level in these rooms. There is no trending of data
- There has been no daily monitoring of aseptic areas and LAF modules for nonviable particulates on a day-of-production basis
- High Efficiency Particulate Air (HEPA) filters have been certified with DOP (di-octyl phthalate) and a particle counter and not with a photometer
- Air velocities have been reported as an average and do not show the individual readings
- HEPA filters need to be DOP tested at least twice a year, not once a year as is currently being done
- Some of the validations of air quality in rooms were done only under static conditions without personnel. Also, no smoke pattern studies on the LAF have been performed to show effects of curtain movements on laminar air flows
- The firm has not set alert and action limits for most environmental samples; it needs to identify all organisms isolated from aseptic area until a database is established for normal flora found in the production environment (with frequency distribution) for use in evaluating sterility test results
- Failure to eliminate objectionable organisms from interior surfaces of transfer carts in which sterilized unsealed containers of drug product are exposed
- No validation studies have been conducted to assure the microbial settle plates are capable of supporting microbial growth after the stated 3-hr exposure time in Class 100 rooms
- The quality control unit did not assure that adequate systems and controls were in place to monitor the functioning, and to detect malfunctions, of the air handling systems used to control and assure aseptic conditions in aseptic manufacturing areas

limitations throughout processing for the primary purpose of preventing the increase of endotoxin (and microbial) contamination that subsequently cannot be destroyed or removed.

Pyrogens can be destroyed by heating at high temperatures. A typical procedure for depyrogenation of glassware and equipment is maintaining a dry-heat temperature of 250°C for 45 minutes. Exposure for 650°C for one minute or 180°C for four hours likewise will destroy pyrogens. The usual autoclaving cycle will not do so. Heating with strong alkali or oxidizing solutions will destroy pyrogens. It has been claimed that thorough washing with detergent will render glassware pyrogen free if subsequently rinsed thoroughly with pyrogen-free water. Rubber stoppers cannot withstand pyrogen-destructive temperatures, so reliance must be placed on an effective sequence of washing, thorough rinsing with WFI, prompt sterilization, and protective storage to ensure adequate pyrogen control. Similarly, plastic containers and devices must be protected from pyrogenic contamination during manufacture and storage, since known ways of destroying pyrogens affect the plastic adversely. It has been reported that anion-exchange resins and positively charged membrane filters will remove pyrogens from water. Also, although reverse osmosis membranes will eliminate them, the most reliable method for their elimination from water is distillation.

A method that has been used for the removal of pyrogens from solutions is adsorption on adsorptive agents. However, since the adsorption phenomenon also may cause selective removal of chemical substances from the solution, this method has limited application. Other in-process methods for their destruction or elimination include selective extraction procedures and careful heating with dilute alkali, dilute acid, or mild oxidizing agents. In each instance, the method must be studied thoroughly to be sure it will not have an adverse effect on the constituents of the product. Although ultrafiltration now makes possible pyrogen separation on a molecular-weight basis and the process of tangential flow is making large-scale processing more practical, use of this technology is limited, except in biotechnological processing.

Sources

By understanding the means through which pyrogens may contaminate parenteral products, their control becomes more achievable. Therefore, it is important to know that water is probably the greatest potential source of pyrogenic contamination, since water is essential for the growth of microorganisms and frequently contaminated with gram-negative organisms. When microorganisms metabolize, pyrogens will be produced. Therefore, raw water can be expected to be pyrogenic and only when it is appropriately treated to render it free from pyrogens, such as WFI, should it be used for compounding the product or rinsing product contact surfaces such as tubing, mixing vessels, and rubber closures. Even when such rinsed equipment and supplies are left wet and improperly exposed to the environment, there is a high risk that they will become pyrogenic. Although proper distillation will provide pyrogen-free water, storage conditions must be such that microorganisms are not introduced and subsequent growth is prevented.

Other potential sources of contamination are containers and equipment. Pyrogenic materials adhere strongly to glass and other surfaces, especially rubber closures. Residues of solutions in used equipment often become bacterial cultures, with subsequent pyrogenic contamination. Since drying does not destroy pyrogens, they may remain in equipment for long periods. Adequate washing will reduce contamination and subsequent dry-heat treatment can render contaminated equipment suitable for use. However, all such processes must be validated to ensure their effectiveness. Aseptic processing guidelines require validation of the depyrogenation process by demonstrating at least 3-log reduction in an applied endotoxin challenge.

Solutes may be a source of pyrogens. For example, the manufacturing of bulk chemicals may involve the use of pyrogenic water for process steps such as crystallization, precipitation, or washing. Bulk drug substances derived from cell culture fermentation will almost certainly be heavily pyrogenic. Therefore, all lots of solutes used to prepare parenteral products should be tested to ensure that they will not contribute unacceptable quantities of endotoxin to the finished product. It is standard practice today to establish valid endotoxin limits on active pharmaceutical ingredients and most solute additives.

The manufacturing process must be carried out with great care and as rapidly as possible, to minimize the risk of microbial contamination. Preferably, no more product should be prepared than can be processed completely within one working day, including sterilization.

SUMMARY OF PRACTICAL APPROACHES TO REDUCE/ELIMINATE RISK OF CONTAMINATION (18)

1. Personnel in the clean room are always to be considered the main source of contamination. Even highly qualified operators will generate hundreds to thousands of organisms per hour. Contamination from people is proportional to their level of activity. The less work people perform near critical work areas and the less rapid and intense their movements, the lower the contamination risk. Adherence to good aseptic practices is always a must. Adherence to good gowning procedures is also a must and any kind of redundancy in gowning, for example, double gloving, sleeve covers, efforts to keep gloves overwrapped on the sleeve are all good practices.
2. Air exchange rates in clean rooms should be sufficient to handle the contamination emitted by the number of workers present during operations. A room that provides a lower number of air exchanges will have a higher level of contamination risk when personnel are present than a room with a much higher number of air exchanges.
3. Always avoid any manual aseptic connections. Rely on clean-in-place and steam-in-place systems to clean and sterilize the connections.
4. Machine setup and adjustment is the most risk-intensive element of aseptic processing. The less the efforts involved in setup and adjustment, the less the potential for contamination. Older equipment tends to have more setup time and manipulations, so investing in newer equipment will be worth it. Setup and adjustment is less of a problem if machines are located in an isolator system that after setup will be decontaminated.
5. Personnel working in clean rooms need to be comfortable. Environments that are too warm and humid will produce discomfort and perspiration, both potentially producing higher levels of contamination. Temperatures in clean rooms should be in the range of 16°C to 18°C.
6. Do not permit the possibility of any moisture in the clean room environment. All clean room surfaces and air must be exceptionally dry. Without moisture, microbial growth is impossible.
7. Employee training in good aseptic practices cannot be guaranteed through didactic training programs and media fills only. How well an operator actually performs can only be ascertained by watching the operator work every day. Thus, the importance of the role of supervision.

REFERENCES

1. Anon. Nosocomial bacteremias associated with intravenous fluid therapy USA. MMWR Morb Mortal Wkly Rep 1971; 20(suppl):2.
2. Phillips I, Eykyn S, Laker M. Outbreak of hospital infection caused by contaminated autoclaved fluids. Lancet 1972; 1:1258–1260.
3. Ringertz O, Ringertz S. The clinical significance of microbial contamination in pharmaceutical and allied products. In: Bean HS, Beckett AH, Carless JE, eds. Advances in Pharmaceutical Sciences. Vol 5. London: Academic Press, 1982:201–226.
4. Turco S, King KE. Sterile Dosage Forms. 3rd ed. Philadelphia: Lea & Febiger, 1987:383–384.
5. Food and Drug Administration. Use of aseptic processing and terminal sterilization, the preparation of sterile pharmaceuticals for human and veterinary use. Fedl Regis, 1991; 56(198):51354–51358.
6. Bloomfield SF. Microbial contamination: Spoilage and hazard, in Guide to Microbiological Control. In: Denyer SP, Baird RM, eds. Pharmaceuticals and Medical Devices. 2nd ed. Boca Raton, FL: CRC Press, 2007:23–50.
7. Avis KE. Summer Parenterals Courses. Memphis, TN: University of Tennessee, 1977–1981.
8. Swenson D. Surface disinfection. Control Sterile Management, April 2006:74–78.
9. Russell AD. Bacterial spores and chemical sporicidal agents. Clin Microbiol Rev 1990; 3(2):99–119.
10. Martinez JE. The rotation of disinfectants principle: True or false? Pharm Tech 2009; 33(2):58–71.
11. Booth A. Environmental monitoring practices and regulations for the sterile manufacturer. Pharm Manuf Packing Sourcer 2003 (Autumn):48–49.

12. Whyte W, Niven L. Airborne bacteria sampling: The effect of dehydration and sampling time. J Parenter Sci Technol 1986; 40:182.
13. Andon BM. Active air vs. passive air (settle plate) monitoring in routine environmental monitoring programs. PDA J Pharm SciTechnol 2006; 60:350–355.
14. Ljungqvist B, Reinmuller B. Cleanroom Clothing Systems: People as a Contamination Source. Bethesda, MD: Parenteral Drug Association, 2004.
15. Akers JE. Environmental monitoring and control: Proposed standards and current practices. PDA J Pharm Sci Technol 1997; 51:36–47.
16. Akers JE, Agalloco J. Environmental monitoring: Myths and misapplications. PDA J Pharm Sci Technol 2001; 55:176–184.
17. Mackler E. Point of view: Environmental monitoring—particle counts are easy. Controlled Environments Magazine, Amherst, NH: Vicon Publishing, Inc. May 2007.
18. Akers JE. Microbiological risk assessment in pharmaceutical production operations. In: Prince R, ed. Microbiology in Pharmaceutical Manufacturing. 2nd ed. Vol 1. 2007:283–304.

BIBLIOGRAPHY

Akers MJ. Parenteral preparations. In: Felton L, et al., eds. Remington's Pharmaceutical Sciences. 21st ed. Philadelphia: Lippincott Williams & Wilkins, 2005:802–836; chap 41.

Austin P. Encyclopedia of Clean Rooms, Bio-Cleanrooms and Aseptic Areas. Contamination Control Seminars Livonia, MI: Cleanroom Consulting Group – A Division of Acorn Industries, 2000. http://www.cleanroom-consulting-group.com/descriptionOfTextBooks.html.

Block SS, ed. Disinfection, Sterilization, and Preservation. 5th ed. Philadelphia: Lippincott, Williams and Wilkins, 2001.

Prince R, ed. Microbiology in Pharmaceutical Manufacturing. 2nd ed. Bethesda, MD: Parenteral Drug Association, 2007.

14 Sterile manufacturing facilities

The manufacturing facility—its floors, walls, ceilings, and all associated equipment and utilities—must be designed, constructed, and operated properly for the production of a sterile product with the excellent quality level required for safety and effectiveness. Materials of construction for sterile product production facilities must be "smooth, cleanable, and impervious to moisture and other damage." All facilities, equipment, and fixtures must be flush fitted to meet the need for smooth and cleanable surfaces. All connections or junctions between ceilings, walls, and floors cannot be 90° angles, but coved for easy cleaning. There is no place for gaps, cracks, recesses, or other defects that can be harbors for microbial contamination to build up.

What might not seem obvious to the average person is the special design of openings for air entrances and returns, doors, windows, light fixtures, and communication systems (speakers, telephones, intercoms) to meet the "smooth, cleanable, and impervious to moisture and other damage" requirements. All of these kinds of essential building items need to have flush fittings and appropriate sealed systems that do not contribute to particle or microbial contamination. Air systems, lighting, and communication systems need to be designed to be accessible from areas that are not part of the clean room. Doors must be designed to be opened and closed easily with minimal disturbance to the normal airflow patterns in the clean room. Of course, rooms are to be built such that air flow moves toward rooms of lesser cleanliness and this is accomplished via room air pressure differentials with air from higher classifications (e.g., Grade A or ISO 5) moving toward rooms with lower classifications (e.g., Grade C or ISO 7). Doors should be self-closing and in rooms where there are two doors (e.g., change room), doors must be interlocked so that only one door at a time can be open. Emergency doors must have workable alarm systems such that if any emergency door is ajar or opened, typically leading to an unclassified environment, personnel are alerted immediately. If such a door is ever opened, purposefully or not, during production, batches risk being rejected because of uncertainty of the classified environment being jeopardized.

Further, the processes used must meet current good manufacturing practices (cGMP) standards. Since the majority of small-volume injectables and topical ophthalmic products are aseptically processed (the finished product not terminally sterilized), adherence to strict cGMP standards with respect to sterility assurance is essential.

The cGMP regulations have several statements regarding facility requirements, but two particularly apply to sterile production facilities:

> Section 211.42: There must be separate or defined areas of operation to prevent contamination, and that for aseptic processing there be, as appropriate, an air supply filtered through HEPA filters under positive pressure, and systems for monitoring the environment and maintaining equipment used to control aseptic conditions.
>
> Section 211.46: Equipment for adequate control over air pressure, microorganisms, dust, humidity, and temperature be provided where appropriate, and that air filtration systems including prefilters and particulate matter air filters, be used when appropriate on air supplies to production areas.

Table 14-1 lists examples of GMP compliance problems (FDA 483 observations) where the above two GMP reference statements were cited as evidence of lack of compliance. It is interesting that most facility problems are a result of poor design to begin with, poor maintenance and repair, and/or concerns about potential facility contributions to lack of sterility assurance of products manufactured in the facility.

Table 14-1 Examples of 483 Observations Related to Manufacturing Facilities for Sterile Product Production

- Capping area not under auspices of a controlled environment
- Loading vials into freeze-dryers not under auspices of a controlled environment
- Evidence of cracks, deterioration of walls, ceilings, floors; debris in clean rooms
- Lack of proper certification of HEPA filters
- Lack of smoke test data and video during operations
- Lack of environmental monitoring data during operations and at specific locations related to actual filling areas
- Inadequate sanitization validation
- Lack of documentation regarding facility maintenance (e.g., cleaning, differential pressure checks, gas filter integrity, improper storage of equipment)
- Inadequate air monitoring (environmental plates, air velocities, air pressure differentials, reactions to out-of-specification limits)
- Facility changes made without change control procedures followed
- Failure to establish a system for maintaining equipment to control aseptic conditions
- Failure to follow appropriate written procedures designed to prevent microbial contamination of drug products in designated facilities
- Deficiencies in the cleaning validation of equipment and facilities and lack of evidence of effectiveness of sanitization procedures
- Inadequate controls to prevent cross-contamination of products within the manufacturing facility
- Inadequate facility design, e.g., porous drywall-like material not easily cleaned
- Rust-like substances seen in several locations in different rooms
- Failure to replace faulty HEPA filters
- No air flow pattern testing in aseptic areas
- Lack of classified environments after vial stoppering

Abbreviation: HEPA, high efficiency particulate air.

GMPs for large-volume injectables (CFR sect. 212) were never formally legalized but several statements in section 212 were adopted and commonly practiced by the parenteral industry at large. Here is one example:

> Section 212.42: Walls, floors, ceilings, fixtures, and partitions in controlled environment area shall (a) have a smooth, cleanable finish that is impervious to water and to cleaning and sanitizing solutions and (b) be constructed of materials that resist chipping, flaking, oxidizing, or other deterioration.

To achieve these specific criteria, the materials of construction specially made and much more expensive than materials used to construct "normal buildings," even nonsterile pharmaceutical manufacturing facilities.

FUNCTIONAL AREAS IN A STERILE PRODUCT MANUFACTURING FACILITY

To achieve the goal of a manufactured sterile product of exceptionally high quality, many functional production areas are involved:

- Warehousing or procurement
- Compounding (or formulation)
- Materials (containers, closures, equipment)
- Preparation
- Filtration
- Sterile receiving
- Aseptic filling
- Stoppering
- Lyophilization (if warranted)
- Packaging
- Labeling
- Quarantine.

Each of these areas has standard design criteria for the acceptable maximum limit of airborne particles greater than or equal to 0.5 μm either per cubic foot (United States classes)

Table 14-2 ISO 14644 Classification of Clean room Particle Limits

ISO Classification	Maximum concentration limits (particles per cubic meter of air) for particles ≥ the sizes per column				
	0.1 μm	0.3 μm	0.5 μm	1 μm	5 μm
1	10	–	–	–	–
2	100	10	4	–	–
3	1000	102	35	8	–
4	10,000	1020	352	83	–
5	100,000	10,200	3520	832	29
6	1000,000	102,000	35,200	8320	290
7	–	–	352,000	83,200	2930
8	–	–	3520,000	832,000	29,300
9	–	–	–	8320,000	293,000

or per cubic meter (EU grades, International Standards Organization classifications). Flow of equipment, materials, and personnel need to move from lower classified environments to the higher classifications (Fig. 12-2). Since ISO classifications are given, even though air is the topic of the next chapter, Table 14-2 provides the ISO classification (ISO 14644) (1) of clean room particulate limits that the Food and Drug Administration adheres to, replacing the old Federal Standard 209 (A, B, C, D, and E) series of clean room classifications (Class 100, 10000, etc.). Chapter 21 (Table 21-1) contains more information about air and microbial classifications of clean rooms according to FDA and European Union standards.

The extra requirements for the aseptic area are designed to provide an environment where a sterile fluid (liquid or dispersed system) or powder may be exposed to the environment for a brief period during subdivision from a bulk container to individual-dose containers without becoming contaminated. Contaminants such as dust, lint, other particles, and microorganisms normally are found floating in the air, lying on counters and other surfaces, on clothing and body surfaces of personnel, in the exhaled breath of personnel, and deposited on the floor. The design and control of an aseptic area is directed toward reducing the presence of these contaminants so that they are no longer a hazard to aseptic filling.

Although the aseptic area must be adjacent to support areas so that an efficient flow of components may be achieved, barriers must be provided to minimize ingress of contaminants to the critical aseptic area. This includes separation of personnel from critical product filling, stoppering, and capping. Such barriers may consist of a variety of forms, including sealed walls, manual or automatic doors, airlock pass throughs, ports of various types, hard plexiglas barriers, plastic curtains, and the like (Figs. 14-1 and 14-2).

FLOW PLAN

In general, the components for a sterile product flow either from the warehouse, after release, to the compounding area, as for ingredients of the formula, or to the materials support area, as for containers and equipment. After proper processing in these areas, the components flow into the security of the aseptic area for filling of the product in appropriate containers. From there the product passes into the quarantine and packaging area where it is held until all necessary tests have been performed. If the product is to be sterilized in its final container, its passage normally is interrupted after leaving the aseptic area for subject to the sterilization process. After the results from all tests are known, the batch records have been reviewed, and the product has been found to comply with its release specifications, it passes to the finishing area for final inspection and final release for shipment. There sometimes are variations from this flow plan to meet the specific needs of an individual product or to conform to existing facilities. Automated operations normally have much larger capacity and convey the components from one area to another with little or no handling by operators.

The key in sterile product facility design is to ensure that movement of equipment, materials, and people is unidirectional, eliminating any crossover of clean and dirty equipment and

Class 100, Grade A, ISO 5 Environment Inside Enclosure

Class 1000-10,000, Grade C, ISO 6-7 Outside Enclosure

Figure 14-1 Hard surface barrier separating internal critical filling process from external rest of room where personnel would be located. *Source*: Courtesy of Baxter Healthcare Corporation.

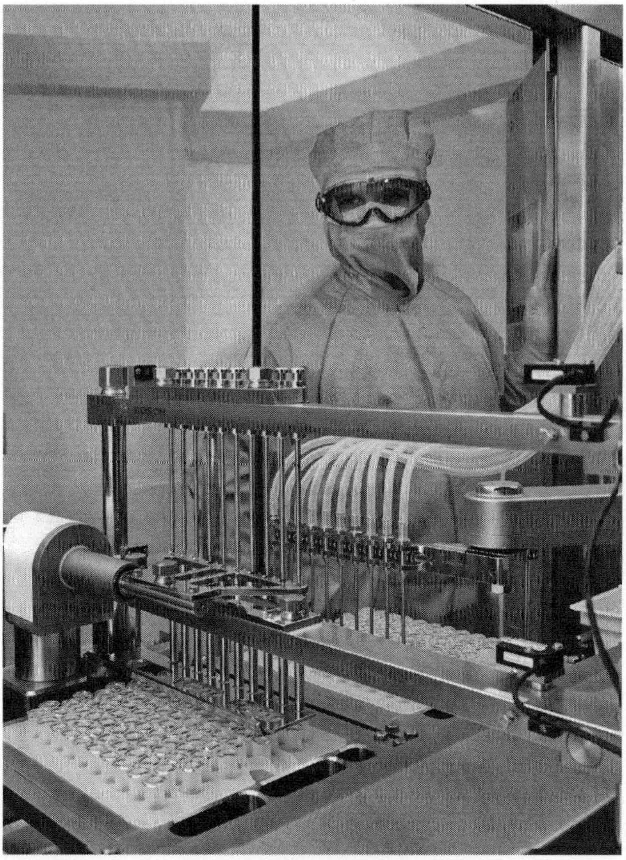

Figure 14-2 Example of barriers erected in filling room to separate personnel from product being filled under Grade A environment. *Source*: Courtesy of Robert Bosch GmbH.

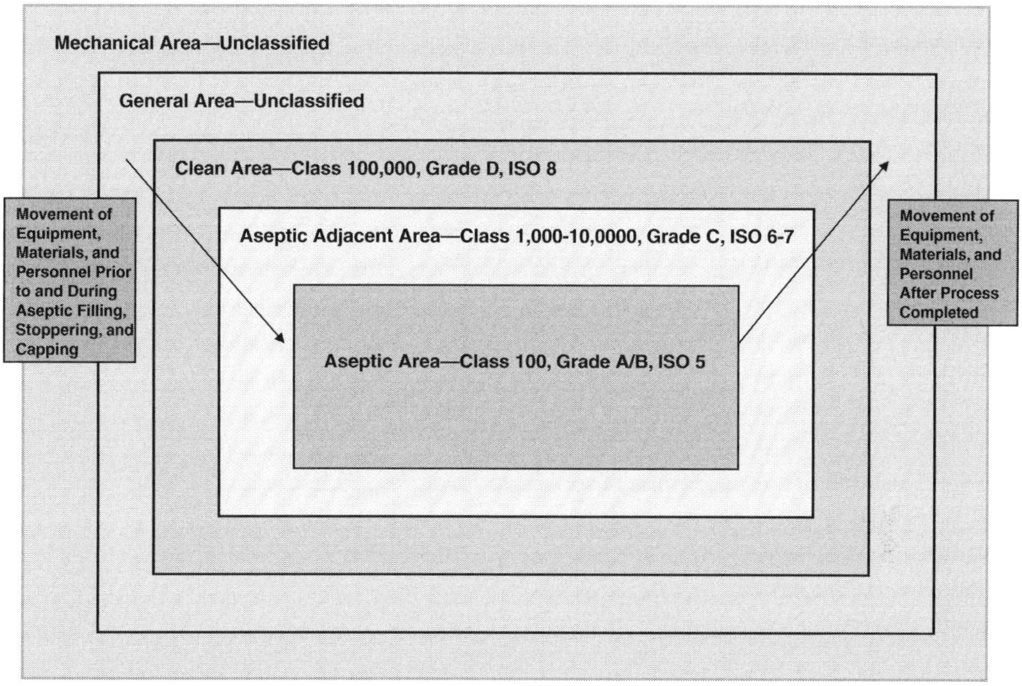

Figure 14-3 Generic floor plan of a sterile dosage form manufacturing facility (refer to Fig. 12-2 for typical activities that take place in each area).

minimizing back and forth movement from uncontrolled to controlled environments (Fig. 14-3). One example of a typical floor plan for sterile manufacturing is shown in Figure 14-4. Note the room classifications and the various entrances for personnel and equipment from lower classified rooms to higher ones.

Appropriate design of a sterile product manufacturing facility and flow of equipment, materials, and people should have the following characteristics (2):

- The exclusion of the surrounding environment (no possibility of surrounding room environment contaminants entering the cleaner environment).
- The removal or dilution of contamination generated during the manufacturing processes, especially potential contamination from personnel.
- A mechanism by which personnel (production operators) are protected from the product and product protected against environmental contamination.
- Optimal working conditions for personnel.
- Achievement of effective monitoring of room conditions at predefined time intervals.

MATERIALS OF CONSTRUCTION

As indicated above, materials of construction for sterile product production facilities must be "smooth, cleanable, and impervious to moisture and other damage" (Table 14-3). Floors start with a concrete slab coated with epoxy terrazzo, urethane, or solid vinyl. Walls are made of solid vinyl or cement plaster with an acrylic–polyvinyl chloride thermoplastic cover (Kydex®) applied for greater durability with the lower sections of the walls. Ceilings are also made of solid vinyl or cement plastic while curtains are typically composed of vinyl or Lexan® (Table 14-4).

Lights in clean rooms should be designed to offer little if any disturbance in airflow and be easily cleaned. In a laminar flow setting, lighting fixtures come in two varieties: the first is called a teardrop and mounts to the "T" grid. Its lens is shaped like an airfoil and contributes to the laminar flow in the room. The second, called a "flowthru," mounts under the high efficiency particulate air (HEPA) filter and allows the clean air to pass through it.

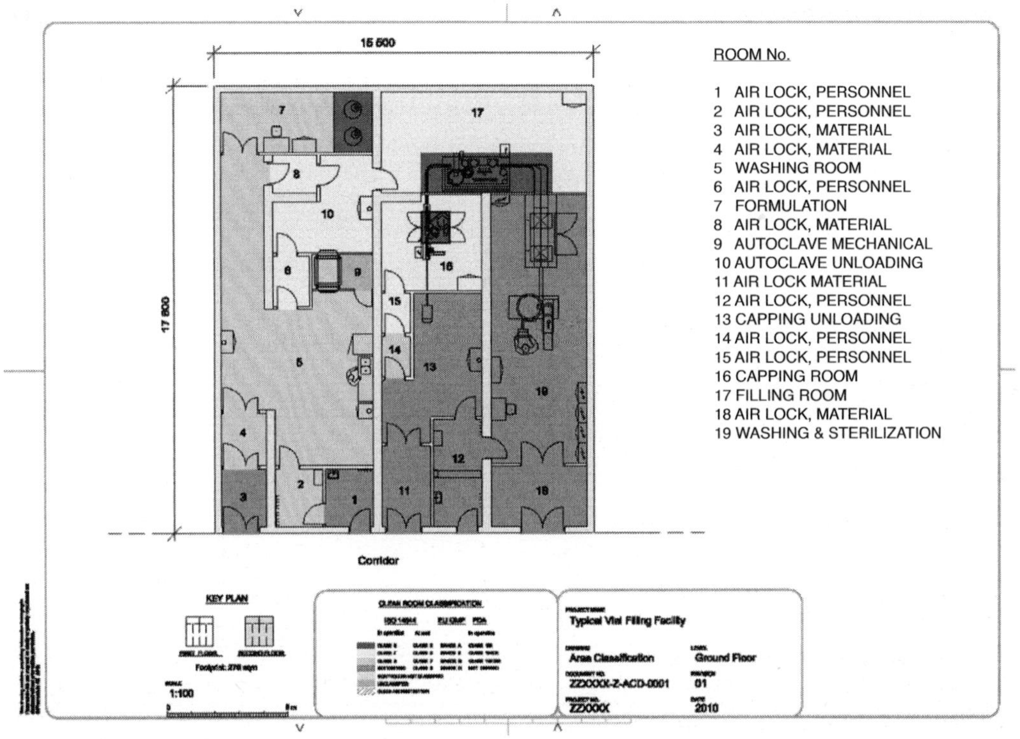

Figure 14-4 Specific floor plan of a sterile dosage form manufacturing facility. *Source*: Courtesy of Pharmadule, Inc.

Table 14-3 General Specifications for Construction Materials for Sterile Product Facility Walls, Floors, Ceilings, Fixtures, and Partitions in Controlled Areas

- Smooth cleanable finish that is impervious to water and to cleaning and sanitizing solutions
- Materials resist chipping, flaking, oxidizing, or other deterioration
- Coved (concave) surface at the floor wall junction to facilitate cleaning
- No horizontal fixed pipes or conduits over exposed components, materials, drug products, and drug product contact surfaces, including containers and closures after final rinse
- Intercompatibility of wall, ceiling, and flooring materials so that they join snugly and exhibit resistance to thermal expansion and contraction
- Avoidance of any surface that will emit or accumulate particular matter

Table 14-4 Construction Materials Used in Parenteral Facilities

Area	Examples of construction materials used
Floors	Concrete base
	Mipolam (solid polyvinyl chloride)
	Terrazo
	Epoxy coating
	Urethane coating
Walls	Cement plaster on gypsum
	Mipolam
	Kydex® (acrylic/polyvinyl chloride) shields
	Epoxy or enamel finish
Ceilings	Seamless plaster
	Gypsum
	Epoxy
	Mipolam
	Lighting and other fixtures recessed
Curtains	Vinyl
	Lexan

What has been described relates to traditional buildings with permanent walls and ceilings typically made of plasterboard dry wall build upon concrete slab floors, then covered with special finishes that give smoothness, cleanability, and durability to the facility. As will be discussed, a more modern approach is the use of modules where the walls are constructed from laminated clean room material mounted on anodized aluminum tracks or attached to joined aluminum extrusions that form the stud and cross members of the room.

CLEAN ROOM CLASSIFIED AREAS

Because of the extremely high standards of cleanliness and purity that must be met by sterile products, it has become standard practice to prescribe specifications for the environments in which these products are manufactured, that is, clean rooms. Because so many sterile products are manufactured at one site for global distribution, air quality standards in aseptic processing areas must meet both United States and European requirements. European standards differ from United States standards in the following ways:

1. Use Grade classifications (A, B, C, and D) rather than Class X (100, 1000, 100,000, etc.).
2. Use particle and microbial limits per cubic meter rather than per cubic foot.
3. Require particle measurements at 5 μm in addition to 0.5 μm in Grade A and B areas.
4. Differentiate area cleanliness dynamically and "at rest" (Grade B).

Air quality is discussed in chapters 13 and 15.

Clean room design traditionally has Class 100 rooms adjacent to Class 100,000 rooms. Regulatory authorities have raised great concerns about this significant change in air quality from critical to controlled areas. It is now preferable to have an area classified from Class 1000 to Class 10,000 in a buffer area between a Class 100 and Class 100,000 area monitored and controlled. Precautions also must be taken to prevent deposition of particles or other contaminants on clean containers and equipment until they have been properly boxed or wrapped preparatory to sterilization and depyrogenation.

COMPOUNDING AREA

In this area the product is prepared, formulated from a "recipe". Although it is not essential that this area be aseptic (unless aseptic formulation using presterilized components/ingredients is to be accomplished), control of microorganisms and particulates should be more stringent than in the materials support area. For example, means may need to be provided to control dust generated from weighing and compounding operations. Cabinets and counters should, preferably, be constructed of stainless steel. They should fit snugly to walls and other furniture so that there are no catch areas where dirt can accumulate. The ceiling, walls, and floor should be similar to those for the materials support area.

ASEPTIC AREA

The aseptic area requires construction features designed for maximum microbial and particulate control. The ceiling, walls, and floor must be sealed so that they may be washed and sanitized with a disinfectant, as needed. All counters should be constructed of stainless steel and hung from the wall so that there are no legs to accumulate dirt where they rest on the floor. All light fixtures, utility service lines, and ventilation fixtures should be recessed in the walls or ceiling to eliminate ledges, joints, and other locations for the accumulation of dust and dirt. As much as possible, tanks containing the compounded product should remain outside the aseptic filling area, and the product fed into the area (Fig. 14-1) through hose lines. Proper sanitization is required if the tanks must be moved in. Large mechanical equipment that is located in the aseptic area should be housed as completely as possible within a stainless steel cabinet to seal the operating parts and their dirt-producing tendencies from the aseptic environment. Further, all such equipment parts should be located below the filling line. Mechanical parts that will contact the parenteral product should be demountable so that they can be cleaned and sterilized.

Comparison of differences in requirements of critical areas and controlled areas for aseptic processing operations is given in Table 14-5. Note that the definition of a classified area includes several criteria besides the number of particles of per cubic foot or meter.

Table 14-5 Requirements of Critical Areas Versus Controlled Areas for Aseptic Processing Operations

Requirement	Critical work area	Controlled work area
Airborne particle content	Class 100 Grades A and B	Class 10,000–100,000 Grades C and D
Velocity of air	100 ft/min $\pm$ 20%	100 ft/min $\pm$ 20%
Changes of air	60–80 times per hour	Not less than 20 times per hour
Airborne microbes	Not more than 1 CFU per 10 cubic ft	Not more than 5 CFU per 10 cubic ft
Surface microbes	$<$1 CFU per contact plate (24–30 cm^2)	5 CFU per contact plate (24–30 cm^2)
Personnel microbes	Gloves: $<$ 1CFU	Gloves: 5 CFU

Personnel entering the aseptic area should enter only through an airlock. They should be attired in sterile coveralls with sterile hats, masks, goggles, and foot covers. Movement within the room should be minimal and in-and-out movement rigidly be restricted during a filling procedure. The requirements for room preparation and the personnel may be relaxed somewhat if the product is to be sterilized terminally in a sealed container. Some are convinced, however, that it is better to have one standard procedure meeting the most rigid requirements.

MODULAR CONSTRUCTION

Modular construction has become a design standard for a number of sterile product companies worldwide. Standardized rooms are constructed to meet strict engineering guidelines while incorporating flexibility in size, classification, and utilization. Figure 14-5 shows an example of a single modular unit being getting ready to be reassembled at the finished product manufacturing site, while Figure 14-6 shows the exterior view of a fill/finish building comprised of modular units.

Figure 14-5 Example of a modular unit for a sterile manufacturing facility. *Source*: Courtesy of Pharmadule, Inc.

Figure 14-6 Exterior view of sterile manufacturing facility comprised of modular units. *Source*: Courtesy of Baxter Healthcare Corporation.

Many sterile facilities today are put together as modular units where each room is built separately, then the entire set of modules put together. Materials of construction are the same as a normal production facility. Modules include process equipment, critical utilities, HVAC (heating, ventilating, and air conditioning), piping, ducting, and electrical installations. The modules are assembled, then tested to ensure that everything is prequalified according to customer approval. The modules are dissembled although equipment and utilities remain within each module. The modules are shipped to the permanent building site, reassembled, and requalified. From start of design until final assembly and qualification, the time required is relatively quick (12–18 months).

Modular construction involves design, construction, testing, and qualification of each module independently. If complexity exists, it should be contained within a module, not between modules. Each module, being independent, has its own supports for utilities, power, instrumentation, piping, and other components. Modules are interconnected at the final site via piping and wall connections.

There are many benefits to modular construction (Table 14-6). Normal delivery time from the modular construction site (e.g., Pharmadule's site is in Nacka, Sweden) to the site for final commissioning is 6 to 12 months from the time the contract is signed. Although costs for modular construction are higher compared with convention construction, the shorter implementation time means faster market introduction of a new product and likely overall greater profitability.

There is so much more that is involved in facility construction that is not covered in this chapter, for example, specifics of the exterior building, specially the roof, fireproofing, caulking, partitions, piping, drains (in lower classified areas), pressurization, temperature and humidity

Table 14-6 Benefits of Modular Construction

- Construction at a single site (e.g., Pharmadule, Sweden) enables better control of weather conditions, labor, and finding building materials in compliance with GMP
- Shorter validation time
- Quicker start-up of production
- Ability to incrementally add to module
- Ability to move modular plants to other locations
- Substantially reduces time to bring new product to market
- Reduced costs overall although initial costs are higher, but time savings from start to completion of installation much faster than traditional construction
- Expansion can occur with minimal interruption to existing production

control, and location of special equipment (sterilizers, washers, water systems, etc.). Interested readers are referred to the bibliography listing for books and articles written by the sterile facility construction experts.

REFERENCES

1. International Organization for Standardization. Standards ISO 14644—Cleanrooms and Associated Controlled Environments; and ISO 14698—Cleanrooms and Associated Controlled Environments—Biocontamination Control. Geneva, Switzerland: ISO, 1999. http://www.iso.org. Accessed June 8, 2010.
2. Cosslett AG. The design of controlled environments. In: Denyer SP, Baird RM, eds. Guide to Microbiological Control of Pharmaceuticals and Medical Devices. 2nd ed. Boca Raton, FL: CRC Press, 2007:69–87.

BIBLIOGRAPHY

Austin PR. Encyclopedia of Clean Rooms, Bio-Cleanrooms and Aseptic Areas. 3rd ed. Contamination Control Seminars, Livonia, MI, 2000.

Avis KE, ed. Sterile Pharmaceutical Products: Process Engineering Applications. Englewood, CO: Interpharm/CRC Press, 1995.

Cole GC. Pharmaceutical Production Facilities: Design and Application. CRC Press, 1998.

Del Ciello R. Buildings and facilities: Subpart C. In: Nally JD, ed. Good Manufacturing Practices for Pharmaceuticals. 6th ed. New York: Informa, 2007:37–50.

Odum JN. Sterile Product Facility Design and Project Management. 2nd ed. Boca Raton, FL: CRC Press, 2004.

Relevant articles in International Society of Pharmaceutical Engineering publications. http://www.ispe.org/. Accessed June 8, 2010.

Whyte W, ed. Cleanroom Technology: Fundamentals of Design, Testing and Operation. New York: John Wiley & Sons, 2001.

15 Water and air quality in sterile manufacturing facilities

This chapter focuses exclusively on the basic highlights of water systems and air handling systems employed in the production of sterile products and the quality requirements of each system. Like almost all other chapters, general references are provided at the end of this chapter for recommended reading to the reader who desires broader and more in-depth coverage of these topics.

WATER

Water is the most commonly used component in sterile product formulations. Like everything else in nature, water has many applications in the sterile product manufacturing industry:

- Solvent in formulations
- Cleaning of components and equipment
- Solvent in cleaning, sanitizing, disinfectant solutions
- Source of clean steam
- Source of cooling water for freeze-dryer compressors
- Source of water for chillers, necessary for
  - Air compressors
  - Rubber closure processors
  - Cooling of depyrogenation tunnels

Water of suitable quality for compounding and rinsing product contact surfaces may be prepared either by distillation or by reverse osmosis (RO) to meet United States Pharmacopeia (USP) and other compendial specifications for Water for Injection (WFI). In active pharmaceutical ingredient manufacturing and in some foreign companies, ultrafiltration (UF) is employed to minimize endotoxins in those drug substances administered parenterally. For some ophthalmic products, such as the ophthalmic irrigating solutions, and some inhalation products, such as Sterile Water for Inhalation, where there are pyrogen specifications, it is expected that WFI be used in their formulation.

Table 15-1 provides a summary of the types of water found in USP monographs. Similar qualities and titles of water exist in other compendia, although this author did not attempt to compare compendial water monographs. Only by distillation or RO is it possible to separate adequately various liquid-, gas-, and solid-contaminating substances from water. With the possible exception of freeze-drying, there is no unit operation more important and none more costly to install and operate than the one for the preparation of WFI.

Preparation

The sources of water used in sterile product manufacture originate from any one of several natural sources—lakes, streams, wells, reservoirs, city water systems, and so forth. Such water, of course, is totally unsuitable for injecting into people and animals because of all the contamination with natural suspended mineral and organic substances, dissolved mineral salts, colloidal material, viable bacteria, bacterial endotoxins, industrial or agricultural chemicals, and other particulate matter. The source water must be pretreated by a combination of the following treatments: chemical softening, filtration, deionization, carbon adsorption, and/or RO purification (Fig. 15-1).

WFI can be prepared by distillation or by membrane technologies (RO or UF). The European Pharmacopeia (EP) only permits distillation as the process for producing WFI. The USP and Japanese Pharmacopeia (JP) allow application of all these technologies.

Distillation is a process of converting water from a liquid to its gaseous form (steam). Since steam is pure gaseous water, all other contaminants in the feed water are removed. Potential

Table 15-1 Water Monographs in the U.S. Pharmacopeia

Water type	Preparation method	Limit for endotoxins	Comments
Purified water, USP	Distillation ion exchange	None	Pharmaceutical solvent
Water for Injection, USP	Distillation reverse osmosis	0.25 EU/mL	Non-sterile, must use within 24 hr or store $< 5^\circ$C or $> 80^\circ$C, used for manufacture of parenteral products going to be sterilized
Sterile Water for Injection, USP	Distillation reverse osmosis	0.25 EU/mL	Single-dose containers same as WFI; also used to reconstitute sterile solids and dilute sterile solutions
Bacteriostatic Water for Injection, USP	Distillation reverse osmosis	0.5 EU/mL	Multiple-dose and single-dose products
Sterile Water for Irrigation, USP	Distillation reverse osmosis	0.25 EU/mL	1 L or larger, wide mouth. Does not meet particulate matter requirements for large volume injections labeled "for irrigation only"
Sterile Water for Inhalation, USP	Distillation reverse osmosis	0.5 EU/mL	Inhalation therapy only
Sterile purified water, USP	Distillation reverse osmosis	None	Used in preparation of non-parenteral compendial dosage forms where sterile form of water is required
Water for hemodialysis, USP	Distillation reverse osmosis	2 EU/mL	Drinking water for patients undergoing hemodialysis. Reduced levels of Al, F, Cl. Bioburden 100 CFU/mL. Not intended for injection

impurities in feed water include bacteria, bacterial endotoxins, particles, electrolytes, organics, colloids, and disinfectants such as chlorine. A distillation system consists of a:

1. Boiler (evaporator) containing feed water (distilland)
2. Source of heat to vaporize the water in the evaporator
3. Headspace above the level of distilland with condensing surfaces for refluxing the vapor, thereby returning nonvolatile impurities to the distilland

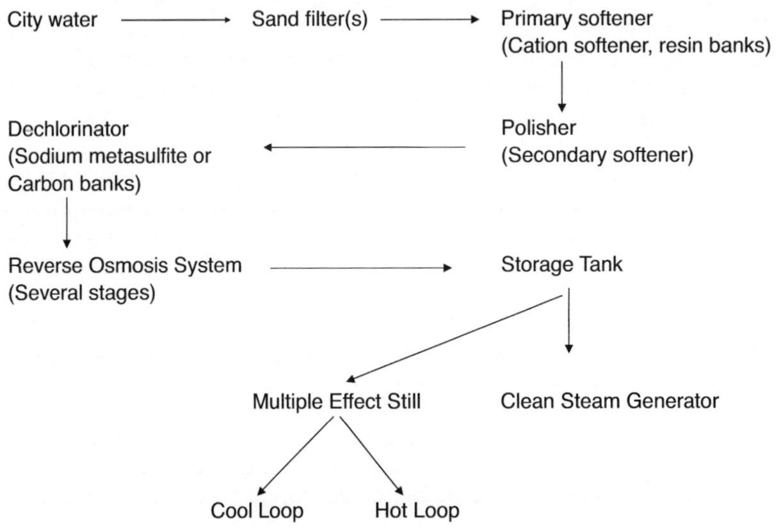

Figure 15-1 WFI system (example of flow from source to end).

4. Means for eliminating volatile impurities (demister/separation device) before the hot water vapor is condensed, and
5. Condenser for removing the heat of vaporization, thereby converting the water vapor to a liquid distillate.

Since a perfect separation never occurs, the distillate must be redistilled to increase its purity and the number of re-distillations defines the process (e.g., triple distilled).

Water is heated in a still until it boils, creating steam. The steam rises, leaving behind bacteria killed in the boiling process, as well as chemicals, heavy minerals, and pollutants found in the water source. The steam moves past a volatile gas vent into the condenser coils where it is cooled by air and condenses to become pure distilled water.

The quality of the feed water will affect the quality of the distillate. Chlorine in water can cause or exacerbate corrosion in distillation units. Silica in water causes scaling within. Controlling the quality of the feed water is essential for meeting the required specifications for the distillate. Pretreatment of feed water is recommended by most manufacturers of distillation equipment and is definitely required for RO units. The incoming feed water quality may fluctuate during the life of the system, depending on seasonal variations and other external factors beyond the control of the pharmaceutical facility. For example, in the spring increases in gram-negative organisms might occur because of all the rainfall. Also, new construction or fires can cause a depletion of water stores in old mains, causing an influx of heavily contaminated water of a different flora. A water system should be designed to operate within these anticipated extremes. Obviously, the only way to know the extremes is to periodically monitor feed water. If the feed water is from a municipal water system, reports from the municipality testing can be used in lieu of in-house testing.

The size of the evaporator will affect the efficiency of the distillation process. It should be sufficiently large to provide a low vapor velocity that reduces the entrainment of the distilland. Baffles (condensing surfaces) within the still determine the effectiveness of refluxing. They should be designed for efficient removal of the entrainment at optimal vapor velocity, collecting and returning the heavier droplets contaminated with the distilland. Redissolving volatile impurities in the distillate reduces its purity. Therefore, they should be separated efficiently from the hot water vapor and eliminated by aspirating them to the drain or venting them to the atmosphere. Contamination of the vapor and distillate from the metal parts of the still can occur. Present standards for high-purity stills are that all parts contacted by the vapor or distillate should be constructed of metal coated with pure tin, 304 or 316 stainless steel, or chemically resistant glass.

The design features of a still also influence its efficiency of operation, relative freedom from maintenance problems, or extent of automatic operation. Stills may be constructed of varying size, rated according to the volume of distillate that can be produced per hour of operation under optimum conditions. Only stills designed to produce high-purity water may be considered for use in the production of WFI. Conventional commercial stills designed for the production of high-purity water are available from several suppliers (examples: *AMSCO, Barnstead, Corning, Getinge, Kuhlman, Vaponics,* and others).

One principal component of the still is the heat exchanger. Because of the similar ionic quality of distilled and deionized water, conductivity meters cannot be used to monitor microbiological quality. Positive pressure, such as in vapor compression or double tubesheet design, should be employed to prevent possible feed water to distillate contamination in a leaky heat exchanger.

An Inspectors Technical Guide[1] of the Food and Drug Administration (FDA) discusses the design and potential problems associated with heat exchangers. The guide points out that there are two methods for preventing contamination by leakage. One is to provide gauges to constantly monitor pressure differentials to ensure that the higher pressure is always on the clean fluid side. The other is to utilize the double-tubesheet type of heat exchanger.

In some systems, heat exchangers are utilized to cool water at use points. For the most part, cooling water is not circulated through them when not in use. In a few situations, pinholes

[1] http://www.pipingnews.com/fdawater.htm

formed in the tubing after they were drained (on the cooling water side) and not in use. It was determined that a small amount of moisture remaining in the tubes, when combined with air, caused a corrosion of the stainless steel tubes on the cooling water side. Thus, it is recommended that when not in use, heat exchangers should not be drained of the cooling water.

There are two basic types of WFI distillation units—the vapor-compression still and the multiple-effect still.

Vapor-Compression Distillation

The vapor-compression still is primarily designed for the production of large volumes of high-purity distillate with low consumption of energy and water. The feed water is heated from an external source in the evaporator to boiling. The vapor produced in the tubes is separated from the entrained distilland in the separator and conveyed to a compressor that compresses the vapor and raises its temperature to approximately 107°C. It then flows to the steam chest where it condenses on the outer surfaces of the tubes containing the distilland; the vapor is thus condensed and drawn off as a distillate, while giving up its heat to bring the distilland in the tubes to the boiling point. Vapor-compression stills are available in capacities from 50 to 2800 gal/hr. They have lost favor in Europe and many other parts of the world, but are still quite popular in the United States.

Multiple-Effect Stills

The multiple-effect still (Fig. 15-2) also is designed to conserve energy and water usage. In principle, it is simply a series of single-effect stills or columns running at differing pressures where phase changes of water take place. A series of up to seven effects may be used, with the first effect operated at the highest pressure and the last effect at atmospheric pressure. Steam from an external source is used in the first effect to generate steam under pressure from feed water; it is used as the power source to drive the second effect. The steam used to drive the second effect condenses as it gives up its heat of vaporization and forms a distillate. This process continues until the last effect, when the steam is at atmospheric pressure, and must be condensed in a heat exchanger.

The capacity of a multiple-effect still can be increased by adding effects. The quantity of the distillate also will be affected by the inlet steam pressure; thus, a 600-gal/hr unit designed to operate at 115 psig steam pressure could be run at approximately 55 psig and would deliver about 400 gal/hr. These stills have no moving parts and operate quietly. They are available in capacities from about 50 to 7000 gal/hr.

Reverse Osmosis (RO)

The principle of osmosis was covered in Chapters 2 and 6. *Osmosis* involves the flow of a solvent through a semipermeable membrane (permeable to the solvent, but impermeable to solutes in the solvent) into a solution of higher solute concentration. Solution flows until concentrations on either side of the membrane are equal. Such concentrations can be measured by osmostic

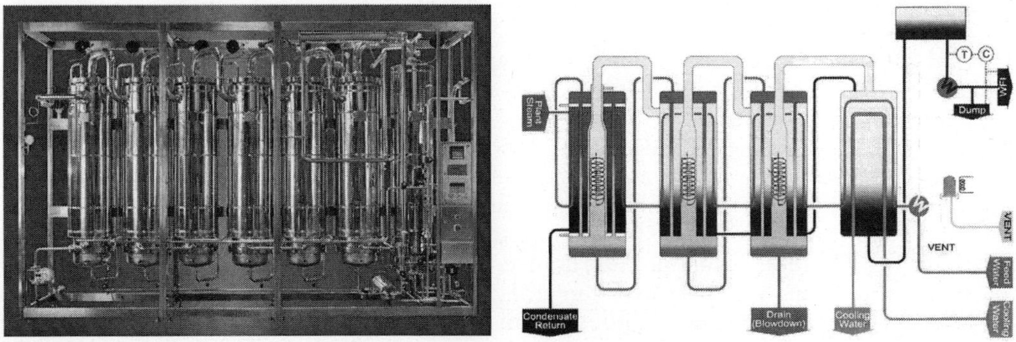

Figure 15-2 Multiple-effect still. *Source:* Courtesy of Getinge Water Systems.

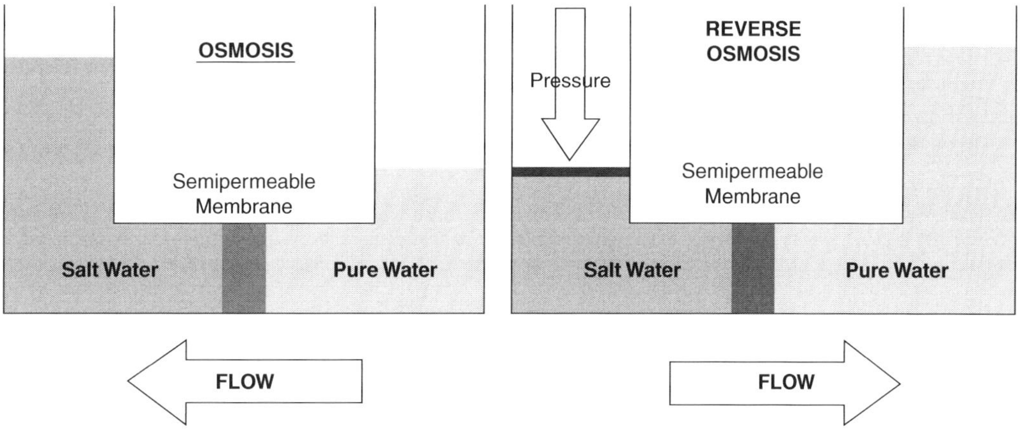

Figure 15-3 Schematic of reverse osmosis compared to osmosis. *Source:* Courtesy of Vertex Hydropore.

pressure instruments (osmometers that typically measure freezing point depression of the solution). In reverse osmosis, pressure, usually between 200 and 400 psig, is applied to overcome natural osmotic flow and force pure water to permeate through the membrane (Fig. 15-3). Membranes, usually composed of cellulose esters or polyamides, are selected to provide an efficient rejection of contaminant molecules (solutes) in raw water. The molecules most difficult to remove are small inorganic ones such as sodium chloride. Passage through two membranes in series is sometimes used to increase the efficiency of removal of these small molecules and to decrease the risk of structural failure of a membrane to remove other contaminants, such as bacteria and pyrogens. With the recognition of microbiological problems, some manufacturers have installed heat exchangers immediately after the RO filters to heat the water to 75–80°C to minimize microbiological contamination.

RO systems can be wall-mounted and fed by a single-pass RO unit that many small biotechnology companies use to produce high-purity water. Most of these systems employ polyvinyl chloride (PVC) or other type of plastic tubing. Because the systems are typically cold, the many joints in the system are subject to contamination. Another potential problem with PVC tubing is the release of extractables. These systems also contain 0.2-μm point-of-use filters to eliminate microbiological contamination and, therefore, reduce the source of endotoxins. However, 0.2-μm filters will not eliminate endotoxins already present. If filters are used in a water system, there should be a stated purpose for the filter, that is, particulate removal or microbial reduction, and a standard operating procedure (SOP) stating the frequency with which the filter is to be changed based on data generated during the validation of the system.

Because of the volume of water actually tested (0.1 mL for endotoxins vs. 100 mL for WFI), the microbiological test offers a good index of the level of contamination in a system. Therefore, unless the water is sampled prior to the final 0.2-μm filter, microbiological testing will have little meaning.

A strong trend in the sterile product manufacturing industry is to utilize both RO and distillation systems for generation of the highest quality water as well as combining highly purified water, RO, and electrodeionization systems (1). Since feed water to distillation units can be heavily contaminated, and, thus, affect the operation of the still, water is first run through RO units to eliminate contaminants. RO systems are available in a range of production and laboratory sizes.

Whichever system is used for the preparation of WFI, validation is required to be sure that the system, consistently and reliably, will produce the chemical, physical, and microbiological quality of water required. Such validation should start with the determined characteristics of the source water and include the pretreatment, production, storage, and distribution systems. All of these systems together, including their proper operation and maintenance, determine the ultimate quality of the WFI.

Storage and Distribution

WFI is either collected in a holding tank or recirculated through facility piping systems (Fig. 15-4). In large operations the holding tanks may have a capacity of several thousand gallons and be a part of a continuously operating system. In such instances the USP requires that the WFI be held at a temperature too high for microbial growth. Normally, this temperature is a constant 80°C. It is possible to use temperatures other than 80°C, but validation of this temperature to maintain water quality will be significantly scrutinized by regulatory authorities.

The USP also permits the WFI to be stored at room temperature but for a maximum of 24 hours. Under such conditions the WFI usually is collected as a batch for a particular use with any unused water being discarded within 24 hours. Such a system requires frequent sanitization to minimize the risk of viable microorganisms being present. The stainless-steel storage tanks in such systems usually are connected to a welded stainless-steel distribution loop supplying the various use sites with a continuously circulating water supply. The tank is provided with a hydrophobic membrane vent filter capable of excluding bacteria and nonviable particulate matter. Such a vent filter is necessary to permit changes in pressure during filling and emptying. The construction material for the tank and connecting lines usually is electropolished 316 L stainless steel with welded pipe. The tanks may also be lined with glass or a coating of pure tin. Such systems are very carefully designed and constructed and often constitute the most costly installation within the plant.

When the water cannot be used at 80°C, heat exchangers must be installed to reduce the temperature at the point of use. Bacterial-retentive filters should not be installed in such systems because of the risk of bacterial buildup on the filters and the consequent release of pyrogenic substances.

The one component of the holding tank that generates the most discussion is the vent filter. It is expected that this filter is integrity tested to assure that it is intact. It is expected, therefore, that the vent filter be located in a position on the holding tank where it is readily accessible.

Typically, filters are now jacketed to prevent condensate or water from blocking the hydrophobic vent filter. If this occurs (the vent filter becomes blocked), either the filter will rupture or the tank will collapse.

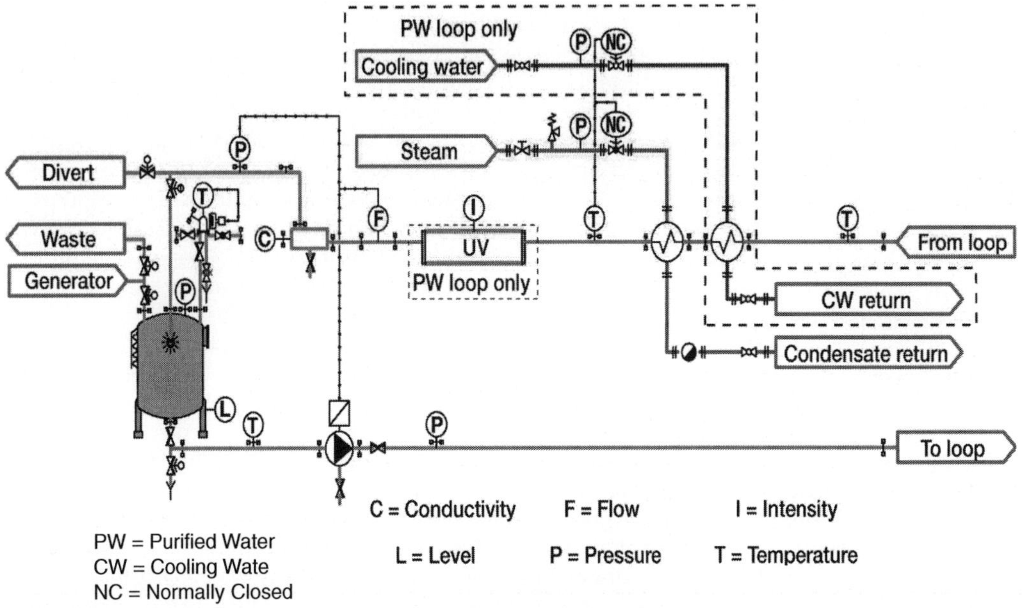

Figure 15-4 Storage and distribution of purified water (PW) and water for injection (WFI). *Source:* Courtesy of Getinge Water Systems. (*Note:* Only difference between PW and WFI is UV system only part of PW loop).

Pumps burn out and parts wear away. Also, if pumps are static and not continuously in operation, their reservoir can be a static area where water will lie. For example, during a FDA inspection some years ago it was noted that a firm had to install a drain from the low point in a pump housing and this eventually resulted in a contamination of *Pseudomonas* species.

Piping in WFI systems usually consist of a high polished stainless steel. In a few cases, manufacturers utilize PVDF (polyvinylidene fluoride) piping. It is purported that this piping can tolerate heat with no extractables being leached. A major problem with PVDF tubing is that it requires considerable support. When this tubing is heated, it tends to sag and may stress the weld (fusion) connection and result in leakage. Additionally, initially at least, fluoride levels are high. This piping is of benefit in product delivery systems where low-level metal contamination may accelerate the degradation of drug product (e.g., biopharmaceuticals).

One common problem with piping is that of "dead-legs." The proposed large volume parenteral (LVP) regulations defined dead-legs as not having an unused portion greater in length than six diameters of the unused pipe measured from the axis of the pipe in use. It should be pointed out that this was developed for hot (75–80°C) circulating systems. With colder systems (65–75°C) any drops or unused portion of any length of piping has the potential for the formation of a biofilm and should be eliminated or have special sanitizing procedures. There should be no threaded fittings in a pharmaceutical water system. All pipe joints must utilize sanitary fittings or be butt-welded. Sanitary fittings will usually be used where the piping meets valves, tanks, and other equipment that must be removed for maintenance or replacement. Therefore, the firm's procedures for sanitization, as well as the actual piping, should be reviewed and evaluated during the inspection.

Water Purity

USP and EP monographs provide the official standards of purity for WFI and Sterile Water for Injection (SWFI). There are four primary quality standards to be met for WFI (Table 15-2). The chemical and physical standards for WFI have changed over the years. The only physical/chemical tests remaining are the new total organic carbon (TOC), with a limit of 500 ppb (0.5 mg/L), and conductivity, with a limit of 1.3 μS/cm at 25°C or 1.1 μS/cm at 20°C. The former is an instrumental method capable of detecting all organic carbon present, and the latter is a three-tiered instrumental test measuring the conductivity contributed by ionized particles (in microSiemens or micromhos) relative to pH. Since conductivity is integrally related to pH, the pH requirement of 5–7 in previous revisions has been eliminated (although with much controversy still remains for USP-packaged SWFI). The TOC and conductivity specifications are now considered to be adequate minimal predictors of the chemical/physical purity of WFI. However, the wet chemistry tests are still used when WFI is packaged for commercial distribution and for SWFI.

Biological requirements continue to be, for WFI, not more than 10 colony-forming units (CFUs)/100 mL and 0.25 USP endotoxin units/mL. The SWFI requirements differ in that since it is a final product, it must pass the USP Sterility Test. The real concern in WFI is endotoxins. Because WFI can pass the Limulus amebocyte lysate (LAL) endotoxin test and still fail the

Table 15-2 Quality Standards for Water for Injection (WFI), USP[a]

Quality standard	How measured	Specification
Inorganic content	Water conductivity at 25°C. USP <645>	$\leq$1.3 μS/cm
Organic content	Total organic carbon, USP <643>	$<$0.5 mg/L
Pyrogen content	Limulus amebocyte lysate test, USP <85>	$<$0.25 EU/mL
Microbial content	Total bacterial count, USP <1231>	$\leq$ 10 CFU/100 mL (generally considered maximum action level for WFI using microbial enumeration methodologies described in USP <1231>)

[a]USP, EP, and JP specifications are harmonized for the above standards. However, EP requires two additional tests—heavy metals (specification NMT 0.1 ppm) and nitrates (specification NMT 0.2 ppm). The JP does not have a requirement for nitrates, but does have a requirement for ammonium (NMT 0.05 mg/L).

above microbial action limit, it is important to monitor WFI systems for both endotoxins and microorganisms.

None of the limits for water are pass/fail limits. All limits are action limits. When action limits are exceeded the firm must investigate the cause of the problem, take action to correct the problem and assess the impact of the microbial contamination on products manufactured with the water, and document the results of their investigation.

WFI and SWFI may not contain added substances. Bacteriostatic Water for Injection (BWFI) may contain one or more suitable antimicrobial agents in containers of 30 mL or less. This restriction is designed to prevent the administration of a large quantity of a bacteriostatic agent that probably would be toxic in the accumulated amount of a large volume of solution, even though the concentration was low.

The USP also provides monographs giving the specifications for Sterile Water for Inhalation and Sterile Water for Irrigation. The USP should be consulted for the minor differences between these specifications and those for SWFI.

With regard to sample size, 100–300 mL is preferred when sampling WFI systems. Sample volumes less than 100 mL are unacceptable.

Organisms exist in a water system either as free floating in the water or attached to the walls of the pipes and tanks. When they are attached to the walls they are known as biofilm, which continuously slough off organisms. Thus, contamination is not uniformly distributed in a system and the sample may not be representative of the type and level of contamination. A count of 10 CFU/mL in one sample and 100 or even 1000 CFU/mL in a subsequent sample would not be unrealistic.

Water System Validation

Validation basically relies on periodic testing for microbiological quality and on the installation of monitoring equipment at specific checkpoints to ensure that the total system is operating properly and continuously, fulfilling its intended function.

Documentation should include a description of the system along with a print. The drawing needs to show all equipment in the system from the water feed to points of use. It should also show all sampling points and their designations. The print should be compared to the actual system annually to ensure its accuracy, detect unreported changes, and confirm reported changes to the system.

After all the equipment and piping have been verified as installed correctly and working as specified, the initial phase of the water system validation can begin. During this phase, the operational parameters and the cleaning/sanitization procedures and frequencies will be developed. Sampling should be done daily after each step in the purification process and at each point of use for 2–4 weeks. The sampling procedure for point-of-use sampling should reflect how the water is to be drawn; for example, if a hose is usually attached, the sample should be taken at the end of the hose. If the SOP calls for the line to be flushed before use of the water from that point, then the sample is taken after the flush. At the end of the 2–4-week time period, SOPs should be finalized for operation of the water system.

The second phase of the system validation is to demonstrate that the system will consistently produce the desired water quality when operated in conformance with the SOPs. The sampling is performed as in the initial phase and for the same time period. At the end of this phase, the data should demonstrate that the system will consistently produce the desired quality of water.

The third phase of validation is designed to demonstrate that when the water system is operated in accordance with the SOPs over a long period of time, it will consistently produce water of the desired quality. Any variations in the quality of the feed water that could affect the operation, and ultimately the water quality, will be picked up during this phase of the validation. Sampling is performed according to routine procedures and frequencies. For WFI systems the samples should be taken daily from a minimum of one point of use, with all points of use tested weekly. The validation of the water system is completed after at least 1 year of data have been accumulated.

While the above validation scheme is not the only way a system can be validated, it contains the necessary elements for validation of a water system. There must be data to support the SOPs. There must be data demonstrating that the SOPs are valid and that the system is capable of consistently producing water that meets the desired specifications. And, there must be data to demonstrate that seasonal variations in the feed water do not adversely affect the operation of the system or the water quality.

The last part of the validation is the compilation of the data, including acceptance criteria, with any conclusions into the final report. The final validation report must be signed by individuals responsible for the operation and quality assurance of the water system.

The FDA Guide to Inspection of Bulk Pharmaceutical Chemicals (July, 1993) comments on the concern for the quality of the water used for the manufacture of drug substances, particularly those drug substances used in parenteral manufacture. Excessive levels of microbiological and/or endotoxin contamination have been found in drug substances, with the source of contamination being the water used in purification. At this time, WFI does not have to be used in the finishing steps of synthesis/purification of drug substances for parenteral use. However, such water systems used in the final stages of processing of drug substances for parenteral use should be validated to assure minimal endotoxin/microbiological contamination.

In the active pharmaceutical ingredient industry, particularly for parenteral grade substances, it is common to see UF and RO systems in use in water systems. While UF may not be as efficient in reducing pyrogens, they will reduce the high-molecular-weight endotoxins that are a contaminant in water systems. As with RO, UF is not absolute, but it will reduce numbers. Additionally, as previously discussed with other cold systems, there is considerable maintenance required to maintain the system.

The FDA Guide to Inspections of Sterile Drug Substance Manufacturers (July, 1994) contains the following paragraph under Section VIII. Water for Injection: "Some manufacturers have attempted to utilize marginal systems, such as single pass Reverse Osmosis (RO) systems. For example, a foreign drug substance manufacturer was using a single pass RO system with post RO sterilizing filters to minimize microbiological contamination. This system was found to be unacceptable. RO filters are not absolute and should therefore be in series. Also, the use of sterilizing filters in a Water for Injection system to mask a microbiological (endotoxin) problem has also been unacceptable."

Typical Problems with Water Systems

FDA and other government or even internal quality inspections for good manufacturing practice (cGMP) compliance often find problems with the tight control of water systems. Examples of problems cited include:

- The water system is not validated with respect to control of quality and purity of WFI.
- Inadequate in-process and routine monitoring for water quality, especially related to microbiological purity.
- Improper responses after failed monitoring results.
- The water system has poor design to begin with.
- Poor system operation and/or maintenance.
- Lack of SOPs for operation, maintenance, and/or monitoring.

If a water system has endotoxin problems, there may be WFI in the condenser at the start-up. Since this water could lie in the condenser for up to several days (i.e., over the weekend), this may produce unacceptable levels of endotoxin.

A common problem is the failure to adequately treat feed water to reduce levels of endotoxins. It must be kept in mind that conductivity meters used to monitor chemical quality have no meaning regarding microbiological quality. Many of the still fabricators will only guarantee a 2.5 log to 3 log reduction in the endotoxin content. Therefore, it is not surprising that in systems where the feed water occasionally spikes to 250 EU/mL, unacceptable levels of endotoxins may occasionally appear in the distillate (WFI). For example, three new stills, including two multi-effect, were found to be periodically yielding WFI with levels greater than 0.25 EU/mL. Pretreatment systems for the stills included only deionization systems with no UF, RO, or distillation. Unless a firm has a satisfactory pretreatment system, it would be extremely difficult for them to demonstrate that the system is validated.

The above examples of problems with distillation units used to produce WFI point to problems with maintenance of the equipment or improper operation of the system. The system likely has not been properly validated or that the initial validation is no longer valid. If you see these types of problems you should look very closely at the system design, any changes that have been made to the system, the validation report, and the routine test data to determine if the system is operating in a state of control.

Since microbiological test results from a water system are not usually obtained until after the drug product is manufactured, results exceeding limits should be reviewed with regard to the drug product formulated from such water. Consideration with regard to the further processing or release of such a product will be dependent on the specific contaminant, the process, and the end use of the product. Such situations are usually evaluated on a case-by-case basis. It is a good practice for such situations to include an investigation report with the logic for release/rejection discussed in the firm's report. End-product microbiological testing, while providing some information, should not be relied on as the sole justification for the release of the drug product. The limitations of microbiological sampling and testing should be recognized.

Manufacturers should also have maintenance records or logs for equipment, such as the still. These logs should also be reviewed so that problems with the system and equipment can be evaluated.

In addition to reviewing test results, summary data, investigation reports and other data, and the print of the system should be reviewed while conducting the actual physical inspection. As pointed out, an accurate description and print of the system is needed in order to demonstrate that the system is validated.

AIR

Chapter 13 discussed standards (limits) for particles and microorganisms for the primary classifications of clean areas in sterile product manufacture. Table 15-3 is an abbreviated summary of air particle standards comparing U.S. and European classifications and clean room designations assigned by the International Society of Pharmaceutical Engineers. The numbers are based on the maximum allowed number of airborne particles/m^3 of 0.5 μm or larger size and, for Europe, 5.0 μm or larger size. The classifications used in pharmaceutical practice normally range from Class 100,000 (Grade D) for materials support areas to Class 100 (Grade A) for aseptic areas. To achieve Class 100 conditions, high-efficiency particulate air (HEPA) filters (Fig. 15-5) are required for the incoming air, with the effluent air sweeping the downstream environment at a uniform velocity, normally 90–100 ft/min $\pm$ 20%, along parallel lines [laminar airflow (LAF)]. HEPA filters are made of densely compacted fiberglass fibers, randomly arranged, that trap particles and other pollutants. HEPA filters are defined as 99.99% or more efficient in removing from the air 0.3-μm particles generated by vaporization of the hydrocarbon Emory 3004. Other characteristics of HEPA filters are given in Table 15-4.

Air Cleaning

Since air is one of the greatest potential sources of contaminants in clean rooms, special attention must be given to air being drawn into clean rooms by the heating, ventilating, and air-conditioning system. This may be done by a series of treatments that will vary somewhat from one installation to another.

Table 15-3 Comparison of Air Cleanliness Classifications

U.S. classification	European grade	ISO room designation	ISPE classification	Particles/m^3 $\geq$ 0.5/5.0 μm
100	A	5	Critical	3,500/0
100	B[a]	6	Clean	3,500/0
10,000	C	7	Controlled	350,200/2,000
100,000	D	8	Pharmaceutical	3,520,000/20,000

[a]Class B is the same as Class A at rest, but during operation, Class B has a limit of 350,200/2,000 particles/m^3 $\geq$ 0.5/5.0 μm. Class C has the same limits as Class D during operation.

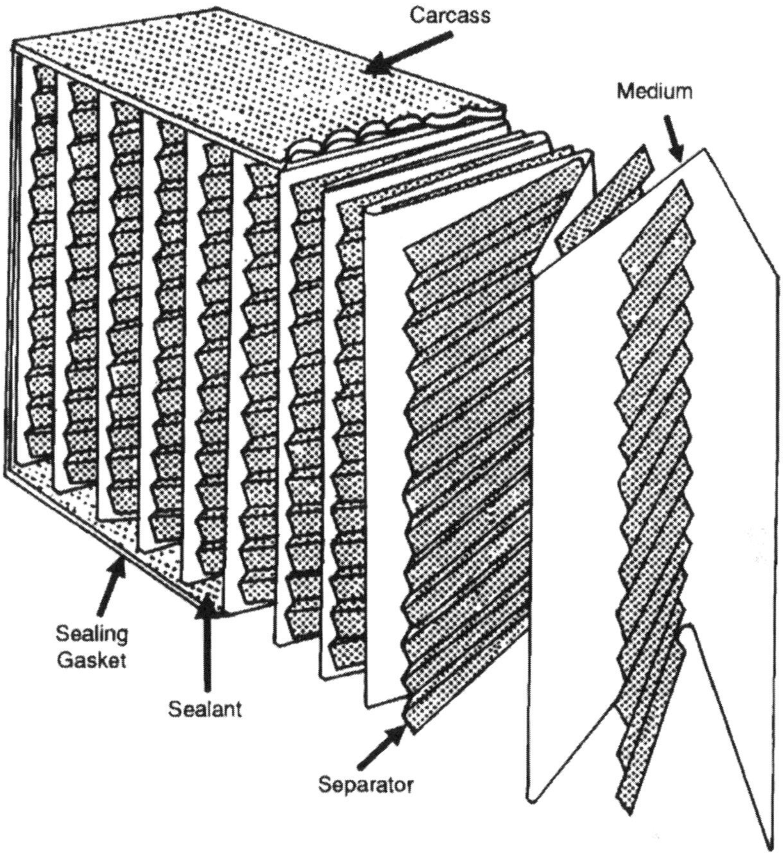

Figure 15-5 HEPA filter construction. *Source:* From Ref. 2.

First, air from the outside is passed through a prefilter, usually of glass wool, cloth, or shredded plastic, to remove large particles (Fig. 15-6). Then it may be treated by passage through an electrostatic precipitator. Such a unit induces an electrical charge on particles in the air and removes them by attraction to oppositely charged plates. The air then passes through the most efficient cleaning device, a HEPA filter.

Figure 15-7 schematically shows clean room air distribution with a LAF unit. Note that the air where the product filling occurs is Class 100 while personnel involved in the filling operation will be separated from the filling equipment by hard plexiglass or soft vinyl barriers with the air classification where people are located usually as Class 1000 or better.

Table 15-4 HEPA Filter Characteristics

- Remove 99.97% of particles $\geq$ 0.3 μm
- Developed initially by the Atomic Energy Commission during World War II
- Made of densely compacted fiberglass fibers randomly arranged into a tightly woven paper
- Pleated filters packed within bonded glass threads, ribbons, or molded media to maintain close, regular packing called "minipleat" design
- Remove particles by
  - Interception (electrostatic retention)—particles stick to fibers
  - Impaction (inertial impaction)—particles embed within the fibers
  - Diffusion (diffusive retention)—smallest particles collide with gas molecules, retarding velocity that enables interception or impaction to occur
  - Note: Particle removal is not dependent on sieving

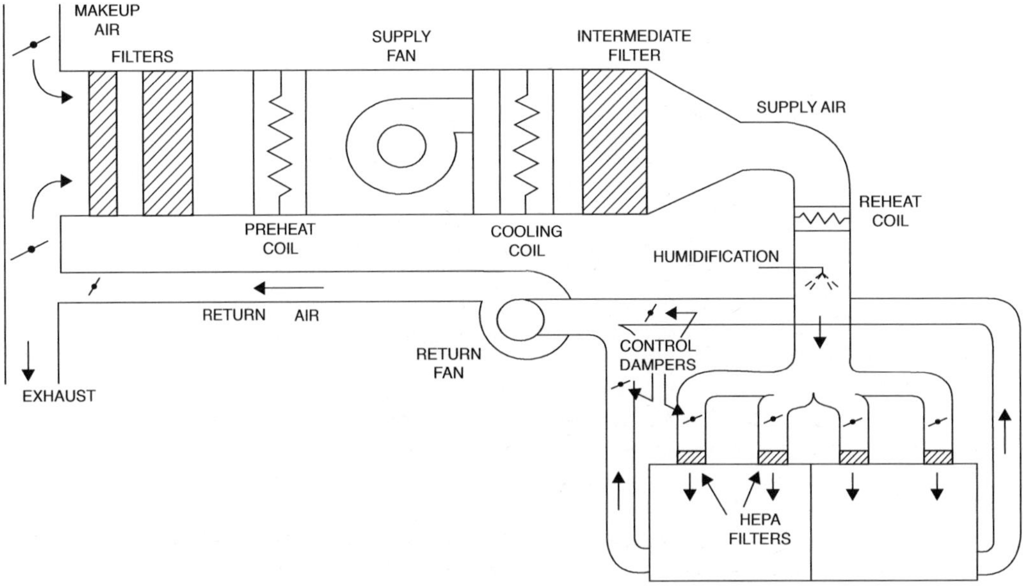

Figure 15-6 Schematic of HEPA filter system. *Source:* From Ref. 3.

For personnel comfort, air conditioning and humidity control should be incorporated into the system. The latter is also important for certain products such as those that must be lyophilized and for the processing of plastic medical devices. The clean, aseptic air is introduced into the Class 100 area and maintained under positive pressure, which prevents outside air from rushing into the aseptic area through cracks, temporarily open doors, or other openings.

LAF rate is usually around 90–100 ft/min, considerably above the rate at which airborne particles will settle. Air at this velocity sweeps suspended matter out of the LAF area, effective in preventing microorganisms from being carried upstream.

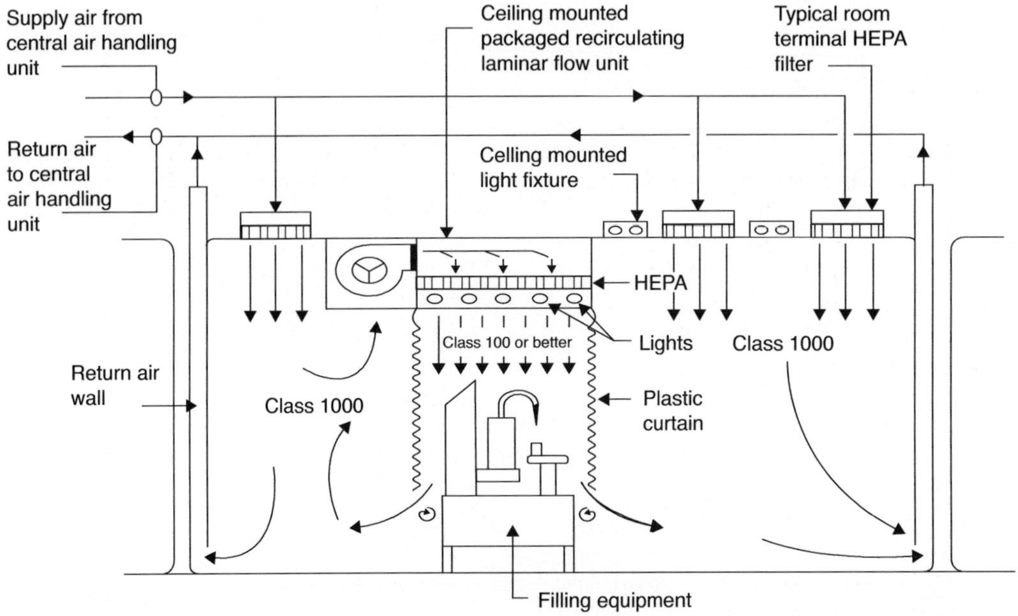

Figure 15-7 Clean room air distribution. *Source:* From Ref. 4.

Laminar-Flow Enclosures

The required environmental control of aseptic areas has been made possible by the use of LAF, originating through a HEPA filter occupying one entire side of the confined space. Therefore, it bathes the total space with very clean air, sweeping away contaminants. The orientation for the direction of airflow can be horizontal or vertical (Fig. 15-8), and may involve a limited area workbench or an entire room. Figure 15-9 shows a vial-filling line protected with vertical LAF from ceiling-hung HEPA filters, a Class 100/Grade A area. Plastic curtains are installed to maintain the unidirectional airflow to below the filling line and to circumscribe the critical filling portion of the line. The area outside the curtains can be maintained at a slightly lower level of cleanliness than that inside, perhaps Class 1000 or 10,000.

Critical areas of processing, wherein the sterile product and sterile product contact surfaces are exposed to the environment, however briefly, such environments must meet Class 100/Grade A/ISO 5 clean room standards.

It must be borne in mind that any contamination introduced upstream by equipment, arms of the operator, or leaks in the filter will be blown downstream. In the instance of horizontal flow this may be to the critical working site, the face of the operator, or across the room. Should the contaminant be, for example, penicillin powder, a biohazard material, or viable microorganisms, the danger to the operator is apparent.

Further, great care must be exercised to prevent cross-contamination from one operation to another, especially with horizontal LAF. For most large-scale operations a vertical system is much more desirable, with the air flowing through perforations in the countertop or through return louvers at floor level. Laminar-flow environments provide well-controlled work areas only if proper precautions are observed. Any reverse air currents or movements exceeding the velocity of the HEPA-filtered airflow may introduce contamination, as may coughing, reaching, or other manipulations of operators. Therefore, laminar-flow work areas should be protected by being located within controlled environments. Personnel should be attired for aseptic processing, as described below. All movements and processes should be planned carefully to avoid the introduction of contamination upstream of the critical work area. Checks of the air stream should be performed initially and at regular intervals (usually every 6 months) to make sure no leaks have developed through or around the HEPA filters.

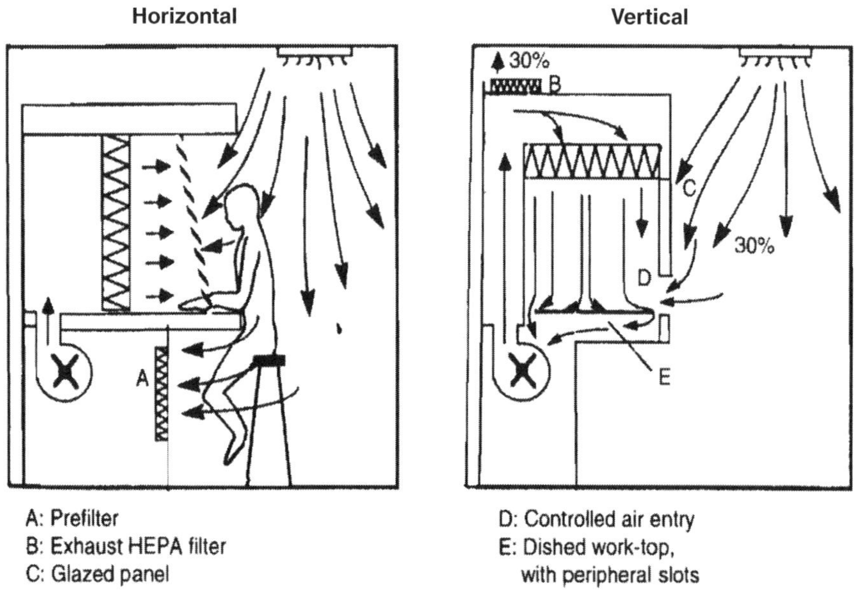

Figure 15-8 Horizontal and vertical laminar airflow. *Source:* From Ref. 2.

Figure 15-9 Filling line under vertical laminar airflow. *Source*: Courtesy of Baxter Healthcare Corporation.

Air Classification in Clean Rooms

The air classification of sterile product work areas generally abides by the following schematic:

> Warehouse (Unclassified) → Preparation of Equipment/Components (Class 100,000) → Compounding of the Product (Class 100,000) → Filling (Class 100) → Capping (Class 100,000) → Sterilization (Unclassified) → Sampling (Unclassified) → Finishing (Unclassified)

Clean room design traditionally has Class 100 rooms adjacent to Class 100,000 rooms. Regulatory authorities have raised serious concerns about this significant change in air quality from critical to controlled areas. It is now preferable to have an area classified from Class 1000 to Class 10,000 in a buffer area between a Class 100 and Class 100,000 area.

Potential Problems

People and equipment, if not positioned properly, will interfere with LAF. When LAF is interrupted, it usually can only be reestablished downstream within a distance equal to three times the diameter of the interfering object. If the interference location is above an open vial, there is usually not sufficient space to reestablish laminarity and turbulent air occurs at the vial opening (Fig. 15-10).

Laminar air filters are fragile and can be easily damaged. Filter material can be punctured easily and chemical splashes can cause filter rupture. This is why filters are usually protected with a screen and good aseptic practices taught so that those working within the confines of these filters realize how easily they are damaged.

An interesting type of problem that is introduced when people work within the confines of LAF is a false sense of security that poor or careless techniques will be compensated by the

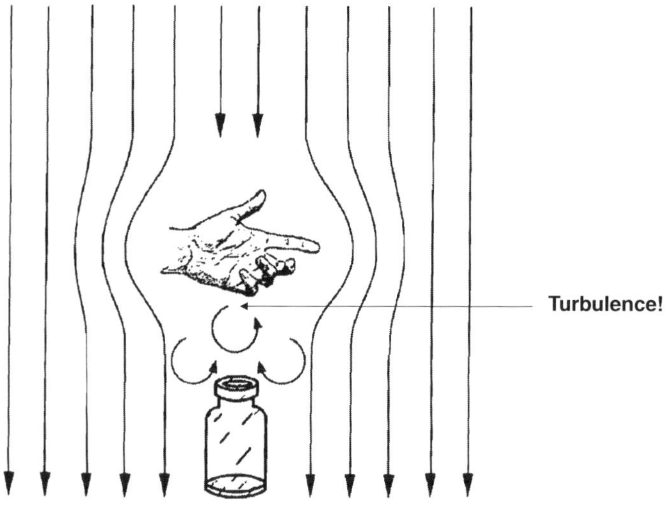

Figure 15-10 Effect of interference with vertical laminar airflow. *Source:* From Ref. 5.

LAF system. Of course, the very best LAF system will never compensate for improper aseptic practices.

REFERENCES

1. Brush H, Zoccolante G. Methods of producing water for injection. Pharm Eng 2009; 29:20–28.
2. Cosslett AG. The design of controlled environments. In: Denyer SP, Baird RM, eds. Guide to Microbiological Control in Pharmaceuticals and Medical Devices. 2nd ed. Boca Raton, FL: CRC Press, Taylor & Francis, 2007:69–88.
3. Keller AM, Lange H. Designing a parenteral production facility. In: Avis KE, Lieberman HA, Lachman L, eds. Pharmaceutical Dosage Forms: Parenteral Medications. Vol. 2. New York, NY: Marcel Dekker, 1993:278.
4. Moore BD. Air handling systems for cleanroom control. In: Avis KE, ed. Sterile Pharmaceutical Products: Process Engineering Applications. Boca Raton, FL: Interpharm/CRC Press, 1996:57.
5. Akers MJ. Good aseptic practices. In: Groves MJ, Murty R, eds. Aseptic Pharmaceutical Manufacturing II. Buffalo Grove, IL: Interpharm/CRC Press, 1996:201.

GENERAL REFERENCES

Akers MJ. Parenteral Preparations, Remington's Pharmaceutical Sciences. 21st ed. Chapter 41. In: Felton L, ed. Philadelphia, PA: Lippincott Williams & Wilkins, 2005:802–836.

Collentro WV. Pharmaceutical Water: System Design, Operation, and Validation. Boca Raton, FL: Interpharm/CRC Press, 1999.

Design Concepts for the Validation of a Water for Injection System. Technical Report No. 4. Bethesda, MD: Parenteral Drug Association, 1983.

Guide to Inspection of High Purity Water Systems, U.S. Food and Drug Adminstration, July, 1993, http://www.fda.gov/ICECI/Inspections/InspectionGuides/ucm074905.htm. Accessed June 9, 2010.

Meltzer TH. The validation of pharmaceutical water systems. In: Carleton FJ, Agalloco JP, eds. Validation of Pharmaceutical Processes. 2nd ed. London, UK: Informa, 1999:279–252.

16 Personnel requirements for sterile manufacturing

Of all the potential sources of microbiological and particle contamination in a sterile product manufacturing facility, people are the worst offenders. Therefore, anything that can be done to reduce contamination levels from people will improve the assurance of sterility of dosage forms prepared in the presence of people. Of course, the ultimate solution to eliminate this major source of contamination is to completely remove any contact of people with product being manufactured. Indeed, the advent of barrier isolators aims to do just that (chap. 23). Perhaps someday all sterile product filling will be accomplished using isolator systems, but until then, control of contamination from people will continue to be the greatest challenge in assuring product sterility.

PERSONNEL CHARACTERISTICS TO WORK IN CLASSIFIED ENVIRONMENTS

Personnel selected to work on the preparation of a sterile product must be neat, orderly, and reliable. They should be in a good health and free from dermatological conditions that might increase the microbial load. If they show symptoms of a head cold, allergies, or similar illness, they should not be permitted in the aseptic area until their recovery is complete. However, a healthy person with the best personal hygiene will still shed large numbers of viable and nonviable particles from body surfaces. This natural phenomenon creates continuing problems when personnel are present in clean rooms; effective training and proper gowning can reduce, but not eliminate, the problem of particle shedding from personnel.

Studies have been published (1–4) showing the huge number of particles emitted from the human body, depending on the type of activity (Tables 16-1 and 16-2). Each adult loses approximately 6–14 grams of dead skin material every day and a complete layer of skin is shed about every 4 days. When these huge numbers are compared against what defines a Class 100/Grade A clean room, it is a small wonder that the presence of human beings in the clean room presents a formidable challenge in maintaining high-quality particle standards. Table 16-3 provides other data on the relative number of particles shed from people and surfaces.

Personnel training is one of three important components that good manufacturing practice (GMP) regulations use to define qualified personnel involved in the preparation and testing of pharmaceutical products. The other two components are education and experience. There might be a fourth component, especially concerning personnel involved in aseptic manufacturing, and that fourth component is attitude. Indeed, in most job functions, personnel are adequately educated, experienced, and trained, but mistakes are still made and such mistakes most of the time can be traced to poor attitudes (lack of discipline, carelessness, apathy) about doing the job right.

There are at least six personnel factors that influence the degree of potential contamination from an individual working in a clean room (2). These are discussed below.

Bathing

Bathing will remove microorganisms, but will increase the number of particles emitted from the body. The washing process removes the outer oily sebum layer of the skin, causing skin scales to dry, curl, and peel off the body. Within 2 hours after bathing, the skin surface will resume its original pattern of microcolonization. Employees working in clean rooms should bathe at least 2 hours before they enter the clean room environment to minimize the extent of skin particle shedding resulting from the bathing process.

Suntan

Suntan dries the skin, causing it to flake and peel more easily. Incidents of contamination occur more frequently during the summer months, partly due to suntan issues. Creams help to reduce

Table 16-1 Particle Generation as a Function of Human Movement

Bodily movement	Number of particles per minute > 0.3 microns
Standing or sitting, no movement	100,000
Sitting with modest movement of head, hand, or lower arm	500,000
Sitting with moderate movement of body, arm, and feet	1,000,000
Standing up	2,000,000
Slow walk (~2.2 mi/hr)	5,000,000
Walking (~3.8 mi/hr)	7,500,000
Walking (~5.6 mi/hr)	10,000,000
Violent exercise	15,000,000–30,000,000

Source: From Ref. 1.

Table 16-2 Particle Generation as a Function of Human Movement and Type of Garment

	Number of particles ≥ 0.3 microns emitted per minute				
Personnel activity	Snap smock	Standard coverall	2-Piece coverall	Tyvek® coverall	Membrane coverall
No movement	100,000	10,000	4,000	1,000	10
Light movement	500,000	50,000	20,000	5,000	50
Heavy movement	1,000,000	100,000	40,000	10,000	100
Change of position	2,500,000	250,000	100,000	25,000	250
Slow walk	5,000,000	500,000	200,000	50,000	500
Fast walk	10,000,000	1,000,000	400,000	100,000	1,000

Note: Light/heavy movements refer to partial body movements (motioning with arm, tapping toes, etc.). Change of position refers to whole body motion (standing up, sitting down, etc.).
Source: From Ref. 5.

skin shedding, but do not entirely solve the problem. The obvious solution is to encourage clean room employees not to expose their skin to excessive sunlight.

Clothing

Friction between clothing and skin will increase the rate of bacterial shedding from the skin. Up to 10 mg of skin particles may get deposited in a person's clothing during a 2-hour period. Hosiery will also increase skin dissemination.

Personal Hygiene

Personal hygiene includes bathing regularly, washing one's hair, trimming facial hair, cleaning the fingernails, and wearing clean clothing and shoes.

Table 16-3 Particle Generation from People and Surfaces

Particle generators	Number of particles > 0.3 microns
Person emits during garmenting process	3,000,000/min
Cleanest skin (hands)	10,000,000/ft^2
Employee street clothes	10,000,000–30,000,000/ft^2
Floor and bench surfaces	>10,000,000/ft^2
Garments supplied by clean room laundry	1,000,000/ft^2

Source: From Ref. 6.

Clean Room Garment

The garment—including the body gown, hood, gloves, booties, masks—must be clean, sterile, and non-shedding. Almost equally important is how the entire garment is put on prior to entering a classified environment.

Traffic Movement and Control

Airborne contamination is directly related to the number of people working in a given area of clean room space and the types of activities they are doing. Therefore, number of personnel and degree of activities must be kept to a minimum. Activities that produce turbulence and increase particle contamination include talking, bodily movements, and anything that interrupts the flow of laminar air.

Human Skin Contamination

Ljungqvisdt and Reinmueller have published many research articles and books on clean room contamination and the problem of people being the main contamination source [(7–9) and references therein]. They have reported that there are >1.2 million/m^2 aerobic bacteria in head and neck regions of both male and female subjects and 0.9–3 million/m^2 in human hands and arms (10). There are even higher numbers of viable anaerobes, primarily *Proprionibacterium acnes*. A fully gowned person sitting in clean room releases approximately 15,000 particles per minute, > 0.5 μm, and a walking person releases approximately 157,000 particles per minute, > 0.5 μm. Therefore, the ratio of total particles > 0.5 μm and viable aerobic organisms encompass a broad range of 600–7000 to 1. People release 600–1300 total particles per hour in > 0.5 μm size range, with approximately 40 colony-forming units (CFUs) of viable aerobic organisms among these. The typical, properly gowned clean room worker will contribute 10–100 CFUs of viable aerobic organisms to the environment per hour (9).

What this reveals is that even the best classified environments and clean rooms with people in them will not approach "sterility." Since release of organisms by gowned personnel is quite normal, it should never be surprising to recover organisms at any location in the clean room. Clean room environments are not sterile as long as people populate them. The only viable approach is to have people gowned properly and strictly adhere to good aseptic practices and techniques.

Gowning

The uniform worn is designed to confine the contaminants discharged from the body of the operator, thereby preventing their entry into the production environment. For use in the aseptic area, uniforms should be sterile. Fresh, sterile uniforms should be used after every break period or whenever the individual returns to the aseptic area. In some plants this is not required if the product is to be sterilized in its final container. The uniform usually consists of coveralls for both men and women, hoods to cover the hair completely, face masks, and Dacron or plastic boots (Fig. 16-1). Sterile rubber or latex-free gloves are also required for aseptic operations, preceded by thorough scrubbing of the hands with a disinfectant soap. Most companies require two pairs of gloves, one pair put on at the beginning of the gowning procedure and the other pair put on after all other apparel has been donned. In addition, goggles are required to complete the coverage of all skin areas.

Dacron or Tyvek® uniforms are usually worn, are effective barriers to discharged body particles (viable and nonviable), are essentially lint-free, and are reasonably comfortable. Air showers are sometimes directed on personnel entering the processing area to blow loose lint from the uniforms.

Gowning rooms should be designed to enhance pre-gowning and gowning procedures by trained operators so that it is possible to ensure the continued sterility of the exterior surfaces of the sterile gowning components. De-gowning should be performed in a separate exit room.

While gowning heroically prevents human particles from grossly contaminating the controlled areas where they work, there are also certain issues with the gown and gowning process. There is no universally accepted procedure for the sequence of gowning steps. Some manufacturers put on boots before masks or hoods; others use the opposite sequence. The use of single versus double gloves still differs among manufacturers, although the great majority of

Figure 16-1 Fully gowned personnel in Class 100/Grade A/B clean rooms. *Source:* Courtesy of Baxter Healthcare Corporation.

manufacturers use double gloves. The type of gown with respect to the quality of stitching at the wrist and ankle cuffs continues to be controversial. Some manufacturers use reusable gowns, while others use disposable gowns. The problem of flogging when goggles are used with personnel requiring corrective glasses continues to be troubling.

The following are the most critical or common mistakes that happen in gowning procedures and working in clean room environments with respect to gowning:

- Failure to follow the proper gowning procedure, for example, incorrect sequence in donning the gowning components.
- Failure to scrub hands and fingernails thoroughly.
- Hair is not completely covered.
- Skin is exposed between the gloved hand and the uniform sleeve.
- The gowned person is able to reach underparts of his or her garment with the gloved hand.
- A part of the garment is dropped on the floor, yet it is not replaced and used as is.
- The gowned person touches parts of his/her face with the gloved hand and fails to sanitize the glove afterward.
- The face mask is not completely covering the face.
- Zippers are not completely zipped and parts of the gown are not completely tucked in or properly overlapped.

Personnel Training

Training is a dynamic process that should/must occur over an entire career of every employee. While training programs exist, the big question is how effective are they? Simply documenting that a person has gone through a training course does not mean that learning actually occurred, even if tests are performed. Training documentation does not verify quality. Systematic training plans should be in place for every job function. These plans should state performance objectives, methods used to achieve these objectives, and an assessment process to measure accomplishment of those objectives.

There are four main methods for measuring training effectiveness:

1. Testing
2. Evaluation of on-the-job error rates
3. Skill-related questioning
4. Employee reports regarding their own assessment of their effectiveness

Personnel involved with manufacturing and testing of sterile products should be given thorough, formal training in the principles of aseptic processing and the techniques to be employed. In fact, personnel training should focus on the following subjects:

1. Minimizing and eliminating sources of contamination
   a. Air quality
   b. Cleaning/disinfection/sterilization
   c. Personal factors
      i. Selection criteria
      ii. Requirements to control contamination
      iii. Gowning procedures
      iv. Aseptic techniques
2. Objective testing
3. Hands-on testing
   a. Gowning test
   b. Broth test
   c. Media fills
4. Remedial training
5. Understanding what the Food and Drug Administration (FDA) evaluates when inspecting for personnel training and aseptic processing

The broth test is a test of aseptic technique while manually filling vials with sterile culture media (broth) and manually stoppering using sterile forceps. The trainee must first be certified on donning sterile gowning correctly. Manual filling involves 250–400 vials at one setting with the exercise repeated one or two more times on separate days. The vials are incubated just like media-filled vials along with positive and negative controls. Also, after each of the three tests the trainee's fingers and chest are sampled with Rodac plates to determine the presence of any contamination while the trainee was performing the test. If any of the 750–1200 vials show contamination after appropriate incubation, the entire broth test must be repeated after remedial training. These tests also are performed in the presence of a trainer who can point out technique errors during the test.

Subsequently, the acquired knowledge and skills should be evaluated to assure that training has been effective, before they are allowed to participate in the preparation of sterile products. Retraining should be performed on a regular schedule to enhance the maintenance of the required level of expertise. An effort should be made to imbue operators with an awareness of the vital role they play in determining the reliability and safety of the final product. This is especially true of supervisors, since they should be individuals who not only understand the unique requirements of aseptic procedures, but are also able to obtain the full participation of other employees in fulfilling these exacting requirements.

Outlines of personnel training curricula are given in Tables 16-4 and 16-5. Table 16-6 provides example test questions. Figures 16-2 and 16-3 are examples of fun exercises to point out some of the "do's" and "don'ts" of aseptic techniques that are listed in Table 16-7.

Role of Management

Management always plays a key role in any organization. Management always needs to be a source of inspiration for all personnel so that every person does his/her best in their job functions. Management should never be a source of problems due to poor leadership, poor decision making, lack of hard work, lack of support, lack of being good examples, incompetent thinking and facilitation, and so forth. So often, employee attitudes about doing the job as best as they can are dictated by attitudes and actions of their management leadership.

With respect to management responsibilities over employees who work in manufacturing and quality environments, there are several points to consider:

- Management themselves must recognize and fully appreciate the need to follow GMPs, including good documentation practices and good aseptic practices.
- Management must hire people who are willing to accept and follow procedures assuring adherence to GMPs.

Table 16-4 Example of a Training Program for Personnel Working in Aseptic Environments

1. Prerequisite training
   a. Safety
   b. Overview of good manufacturing practice regulations
   c. Good documentation practices
   d. Garment/gowning practices
   e. Good hygiene practices
2. General aseptic practices
   a. Basics of laminar airflow hoods/environments
   b. General aseptic techniques and procedures
   c. Understanding the aseptic environment (see Table 16-5)
   d. Aseptic connections
   e. Introducing equipment through air locks
   f. Preparation of areas for aseptic processing
   g. Use of goggles
   h. Sanitization of aseptic areas
   i. Environmental monitoring of aseptic areas
   j. Recent internal audit findings
   k. Review of FDA and government inspections
3. Specific aseptic practices
   a. Introduction to aseptic gowning
   b. Aseptic gowning procedure
   c. Aseptic operator monitoring
   d. Gowning certification
   e. Broth test procedure
   f. Broth test certification
   g. Participation in line media fills

- Management must effectively communicate and exemplify the importance of GMPs without breeding negative feelings among employees.
- Management must realize that employees' attitudes are extremely important and attitudes are markedly influenced by how employees are treated.
- Management must support thorough and ongoing training programs.
- Management should strive to be teachers and leaders in enabling their employees to want to learn and keep learning and follow all procedures correctly.

Table 16-5 Outline of Course on Understanding the Aseptic Environment

1. Purpose
   a. To increase knowledge of basic microbiology
   b. To increase awareness of contamination
   c. To understand how the employee can help control contamination through proper behavior and technique
2. Part I: Microbiology
   a. Definitions (e.g., parenteral, sterile, clean, asepsis, etc.)
   b. Viewing and discussion of videotape Basic Microbiology[a]
   c. Origin of microbial contamination including human contamination
   d. Growth requirements for microorganisms
3. Part II: Contamination Control
   a. Viewing and discussion of videotape Basic Contamination Control[a]
   b. Definition, sources and elimination of contamination
   c. Laminar airflow, understanding of HEPA filters, and general clean room technology
4. Part III: Behavior in the Clean Room
   a. Viewing and discussion of videotape Behavior in the Clean Room[a]
   b. Demonstrate or discuss video of "Do's and Don'ts" in the clean room (see Table 16-7)
      i. Dress
      ii. Work
      iii. Act
      iv. Move

[a]Videotapes obtained from Micron Video International, Inc., http://www.mvitraining.com

Table 16-6 Examples of Test Questions Related to Aseptic Manufacturing Practices

Objective Questions

1. Define basic terms: sterile, sterility assurance, aseptic, aseptic processing, terminal sterilization, endotoxin, pyrogen, bioburden, clean room grad and class, HEPA filter, laminar flow, media fill.
2. Identify the criteria required for facilities and type of equipment used in an aseptic environment.
3. Describe the value/limitations of laminar flow/sanitization in promoting sterility.
4. Define environmental monitoring and identify what it can and cannot do to assure sterility.
5. Identify the steps required for vial/stopper preparation.
6. Define and explain six sterilization methods.
7. Describe the materials used in container/closure systems and the pros/cons of each type.
8. Identify the factors that contribute to sterility assurance.
9. Describe the unit operation steps in aseptic processing.
10. Identify five major sterility issues that affect pharmaceutical companies today.
11. Identify the topics covered within FDA guidelines on aseptic processing and explain key points associated with each topic.
12. Evaluate case studies related to sterility assurance/aseptic process validation issues. Determine the appropriate course of action to take based on your understanding of FDA guidelines and GMPs.

True or False Questions

1. Air samples are quantitatively accurate.
2. Microbiology is an exact science.
3. Contamination detected by active air samplers means that the product made at the same time is contaminated.
4. Air sampling devices are generally equal in their ability to detect contamination.
5. Statistics plus sampling plan analysis is sufficient to enable you to create a formula for determining accept/reject of a product lot.
6. Microbes will survive forever in a clean room unless killed by a disinfectant.
7. Microbes develop resistance to chemical disinfectants over time; therefore, disinfectants must be rotated.
8. Once released into a clean room environment, microbes will proliferate.
9. Microbes are highly motile and can easily float and fall into a product.
10. RODAC plates give the best data when used at the END of a day because they can find all contamination that might have fallen out of the air.
11. Sampling of product contact surfaces after completion of an aseptic process can give excellent indication of the environmental conditions.
12. Microbiological data can be trended and evaluated using ordinary statistical methods.
13. Detection of a number of CFUs (colony forming units) higher than expected is cause for immediate concern.
14. Anaerobes, molds, and yeast are common contaminants in aseptic processing areas and monitoring for their presence is essential.
15. The aseptic environment is full of organisms that cannot be detected in our EM (environmental monitoring) programs. These organisms pose a serious health threat to consumers.
16. Formaldehyde and UV light are not effective antimicrobial agents.
17. Lack of sterility assurance has been the no. 1 reason for recalls for the past 4 years. The greatest number of these recalls occurred in 2001.
18. In certain situations, it can be acceptable to use a non-sterilized tool for an intervention during aseptic processing.
19. About half of the drug products recalled due to nonsterility over the past 10 years were produced by aseptic processing.
20. Data indicating loss of environmental control may not always need to be treated seriously.
21. According to the FDA, there may not be any level of microcontamination in aseptic processing rooms.
22. FDA becomes very concerned about EM data when they show an adverse trend. A single atypical result is not cause for alarm.
23. FDA is very concerned about temperature differences inside and outside a freeze dryer that result from air flowing into the chamber when the chamber door is open. This air must be HEPA-filtered.
24. The FDA has not identified any concerns related to barrier/isolator technology, which is why so many pharmaceutical companies are interested in using it.
25. Loss of GMP control in aseptic rooms is usually the result of poor equipment design.
26. Equipment design issues outweigh poor personnel practices in causing deviations in acceptable environmental monitoring data.
27. Invalidation of a media fill is acceptable if the deviation would also be cause for aborting a commercial run.
28. You can justify removing a unit of media if the unit legitimately would be removed as part of the aseptic process during an intervention.

1 Excess paper
2 Tweezer on work surface
3 Contact with open fill port
4 Bare wrist
5 Mask low on nose
6 Goggles on forehead
7 Gown unzipped
8 Nonsterile supplies in aseptic area
9 Arm resting on equipment
10 Adjusting goggles
11 Hood out of gown
12 Second hood out of gown
13 Rip in uniform
14 Bare wrist

Figure 16-2 Example of a training cartoon showing 14 aseptic practice errors.

Other responsibilities of management that have great impact on employees, especially those working in classified air environments, are listed in Table 16-8.

FDA Audits

It must be clearly appreciated that employee training occurs not to satisfy FDA and other regulatory group expectations, but rather to ensure that safe and effective drug products are manufactured and tested properly. When FDA GMP inspections focus on personnel training, they basically question what kind of training is performed, how is it documented, how is

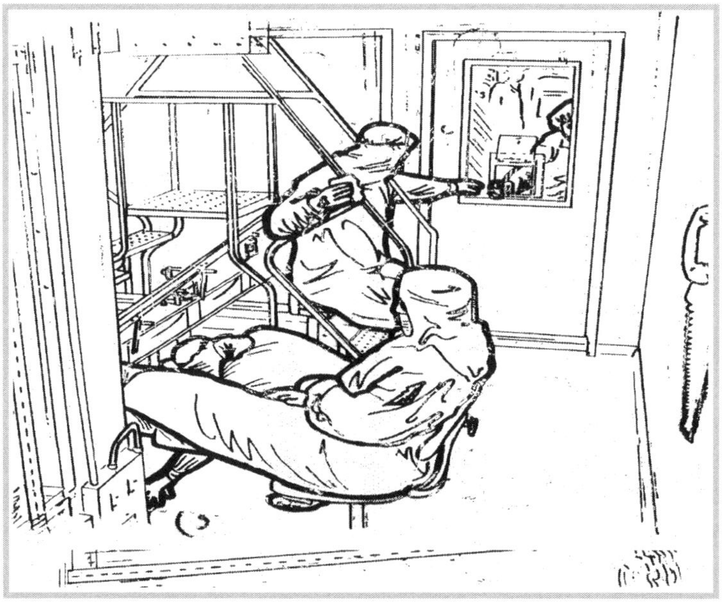

1 Transfer from less clean environment
2 Gloved hand about to touch non-sterile object
3 Goggles on head
4 Hood outside gown
5 Hand holding potential non-sterile item
6 Tear in glove
7 Sitting on ladder
8 Face mask below nose
9 Torn uniform
10 Hand touching lower body
11 Body posture shows poor attitude
12 Third hood outside gown
13 Picking up item on floor
14 Knee touching wall

Figure 16-3 Another example of a training cartoon showing 14 aseptic practice errors.

Table 16-7 "Do's and Don'ts" of Aseptic Practices

Dress Correctly
- Understand rules of gowning.
- Proper gowning procedures and protection at all times.

Work Correctly
- Proper cleaning and sanitization of room and all work surfaces.
- Never store sanitizing solution in a critical area.
- Know when to re-sanitize or change gloves.
- Sanitize gloved hands each time before entering a critical area.
- No paper except sterile bioshield paper is allowed in a critical area.
- Pens, calculators, etc. must never be placed inside a critical area.
- Avoid of any particle shedding object—pencils, paper, exposed hair, and skin.
- Keep laminar hood doors closed as much as possible.
- When entering a critical area make sure that all other doors are closed.
- Don't interrupt the laminar airflow pattern above or around any sterile opening or object.
- Do be aware of a false sense of security when working in laminar airflow areas.
- Follow strict personal hygiene procedures.
- Always be aware of hands and fingers with respect to source of HEPA-filtered air.
- During setup bring sterile equipment as close to the critical area as possible before transferring the equipment.
- All wrapped sterile equipment must be unwrapped in the critical area. Utmost care must be taken to protect the sterility of this equipment.
- Never touch any product contact part with gloved hands.
- Never touch the floor.
- Nothing should be placed between the product or contact part and the source of HEPA-filtered air.
- If it is absolutely necessary to reach over a sterile opening of a container, that product unit must be discarded.
- Once unwrapped, sterile forceps, tweezers, and/or hemostats must remain in the critical area.
- Never pick up anything off the floor unless it is determined that a safety hazard exists (e.g., spilled cytotoxic agent). If anything is picked up, gloves and perhaps entire gown must be replaced.

Move and Act Correctly
- Minimize talking.
- Minimize body movement.
- Never eat or have anything in your mouth while in any classified area.
- Never touch exposed skin and avoid touching others' clothing as well as your own. No scratching or rubbing.
- Never sit on tables, ladders, waste receptacles, etc. Only approved chairs suitable for aseptic areas can be used.
- Never put feet on anything that could come into contact with your hands or gown.
- Never open sterile gown unless degowning when exiting the critical area.
- Avoid unnecessary motions in any critical area, especially Grade A/B areas.
- Movements should be slow and deliberate.

Table 16-8 Responsibilities of Management That Impact Quality of Training and Conduct of Employees in Clean Room Operations

- Impart belief that good aseptic practices are essential for the manufacture and control of sterile products possessing the GMP values of safety, identity, strength, purity, and quality.
- Instill feelings of pride and confidence in clean room operations.
- Stress the concept of teamwork.
- Help employees feel honored to be chosen for such critical job functions.
- Keep employees informed continuously about what is going on internally and externally; e.g.,
    - New or revised GMP regulations
    - Learning points from QA and FDA/other government GMP inspections
    - Advances made from reading literature, attending conferences
- Involve employees in goal setting, problem solving, and decision making.
- Actively listen and respond to employee feedback.
- Creatively recognize conscientious and outstanding performance.

Table 16-9 FDA Audit Findings Related to Personnel Practices in Clean Rooms

- Inappropriate techniques were observed within aseptic areas.
- Different degrees of proper aseptic gowning were widely observed.
- Not all personnel observed in the aseptic areas were wearing goggles as required.
- Operator observed leaning over the accumulator for no apparent reason.
- Exaggerated movements (dancing) were observed.
- Plexi-panels were open on both sides of critical area so that operators could talk to one another.
- Too may people located within aseptic areas.
- One operator noted to run up to the filling line, arms waving.
- A group of five operators congregated inside the Class 100 critical area.
- Too much leaning over exposed vials observed.
- Operator appeared to be touching sterile tweezers while hand stoppering.
- Operator went into critical area three times without sanitizing their hands.
- Operator not correctly using tweezers to remove overturned bottles on accumulator.
- Hands were sanitized using dirty LPH, which was used to sanitize several stopper torpedoes.
- Cleaning/sanitizing of aseptic areas not unidirectional.
- Head covers did not always cover the face.
- Beard covers did not always cover beards.
- Operator was observed in Class 100 area with regular glasses, not goggles.
- Operator observed with goggles up, resting on forehead while working inside the critical adjacent area.

training evaluated to assure that learning actually occurred, and how remedial or retraining procedures are done in cases involving personnel failures. The FDA investigator might request to observe personnel gowning procedures, hand washing and sanitation techniques, and how an operator works in the clean room. The following are general expectations of the FDA and other regulatory bodies regarding GMP training programs:

- Training is a dynamic process that should keep occurring over the entire career.
- It must not be assumed that all training is effective or that documentation that training occurred assures that learning took place.
- Each job function must have a training plan that contains performance objectives and assessments to measure accomplishment of those objectives.
- The main methods for measuring training effectiveness include testing, demonstration of skills learned, and evaluation of on-the-job error rates.

Table 16-9 lists examples of 483 citations resulting from audits involving personnel training and observations as they work.

REFERENCES

1. Howorth FH. Movement of airflow, peripheral entrainment, and dispersion of contaminants. J Parenteral Sci Tech 1998; 42:14–19.
2. Luna CJ. Personnel: The key factor in clean room operations. In: Avis KE, Lieberman HA, Lachman L, eds. Pharmaceutical Dosage Forms: Parenteral Medications. Vol 2. 2nd ed. Marcel Dekker, 1993:367–410.
3. Akers MJ. Good aseptic practices: Education and training of personnel involved in aseptic processing. In: Groves MJ, Murty R, eds. Aseptic Pharmaceutical Manufacturing II. Buffalo Grove, IL: Interpharm Press, 1995:181–221.
4. Whyte W. Cleanroom clothing. In: Whyte W, ed. Cleanroom Technology: Fundamentals of Design, Testing, and Operation. New York, NY: John Wiley & Sons, 2001:237–262.
5. Dr. Philip Austin, Austin Contamination Index, Encyclopedia of Clean Rooms, Bio-Cleanrooms and Aseptic Areas, 2000, Contamination Control Seminars, Livonia, MI, p. 16.
6. Dr. Philip Austin, Cleanroom Garments, Encyclopedia of Clean Rooms, BioCleanrooms and Aseptic Areas, 2000, Contamination Control Seminars, Livonia, MI, p. 16.
7. Ljungqvist B, Reinmuller B. Cleanroom design: Minimizing contamination through proper design. Boca Raton, FL: Interpharm/CRC Press, 2002.

8. Ljungqvist B, Reinmuller B. People as a contamination source: Cleanroom clothing systems after 1, 25, and 50 washing/sterilizing cycles. Eur J Parenteral Pharm Sci 2003; 8:75–80.
9. Ljungqvist B, Reinmuller B. Predicted contamination levels in clean rooms when cleanroom dressed people are the contamination source. Pharm Tech (special supplement on aseptic processing), 2006.
10. Ljungqvist B, Reinmuller B. Hazard analyses of airborne contamination in clean rooms—Application of a method for limitation of risks. PDA J Pharm Sci Tech 1995; 49:239–243.

17 Sterilization methods in sterile product manufacturing

The entire field of discipline of sterile product science and technology is based on the ability to render finished dosage forms sterile. Sterility is defined theoretically as the complete absence of microbial life. Achieving sterility is the subject of this chapter.

There are four main methods that sterilize items used in parenteral manufacturing, testing, and administration.

1. Heat
2. Gas
3. Radiation
4. Filtration

There is a fifth possibility—bright light—but at the time of writing this text, light sterilization had not yet reached a status where it can be considered a standard sterilization technique. More discussion of bright light sterilization will be given at the end of this chapter. This chapter will cover heat, gas, and radiation sterilization while chapter 18 will cover filtration sterilization.

Another way to classify sterilization methods can be the following:

1. Thermal
   (a) Moist
   (b) Dry
2. Nonthermal
   (a) Filtration
   (b) Radiation
3. Chemical
   (a) Gaseous
   (b) Liquid

Before describing each of these sterilization methods, basic microbiology principles are presented. Both the United States aseptic processing guidelines and the European Union manufacture of sterile medicinal products documents require all personnel working with sterile products to have formal training on basic microbiology principles. In addition, prior to providing the basics of sterilization methods, the basics of microbial death kinetics will be covered.

SOME BASIC MICROBIOLOGY PRINCIPLES

Terms used frequently in discussing sterilization procedures in sterile product manufacturing include the following:

- Sterility—Absolute freedom from biological contamination
- Asepsis—Freedom from microbial infection potential (sepsis)
- Sterilization—Elimination of all viable microorganisms
- Disinfection—Renders objects noninfectious
- Sanitizing agent—Reduces the microbial population
- Spore—Resistant hibernation state of microorganisms
- Vegetative cell—capable of multiplication.

Most information developed on the growth, survival, and death of microorganisms comes from work performed under ideal conditions of a laboratory. Microorganisms found in a sterile production area are typically under nutritional, chemical, dehydration of other form of stress. Therefore, what is known ideally about microorganisms may not predict actual situations.

Sterilization procedures destroy or eliminate bacterial, fungal (yeast and mold), and viral contamination. Some brief instructions about each of these life forms: Bacteria can be gram positive or gram negative. Whether a bacterial life form is gram positive or negative depends

Table 17-1 Biological Indicators and *D* Values

Sterilization process	Biological indicator	ATCC number	Typical *D* value range
Steam	*Geobacillus* (formerly *Bacillus*) *stearothermophilus*	7953	1.5 min @ 121°C
Dry Heat	*Bacillus subtilis* var. *niger*	9372	1.0 min @ 180°C
Ionizing radiation	*Bacillus pumulis*	14884	3.0 kG
Ethylene oxide	*Bacillus subtilis* var. *niger*	9372	5.8 min @ 600 mg/L, 54°C, 60% RH
Vapor phase hydrogen peroxide	*Geobacillus* (formerly *Bacillus*) *stearothermophilus*	7953	NA
Peracetic acid	*Bacillus subtilis* var. *niger*	9372	NA

Abbreviation: RH, relative humidity.
Source: From Ref. 9.

on the presence of a cellular envelope. Gram-positive bacteria do not contain an outer cell wall while gram-negative bacteria do. It is this outer cell wall of gram-negative bacteria that contains layer(s) of lipopolysaccharide that produces endotoxins when such bacteria grow and die. Major examples of gram-positive bacteria are *Staphylococcus, Streptococcus, Bacillus, and Clostridium.* Major examples (genus) of gram-negative bacteria are *Pseudomonas, Escherichia coli, Salmonella, Klebsiella, and Serratia.*

Bacteria can also be classified as aerobic (requiring oxygen to grow), anaerobic (can grow in nonoxygen environments, e.g., nitrogen saturated solutions), or facultative (can grow in either environment). Bacteria are pathogenic, nonpathogenic, or opportunistic. Some, but not all, bacteria can form spore forms. Spore formation results where the bacterial cell, in order to continue to survive, develops a sort of outer shell that protects it from adverse environmental conditions such as heat, chemicals, and nutrient depletion. The most common spore formers are *Bacillus* and *Clostridium* species. Both are gram-positive bacteria and are commonly used as biological indicators (Table 17-1) since spores are hundreds of times more resistant than vegetative bacteria to the effects of sterilization treatments. Biological indicators measure the effectiveness of sterilization methods.

Fungi are cellular forms that are very similar to human cells. While bacterial cells are called prokaryotic cells, fungal cells are called eukaryotic cells, the same classification as human cells. Fungal cells are much more difficult to destroy plus attempts to kill fungal cells may also kill human cells because of their similar cell types. About 10% of all known fungi are pathogenic. *Candida* species and some dermatophytes are the only known fungi transmitted from person to person.

Viruses are intracellular parasites that do not need food to survive. Viruses are extremely small and will easily pass through bacterial retentive filters. Viruses are readily inactivated by heat at relative low (~ 65°C or above) temperatures. They are very susceptible to surface disinfectants. The environmental detection of viral contaminations can be very costly. In light of the fact that sterile manufacturing environments are extremely harsh for viral survival, viral monitoring and concerns about viral contamination in finished product manufacture are practically nonexistent.

MICROBIAL DEATH KINETICS

Figure 17-1 displays the ideal growth and death phases of microorganisms. Since microbial growth is cellular duplication and multiplication (geometric progression), growth is plotted logarithmic (exponential). When contamination first occurs, assuming that it is very low level (a few cells), there is a lag time before sufficient cells can be measured. This lag phase can be minutes to much longer times (even years). In sterilization microbiology, lag times are considered to be minutes to hours. Once growth starts, it progresses quite rapidly. For example, a typical bacterial cell might duplicate itself every 20 minutes under ideal growth conditions. Therefore, after one-hour incubation, one cell has grown to eight cells. After two hours 8 cells have grown to 64 cells. After 8 hours of duplication every 20 minutes that original single cell has produced over 1,300,000 cells!

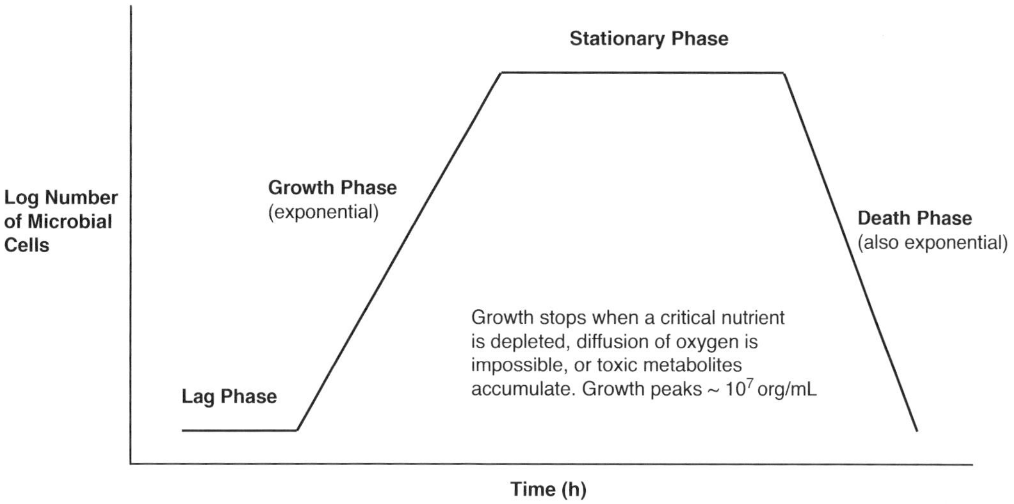

Figure 17-1 Microbial growth and death.

Eventually microbial growth will plateau (stationary phase in Fig. 17-1) when a critical nutrient is depleted or oxygen diffusion cannot go on, or toxic metabolites accumulate. Although variable, the approximate population where stationary phases exist are 10,000,000 cells/mL.

Microbial death kinetics is also exponential. Logarithmic plots of microbial population versus time for heat or gas sterilization or versus dose for radiation sterilization allow microbiologists to develop rate constants for sterilization conditions just like chemical kinetic plots are used to determine drug stability profiles. An example of a microbial death kinetic plot is given in Figure 17-2.

There are several common terms used in microbial death kinetic studies—initial microbial population or bioburden, *D* value, *Z* value, and *F* value.

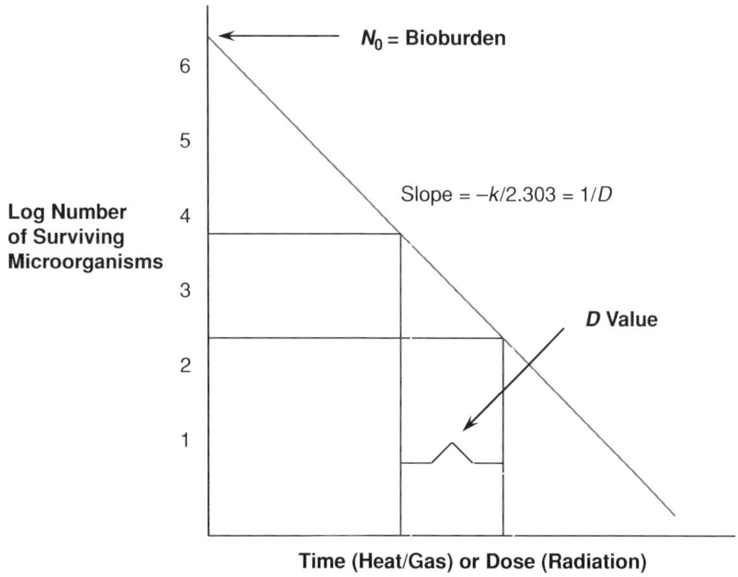

Figure 17-2 Microbial death kinetics. *D* value is defined as the time in minutes required for a one-log cycle or 90% reduction of a microbial population under specified lethal conditions. It is a fundamental biological parameter in sterilization process analysis and can be determined by the survivor curve method or the fraction-negative method.

Bioburden

Bioburden is a measure of microorganism recovery in a unit of product or substance (raw material, solution, surface, etc.). It is the initial population of microorganisms prior to being subjected to a sterilization procedure; thus, bioburden is the Y-intercept of the microbial death kinetic plot. Bioburden determination in a laboratory for *D* value determination is relatively reproducible compared with bioburden determination in a compounded product solution prior to filtration. In the manufacturing environment, most microorganisms are difficult to recover (grow). Bioburden recovery on surfaces using swabs or Rodac plates is usually very poor; on the order of 30% to 40% of what might actually present.

In manufacturing environments, the expected maximum level of microorganisms prior to filtration is not more than 10 CFUs/100 mL solution. United States Pharmacopeia (USP) General Chapter <1111> provides guidelines for raw materials, excipients, and bulk drug substances with a proposed bioburden limit of not more than 1000 CFU/g or mL.

D Value

The *D* value (decimal point reduction value) is the time or dose required for a one log reduction in the microbial population under specific conditions. An example of a plot of microbial population versus time or dose is shown in Figure 17-2 where it is demonstrated how the *D* value is obtained. In reality, a linear regression line is calculated, as microbial population versus time data is rarely linear. The *D* value is calculated from this best fit line from the microbial reduction data over time or dose. Several complicated mathematical approaches are available for accurately calculating the *D* value (1). If the *D* value is one minute, this means that it required one minute at a given temperature for the microbial population to be reduced by one log unit. The *D* value is dependent on many factors including type of microorganism, temperature or dose, and the medium/substance containing the microorganism. *D* values for a variety of microorganisms at 121°C are given in Table 17-2. Note the differences in the time required to reduce different species at the same temperature. Table 17-2 also shows the effect of type of sterilization method on the *D* value of the same microorganism (*B. subtilis var. niger*) showing the effectiveness of different treatments and different temperatures on microbial level reduction.

D values for some biological indicator organisms used currently are shown in Table 17-1. Over the years these values have increased indicating the natural tendency of microbial life to develop resistance to methods used to destroy them.

Z Value

The *Z* value is the number of degrees or dosage units required for a one log reduction in the *D* value. The *Z* value measures resistance of the microorganism to the sterilization source. Figure 17-3 shows a logarithmic plot of *D* value for a particular microorganism versus temperature for a heat sterilization process. The steeper the slope the more resistant is the indicator organism. The conventional *Z* value used for steam sterilization is 10°C (2).

Table 17-2A *D* Values for Different Bacterial Spores by Steam Sterilization

Spore	*D* value range (minutes at 121°C)
Bacillus stearothermophilus	1.5–3.0
Bacillus subtilis ATC 5230	0.3–0.7
Bacillus coagulans	0.4–0.8
Clostridium sporogenes	0.4–0.8

Original source unknown. Information obtained from Kenneth E. Avis course notes, University of Tennessee, 1980. *D* Values may no longer be accurate, but purpose of the table is to point out relative differences among bacterial spores to steam sterilization.

Table 17-2B *D* Values for *Bacillus subtilis var. niger* Exposed to Different Sterilization Treatments

Process	Parameters	*D* value (min)
Steam	121°C	0.5
	118°C	5.0
Dry heat	250°C	5.0
Ethylene oxide	600 mg/L gas; 54°C, 60% RH	3.9
Gamma radiation	NA	0.6 kG

Original source unknown. Information obtained from Kenneth E. Avis course notes, University of Tennessee, 1980. *D* Values may no longer be accurate, but purpose of the table is to point out relative differences among sterilization treatments to the same bacterial spore destruction.

F Value

The *F* value is the equivalent time at a given temperature that a lethal amount of sterilization is delivered to a unit of product. The *F* value is the sterilization process equivalent time and applies to steam sterilization primarily. It has been applied to dry heat sterilization kinetics, but the main emphasis of *F* value in the pharmaceutical industry has been to determine minimum and overkill cycles for terminal sterilization processes used steam.

The *F* value is a convenient measure of the lethality delivered per unit time. Unlike *D* value, the *F* value term is not clock time, but "equivalent time." *F* value is a single quantitative value that relates the microbial death efficiency of a given temperature to a standard temperature known to produce microbial kill. For example, an *F* value of eight minutes means that the item being sterilized was exposed to the equivalent of eight minutes at the reference temperature (e.g., 121°C) regardless of actual temperatures attained.

F values are calculated according to the following thermal algorithm:

$$F = \Delta t \sum 10^{(T-T_0)/Z} \qquad \text{(Equation 1)}$$

where T is the measured temperature, T_0 is the reference temperature (e.g., 121°C), Z is the thermal resistance value calculated from D values at different temperatures, and Δt is the time interval between temperature determinations.

Table 17-3 and Figure 17-4 exemplify the calculation of the *F* value. Table 17-3 gives real time and temperature data with the last column being the calculation of the exponential

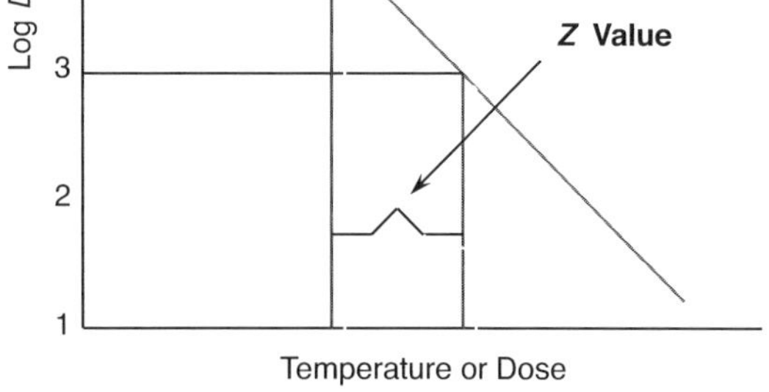

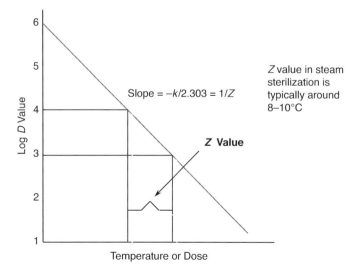

Figure 17-3 Microbial resistance value (*Z* value). The *Z* value is the change in sterilization condition that affects a 10-fold (1 log) reduction in the *D* value.

Table 17-3 Manual Calculation of Lethality Value

Time (min)	Temp (°C)	Lethal value $[10^{(T-T_0)/z}]$
5	94	0.0017
6	100	0.0100
7	106	0.0359
8	111	0.0880
9	114	0.1668
10	118	0.2615
11	118.5	0.3831
12	119	0.4948
13	119.5	0.5995
14	120	0.6813
15	120.2	0.7743
16	120.5	0.8254
17	120.6	0.8799
18	120.7	0.9031
19	120.8	0.9353
20	120.9	0.9504
21	119	0.5275
22	97	0.0036

relationship among actual temperature, reference temperature (121°C), and a Z value of 10°C, called lethality value. Adding the lethality values $[\sum 10^{(T-T_0)/Z}]$ gives a total of 8.522 minutes. Since Δt is one minute then the F value for this particular cycle is 8.522. Figure 17-4 compares the actual temperature versus time data during a sterilization cycle and the calculated exponential term at a given time interval. This final F value result is actually the area under the time–temperature curve (darkened area).

The earlier method of calculating F values is a physical approach where data required are time and temperature data. F values may also be calculated by what is called a biological equation:

$$F_T = (\mathrm{Log}N_0 - \log N_T) \times D_T \qquad \text{(Equation 2)}$$

The biological F value calculation is used to determine what F value is required to obtain the appropriate spore log reduction value (Log N_0 – log N_T) as a function of the D value of the specific spore. For example, if the D value is known to be two minutes and a 12-log reduction in that spore indicator organism is required for sterilization validation, then the minimum F value required is 24 minutes. To obtain this F value of 24, the physical F value equation is used to

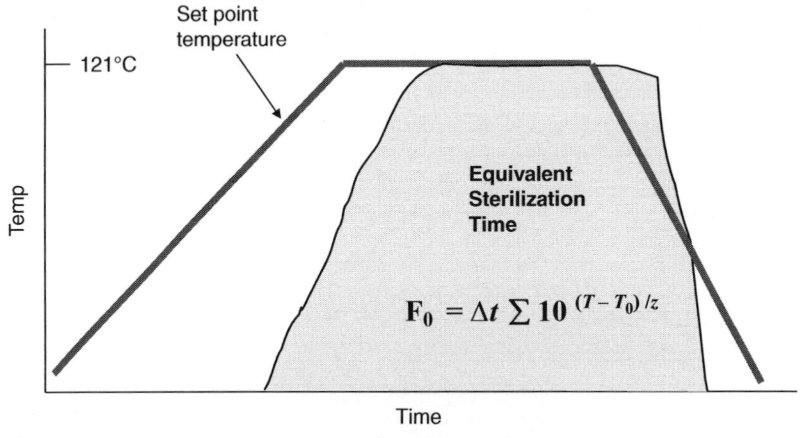

Figure 17-4 Comparison of sterilizer temperature–time curve and equivalent sterilization time.

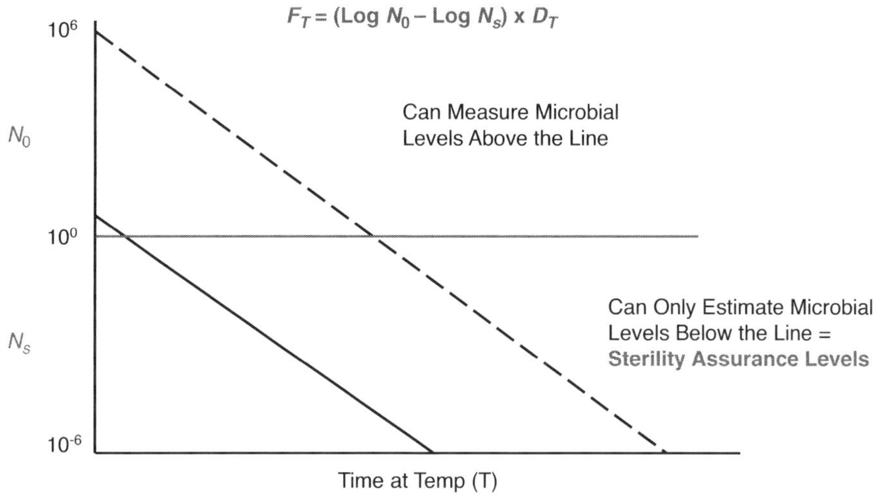

Figure 17-5 Plot of microbial population versus time at a given temperature.

calculate the F value according to actual temperature versus time profiles. Thus, manipulation of temperature and time in designing the sterilization cycle will ultimately produce a cycle with a calculated F value of 24.

Figure 17-5 shows a plot of microbial population versus time at a given temperature, the same kind of plot as a D value graph. This figure shows the fact that because microbial death is logarithmic, it never hits zero (i.e., $10^0 = 1$; $10^{-1} = 0.1$, and so forth). Therefore, below measurable levels of microbial growth, the term "probability of nonsterility" or "sterility assurance level (SAL)" is introduced. It can be easily seen that the higher the initial microbial population (bioburden), the longer it will take to reach a certain SAL, for example, 10^{-6} or a probability of nonsterility of one in one million. By reducing the initial bioburden, the 10^{-6} SAL can be achieved much faster. The "overkill" sterilization cycles use this kind of plot to determine the time required to achieve a certain SAL or microbial log reduction. For example, a cycle that provides more than a 12-log reduction of a resistant biological indicator with a known D value of not less that one minute is considered an overkill cycle. If the initial bioburden of this sample were 10 and the sterilization cycle produced a 12-log reduction of that organism, then the SAL would be 10^{-13} or the probability that only 1 unit out of 10 billion units sterilized with this cycle would be contaminated.

Table 17-4 shows the relationship of F values as a function of temperature. For example, the lethal effect of one-minute exposure at 118°C is 50% that of a one-minute exposure at 121°C. Obviously, as temperature falls away from 121°C, the lethal effect exponentially decreases while if temperature increases from 121°C, the lethal effect exponentially increases. These

Table 17-4 Lethality Ratios As a Function of Temperature During a Steam Sterilization Cycle (Reference Temperature = 121°C and Z Value = 10°C)

Product temperature (°C)	Lethality ratio
110	0.08
112	0.13
114	0.19
116	0.32
118	0.50
120	0.79
121	1.00
122	1.26

relationships obtained using *F* value calculations can design optimal sterilization cycles using any temperature profile usually within $\pm$ 10°C of 121°C.

HEAT STERILIZATION

Heat used to sterilize items is either wet heat or dry heat. Wet heat is also known as steam sterilization, steam under pressure sterilization, or autoclaving. Items traditionally sterilized by steam under pressure include rubber and durable plastic materials (e.g., filtration and tubing materials are durable whereas flexible plastic containers are not), mixing tanks, other equipment parts, filling equipment, freeze-dryer chambers, and, if possible, filled containers with product if the product can withstand high-temperature exposure.

Items sterilized by dry heat include glass containers, stainless steel equipment, and dry powders, again if the powder can withstand high-temperature exposure.

Lethality of microorganisms depends on the degree of heat exposure, duration of heat exposure, and moisture. Heat destruction of microorganisms occurs by coagulation of the proteins in the cell. Moist heat is much more effective as a sterilization method than dry heat. Moist heat involves raising the boiling point of water from 100°C to 121°C by applying 15 pounds per square inch pressure above atmospheric pressure. At this pressure and temperature water becomes saturated steam. Sterilization by moist heat means that liquid water is essential for denaturation of proteins in the bacterial cell wall. When saturated steam hits a surface cooler than itself, temperature of the surface increases and with this increase there is a release of what is called heat of condensation. The heat of condensation results from the need for a phase change to maintain the balance of liquid water in saturated steam. Heat of condensation releases a huge amount of energy (hundreds of calories) and this energy is what kills microorganisms so effectively. By contrast, dry heat at the same temperature will only release about 1 calorie per gram. Therefore, the presence of liquid water is essential for the effective microbial destructive effects of moist heat sterilization; therefore, oils or enclosed dry systems cannot effectively be terminally sterilized by wet heat. Dry heat will work but must use much higher temperatures to compensate for the lack of the heat of condensation energy effect.

Steam (Wet Heat) Sterilization

Steam sterilization is conducted in a pressurized vessel called an autoclave. Steam must be pure and saturated with no air or other noncondensable gases. The problem of removing air from the chamber and replacing it with pure saturated steam is the greatest challenge in effective steam sterilization. Keep in mind that thermocouples used to monitor temperature at various locations in the chamber and within the items being sterilized cannot detect whether the atmosphere is wet or dry heat. Other indicators called Bowie–Dick (Fig. 17-6) or Dart indicators are used to verify that the temperature measured is steam heat, not dry heat. If adequate steam penetration has occurred, dark brown stripes will appear across the Bowie–Dick tape. If the stripes are pale brown, steam penetration is inadequate and the heat in that area has been dry not wet. These indicators are used as part of other measurements—temperature, pressure, biological indicators—to validate a successful sterilization cycle.

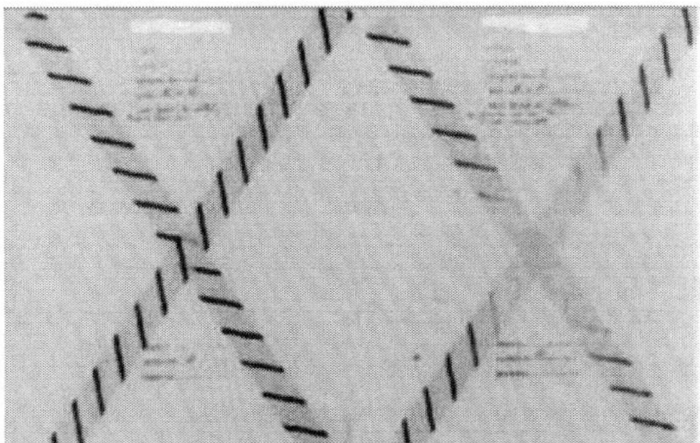

Figure 17-6 Bowie–Dick physical indicators of exposure to steam or dry heat. Bowie–Dick tape showing dark brown stripes (left) and stripes not all dark brown (right) indicating that the item on the right has not been adequately steam sterilized.

The basic steam sterilization cycle involves three primary steps:

1. Preconditioning the chamber and load within the chamber to remove air and replace it with saturated steam
2. The sterilization cycle
3. Removal of steam and release of pressure.

The removal of air and replacement with saturated steam is the most important technical aspect of steam sterilization. Removal of air depends on (*i*) the availability of moisture to displace air, (*ii*) the system used (e.g., vacuum) to displace air in an autoclave, (*iii*) the configuration of the load being sterilized, and (*iv*) the absence of air leaks.

The sterilization cycle is dependent on the ability of saturated steam to reach the innermost recesses of the load/materials being sterilized. This will dictate how much lag time is required before temperatures will reach levels where microbial destruction can occur. If it is finished, product being sterilized, for example, solutions in containers, is being terminally sterilized, the walls of each container must be heated to raise the temperature of the solution inside the container and this, in turn, generates steam inside the container. This points out that any container that is sealed or covered (e.g., rubber stoppers in cans, small pressure vessels, tubing sealed on both ends, any covered container) must have some degree of moisture inside the sealed/covered system. Otherwise, steam cannot penetrate container or vessel and temperature inside will only reach 100°C and not be saturated steam.

For solutions being sterilized inside containers, the pressure inside the container will be higher than the pressure outside. During sterilization, the vapor pressure from the solution in the container will increase to the same as the pressure in the chamber, but the partial pressure of the airspace in the container will increase and thermal expansion of the solution also will contribute to the increase in internal pressure. This increased internal pressure is safe for all glass containers (e.g., ampoules) and rubber-closed glass containers provided that the seal force torque of the rubber closure is adequate. However, for plastic containers, syringes, or any kind of container without a firm closure or cap, traditional steam sterilization is unsafe and must be replaced by using counterpressure steam sterilization methods.

The third stage of a steam sterilization process is the poststerilization stage where steam is replaced by air and pressure is reduced. There are several different designs of autoclaves differing primarily in how the poststerilization stage is accomplished (Table 17-5).

Autoclave with Vacuum and Time-Controlled Vacuum Maintenance

The batch or load after the sterilization cycle is dried and cooled by vacuum purges. Solid materials, both porous and nonporous, can be sterilized with this kind of autoclave.

Autoclave with Circulating Cold Water in the Jacket

With cold water circulating within the jacket of the autoclave, steam is removed through the introduction of compressed sterile air at pressures equal to the sterilization pressure. This prevents solutions from boiling and improves the heat exchange between the load in the autoclave and the autoclave jacket. Culture media used for sterility testing and media fills are sterilized using this type of autoclave.

Autoclave with Nebulized Spray Water

Cool water is nebulized and sprayed onto the load, producing rapid condensation of steam and sudden pressure drops. Pressure inside the containers remains high because the solution

Table 17-5 Poststeam Sterilization Possibilities

1. Autoclave with vacuum and time-controlled vacuum maintenance
2. Autoclave with circulating cold water in the jacket
3. Autoclave with nebulized spray water
4. Autoclave with superheated water spray (water cascade)
5. Autoclave with air over steam counter pressure

will cool slowly. Glass-sealed ampoules and plastic containers can be sterilized in this type of autoclave. Cooling stops when the solution inside the container reaches around 75°C, thus helping to dry the exterior of the container when stored outside the autoclave.

Autoclave with Superheated Water Spray (Water Cascade)

After loading this type of autoclave, the lower circular sector of the autoclave is filled with purified water. At the beginning of the process, no air in the chamber is removed. Water is circulated in the heat exchanger, then sprayed onto the load. This process provides excellent temperature uniformity and very small F_0 excursions, thus minimizing the sterilization time. The circulation water continues to circulate after the sterilization phase. Cold tap water flow into the plates of the heat exchanger to replace the steam and then cools the load. During all phases of the sterilization cycle, sterile air counterpressure is maintained inside the chamber so that no thermal or pressure shock occurs. This autoclave is used to sterilize flexible containers that cannot withstand sudden changes in temperature and pressure together. One major disadvantage of this process is the obvious fact that the load cannot be dried inside the chamber.

Autoclave with Air Over Steam Counter Pressure

This autoclave is similar to the water cascade autoclave in many respects. For example, the air in the chamber is not initially removed before steam enters the chamber. Partial air pressure of this mixture of air and steam is adjusted during the entire process with fans and flow deflectors in the chamber assuring a homogeneous steam and air mixture. Pressures inside the chamber of this kind of autoclaves are much higher than conventional pure saturated steam autoclaves. The cooling phase consists of air feeding into the chamber to condense the steam while maintaining the sterilization phase pressure. Cold tap water is then fed into the heat exchanges. The load is cooled while maintaining a constant controlled pressure. This autoclave also is used to sterilize flexible containers with the advantage of being able to dry the containers during the cycle. However, this type of autoclave has a cooling phase that takes much longer than the superheated water spray autoclave.

Sterilization-in-Place (SIP)

When this term is used, it is always referring to steam sterilization of large equipment items such as mixing tanks, vessel-filter-filler systems, and even complete isolator units. The same steam sterilization principles apply in that effective air removal must occur first, followed by adequate time–temperature exposure of all surfaces for overkill sterilization to take place. SIP is preceded by effective clean-in-place (CIP) procedures, so for both effective and successful CIP and SIP, equipment design and construction must be able to achieve the following (3):

- Withstand pressures required for steam sterilization
- Adequate air venting using microbial retentive filters
- Condensate must be trapped
- No leaks
- No inner surfaces that cannot be exposed to water and steam.

Dry Heat Sterilization

Dry heat destroys microorganisms by oxidation (basically exploding the cells) because of the very high temperatures employed, at least 170°C. Materials typically sterilized by dry heat include glassware, metal parts, oils, and dry powders. The process of dry heat sterilization is quite simple—heat with filtered air with blower fans enabling heat to be uniformly distributed in the sterilizer. Besides simplicity, the main advantage of dry heat sterilization is its effectiveness in destroyed endotoxins. In fact, dry heat is perhaps the most effective method to destroy endotoxins although temperatures required to validate the depyrogenation process for glass containers are a minimum of 250°C. Thus, dry heat depyrogenation requires higher temperatures and longer exposure times than that required by sterilization. Other advantages of dry heat sterilization are materials being dry at the end of the cycle and corrosion of materials is not an issue.

Disadvantages of dry heat sterilization include the fact that the process is difficult to control within precise temperature limits. The USP states that "a typical acceptable range in temperature in the empty chamber is ± 15°C when that unit is operating at not less that 250°C." Dry heat sterilization is very slow because heat transfer occurs by convection in air, a poor heat conductor. Heat penetration is slower than steam heat because of the long exposure times required to kill resistant spore organisms. The high temperatures required may cause degradation of materials, this being a major limitation of the wide applicability of dry heat as a sterilization method. Heat penetration through steel is faster than penetration through glass. Heat must penetrate to interior surfaces of items via conduction. Reflectance from shiny surfaces and differences in air density with temperature will have significant effects on the rate and extent dry heat sterilization. Of course, air tends to stratify, so fans or blowers must be used to aid heat circulation. One final limitation—materials will expand during heating and contract during cooling. Contraction could draw in microorganisms; therefore, all openings must be covered securely.

Dry heat sterilization is accomplished using either cabinet ovens or conveyer tunnels. With cabinet ovens, filtered air flows across the load, moved by a blower. High efficiency particulate air (HEPA) vent filters are used for the air inlets and outlets. There is always a concern about particulate matter being generated from the heat source. The door opening to the dryer must be sealed adequately. Temperature, time, and blower speed are controlled. The size of the chamber is limited, and manual loading and unloading are required that limit the rate of processing.

Tunnel sterilizers are conveyor systems where primarily glass containers are sterilized and depyrogenated while moving from a heat zone through a cooling zone. A schematic example of a dry heat sterilization and depyrogenation tunnel is shown in Figure 17-7. The source of heat is either convection or radiant heat while the cooling zone contains HEPA-filtered, vertical laminar air flow units. The conveyor belt is stainless steel and containers are moved from a nonsterile area after washing to critical work area where they exist typically right onto a collection/accumulation table for immediate filling of sterile product. Tunnel dry heat sterilizers are used for products having large volumes of glass containers to be filled. Tunnel sterilizers are more difficult to validate (than cabinet ovens) are more difficult to control uniform heating throughout the conveyor system, and, like any dry heat system, may generate particles from the heating zone.

Gas Sterilization

Many gases have been tried and used over the years to sterilize pharmaceutical materials. Most prominent of these gases have been ethylene oxide, peracetic acid, chlorine dioxide, and vapor phase hydrogen peroxide. Other gases used less frequently or not at all any more include formaldehyde, propylene oxide, beta propiolactone, and ozone. Items traditionally sterilized by gases include plastic containers, gowning materials, plastic devices, and other heat-labile equipment and materials.

Ethylene oxide (EtO) (Fig. 17-8) has been the classic sterilization gas. It is an alkylating agent that is very potent and highly penetrating. It is also a carcinogen. For many years it was used in a mixture with Freon, normally 12 parts EtO and 88 parts Freon. When the use of Freon was banned in 1996, EtO had to be used either as 100% (relatively dangerous because of its flammability) or in combination with carbon dioxide.

EtO lethality is influenced by four main factors:

1. Gas concentration—Ranges used are 400 to 1200 mg/L
2. Temperature—Temperature used depends on gas concentration used. Normal temperature range is 50°C to 60°C
3. Relative humidity—The normal range used is 35% to 80% RH. The *D* value of biological indicators used for EtO sterilization validation may range up to 10-fold over the range of RH values
4. Exposure time.

One of the major drawbacks of using EtO as a sterilization gas is its reacting with water or other components of the item(s) being sterilized and forming EtO residual compounds. These residual compounds at certain levels are hazardous to people and to the environment. For years, there have been threats both by the occupational safety and health organization (OSHA) and the environmental protection agency (EPA) to ban the use of EtO because of residuals. Typical EtO residuals are ethylene glycol from the reaction of EtO and water, and ethylene chlorhydrin from the interaction of EtO and chloride compounds. Each residual level has an upper limit that is usually achievable through aeration of the material after EtO exposure.

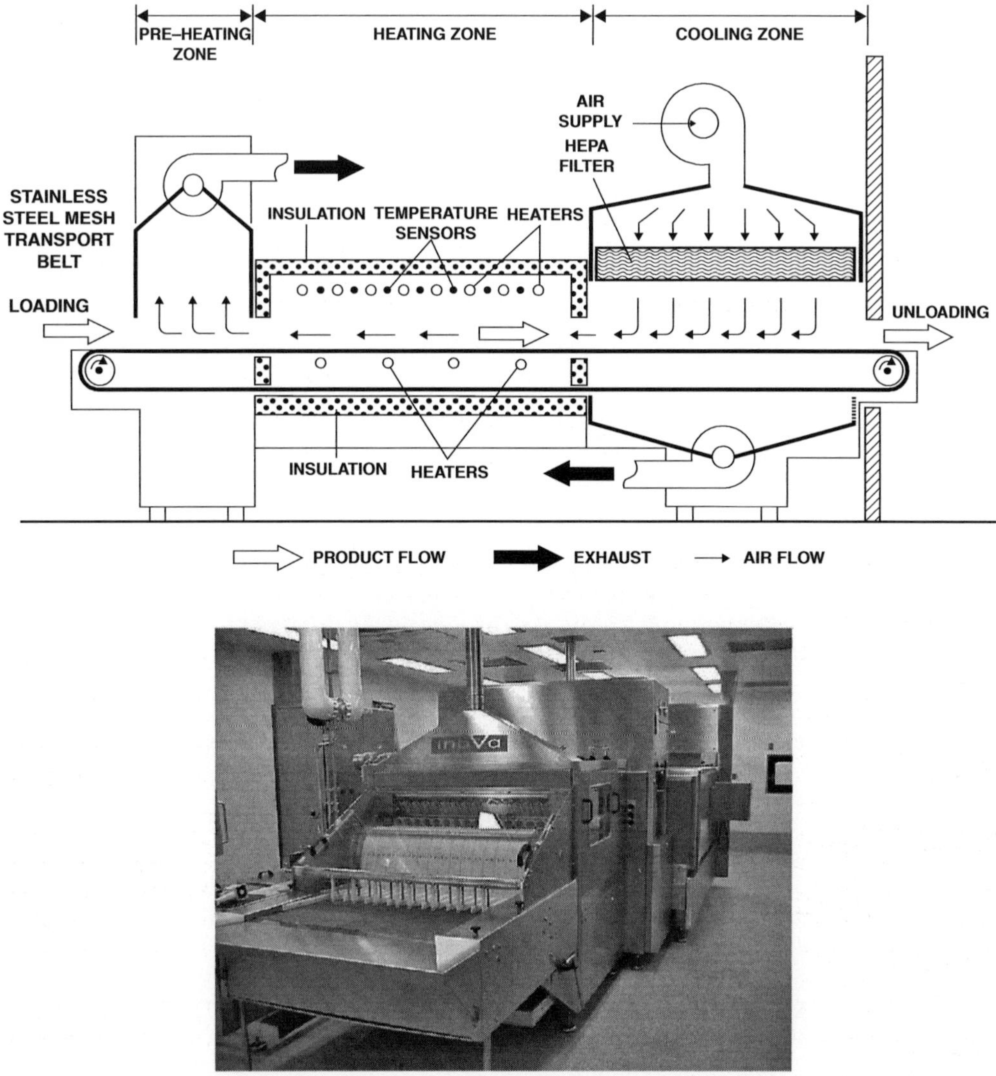

Figure 17-7 Dry heat sterilization/depyrogenation tunnel. Schematic (*Source*: From Ref 8 and courtesy of Fedegari Autoclavi) and photo (courtesy of Baxter Healthcare Corporation) of a dry heat sterilization/depyrogenation tunnel.

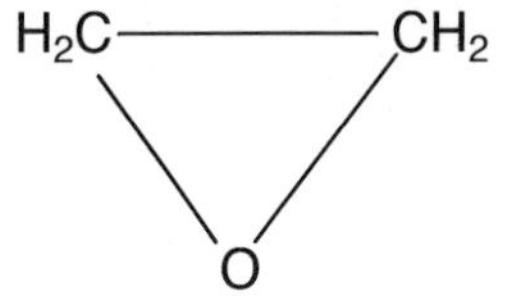

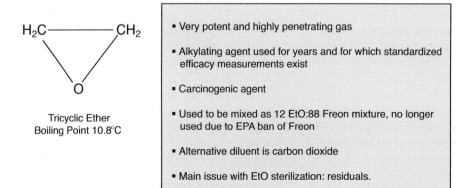

Tricyclic Ether
Boiling Point 10.8°C

- Very potent and highly penetrating gas
- Alkylating agent used for years and for which standardized efficacy measurements exist
- Carcinogenic agent
- Used to be mixed as 12 EtO:88 Freon mixture, no longer used due to EPA ban of Freon
- Alternative diluent is carbon dioxide
- Main issue with EtO sterilization: residuals.

Figure 17-8 Ethylene oxide gas.

Table 17-6 ISO 10933-7 Requirements for Ethylene Oxide Residuals

Exposure	Patient exposure	Average daily dose to patient	Dose in first 24 hours	Dose in first 30 days	Dose over lifetime
Limited	Contact up to 24 hours	EtO < 20 mg ECH < 12 mg ECH < 12 mg	EtO < 20 mg ECH < 12 mg ECH < 12 mg	Not applicable	Not applicable
Prolonged	24 hours to 30 days	EtO < 20 mg ECH < 12 mg	EtO < 2 mg/day ECH < 2 mg/day	EtO < 60 mg ECH < 60 mg	Not applicable
Permanent	Greater than 30 days	EtO < 20 mg ECH < 12 mg	EtO < 0.1 mg/day ECH < 2 mg/day	EtO < 60 mg ECH < 60 mg	EtO < 2.5 mg ECH < 50 mg

The International Organization for Standardization (ISO) developed ISO10993-7 entitled "Biological Evaluation for Medical Devices—Part 7: Ethylene Oxide (EtO) Sterilization Residuals" (1995) that specified requirements for establishing allowable limits of EtO residuals on medical devices (Table 17-6). Formerly, the FDA in 1978 established limits as:

Ethylene oxide	250 ppm
Ethylene chlorhydrin	250 ppm
Ethylene glycol	500 ppm

A typical EtO sterilization cycle is shown in Figure 17-9. Note that the sterilization exposure period is relatively long and, although not shown, the aeration period also can be very long, up to 24 hours or more.

Chlorine dioxide and vapor phase hydrogen peroxide (VPHP) are alternative gas sterilization agents, both used primarily to sterilize surfaces of flexible and rigid barrier isolation systems. Ozone and peracetic acid are other alternative agents with some applications in the parenteral industry, ozone for deionized water treatment, peracetic acid for isolator sterilization. Chlorine dioxide typically uses a mixture of 2% chlorine gas and 98% nitrogen, and employs cycle parameters (vacuum, humidity, gas exposure time, aeration time) similar to EtO cycles although chloride dioxide sterilization cycles are shorter. Penetration of materials is better with chloride dioxide that with VPHP, but neither gas can penetrate as readily as EtO. Chlorine dioxide gas has a greenish color that is easily seen during a sterilization cycle.

VPHP has been the standard sterilant of choice for isolators. It is a very effective sporicidal agent, relatively safe, and environmentally friendly. The vapor form of peroxide, unlike the

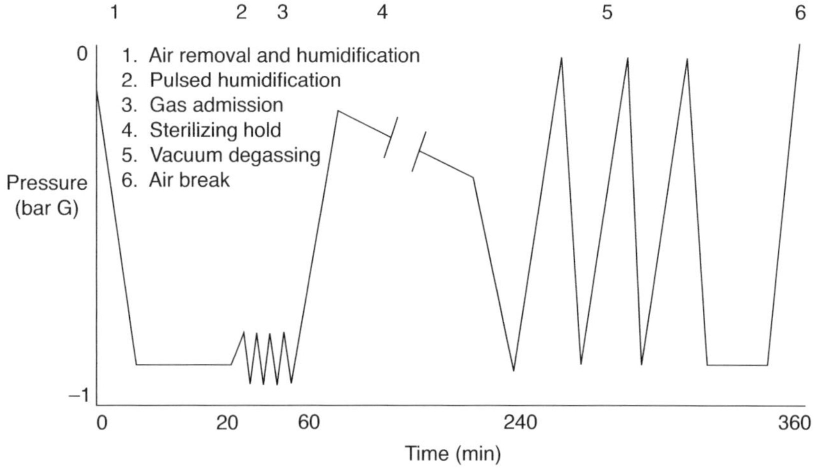

Figure 17-9 Schematic diagram of a 100% ethylene oxide sterilization cycle. *Source*: From Ref. 10.

liquid form, is noncorrosive and requires relatively low temperatures (20–35°C). The concentration used for VPHP sterilization of isolators depends on the internal surface area of the isolator. Typically, a 30% solution of liquid peroxide is pressurized in a VPHP generator (example shown in Figure 27-3 with the generator next to the sterility test isolator) to produce vapor peroxide at a concentration of 1 to 2 mg/L. The fact that VPHP sterilization occurs at ambient temperature is a major advantage of this sterilant over other gaseous sterilants (chlorine dioxide, peracetic acid). Among the disadvantages of using VPHP is its tendency to be absorbed by plastic and other types of materials.

RADIATION STERILIZATION

Radiation sterilization may be achieved by gamma radiation, beta particle (electron beam; accelerated electrons), or ultraviolet light. Microwave radiation has been the studied as a sterilization method for empty vials, in-line glass ampoule solution products, and hydrophilic contact lenses, but is not considered a major radiation sterilization method. Gamma radiation, typically cobalt 60 high-energy photons, is the most penetrative and effective radiation sterilization method. Beta particles are ionizing radiations, not electromagnetic like gamma rays, and, thus, are less penetrative. Electrons are generated from a radioactive element, for example, Strontium 90, and accelerated mechanically to extremely high energy levels in the range of 5 to 30 kilogray (kG) (0.5–3.0 mRad). Ultraviolet light is considered a surface sterilizing method only, as its energy level is insufficient to penetrate materials. Gamma sterilization is overall the method of choice although beta particle radiation being more closely evaluated, especially as a possible terminal sterilization method for finished products.

Items sterilized by radiation sterilization are essentially the same items that can be sterilized by gaseous methods—plastic materials, heat-labile materials, powders.

Radiation will damage the nucleoproteins of microorganisms. Effectiveness of radiation is dependent on the dose of radiation and time.

The 12 *D* overkill approach is always used in radiation sterilization. What this means is that whatever the *D* value of the most resistance microbial spore form in the material to be sterilized, the typical radiation dose is sufficient to produce a 12-log reduction in the spore population. For example, typical *D* values for the most resistant bacterial spore (*Bacillus pumulis*) to radiation is 1.7 to 2.0 kG [or megarad (mRad)] (remember that the *D* value determination for radiation uses dose rather than time). The typical radiation dose is 25 kG, greater than the 12-fold the *D* value of *B. pumulis*. In addition, during the radiation sterilization treatment, dosimeters are placed at strategic locations in order to monitor radiation doses received throughout the load.

A schematic depiction of a radiation sterilization conveyor system is seen in Figure 17-10. The product enters the entrance to the conveyor, then travels through different sections where different doses of radiation are given. The total dose may be 25 kG, but that dose is distributed throughout the conveyor system in order not to apply overwhelming and perhaps damaging dose of radiation at any given segment. A major concern when attempting to sterilize finished products or active pharmaceutical ingredients is the formation of radiolytic byproducts (e.g., *OH) that in turn may cause damage to the raw material and/or the packaging system.

Validation of radiation sterilization follows the guidelines of the Association for the Advancement of Medical Instrumentation (AAMI). Validation involves the radiation dose required to destroy 10^8 spores of the biological indicator in a maximum load size. Validation also requires determination of the radiation absorbed (using dosimeters) in the material being sterilized. Factors that affect radiation sterilization validation include the *D* value of the biological indicator in the item being sterilized, the radiation strength, the radiation dose rate, and the conveyor speed.

BRIGHT LIGHT (OR PULSED LIGHT) STERILIZATION

Bright light has been developed as a possible terminal sterilization method for drug products in final containers, specifically polyethylene containers produced from form-fill-finish operations. Maxwell Technologies developed pure pulsed light to inactivate all microbial life. Light is produced by ionizing xenon gas in a quartz lamp using a high voltage pulse of short duration (specifically, a few hundred millionths of a second). Light produced is claimed to be 20,000 times more intense (brighter) than sunlight (initially estimated to be 90,000 times brighter). A single

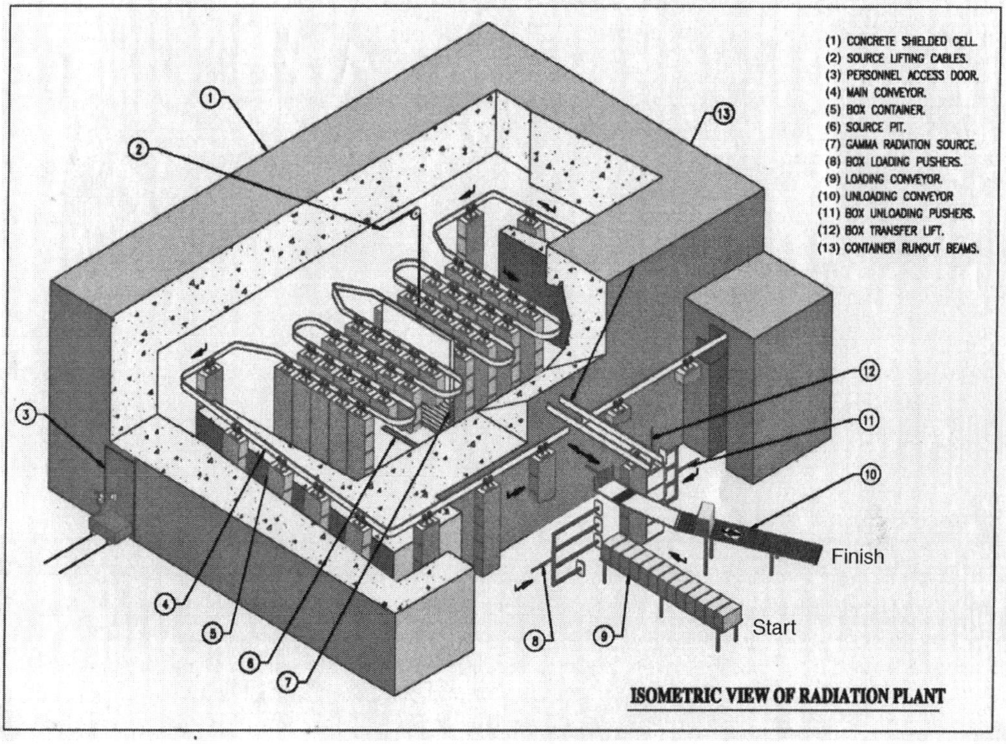

Figure 17-10 Schematic of radiation sterilization conveyor. Percentage of total radiation dose is distributed in varying amounts as items being sterilized are conveyed from start to finish. Percentages of total dose are greatest along conveyor closest to the source of gamma radiation (#7). *Source*: Courtesy of Dr. AK Kohli of BRIT/BARC Vashi Complex, Mumbai, India.

flash of pulsed light at 1 to 2 J/cm^2 will kill greater than one million colony-forming units of bacterial spores. Such brief light exposure does not affect the temperature of the product. Pulsed light can sterilize products in containers that can transmit light–polyethylene, polypropylene, nylon, and ethylene vinyl acetate. Polyvinyl chloride and polystyrene are examples of plastics that will not readily transmit pulsed light. Pulsed light is readily transmitted through water and most pharmaceutical solutions.

Filtration Sterilization

Filtration sterilization is covered in chapter 18.

TERMINAL STERILIZATION

Whenever possible, the parenteral product should be sterilized after being sealed in its final container (terminal sterilization) and within as short a time as possible after the filling and sealing have been completed. Since this usually involves a thermal process (although there is a trend in applying radiation sterilization to finished products), due consideration must be given to the effect of the elevated temperature upon the stability of the product. Many products, both pharmaceutical and biological, will be affected adversely by the elevated temperatures required for thermal sterilization. Heat-labile products must, therefore, be sterilized by a nonthermal method, usually by filtration through bacteria-retaining filters. Subsequently, all operations must be carried out in an aseptic manner so that contamination will not be introduced into the filtrate. Colloids, oleaginous solutions, suspensions, and emulsions that are thermolabile may require a process in which each component is sterilized separately and the product is formulated and processed under aseptic conditions.

The performance of an aseptic process is challenging, but technical advances in aseptic processing, including improved automation, use of isolator systems, formulations to include antimicrobial effects, and combinations of limited sterilization with aseptic processing, have decreased the risk of contamination. Therefore, the successes realized should encourage continued efforts to improve the assurance of sterility achievable with aseptic processing. The importance of this is that for many drug solutions and essentially all biopharmaceutical products, aseptic processing is the only method that can be considered for preparing a sterile product.

Interaction among environmental conditions, the constituents in the closure, and the product may result in undesirable closure changes such as increased brittleness or stickiness, which may cause loss of container–closure seal integrity. Thus, shelf life integrity is an important consideration in closure selection and evaluation.

The assessment of aseptic-processing performance is based on the contamination rate resulting from periodic process simulations using media filling instead of product filling of containers. A contamination rate no greater than 0.1% at 95% confidence has generally been considered as indicative of satisfactory performance in the industry. However, with current advances in aseptic-processing capabilities, lower contamination rates may be achievable.

Radiation sterilization, as mentioned, is gaining some momentum as an alternative terminal sterilization method. There has been limited understanding of the molecular transformations that may occur in drug molecules and excipients under exposure to the high-energy gamma radiation levels of the process. However, lower energy beta particle (electron beam) radiation has seen some success. There is still significant research that must be accomplished before radiation sterilization is used as a terminal sterilization process. The use of radiation for the sterilization of materials such as plastic medical devices is well established.

Dry heat sterilization may be employed for a few dry solids that are not affected adversely by the high temperatures and for the relatively long heating period required. This method is applied most effectively to the sterilization of glassware and metalware. After sterilization, the equipment will be sterile, dry, and, if the sterilization period is long enough, pyrogen free.

Saturated steam under pressure (autoclaving) is the most commonly used and the most effective method for the sterilization of aqueous liquids or substances that can be reached or penetrated by steam. A survival probability of at least 10^{-6} is readily achievable with terminal autoclaving of a thermally stable product. However, it needs to be noted that for terminal sterilization, the assurance of sterility is based on an evaluation of the lethality of the process, that is, of the probable number of viable microorganisms remaining in product units. However, for aseptic processing, where the components used have been sterilized separately by validated processes and aseptically put together, the level of sterility assurance is based on an evaluation of the *probable* number of product units that were contaminated during the process.

Because the temperature employed in an autoclave is lower than that for dry heat sterilization, equipment made of materials such as rubber and polypropylene may be sterilized if the time and temperature are controlled carefully. As mentioned previously, some injections will be affected adversely by the elevated temperature required for autoclaving. For some products, such as Dextrose Injection, a shortened cycle using an autoclave designed to permit a rapid temperature rise and rapid cooling with water spray or other cooling methods will make it possible to use this method. It is ineffective in anhydrous conditions, such as within a sealed ampoule containing a dry solid or an anhydrous oil. Other products that will not withstand autoclaving temperatures may withstand marginal thermal methods, such as tyndallization or pasteurization. This describes a practice where a product is heated to approximately 100°C in order to activate any bacterial spores present to revert to their vegetative forms, thus much easier to destroy. A second 100°C exposure will destroy the vegetative form and a third 100°C exposure will add assurance that complete bacterial destruction has occurred. Theoretically, this practice could be applied as a terminal sterilization method for products containing heat sensitive active pharmaceutical ingredients. However, since the process cannot be validated using conventional biological indicators (remember, BIs are bacterial spores), it is not considered as a viable option for terminal sterilization.

Table 17-7 International Standards Organization (ISO) Technical Committee (TC) 198 Standards for Sterilization of Health Care Products

ISO document reference	Brief description of reference
Standards	
ISO 11134:1994	Sterilization of health care products—requirements for validation and routine control—industrial moist heat sterilization
ISO 11135:1994	Medical devices—validation and routine control of ethylene oxide sterilization
ISO 11135:1994/Cor 1:1994	
ISO 11137:1995	Sterilization of health care products—requirements for validation and routine control—radiation sterilization
ISO 11137:1995/Cor 1:1997	
ISO 11137:1995/Amd 1:2001	Selection of items for dose setting
ISO 11138–1:1994	Sterilization of health care products—biological indicators–Part 1: General
ISO 11138–2:1994	Sterilization of health care products—biological indicators–Part 2: biological indicators for ethylene oxide sterilization
ISO 11138–3:1995	Sterilization of health care products—biological indicators—part 3: biological indicators for moist heat sterilization
ISO/TS 11139:2001	Sterilization of health care products—vocabulary
ISO 11140–1:1995	Sterilization of health care products—chemical indicators—Part 1: general requirements
ISO 11140–1:1995/Amd 1:1998	
ISO 11140–2:1998	Sterilization of health care products—chemical indicators—Part 2: test equipment and methods
ISO 11140–3:2000	Sterilization of health care products—chemical indicators—Part 3: class 2 indicators for steam penetration test sheets
ISO 11140–4:2001	Sterilization of health care products—chemical indicators—Part 4: class 2 indicators for steam penetration test packs
ISO 11140–5:2000	Sterilization of health care products—chemical indicators—Part 5: class 2 indicators for air removal test sheets and packs
ISO 11607:2003	Packaging for terminally sterilized medical devices
ISO 11737–1:1995	Sterilization of medical devices—microbiological methods–Part 1: estimation of population of microorganisms on products
ISO 11737–2:1998	Sterilization of medical devices—microbiological methods—Part 2: tests of sterility performed in the validation of a sterilization process
ISO 11737–3:2004	Sterilization of medical devices—microbiological methods—Part 3: guidance on evaluation and interpretation of bioburden data
ISO 13408–1:1998	Aseptic processing of health care products—Part 1: general requirements
ISO 13408–2:2003	Aseptic processing of health care products—Part 2: filtration
ISO 13408–6:2005	Aseptic processing of health care products—Part 6: isolator systems
ISO/TS 13409:2002	Sterilization of health care products—radiation sterilization—substantiation of 25 kG as a sterilization dose for small or infrequent production batches
ISO 13683:1997	Sterilization of health care products—requirements for validation and routine control of moist heat sterilization in health care facilities
ISO 14160:1998	Sterilization of single-use medical devices incorporating materials of animal origin—validation and routine control of sterilization by liquid chemical sterilants
ISO 14161:2000	Sterilization of health care products—biological indicators—guidance for the selection, use and interpretation of results
ISO 14937:2000	Sterilization of health care products—general requirements for characterization of a sterilizing agent and the development, validation, and routine control of a sterilization process for medical devices
ISO/TS 15843:2000	Sterilization of health care products—radiation sterilization—product families and sampling plans for verification dose experiments and sterilization dose audits, and frequency of sterilization dose audits
ISO/TR 15844:1998	Sterilization of health care products—radiation sterilization—selection of sterilization dose for a single production batch
ISO 15882:2003	Sterilization of health care products—chemical indicators—guidance for selection, use and interpretation of results
ISO 17664:2004	Sterilization of medical devices—information to be provided by the manufacturer for the processing of resterilizable medical

Source: From Ref. 11.

Table 17-8 FDA Guidance for Industry for Submission Documentation for Sterilization Process Validation in Applications for Human and Veterinary Drug Products (November, 1994) (Ref. 7)

Information for terminal moist heat sterilization

A Description of process and product
  1. Container/closure system
  2. Sterilization process
  3. Performance specs of autoclave
  4. Loading patterns
  5. Monitoring controls
  6. Requalification
  7. Reprocessing

B Thermal qualification of the cycle
  1. Heat distribution and penetration
  2. Monitors
  3. Effects of loading
  4. Batch record information

C Microbiological efficacy of the cycle
  1. Identification and characterization of bioburden organisms
  2. Bioburden specification
  3. Identification, resistance, and stability of biological indicators
  4. Comparison of bioburden organisms and biological indicator challenge

D Microbiological monitoring of environment

E Container/closure and package integrity
  1. Simulation of stresses from processing
  2. Demonstration of integrity following maximum exposure
  3. Multiple barriers
  4. Test sensitivity
  5. Integrity over product shelf-life

F Bacterial endotoxins test and method

G Sterility testing methods and release criteria

H Evidence of formal, written SOPs

Information for ethylene oxide sterilization

A Description
B Cycle parameters
C Microbiological methods
D Stability

Information for radiation sterilization

A Facility
B Process
C Packaging
D Multiple dose mapping
E Microbiological methods
F Stability

Information for aseptic fill manufacturing processes

A Buildings and facilities—floor plan and equipment location
B Overall manufacturing operation—filtration, hold period specs, critical operations
C Sterilization and depyrogenation of containers, closures, equipment, components
D Procedures and specs for media fills
E Actions concerning product when media fills fail
F Micro monitoring of environment—micro methods, yeast, molds, anaerobes, exceeded limits
G Container–closure and package integrity
H Sterility testing methods and release criteria
I Bacterial endotoxins test and method
J Evidence of formal standard operating procedures

Maintenance of microbiological control and quality; stability considerations

A Container–closure integrity
B Preservative effectiveness
C Pyrogen or endotoxin testing

All additional information

Abbreviation: SOP, standard operating procedures.

Some manufacturers have incorporate antimicrobial preservative agents in the formulation to aid in the sterilization of marginal sterilization practices like tyndallization, but again, these practices today are unacceptable from a sterilization validation standpoint.

Articles to be sterilized must be properly wrapped or placed in suitable containers to permit penetration of sterilants and provide protection from contamination after sterilization. Sheets or bags made of special steam-penetrating paper or polymeric materials are available for this purpose. Further, containers or bags impervious to steam can be equipped with a microbe-excluding vent filter to permit adequate steam penetration and air exit. Multiple wrapping permits sequential removal of outer layers, as articles are transferred from zones of lower to higher environmental quality. The openings of equipment subjected to dry heat sterilization often are covered with metal or glass covers. Laboratories often used silver-aluminum foil for covering glassware to be used for endotoxin testing. Wrapping materials commonly used for steam sterilization may be combustible or otherwise become degraded under dry heat sterilization conditions.

The effectiveness of any sterilization technique must be proved (validated) before it is employed in practice. Since the goal of sterilization is to kill microorganisms, the ideal indicator to prove the effectiveness of the process is a resistant form of an appropriate microorganism, normally resistant spores (a biological indicator, or BI). Therefore, during validation of a sterilization process, BIs of known resistance and numbers are used in association with physical-parameter indicators, such as recording thermocouples. Once the lethality of the process is established in association with the physical measurements, the physical measurements can be used for subsequent monitoring of in-use processes without the BIs. Eliminating the use of BIs in direct association with human-use products is appropriate because of the ever-present risk of an undetected, inadvertent contamination of a product or the environment. The number of spores and their resistance in BIs used for validation studies must be accurately known or determined. Additionally, the manner in which BIs are used in validation is critical and must be controlled carefully.

In addition to the data printout from thermocouples, sometimes other physical indicators are used, such as color change and melting indicators, to give visual indication that a package or truckload has been subjected to a sterilization process. Such evidence can become a part of the batch record to confirm that sterilization was accomplished.

Table 17-7 provides a comprehensive listing of all the ISO standard documents for sterilization of health care products.

PARAMETRIC RELEASE

Many products, especially large-volume injectables, that are terminally sterilized using overkill cycles, can be released to the market without the need to perform compendial sterility testing. Parametric release must be approved by the FDA or other appropriate regulatory body. Parametric release requires well-defined and validated sterilization cycles where the physical parameters of processing are well defined, predictable, measurable, and the lethality of the cycle has been microbiologically validated (4). Both FDA and EU guidelines permit parametric release of products with prior approval (5,6).

Sterilization Information Required for Commercialization of Sterile Products

The FDA published a guidance document that is still used today for submitting documentation related to the sterilization and validation of that sterilization process for a sterile drug product (7). Any new or abbreviated drug application for marketing of a sterile drug product must submit documentation that clearly details how the product is rendered sterile and sterility maintained from preparation, through release and throughout the shelf-life period of every batch of the product. Table 17-8 summarizes what is contained in this guidance document.

REFERENCES

1. Shirtz J. F, D and z values. In: Agalloco JP, Carleton FJ, eds. Validation of Pharmaceutical Processes. 3rd ed. Informa, New York: CRC Press, 2007:159–173.
2. Pflug IJ. Heat sterilization. In: Phillips GB, Miller WS, eds. Industrial Sterilization. Durham, NC: Duke University Press, 1973:239–282.

3. Seiberling DA, Ratz AJ. Engineering considerations for CIP/SIP systems. In: Avis KE, ed. Sterile Pharmaceutical Products: Process Engineering Applications. Prairie View, IL: Interpharm Press, 1995:135–220.
4. Tirumalai R, Porter D. Terminal sterilization and potential for parametric release. Am Pharm Rev 2005; 8(4):26–31.
5. United States Pharmacopeia. General Chapter <1222>. Terminally Sterilized. Pharmaceutical Products-Parametric Release, 2010.
6. The European Agency for the Evaluation of Medicinal Product. Committee for Proprietary Medicinal Products, Note for Guidance for Parametric Release. London: EMEA, 2001.
7. Food and Drug Administration. Guidance for Industry for the Submission Documentation for Sterilization Process Validation in Applications for Human and Veterinary Drug Products. Silver Spring, MD: FDA, 1994. http://www.fda.gov/downloads/Drugs/GuidanceComplianceRegulatoryInformation/Guidances/ucm072171.pdf.
8. Hagman DE. Sterilization. In: Remington: The Science and Practice of Pharmacy. 21st ed. Lippincott Williams & Wilkins, 2005:787.
9. Hoxey EV, Thomas N, Davies DJG. Principles of sterilization. In: Denyer SP, Baird RM, eds. Guide to Microbiological Control in Pharmaceuticals and Medical Devices. 2nd ed. Boca Raton, FL: CRC Press, 2007:229–271.
10. Dewhurst EL, Hoxey EV. Sterilization methods. In: Denyer SP, Baird RM, eds. Guide to Microbiological Control in Pharmaceuticals and Medical Devices. 2nd ed. CRC Press, 2007.
11. http://www.iso.org/iso/iso_catalogue/catalogue_tc/catalogue_tc_browse.htm?commid = 54576.

BIBLOGRAPHY

Block SS. Disinfection, Sterilization, and Preservation. 5th ed. Philadelphia: Lippincott, Williams & Wilkins, 2001:695–864.

Booth AF. Radiation sterilization: Validation and routine operations handbook, PDA-DHI, Bethesda, 2007.

Booth AF. Ethylene oxide sterilization validation and routine operations handbook, PDA-DHI, Bethesda, 2008.

Dewhurst EL, Hoxey EV. Sterilization methods. In: Denyer SP, Baird RM, eds. Guide to Microbiological Control in Pharmaceuticals and Medical Devices. Boca Raton, FL: CRC Press, 2007:229–271.

Groves MJ. Sterilization of pharmaceutical solutions. In: Nordhauser F, Olson WP, eds. Sterilization of Drugs and Devices. Prairie View, IL: Interpharm Press, 1998:5–44.

Hagman DE. Sterilization. In: Felton L, ed. Remington's: The Science and Practice of Pharmacy. 21st ed. Lippincott, Williams & Wilkins, 2005:776–801.

Jacobs GP. Radiation in the sterilization of pharmaceuticals. In: Groves MJ, Olson WP, Anisfeld MH, eds. Sterile Pharmaceutical Manufacturing. Prairie View, IL: Interpharm Press, 1991:57–78.

Pflug IJ. Selected Papers on the Microbiology and Engineering of Sterilization Processes. 5th ed. Minneapolis, MN: Environmental Sterilization Laboratory, 1988.

United States Pharmacopeia General Chapters

- <55> Biological Indicators—Resistance Performance Tests
- <1208> Sterility Testing—Validation of Isolator Systems
- <1209> Sterilization—Chemical and Physicochemical Indicators and Integrators
- <1211> Sterilization and Sterilility Assurance of Compendial Articles
- <1222> Terminally Sterilized Pharmaceutical Products—Parametric Release.

Parenteral Drug Association, Bethesda, MD, Technical Reports

- No. 1—2007 Revision—Validation of Moist Heat Sterilization Processes
- No. 3—1981—Validation of Dry Heat Processes Used for Sterilization and Depyrogenation
- No. 11—1988—Sterilization of Parenterals by Gamma Radiation

18 Sterile filtration

Sterilization by filtration is called the "cold" method of sterilization since it is the only method that does not rely on either elevated temperature or some other form of energy to destroy microorganisms. Sterile filtration, of course, does not destroy microbial life; rather, it removes or separates microbial life from the rest of the product. The first sterile filters used in the late 19th century were composed of asbestos. The technology has come a long way since.

TYPES OF FILTERS

There are four primary types of filters used in the parenteral and biopharmaceutical industry (Table 18-1).

Particle Filters

Porosity of particle filters range from 10 to 200 μm and are used as depth or prefilters to remove dirt, pollen, some bacteria, and most particles. Examples of materials used as depth filters include cellulose fibers, diatomaceous earth, glass fibers, sand and gravel, and polypropylene yarn. Asbestos also was used for decades as a depth filter until banned because of its carcinogenic properties. Prefilters or surface filters are commonly used simply to protect the membrane microfilter from clogging too quickly. Examples of materials used as prefilters include cellulose ester and heat-bonded polypropylene fibers.

Microfilters

Porosity of microfilters ranges from 0.1 to 10 μm and is used to remove all bacteria, yeast, and colloidal forms. Microfilters are the classic sterilizing filters used in the industry. These filters have very narrow pore distribution because of controlled polymeric structures and can be integrity testing. Examples of materials used in microfilters are listed in Table 18-1. There also exist combination filters combining different membrane pore sizes or combining depth media and a membrane filter. Typically combination filters are final filters used with syringes prior to product administration.

Ultrafilters

Porosity of ultrafilters ranges from 0.001 to 0.1 μm and is used to remove most viral life forms and large organic compounds (>10,000 Da).

Nanofilters

Porosity of nanofilters is less than 0.001 μm and is used to remove small organic compounds and ionic forms. These filters are used in reverse osmosis systems. Nanofilters are composed of a variety of materials including nano-sized activated alumina particles bonded onto glass fiber matrices, electrospun Nylon 6 fibers, polycarbonate, polyethersulfone, and other polymers.

Polymeric filter materials are broadly classified as either hydrophilic or hydrophobic. Hydrophilic filters wet spontaneously and are the filters used in sterile filtration of solutions. Hydrophobic filters do not wet spontaneously and are the filters used in the sterile filtration of gases or solvents or strongly acidic or alkaline solutions.

Microfilters have a rated porosity of 0.22 μm or smaller. Filters can be rated either nominally or absolutely. Nominal ratings describe the weight percent removal of particles/bacteria at a particular size. Filter manufacturers assign arbitrary micron values based on data obtained in the removal of samples of known particle sizes at different weights. For example, a filter having a nominal rating of 67% for 1 μm or greater means that such a filter will remove 67% of all particles $\geq$ 1 μm.

Table 18-1 General Types of Filters Used in the Sterile Product Industry

Filter type	Size range (μm)	Examples of what is removed by this filter type
Particle	10–200	Pollens Particles Some bacteria
Microfilter	0.1–10	All bacteria Yeasts Colloids
Ultrafilter	0.001–0.1	Most viruses Large organic compounds (>10,000 Da)
Nanofilter (reverse osmosis)	Less than 0.001	Small organic compounds Ions

Absolute ratings, much more commonly used in the sterile filtration industry, define the diameter of the largest particle that will pass through the filter. Therefore, using the 0.22 μm filter means that no particle larger than 0.22 μm will pass through that filter, unless, of course, the filter is damaged.

After a product has been compounded, it must be filtered if it is a solution. The primary objective of filtration is to clarify a solution. A further step, removing particulate matter down to 0.2 μm in size, would eliminate microorganisms and would accomplish cold sterilization. A solution with a high degree of clarity conveys the impression of high quality and purity, desirable characteristics for a parenteral solution.

MECHANISMS OF AND FACTORS AFFECTING FILTER REMOVAL OF PARTICLES AND MICROORGANISMS

Filters are thought to function by one or, usually, a combination of the following: (*i*) sieving or screening, (*ii*) entrapment or impaction, and (*iii*) electrostatic attraction (Fig. 18-1). When a filter retains particles by sieving, they are retained on the surface of the filter. Entrapment occurs when a particle smaller than the dimensions of the passageway (pore) becomes lodged in a turn or impacted on the surface of the passageway. Electrostatic attraction causes particles opposite in charge to that of the surface of the filter pore to be held or adsorbed to the surface. It should be noted that increasing, prolonging, or varying the force behind the solution may tend to sweep particles initially held by entrapment or electrostatic charge through the pores and into the filtrate.

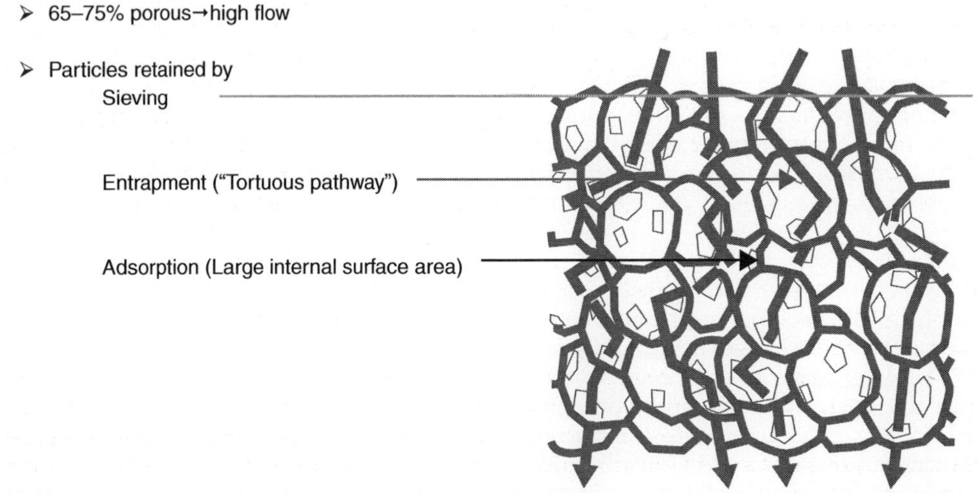
➢ 65–75% porous→high flow
➢ Particles retained by
Sieving
Entrapment ("Tortuous pathway")
Adsorption (Large internal surface area)

Figure 18-1 Membrane filter characteristics. *Source*: Courtesy of Millipore Corporation.

Several factors affect filter efficiency in microbial and particle retention (1).

1. Type of particle—Source, shape, charge, size.
2. Filter material (type of polymer)—Filter composition plays a role in charge-related attraction of particles, including microorganisms, with zeta potential van der Waals forces, hydrogen bonding, and hydrophobic attraction properties of the filter all involved.
3. Filter membrane thickness—Filter thickness slows the flow characteristics and affects the particle adsorption mechanism although a coarse and thick membrane can be just as efficient as a fine, thin one.
4. Filter porosity—Obviously the smaller the porosity, the greater the retention of microorganisms and particles, but flow rates are retarded. Also potential incompatibilities with liquids being filtered are greater with smaller porosities.
5. Temperature—influences microbial proliferation and viability and affects the Brownian motion of suspended organisms, increasing possibility of their adsorptive contact with the pore walls. The smaller the organism or particle, the greater the Brownian-motion effect upon it.
6. Type of fluid/solution being filtered—Increasing the viscosity of the solution will require some increase in applied pressure that, in turn, will increase the shear force on any bacterial cells present. Increased viscosity will disrupt adsorption interactions on the filter membrane, but will have no effect on size exclusion properties of the membrane. Surface-active agents in the solution formulation will decrease the surface tension of the solution and lower the bubble point of the filter (explained later). Surface-active agents normally will bind to solid surfaces and may reduce or eliminate bacterial adsorption in the filter, but will have no effect on membrane structure or changing the size of the bacterial cell.
7. Applied pressure, flow rate, and time—While it seems intuitive that increasing pressure or flow rate or time will have adverse effects on the integrity of the filter and perhaps affect microbial cell size, there are no compelling data to support this. Commercially used membrane sterilizing filters can be used for up to one week without changes in retention characteristics according to filter manufacturers' technical literature.

Filter manufacturers publish many technical articles and data sheets (most available on their web sites) describing and explaining the properties and functionalities of all their commercially available filters.

APPLICATIONS

Membrane filters are used exclusively for sterilizing solutions because of their particle-retention effectiveness, nonshedding property, nonreactivity, and disposable characteristics. However, it should be noted that nonreactivity does not apply in all cases. For example, polypeptide products may show considerable adsorption through some membrane filters, but those composed of polysulfone and polyvinylidine difluoride (PVDF) have been developed to be essentially nonadsorptive for these products. The most common membranes are composed of cellulose esters, nylon, polysulfone, polycarbonate, PVDF, or polytetrafluoroethylene (Teflon) (Table 18-2).

Filters are available as flat membranes or pleated into cylinders or cartridge filters (Fig. 18-2) to increase surface area and, thus, flow rate (suppliers: *Cuno, Gelman, Meissner, Millipore, Pall, Sartorius, Schleicher, perhaps others*). Fluid enters the outside of the filter cartridge with applied positive pressure forcing the fluid inward through the filter with the sterile effluent exiting from the center of the cartridge (Fig. 18-3).

The filter is assembled in the stainless steel housing by first wetting the O-rings, making certain that the filter is oriented properly within the housing, then the clamps are hand tightened. There must be no direct hand contact with the cartridge during assembly. The filter and housing are then steam sterilized, typically by steam-in-place (SIP) systems. Inlet pressure must be matched to the maximum cartridge temperature with differential pressure controlled to ensure filter integrity. Both pressurization before and depressurization after sterilization must be gradual to protect filter integrity. Like other steam sterilization processes, there must be assurance that air is replaced by steam. Validation of the sterilization of filters occurs with thermocouples and spore strips located at identified "cold spots" within the filter assembly. After sterilization the filter cartridges are dried with filtered compressed gas.

Table 18-2 Microfilter and Ultrafilter Polymers

Membrane polymer	Advantages	Disadvantages
Cellulose acetate	Very low adsorption High flow rates	Limited pH compatibility
Cellulose nitrate	Good flow rate	High adsorption Limited pH compatibility
Regenerated cellulose	Very low adsorption Very high flow rates	Limited pH compatibility
Modified regenerated cellulose	Very low adsorption Broad pH compatibility	Moderate flow rates
Polyamide (nylon)	Good solvent compatibility Good mechanical strength Broad pH compatibility	High protein adsorption Moderate flow rates
Polycarbonate	Good chemical compatibility	Moderate flow rates Difficult to produce
Polyethersulfone	High flow rates Broad pH compatibility	Moderate-to-low adsorption Limited solvent compatibility
Polysulfone	High flow rates Broad pH compatibility	Moderate-to-high adsorption Limited solvent compatibility
Polypropylene	Excellent chemical and mechanical resistance	Hydrophobic material High adsorption
Polyvinylidenedifluoride (PVF)	Low adsorption Good solvent compatibility	Moderate flow rate Hydrophobic base, made hydrophilic by chemical surface treatment High cost
Polytetrafluoroethylene (PTFE)	Excellent chemical and mechanical resistance	Hydrophobic material High adsorption High cost

Source: From Ref. 2.

FILTER VALIDATION

Filter validation includes both destructive testing to qualify the filter initially and nondestructive testing that is performed prior to and after using the filter in batch production. Destructive testing includes three main tests—(*i*) bacterial retention using actual final formulation of drug product, (*ii*) filter extractables/leachables, and (*iii*) compatibility of filter with drug product.

Bacterial Retention

In this testing phase, the filter is challenged with a known population of microorganisms using conditions that simulate the actual process. It is important that the microbial challenge involves the final product formulation. Formerly, it was acceptable to use a placebo form of the final product where critical attributes like pH, viscosity, osmolarity, ionic strength, and surface tension were simulated, but today the actual final product must be used as the solution

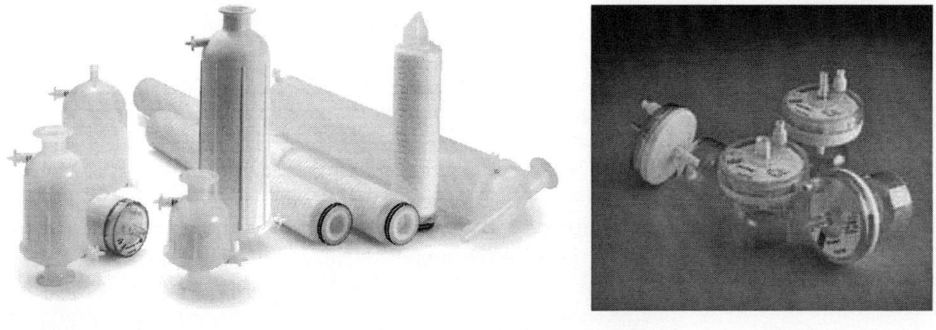

Figure 18-2 Cartridge and disc filters. *Source*: Courtesy of Millipore Corporation.

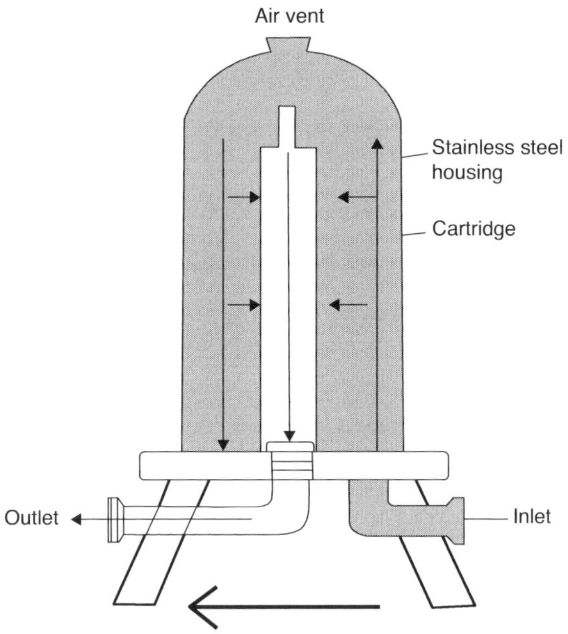

Figure 18-3 Schematic of filter assembly and fluid flow. Note that fluid flow enters outside of filter cartridge, filtration proceeds inward, and sterile effluent exits from center of cartridge.

containing the microbial challenge. The antimicrobial properties of the formulation must be neutralized before adding the microbial challenge. The microorganisms used are *Brevundimonas diminuta*, selected because these cells are approximately 0.3 μm in size, barely above the absolute rating of the 0.22 μm sterilizing filter. The concentration of these cells is 10^7 cells per cm^2 filter surface area. Processing conditions that are simulated include the following:

- Filtration pressure and flow rate
- Duration of filtration process
- Using the same filter type and configuration (disc or cartridge) that will be used in actual process
- Temperature.

Figure 18-4 depicts a schematic of the filter validation retention test apparatus used at filter companies. Note that the apparatus uses both a 0.22 μm filter and a 0.45 μm filter with the latter used as a positive control.

How much volume of product is required to perform the bacterial retention test? The following is a hypothetical example.

- If a 293 mm filter disc is used, this filter has a surface area of 530 cm^2 that, in turn, contains 10^{12} pores.
- The requirement for the bacterial challenge is 10^7 organisms or cells per cm^2 filter area.
- Therefore, a filter with a surface area of 530 cm^2 will require a microbial challenge of 5.3×10^9 organisms.
- It is assumed that one organism will cover approximately 200 pores on a filter surface.
- The standard suspension concentration containing the microbial challenge is 10^6 organisms or cells per mL.
- Therefore, a filter requiring 5.3×10^9 organisms from a stock standard of 10^6 organisms or cells per mL will, in turn, require a product volume of at least 5.3×10^3 mL (5.3×10^9 organisms/10^6 organisms per mL).

Similar calculations can be applied to other filters of different surface areas.

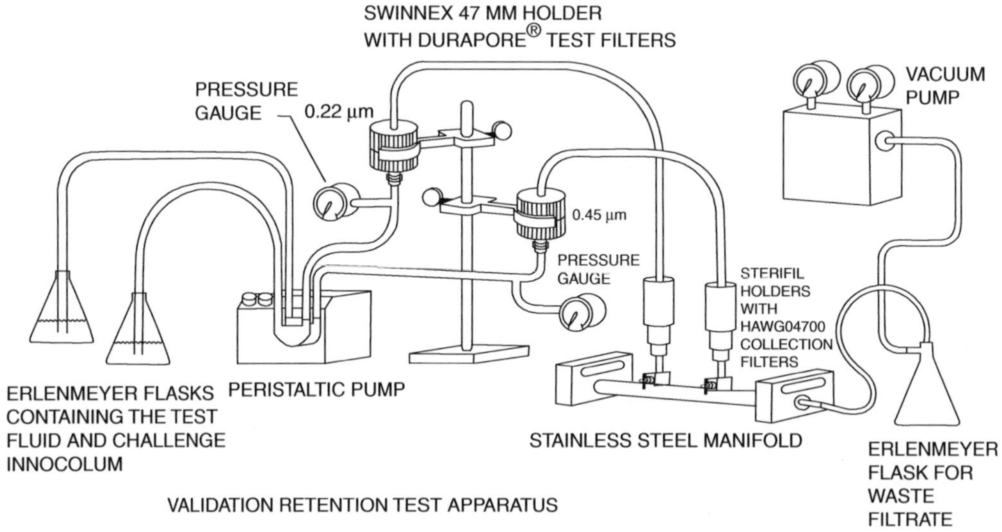

Figure 18-4 Filter validation retention test apparatus. *Source*: From Ref. 3.

Product-Filter Compatibility

Tests must be performed to demonstrate that (1) the product does not adversely affect the retention properties of the filter, as is accomplished in most cases with the bacterial retention studies discussed above, and (2) does not cause the filter to leach materials into the product. Compatibility and extractable studies are performed by the filter manufacturer, although like bacterial retention studies, the product manufacturer is ultimately responsible for the validity of the data. The filter manufacturer will provide information on the flush volume required to yield negative oxidizable substances and provide the data on the level of extractables obtained with different solvent exposures. Potential filter extractables include oligomers, mold release agents, antioxidants, wetting agents, manufacturing debris, plasticizers, membrane backing, cartridge body, and 0-ring material.

There are a few examples, almost all involving protein drug products, where the protein will bind to the filter material with most studies involving in-line membrane filters, not large surface area filters used in commercial manufacturing (4–6). Typically, an insignificant amount of drug will adsorb on the filter surface and occupy all the binding sites. One purpose of a preflush step prior to filtration is to occupy available binding sites as well as remove potential extractables. Polyethersulfone (PES) and PVDF filters are low protein-binding filters.

Other data provided by the filter manufacturer in performing qualification studies on the filter to be used with the finished product include:

- Limits for flow rate, temperature, and pressure
- Ensure that the filter meets the nonfiber releasing criteria from 21CFR 210.3b(6)
- Procedures for filter sterilization
- The filter bubble point or diffusion rate for the in-process integrity tests
- Correlation of the integrity test value and the amount of *B. diminuta* retained
- Written instructions and specifications for the filter integrity test.

IN-PROCESS FILTER INTEGRITY TESTING

Prior to actual filtration of the product, the filter should be flushed either with product or with water for injection to reduce potential extractables and downstream particles. The filter is then subjected to a filter integrity test (prefiltration filter integrity test) and after the solution is filtered, the filter is again subjected to a second filter integrity test (postfiltration filter integrity test). This integrity test usually is performed either as the *bubble-point test* or as the *diffusion or forward flow* test. The bubble point test is commonly used on smaller filters. As the surface area of filters becomes large, diffusion of air through the water-filled pores tends to obscure the bubble point. Therefore, the diffusion test has been developed as an integrity test for filters with large surface areas. A *pressure hold test* also can be applied to large surface area filters. The filter manufacturer will recommend the best integrity test for the filter system in question.

These are tests to detect the largest pore or other opening through the membrane. The basic test is performed by gradually raising air pressure on the upstream side of a water-wet filter. It is imperative that the filter be completely wetted or else the integrity test will fail because applied pressure gas will easily pass through pores not filled with liquid. Either water for injection or actual product is used to wet the filter prior to performing the prefiltration filter integrity test. The bubble point test keeps raising pressure until a pressure is obtained where air bubbles first appear downstream from the filter.

The principle of the bubble point test follows the fact that a fully wetted membrane filter of very small pore size will hold liquid in the pores by surface tension and capillary force. The pressure of a gas required to force the entrapped liquid through and out of the fully wetted pore capillary is called the bubble point because after the liquid is forced out, air bubbles will appear (Fig. 18-5). The bubble point is a function of the type and pore size of the filter membrane, the surface tension of the liquid and temperature. The equation for bubble point pressure is

$$P = 4\,k\,\gamma\,\cos\Theta/d \qquad \text{(Equation 1)}$$

where P is the bubble point pressure that is directly proportional to the shape correction factor of the filter, k, the liquid surface tension, γ, and the liquid contact angle, Θ, and inversely proportional to the pore diameter, d. The bubble point pressure correlates to the microbial log reduction value as shown in Figure 18-6. Table 18-3 provides the standard bubble point ratings for various types of membrane filters (although these ratings are subject to change).

The diffusion or forward flow test raises pressure to some point below the known bubble point pressure, then diffusion flow (usually in mL/min) is measured. The principle of the

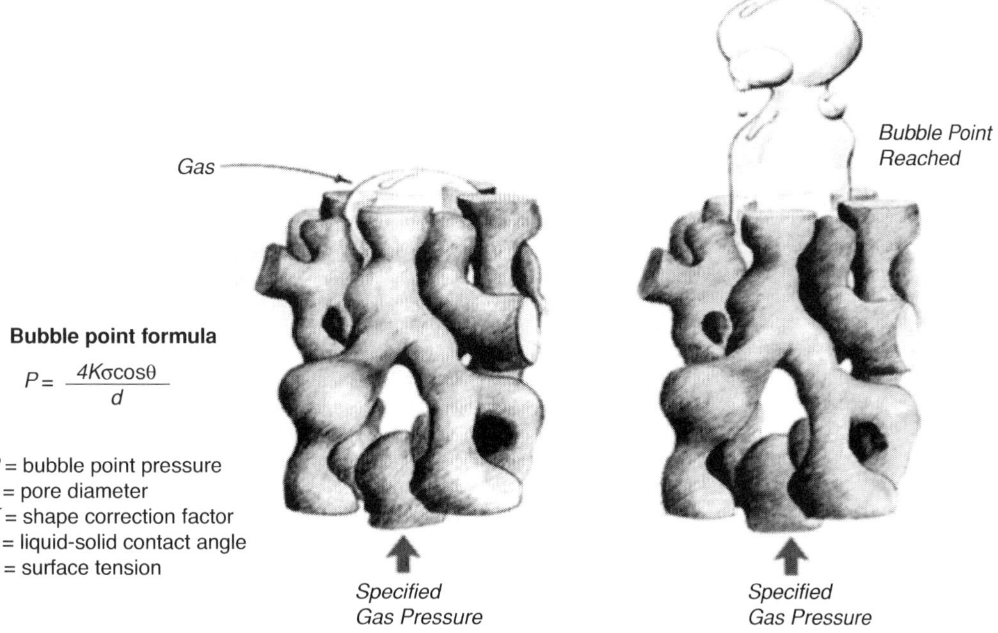

Figure 18-5 Bubble point filter integrity test. The bubble point test is based on the fact that liquid is held in a capillary tube by surface tension. The minimum gas pressure required to force liquid out of the tube is a direct function of tube diameter. The pressure required to force liquid out of a liquid-filled capillary must be sufficient to overcome surface tension and is a direct measure of effective. *Source*: Courtesy of Millipore Corporation.

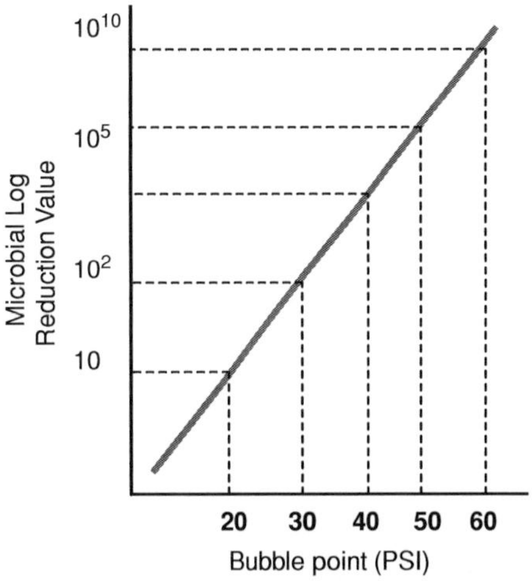

Figure 18-6 Correlation estimate of bubble point pressure and microbial reduction. *Source*: From Ref. 7.

diffusion test is very similar to that of the bubble point test. A gas dissolved in a liquid held in the pores of a fully wetted filter will slowly diffuse out of the filter pores as a pressure differential is applied to the filter that results in a concentration gradient across the filter. This is described by the following equation:

$$Q = K\,(P_1 - P_2)\,A\,\varphi/L \qquad \text{(Equation 2)}$$

where the diffusional flowrate, Q, is dependent on the diffusivity coefficient, K, the pressure differential, $(P_1 - P_2)$, the area of the membrane, A, and the membrane porosity, φ, and inversely proportional to the effective path length, L, of the gas flow in the filter. Figure 18-7 shows the setup of a diffusion flow test apparatus and Table 18-3 provides acceptance criteria for diffusion flow as a function of the type of cartridge filter.

These pressures are characteristic for each pore size of a filter and are provided by the filter manufacturer. For example, a 0.2 μm cellulose ester filter will bubble at about 50 psig or a diffusive flow rating of no greater than 13 mL/min at a pressure of 40 psig. If the filter is wetted with other liquids, such as a product, the bubble point will differ and must be determined experimentally. If the bubble point is lower than the rated pressure, the filter is defective, probably because of a puncture or tear, and should not be used.

Table 18-3 Examples of Bubble Point and Diffusion Test Standards for Filter Integrity Criteria of Different Filters

Bubble points	
Filter	**Bubble point (≥ pounds/square inch)**
0.1 μm PVDF	70
0.22 μm PVDF	50
0.22 μm MCE	55
0.45 μm MCE	30
0.65 μm MCE	17
Diffusion rates	
Filter	**Diffusion rate (mL/min@ × psi)**
0.1 μm PVDF cartridge	<20 @ 56
0.22 μm PVDF cartridge	<13 @ 40
0.45 μm PVDF cartridge	<15 @ 15

Abbreviations: MCE, mixed cellulose ester; PVDF, polysulfone and polyvinylidine difluoride.
Source: From Millipore Corporation Technical Literature.

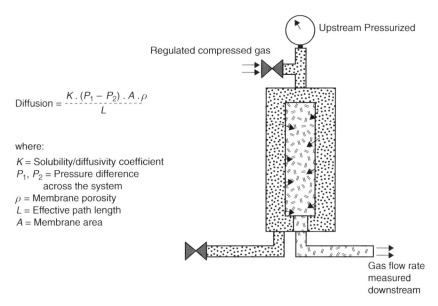

$$\text{Diffusion} = \frac{K \cdot (P_1 - P_2) \cdot A \cdot \rho}{L}$$

where:

K = Solubility/diffusivity coefficient
P_1, P_2 = Pressure difference across the system
ρ = Membrane porosity
L = Effective path length
A = Membrane area

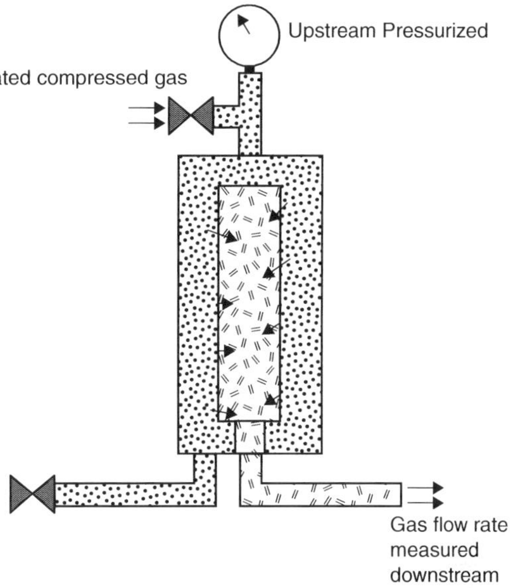

Figure 18-7 Gas diffusion (forward flow). *Source*: Courtesy of Millipore Corporation.

Figure 18-8 shows a comparison of the bubble point and diffusive flow test. The bubble point test is best applied for small disc filters while the diffusive flow test is best for cartridge filters. The bubble point test is quick and easy and relates to the largest pore size of the filter, but errors in estimating the bubble point occur, especially when manually done. The diffusion test is more quantitative, confirms the absence of large pores, but is not applicable for small area filters because of insufficient diffusive gas flow. Neither test will measure pore size.

While membrane filters are disposable and thus discarded after use, the holders must be cleaned thoroughly between uses. Today, clean, sterile, pretested, disposable assemblies for small as well as large volumes of solutions are available commercially.

MEMBRANE FILTER INTEGRITY TESTING WITH PRODUCT THAT LOWERS THE BUBBLE POINT

Again, the bubble point equation is

$$P = \frac{4k\gamma \cos ø}{d} \qquad \text{(Equation 3)}$$

P = bubble point pressure
k = shape correction factor
γ = surface tension
ø = contact angle
d = pore diameter

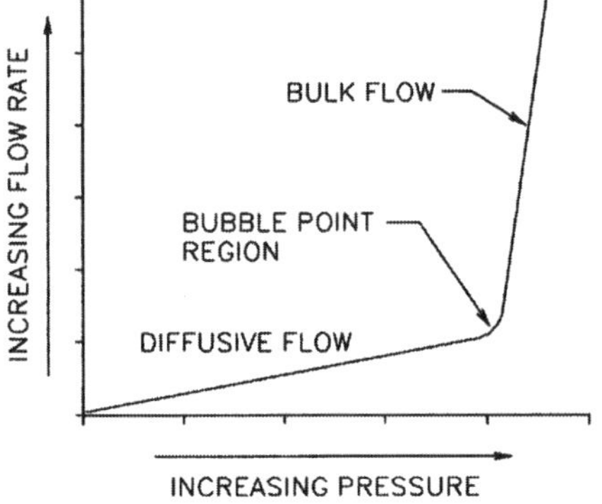

Figure 18-8 Diffusive flow rate as a function of applied pressure on a filter membrane. *Source*: From Ref. 8.

where the bubble point pressure, P, is directly proportional to the surface tension, γ. The surface tension of water is 72 dyn/cm (contrasted with 21.3 dyn/cm for isopropanol) meaning that water has very high cohesive (attraction) forces for the capillary walls of a membrane filter. That is why it requires a very high amount of pressure to overcome this cohesive force and drive water out of the pores of a membrane filter.

Surface-active agents are amphiphilic wetting agents that lower the surface tension of water and are positively adsorbed at the liquid/air interface, thus preventing proteins from adsorbing at this interface and minimizing protein denaturation (aggregation) due to hydrophobic (air) interactions. Thus, the presence of surface-active agents in aqueous solutions will depress the bubble point compared with water alone.

Sometimes, the depressed bubble point can be restored by copious amounts of rinsing the filter with water. However, this is not always successful. If not with the product in question, then filter validation studies need to be conducted using the final product formulation containing the required 10^7 organisms (*B. diminuta*) per cm^2 filter surface area.

A simple relationship can be applied between the minimum acceptable product bubble point pressure and the manufacturer's stated minimum allowable bubble point:

$$P_p = \frac{P_o}{P_w} P_m \qquad \text{(Equation 4)}$$

P_p = minimum acceptable product bubble point
P_o = observed bubble point using product
P_w = water bubble point
P_m = filter manufacturer's stated minimum allowable bubble point

AUDITING FILTRATION PROCESSING AND FILTER VALIDATION

When the FDA or other current good manufacturing practice (cGMP) compliance auditor inspects a filtration process, they will expect to see validated data on the effect of product and process properties such as flow rate, pressure drop, viscosity, pH, temperature, organic content, and ionic strength on the bacterial retention properties of the filter. They also will ask for data supporting the compatibility of the product and the filter and how the bacterial challenge test was done and validated. Other information reviewed will be assembly and sterilization of the filter, validated filtration time limits, and integrity test methods and specifications.

ONGOING CONTROVERSIES

Sometimes, evidence is reported that 0.2 μm filters do not remove all possible microbial contamination (9,10), potentially necessitating the need to use certain types of 0.1 μm membrane filters (11). However, most of the parenteral pharmaceutical industry continues to use 0.2 μm filters although now employing redundant (two 0.2 μm filters side-by-side) filtration systems. Double filtration indeed increases the probability of adsorptive organism capture of organisms, such as "L-forms" smaller than 0.2 μm filter pore size (12). Pre and postfiltration integrity tests must be done on both filters and, although some exceptions exist, all four integrity tests must pass.

Ongoing technical issues or controversies in sterile filtration technology include the following:

1. Defining "worst-case" conditions for filter validation studies
2. The need for validating the removal of the smallest types of organisms, for example, mycoplasma and viruses
3. Effects of the filter and the process of bacterial deformation
4. The potential for air entrapment during filter sterilization by steam
5. Is there really a significant trend toward replacing 0.2 μm filters with 0.1 μm filters as final sterilization filters?

REFERENCES

1. Meltzer TH, Jornitz MW, Johnston PR. Relative efficiencies of double filters or tighter filters for small-organism removals. Pharm Technol 1999; 23:98–106.
2. Jornitz MW. Filters and Filtration, Encyclopedia of Pharmaceutical Technology. London: Informa, 2003:1754.
3. MacDonald WD, Pelletier CA, Gasper CA. DL practical methods for the microbial validation of sterilizing grade filters used in aseptic processing. J Paren Sci Technol 1989; 43:266–270.
4. Turco S, King RE. Sterile Dosage Forms. 3rd ed. Philadelphia: Lea & Febiger, 1987:312–316.
5. Brophy RT, Lambert WJ. The adsorption of insulinotropin to polymeric sterilizing filters. J Parenter Sci Technol 1994; 48:92–94.
6. Brose DJ, Waibel P. Adsorption of proteins in commercial microfiltration capsule. Pharm Technol 1996; 20:48–52.
7. Leahy TJ, Sullivan MJ. Validation of bacterial retention capabilities of membrane filters. Pharm Technol 1978; 2:65–75.
8. Sterilizing Filtration of Liquids. PDA J Pharm Sci Technol Technical Report No. 26, 1998 Supplement, p. 13, 2008; 62:S-5.
9. Sundaram S, Eisenhuth J, Howard Jr, et al. Retention of water-borne bacteria by membrane filters, Part I. Bacterial challenge tests on 0.2 and 0.22 micron rated filters. PDA J Pharm Sci Technol 2001; 55:65–86.
10. Sundaram S, Mallick S, Eisenhuth J, et al. Retention of water-borne bacteria by membrane filters, Part II. Scanning electron microscopy (SEM) and fatty acid ester (FAME) characterization of bacterial species recovered downstream of 0.2/0.22 micron rated filters. PDA J Pharm Sci Technol 2001; 55:87–113.
11. Sundaram S, Eisenhuth J, Howard Jr, et al. Retention of water-borne bacteria by membrane filters, Part III. Bacterial challenge tests on 0.1 micron rated filters. PDA J Pharm Sci Technol 2001; 55:114–126.
12. Jornitz MW, Meltzer TH. Sterile double filtration. Pharm Technol 1998; 22:92–100.

BIBLIOGRAPHY

Antonsen HR, Awafo V, Bender JL, et al. Sterilizing filtration of liquids. PDA Technical Report No. 26 (revised 2008). PDA J Pharm Sci Technol 2008; 62(Suppl. 5):2–60.

Johnston PR. Fluid Sterilization by Filtration. 3rd ed. Boca Raton, FL: Interpharm/CRC, 2003.

Jornitz MW, ed. Sterile filtration. In: Scheper T, series ed. Advances in Biochemical Engineering/Biotechnology. Vol 98. Berlin: Springer, 2006.

Jornitz MW, Meltzer T. Sterile Filtration—A Practical Approach. New York: Informa, 2001.

Meltzer TH, Jornitz MW. Anatomy of a Pharmaceutical Filtration Differential Pressures, Flow Rates, Filter Areas, Throughputs and Filter Sizing. Bethesda, MD: PDA/DHI, 2009.

19 Sterile product filling, stoppering, and sealing

*Mark A. Kruszynski**

During the filling of containers with a product, the most stringent requirements must be exercised to prevent contamination, particularly if the product has been sterilized by filtration and will not be sterilized in the final container. This process is called an *aseptic fill* and is validated with media fills (see chap. 21). The assurance of product sterility is supported by the design of the filling complex, the training of the operators, environmental monitoring, filter validation, and by mimicking the manufacturing process through media fills. During the filling operation, the product must be transferred from a bulk container or tank and subdivided into dose containers. This operation exposes the sterile product to the environment, equipment, and manipulative technique of the operators until it can be sealed in the dose container. Therefore, this operation is carried out with a minimum exposure time, even though maximum protection is provided by filling under a blanket of high efficiency particulate air (HEPA)-filtered laminar flow air within the aseptic area.

Most frequently, the compounded product is in the form of a liquid. However, products are also compounded as suspensions or emulsions and as powders. A liquid is more readily subdivided uniformly and introduced into a container having a narrow mouth than is a solid. Mobile liquids are considerably easier to transfer and subdivide than viscous, sticky liquids, which require heavy-duty machinery for rapid production filling.

Although many devices are available for filling containers with liquids, certain characteristics are fundamental to them all. A mechanism is provided for repetitively forcing a measured volume of the liquid through the orifice of a delivery tube that is introduced into the container. The size of the delivery tube will vary from that of about a 20-G hypodermic needle to a tube $^1/_2$ in. or more in diameter. The size required is determined by the physical characteristics of the liquid, the desired delivery speed, and the inside diameter of the neck of the container. The tube must enter the neck and deliver the liquid well into the neck to eliminate spillage, allowing sufficient clearance to permit air to leave the container as the liquid enters.

The delivery tube should be as large in diameter as possible to reduce the resistance and decrease the velocity of flow of the liquid. Product surface tension, viscosity, and temperature dictate the potential of product dripping or the formation of "threads" of product on the sealing surface of the vial or syringe wall. To reduce the possibility of the product splashing out of the container, most automated filling systems fill "bottom up" with the filling tube inserted to its greatest depth at the start of the filling cycle and withdrawing the tube as the product is dosed into the container.

For smaller volumes of liquids, the delivery usually is obtained from the stroke of the plunger of a syringe, forcing the liquid through a two-way valve providing for alternate filling of the syringe and delivery of mobile liquids. For heavy, viscous liquids, a sliding piston valve, the turn of an auger in the neck of a funnel, or the oscillation of a rubber diaphragm may be used. For large volumes, the quantity delivered usually is measured in the container by the level of fill in the container, the force required to transfer the liquid being provided by gravity, a pressure pump, or a vacuum pump.

The narrow neck of an ampoule limits the clearance possible between the delivery tube and the inside of the neck. Since a drop of liquid normally hangs at the tip of the delivery tube after a delivery, the neck of an ampoule will be wet as the delivery tube is withdrawn, unless the drop is retracted. Therefore, filling machines should have a mechanism by which this drop can be drawn back into the lumen of the tube, called a "suck-back feature." Since the liquid will be in intimate contact with the parts of the machine through which it flows, these must be constructed

* This chapter coauthored by Mark A. Kruszynski of Baxter Biopharma Solutions.

of nonreactive materials such as borosilicate glass or stainless steel. Modern coatings, such as AMCX2286, are used to coat stainless steel needles for products that are affected by contact with metals, for example, formulations containing chelating agents or having very acidic or alkaline pH values. In addition, they should easily be demountable for cleaning and sterilization.

FILLING MECHANISMS

Filling machines are classified by the type of driving device or filling mechanism used to deliver the drug-containing formulation into the primary package. There are at least four driving devices and four filling mechanisms:

Driving device	Filling mechanism(s)
Gravity (solids and liquids)	Gravimetric, time–pressure, fill-by-weight
Piston (liquids and gases)	Rotary piston
Rotary pump (liquids and gases)	Rotary peristaltic
Auger screw or vibrator (solids)	Vibratory/mechanical force

Gravity/Time Pressure Filling

The gravity-based filling machine is the oldest type and most economical. The filling principle is simple; the amount of product flowing through the filling nozzle is driven by gravity and will always be the same for a fixed amount of time. The finished bulk solution is pumped into a holding tank above a set of pneumatically operated valves. Each valve is independently timed by a master computer for the filling machine so that precise amounts of liquid will flow by gravity into the container. The amount of product dispensed is controlled by adjusting the time for closing the valve. In more precise systems, weight feedback is used to control the volume of dispensed product. Independent timing of each filling valve/nozzle corrects for minor variations in flow rates so that each container is filled accurately and uniformly. Improvements in holding tank headspace pressure control and feedback control have made time pressure filling machines more accurate than pump systems.

The disadvantage of this type of technology is that the dynamics of the fluid path and nozzle actuation characteristics continuously change over time. This requires the operator to make adjustments to the machine's stored parameters more frequently than other filling mechanisms.

Fill-by-Weight

This is a very simple system where the bulk solution tank is controlled by a valve to release product through a filling nozzle into the container that sits on a balance that controls the volume of product actually filled. This filling method is not used much anymore because of significant disadvantages using a balance or load cell (Table 19-1). However, there is better accuracy and control of the fill volume per container compared with other methods.

Piston Filling

Piston filling includes pumps with lapped rotary or check valves and pumps that use a rolling diaphragm. Lapped rotary pumps involve a cylinder that is lapped by both the piston and the rotary valve to produce an exceedingly tight fit. Pumps with check valves are not used for injectable filling because the valves are difficult to clean. Pumps with the rolling diaphragm use a flexible membrane attached to the pump at its outside diameter and to the piston at its inside diameter (1). A space between the piston and the body internal cylinder allows the diaphragm to be doubled and to roll as the piston moves up and down. Vacuum is required to maintain the shape of the diaphragm and to pull the piston downward on the refill part of the filling cycle.

Piston pumping machines are the most commonly used filling machines for liquids. They are not the best choice as a filling mechanism for shear-sensitive liquids and suspensions because of the tight clearances between the piston and the cylinder. In piston-driven filling machines, the product enters the dispensing cylinder by opening an infeed valve moving the piston in a reverse direction, closing the infeed valve, opening a discharge valve, and driving the piston in the opposite direction so that the product is propelled to the nozzle and into the collection

Table 19-1 Advantages and Disadvantages of Sterile Product Filling Methods

Filling mechanism	Advantages	Disadvantages
Time–pressure	• Clean—few parts in the product path • Easy to maintain and change over • Product path easy to CIP/SIP • No leakage in product path • Can handle sensitive products • Can run dry • Accurate for small or large fill volumes • Possible self-adjustment of fill volumes between fill cycles at line speed	• Pressure change sensitivity • Pressure control and monitoring required • Temperature sensitivity • Viscosity sensitivity • Expensive parts for change over of fill volume
Fill-by-weight	• Clean • Easy to maintain and change over • Easy to clean and to CIP/SIP • No leakage in product path • Can handle sensitive products • Can run dry • Real-time fill volume control • 100% documentation of fill volumes • Accurate for small and large fill volumes	• High cost of scales and control system • Maintenance (potential of spillage on scales) • Complex container handling • Longer fill times • Accuracy for small fills decreases with fill volume reduction • Dripping or spills can damage scales—protection of scales complicates the system
Piston pumps with lapped rotary valves	• Simplicity-3 parts—no consumable parts such as a rubber seal or diaphragm • Simplicity of motion • No sophisticated controls required • Accurate for small or large fill volumes • Reasonably easy to CIP/SIP	• May damage shear-sensitive products • Greater source of metallic particles • Push–pull actuation—actuating mechanism must be backlash free • Thorough cleaning required between filling campaigns • Must be located in clean environment • Costs • Handling issues during cleaning—nicking, cannot interchange piston and cylinder • Cannot run dry • Potential for seizing • Leakage varies with input pressure
Piston pumps with rolling diaphragm	• Clean—few parts in product path • No leakage in product path • Gentle to sensitive products since no shear is involved • Can run sugar-based products without seizing • Can handle slurries • Can run dry • Accurate for small or large fill volumes • Pump loads the actuating mechanism, eliminating backlash	• Special assembly requirements, need highly trained people to assemble • Vacuum source required • Diaphragm must be discarded and replaced • Number of components • Pump must be horizontally oriented for CIP/SIP that can affect drainage
Peristaltic	• Clean—very few parts in product path • Easy to maintain and change over • Product path easy to CIP/SIP • No leakage in the product path • Can handle sensitive products • Can handle suspensions and slurries • Can run dry • Easy cleanup for potent products—best of all filling systems	• Pulsating flow • Accuracy issues due to tubing tolerances, angular position of rotor at start and stop, change of tubing size and shape over time, and check weigh and adjustment must compensate for volume drift

Abbreviations: CIP, clean-in-place; SIP, sterilize-in-place. *Source*: From Ref. 1.

container. The volume of the filled product is controlled by adjusting the stroke of the piston. The steps of a piston filling machine are as follows:

- Suck back
- Rotary valve change position
- Nozzle open
- Piston forward to discharge solution
- Nozzle close
- Rotary valve change position.

Syringe filling machines typically are valve-less rotary piston fillers, although peristaltic and time pressure syringe fillers do exist. Instead of the existence of a solid piston, a portion of the piston body is removed. On the infeed stroke, the side of the piston with the cavity is rotated to the inlet. The down stroke creates a vacuum and product enters the pump body. The piston rotates 180°, and the liquid-filled cavity faces the outlet. The pump upstroke occurs and the product is forced out of the pump. The rotation continues another 180° and the cycle repeated.

Peristaltic Filling

Peristalsis describes movement of ingested food in the gastrointestinal tract. The same principle is used for filling machines. Peristaltic filling involves positive displacement where the solution contained within a flexible tube that is fitted inside a circular (rotary) or elongated (linear) (Fig. 19-1) pump casing. A rotor with a number of "rollers," "shoes," or "wipers" attached to the external circumference compresses the flexible tube. As the rotor turns or moves, the part of tube under compression closes (or "occludes") thus forcing the fluid to be pumped to move through the tube. Additionally, as the tube opens to its natural state after the passing of the cam ("restitution") fluid flow is induced to the pump.

Since there are no moving parts in contact with the fluid, peristaltic pumps are inexpensive to manufacture. Their lack of valves, seals, and glands makes them comparatively inexpensive to maintain, and the use of a hose or tube makes for a relatively low-cost maintenance item compared with other pump types. Peristaltic pumps also minimize shear forces experienced by the product solution, good for shear-sensitive protein products. However, they are not as good for high viscosity liquids and cannot match rotary piston machines for small-volume filling precision.

Typical tubing systems used for filling machines, regardless of mechanism, are silicone rubber, polyvinyl chloride, and fluoropolymer.

Advantages and disadvantages of each filling mechanism are summarized in Table 19-1 (1).

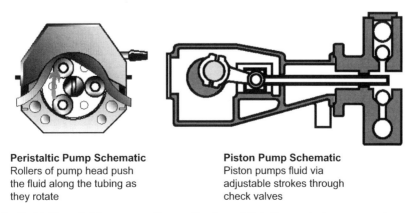

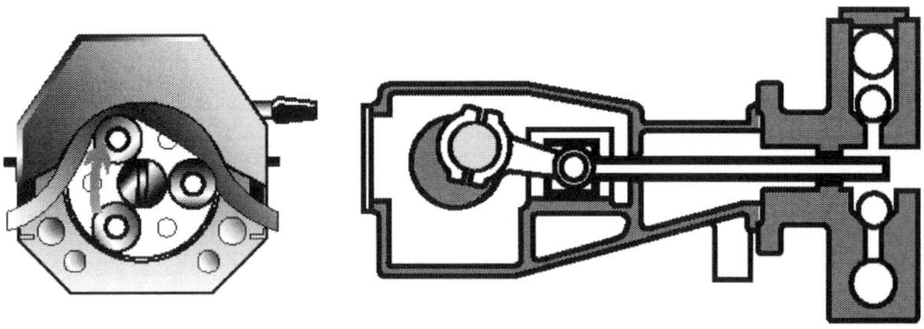

Peristaltic Pump Schematic
Rollers of pump head push the fluid along the tubing as they rotate

Piston Pump Schematic
Piston pumps fluid via adjustable strokes through check valves

Figure 19-1 Peristaltic and piston pumps. *Source*: Courtesy of Cole-Parmer (coleparmer.com).

LIQUID FILLING

There are three main methods for filling liquids into containers with high accuracy: volumetric filling, time/pressure dosing, and net weight filling. Volumetric filling machines employing pistons or peristaltic pumps are most commonly used, for example, the Chase-Logeman filling machine (Fig. 19-2). This filler is best suited for small batch filling of 2 mL vials (13 mm openings) to 100 mL vials or bottles with 20 mm openings. Filling speeds for 2 to 20 mL vials are usually 80 to 120 vials filled per minute.

Stainless steel syringes are required with viscous liquids because glass syringes are not strong enough to withstand the high pressures developed during delivery.

When high-speed filling rates (now up to 800 per minute for vials and 60,000 per minute for 0.5–1.0 mL long syringes) are desired, but accuracy and precision must be maintained, multiple filling units can be joined together and electronically coordinated. When the product is sensitive to metals, a peristaltic pump filler may be used because the product comes in contact only with silicone rubber tubing. While there might be some sacrifice of filling accuracy ($\pm$ 3% with <0.5 mL fill volumes), technology now uses 100% check weighing of filled containers so that filling accuracy is still quite good.

Time–pressure (or time–gravity) filling machines are gaining in popularity in filling sterile liquids. A product tank is connected to the filling system that is equipped with a pressure sensor. The sensor continuously measures pressure and transmits values to the programmable logic control (PLC) system that controls the flow of product from tank to filling manifold. Product flow occurs when tubing is mechanically unpinched and stops when tubing is mechanically pinched. The main advantage of time/pressure filling operations is that these filling apparatuses do not

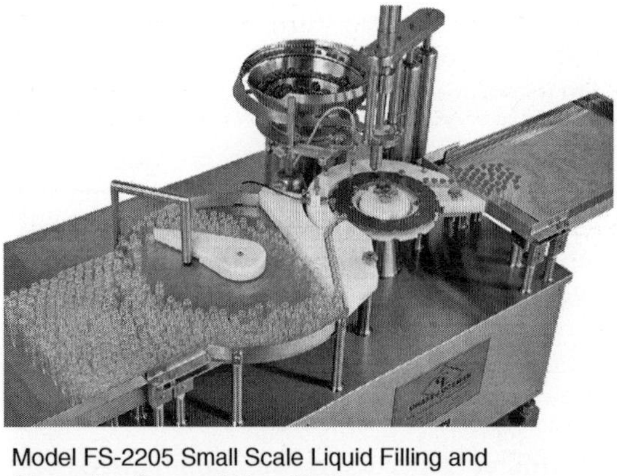

Model FS-2205 Small Scale Liquid Filling and Stoppering Compact Monoblock

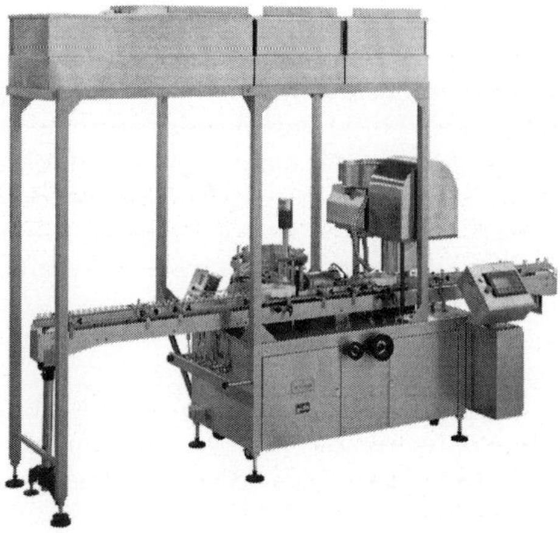

Model NwFSAS12 Liquid Filling, Stoppering, and Crimp Sealing Monoblock with HEPA Filtration Unit

Model FSTS Production Liquid Filling, Stoppering, and Automatic Tray Loading

Figure 19-2 Examples of a commonly used sterile filling machines. *Source*: Courtesy of Chase-Logeman Corporation.

Figure 19-3 Time–pressure filling machine with close-up view. *Source*: Courtesy of Baxter Healthcare Corporation.

contain mechanical moving parts in the product stream. The product is driven by pressure (usually nitrogen) with no pumping mechanism involved. Thus, especially for proteins that are quite sensitive to shear forces, time/pressure filling is preferable.

The Inova VFVM2428 model (Fig. 19-3) is an example of a time–pressure filling machine designed for fast filling of large batches. It can fill vial sizes from 2 to 100 mL at filing rates of 60 to 300 per minute.

Most high-speed fillers for large-volume solutions use the bottle as the measuring device, transferring the liquid either by vacuum or positive pressure from the bulk reservoir to the individual unit containers. Therefore, a high accuracy of fill is not achievable.

The United States Pharmacopeia (USP) requires that each container be filled with a sufficient volume in excess of the labeled volume to ensure withdrawal of the labeled volume and provides a table of suggested fill volumes (Table 2-2).

The filling of a small number of containers may be accomplished with a hypodermic syringe and needle, the liquid being drawn into the syringe and forced through the needle into the container. A device for providing greater speed of filling is the Cornwall Pipet (*Becton Dickinson*). This has a two-way valve between the syringe and the needle and means for setting

Table 19-2 Possible Problems Encountered During Filling Operations

- Product splashing
- Product spills
- Product foaming and effect on dose accuracy
- Viscous product and potential problems with dose accuracy and uniformity
- Out of tolerance fill volumes/weight
- Receiving vessel overflows
- Receiving vessel overpressurized
- Filling needles are bent
- Filling needles are plugged with product
- Control of dose from container-to-container
- Adsorption of active ingredient on the surface of the tubing used with the filling machine
- Protein aggregation due to tubing surface interactions
- Leachables from tubing
- Fill pump leak
- Power outage

the stroke of the syringe so that the same volume will be delivered each time. Clean, sterile, disposable assemblies (suppliers: *Burron, Pharmaseal*) operating on the same principle have particular usefulness in hospital pharmacy or experimental operations.

PREFILLED SYRINGE PROCESSING AND FILLING

Syringes are cleaned, sterilized (by ethylene oxide or radiation), and sealed with a puncture proof cover by the syringe manufacturer before delivering to the finished product manufacturer. Syringes are contained in a plastic tub system double wrapped that maintains sterility of the syringes [e.g., Becton Dickinson's Hypak™ syringes (Fig. 7-7)]. The transfer of these tubs containing sterile syringes from a receiving area into the aseptic filling area presents a challenge with respect to maintaining sterility. Typically, the outer bag wrap is removed within a Grade C/ISO 8 area and the inner bag wrap is sanitized (alcohol or hydrogen peroxide vapor) before moving into the aseptic area. At the time of the publication of this book, low energy e-beam radiation was becoming a new alternative as a surface decontamination process that increases the level of sterility assurance in the transfer of presterilized syringe tubs into the aseptic area. In the aseptic area, an operator removes the lid of the tub and the tub is placed on the filling line. Syringes are filled row by row with precise filling volumes (can be accurate within 0.1 mL) and then the rubber plunger is accurately inserted at the predetermined location within the syringe barrel to ensure accurate delivery volume.

An example of a common syringe filling machine (Inova) is seen in Figure 19-4. Syringe fillers are designed to first fill the product into sterile syringes, then the sterile stopper is inserted. If the stopper insertion rods or tubes are not properly aligned then the product could potentially contact the rods and tubes and glass will break. Syringe fillers typically can fill 0.5 to 20 mL syringe at rates between 60 to 600 syringes per minute.

CARTRIDGE FILLING

Example of a common cartridge filling machine is seen in Figure 19-5. This is a Bausch + Stroebel machine that fills up to 3 mL cartridges at rates of 300 per minute. With cartridges, the rubber plunger is first inserted to a predetermined place within the barrel of the siliconized cartridge. The product is filled, typically with a two, even three-shot fill so that there is no significant headspace; then the cartridge is sealed with a sterile, rubber septum within a aluminum cap. Excessive air space in a cartridge will affect dose accuracy when the contents of the cartridges are ejected through a pen delivery system.

Issues with Liquid Filling

Table 19-2 lists examples of potential problems that may occur with filling of liquid products. These potential problems illustrate the extreme importance of process research and development with the ultimate goal of process validation with respect to filling accuracy and effect of filling phenomena on product quality. While many of these studies can be conducted in a laboratory, final verification and validation must be conducted in a pilot or production filling facility at scale.

Figure 19-4 Syringe filling machine (Inova). *Source*: Courtesy of Baxter BioPharma Healthcare Corporation.

Nominal and exact dose filling

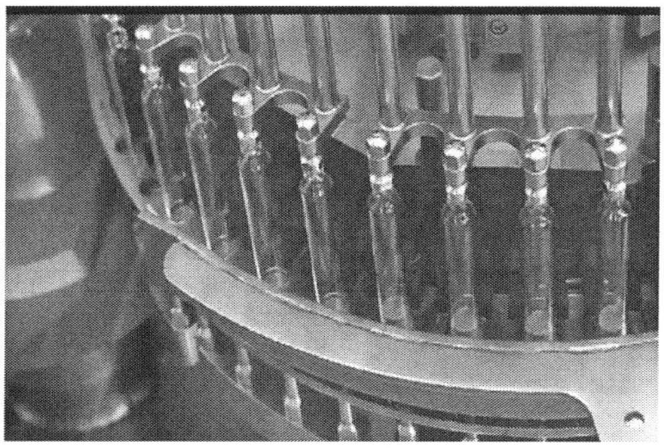

1.5–3.0mL volumes

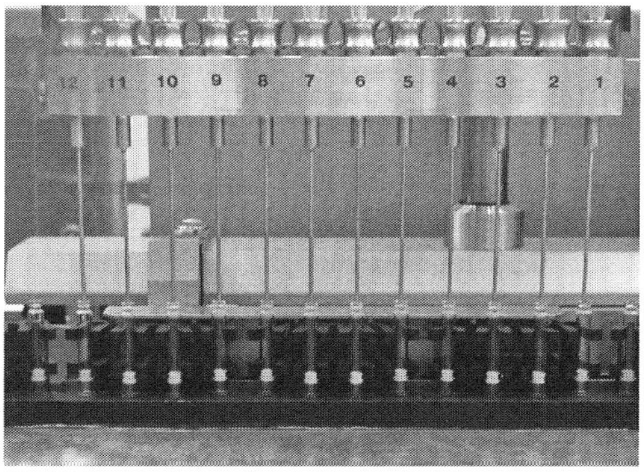

12 (24) Two-stage filling

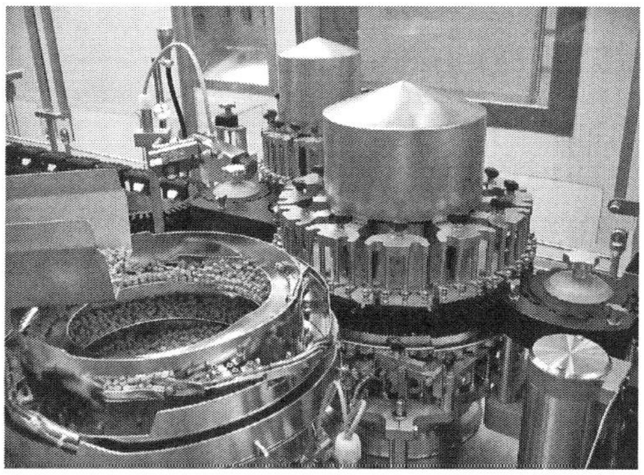

Time/pressure and Sensor Filling

Figure 19-5 Cartridge filling machine (Bausch & Strobel KFM-6024). *Source*: Courtesy of Baxter Healthcare Corporation.

SOLID FILLING

Sterile solids, such as antibiotics, are more difficult to subdivide evenly into containers than are liquids. The rate of flow of solid material is slow and often irregular. Even though a container with a larger diameter opening is used to facilitate filling, it is difficult to introduce the solid particles, and the risk of spillage is ever present. The accuracy of the quantity delivered cannot be controlled as well as with liquids. Because of these factors, the tolerances permitted for the content of such containers must be relatively large.

Powder filling also must control dose uniformity and accuracy that is a function both of the engineering of the powder filling machine and the particle size characteristics dictated by methods used to produce the solid product. Control of relative humidity during filling and minimizing foreign particle contamination also are challenges with powder filling. Primary vendors of powder filling machines are Perry and Chase-Logeman.

Some sterile solids are subdivided into containers by individual weighing. A scoop usually is provided to aid in approximating the quantity required, but the quantity filled into the container finally is weighed on a balance. This is a slow process. When the solid is obtainable in a granular form so that it will flow more freely, other methods of filling may be employed. In general, these involve the measurement and delivery of a volume of the granular material that has been calibrated in terms of the weight desired. An adjustable cavity in the rim of a wheel is filled by vacuum and the contents held by vacuum until the cavity is inverted over the container. The solid material then is discharged into the container by a puff of sterile air.

The Perry Accofil® system was developed as a solution to the problem encountered by the pharmaceutical industry with the introduction of penicillin many years ago. Until the availability of this machine, powders had to be manually weighed, causing lots of potential problems with sterility assurance and operator exposure to powder particles. The principle of

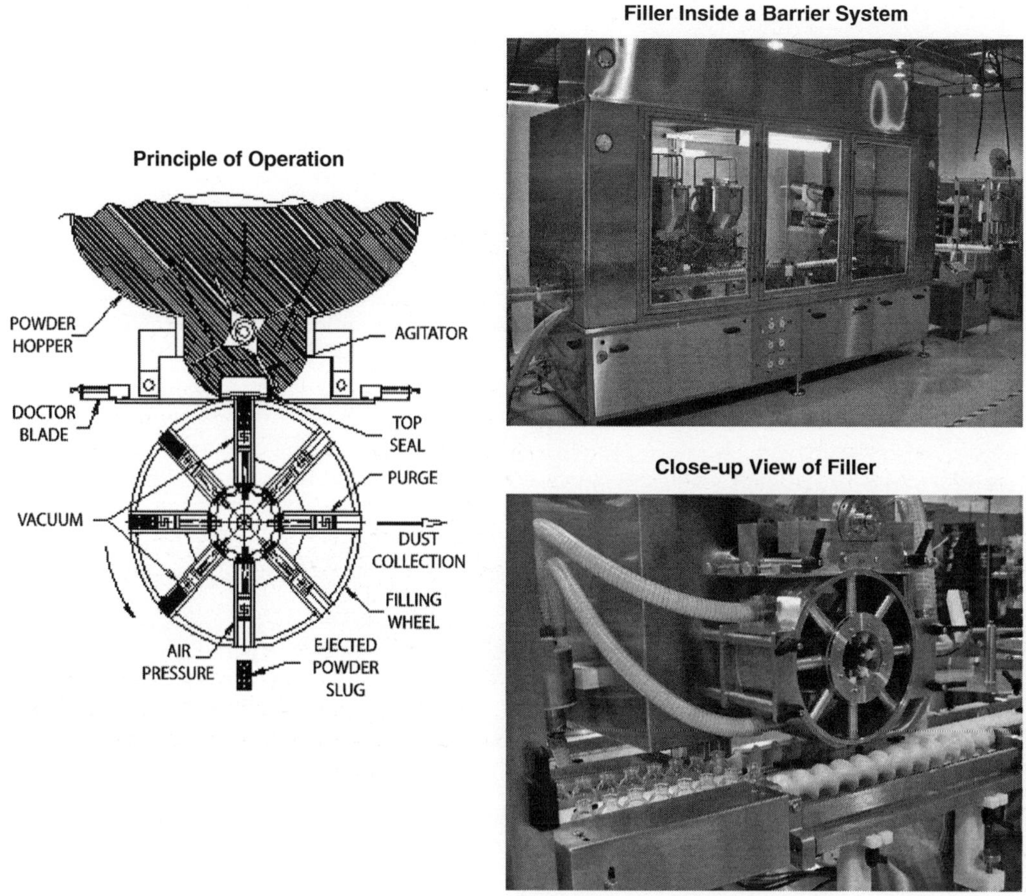

Figure 19-6 Perry Accofil® sterile powder filling machine. *Source*: Courtesy of M&O Perry Industries, Inc.

the Accofil system is shown in Figure 19-6 along with photos of the filling machine. A metering cylinder contains an adjustable piston with a porous filter head that is impervious to powder, but will pass air. The piston head forms the bottom of the cylinder and can be adjusted to provide a desired powder volume. The vacuum is applied through the piston with filter, which causes the powder to be drawn into the cylinder from a bulk supply hopper. Since the filter material of the piston head passes air but not powder, a compact slug of powder material is formed in the cylinder by the vacuum. When the cylinder is withdrawn from the bulk hopper, a mushroom of powder will come up with the filled cylinder. The excess powder is doctored off the end of the cylinder and remains in the hopper broken down into its original powder form, since the vacuum is no longer applied to it. The powder slug formed is then discharged by replacing the vacuum behind the porous filter with a pulse of low-pressure air. Perry claims a fill accuracy of $\pm 0.5\%$ to 2.0%.

Filling of sterile powders will always offer more challenges than filling of liquids. The issues or problems that may occur in the filling of solids include the following:

1. Dose accuracy container-to-container
2. Content uniformity of the solid has more than one component
3. Environmental humidity not controlled
4. Maintaining aseptic conditions, especially with particulate controls
5. Increased probability of particulate matter in the product.

SUSPENSIONS AND OTHER DISPERSED SYSTEM FILLING (see also pp. 129–131)

The main issues or potential problems that may occur in the filling of dispersed systems include the following:

1. Maintaining dose homogeneity container-to-container
2. Validation of dose homogeneity especially with higher product viscosities
3. Clogging of filling needles/nozzles
4. Batch size
5. Aseptic additions
6. Particle size reduction under aseptic conditions.

Maintaining dose homogeneity during filling operations is a huge challenge. Dose homogeneity is a function of the ability of recirculation system supporting the filling system to prevent suspension particle settling or emulsion globule interaction and growth.

Suspension products are filled in two ways. The primary way is filling of the recirculated suspension; an alternative although not performed much at all is a two step solution filling where the suspension is formed in situ once the second solution is added to the first. Insulin NPH suspensions are approved to be filled this way where the first solution filled into the container is the insulin solution and the second solution filled contains the complexing agent, protamine, that immediately interacts with the previously filled insulin solution to form a suspension. The amount of protamine in the second fill is precalculated to stoichiometrically bind all of the insulin from the first fill.

CHECK WEIGHING

All filling operations must be checked for accurate dose filling, both prior to the start of the filling operation to make proper initial adjustments and during filling by checking fill volumes periodically to ensure that predetermined volumes or weights are within specifications.

There are a number of check weighing methods (focus on vials) (2).

- Manual check weighing
- Vacuum starwheel check weighing of a full vial set
- Robotic check weighing of a single container
- Robotic check weighing of a full container set
- 100% noncontact check weighing.

Whatever check weighing method is used, control charts are established and monitored during a filling operation (Fig. 19-7). Each filling operation has a target fill volume or weight with upper and lower acceptance limits. Typical fill requirements are $\pm 0.5\%$ of the target fill volume for each and every filling nozzle (1). For example, a target fill weight might be 5.0 g with the upper limit being 5.1 g and the lower limit being 4.9 g. Obviously, for liquid-filled products, the product density (or specific gravity) must be accurately known so that a conversion to weight can be determined. Periodic weight checking is performed and the data recorded on a control chart. Filling precision is calculated using the smaller of the following two calculations:

1. (Upper specification limit—average weight)/3σ or
2. (Average weight—lower specification)/3σ.

where 3σ is three standard deviations from the average (mean) weight value, where 99.73% of all data fall within this range.

STOPPERING

These operations must occur under Grade A/B (ISO 5) clean room conditions.

Ampoules, of course, do not require rubber closures and are sealed with a flame. Vials are closed with rubber stoppers (or, for vials containing solution to be freeze-dried, the stopper is partially inserted into the vial opening) and syringes and cartridges closed with rubber plungers at the distal end (with rubber septa sealing the proximal end except for staked-needle syringes). Rubber stoppers and plungers need to be lubricated either with applied silicone oil or emulsion or with special coatings (see chap. 7) that permit and facilitate rubber units to move easily from the hopper along stainless steel tracks or rails to the openings of the primary

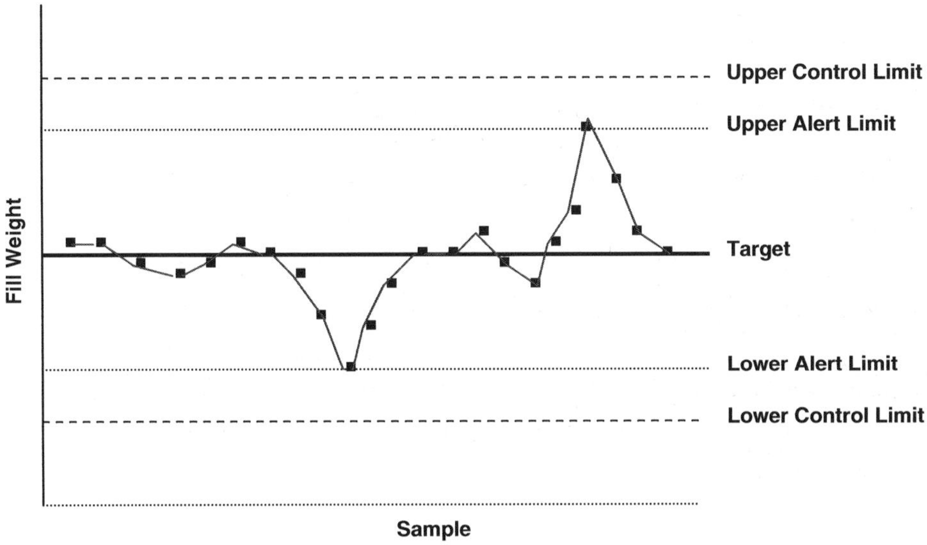

Figure 19-7 Example of a fill weight control chart.

containers (Fig. 19-8). For vial openings, the closure must fit snugly, not "pop out." Often, filling efficiencies are dependent more on the stoppering process than on the actual filling process, as there are tendencies for rubber closures to slip or pop off the openings of vials. For syringes and cartridges, the placement of the rubber plunger is dictated by the desire position of the plunger within the barrel of the syringe or cartridge to deliver the claimed volume of product.

The closing of primary containers will affect the final integrity of the container/closure interface. For syringes and cartridges, no further sealing is done although units are either placed in secondary packaging for unit dosing or part of a tray system, for example, Hypak™ (Becton-Dickinson). For vials and bottles, aluminum seals (Fig. 19-9) are crimped around the rubber closure and top of the container. Seal force integrity is measured by a torque-testing device.

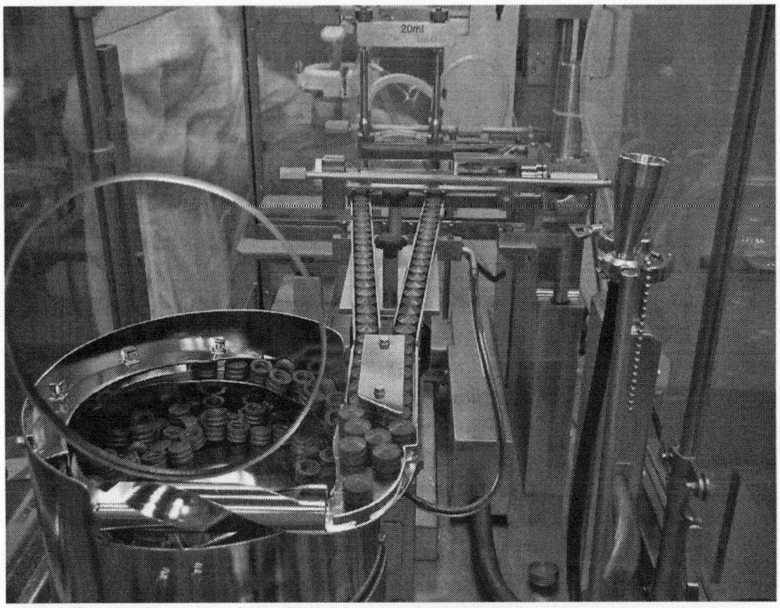

Figure 19-8 Rubber closure hoppers. *Source*: Courtesy of Baxter Healthcare Corporation.

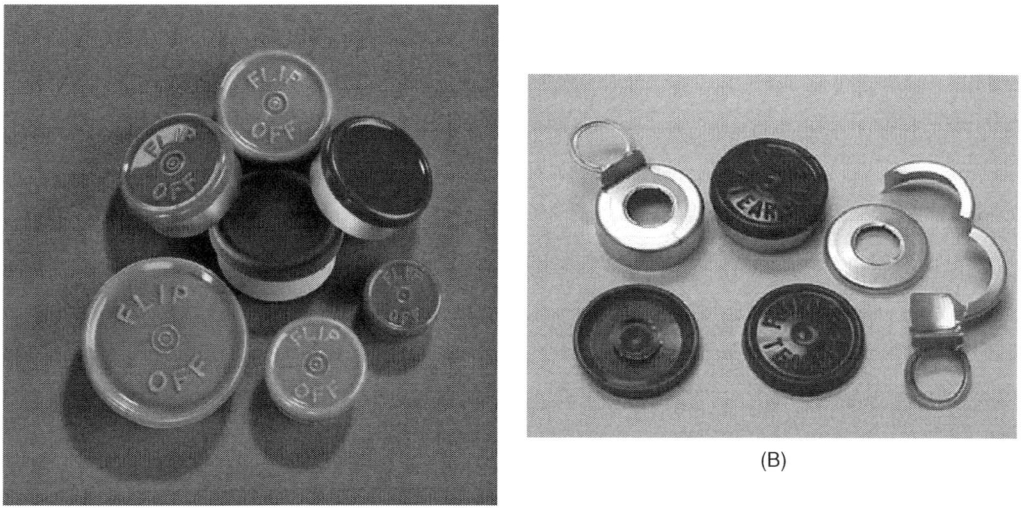

(B)

(A)

Figure 19-9 Examples of aluminum–plastic seals. (A) Flip-Off® (Flip-Off® is a registered trademark of West Pharmaceutical Services in the United States and other jurisdictions) seals: aluminum shell with a removable plastic button in order to access stopper surface. (B) Flip-Tear seals: aluminum shell is completely removed from container by flipping off the plastic button that allows stopper removal. *Source*: Courtesy of West Pharmaceutical Services.

Problems encountered during stoppering include the following:

- Too little or too much silicone on stoppers
- Misaligned or bent syringe stopper insertion rods or tubes
- Stoppers become jammed on the track
- Improper headspace (syringes)
- Stoppers are not completely seated.

SEALING

Ampoule-filled containers should be sealed as soon as possible to prevent the contents from being contaminated by the environment. Ampoules are sealed by melting a portion of the glass neck. Two types of seals are employed normally: tip seals (bead seals) or pull seals (Fig. 19-10).

Tip seals are made by melting enough glass at the tip of the neck of an ampoule to form a bead and close the opening. These can be made rapidly in a high-temperature gas-oxygen flame. To produce a uniform bead, the ampoule neck must be heated evenly on all sides, such as by burners on opposite sides of stationary ampoules or by rotating the ampoule in a single flame. Care must be taken to adjust the flame temperature and the interval of heating properly to completely close the opening with a bead of glass. Excessive heating will result in the expansion of the gases within the ampoule against the soft bead seal and cause a bubble to form. If it bursts, the ampoule is no longer sealed; if it does not, the wall of the bubble will be thin and fragile. Insufficient heating will leave an open capillary through the center of the bead. An incompletely sealed ampoule is called a leaker.

Pull seals are made by heating the neck of the ampoule below the tip, leaving enough of the tip for grasping with forceps or other mechanical devices. The ampoule is rotated in the flame from a single burner. When the glass has softened, the tip is grasped firmly and pulled quickly away from the body of the ampoule, which continues to rotate. The small capillary tube thus formed is twisted closed. Pull sealing is slower, but the seals are more sure than tip sealing.

Powder ampoules or other types having a wide opening must be sealed by pull sealing. Fracture of the neck of ampoules during sealing may occur if wetting of the necks occurred at the time of filling. Also, wet necks increase the frequency of bubble formation and unsightly carbon deposits if the product is organic.

Figure 19-10 FPS1 automatic monoblock ampoule filling and sealing machine. *Source*: Courtesy of Cozzoli Machine Company.

To prevent decomposition of a product, it is sometimes necessary to displace the air in the space above the product in the ampoule with an inert gas. This is done by introducing a stream of the gas, such as nitrogen or carbon dioxide, during or after filling with the product. Immediately thereafter the ampoule is sealed before the gas can diffuse to the outside. This process should be validated to ensure adequate displacement of air by the inert gas in each container.

Vials and bottles are sealed by closing the opening with a rubber closure (stopper). This must be accomplished as rapidly as possible after filling and with reasoned care to prevent contamination of the contents. The large opening makes the introduction of contamination much easier than with ampoules. Therefore, during the critical exposure time the open containers should be protected from the ingress of contamination, preferably with a blanket of HEPA-filtered laminar airflow.

The closure must fit the mouth of the container snugly enough so that its elasticity will seal rigid to slight irregularities in the lip and neck of the container. However, it must not fit so snugly that it is difficult to introduce into the neck of the container. Closures preferably are inserted mechanically using an automated process, especially with high-speed processing. To reduce friction so that the closure may slide more easily through a chute and into the container opening, the closure surfaces are halogenated or treated with silicone. When the closure is positioned at the insertion site, it is pushed mechanically into the container opening. When small lots are encountered, manual stoppering with forceps may be used, but such a process poses greater risk of introducing contamination than automated processes. This is a good test for evaluation aseptic operator aseptic techniques, but not recommended for any product filling and stoppering.

Container–closure integrity testing has become a major focus for the industry because of emphasis by regulatory agencies. Container–closure integrity measures the ability of the seal between the glass or plastic container opening and the rubber closure to remain tight and fit and to resist any ingress of microbial contamination during product shelf life. This topic will be thoroughly discussed in chapter 30.

Rubber closures are held in place by means of aluminum caps. The caps cover the closure and are crimped under the lip of the vial or bottle to hold them in place. The closure cannot be removed without destroying the aluminum cap; it is tamperproof. Therefore, an intact aluminum cap is proof that the closure has not been removed intentionally or unintentionally. Such confirmation is necessary to ensure the integrity of the contents as to sterility and other aspects of quality.

Figure 19-11 CM200 continuous motion crimping machine. *Source*: Courtesy of Cozzoli Machine Company.

The aluminum caps are so designed that the outer layer of double-layered caps, or the center of single-layered caps, can be removed to expose the center of the rubber closure without disturbing the band that holds the closure in the container. Rubber closures for use with intravenous administration sets often have a permanent hole through the closure. In such cases, a thin rubber disk overlayed with a solid aluminum disk is placed between an inner and outer aluminum cap, thereby providing a seal of the hole through the closure.

Single-layered aluminum caps may be applied by means of a hand crimper known as the Fermpress (suppliers: *West, Wheaton*). Double- or triple-layered caps require greater force for crimping; therefore, heavy-duty mechanical crimpers (Fig. 19-11) are required (suppliers: *Bosch, Cozzoli, Perry, West, Wheaton*).

A relatively recent trend, although now standard practice, is the requirement that sealing of vials and other containers be accomplished in Class 100/Grade A/ISO 5 clean room environments. Formerly such sealing occurred in unclassified environments.

ADVANCES IN VIAL AND SYRINGE FILLING

While the emphasis of this entire book is sticking to the basics, some discussion of advances (3,4) need to be mentioned although at the time of finishing this writing, it is unclear how routine these advances will become.

Flexible Lines

Because of extremely high costs of some new drugs, especially biopharmaceuticals, it is preferable to fill small batches to reduce the risk of unacceptable monetary losses in the event of a manufacturing deviation that results in batch rejection. The move toward smaller batch filling has necessitated the requirement for more accurate fills and faster line change overs. One way that this is accomplished is through the use of single-use, disposable closing systems[2] in which the entire product path is discarded after use. Another approach is to modify filling designs so that only one change part is required for a vial diameter change. Filling machines are available that have more than one dosing system to increase flexibility for filling a variety of products.

[2] For example, the Acerta® DS1 dispensing system.

Reduced Customization

Many companies have experienced too many problems with highly customized filling equipment; thus, standardization of filling machines has made a comeback. Standardization includes vendor selection, PLCs, human machine interfaces (HMI, touch screens), component transfer systems, filling method, and design of rapid access barrier systems (RABS) or isolator enclosures. Reduced customization has resulted in faster line fabrication, shorter factory acceptance testing (FAT), and reduced risk associated with startup, site acceptance testing (SAT), installation qualification (IQ), operational qualification (OQ), and performance qualification (PQ). Also, to be expected, maintenance is simpler and there is reduced need for space parts.

Integrated and Compact Lines

The pharmaceutical industry is moving toward single-sourced, integrated filling lines. The BOC Edwards production freeze-dryers and associated automatic loading and unloading systems are a good example. For low-to-intermediate production volumes, compact lines such as IMA's Modular Aseptic Compact System have been implemented that includes vial washing, depyrogenation oven, and filling machine integrated as one complete unit.

Filling Machines for Integration with Barrier Isolators or Rapid Access Barrier Systems

Streamlined filling machines have been produced to fit precisely into these isolator systems to optimize airflow, aid in sterilant distribution, be ergonomic with the gloveports, facilitate removal of waste, and making it easier to remove the source of jams. Such filling machines are linear fillers with small widths. Vial transport systems to these isolator filling machines have been improved to allow complete exposure to sterilizing gases, typically vapor phase hydrogen peroxide. Electron beam tunnels are available to surface sterilize tubs of prefilled syringes directly feeding a syringe filler. Automated bag opening has been integrated upstream of these tunnels and automated tub lid removal downstream to provide greater separation of operators from the process.

Higher Grade Vial Capping

Because of European Union requirements for Grade A air supply over capping operations, capping machines are available with RABS enclosures that target unidirectional downward airflow over the capping head, sorting bowl, and chute.

Integration of External Vial Washing

Vial washing machines can be purchased to wash the vial exterior after filling to remove potent compounds on the exterior surface for added operator and user safety. Such machines aim water rinses so that the vial caps are not wetted and filtered compressed air is used to dry the vials. External vial washing also can help to remove cosmetic defects.

Closed Vial Filling Systems

Aseptic Technologies developed the Crystal® Closed Vial Filling System (CVFS) where a ready-to-fill plastic (cyclo-olefin copolymer) vial and thermoplastic elastomer are molded in a Grade A clean room, assembled robotically, then gamma irradiated prior to delivery to the manufacturer. The specialized filling machine needle pierces the stopper, liquid is filled into the vial, the needle is withdrawn, and the piercing trace is laser resealed to restore closure integrity. A cap, designed to keep the stopper surface protected until use, is placed by snap fit. All these operations are conducted inside a CVFS that ensures Grade A environmental control. More discussion of this technology is presented in Chapter 23, page 360.

REFERENCES

1. Peterson A. Filling methods as they apply to parenteral product quality and biopharmaceutical microdosing. In: Lysfjord J, ed. Practical Aseptic Processing: Fill and Finish. Vol 1. Bethesda, MD: Parenteral Drug Association, 2009:145–165.
2. Peterson A. Checkweighing fill weight of parenteral product is the heart of process quality. In: Lysfjord J, ed. Practical Aseptic Processing: Fill and Finish. Vol 1. Bethesda, MD: Parenteral Drug Association, 2009:135–144.

3. Heyman P. Recent trends in vial and syringe filling. Pharmaceutical Processing July 2009; 18–23.
4. Auerbach M. Aseptic technologies process revolutionizes closed vial filling. Pharmaceutical Processing May 2009; 8–11.

BIBLIOGRAPHY

Akers MJ. Parenteral preparations. In: Felton, L., ed. Remington's Pharmaceutical Sciences. 21st ed. Philadelphia: Lippincott Williams & Wilkins, 2005:802–836.

20 | Freeze-dry (lyophilization) processing

Many parenteral drugs, particularly biopharmaceuticals, are too unstable in solution to be available as ready-to-use liquid dosage forms. Such drugs can still be filled as solutions and placed in a chamber where the combined effects of freezing and drying under low pressure will remove the solvent and residual moisture from the solute components, resulting in a dry powder that has sufficient long-term stability. The process of freeze-drying has taken on greater prominence in the parenteral industry because of the advent of recombinant DNA technology. Proteins and peptides and other active biological compounds generally must be freeze-dried for clinical and commercial use. There are other technologies available to produce sterile dry powder drug products besides freeze-drying, such as sterile crystallization, film drying, or spray drying. However, all these technologies require powder filling of product in the final container while freeze-drying allows product filling as liquids, far more preferable from environmental control considerations. Freeze-drying is by far the most common unit process for manufacturing drug products too unstable to be marketed as solutions.

This chapter focuses on the process of freeze-drying or lyophilization (1–6), whereas chapter 10 focused on formulation and packaging of lyophilization products.

The term "lyophilization" describes a process to produce a product that "loves the dry state." However, this term does not include the freezing process. Therefore, although lyophilization and freeze-drying are used interchangeably, freeze-drying is a more descriptive term. Equipment used to freeze-dry products are called freeze-dryers or lyophilizers. Table 20-1 lists the advantages, features, and disadvantages of freeze-drying.

Prior to placing primary units of product into a freeze-dryer, the manufacturing process is identical to a ready-to-use solution process with the exception of the placement of the rubber closure. After filling the solution into the primary container, the specially designed rubber closure is partially fitted on top of the container (Fig. 10-1), not fully seated, so that there is sufficient opening for the sublimation process (frozen ice to vapor) to take place.

FREEZE-DRYING STAGES

Cooling/Freezing Stage

Cooling starts with a true solution containing, on average, around 5% solute (therefore, 95% solvent is usually water). This stage cools the product solution at a temperature below the product eutectic (crystalline) or glass transition temperature where the solution is completely in the frozen state.

Primary Drying Stage

At the start of primary drying the product container, on average, 95% is water. At the end of primary drying, the amount of water remaining is on the order of 5% to 10%. This stage removes the solvent (ice) from the product by evacuating the chamber, usually below 0.1 torr (100 μm Hg) and subliming the ice onto a cold, condensing surface at a temperature below that of the product, the condensing surface being within the chamber or in a connecting chamber. During primary drying, the temperature of the product must remain slightly below its critical temperature, called "collapse temperature." Collapse temperature is best measured by visual observation using a freeze-dry microscope that simulates the freeze-drying process. Generally, collapse temperature is similar to the eutectic or glass transition temperature of the product.

Secondary Drying Stage

Secondary drying stage starts with 5% to 10% water and ends when the residual water (moisture) content in the lyophilized cake is somewhere between 0.1% and 2.0% (the allowed residual moisture range must be determined via experimentation in order to assure achievement of

Table 20-1 Advantages, Features, and Disadvantages of Lyophilization

Advantages	Features	Disadvantages
Product is stored in dry state-few stability problems	Pharmaceutically elegant-appearing solid product (cake)	Volatile compounds may be removed by high vacuum
Product is dried without elevated temperatures	Active ingredient is maintained at sufficient strength/potency	The drug may not be stable as a freeze-dried solid, e.g., cephalosporins
Good for oxygen and/or air-sensitive drugs	Uniform color of the solid powder	Many biological molecules are damaged by the stresses associated with freezing, freeze-drying, or both
Rapid reconstitution time	Sufficiently dry to maintain acceptable stability throughout shelf life	Not all solutes can be freeze-dried to form a pharmaceutically acceptable cake
Constituents of the dried material remain homogeneously dispersed	Cake sufficiently porous for rapid dissolution	Cost may be an issue, depending on the product
Product is processed in the liquid form	Sterile, pyrogen-free, and particulate-free after in solid state and after reconstitution	Some issues associated with sterilization and sterility assurance of the dryer chamber and aseptic loading of vials into the chamber
Sterility of product can be achieved and maintained		

desire product shelf-life), thereby removing bound water from solute(s) to a level that assures long-term stability of the product. This is accomplished by introducing heat to the product under controlled conditions, thereby providing additional energy to the product to remove adsorbed water. The temperature for secondary drying should be as high as possible without causing any chemical degradation of the active ingredient. Generally, for small molecules, the highest secondary drying temperature used is 40°C whereas for proteins it is no more than 30°C.

Figure 20-1 provides a diagram of the different stages of the freeze-drying process and shows the relationship between shelf temperature and product temperature. The figure also

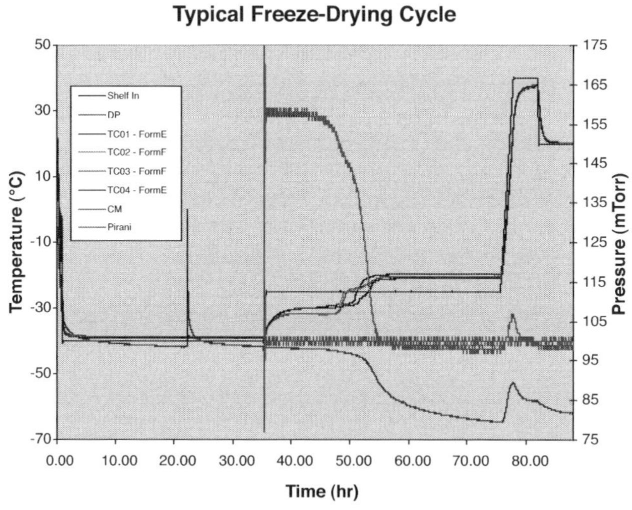

Freezing

Need to know eutectic or glass transition temperature of product and freeze below this temperature

Rate of freezing affects size of ice crystals and subsequent rate of drying

Primary Drying

Removal of solvent (ice) water.

Critical that product temperature remain below the collapse temperature (or T_g' or T_e) so that all ice is sublimed (solid to vapor transition)

Secondary Drying

Removal of "bound" water; water from solute

Because ice is gone, the temperature of the product does not have to be held below the collapse temperature

Shelf temperature often raised > 40°C to accelerate desorption process.

Figure 20-1 Stages of lyophilization cycle.

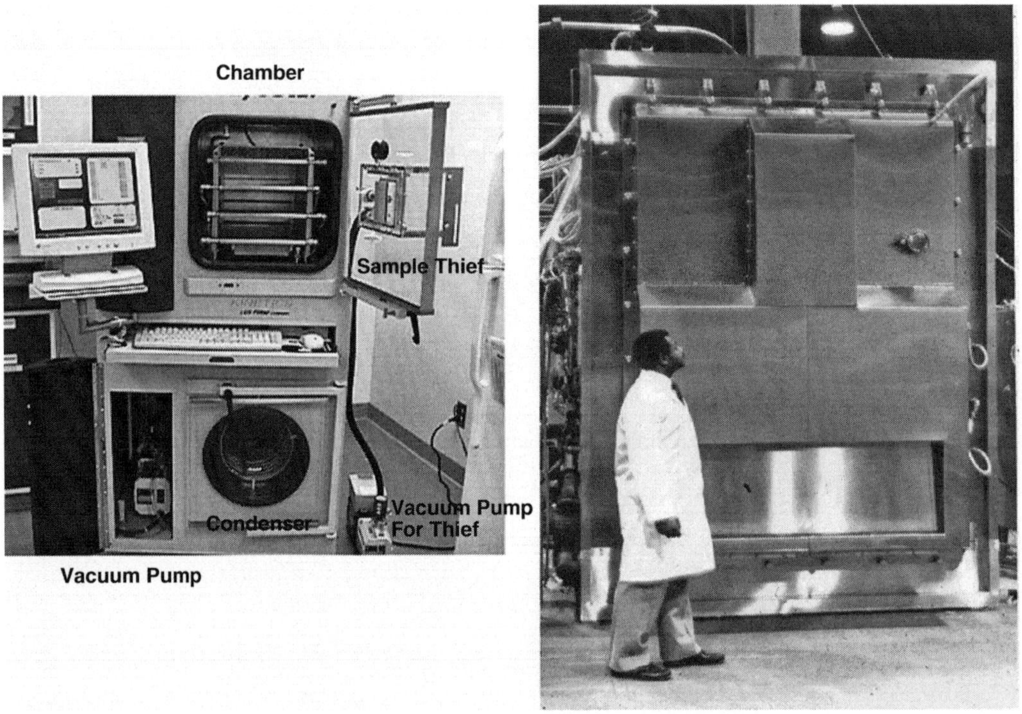

Figure 20-2 Small-scale and large-scale freeze-dryers.

points out on the side bar other common phenomena that occur and should be considered during a typical freeze-dry cycle.

Figure 20-2 shows a photo of a small-scale lyophilization system and its functional components (and a production-size freeze-dryer to show a comparison of size). The cycle begins with the product in its container being cooled and frozen on the shelf in the chamber by circulating refrigerant (usually silicone) from the compressor through pipes within the shelf. After freezing is complete (although with amorphous components there likely is unfrozen water in the freeze concentrate), which may require several hours, the chamber and condenser are evacuated by the vacuum pump, the condenser surface having been chilled previously by circulating refrigerant from the large compressor.

Heat then is introduced from the shelf to the product under graded control by electric resistance coils or by circulating silicone or glycol. Heat transfer proceeds from the shelf into the product vial and mass transfer (ice) proceeds from the product vial by sublimation through the chamber and onto the condenser. The process continues until the product is dry (usually 1% or less moisture except for some proteins that require a minimum amount of water for conformational stability), leaving a sponge-like matrix of the solids originally present in the product, the input of heat being controlled so as not to degrade the product.

For most pharmaceuticals and biologicals the liquid product is sterilized by filtration before being filled into the dosage container aseptically. The containers must remain open during the drying process to allow water vapor to escape; therefore, they must be protected from contamination during transfer from the filling area to the freeze-drying chamber, while in the freeze-drying chamber, and at the end of the drying process until sealed. Automated loading and unloading of product to and from the freeze-dryer shelves is now state-of-the-art where partially open vials are always under the auspices of Class 100 air and human intervention is eliminated.

Freeze-dryers are equipped with hydraulic or pneumatic internal-stoppering devices designed to push slotted rubber closures into the vials to be sealed while the chamber is still evacuated, the closures having been partially inserted immediately after filling, so that the slots

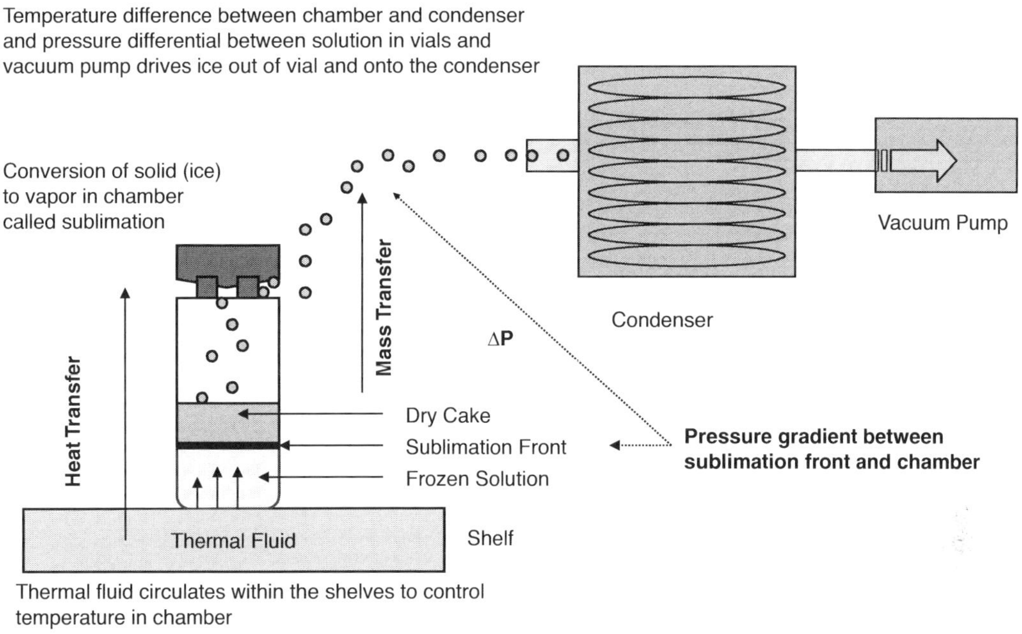

Figure 20-3 Schematic of Heat and mass transfer in the freeze-dryer.

were open to the outside. If internal stoppering is not available or containers such as ampoules are used, filtered dry air or nitrogen should be introduced into the chamber at the end of the process to establish atmospheric pressure.

Freeze-drying essential involves the manipulation of both temperature and pressure differences between the chamber and condenser to establish a pressure gradient between the surface of the product in the vial and the vapor pressure of the chamber. This pressure gradient drives ice from the product onto the condenser, schematically presented in Figure 20-3.

Freeze-dryer compressors require large amounts of water. For example, a 220-square foot dryer needs a minimum of 170 gallons of water per minute for its compressors. Typically, production freeze-dryers have their own water systems, separate from water systems that supply water for cleaning, compounding, and other process operations.

COOLING/FREEZING STAGE

Water freezes at 0°C at atmospheric pressure, but when solute is dissolved in water, the freezing point of water decreases. The main question that must be answered for the freezing step in the freeze-dry cycle for a sterile product is at what temperature will the product truly freeze, that is, ice completely crystallizes and solute either crystallizes too or remains in an immobile state.

Freezing rate typically is the term used when, in fact, the more correct term is cooling rate. Cooling rates range from <1°C/min to >10°C/min in ramped freeze dryers or much more rapidly (e.g., perhaps 900°C/min) if solution containers are placed in liquid nitrogen. Rapid cooling is very rare in commercial sterile product freeze-drying. Freezing occurs at the temperature when ice crystallizes and postnucleation of ice occurs. When ice crystallizes, heat is released (exothermic event) and an abrupt temperature increase occurs (Fig. 20-4).

Supercooling is the phenomenon where the solution does not freeze (crystallize) until at some temperature far below the expected (normal) freezing temperature.

Water supercools to −15°C. Sterile solutions rely on supercooling because of lack of nucleation centers since there are no particles in solution and no imperfections on vial surfaces. Once freezing starts with supercooled water, the entire solution cannot freeze immediately, because supercooled water cannot absorb all of the heat at once yielded by ice formation. Ice propagates from nucleation site and crystallizes in multibranching, tortuous paths. The remaining water freezes when the previously formed ice crystals keep growing.

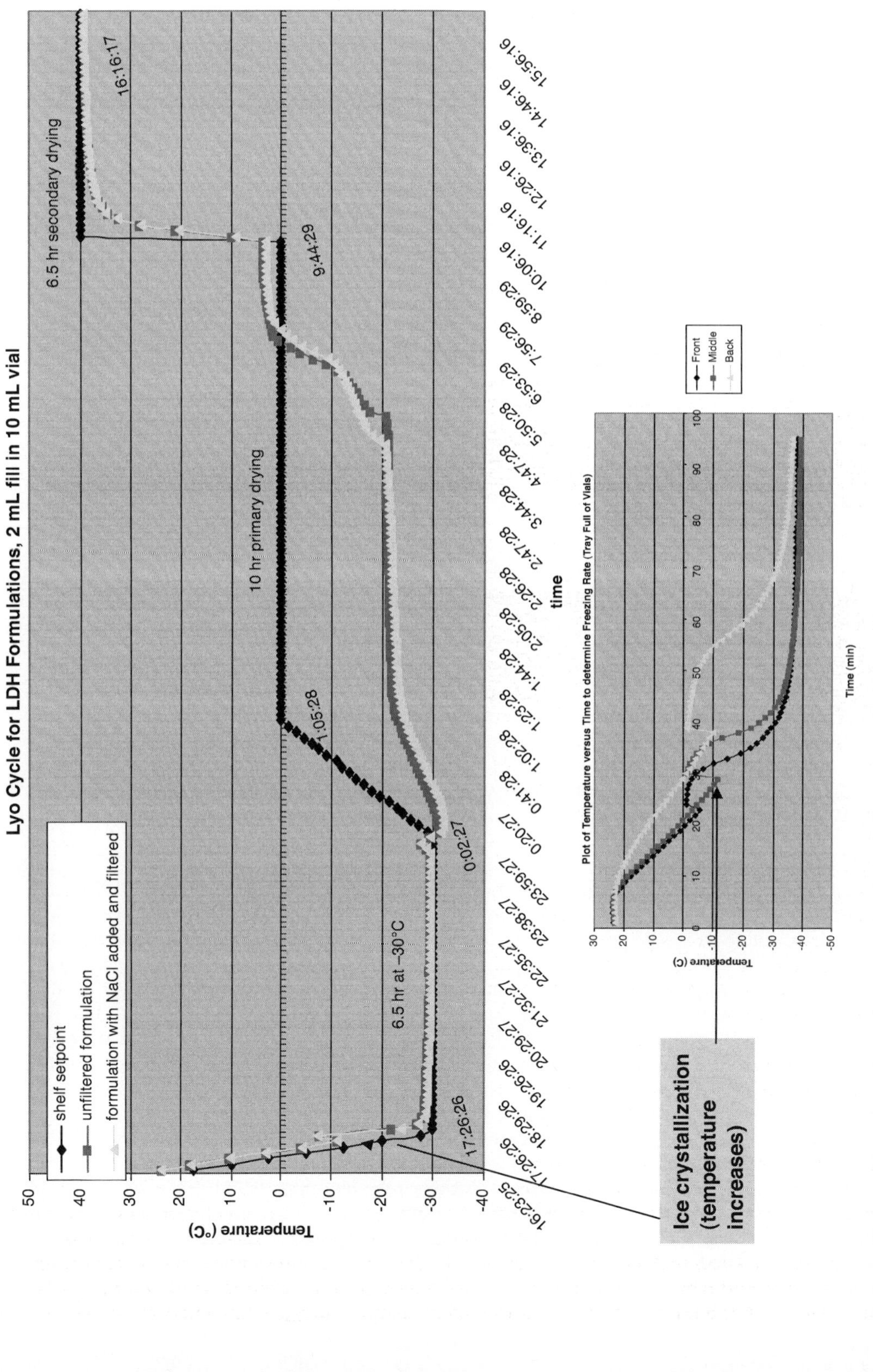

Figure 20-4 Temperature-time profiles of sample showing the ice crystallization exothermic event.

Table 20-2 Importance of the Rate of Cooling and Freezing

Cooling Rate	Degree of supercooling	Rate of ice crystallization	Ice crystal size	Duration of primary drying	Recovery of protein activity
Fast	High	Fast	Small (large number of small ice crystals)	Slow (smaller pores; increased resistance to mass flow)	Less (Due to larger ice–water interfacial area)
Slow	Low	Slow	Large (small number of large ice crystals)	Fast (larger pores; decreased resistance to mass flow)	More (Less ice–water interfacial area)

The degree of supercooling and the rate of cooling and freezing have a potentially major impact on the degree of protein degradation during this stage. Table 20-2 summarizes the effects of fast (>10°C/min) versus slow (<1°C/min) cooling rates. The information in this table is intended to be relative, not absolute, and general, not necessarily true for every kind of active ingredient that may be freeze-dried.

Issues associated with freezing are:

- Thermocouples in vials will influence the size of ice crystal formed during freezing so the rate of drying during primary drying will be faster for thermocoupled vials than all other vials, because thermocouples serve as nucleation centers and ice crystals will grow and be larger than ice crystals in nonthermocoupled vials.
- The greater the fill depth in a vial, the more likely that freezing of solution is nonuniform. Cooling the solution in segments will minimize the differences in ice crystal size at different parts of the product that would be manifested in nonuniform cake appearance.
- Crystallization of some components might cause problems. Dibasic sodium phosphate crystallization will lower pH and this might affect active ingredient stability although not likely a major factor because of the extremely cold temperatures. Sodium chloride crystallization may change solution ionic strength and again might have stability consequences. It is well known that mannitol and glycine are prone to crystallization and if such crystallization is not well controlled, later crystallization of these components (either during primary drying or even over shelf life of the product) could affect product stability and/or rate of dissolution. Also late mannitol crystallization, for example, crystallization during primary drying is known to be a cause of vial breakage. This is why an annealing step might be a wise safeguard against incomplete crystallization unless the cycle is designed to produce predominantly amorphous cakes.
- Because freezing affects the thickness of channel walls and size of pores, this can eventually affect rate of reconstitution of dry powder.
- Freezing can denature (unfold) proteins during formation of the ice–water interface.
- Rate of cooling and freezing can have a significant effect on ice crystal size and on the resultant rate of solution and chemical/biological activity of the finished product (Table 20-2).

The freezing stage must achieve a sufficiently low temperature such that each solution in every single vial is completely frozen. Ice will crystallize and concentrate, solutes that are capable of crystallization when frozen will crystallize and concentrate, but solutes that do not readily crystallize (e.g., sugars and proteins) will remain amorphous and may not completely freeze. This unfrozen phase is called the freeze concentrate and it is within this concentrate that buffers can crystallize and incompatible solutes (excipients) can undergo liquid–liquid phase separation. Phase separation between two or more excipients or between a protein and excipients, can adversely affect the molecular interactions required to stabilize the protein.

Even if not completely frozen, it is very important that amorphous components reach a temperature and stay below that temperature where they are rigid (glassy) and immobile. This "critical" temperature is the glass transition temperature. As discussed in chapter 10, the glass transition temperature is depending on the type and concentration of formulation components and can be determined by thermal analytical techniques (e.g., differential scanning calorimetry or thermoelectric analysis) or by freeze-dry microscopy. For crystalline solutes, the "critical" temperature at which the solute and ice crystallize is called the eutectic temperature. Solutions being freeze-dried must be cooled and achieve complete freezing (or remain rigid for amorphous components) and remain at temperatures below their critical temperatures prior to the beginning of primary drying.

Annealing

Annealing is a possible step performed during the freezing stage[1] where, after the product solution is frozen, the shelf temperature is raised to allow the product temperature to rise above its glass transition temperature, but, if applicable, below its eutectic melting temperature. Above the glass transition temperature, amorphous components will crystallize. The shelf temperature is then lowered to its original freezing stage temperature prior to the beginning of primary drying. Forced crystallization during freezing via annealing assures that complete crystallization occurs that will accelerate the primary drying stage and help to assure that any amorphous solute that tends to crystallize will be forced to crystallize at this stage and not slowly crystallize later either in the drying process or in the dry state during shelf life. Annealing can prevent partial collapse of the cake, reduce intravial heterogeneity, relieve stress within the glassy formulation that reduces protein damage, and reduce cake cracking and air bubble formation during rehydration (7).

An example of the positive effects of annealing on product stability involves recombinant human interferon-γ (rhIFN-γ) (8). This protein adsorbs to both air–liquid and ice–liquid interfaces, resulting in protein aggregation after air–liquid interfacial adsorption. Annealing was found to offer a significant advantage, as it acted to minimize air bubble formation during reconstitution, thus avoiding damage at air–liquid surfaces.

Potential limitations of annealing frozen samples include increased secondary drying times and potential greater residual moisture in the final product, decreased reconstitution times, and its relatively unknown effect on protein structure and stability.

Rate of Cooling

The rate of product cooling or freezing[2] can potentially affect protein stability, primarily physical, and ice crystal morphology. The more rapid the rate of freezing (that is, the actual growth of ice crystals as opposed to solution cooling prior to nucleation of the ice), the larger is the ice-liquid interfacial area where aggregation of proteins has been suggested to occur (9). This observation was corroborated with respect to storage stability of recombinant tissue-type plasminogen activator (10), recovery of enzymatic activity of lactate dehydrogenase (LDH) and β-galactosidase after freezing (11), and degree of insoluble aggregate formation in bovine IgG after freeze-drying (12). In all of these studies, fast freezing was either achieved by quench freezing (dipping vials in liquid nitrogen) or placing vials on shelves already precooled to −45°C temperatures.

The rate of cooling can have a significant effect on the morphology of the ice crystals that, in turn, can affect both resistance to vapor flow during sublimation and the quality of the final product when collapse is a problem (13). An approximate 3% increase in primary drying time occurred for every 1°C decrease in ice nucleation temperature (14). The temperature at which nucleation occurs is dependent on both formulation and processing conditions. Nucleation temperatures could be somewhat controlled, and heterogeneities both within batches and

[1] Annealing potentially could also be applied during the secondary drying stage if annealing during the freezing stage for whatever reason cannot completely crystallize the crystalline components.

[2] Cooling is the more correct term than freezing, but both are used in the literature to describe the first step of the freeze-drying process. Rate of cooling or freezing can be easily misinterpreted. There is a difference between actual cooling/freezing rate and the rate of temperature ramping. The freezing process first involves cooling the solution prior to ice nucleation, then ice crystals begin to grow at a certain rate. Freezing rate is determined largely by the amount of supercooling prior to ice nucleation. The amount or rate of supercooling cannot be directly controlled.

during batch scale-ups could be minimized, with the addition of nucleating agents (particles, silver iodide, protrusions in the inner walls of the vials). However, such nucleating agents are impractical. One practical technique to eliminating intravial heterogeneity due to differences in nucleating temperatures is to incorporate an annealing step as previously introduced (15). Realize that annealing could be detrimental to the long-term stability of the protein product for formulations that must maintain amorphous solutes for stabilization, or for formulations that might experience a pH shift due to crystallization of buffer components.

Damage to hemoglobin during freeze-drying could be avoided by rapidly freezing samples using liquid nitrogen (16). This minimizes the time the protein spends in temperature ranges between −3°C and −23°C where the formulation containing mannitol undergoes phase separation. Even adding noncrystallizing sugars (sucrose or trehalose) or nonionic surfactants (Tween 80 or Triton X-100) does not protect against phase separation-induced damage during freeze-drying.

Other Aspects of the Freezing Stage

Cold denaturation of L-asparaginase, both when frozen to −40°C and, using liquid nitrogen, to −190°C, can be minimized by the addition of hetastarch (17). The cryoprotective effect was hypothesized to be the highly viscous environment created by hetastarch, because other commonly used additives such as glucose and lactose, having significantly lower viscosities, had no effect.

In fact, cold denaturation itself may be only part of the reason for loss of activity. Other possible mechanisms involved in loss of activity during freeze-drying include denaturation at the ice–freeze interface, changes in microenvironmental pH due to component crystallization, and loss of native structure by pulling away unfrozen water during secondary drying. Cochran and Nail found an inverse relationship between the extent of supercooling during freezing and recovery of protein activity with protein inactivation—the result of adsorption of protein at the ice–freeze concentrate interface during the freezing process (18).

The impact of the freezing process on the primary drying rate was shown for a concentrated (15%) formulation with a high-fill (15 mL in a 30-mL vial) depth (19). Annealing, vacuum-induced freezing, and the addition of tertiary butyl alcohol or silver iodide as an ice nucleation accelerant, all accelerated the sublimation process. The authors recommended the combination of two-step freezing with an annealing step as the optimal method for a high-fill depth product.

Because phosphate buffers are commonly used in freeze-dried formulations, crystallization of dibasic sodium phosphate and resultant change in solution pH during the freezing stage has drawn much attention. Changes in pH associated with crystallization of sodium phosphate buffer were found to be directly related to the initial concentration of buffer (e.g., the higher the concentration, the greater the pH change) and that the lower the initial pH of the buffer, the higher the observed pH upon crystallization (20). Addition of solutes such as sucrose and mannitol inhibited the crystallization of dibasic sodium phosphate with subsequent low-level change in pH upon freezing. A shift in pH during freezing can have a negative effect on protein activity, as in the case of LDH, which dissociates as the pH shifts from 7.5 to 4.5 during freezing (21).

LDH dissociation during freezing was later reported to be prevented by either polymers [e.g., dextran, Ficoll, polyethylene glycol (PEG)] and/or sugars (e.g., sucrose, trehalose, glucose) added to the formulation (22). Surface-active agents did not protect LDH during freezing (or, for that matter, during freeze-drying either). The authors proposed that polymers or sugars offer freeze protection because they thermodynamically inhibit freeze-induced dissociation of the protein. The polymers are preferentially excluded from the surface of LDH, thus increasing the free energy of the dissociation state. These stabilization effects are true not only for freeze thawing, but also for freeze-drying of LDH. The drying-related protection was hypothesized to result from protein forming hydrogen bonds with a lyoprotectant.

Large molecules such as dextran have been reported to be too bulky (steric hindrance) to form hydrogen bonds with proteins in the dry state (23–25). This was also true for dextran at concentrations below ≤1% in the stabilization of LDH during freezing and freeze-drying (22). Ten percent dextran offered better protection, although not as effective as PEG or Ficoll at a concentration of at least 1%.

In another report, maltodextrins were found to protect LDH against inactivation during freeze-drying, with lyoprotection improved by the addition of PEG 8000 (26).

PRIMARY DRYING STAGE

The primary drying stage starts after all solutions in all product containers on all the shelves in a freeze dryer are completely frozen. Of course, it is possible that not every solution is completely frozen, but this is the assumption made on the basis of previous studies. The start of primary drying involves the increase in shelf temperature to some predetermined temperature that still keeps the product temperature below its collapse temperature. The start of primary drying also involves the introduction of vacuum (low pressure) to a point where there is a sufficient pressure differential between the product at its sublimation front and the pressure in the chamber (Fig. 20-3). Primary drying involves the sublimation of ice from the frozen solution that produces a dried layer of solute. The solute must form a rigid structure that will support its weight after the removal of ice. That is why it is essential that the product temperature is maintained below its collapse temperature during primary drying to maintain solute rigidity/immobility in order that the finished product both looks pharmaceutically elegant and does not have excessive residual moisture that could cause degradation of the active ingredient.

From the diagram in Figure 20-3, it can be seen that the direction of heat and mass transfer causes the top of the product to dry first with drying proceeding downward to the bottom of the vial. Therefore, as drying proceeds, there exists a three component or layer system in each vial—the upper dry product, the middle sublimation front, and the lower frozen liquid product. As the dried layer increases, it becomes a greater barrier or the source of greatest resistance to the transfer of mass out of the vials. This points out the importance of vial dimensions and volume of product per vial on the efficiency of the freeze-drying process. If large volumes of solution must be processed, the surface area relative to the depth may be increased utilizing larger vials or by using such devices as freezing the container in a slanted position to increase the surface area.

The actual driving force for the process is the vapor pressure differential (ΔP) between the vapor at the surface where drying of the product is occurring [the drying boundary or ice vapor pressure (ice VP)] and that at the surface of the ice on the condenser [also called partial pressure (PP) of water vapor in the chamber]. The latter, PP, is determined by the temperature of the condenser as modified by the insulating effect of the accumulated ice. The former is determined by a number of factors, including:

- The rate of heat conduction through the container and the frozen material, both usually relatively poor thermal conductors, to the drying boundary while maintaining the product below its eutectic temperature.
- The impeding effect of the increasing depth of dried, porous product above the drying boundary.
- The temperature and heat capacity of the shelf itself.

The ΔP between the sublimation front (ice VP) and the chamber/condenser (PP water vapor) follows certain general rules:

- Increasing shelf temperature increases the VP of ice at the sublimation front; therefore not as much vacuum is required in the chamber to have an adequate ΔP.
- Increasing the vacuum (decreasing chamber pressure) decreases PP.
- It is desirable to have PP to be no greater than 30% than that of ice VP.
- It is always desirable to have ice VP to be as high as possible over PP.
- PP greater than 200 mTorr (microns) is not very effective in facilitating the rate of secondary drying because heat does not transfer easily in a vacuum.

The optimal primary drying cycle involves a trade-off between heat transfer and mass transfer considerations. On one hand, chamber pressure must be lower than vapor pressure of ice in the product for drying to occur. On the other hand, if pressure is too low, the sublimation process is too slow. Chamber pressure should be no more than half and not less than quarter the vapor pressure of ice at the desire product temperature.

The passageways between the product surface and the condenser surface must be wide open and direct for effective operation. The condensing surfaces in large freeze-dryers may be in the same chamber as the product or located in a separate chamber connected by a duct to the drying chamber. Evacuation of the system is necessary to reduce the impeding effect that collisions with air molecules would have on the passage of water molecules. However, the residual pressure in the system must be greater than the vapor pressure of the ice on the condenser or the ice will be vaporized and pulled into the pump, an event detrimental to most pumps.

The amount of solids in the product, the ice crystal size, and their thermal conductance will affect the rate of drying. The more solids are present, the more impediment will be provided to the escape of the water vapor. The degree of supercooling (how much lower the product temperature goes below its equilibrium freezing point before ice crystals first form) and the rate of ice crystallization affect the freezing process and efficiency of primary drying (again refer to Table 20-2). The larger the size of ice crystals formed, usually as a result of slow freezing, the larger the pore sizes are when the ice sublimes and, consequently, the faster will be the rate of drying. A high degree of supercooling will produce a large number of small ice crystals, a small pore size when the ice sublimes in the dried layer, and a greater resistance to water vapor transport during primary drying. The poorer the thermal conducting properties of the solids in the product, the slower will be the rate of heat transfer through the frozen material to the drying boundary.

The rate of drying is slow, most often requiring 24 hours or longer for completion. The actual time required, the rate of heat input, and the product temperatures that may be used must be determined for each product and then reproduced carefully with successive processes.

Drying Steps

The primary drying segment of lyophilization was investigated to evaluate the relationship between resistance to water vapor flow through the dried layer and the microstructure of the cake (27). Mass transfer resistance was calculated from the following equation, derived by these authors from original work published by Pikal (28–30):

$$R_{\mathrm{p}} = \frac{A_{\mathrm{p}}}{m} \times \left(2.7 \times 10^{13} \times \exp\left|\frac{-6145}{T_{\mathrm{p}} - (\Delta H_{\mathrm{s}} \times m)/(A_{\mathrm{v}} K_{\mathrm{I}}) \times (L - l) + 273.15}\right| - P_{\mathrm{c}}\right)$$

where resistance, R_{p}, is a function of the dried layer thickness (l), sublimation rate (m), product temperature (T_{p}), and vial dimensions (A_{p} A_{v}). Readers need to consult the original articles for more details on the development of this equation.

Mass transfer resistance was found to decrease with increases in temperature and was also dependent on formulation composition, with trehalose and sucrose offering less resistance than the recombinant antibody studied (Fig. 20-5). Lower resistance to water vapor flow in primary drying near a collapse or eutectic melt was a result of small-scale collapse that was

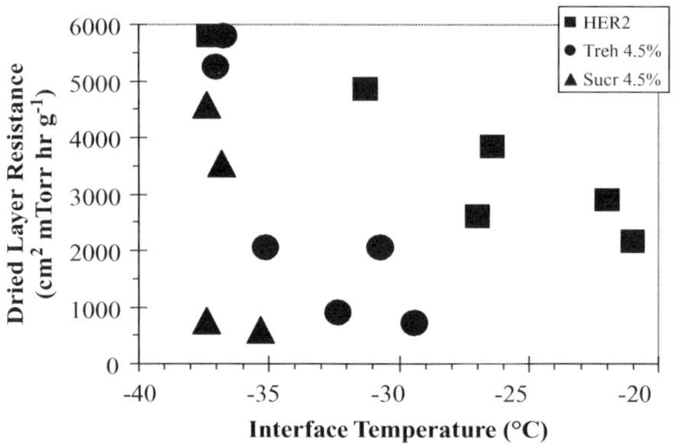

Figure 20-5 Mass transfer resistance for formulated rhuMAb HER2 (■), trehalose (●), and sucrose (▲) versus interface temperature. Resistance followed the series rhuMAb HER2 > trehalose > sucrose and decreased with increased interface temperature. *Source*: From Ref. 31.

dependent on the formulation. Microscopic structure of the lyophilized material was evaluated by scanning electron microscopy (SEM) and fluorescence microscopy with the presence or absence of collapse during freeze-drying assessed using a freeze-drying microscope.

Holes observed in the plate-like structure of freeze-dried lactose using SEM were defined as "small-scale collapse" or "microcollapse" (32), formed due to thin spots in the dried material as surface tension forces became active as the dried material viscosity decreased. Such holes are clearly observable microscopically without visible evidence of cake shrinkage or gross collapse of the dried structure.

Sublimation rates were found to be significantly higher for vials located in the front of the freeze-dryer compared to vials in the center (33). This is due to atypical radiation heat transfer phenomena experienced by edge vials. This effect can be minimized by using appropriate radiation shields (e.g., aluminum foil) or coating vials with gold. Absence of a guard rail will cause higher sublimation rates among edge vials because the metal band of the rail acts as a thermal shield. These observations have a practical value in that heat transfer and edge vial effects that are present in a laboratory dryer might impact scale-up results when transferring processes to production scale. Such variables can be minimized by using radiation shields in laboratory dryers.

Sublimation rates are influenced by design features of the vial (34). The type of glass, vial diameter, bottom radius, and fill volume will affect sublimation. One surprising result of this study was that the concavity of the vial bottom did not have a significant effect on sublimation rate. This was already discussed in chapter 10 (see Fig. 10-3) that drying within a vial occurs along the sides of the container first then proceeds inward such that the last part of the product to sublime is the center bottom of the container (35).

A freeze-dry cycle developed for the antileukemia enzyme, *Erwinia* L-asparaginase, maintained product temperature very close to but not exceeding the collapse temperature during all of primary drying (36). At the beginning of primary drying, vacuum control, via a controlled air bleed into the chamber, maintained the increased chamber pressure needed to maximize heat transfer from the shelf to the product. As drying progressed, and vapor diffusion was impeded, the product gradually warmed to near the collapse temperature. The chamber pressure was subsequently lowered by turning off the air bleed and allowing the vacuum to pull to maximum low. Product temperature dropped by 5°C, and again gradually climbed as impedance to vapor diffusion continued to increase. When the product again neared the collapse temperature, the shelf temperature was gradually lowered to maintain the product temperature below the collapse temperature. These conditions maintained the sublimation interface slightly below the product collapse temperature, reduced the cycle time, and prevented product collapse during primary drying.

An example of efficient freeze-dry cycle optimization was reported that took advantage of high shelf temperatures (30°C, still below the Tg′ of the product) and low chamber pressure (325 μm) for both primary and secondary drying steps (37). This resulted in the development of a single-step drying cycle (Fig. 20-6) for the product (recombinant human interleuken-1 receptor antagonist).

Drying rates can be limited by what refers to as sonic "choked" flow[3]. The flow of water vapor from the product in the chamber to the condenser can be restricted depending on the cross-sectional area of the opening connecting the chamber and the condenser and the aerodynamic properties of the gas flow path between the product vials and the condenser coil surfaces (38). Figure 20-7 shows a plot of sublimation rate versus chamber pressure with a red line superimposed that represents the maximum "choked flow" drying rate limit of a given lyophilizer (37,38). The combination of shelf temperature and chamber pressure necessary to maintain a given product temperature can be manipulated to maximize the sublimation rate. Generally, decreasing the chamber pressure while correspondingly increasing the shelf temperature will increase the sublimation rate, but such increases will be limited if pressures are below the choked flow limit.

[3] Sonic flow is the speed of sound, the maximum possible velocity of a gas flowing in a cylinder. Sonic choked flow occurs when downstream pressure is reduced and gas flow reaches sonic flow at the exit point. Any further reduction in downstream pressure will not affect gas flow and the system is said to be "choked."

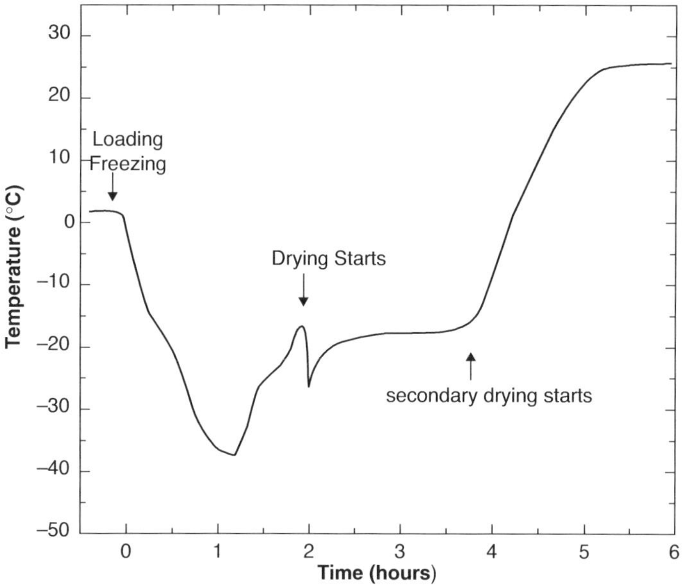

Figure 20-6 Example of a single-step freeze-drying cycle. The chamber pressure was maintained at 325 μm Hg and the shelf temperature was maintained at 30°C for both primary and secondary drying. *Source*: From Ref. 37.

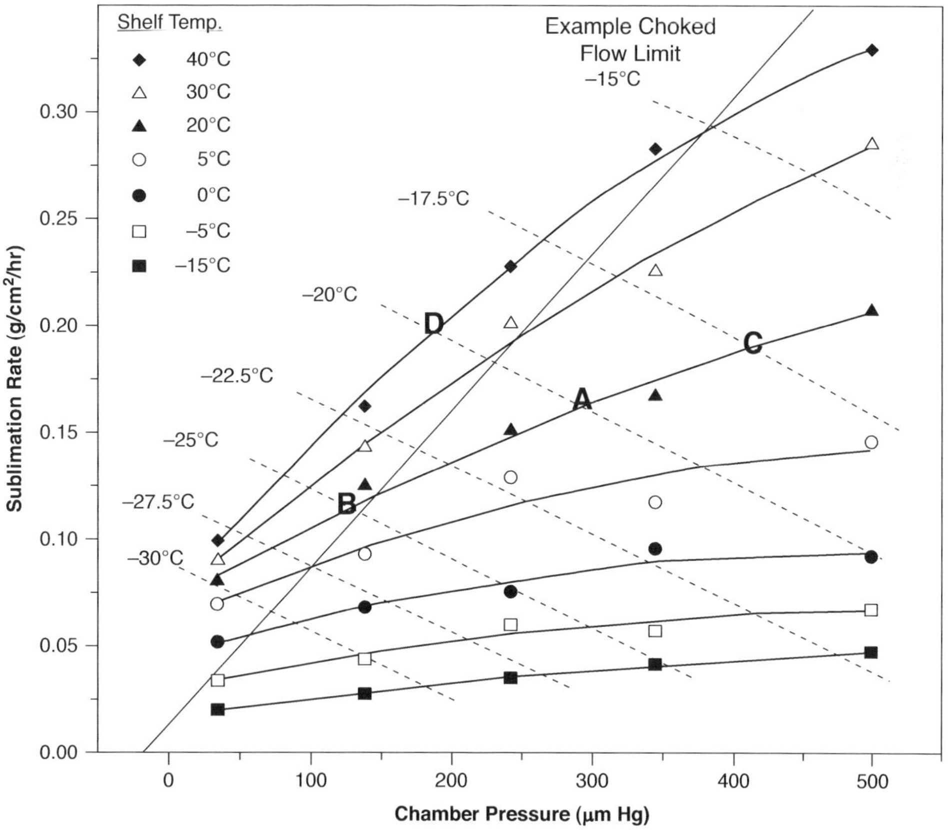

Figure 20-7 Plot of sublimation rate versus chamber pressure for different shelf temperatures with the temperatures associated with the dotted lines being product temperatures. *Source*: From Ref. 37. Furthermore, the diagonal line drawn representing the maximum "choked flow" drying rate limit taken with permission from Ref. 38.

Consequences of Inadequate Primary Drying

Inadequate temperature control during primary drying may lead to a variety of problems including complete collapse, partial collapse, or meltback of the dried finished product. These and other potential problems are listed in Table 20-3 along with some example photos.

The safest way to prevent or minimize these problems is to establish the shelf temperature and chamber pressure so that the product temperature always remains below its critical temperature, that is, its collapse temperature. Remaining below the critical temperature, the solute remains immobile and will not interact with the ice crystals. However, above the collapse temperature, solute molecules develop mobility and will lose their ability to remain rigid even after most of the ice is sublimed. Basically, inadequate primary drying allows excessive solvent to remain associated with the solute and once the product temperature is increased to some temperature above the freezing point, the ice (solvent) remaining will melt and solute molecules associated with the ice will collapse, partially collapse, or even melt.

To maintain product temperature below its collapse temperature seems simple, but keep in mind that the shelf temperature is not the same as the product temperature. As long as ice remains in the product container, the product temperature will always be lower than the shelf temperature. The intricacies of heat transfer not only from the shelf but also from the product container surroundings (other containers, walls, dryer door, chamber space) plus differences in effectiveness of heat transfer into the product, all result in a variable range of product temperatures from one vial to the next. When you consider temperature variations from front to back, side to side, and top to bottom of a multishelved, large freeze-dryer, it is practically impossible to achieve the same temperatures throughout the product load. What this all means is that you cannot assume that once a shelf temperature is established that the temperature differential between the product and the shelf is the same for every unit container in the entire load. This is added reason why the shelf temperature needs to be set to assure that even the unit container having the highest product temperature, wherever that unit is located in the freeze-dryer, is still lower than the critical temperature for that formulation.

While it is always wise to primary dry products at product temperatures below the critical temperature of the product formulation (typically 2–5°C below the glass transition or collapse temperature), there are exceptions to this "rule" where some products can been successfully freeze-dried at temperatures above collapse temperature (39–41). The need for reducing lengthy freeze-drying cycles sometimes requires more aggressive parameters, primarily applying shelf temperatures as high as possible that border on or exceed product critical temperatures, but this approach always will present risks to potential product collapse, even meltback, problems and result in product instability.

In general, on the basis of this author's experience, the shelf temperature needs to be somewhere in the vicinity of 18°C to 20°C higher than the product temperature to keep the product below its critical temperature.

SECONDARY DRYING

The secondary drying stage begins as soon as it is determined that the ice (solvent) has been removed (sublimed) from every unit container in the batch load. The time within the freeze-drying cycle that this occurs can be determined by several measurements (Fig. 20-8)

- Increase in product temperature
- Increase in chamber pressure
- Decrease in chamber moisture

During secondary drying, residual water associated with the remaining solute is removed. This is accomplished by raising the product temperature via increasing shelf temperature. Product temperature is raised to drive off adsorbed water (crystalline systems) or nonfrozen water in the glassy phase (amorphous systems). Chamber pressure can be increased but not much (usually $\leq$200 mT) and should not be decreased. Low chamber pressure during secondary drying will slow the rate of drying by inhibiting heat transfer from shelf to product. Also, slow pressure during secondary drying can accelerate the leaching of components of the rubber closure (although coated rubber closures protect largely against this).

Table 20-3 Potential Defects with Freeze-Dried Solids

- Collapse
  - Product temp during primary drying rises above its true freezing temperature (Tc), high viscosity lost, dry powder cannot hold up under its own weight

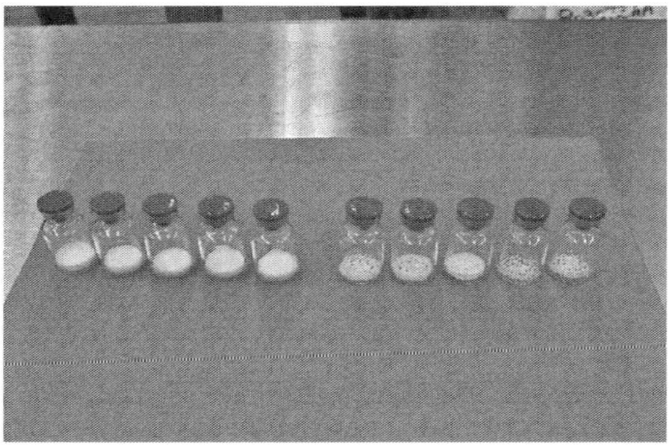

The five right side vials show collapse

- Partial Collapse
  - Product temp approaches but does not exceed collapse temp, some product in liquid state when heat transfer started
- Meltback
  - Insufficient drying/removal of water, solute dissolves
- Holes/pores
  - Microcollapse of solute
- Crust/glaze

Part of product still in liquid state when heat transfer began

- Puffing
  - Dryer evacuation occurs before freezing completed
- "Chimney effect" (2)

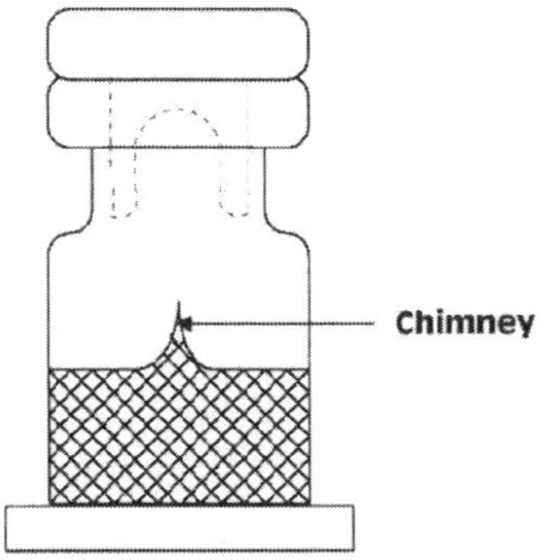

Product does not completely freeze ("mush"). Liquid layer above mush forced upward a bit due to convection currents in chamber.

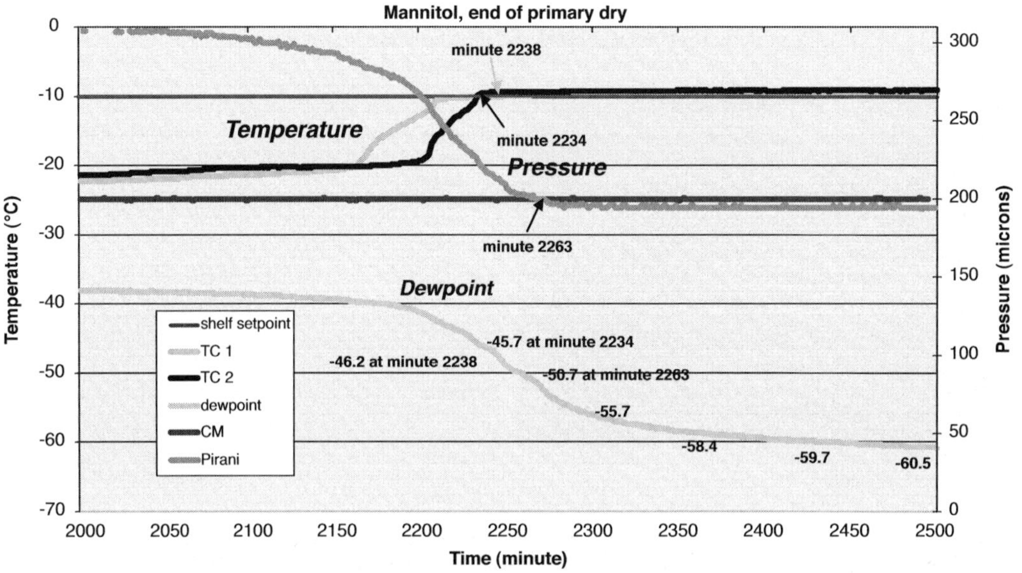

Figure 20-8 Comparison of temperature, pressure, and moisture data indicative of the end of primary drying. *Source*: Courtesy of Ms. Lisa Hardwick, Baxter BioPharma Solutions.

Preliminary experimentation should determine how high secondary drying temperatures can be achieved for a given drug product without causing any chemical degradation. In general, small molecule freeze-dried products can withstand secondary drying temperatures between 40°C and 50°C whereas secondary drying temperatures for large molecules cannot exceed 30°C to 35°C. Using lower temperatures during secondary drying, while taking more time to achieve the desired residual moisture level, will allow better product uniformity of water content.

Generally speaking, secondary drying requires whatever length of time required to achieve residual moisture levels of less than 1%. Time and temperature conditions for secondary drying should be determined experimentally (not blindly) using a sample thief to pull product samples at different times during secondary drying and product measured for residual moisture by Karl Fischer or other water determination measurement method as well as correlated water content with drug potency, purity, and/or activity

Product Temperature

In looking at the product temperature profile in Figure 20-8, point A is the time where you can see product temperature starting to increase. This indicates the end of primary drying for that monitored unit container. However, this does not mean that all the ice has been removed from every other unit container in the batch. Indeed, an earlier point deserves to be reiterated here in that unit containers having thermocouples in them will dry faster than all other containers without thermocouples. State-of-the-art practice requires that the primary drying cycle continue for several more hours to provide the "fudge factor" for all the other vials to "catch up" to the temperature of the thermocoupled vial(s).

Chamber Pressure

For modern automated loading/unloading freeze-drying systems, thermocouples are not used because this would defeat the purpose of automation to maintain ISO 5 environments throughout the filling/loading/lyophilization/unloading process without intervention. Therefore, monitoring of chamber pressure (or moisture) dictates the end of primary drying.

At the end of secondary drying, a pressure rise in the chamber and a decrease in water vapor pressure in the chamber will be observed. Chamber pressures are measured either by a Pirani gauge that is a thermal conductivity gauge or a capacitance manometer (CM). The

Pirani gauge responds to differences in thermal conductivity of individual gases, and will read differently as the total gas composition changes over the course of the cycle.

Thermal conductivity of water vapor is 50% greater than that for nitrogen. Thus, as chamber moisture levels decline, chamber inert gas will become more predominant and the Pirani gauge is more sensitive to detect this change in chamber gas content. The CM provides a true measure of force per unit area and measures pressure independently of gas phase composition. Pressure is controlled at a constant level using the CM as the sensor. As drying slows, the decrease in partial pressure of water vapor in the chamber is reflected in the decreased output of the Pirani gauge.

Chamber Moisture

As seen in Figure 20-8, chamber moisture is measured by the dewpoint with dewpoint being the measure of the partial pressure of water vapor, or the relative humidity in the chamber. Dewpoint is measured by a capacitive hygrometer that typically is installed by the freeze-dryer vendor. The hygrometer measures the electrical conductivity of water vapor molecules and converts these values into dewpoint temperature.

Sublimation Rate

The rate of water flow through the duct between the chamber and condenser can be measured by laser technology (Tunable Diode Laser Absorption Spectroscopy—TDLAS, LyoFlux, Physical Sciences, Inc) (42,43). A laser beam is generated by a near infrared-laser and launched through the sample at an angle about 45 degrees to the gas-flow axis between chamber and condenser (Figure 20-9). The absorption spectrum is measured as a function of the wavelength continuously, and the area under the absorption profile is integrated to determine the concentration (molecules/cm^3) of the target gas. Because the TDLAS unit is installed at an angle relative to the duct, the beam is subjected to a Doppler shift resulting in a wavelength shift of the absorption spectrum. The higher the Doppler shift, the greater the velocity of water vapor (note: speed of water vapor hits a limit at around 400 m/sec) and the faster the rate of sublimation. The comparison of this spectrum to a reference (measurement is made at an angle normal to the gas-flow axis) allows calculation of flow velocity and of mass flow rates through the spool piece. A combination with MTM technology appears desirable for the design and optimization of lyophilization cycles and calculation of process parameters in pre-existing cycles.

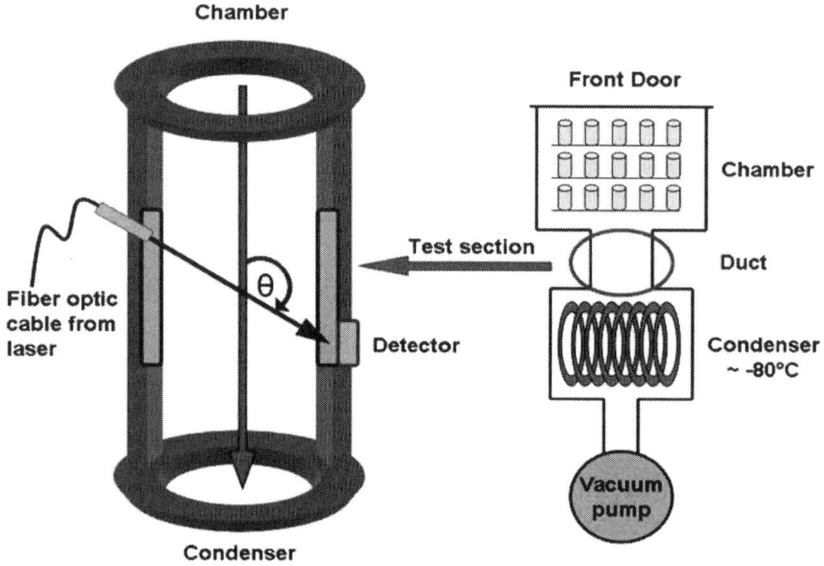

Figure 20-9 Schematic of tunable diode laser absorption spectroscopy (TDLAS). *Source*: Courtesy of Bill Kessler and Physical Sciences, Inc.

EQUIPMENT

In some instances a product may be frozen in a bulk container or in trays rather than in the final container and then handled as a bulk solid. Such a state requires a continuation of aseptic processing conditions as long as the product is exposed to the environment.

When large quantities of material are processed, it may be desirable to use ejection pumps in the equipment system. These draw the vapor into the pump and eject it to the outside, thereby eliminating the need for a condensing surface. Such pumps are expensive and usually practical only in large installations.

Available freeze-dryers (suppliers: BOC Edwards, FTS, Hull, Serail, Stokes, Usifroid, Virtis, others) range in size from small laboratory units to large industrial models. Their selection requires consideration of such factors as:

- The tray area required
- The volume of water to be removed
- How the chamber will be sterilized
- Whether internal stoppering is required
- Whether separate freezers will be used for initial freezing and condensation of the product
- The degree of automatic operation desired

Other factors involved in the selection and use of freeze-drying equipment are beyond the scope of this chapter, but references (2,3), and (5) can be consulted for more information as well as, of course, technical discussions with freeze-drying equipment manufacturers.

Freeze-drying is being used now for research in the preservation of human tissue and is finding increasing application in the food industry. Most biopharmaceuticals require lyophilization to stabilize their protein content effectively. Therefore, many newer developments in the lyophilization process focus on the requirements of this new class of drug products.

Aseptic Technologies (Gembloux, Belgium) has introduced the concept of "closed-vial" technology (discussed in chap. 23) and this includes the ability to freeze-dry vials (44). Vials are stoppered prior to filling with filling occurring through a needle piercing the rubber closure. The needle gauge for filling of vials for freeze-drying is 11G, slightly larger than the 13G used for vials that are not freeze-dried. A device called a "penetrator" is placed on top of the vial, then the vial, still closed, is conveyed to the freeze-dryer shelf. Once all vials are located within the freeze-dryer and the door closed, the shelves are moved downward, pushing on the penetrator's cone that reopens the piercing trace made by the 11-G filling needle. The lyophilization cycle is started with the shelves kept at the low position to keep the stoppers open via the penetrator. At the conclusion of the cycle, the shelves are lifted, the stopper assumes its original shape, the penetrator is lifted, and the rubber stopper re-seals. Vials are removed from the dryers, penetrators removed, and laser re-sealing of the stopper occurs (see chap. 23 for further explanation). At the time of this publication, the author was not aware of any company adopting this technology, but perhaps it is only a matter of time before such technology becomes state-of-the-art. There are still many challenges and limitations with the closed vial concept, as will be discussed in chapter 23.

REFERENCES

1. Nail SL, Gatlin LA. Freeze drying: Principles and practices. In: Avis, et al, eds. Pharmaceutical Dosage Forms: Parenteral Medications. Vol 2. New York, NY: Marcel Dekker, 1993; also see new revision published 2010.
2. Jennings TA. Lyophilization: Introduction and Basic Principles. Boca Raton, FL: CRC Press, 1999.
3. Nail SL, Jiang S, Chongprasert S, et al. Fundamentals of freeze-drying. In: Nail SL, Akers MJ, eds. Development and Manufacture of Protein Pharmaceuticals. New York, NY: Plenum, 2002: 281–360.
4. Rey L, May JC, eds. Freeze-Drying/Lyophilization of Pharmaceutical and Biological Products. 2nd ed. New York, NY: Marcel Dekker, 2004; revised and expanded.
5. Costantino, HR. and Pikal, MJ. Lyophilization of Biopharmaceuticals, AAPS Press, 2005.
6. Schwegman JJ, Hardwick LM, Akers MJ. Formulation and process development of freeze-dried biopharmaceuticals. Pharm Dev Technol 2005; 10:151–173.
7. Randolph TW, Searles JA. Freezing and annealing phenomena in lyophilization: Effects upon primary drying rate, morphology and heterogeneity. Am Pharm Rev 2002; 4:40–46.

8. Webb SD, Cleland JL, Carpenter JF, et al. Effects of annealing lyophilized and spray-lyophilized formulations of recombinant human interferon-γ. J Pharm Sci 2002; 92:715–729.
9. Chang BS, Kendrick BS, Carpenter JF. Surface-induced denaturation of proteins during freezing and its inhibition by surfactants. J Pharm Sci 1996; 85:1325–1330.
10. Hsu CC, Nguyen HM, Yeung DA, et al. Surface denaturation at solid-void interface—A possible pathway by which opalescent particulates form during the storage of lyophilized tissue-type plasminogen activator at high temperature. Pharm Res 1995; 12:69–77.
11. Jiang S, Nail SL. Effect of process conditions on recovery of protein activity after freezing and freeze-drying. Eur J Pharm Biopharm 1998; 45:249–257.
12. Sarciaux JM, Mansour S, Hageman MJ, et al. Effects of buffer composition and processing conditions on aggregation of bovine IgG during freeze-drying. J Pharm Sci 1999; 88:1354–1361.
13. Patahoff TA, Overcashier DE. The importance of freezing on lyophilization cycle development. Biopharm 2002; 3:16–21.
14. Searles JA, Carpenter JF, Randolph TW. The ice nucleation temperature determines the primary drying rate of lyophilization for samples frozen on a temperature-controlled shelf. J Pharm Sci 2001; 90:860–871.
15. Searles JA, Carpenter JF, Randolph TW. Annealing to optimize the primary drying rate, reduce freezing-induced drying rate heterogeneity, and determine T(g)′ in pharmaceutical lyophilization. J Pharm Sci 2001; 90:872–822.
16. Heller MC, Carpenter JF, Randolph TW. Protein formulation and lyophilization cycle design: Prevention of damage due to freeze-concentration induced phase separation. J Pharm Sci 63; 166–174.
17. Jameel F, Kalonia D, Bogner R. The effect of hetastarch on the stability of L-asparaginase during freeze-thaw cycling. PDA J Pharm Sci Technol 1995; 49:127–131.
18. Cochran T, Nail SL. Ice nucleation temperature influences recovery of activity of a model protein after freeze-drying. J Pharm Sci 2009; 98:3495–3498.
19. Liu J, Viverette T, Virgin M, et al. A study of the impact of freezing on the lyophilization of a concentrated formulation with a high fill-depth. Pharm Dev Technol 2004.
20. Gomez G, Pikal MJ, Rodriguez-Hornedo N. Effect of initial buffer composition on pH changes during far from equilibrium freezing of sodium phosphate buffer solutions. Pharm Res 2001; 18:90–97.
21. Anchordoquy TJ, Carpenter, JF. Polymers protect lactate dehydrogenase during freeze-drying by inhibiting dissociation in the frozen state. Arch Biochem Biophys 1996; 332:231–238.
22. Anchordoquy TJ, Izutsu KI, Randolph TW, et al. Maintenance of quarternary structure in the frozen state stabilizes lactate dehydrogenase during freeze-drying. Arch Biochem Biophys 2001; 390:35–41.
23. Pikal MJ. Freeze drying of proteins. In: Cleland JL, Langer R, eds. Formulation and Delivery of Proteins and Peptides. Vol. 567. Washington, DC: ACS Symposium Series, 1994:120–133.
24. Carpenter JF, Pikal MJ, Chang BS, et al. Rational design of stable lyophilized protein formulations: Some practical advice. Pharm Res 1997; 14:969–975.
25. Allison SD, Chang B, Randolph TW, et al. Hydrogen bonding between sugar and protein is responsible for inhibition of dehydration-induced protein unfolding. Arch Biochem Biophys 1999; 365:289–298.
26. Corvelyn S, Remon JP. Maltodextrins as lyoprotectants in the lyophilization of a model protein. LDH Pharm Res 1996; 13:146–150.
27. Overcashier DE, Patapoff TW, Hsu CC. Lyophilization of protein formulations in vials: Investigation of the relationship between resistance to vapor flow during primary drying and small scale product collapse. J Pharm Sci 1999; 88:688–695.
28. Pikal MJ, Shah S, Senior D, et al. Physical chemistry of freeze-drying: Measurement of sublimation rates for frozen aqueous solutions by a microbalance technique. J Pharm Sci 1983; 72:635–650.
29. Pikal MJ, Roy ML, Shah S. Mass and heat transfer in vial freeze-drying of pharmaceuticals: Role of vial. J Pharm Sci 1984; 73:1224–1237.
30. Pikal MJ. Use of laboratory data in freeze drying process design: Heat and mass transfer coefficients and the computer simulation of freeze drying. J Parenter Sci Technol 1985; 39:115–138.
31. Overcashier, D.E.; Patapoff, T.W.; Hsu, C.C., Lyophilization of protein formulations in vials: Investigation of the relationship between resistance to vapor flow during primary drying and small scale product collapse, J Pharm Sci 1999, 88, 688–695.
32. Milton N, Pikal MJ, Roy ML, et al. Evaluation of manometric temperature measurement as a method of monitoring product temperature during lyophilization. PDA J Pharm Sci Technol 1997; 51:7–15.
33. Rambhatla S, Pikal MJ. Heat and mass transfer scale-up issues during freeze drying, I: Atypical radiation and the edge vial effect. AAPS PharmSciTech 2003; 4:111–120.
34. Cannon AJ, Shemely K. Statistical evaluation of vial design features that influence sublimation rates during primary drying. Pharm Res 2004; 21:536–542.
35. Pikal MJ, Shah S. Intravial distribution of moisture during the secondary drying state of freeze drying. PDA J Pharm Sci Technol 1997; 51:17–24.

36. Adams GDJ, Ramsay JR. Optimizing the lyophilization cycle and the consequences of collapse on the pharmaceutical acceptability of *Erwinia* L-asparaginase. J Pharm Sci 1996; 86:1301–1305.
37. Chang BS, Fischer NL. Development of an efficient single-step freeze-drying cycle for protein formulations. Pharm Res 1995; 12:831–837.
38. Searles JA. Observation and implications of sonic water vapor flow during freeze-drying. Am Pharm Rev 2004; 7:58–68, 75.
39. Colandene JD, Maldonado LM, Creagh AT, et al. Lyophilization cycle development for a high-concentration monoclonal antibody formulation lacking a crystalline bulking agent. J Pharm Sci 2006; 96:1598–1608.
40. Luthra S, Obert JP, Kalonia DS, et al. Investigation of drying stresses on proteins during lyophilization: Differentiation between primary and secondary-drying stresses on lactate dehydrogenase using a humidity controlled mini freeze-dryer. J Pharm Sci 2007; 96:61–70.
41. Schersch K, Betz O, Garidel P, et al. Systematic investigation of the effect of lyophilzate collapse on pharmaceutically relevant proteins I: Stability after freeze-drying, J Pharm Sci 2010; 99:2256–2278.
42. Gieseler H, Kessler WJ, Finson M, et al. Evaluation of tunable diode laser absorption spectroscopy for in-process water vapor mass flux measurements during freeze-drying. J Pharm Sci 2007; 96:1776–1793.
43. Kuu WY, Nail SL. Rapid freeze-drying cycle optimization using computer programs developed based on heat and mass transfer models and facilitated by tunable diode laser absorption spectroscopy (TDLAS). J Pharm Sci 2009; 98:3469–3482.
44. Thilly J, Mayeresse Y. Freeze-drying with closed vials [supplement on Advances in Sterile Manufacturing and Aseptic Processing]. Pharm Tech, 2008; s38–s42.

21 Aseptic processing

Sterile products are sterilized either by terminal sterilization technologies or by filtration, followed by aseptic processing. The majority of small-volume injectables are aseptically processed. What aseptic processing means is that components of the final product are separately sterilized, and then put together under aseptic conditions with no terminal sterilization step after the product is filled, stoppered, and sealed. While significant advances have been made in sterility assurance of aseptic production processes, such assurance will never be greater than that achieved with terminal sterilization processes. At the time of the publication of this book, almost all drug products recalled by the Food and Drug Administration (FDA) with either confirmed or suspected issues with sterility assurance were products produced by aseptic processes. Therefore, significant scrutiny is placed upon all aspects of the production facility, processes, personnel, procedures, and documentation with regard to validation of aseptic processing.

Successful aseptic processing relies on validated cleaning, sanitization, and sterilization procedures for all facilities and components involved in the process. The facility must be adequately cleaned and sanitized and then maintained in order to meet classified work area requirements. High-efficiency particulate air (HEPA) filters and laminar air stations must be certified and maintained. All equipment must be sterilized by validated procedures and then sterility maintained through usage. Think about all the equipment that must be cleaned, sterilized, and maintained—tanks, tubing, filling nozzles, stopper bowls, freeze-dryers, utensils, whatever is to be used in a filling and stopper operation. Once equipment are ready, then packaging components must be cleaned, sterilized, and depyrogenated. Glass containers are sterilized by dry heat and rubber components sterilized by steam. Finally, the formulation itself must be sterilized by filtration, then the sterile formulation is filled into the sterile primary packaging, and stoppered by sterile rubber stoppers, all this performed under aseptic conditions. For products that are freeze-dried, maintenance of aseptic conditions takes on greater vulnerability because of the significant added time and movement of filled product from the filling line to the freeze dryer, with partially stoppered vials exposed to the environment until the freeze-drying process is completed.

Then add people who are part of the manufacturing process. After doing so, one can easily see how difficult it can be to validate an aseptic process. So how does a company validate aseptic processing with so many potential contributors to rendering the process non-sterile? Contamination control aspects, as described in Chapter 13 and other previous chapters, certainly come into play here. All of the components that contribute to sterility assurance come into play here. The facility must be cleaned and sanitized by valid procedures and then consistently evaluated by valid environmental monitoring procedures. All personnel involved must be adequately trained and certified for correct gowning procedures and aseptic techniques. Air handling systems must be certified and maintained. All components—equipment and packaging—must be cleaned and sterilized by valid procedures. The product itself must be sterile filtered by valid procedures. Finally, everything comes together where all aspects of an aseptic processing operation are validated by a process simulation procedure, commonly called the "media fill," to be covered extensively later in this chapter.

Table 21-1 lists all the processing steps and support systems that are components of an aseptic process, each of which can affect the sterility and other quality parameters of an aseptically processed product and, therefore, requires qualification or validation (1). Most of the remainder of this chapter highlights the requirements of the FDA and European Union (EU) guidelines for aseptic manufacture of sterile products (2,3).

Table 21-1 Aseptic Process Operations and Support Systems That Should Be Part of a Qualification and/or Validation Study

- Cleaning and disinfection of tanks, mixing vessels, and transfer lines
- Raw material bioburden and pyroburden
- Raw material addition
- Compounding and mixing
- All product transfer steps
- Product filtration (sterilization step) including microbial retention, product compatibility, and filter extractables
- Sterilization of all filters (liquid and gas)
- Cleaning and sterilization of product contact surfaces, parts, vessels, lines, housings, all accessories
- Cleaning of all packaging components
- Cleaning and sanitization of clean room equipment, walls, floors, surfaces
- Operation of component handling and transport equipment, unscramblers, hoppers, bottle orienters, star wheels, component bowls and tracks, conveyors, turntables
- Operation of filling equipment, inert gas overlay systems, stopper inserters, cappers
- Operation of product removal systems, check weighers, volume detectors, leak detectors, inspection systems, vision systems
- Operation of labelers, sealers, cartoners, all packaging equipment
- Utilities generation and transport systems for air, water, cooling medium, vacuum, dust collection, nitrogen, plant and clean steam
- Monitoring systems, building automation, facility monitoring, distributed control systems, PLCs, LIMS data collection, all electronic record generation and storage systems
- Warehouse, cold storage, handling
- Disinfectant (sanitizing) and cleaning effectiveness
- Gown and glove sterilization
- Effectiveness of clean room HEPA filters
- Operation of clean room air handling systems
- Clean room airflow in and around exposed product and product contact surfaces in relation to the aseptic process and interventions
- Cleaning and disinfection of isolators or RABS interior
- Operator gowning
- Operator hygiene, aseptic techniques and practices

Abbreviations: PLC, programmable logic processor; LIMS, laboratory information management system; RABS, restricted access barrier system. *Source*: From Ref. 1.

BUILDINGS AND FACILITIES

Air classification is in accordance with both the FDA and EU guidelines, as seen in Table 21-2A and 21-2B, with surface quality standards given in Table 21-2C. Air particle counts must be measured not more than 1 foot away from the actual work site and should be measured during actual filling and closing operations. Air quality of powder filling operations must be certified under dynamic conditions (machinery running) without filling of actual powder. Air particle counts must be measured frequently during each shift, bracketing the beginning and end of the filling operation.

The air supplied to the Grade A/B or Class 100 clean room must have a velocity of 90–100 ft/min with a range of ± 20%. Airflow patterns must be determined using smoke tests and videotaping the smoke test results. Smoke tests verify the unidirectional flow of air. The

Table 21-2A Air Classifications According to FDA Aseptic Processing Guidelines

Clean area classification	Particles $\geq 0.5 \mu m/ft^3$	Particles $\geq 0.5 \mu m/m^3$	Air microbial action level (CFU/m^3)	Settle plate action level (CFU/4 hr)
100 ISO 5	100	3520	1 (Expect zero)	1 (Expect 0)
1000 ISO 6	1000	35,200	7	3
10,000 ISO 7	10,000	352,000	10	5
100,000 ISO 8	100,000	3,520,000	100	50

Table 21-2B Air Classifications According to European Grade

	At rest		In operation	
Grade	**Particles ≥ 0.5 μm/m³**	**Particles ≥ 5 μm/m³**	**Particles ≥ 0.5 μm/m³**	**Particles ≥ 5 μm/m³**
A	3520	0	3500	0
B	35,200	0	350,000	2,000
C	352,000	2000	3,500,000	20,000
D	3,520,000	20,000	Not defined	Not defined

Table 21-2C Recommended Limits for Microbial Contamination According to European Grade Clean room Classification

	Recommended limits for microbial contamination			
Grade	**Air sample (CFU/m³)**	**Settle plates (CFU/4 hr)**	**Contact plates (CFU/Plate)**	**Glove print (CFU/Glove)**
A	<1	<1	<1	<1
B	10	3	5	5
C	100	50	25	–
D	200	100	50	–

source of smoke typically is a glycol-based[1] fog generator that clearly shows if there is any turbulence in the room(s). If there is any suspicion of a breach in the security of a clean room (e.g., an emergency door opened where the alarm did not work), application of a smoke test will determine if such a breach caused turbulence.

Air pressure differentials between rooms must be different by at least 12.5 Pascals (0.05 inches of water). At least 20 changes of air per hour are required for all clean rooms with no specific minimum given for Grade A/B or Class 100 rooms. Microbiological monitoring is performed with the expectation of zero growth of any plate at any time.

The area immediately adjacent to the aseptic processing line should meet a minimum of Grade C or Class 10,000 conditions under dynamic operations, although the current preference is Class 1000 or maintaining the entire aseptic filling room at Class 100.

All compressed gases used in aseptic processing must be free from demonstrable oil vapor, sterilized tanks must be held under continuous overpressure, and all gas filters must be periodically integrity tested.

HEPA filters are to be integrity tested twice a year using poly-alpha-olefin (PAO) aerosol. This aerosol is a polydisperse, nontoxic liquid that possesses a light scattering mean droplet diameter of 0.7 micrometers. Dioctyl phthalate (DOP) is another aerosol used, although PAO is more widely used. The starting liquid is heated to the point of vaporization and reconstituted into 0.3-μm particles to form a monodisperse aerosol. These single-size particles are diluted with air until a concentration of 100 μg/L is reached, and the aerosol-air mixture is passed through the filter. The sample rate should be at least 1 ft^3/min with the probe 1–2 inches from the face of the filter. Since the upstream aerosol concentration is known, and the photometer is linear, the downstream samples may be read out in percent of concentration. Typical readings at the filter face range from 0.004% to 0.008%. Any leak greater than 0.01% of the upstream concentration is considered a significant leak and the location of the leak needs to be repaired (patched). There should be no leaks around the filter seals. If 10% or more of the filter face fails this challenge test, the entire filter must be replaced.

The design of the building and facilities has been covered in Chapter 14. A summary of the requirements of facilities according to the FDA aseptic processing guidelines include proper ergonomics of all equipment used, minimization of entries and exits, proper design of airlocks,

[1] Water-based fog generators (e.g., carbon dioxide or liquid nitrogen) create an effluent that is heavier than air that may not accurately demonstrate actual air patterns.

seamless construction, sanitary fittings and valves, no drains in Class 100 rooms and drains in other rooms that have air breaks, and equipment designed to be cleanable and not block HEPA filter airflow.

PERSONNEL TRAINING AND QUALIFICATION

FDA aseptic processing guidelines stress vigilant adherence to fundamental principles of aseptic techniques (see chap. 16). Personnel training should include didactic teaching and hands-on performance of proper aseptic techniques, basics of microbiology, personal hygiene, gowning, and all appropriate standard operating procedures (SOPs). All personnel are required to participate in at least one media fill per year. Between media fills, all personnel should participate in regular training updates supplemented by routine evaluations by supervisory personnel of each operator's conformance to written procedures and basic aseptic techniques during actual operations.

Personnel working within a defined classified work area should be kept to the absolute minimum number. Gowning requirements and training thereof are elaborated in Chapter 16. On a daily basis, all personnel are monitored by surface sampling plates, which includes daily glove monitoring plus regular sampling of one or more strategically selected locations of the gown. It is somewhat of a common, albeit wrong, practice for operators to sanitize one's gloves prior to surface sampling.

Laboratory personnel working on the microbiological testing of sterile products must have the same type of training as that of manufacturing personnel.

COMPONENTS (ACTIVE INGREDIENTS AND EXCIPIENTS)

FDA aseptic processing guidelines require that the microbial content of each component liable to contamination be characterized and appropriate acceptance limits established based on these data. Established specifications for acceptance or rejection of each component for the presence of endotoxins also are required.

CONTAINERS AND CLOSURES

FDA aseptic processing guidelines require containers and closures to undergo a final rinse with water that meets USP (United States Pharmacopeia) Water for Injection (WFI) specifications. Containers and closures must be depyrogenated by whatever means necessary that can be validated to reduce the endotoxin content by at least 3 log units. Typically glass containers are depyrogenated by dry heat and plastic containers depyrogenated by ethylene oxide. Rubber closures are depyrogenated by copious amounts of USP WFI rinsing. The greatest challenge is not endotoxin removal per se by any of the procedures, but the validation of removal because endotoxin spikes on glass, plastic, and rubber are not easily recoverable. The risk of false negative results is a concern because endotoxin will bind to surfaces. Valid techniques (e.g., sonification) for quantitative removal of spiked endotoxin as controls must be demonstrated.

There must be time limits established between washing and sterilizing of containers and closures so that microbial buildup does not occur. If silicone is used, it must be rendered sterile and show no adverse effect on safety, quality, or purity of the drug product. Contractors who sterilize and depyrogenate containers and closures are subject to the same good manufacturing practice (GMP) compliance requirements as the sterile drug manufacturers.

The manufacturer must detect and remove via a dependable inspection process any product lacking in container/closure integrity (see chap. 30). Also for delivery devices such as syringes and cartridges, there should be in-process tests that can detect any functionality problems, for example, poor syringeability or lack of control of delivery volume.

ENDOTOXIN CONTROL

The focus on endotoxin control relates to the formulation components, containers, closures, equipment, and storage time limits. All depyrogenation processes must be validated for the reduction/removal of endotoxin. Some depyrogenation processes involve clean-in-place procedures and final rinsing procedures, all of which require proof of reducing by at least 3 log units an applied endotoxin challenge. Endotoxin control must be practiced for all product contact surfaces prior to and after sterile filtration.

Table 21-3 Examples of 483 Observations Related To Media Fill Processes and Related Documentation

- Inadequate investigation of media fill failure.
- Inadequate training of employees after media fill failure.
- Media fills did not follow SOP.
- Media fill aborted due to high particulate counts, but inadequate investigation into reasons for high counts.
- Media fill did not start at point after product had been sterilized.
- Defective vials discarded prior to incubation and not counted as failures.
- Number of units filled too small.
- Media fills did not simulate what was documented in batch records.
- Certain environmental data not collected during fill.

TIME LIMITATIONS

Aseptic processing guidelines require that maximum hold times be established through reproducible studies for:

- Filtration processes
- How long a non-sterilized bulk solution can be held prior to filtration
- How long a sterilized solution can be held prior to filling
- How long sterilized equipment can be held prior to using them
- How long sterilized containers and closures can be held prior to using them

PROCESS VALIDATION AND EQUIPMENT QUALIFICATION

There are three main aspects to aseptic process validation and equipment qualification for aseptic processing—process simulation testing (media fills), filtration efficiency, and sterilization of equipment and materials.

Process Simulations (Media Fills)

The FDA aseptic processing guidelines and the EU guidelines for sterile drug manufacturing contain a large number of specific guidances for the sterile drug industry to abide by. FDA inspections have increasingly focused on media fill studies that truly simulate the production process. Table 21-3 lists some examples of 483 observations issued by FDA inspectors related to media fill operations and documentation.

Because so many factors affect the assurance of sterility of an aseptic process operation, the use of sterile culture media has become the best determinant to validate the fact that all these factors are in place. Basically, culture media replaces the product that is prepared, filtered, and filled into the final container. Since culture media will support microbial growth, the presence of microorganisms due to any breach of asepsis in the manufacturing area, components and equipment used, the entire process, and personnel involved will show up as positive growth in culture media filled and stoppered into final containers. There are many factors that must be considered in designing a valid simulation of the actual process (Table 21-4).

The *media fill* or *process simulation test* involves preparation and sterilization (often by filtration) of sterile trypticase soy broth and filling this broth into sterile containers under conditions simulating as closely as possible those characteristics of a filling process for a product.[2] The key is designing these studies that simulate all factors that occur during the normal production of a lot. The entire lot, normally at least 4750 units, is incubated at temperatures verified to support microbial growth, usually rotating 20–25°C storage and 30–35°C storage, for at least 14 days and examined for the appearance of growth of microorganisms. It must be verified that the media used is capable of supporting microbial growth. If growth occurs, contamination has entered the container(s) during the processing. To pass the test at 95% confidence, not more than 0.1% of the challenged units may show growth, although the current expectation of regulatory agencies is "approaching zero." This evaluation also has been used as a measure of the proficiency of an individual or team of operators. This test is a very stringent evaluation of the

[2] For sterile powder filling, sterile lactose is used to simulate the filling process followed by dissolving with trypticase soy broth under aseptic conditions prior to incubation.

Table 21-4 Factors to Consider In the Design of Media Fill Studies

- Duration of longest run
- Worst-case environmental conditions
- Number and type of interventions, stoppages, adjustments, transfers
- Aseptic assembly of equipment
- Number and activities of personnel
- Number of aseptic additions
- Shift breaks, changes, multiple gownings
- Number/type of aseptic equipment disconnections and connections
- Aseptic samples
- Line speed/configuration
- Manual weight checks
- Operator fatigue
- Container/closure types run on the line
- Temperature/relative humidity extremes
- Conditions permitted before line clearance
- Container/closure surfaces which contact formulation during aseptic process

efficiency of an aseptic filling process and, by many, is considered to be the most evaluative test available.

The media fill provides a "one-time" representation of the capabilities of an aseptic processing operation. Media fills are conducted when a new filling line or new product container is introduced. For initial qualification of a line or product, three consecutive, separate, and successful media fill runs must take place. The FDA stresses that three is a minimum number of runs. Today, the term "successful" means that there is no growth in any of the units filled with sterile broth. All activities and interventions representative of each shift on each line must be simulated during the media fill. All personnel involved in the aseptic filling of a product (operators, maintenance personnel, microbiology support personnel) must participate in at least one media fill run per year. Typically, for each filling line and process, the filling operation will be validated for the smallest and largest container size that will be used.

After initial qualification, media fills are then conducted on a periodic basis, usually twice a year on the same filling line, to assure that conditions that existed during the initial qualification have been maintained. For periodic qualification, only one successful media fill run is required. If any media fill run fails or significant changes occur with the line, facility, or personnel, then the initial qualification media fill (three consecutive successful runs) must be conducted. Any changes in the process must be evaluated for its level of significance (change control quality system) that would necessitate a media fill validation run. Any media fill failure must be thoroughly investigated and followed by multiple repeat media fill runs. It is generally considered inappropriate to "invalidate" a media fill run.

The number of containers filled with media ideally should be the same as the actual number filled according to the batch record for the product being validated. Of course, this is unrealistic for large batch sizes. Therefore, the number of units filled must be sufficient to reflect the effects of all worst-case filling rates. For example, operator fatigue and the maximum number of interventions and stoppages must be incorporated into the media fill protocol. When media filling first started, the acceptable rate of positives (number of containers that showed contamination after incubating the culture media) was 1 out of 1000 (0.1%). Later that number became 1 out of 3000 to account for 95% confidence of a contamination rate of 0.1%. Today, 1 positive out of 3000 is no longer acceptable. Table 21-5 presents that International Standards Organization (ISO) standard used to determine the minimum number of containers filled with media and the acceptable number of positives. The most common number of containers filled with media in the industry is 4750, with three consecutive runs of 4750 used for initial performance qualification of a new product and/or filling/closing line. This same number of units filled—4750—is also used for the routine semiannual re-qualification media fills. The expected number of positive media fills (growth seen upon incubation) is zero. One

Table 21-5 ISO 13408-1 Standards for Minimum Number of Containers Filled with Media and the Acceptable Number of Positives

Number of media fill units	Allowable number of failed units (95% C. L.) by ISO	Allowable number of failed units by simple math
3,000	1	3
4,750	2	4
6,300	3	6
7,760	4	7
9,160	5	9
10,520	6	10
11,850	7	11
13,150	8	13
14,440	9	13
15,710	10	15
16,970	11	17

Abbreviation: CL, confidence limit.

or more failures likely means that there is a significant breach in the aseptic manufacturing process and the ensuing investigation must do everything possible to find the assignable cause.

After filling with culture media, but prior to incubation, all units should be inverted or swirled to enable the media to make contact with all internal surfaces of the container/closure system.

The culture media used for each media fill exercise must be tested to ensure that it will support the growth of microorganisms if they are present. Challenge organisms used in the media challenge pre-testing should include those isolated from environmental/personnel monitoring, those isolated from positive sterility test results, and USP growth promotion microorganisms. The positive control units inoculated with approximately 100 colony-forming units (CFUs) of these challenge organisms are incubated at temperatures and times validated to show microbial growth, if present. After the 14-day incubation period of the media fill containers, negative control units should then be inoculated with challenge organisms to prove that the media will still support growth, if present.

Inspection of media-filled units before and after incubation is conducted by individuals trained as qualified inspectors and certified by the quality control (QC) unit. It is permissible that any unit that is found to lack integrity after filling be rejected from being part of the media fill incubation just as a product vial would be rejected if a critical defect were found. However, if a media fill unit is found damaged after incubation is underway, it must remain incubated and counted in the data for the media fill batch. Procedures must be very clear and specific regarding samples taken during the media fill that simulate the actual sampling process and why these units are not part of those incubated.

Other requirements of a valid media fill experiment include:

- An appropriate criteria for batch yield and accountability just like a product batch.
- Identifying any contaminant to the species level and performing complete investigations of failed media fills.
- FDA advocates videotaping media fills to identify personnel practices that could negatively impact the aseptic process.
- Media fill duration, according to FDA, EU, ISO, CEN (European Committee of Standardization), and PIC, must be sufficiently long to include all required manipulations and cover the same length of time that is normally consumed by the commercial process. Most media fills are a minimum of 3 hours; some may be as long as 24 hours.

Filtration Efficiency (see also chap. 18)

The challenge organism used to validate the retention capability of a filter is *Brevundimonas diminuta* (ATCC 19146) because of the small mean diameter of this microorganism (~0.3 μm).

The challenge concentration is at least 10^7 organisms per cm^2 filter surface area. It might also be wise to conduct bacterial retention studies with microorganisms known to be bioburden isolates. The pre-filtered bulk solution used in this challenge study is sampled for bioburden in order to track potential contaminant organisms.

In addition to filter retention studies, filter validation also must include the determination of the effect of the product formulation and filter process on filter efficiency. Properties of the product, for example, pH, viscosity, ionic strength, and osmolality, could affect the ability of the filter to retain microbial challenges. Process variables such as pressure, flow rate, maximum time of filtration, temperature, and hydraulic shock need to be studied for effects on filtration efficiency. The actual product, not a simulated product, must be used except in cases where the product has bactericidal activity against *B. diminuta* or products that are oil-based formulations.

Filter validation studies are usually performed by the filter manufacturers, but the filter user is responsible for the data.

Sterilization of Equipment and Materials (see also chap. 17)

A main emphasis in the qualification and process validation of steam sterilizers is the ability to remove air and replace with steam. It is incumbent on the manufacturer to locate the most difficult area for heat to penetrate in the batch to be sterilized. Heat distribution studies are conducted as part of the qualification of the empty sterilization chamber. Validation of the loaded chamber usually focuses on a 6 log safety factor, that is, the cycle is extended to add an additional 6 logs of lethality to the product. Loading patterns are validated by identifying the cold spots within the product load using thermocouples and biological indicators. Subsequent batches must use the same loading patterns with thermal (thermocouple) and microbial (biological indicator) monitoring occurring at the previously identified cold spots. Any changes in the loading pattern must be revalidated as the cold spots might change.

It is expected that proper calibration of equipment controls and instrumentation be implemented. This would include controls for temperature, pressure, and quality of steam.

Laboratory Controls

FDA guidelines have specific requirements for the following laboratory control functions: environmental monitoring, microbiological media identification and trending, pre-filtration bioburden, and particulate matter testing.

Environmental Monitoring

Air, floors, walls, and equipment surfaces are to be monitored on every shift. Written procedures should include a list of locations to be sampled and when, how long, and how frequent the sampling will occur, the surface area and air volume monitoring, and what are the alert and action limits. Critical surface sampling may be performed at the conclusion of the batch process. Air and surface samples should be taken at the actual working level or actual surface. Daily surface samples of each aseptic operator's gown and finger pads must be taken, employing random intervals. The personnel monitoring program should be considered a separate program from the air and surface environmental monitoring in order to accommodate different types of follow-up actions, for example, increased scrutiny, retraining, and/or re-qualification.

Low-level contamination is not always detected. Because of the possible existence of false negatives during monitoring, consecutive growth results should not be considered the only type of adverse trend. It is advised to look for increased evidence of contamination over a given period in comparison to that normally detected.

The environmental monitoring program must have SOPs describing how the data are reviewed, isolates identified, and how responses to trends are conducted by the QC unit and regular updates given to the responsible management. Trend reports should be generated as a function of location, shift, lot, room, operator, or other search parameters.

Environmental monitoring alert and action limits must be established based on the relationship of the monitoring location and the critical operation. Individual results that exceed the alert limits should focus on trend analysis of historical data and associated manufacturing deviation records and actions taken as a result of the deviation. Individual results that exceed action limits must prompt more thorough investigations and documented results of those investigations. Any atypical microorganisms isolated definitely must be investigated to determine their source.

Sanitization efficiency is assessed by environmental monitoring. Prior to implementing a sanitization procedure, environmental data must be collected to verify the effectiveness of the sanitization agent and procedure. After implementation, monitoring should occur frequently to prove that the agent and procedure are still effective. Sanitization agents must be rendered sterile prior to use. SOPs must state the validated time limits for use of sanitization agents. These agents should retain efficacy against normal microbiological flora and against spore-forming bacteria. It has been realized that isopropyl alcohol does not kill spores so if this agent is used as a sanitization agent, another agent that is known to be sporicidal must be rotated with isopropyl alcohol.

Environmental monitoring systems recognized/accepted by the FDA include:

Surface sampling	**Air monitoring**
Touch plates	Slit-to-agar samplers
Swabs	Centrifugal samplers
Contact plates	Liquid impingement
	Membrane filtration
	Settle plates

More information on use of these methods can be found in Chapter 13.

Microbiological Media and Identification

Any organism isolated on an environmental monitoring plate should be identified to the species level. Sufficient characterization of isolates should be able to establish the relationship between the environmental isolate and any isolate found in a non-sterile unit during sterile media fill runs and/or any failure in product sterility testing. The culture media used in environmental monitoring programs must be validated to support the growth of bacteria and fungi. Incubation conditions must be validated. Typical incubation conditions for aerobic bacteria are storage at 30–35°C for 48–72 hours. Total combined yeast and mold incubation requires storage at 20–25°C for 5–7 days.

Pre-filtration Bioburden and Particulate Matter Testing

Limits for bioburden level for each formulated drug product should be established. Critical areas should be monitored for particulate matter frequently throughout daily operations at predetermined locations during production activities. Any result outside of qualified processing norms must be investigated to be consistent with the severity of the excursion.

Sterility Testing

The FDA still affirms the importance of conducting sterility testing despite severe limitations of the test (see chap. 27). The sterility testing laboratory should employ facilities and controls comparable to those used for filling operations. The use of isolators in sterility testing laboratories is widely recognized and is now state-of-the-art technology. Conducting sterility testing in isolator environments has been proven to be a better environment for minimizing the incidence of false positive tests.

The membrane filtration sterility test, as opposed to the direct inoculation method, is the preferred test method, whenever feasible. Samples for sterility testing should be taken from the beginning, middle, and end of the batch process and in coordination with any intervention or excursion during the process. Because microbial contamination, if present, will be quite small and such microorganisms likely will be nutrient-depleted, the sterility test may require longer incubation times to allow microbial growth to occur.

No amount of retesting can overcome a valid initial positive sterility test result. FDA guidelines specifically stress that persuasive evidence must be found to show absence of laboratory error. If laboratory error is absent or inconclusive, then the manufacturer should err on the side of safety and reject a batch that has experienced an initial sterility test failure, regardless of the results of the retest. A finding of "no-growth" in the retest should be accorded less weight than other parts of the investigation. The identity of the organism causing the sterility test failure should be known to the species level.

Trend analyses of sterility testing results should be reviewed periodically. Manufacturers must react to any trend showing an increase in false positive results. Trends should be separated by product, container type, filling line, and type of aseptic process. Environmental monitoring data trends in the production area should also be monitored for any correlation to the failed sterility tests. Chapter 27 lists many other records and documents that must be evaluated during an investigation following a sterility test failure.

ASEPTIC PROCESSING ISOLATORS

FDA aseptic processing guidelines state that the following isolator systems require daily attention and preventative maintenance:

- Gloves
- Half suits
- Seams
- Gaskets
- Seals
- HEPA filters

A major weakness of isolators is glove integrity. Durable glove materials must be used and aggressive replacement frequency must be practiced. Glove integrity batch-by-batch or daily evaluation should be performed using both visual and mechanical integrity tests. Evidence of any leakage terminates the operation.

Airflow in the isolator must be either HEPA- or ultra low particulate air-filtered. FDA prefers rigid wall construction over flexible materials. The air pressure differential should range from 0.075 to 0.2 inches of water gauge. Where any opening exists, for example, the exit, Class 100 protection should exist. Class 100 environment should be present in the interior with the isolator background (where it is located) being at least Class 10,000, especially for applications with multiple transfers, mouseholes, and sanitizing transfer ports. Rapid transfer ports (RTPs) are considered effective as transfer systems, but still should be kept to a minimum in the isolator design. Ultraviolet light and localized, HEPA-filtered air can be used in transfer ports.

All surfaces within the isolator must be exposed to a chemical sterilant. Surface sterilization validation studies should include a thorough determination of the limitations of the sterilization cycle using biological indicators at various locations, especially "tough-to-reach" locations. The entire path of the sterilized liquid stream must be steam sterilized. An environmental monitoring program must be established where within the isolator, the air quality is evaluated periodically during actual operations. Even with isolator technology, the human factor still remains an integral component and concern, especially considering the fatigue factor and its effect on practicing inerrant aseptic technique.

Validation of barrier isolators is covered in more detail in Chapter 23.

ASEPTIC CONNECTIONS AND SAMPLING METHODS

A single pharmaceutical company will make anywhere from 25,000 to over 100,000 aseptic connections in a single year (4). Each connection or sample runs the risk of introducing contamination. In response, several vendors have made available new devices for making connections and taking samples more easily and quickly. Description of the various aseptic connectors and samplers is beyond the scope of this chapter, although some additional brief coverage can be found in Chapter 23. Major vendors such as Pall (Kleenpak™), Millipore (NovaSeptum®),

BioQuate, Asepco, Stedim, and others provide these devices and their websites can be easily accessed for more information and technical literature.

Comparison of FDA and EU Guidelines—Questions on Aseptic Processing

When teaching on aseptic processing, this author would give his class the list of questions given in Appendix A, requiring participants to read and compare the FDA and the EU aseptic processing guidelines and requirements. Answers are given following the questions.

REFERENCES

1. Baseman H. Aseptic process validation and aseptic process simulation studies. In: Lysfjord J. ed. Practical Aseptic Processing, Fill and Finish. Vol 2. Bethesda, MD: Parenteral Drug Assocation/Davis Healthcare, 2009:49–79.
2. Guidance for Industry: Sterile Drug Products Produced by Aseptic Processing—Current Good Manufacturing Practice, FDA, September, 2004. http://www.fda.gov/downloads/Drugs/Guidance ComplianceRegulatoryInformation/Guidances/ucm070342.pdf. Accessed June 12, 2010.
3. EudraLex—Vol 4, Good manufacturing practice (GMP) guidelines, part 1—Basic requirements for medicinal products, Annex 1, Manufacture of sterile medicinal products, revision November, 2008. http://ec.europa.eu/enterprise/sectors/pharmaceuticals/files/eudralex/vol-4/2008_11_25_gmp-an1_en.pdf. Accessed June 12, 2010.
4. http://www.hospital-int.net/categories/aseptic-connection-devices/aseptic-connection-devices.asp. Accessed November, 2009. Accessed June 12, 2010.
5. Knapp JZ, Kushner HK. Generalized methodology for evaluation of parenteral inspection procedures. J Parenter Drug Assoc 1980; 34:14.
6. Akers et al. Parenteral Quality Control. New York: Marcel Dekker, 2003:231.

BIBLIOGRAPHY

Carleton FJ, Agalloco JP, eds. Validation of Aseptic Pharmaceutical Processes. 3rd ed. New York, NY: Informa Healthcare, Inc, 2008.

Current practices in the validation of aseptic processing–2001, Technical Report No. 36, PDA J Pharm Sci Tech, 56 (3), May–June, 2002.

Process Simulation Testing for Aseptically Filled Products. Technical Report No. 22. PDA J Pharm Sci Tech 1996; 50:S1–S16. (Note: A draft revision was published in March 2009).

Quality risk management for aseptic processes. Technical report No. 44. PDA J Pharm Sci Tech 2008; 62(suppl 1):2–39.

APPENDIX A

QUESTIONS ON ASEPTIC PROCESSING AND ASEPTIC PROCESS VALIDATION

1. What are the two general categories of manufacturing operations for sterile products (FDA)?
2. With respect to airborne particulate classifications, what are the differences between the FDA guidelines and the EU guidelines? Extra credit: How long should "in operation" state return to "at rest" state according to EU?
3. What are differences between FDA and EU guidelines with respect to limits for microbial contamination in the different clean areas?
4. What is recommended by FDA that will determine the absence or presence of turbulence in the aseptic processing line or clean zone?
5. What air classification should solutions being prepared prior to filtration be located (EU)?
6. What is the minimal air classification required for location of isolators (FDA and EU)?
7. What is the maximum number of hours that a settle (fallout) plate should be exposed (EU)?
8. What environmental grade should products intended for terminal sterilization be located according to EU?
   a. "Slow filling operations"?
   b. Products that support microbial growth?
   c. All other products?

9. Besides filter leak testing semiannually, what must be done more periodically to monitor performance of HEPA filters (FDA)?
10. What basic sciences should personnel involved in aseptic processing be trained on (EU and FDA)?
11. What are the two main requirements for clean room construction of floors, walls, and ceilings (FDA and EU)?
12. What should be fitted between a drain and the machine or sinks it serves (EU)? Are drains permitted in Grade A environments? Or Grade B environments?
13. What is the pressure differential guidance (both in U.S. units and European units) between adjacent rooms of different grades?
14. What should happen after equipment maintenance has occurred within a clean room (EU)?
15. What does FDA state is improper to do prior to environmental sampling of gloves on operators?
16. What is a unique requirement for disinfectants and detergents used in Grades A and B areas (EU)?
17. What is the requirement for validation of glass container depyrogenation (FDA)?
18. Why is there a time limitation between washing and sterilization of components and equipment (two answers) (FDA and EU)?
19. What is the minimum requirement for (a) the number of separate media fills required initially to qualify a new filling line or process; (b) the number of revalidation runs per shift and processing line; and (c) the number of media fills that each person involved in aseptic processing should be part of (FDA)?
20. What is the minimum number of container units to be filled during a media fill (FDA)?
21. What is to be done with a media-filled container if it is found to be defective (a) prior to incubation; (b) during or after incubation (FDA)?
22. What is the minimum acceptable contamination rate for a media fill (FDA and EU)?
23. What tests should be conducted prior to filtration sterilization (EU)?
24. What is the challenge a filter must pass in order to be a validated filter system for a given product? Can you name the three expectations FDA and other regulatory authorities now expect to review when reviewing filter system validation?
25. True or False. FDA will not allow filter validation to be done without filtering the actual product and actual process conditions?
26. What is the minimum sterility assurance level required by FDA for sterilization of a "load"?
27. What items does FDA indicate are most difficult to sterilize?
28. What two physical parameters must be monitored during moist heat sterilization (EU)?
29. When should critical surface sampling be performed and where should samples be taken during production activities (FDA)?
30. True or False. The filtration process should be complemented by some degree of heat treatment (EU)?
31. True or False. Double filtration is "advisable" (EU)?
32. True or False. Parametric release is recognized in the United States, but not in Europe?
33. For sampling of aseptically filled products for sterility testing, when during the process should samples be taken (FDA and EU)?

ANSWERS TO QUESTIONS ON ASEPTIC PROCESSING AND ASEPTIC PROCESS VALIDATION (From Refs. 2 and 3)

1. What are the two general categories of manufacturing operations for sterile products? (FDA)

   Terminal sterilization and aseptic processing

2. With respect to airborne particulate classifications, what are the differences between the FDA guidelines and the EU guidelines?

   FDA units in cubic feet while EU units in cubic meters
   EU requires limits for particles ≥5 microns while FDA does not
   EU differentiates rooms "at rest" and "in operation"

FDA has four separate classifications of rooms based on particle limits while EU has only three.

Extra credit 15–20 minutes:

3. What are differences between FDA and EU guidelines with respect to limits for microbial contamination in the different clean areas?

   FDA Class 100 value in CFU/cubic meter is "1" (expect 0)

   EU Grade A value in CFU/cubic meter is <1

   EU requires four different measurements—quantitative air, settle plate, contact plate on surface, and contact plate on fingers—while FDA has two measurement requirements—quantitative air and settle plate.
4. What is recommended by FDA that will determine the absence or presence of turbulence in the aseptic processing line or clean zone?

   Smoke studies documented by videotaping the area in question
5. What air classification should solutions being prepared prior to filtration be located (EU)?

   Grade C
6. What is the minimal air classification required for location of isolators (FDA and EU)?

   FDA: Class 10,000 to Class 100,000; EU: Grade D
7. What is the maximum number of hours that a settle (fallout) plate should be exposed (EU)?

   Less than 4 hours. FDA used to state 4 hours, but now no longer gives a value.
8. What environmental grade should products intended for terminal sterilization be located according to EU?
   a. "Slow filling operations"?—*Grade A with a Grade C background*
   b. Products that support microbial growth?—*Grade C*
   c. All other products? *Grade C*
9. Besides filter leak testing semiannually, what must be done more periodically to monitor performance of HEPA filters (FDA)?

   Uniformity of velocity across filter face and relative to other filters
10. What basic sciences should personnel involved in aseptic processing be trained on (EU and FDA)?

    Microbiology and hygiene. FDA also identifies "aseptic technique," clean room behavior, gowning, patient safety hazards, and SOPs.
11. What are the two main requirements for clean room construction of floors, walls, and ceilings (FDA and EU)?

    FDA: Smooth, hard, easily cleaned, and sanitized

    EU: Smooth, impervious, and unbroken
12. What should be fitted between a drain and the machine or sinks it serves (EU)? Are drains permitted in Grade A environments? Or Grade B environments?

    An air break; drains are not permitted in either Grade A or B environments
13. What is the pressure differential guidance (both in U.S. units and European units) between adjacent rooms of different grades?

    US; 12.5 Pascals (0.05 inches of water) with the doors closed

    EU: 10–15 Pascals
14. What should happen after equipment maintenance has occurred within a clean room (EU)?

    Clean, disinfect, and/or sterilize the room
15. What does FDA state is improper to do prior to environmental sampling of gloves on operators?

    Sanitizing them
16. What is a unique requirement for disinfectants and detergents used in Grades A and B areas (EU)?

    Should be sterile prior to use (FDA also adopted this)

17. What is the requirement for validation of glass container depyrogenation (FDA)?

    3-log reduction in endotoxin challenge

18. Why is there a time limitation between washing and sterilization of components and equipment (two answers) (FDA and EU)?

    FDA: Concerns both about bioburden and endotoxin
    EU: No reason given

19. What is the minimum requirement for (a) the number of separate media fills required initially to qualify a new filling line or process; (b) the number of revalidation runs per shift and processing line; and (c) the number of media fills that each person involved in aseptic processing should be part of (FDA)?
    (a) *Three*
    (b) *Semiannual*
    (c) *Once a year*

20. What is the minimum number of container units to be filled during a media fill (FDA)?

    Starting point range is 5,000 to 10,000 units unless batch size lower than 5,000, in which case the media fill units are the same as the batch size

21. What is to be done with a media-filled container if it is found to be defective (a) prior to incubation; (b) during or after incubation (FDA)?

    Defects not related to integrity (cosmetic) should be incubated; units lacking integrity should be rejected. Damage during incubation should be included in final data analysis.

22. What is the minimum acceptable contamination rate for a media fill (FDA and EU)?

    FDA: "Approaching zero"; 0 out of <5,000 units; 1 out of 5,000–10,000 units; 1 out of >10,000 units. Any failure should be investigated for assignable cause.
    EU: "<0.1% with 95% confidence"

23. What tests should be conducted prior to filtration sterilization (EU)?

    Bioburden and filter integrity testing

24. What is the challenge a filter must pass in order to be a validated filter system for a given product? Can you name the three expectations FDA and other regulatory authorities now expect to review when reviewing filter system validation?

    Microbial retention of 10^7 Brevidumonas diminuta cells per cm^2 filter surface area plus proof that properties of drug product do not affect microbial retention and filter does not leach extractables into product

25. True or False. FDA will not allow filter validation to be done without filtering the actual product and actual process conditions?

    False. Must justify any divergence from a simulation using the actual product and process

26. What is the minimum sterility assurance level required by FDA for sterilization of a "load"?

    1×10^{-6} (no higher probability of contamination than 1 in 1 million)

27. What items does FDA indicate are most difficult to sterilize?

    Filter installations in piping, tightly wrapped or densely packed supplies, securely fastened load articles, lengthy tubing, sterile filter apparatuses, hydrophobic filters, and rubber stoppers

28. What two physical parameters must be monitored during moist heat sterilization (EU)?

    Temperature and pressure

29. *When should critical surface sampling be performed and where should samples be taken during production activities (FDA)?*

    At conclusion of operation and taken at actual working site and at locations where significant activity or product exposure occurs. Consider where contamination risk is, for example, difficult setups, length of processing time, and impact of interventions

30. True or False. The filtration process should be complemented by some degree of heat treatment (EU)?

    True

31. True or False. Double filtration is "advisable" (EU)?
    True. FDA also supports this.
32. True or False. Parametric release is recognized in the United States, but not in Europe?
    False. EU also recognizes possibility of parametric release
33. For sampling of aseptically filled products for sterility testing, when during the process should samples be taken (FDA and EU)?
    Both state that samples must be taken at beginning, middle, and end of process and after any intervention or excursion during process.

22 | Inspection, labeling, and secondary packaging

After a product is filled into a primary container, if the container is a vial, it is stoppered (partially inserted for lyophilized products) and crimped with an aluminum seal. If the container is a syringe, it is stoppered with a rubber plunger. If the container is an ampoule, it is heat sealed. Products filled into glass or plastic bottles are rubber stoppered and sealed and products filled into plastic bags are heat sealed, usually employing form-fill-finish technology. However, for all these products, before final labeling and placement in secondary packaging for distribution, inspection for product defects must be conducted. This is part of the quality control and assurance program to ensure that only product dosage forms meeting all current good manufacturing practice (cGMP) attributes are released and available to the marketplace consumer.

The United States Pharmacopeia (USP) requirement for injectable products specifies that (1).[1]

> Each final container of all parenteral preparations shall be inspected to the extent possible for the presence of observable foreign and particulate matter ("visible particulates") in its contents. The inspection process shall be designed and qualified to ensure that every lot of all parenteral preparations is essentially free from visible particulates. Qualification of the inspection process shall be performed with reference to particulates in the visible range of a type that might emanate from the manufacturing or filling process. Every container whose contents shows evidence of visible particulates shall be rejected.

Inspection for visible foreign particulate matter is an extremely important function of final product inspection of sterile product solutions, but it is not the only potential product defect to look for. Detection and measurement of particulate matter in sterile products is covered in chapter 29.

Each company has its own specific criteria for what is classified as a defect and the degree of severity for each defect. For example, some companies use defect classifications such as critical, major, and minor. Some companies use subcategories of major and minor defect types. There is general agreement on what can be classified as a critical defect. Critical defects universally include missing closures, severely cracked glass, evidence of microbial contamination, and obviously bad seals.

Table 22-1 summarizes defect classifications, criteria, examples, and a general acceptable quality limit. Tables 22-2, 22-3, and 22-4 provide examples of classifications of product defects for vials and cartridges, for syringes, and for lyophilized powders, respectively. These are not one company's classifications, but a compilation by the author based on many years of teaching and interaction with industry experts.

The body of published literature on visual inspection of sterile products focuses on inspection in a production environment, where the objective is valid accept/reject decisions on individual units of product, where defective units are discarded. In a product development environment, the objective of inspection is not making accept/reject decisions, but rather on gathering information on the suitability of parenteral formulations, container/closure systems, and processing methods. Also, in a production environment, there is no concern over the development of particulate matter during storage of the product, whereas this is very much a concern in a product development environment (2).

The probability of detecting particulate matter by visual inspection depends on the conditions used to observe product; that is, the lighting source, lighting intensity, lighting uniformity, background, and pacing of the inspection process. The survey cited earlier (2) pointed out large discrepancies in visual inspection conditions used during product development.

[1] See discussion later in this chapter about a potential revision in the wording of this section of USP General Chapter Injections <1>.

Table 22-1 Example of Defect Classifications and Inspection Recommendations

Defect classification	Criteria	Examples of defects	Acceptance quality limit	Accept/reject number/315 samples
Critical	Life threatening when used as directed. Would trigger a product recall	Cracks, bad seals, missing stoppers, microbial contamination	Zero	0/0
Major A	Nonlife threatening. Would trigger a product complaint	Visible foreign particulate matter	0.25%	1/1
Major B	Could trigger a product complaint	Discoloration	0.65%	2/3
Minor	Could lower customer perception of quality if noticed	Scratches, crooked seals	2.5%	8/9

There is a need for a guidance document—perhaps an informational chapter in USP—aimed at establishing more consistent, and more scientifically defensible, practices for visual inspection, particularly in a product development setting. There is also need, at the time of this publication, for other questions to be answered and conflicts to be resolved; for example:

- On-line inspections require 100% review of each product container with every container seen with visible particles being rejected. Release testing relies on a relatively small sample of product units from the entire batch that are inspected with allowance for a certain number of product units (albeit very small) containing visible particles for the batch to be released. This seems to be a contradictory situation.
- How are the compendial requirements "essentially free from particles" to be interpreted?
- Is there any possibility that inspection methodology for visual particulates will be harmonized worldwide?
- Is there any possibility that "haze" testing can be harmonized and acceptable to all compendia?

USP Chapter <788>, dealing with subvisible particulate matter in parenterals, allows up to 6000 and 600 particles per container for particles greater than or equal to 10 μm and greater than or equal to 25 μm, respectively. In time, it is likely that these compendial acceptance limits will be reduced. Industrial production environments, processes, and personnel training and practices all have improved significantly such that typical particle counts rarely exceed 1000 $\geq$ 10 μm and 100 $\geq$ 25 μm per container.

CURRENT PERSPECTIVES IN VISUAL INSPECTION PRACTICES FOR STERILE PRODUCTS

Inspection of solutions for visible particles has reached new heights of regulatory priority and scrutiny. The FDA and other regulatory compliance groups demand higher levels of product quality with respect to visible particles. The industry itself struggles to define quality policies related to visible particles. Particles are a ubiquitous problem and exist in all sterile products. Ideally, no sterile solution containing visible particles is ever released for commercial sale, yet this does happen (otherwise the acceptance quality limit would always be zero). Subvisible particles ($\leq$25 μm) exist, but there are regulatory limits to control how many exist in a sterile product administered to humans and animals. Where is the line drawn with respect to acceptable or unacceptable product quality with respect to particulate matter?

There are significant struggles with interpreting the pharmacopeias with respect to visible particle testing for parenterals. All sterile products are subjected to a 100% inspection, typically automated to some degree. However, each unit of product is viewed/analyzed by a human inspection for only a second or two (at most maybe 5 seconds, with the exception of Japan which may inspect for up to 10 seconds per unit of product!). Following 100% inspection,

Table 22-2 Example Descriptions of Defects in Vial and Cartridge Primary Containers

General classification	Defect category	Description
Container defects	Bruises	Small but noticeable defects in/on glass
	Stones	Stones or foreign embedded material in glass
	Blisters	Air bubbles in glass
	Broken/fractured	Broken or fractured glass
	Discolored	Discoloration of container
	Leaking	Product leaking from broken or defective container
	Cosmetic	Heavy seams, lines rough surface, or other appearance discrepancies
Vial cap defects	Cracked	Cap is cracked
	Scratched	Cap has excessive scratches
	Loose	Cap is loose
	Missing	Cap is missing
	Incorrect color	Cap of a color other than that specified by the batch record
Vial seal defects	Dented	Significant dents in seal
	Crimp	Improper crimp; no crimp; partial crimp; loose fit, jagged edges
	Dirt	Excessive dirt or residue on or around seal. Dried solution under the metal ring may indicate leakage
	Imprint	Illegible printing on seal
Cartridge seal defect	Defective crimp	Improper crimp; no crimp; partial crimp; loose fit, jagged edges
Rubber closure defects	Embedded material	Foreign material embedded in closure
	Damaged	Incomplete stopper or other functional defect
	Missing	No stopper present
	Particles	Visible particulate matter on the surface of the stopper (in contact with product)
	Smudges/streaks	Streaks or smudges on closure
	Discolored	Stains or variation in normal color
Cartridge disc seal defects	Septum	Flat septum, square septum, bulging septum
Liquid product defects for vial/cartridge	Fibers	Visible particulate contaminant
	White particles	White or light-colored particles visible in the product (if product is monoclonal antibody, small amounts of particles might be acceptable, depends on product
	Dark particles	Dark colored or black particles visible in product
	Metallic particles	Visible particles that appear to be metallic
	Glass particles	Visible particles that appear to be glass fragments
	Cloudy/hazy	Cloudiness or a "tornado" effect caused by many small particles in solution
	Precipitated/ crystallized	Product contains insoluble material
	Discolored	Color of product not within specifications
	High fill	Level of product in container higher than specification
	Low fill	Level of product in container lower than specification
Liquid product defects for cartridge only	Bubble	Air bubble in the cartridge that exceeds limit defined by ISO 11608
	Partial fill/fill volume low	Cartridge "shoulder" empty; fill volume too low
	Empty	Cartridge without product
Powder product defects	Discolored	Discoloration of product
	Particles	Fibers or other solid different from the bulk of the powder, clumps, or particles larger than the bulk powder particles
	Dark particles	Dark colored or black particles visible in product
	Metallic particles	Visible particles that appear to be metallic
	Glass particles	Visible particles that appear to be glass fragments
	High fill	Level of product in container higher than specification
	Low fill	Level of product in container lower than specification
	Empty	No product in vial

Table 22-3 Example Descriptions of Syringe Product Defects

General classification	Defect category	Description
General	Broken or cracked syringe barrel, broken or cracked tip	Crack on syringe barrel or on syringe's tip
	Broken flange	
	Cracked flange	Crack on syringe flange
	Foreign particle	
	Leak	Leaking needle or plunger head
Dose	Empty syringe	No product in the syringe
	Low dose	Volume lower than specification
	High dose	Volume greater than specification
Clarity	Fiber product	Visible fibers in product
	Particle product	Visible black, metallic or other particles
	Glass particles	Visible glass particles in product
Needle guard	Needle guard incorrectly installed or missing	Needle guard incorrectly installed or torn, compromising product integrity
	Twisted needle guard	Twisted or bent needle compromising use
	Needle guard incorrectly molded	Defect of molding or incorrect trimming
	Particle or stain	Presence of particle or stain that cannot be removed
Plunger head (rubber)	Plunger head assembly misplaced	Wrong position in the barrel compromising packaging in a combination package
	Stained plunger head (nondetachable)	Stains can be either product contact or non-product contact
	Particle in plunger head (detachable)	Particle can be either in contact with product or on first lip of the plunger head
	Product in plunger head	Product bridging the ribs of the stopper/plunger head
	Creased plunger head (Flurortec rubber only)	A crease that extends above the Flurotec coating that may affect container integrity
Syringe	Scratched syringe	Deep scratch (detectable by feel)
	Defective glass	Detectable by feel
	Excess of material	Excess of glass material, irregular cut
	Stained syringe	Colored stain embedded in glass barrel or on internal surface of glass barrel
		Colored stain on external side of glass barrel that cannot be removed
Plunger rod	Particle/stain plunger rod	Presence of particle or stain between plunger rod and syringe barrel
	Colored fiber plunger rod	Presence of colored fiber between plunger rod and syringe barrel
	Scratched plunger rod	Scratch on plunger rod

product lots are subjected to a more demanding release test, described in the compendia. In fact the compendia have the following definitions:

USP: "essentially free from particles that can be observed on visual inspection" (1).
EP: "clear and practically free from particles" (3).
JP: "clear and free from readily detectable foreign insoluble matter" (4).

Here are some examples of questions that the sterile product industry as a whole has to deal with:

1. How are 100% on-line inspections separated from release testing where on-line inspections require rejection of all product units that contain particles while release testing require that product units "be essentially free from particles"?
2. How are the compendial requirements "essentially free from particles" to be interpreted?

Table 22-4 Example Description of Lyophilized Product Defects

General classification	Defect category	Description
Container defects	Bruises	Small nicks in glass caused by impact
	Stones	Stones or foreign embedded material in glass
	Blisters	Air bubbles in glass
	Broken/fractured	Broken or fractured glass, cracks
	Discolored	Discoloration of container
	Cosmetic	Heavy seams, lines rough surface, or other appearance discrepancies
Stopper defects	Embedded material	Foreign material embedded in closure
	Damaged	Incomplete stopper or other functional defect
	Missing	No stopper present
	Particles	Visible particulate matter on the surface of the stopper (in contact with product)
	Smudges/streaks	Streaks or smudges on closure
	Discolored	Stains or variation in normal color
	Product on stopper	Loose, dried product on stopper or neck of vial above the vent of the stopper
	Cosmetic product on stopper	Loose, dried product on stopper or neck of vial above the vent of the stopper—rejection depends on specific product
Seal defects	Dented	Significant dents in seal
	Crimp	Improper crimp; no crimp; partial crimp; loose fit, jagged edges
	Dirt	Excessive dirt or residue on or around seal. Dried solution under the metal ring may indicate leakage
	Imprint	Illegible printing on seal
Product defects	Fibers	Visible particle contamination
	White particles	White or light colored particles visible in product, different from the bulk of the powder
	Dark particles	Dark colored or black particles visible in product, different from the bulk
	Metallic particles	Visible particles that appear to be metallic, different from the bulk of the powder
	Glass particles	Visible particles that appear to be glass, different from the bulk of the powder
	Discolored	Color of product does not match description in batch record
	High fill	Cake height is obviously larger than the other cakes in the batch, with no other cake defects (puffing, amorphous, or layered)
	Low fill	Cake height is obviously smaller than the other cakes in the batch, with no other cake defects (puffing, amorphous, or layered)
Cake defects	Collapse	Cake is smaller than usual and may have rounded or raised edges at the top of the cake
	Meltback	Cake appears as a small glob or bubbles and may appear moist. Meltback is a severe form of collapse
	Partial meltback	Part of cake may appear normal while other parts may seem to be missing or melted. Partial meltback will most often be seen at the bottom of the vial
	Amorphous or incomplete	Cake does not have typical cylindrical shape. Product may be dried on the shoulders of the vial with or without a pit in the center of the cake
	Shrunken plug	Cake has been abnormally pulled from the sides of the vial, but has a cylindrical shape and appears to be dry
	Puffing	Cake appears swollen, top surface may be grossly uneven
	Discolored	Cake color is spotty, uneven, or inconsistent
	Layered	Cake has clearly different horizontal layers. Differences may be in shade of color or structure of cake

Table 22-5 Examples of Responses To The Question "What is the Meaning of 'Essentially free' of Visible Foreign Particulate Matter?"

- "Essentially free" means free from product-related or particulate matter from interactions with the container/closure system. There are no numerically based criteria
- We have used the British Pharmacopeia Supplemental Chapter 1 N, suggesting that $\geq$ 19 units of each 20 should be free of visible particles
- No more than 3 particles per 5 mL of product
- Less than or equal to three particles per vial
- If the product passes our standard visual inspection, then it is considered to be essentially free of particulates
- The material is considered essentially free of visible particulate matter if we cannot see particles without the aid of a magnifying lens
- "Essentially free" = "free"
- "Essentially free" means that no particles are detected under a defined set of observation conditions; that is, using a defined light source, light intensity, background, and inspection rate, with no magnification

Source: From Ref. 2.

3. Is there any possibility that inspection methodology for visual particulates will be harmonized worldwide?
4. Is there any possibility that "haze" testing can be harmonized and acceptable to all compendia?

There is considerable diversity of opinion of what "essentially free" means. Table 22-5 provides a sampling of responses from a recent survey of pharmaceutical scientists involved in development of parenteral products (some responses, e.g., no more than three particles per 5 mL do not make sense for visible particle inspections, but the responses were anonymous) (2). These responses further exemplify how confusing and diverse are the interpretations of this USP statement "essentially free." Nevertheless, FDA presentations have indicated that quality control of visible particulate matter is of great concern to inspectors. Industry and government need to continue to work together to agree on acceptable practices and standards in inspection practices for visible particulate matter.

Warning Letters from the FDA to manufacturers of injectable products reveal the following specific observations with respect to concerns about visible particulates:

- For the same lot there is sometimes a high rejection rate by certain personnel while other personnel have low rejection rates.
- The results from statistical sampling at the beginning, middle, and end of a filling operation found no defects whereas the entire lot was rejected for release because of visual defects during 100% inspection.
- Failure to establish a maximum acceptable level of vials rejected during 100% inspection.
- Failure to take adequate action to ensure the quality of released product when the particulate reject limit was exceeded.
- Personnel responsible for detection of visible particulate matter must be thoroughly trained for this important quality evaluation. Training is not an easy task because of a variety of reasons: vision capabilities, concentration, sample standards of particulate types, inspection environment, and qualifications of the trainer(s).

PROPOSAL TO REVISE USP GENERAL CHAPTER <1>

In response to all the concerns raised earlier about inspections for visible particulate matter in injectable products, a proposal was published in *Pharmacopeial Forum* at the end of 2009 (5). The authors stated that a more precise definition of the term "essentially free" was desirable to prevent misunderstanding of what this term really means and how it applies globally. Here is the proposal definition: "Where used in this Chapter (USP <1>) the term essentially free means that when the batch of Injection is inspected as described (within a subheading under Foreign and Particulate Matter under the heading Visible Particulates in Injections) no more than the specified number of units may be observed to contain visible particulates."

The proposed wording of the section "Visible Particulates in Injections" is as follows:

> This test is intended to be applied to product that has been 100% inspected as part of the manufacturing process; it is not sufficient for batch release testing alone, and a complete program for the control and monitoring of particulate matter remains an essential prerequisite. This includes dry sterile solids for injection when reconstituted as directed in the labeling. Other methods that have been demonstrated to achieve the same or better sensitivity for visible particulates may be used as an alternative to the one described below. Injections shall be clear and free from visible particulates when examined without magnification (except for optical correction as may be required to establish normal vision) against a black background and against a white background with illumination that at the inspection point has an intensity between 2000 and 3750 lux. This may be achieved through the use of two 15-W fluorescent lamps (e.g. F15/T8). The use of a high-frequency ballast to reduce flicker from the fluorescent lamps is recommended. Higher illumination intensity is recommended for examination of product in containers other than those made from clear glass. Before performing the inspection, remove any adherent labels from the container and wash and dry the outside. The unit to be inspected shall be gently swirled, ensuring that no air bubbles are produced, and inspected for approximately 5 s against each of the backgrounds. The presence of any particles should be recorded. For batch-release purposes, sample and inspect the batch using ANSI/ASQ Z1.4 (2008)[2] General Inspection Level II single sampling plans for normal inspection, AQL 0.65. Not more than the specified number of units contains visible particulates. For product in distribution, sample and inspect 60 units. Not more than one unit contains visible particulates.

PERSONNEL

Personnel responsible for detection of visible particulate matter must be thoroughly trained for this important quality evaluation. Training is not an easy task because of a variety of reasons: vision capabilities, concentration, sample standards of particulate types, inspection environment, and qualifications of the trainer(s).

The following sections will discuss current practices, procedures, issues and trends with respect to personnel training and qualification, inspection of different products, and establishment of inspection criteria and limits.

Training/Qualification of Inspectors

All inspectors should be trained and evaluated based on objective standards. Examples of acceptable and defective containers are very useful for training inspectors especially for defect types such as minor blemishes on a glass vial, which are very subjective. Regular inspectors as well as production representative or Quality Assurance personnel performing the inspection of statistical samples to verify inspection effectiveness should be trained in the same manner. Qualification of inspectors should be conducted at the same speed at which regular inspections will be carried out.

The human inspector determines the quality and success of the manual inspection process. Since the inspection process is subjective in nature, the main limitation of the process lies with restriction in the vision, attitude, and training of the individual inspector.

As a minimum standard, personnel assigned as inspectors should have good vision, corrected, if necessary, to acceptable standards. Inspectors should not be color-blind. Visual acuity should be tested at least on an annual basis.

Since the number and size of particles in parenteral solutions have become important characteristics to evaluate, it has been assumed that particles larger than 40 or 50 μm are detectable by the unaided eye. Thus, in complying with the USP requirements that any container showing visible evidence of particulate matter be rejected, it must be assumed that the average inspector will pass those solutions containing particles with a size > 40 μm. This, of course, presents some discomfort for those who believe that particulate matter, especially in the size range of 10 to 40 μm, is clinically hazardous.

[2] Two other inspection standards are MIL-STD-105E and ISO 2859-1 (1999).

Table 22-6 Size of Particles of Varying Probability Levels[a]

Particle concentration	Particle size (μm) 50% chance	Particle size (μm) 100% chance
USP limit 50 particles/mL[b]	18.82	51.45
USP limit 5 particles/mL[c]	19.96	54.88
1 mL ampul, 1 particle	20.07	55.21
2 mL ampul, 1 particle	20.08	55.25
5 mL ampul, l particle	20.09	55.28
10 mL ampul, 1 particle	20.10	55.29
20 mL ampul, 1 particle	20.10	55.29
50 mL vial, 1 particle	20.10	55.29
1 L large volume, 1 particle	20.10	55.29

[a]Arcsine $P_1 = 0.33689252 + 0.02231515$ size $+ 0.000035$ size versus concentration -0.00008694 concentration.
[b]Not more than 50 particles/mL equal to or larger than 10 μm.
[c]Not more than 5 particles/mL equal to or larger than 10 μm.
Source: From Ref. 6.

It is not only the size, but also, and probably more importantly, the number of large particles injected into man intravenously that is considered dangerous. Thus, official standards have been enforced for maximum allowable numbers of certain-sized particles in parenteral solutions.

At least one attempt has been made to quantify the size and concentration of particles that can be detected by the unaided eye (6). Five-milliliter ampuls containing 10 to 500 particles per mL of particle sizes between 5 and 40 μm (using polystyrene beads) were inspected by 17 inspectors in a standard booth. Based on a multiple linear analysis model that calculated the probability of rejecting an ampul as a function of particle size and concentration, sizes of particles detected at various concentration levels at 50% and 100% probability of rejection rates were predicted. These data are reproduced in Table 22-6. The authors concluded that a 50% probability of rejection rate be achieved with 20 μm particles in sample solutions in order for potential inspectors to be qualified for in-line inspection. However, it is interesting to note that a minimum particle size of 55 μm was required for all inspectors to reject all solutions containing this size of particle.

Good attitude and concentration cannot be overemphasized. One of the major limitations of human inspection for particulate matter is reduced efficiency of the individual because of a lack of concentration. This can easily occur if the inspector suffers from extreme worry or other distraction resulting from outside personal pressures. Obviously, emotional stability is an important criterion in selecting inspectors.

Fatigue also becomes a major limitation of human inspection. Personnel should be provided appropriate relief from the inspection function by rotating jobs and allowing for rest periods.

Formal training programs must precede the acceptance of an individual as a qualified inspector. The training program should include samples of both acceptable and unacceptable product containers that must be distinguished by the trainee. During the training period, all units inspected by the trainee should be reinspected by qualified inspectors to ensure the quality of the inspection and the development of the trainee. After the inspector has passed his/her training period, performance tests should be done at random intervals to ensure that quality standards are being maintained. Personal experience plus some older literature reports (not cited) supports the logical conclusion that the more training and experience an inspector accumulates, the better the discrimination ability of the inspector to detect particles and other defects in finished products.

Methodology

Most inspection processes are referred to as off-line inspections, in which the inspection procedure occurs at the completion of the manufacturing, filling, and sealing process. In-line

Table 22-7 Basic Procedure for Manually Inspecting Clear Solutions for Visible Evidence of Particulate Matter and Other Defects

1. Container of parenteral solution must be free of attached labels and thoroughly cleaned. Use a dampened nonlinting cloth or sponge to remove external particles
2. Hold container by its top and carefully swirl contents by rotating the wrist to start contents of the container moving in a circular motion. Vigorous swirling will create air bubbles, which should be avoided. Air bubbles will rise to the surface of the liquid; this helps to differentiate them from particulate matter
3. Hold the container horizontally about 4 in. below the light source against a white and black background. Light should be directed away from the eyes of the inspector and hands should be kept under the light source to prevent glare
4. If no particles are seen, invert the container slowly and observe for heavy particles that may not have been suspended by swirling
5. Observation should last for about five seconds each for the black and white background
6. Reject any container having visible particles at any time during the inspection process

inspection of container components can also be done, especially if the production process can be suitably adapted to achieve the desired results without increasing the risk of microbial and particulate contamination. Obviously, the removal of defective containers, such as those showing cracks or the presence of particles, prior to the filling of the product ensures product quality and minimizes loss of expensive drug products.

Standard operating procedures for inspection of parenteral containers depend on the kind of container inspected, that is, procedures will be slightly different for ampuls than for large-volume glass bottles, for amber vials than for flint vials, and for plastic bags than for glass containers. However, a basic procedure can be followed regardless of the type or size of container, and an example of such a procedure is given in Table 22-7.

PRODUCT

Tubing Versus Molded Vials

The type of container and closure used can have a profound impact on the quantity of product rejected in a visual inspection. Molded vials tend to have a wide variety of cosmetic flaws thus making it more difficult to train inspectors to identify each type and classify as acceptable or defective during manual inspection. To complicate matters, different glass manufacturers do not always use consistent terminology to describe each defect type, so it is difficult to train operators to classify defects consistently for trending purposes. Molded vials generally have a much higher reject rate for cosmetic flaws than tubing vials. Therefore, it is recommended that products being marketed in Japan or other markets that require an excellent appearance should be filled into tubing vials whenever possible.

Preinspection of Containers

If molded vials are used, and the expected reject rate in the final product is unacceptable due to extremely valuable or scarce products, the containers can be inspected prefilling to reduce finished product rejects. Preinspection should ideally be conducted under the same conditions and by the same personnel who will be inspecting the finished product containers. This ensures that the same criteria are applied for identifying reject containers. There are also psychological factors to consider when preinspecting vials—one is that post-filling, if the inspectors are not informed that the batch they are inspecting was filled into preinspected containers, there may be a tendency to be more critical—if they always have a 1% rejection rate for container defects, there may be a tendency to reject at this level because of a fear of "missing" something. If the same operators are employed postfilling, and are informed that the batch to be inspected was filled into the previously inspected containers, the container defect reject rate may be much lower due to the inspectors' expectations.

The costs of preinspecting the glass must be carefully weighed against its benefits. The additional handling and repackaging of the containers can create additional blemishes and scratches as well as increasing the particle and bioburden levels. If the preinspection is to take place at a location distant from the glass washing and depyrogenation area, the postinspection repackaging, storage, and transportation must be considered. Another critical consideration is how the preinspected glass units will be differentiated by warehouse personnel from noninspected containers of the same type. An alternative to preinspection is to make special arrangements with the glass manufacturer to have the glass meet a stricter standard of quality.

It is also important to have realistic expectations that even with preinspection, some additional defects may be detected in the finished products due to scratches or breakage generated during the process or flaws that were not rejected in the initial inspection. Preinspection can sometimes be avoided if procedures are in place specifying that certain cosmetic container defects can be accepted in situations such as clinical trial materials where marketability is not an issue. One then must specifically decide which defects are acceptable and convey this to the inspection operators. This then can become problematic, especially in a multiproduct facility, as the inspectors must then be informed as to which set of criteria to apply to any given batch and to essentially disregard their normal procedures and training in these special cases—not an easy thing for a trained and experienced inspector to do.

Particles and Other Defects

Anything that directly or indirectly comes in contact with a parenteral solution, including the solvent and solutes composing the solution itself, represents a potential source of particle contamination. Table 22-8 lists common sources of particulate matter found in parenteral solutions.

The smallest capillary blood vessels are considered to have a diameter of approximately 7 μm. Thus, all particles having a size equal to or greater than 7 μm can conceivably become entrapped in and occlude a blood capillary. Most particulates, as seen in Table 22-8, potentially can be this size and, obviously, represent a hazard to the health of a patient administered parenteral medications containing these contaminants.

It seems that regardless of whatever painstaking procedures are undertaken to eliminate particle contamination, parenteral solutions always contain a certain degree of particulate matter. It is always an uncertainty whether the particles originated during the manufacturing and packaging process or were introduced during the analysis of the solution for the presence of

Table 22-8 Common Sources and Types of Particulate Matter

Source	Type or example
Chemical	Undissolved substances—starch, zinc oxide, crystalline substances
	Trace contaminants
Solvent impurities	Insoluble forms
Packaging components	Glass—alkali leachables, glass particles
	Plastic—plasticizers, other leachables and extractables
	Rubber—zinc oxide, carbon black, talc
	IV administration sets
Environmental contaminations	Air
	Surfaces
	Insect parts
	Microorganisms
Processing equipment	Glass
	Stainless steel
	Rubber
	Plastic components
	Rust
Filters	Fibers–e.g. cellulose, other polymer sources
People	Skin
	Hair
	Gowning

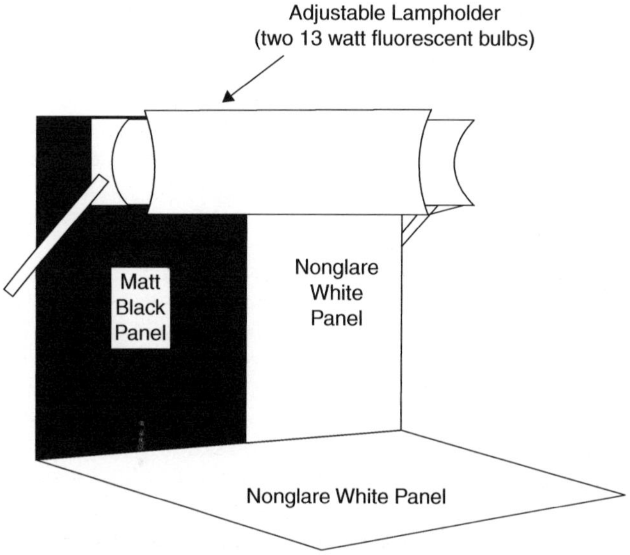

Figure 22-1 European pharmacopeia apparatus for visual inspections.

particulates. It is imperative that particles seen in solutions not have originated during the particle measurement and identification procedures.

PROCEDURES

Manual Inspections

Manual inspection by human inspectors for the presence of visible particulate matter in parenteral solutions remains the standard 100% inspection method both in-line and inspection of statistical samples.

A white and black background lighted with nonglaring light is the standard environment used for visual inspection of product containers. The white background aids in the detection of dark-colored particles. Light or refractile particles will appear against the black background.

Lighting may be fluorescent, incandescent, spot, and/or polarized. The most common source of light is fluorescent. The light source may be positioned above, below, or behind the units being inspected. The range of light intensity may vary between 100 and 350 foot-candles.[3] This intensity can be achieved either with one 100-watt, inside-frosted incandescent light bulb, or with three 15-watt fluorescent bulbs with the container held 10 in. from the light source. Certain types of products (e.g., colored solutions) or certain types of containers (e.g., amber) require increased light intensity over that normally used. As light intensity begins to weaken, due to age or usage, lamps should be replaced. Good practice demands that inspection lamps be monitored periodically.

The European Pharmacopeia provides a figure of the type of apparatus to be used in visible inspection of particles and other defects (Fig. 22-1). The apparatus consists of the following:

- A matt black panel of appropriate size held in a vertical position
- A nonglare white panel of appropriate size held in a vertical position next to the black panel
- An adjustable lampholder fitted with a suitable, shaded, white-light source, a suitable light diffuser (details provided in the EP), and illumination intensity maintained between 2000 and 3750 lux with higher intensities preferred for colored glass and plastic containers.

[3] The European Pharmacopeia 2.9.20 specifies light intensity of 2000–3750 lux (185–350 foot-candles) while the Japanese Pharmacopeia (General Tests 6.06) specifies approximately 1000 lux (93 foot-candles) for glass containers, 8000 lux (744 foot-candles) to 10000 lux (929 foot-candles) for plastic containers.

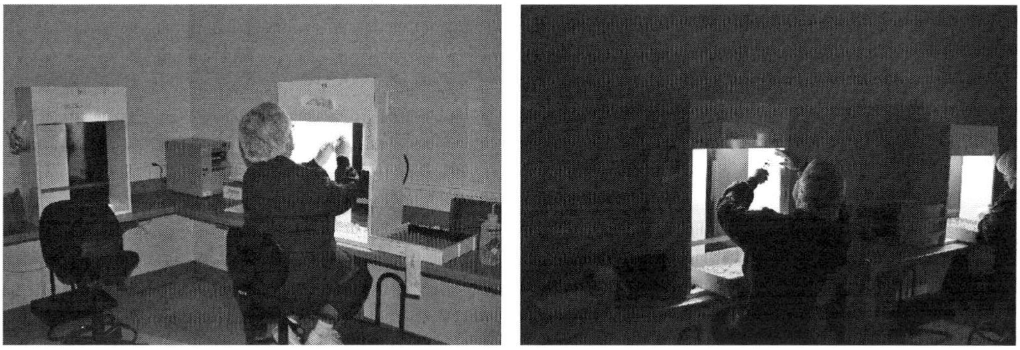

Figure 22-2 Manual visual inspections. *Source*: Courtesy of Baxter Healthcare Corporation.

A standard inspection booth contains an all-black interior except for the front entrance for the inspector (Fig. 22-2). A vertical screen in the back of the booth is half black and half white. Light usually is projected vertically with frontal blockage to protect the observer's eyes from direct illumination. A magnifying lens at 2.5× magnification may be set at eye level to aid the inspector in viewing the container in front of the white/black background. Excellent viewing is provided without distraction, and acuteness of vision is increased to improve the level of discrimination. It could be argued that the level of discrimination becomes too high, that is, containers are rejected that would not have been rejected had no magnification been used.

Inspection cabinets should have black sidewalls with a baffle to prevent the light source from impinging on the inspector's eye. Fluorescent lamps provide a better light source because these are more diffuse than incandescent lamps.

Standard operating procedures for inspection of parenteral containers depend on the kind of container inspected, that is, procedures will be slightly different for ampuls than for large-volume glass bottles, for amber vials than for flint vials, and for plastic bags than for glass containers. However, a basic procedure can be followed regardless of the type or size of container, and an example of such a procedure is given in Table 22-7.

Semiautomated to Automated Inspections

High technology strives for sophisticated automatic methodology to replace the dependency on human manual inspection. The area of technology that offers the greatest potential in replacing human examination in 100% container inspection requirements is the area of computer-controlled, automatic electro-optic systems. Such systems are rapid, nondestructive, and reproducible in their inspection of parenteral products.

Technology has made significant improvements in fully automated parenteral product inspection procedures. Disadvantages of earlier automated systems, such as lack of standardization of performance, separating marks on the outer container surface from particles inside, failures to detect underfills or empty containers, and machine variabilities, have largely been eliminated with the automated systems available today.

Video inspection employs one of two basic mechanisms for automated container inspection.

1. Using imaging optics in which the particles suspended in the solution are illuminated by a fiber optic light system and imaged on a video display. Brightwell's Micro-Flow Imaging technology is one of the most popular systems for counting, sizing, and classifying particles in liquids although at the time of this publication this technology was only being used as a research tool (7).
2. Using light scattering techniques where particles scatter light that is received by a detection system and projected onto a television camera system. The Eisai AIM system is the most widely used automated detection system (8,9).

The fact that the liquid contents are swirling while the container itself is motionless during the inspection process has a very important implication. The master picture is based on a motionless container. All scratches, printing, or other marks on either the outer or inner surface of the container are part of the master picture. Any difference between the master and any one of the subsequent comparison pictures of the single container, therefore, would be caused only by particulate matter moving within the liquid contents, reflecting light back to the camera.

Several companies offer automated inspection equipment with the two most widely used at the time of this book publication being Eisai and Seidenader.

Eisai's inspection for particles uses static diode array sensors while Seidenader used a vision (camera) sensor system. The camera takes still images and looks for patterns that are unexpected.

Eisai System

The Eisai system uses a static diode array sensor system white light as the source of detection of particles. Its sensors detect moving shadows produced by foreign material in a container of solution and cast these images onto the diode array. Each container is spun around (thousands of revolutions per minute) and stopped so that only the liquid in the container is still rotating when the container enters the light. If any foreign matter is floating and rotating in the liquid, the light transmitted through the liquid is blocked and a shadow is cast by the moving particles. Eisai systems employ a phototransistor that converts moving shadows into electrical signals. There are 24 images obtained per inspected unit. These signals are compared with preset detection sensitivity signal standards and if the standard sensitivity is exceeded, the container is rejected. The Eisai detector does not react to scratches, stains, and colors of the container or the color of the liquid contents since these are all perceived as stationary objects.

The Eisai system checks the volume of liquid in the container and can reject overfilled, underfilled, and empty containers. The shadow cast by the liquid meniscus of a properly filled container is expected to fall within a certain preset range within the inspection field. If it falls above or below this range the container is rejected. Adjustments in the Eisai system can be easily made for different sizes, color, and viscosity of the liquid contents.

The typical number of units inspected per minute with an Eisai machine is 300 to 600. Contrast this with the average experienced human inspector inspecting four units per minute for the same defects.

The conveyance and inspection mechanism of the Eisai system is shown in Figure 22-3. Containers are conveyed by the star wheel onto the inspection table, spun at a high speed, and stopped before reaching the light beam. When the container enters the light beam, the light projector and detector follow the container while liquid is still rotating inside. After one container is inspected by two sets of projectors and receptors (thus, a double inspection system) the next container is carried through the same process. Containers are moved by the screw conveyor to the sorting pendulum, where rejected and accepted units are separated. The system automatically keeps count of the number of accepted and rejected containers and displays these numbers on the display panel.

Advantages of the Eisai system include (*i*) versatility, that is, ability to handle a large variety of vial sizes, products, and viscosities, (*ii*) the adjustable sensitivity level, (*iii*) attainable speeds, (*iv*) results of performance studies, and (*v*) cost. One main disadvantage is that it cannot inspect molded vials due to imperfections in the glass that inherently occur during the molding process. Only tubing vials can be inspected by the Eisai system. Also, it takes a lot of expertise to set up the Eisai system and its many "tools" (e.g., edges, regions, blob, and pixels) for proper inspection of the particular type of primary package. Syringes commonly inspected by Eisai machines.

De la Montaigne et al. (10) provide additional coverage of the sensing mechanism, capabilities, cameras, and validation of the Eisai (and other) automated inspection system.

Probabilistic Particulate Detection Model

The probabilistic inspection model is based on the finding that particulate inspection methodologies, human or robotic, are probabilistic rather than deterministic in nature. In other words, no final container of solution is acceptable or unacceptable; rather, each final container of solution possesses a probability of being rejected for whatever inspection process is being evaluated. Rejection probabilities are determined simply by recording the number of times a numbered

Figure 22-3 Eisai automated inspection machine. *Source*: Courtesy of Eisai, Inc. and Baxter Healthcare Corporation.

container is passed and the number of times that same container is rejected during a manual or automatic inspection process. Each container accumulates an accept/reject record. If 1000 containers are inspected several times and each of the 1000 containers yields an accept/reject ratio, a histogram can be constructed plotting the number of containers in each probability group against an empirically determined rejection probability. Such a histogram is shown in Figure 22-4 and represents the cornerstone for the conversion by Knapp et al. (11–16) of particulate inspection from a craft to a science.

The abscissa in Figure 22-4 represents rejection probabilities grouped arbitrarily into 11 intervals. The ordinate represents the logarithmic number of containers (vials) within each of the 11 probability groups. For example, of the 1000 vials inspected for particulate contamination, 805 vials were found to be particulate free in each of the 50 inspections while 2 vials contained particulates that were detected in each of the 50 inspections.

The dashed lines on the lower half of the histogram show the average number of vials rejected in a single inspection or two sequential inspections in each probability group. These values are obtained from the relationship

$$P(\mathrm{M}n)_{\mathrm{i}} = P(\mathrm{MI})_{\mathrm{i}}$$

where $P(\mathrm{M}n)_{\mathrm{i}}$ is the rejection probability associated with the nth manual inspection in a probability group, $P(\mathrm{MI})_{\mathrm{i}}$ is the quantity of vials rejected in a rejection probability group in a single inspection, and n is the number of inspections of rejected material. For example, of the eight vials located in the 0.6 rejection probability $(P\ \mathrm{MI})_{\mathrm{i}}$ group, five were rejected following a single inspection while only three were rejected following two sequential inspections. This indicates that improved discrimination occurs following a reinspection of initial rejects. The reinspection was utilized as a practical response to the existence of particulates even in well-controlled parenteral manufacturing areas. From the information contained in the reinspection histogram of

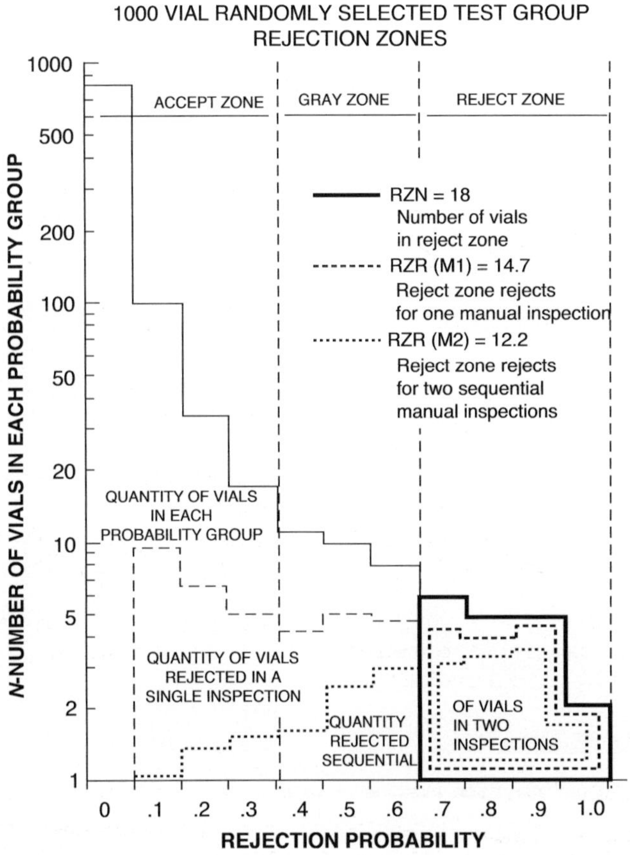

Figure 22-4 Histogram plotting number of vials per each probability of rejection group. *Source*: From Ref. 12.

Figure 22-4, Knapp and Kushner (12) defined three zones within the rejection probability limits of 0 and 1.

The accept zone contains all vials that have less than one chance in 10 of rejection in two sequential inspections. The reject zones contain all vials that have at least one chance in two of being rejected in two sequential inspections. The gray zone exists between the accept and reject zones. For single inspections, the probability limits for the three zones are seen in Figure 3.6 where

Accept zone $P \leq 0.3$
Gray zone $0.3 \leq P \leq 0.7$
Reject zone $P \geq 0.7$

Figure 22-4 also shows three terms abbreviated RZN, RZR (M1), and RZR (M2). The definitions of these terms are given in the figure (8). Using these terms a variety of parameters can be measured, including reject zone efficiency (RZE) and undesired reject rate (RAG). By definition, RZE = RZR/RZN. In the example given in Figure 22-4, the RZE after a single inspection is 81.7%. This means an 81.7% probability exists for a manual single inspection method to reject those vials known to exist in the reject zone. Matching or exceeding this objective measure of the security achieved by a manual parenteral inspection procedure should be the only GMP requirement for validation of any alternative inspection technique or process (13).

SETTING LIMITS

Defect limits define the quality of product that the manufacturer is willing to release into commercialization. Limits are defined by safety factors (e.g., potential for contamination), consumer acceptability (potential for product complaints), industry standards (state of the technology, expectations, FDA feedback, survey information, etc.), and production capabilities (clean room technology, container quality, current GMP applications, etc.). Each type of defect has a defined "acceptable quality level" (AQL). AQL is the highest percentage of a particulate defect at or below which the batch is acceptable for release to the marketplace. AQL is based on a thorough inspection of a statistically valid sample of finished product units from throughout the completed batch. It is beyond the scope of discussion in this chapter, but the probability of wrongly accepting a defective lot (Type II error) or wrongly rejecting an acceptable lot (Type I error) can be estimated for each sampling plan[4] using the operating characteristics (OC) curve (17).

Examples of AQLs for many sterile product defects are given in Table 22-1. For example, with a sample of 315 vials, and an AQL of 0.25%, finding 2 vials with visible foreign particulates will be acceptable, but finding 3 vials with particulates will cause rejection or resorting of the batch. If an AQL value for visible foreign particles is established for a manufacturer that indicates that the product itself, its package, and the manufacturing environment and process can produce such a low level of potential defective units and that this level is consistent with safety, industry practices, and with expectations of the consumer market and regulatory bodies.

However, it can be seen in Table 22-1 that for what are considered critical defects (likely to cause contamination and/or loss of potency) that the AQL is zero so that of the sample size inspection, not one product with a critical defect is acceptable.

Reinspections and Investigations

Procedures for reinspections when AQLs are exceeded vary among manufacturers. The sample number might be doubled and the inspection process repeated. The entire batch may have to be thoroughly inspected especially if more than one critical defect (e.g., cracked glass) was found. If the number of defects found is higher than usual, an investigation should be initiated to determine assignable causes. Again, procedures should be written and followed concerning actions upon failing initial inspection criteria for finished product defects.

Identification of Particles

It is useful to identify particles to isolate and eliminate their source. Polarized light microscopy and fourier transform infrared spectroscopy (FTIR) microscopy are valuable in identifying isolated particles. Establishing a particle library of known materials in your facility is extremely helpful in identifying particles found in the finished products. Keeping the physical samples for a visual comparison and/or generating photomicrographs are as valuable as the spectral data obtained from FTIR microscopy when comparing unknown samples.

LABELING

Labeling requirements were described in chapter 2 with some reiteration here. The labeling of an injection must provide the physician or other user with all of the information needed to ensure the safe and proper use of the product. Since all of this information cannot be placed on the immediate container and be legible, it may be provided on accompanying printed information sheets.

A restatement of the labeling definitions and requirements of the USP for injections is as follows: The term *labeling* designates all labels and other written, printed, or graphic matter upon an immediate container or upon, or in, any package or wrapper in which it is enclosed, with the exception of the outer shipping container. The term *label* designates that part of the labeling upon the immediate container.

[4] There are standard sampling plans for attributes (ANSI/ASQC Z1.4 or MIL STD 105) and variables (ANSI/ASQC Z1.9 or MIL STD 414) that are based on acceptable quality levels (AQLs).

The label states the following information:

- The name of the preparation
- The percentage content of drug of a liquid preparation
- The amount of active ingredient of a dry preparation
- The volume of liquid to be added to prepare an injection or suspension from a dry preparation
- The route of administration
- A statement of storage conditions
- An expiration date
- The name of the vehicle and the proportions of each constituent, if it is a mixture
- The names and proportions of all substances added to increase stability or usefulness
- The name of the manufacturer or distributor
- An identifying lot number.

The lot number is capable of providing access to the complete manufacturing history of the specific package, including each single manufacturing step. The container label is so arranged that a sufficient area of the container remains uncovered for its full length or circumference to permit inspection of the contents.

Preparations labeled for use as dialysis, hemofiltration, or irrigation solutions must meet the requirements for injections other than those relating to volume and also must bear on the label statements that they are not intended for intravenous injection. Injections intended for veterinary use are so labeled.

Labeling of drug products is moving toward authentication and "track and trace" technologies intended to provide authenticity of drug products and to identify counterfeit products. Authentication technologies include overt (tamper evidence, security graphics and inks, optical variable devices, color shifting films, etc.) and covert systems (invisible printing, laser coding, security inks, etc.). A familiar example of a security ink is UV light detected ink. Track and trace technologies include 2D bar codes and radio frequency identification (RFID). Any added authentication technology always runs the risk of added leachable or extractable problems. One major problem in incorporating these technologies on labels is the practical fact that so many products are contained in very small packaging systems (1–5 mL) that has limited space because of all the other information required (see in the preceding text). Track and trace technologies are being heavily investigated for their effects on drug stability and proper placement on the label or on the package itself.

A web site featuring a 2008 USP presentation by Michael Eakins (18) contains examples of 2D bar codes and RFID placements on labels or containers. This is a subject that will continue to draw much attention as experts determine how to apply these technologies on all drug products

Uhlmann Packaging Machine places prefilled syringes into plastic trays, heat seals a lid on the trays, then cuts into desired formats for cartoning

Broad view of Uhlmann Packaging Machine | Close-up view of Uhlmann Packaging Machine

Figure 22-5 Finishing apparatuses for secondary packaging. *Source*: Courtesy of Baxter Healthcare Corporation.

that will accomplish the primary goals of drug product authentication without distorting current labeling features and producing any adverse effects on the product.

SECONDARY PACKAGING

Secondary packaging, while perhaps not important from a stability and sterility assurance viewpoint, has a major effect on marketing and consumer image of the product. Secondary packaging is also useful, even essential, for protecting certain drug products from excessive chemical degradation; for example, adverse effects of light exposure. Examples of secondary packaging include outer individual boxes, boxes containing multiple unit packages, blisters, and cartons. For the majority of biopharmaceutical products that must be distributed and stored in containers that maintain cold temperatures for the primary packages, secondary, even tertiary packaging (e.g., large cold storage containers, palletized large boxes) is essential for maintaining quality parameters, including long-term stability of the product. Examples of secondary packaging and labeling equipment are shown in Figure 22-5.

REFERENCES

1. USP. United States Pharmacopeia-National Formulary. Rockville, MD: The United States Pharmacopeial Convention. General Chapter <1>, 2010.
2. Nail SL, Guazzo D, Rajagopalan N, et al. Visual inspection of parenteral products in a development environment. Am Pharm Rev 2006; 9:96–101.
3. European Medicines Evaluation Agency. European Pharmacopeia 6.0, Parenteral Preparations, Injections, European Directorate for the Quality of Medicines & HealthCare. Strasbourg, France: EMEA, 2008:736.
4. Japanese Pharmacopoeia, Fifteenth ed. General Rules for Preparations, 11, Injections. Tokyo, Japan. Ministry of Health, Labour, and Welfare 2006:11.
5. Madsen RE, Cherris RT, Shabushnig JG, et al. Visible particulates in injections—A history and a proposal to revise USP general chapter injections <1>. Pharm Forum 2009; 35(5):1383–1387.
6. Leelarasamee N, Howard SA, Baldwin HJ. Visible particle limits in small-volume parenterals. J Parenter Drug Assoc 1980; 34:167–174.
7. Huang CT, Sharm D, Oma P, et al. Quantitation of protein particles in parenteral solutions using micro-flow imaging. J Pharm Sci 2009; 98:3058–3071.
8. Knapp JZ, Abramson LR. Evaluation and validation of nondestructive particle inspection methods and systems. In: Knapp JZ, Barber TA, Lieberman A, eds. Liquid- and Surface-Borne Particle Measurement Handbook. New York, NY: Marcel Dekker, 1996:295–450.
9. Barber TA. Control of particulate matter contamination in healthcare manufacturing, 2000. Interpharm Press, 2000:257–259.
10. De la Montaigne M, Mendez PJ, Tagaya R. Inspection of parenteral products in vials, ampoules, syringes, and cartridges. In: Lysfjord J, ed. Practical Aseptic Processing. Vol 1. Bethesda, MD: Parenteral Drug Association, 2009:275–304.
11. Knapp JZ, Kushner HK. Implementation and automation of a particle detection system for parenteral products. J Parenter Drug Assoc 1980; 34:369.
12. Knapp JZ, Kushner HK. Generalized methodology for evaluation of parenteral inspection procedures. J Parenter Drug Assoc 1980; 34:14.
13. Knapp JZ, Kushner HK, Abramson LR. Automated particulate detection for ampuls with the use of the probabilistic particulate detection model. J Parenter Sci Technol 1981; 35:21.
14. Knapp JZ, Kushner HK, Abramson LR. Particulate inspection of parenteral products: An assessment. J Parenter Sci Technol 1981; 35:176.
15. Knapp JZ, Kushner HK. Particulate inspection of parenteral products: From biophysics to automation. J Parenter Sci Technol 1982; 36:121.
16. Knapp JZ. The scientific basis for visible particle inspection. PDA J Pharm Sci Technol 1999; 53:291–301.
17. Sower JM, Savoie M. Are acceptance sampling and statistical process control complementary or incompatible? Qual Prog 1993; 26:85–89.
18. Eakins, EN. Progress towards establishing e-pedigree for drug products, United States Pharmacopeia annual meeting, September 24, 2009, Toronto, http://www.usp.org/pdf/EN/meetings/ASM2009/S3T1MichaelEakins.pdf. Accessed June 13, 2010.

BIBLIOGRAPHY

Knapp JZ, Barber TA, Lieberman A, eds. Liquid and surface-borne particle measurement handbook, 1999, Marcel Dekker, New York.

23 Barrier and other advanced technologies in aseptic processing

The rough estimate that 80% of all small-volume injectable and ophthalmic products are aseptically filled has led to a number of innovations designed to increase the level of sterility assurance during aseptic operations. Barrier isolator technology is designed to isolate aseptic operations from personnel and the surrounding environment. Considerable experience has been gained in its use for sterility testing, essentially eliminating false-positive test results. Efforts in adapting automated, large-scale, aseptic filling operations to isolators have steadily gained momentum and will continue to grow in application in sterile product manufacturing (1,2).

Because the large majority of small-volume injectable products are aseptically processed and not terminally sterilized, sterility assurance is always a significant hot topic in sterile product processing. An almost endless stream of external conferences and publications and internal meetings in companies that manufacture sterile products are necessitated over sterility assurance concerns with aseptically produced dosage forms. These concerns were highly reflected in the 2004 FDA Aseptic Processing Guidelines (3). Compared to the 1987 guidelines, the 2004 revised guidelines had many more requirements, particularly with respect to requirements for media fills, personnel training, facility design and control, endotoxin control, prefiltration bioburden, and use of aseptic processing isolators.

Several initiatives, most not that new, but ever improving, have taken place in the attempt to increase the level of sterility assurance in aseptic processing (Table 23-1) (4–7). Many of these advances basically minimize or remove the need for human intervention.

Several factors have contributed to the increased importance and utilization of barrier isolator technology.

1. The high level of concern from manufacturers and regulatory agencies over the level of sterility assurance in aseptic processing.
2. Continued relatively high level of product recalls due to concerns—proven or suspected—over contamination potential.
3. The surge of potential heat-labile products from biotechnology, and the inability to terminally sterilize these molecules. There are needs to control the environment not only from contamination but also with respect to stability considerations—temperature, humidity, and, if necessary, anaerobics.
4. Many new drug compounds are cytotoxic or otherwise highly potent where safety considerations demand separation of these drugs from human operators.
5. Because so many biopharmaceutical drugs are so expensive, there is a trend toward smaller batch production. Smaller batch production makes construction of large manufacturing facilities unnecessary, yet there is still the need to manufacture in Class 100/Grade A/ISO 5 clean rooms. Isolators are ideal for smaller facilities plus are much more economical from the standpoint of capital, labor and maintenance, and operator (e.g., number of employees, gowning) costs.

The main features of barrier/isolator technology are the ability to sterilize (more than sanitize) the environment to which sterile solution is exposed during filling and stoppering and the removal of direct human contact with the exposed sterile product. Isolators not only protect the product from potential human contamination, but also protect the human from potential toxic effects of direct exposure to the drug product, especially important for cytotoxic drugs.

Barrier technology has long been used in the pharmaceutical industry and ranges from simple screens to restricted access barrier systems (RABS). Figures 23-1 and 23-2 show schematic comparisons of the conventional clean room filling operation versus a true isolator (closed RABS) versus a passive RABS and an active RABS (8). Figure 23-3 shows schematic comparisons of a

Table 23-1 Examples of Advances in Aseptic Processing

- Barrier isolation technologies, both isolators and restricted access barriers
- Improved clean room designs and operational performance
- Increased automation, e.g., in-process control, environmental monitoring, filter-integrity testing, robotics for component feeding, elimination of unfilled containers, and containers without stoppers
- Improvements in depyrogenation tunnels that do not allow in-feed of containers until space; more accurate fill-dose systems, easy to clean and sterilize, set up remotely, require no in-process adjustments, and minimize the need for manual weight checking
- Improved aseptic processing equipment requiring fewer line interventions and more rigorous procedures for performing interventions during processing
- Complete elimination of aseptic connections, instead using automated CIP and SIP systems
- Increased level and analysis of environmental monitoring
- More rigorous cleaning and sanitizing validation
- Improved clean room garments
- More rigorous training in gowning and aseptic technique practices
- High-quality glassware and rubber closures that eliminate equipment misfeeds
- Cappers that use negative pressure relative to a filling-capping isolator that prevents aluminum particles from entering the classified (e.g., ISO 5) environment
- More accurate seal integrity measuring systems

Abbreviations: CIP, clean-in-place; SIP, steam-in-place.

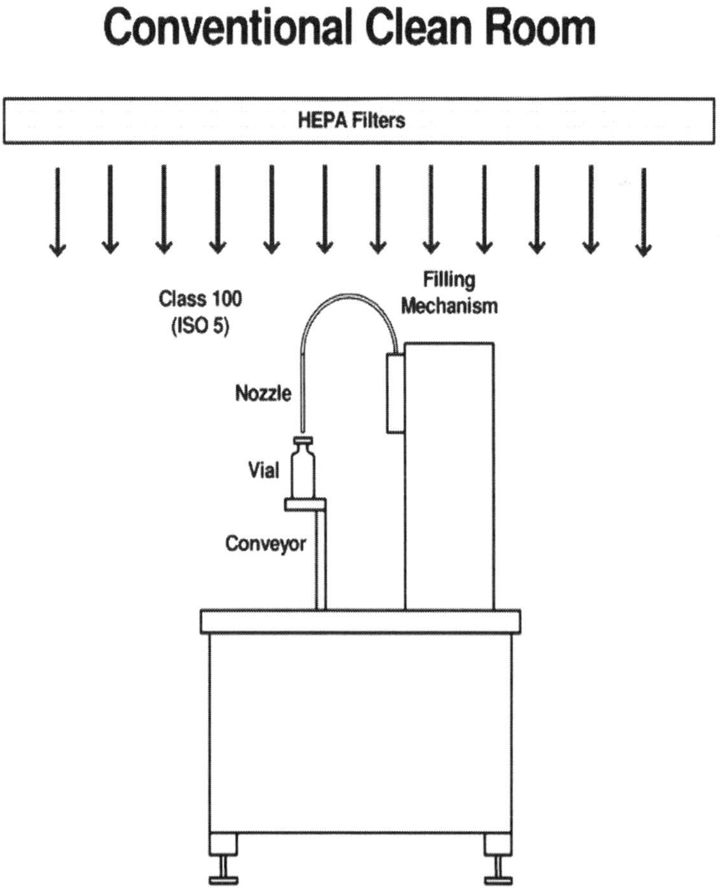

Figure 23-1 Schematic of a conventional clean room. *Source*: From Ref. 8. Parenteral Drug Association and Davis Healthcare Int'l Pub, LLC.

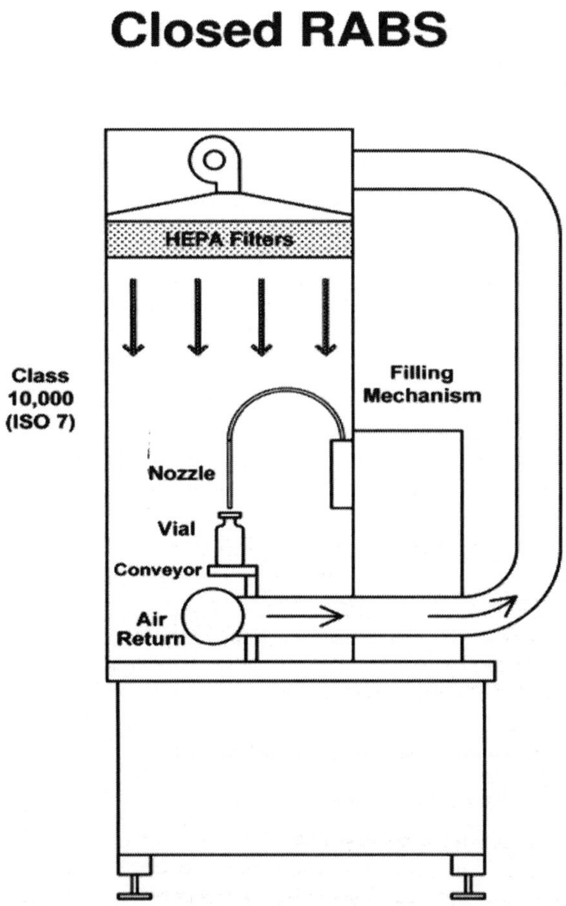

Figure 23-2 Schematic of a closed restrictive access barrier system (RABS) clean room. *Source*: From Ref. 8. Parenteral Drug Association and Davis Healthcare Int'l Pub, LLC.

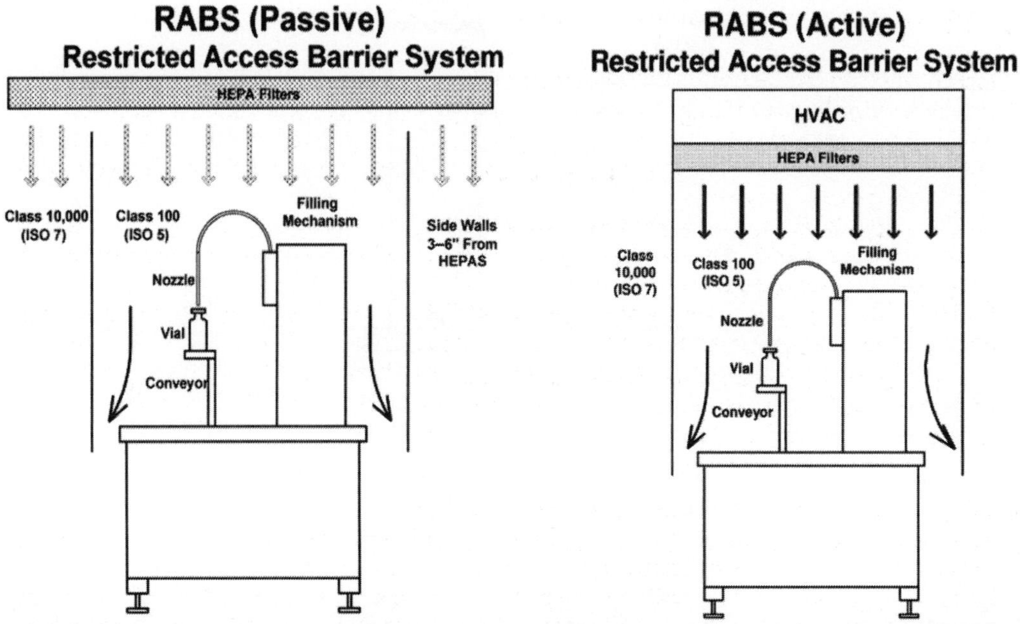

Figure 23-3 Schematic comparison of passive versus active RABS. *Source*: From Ref. 8. Parenteral Drug Association and Davis Healthcare Int'l Pub, LLC.

passive and active RABS. A closed RABS has all the appearances of a true isolator, but requires manual cleaning and manual decontamination. A passive RABS has its HEPA (high-efficiency particulate air)-filtered air provided by the facility, not by the barrier itself. An active RABS has its own integral heating, ventilating, and air conditioning (HVAC) and HEPA system with an interconnected control system with the processing line. All RABS, like the true isolator systems, use glove ports and rapid transfer ports (RTPs).

Recent isolator development has been significant and such systems are now better specified than ever. In advanced aseptic processing facilities, it has been proven that isolators can provide zero colony-forming unit (0 CFU) contamination in process operations, while the background environment is only at ISO 8-EC grade D level. However, cost savings in clean room construction and operation may be offset by the construction and validation costs of the isolator system.

Isolators are enclosed, usually positively pressurized units with HEPA filters, supplying ISO 5 airflow in a unidirectional manner to the interior. Air recirculates by returning it to the air handlers through sealed ductwork. Cleaning can be manual or automated (clean-in-place). Bio-decontamination occurs through an automated cycle typically using vaporized hydrogen peroxide. Access to an isolator is through glove ports and sterile transfer systems. Isolators can be located in an ISO 8 or better environment.

Figures 23-4(A) and 23-4(B) illustrate the adaptation of a large-scale filling line to isolator technology. A smaller version of an isolator used for sterility testing is shown later in Figure 27-3. The operations are performed within windowed, sealed walls with operators working through glove ports. An example of the interior of an isolator is shown in Figure 23-5. The sealed enclosures are presterilized, usually with peracetic acid, hydrogen peroxide vapor, or steam. Sterile supplies are introduced from sterilizable movable modules through uniquely engineered transfer ports or directly from attached sterilizers, including autoclaves and hot-air sterilizing tunnels. Results have been very promising, giving expectation of significantly enhanced control of the aseptic processing environment.

Traditionally RABS operate in clean room environments of ISO 7-EC grade B. They can provide further zoning via screened barriers and HEPA clean-air filtration, such that an ISO 5-grade A critical zone can be established. One key consideration, particularly where process operations require "open door" access and manual intervention, is that RABS provide a high level of protection against contamination from operator intervention.

Some of the key differences between RABS and isolators are (6):

1. RABS offer a combined physical and aerodynamic barrier, ideally controlled by positive pressure with clean-air filtration providing air exchanges and particulate cleanup for an ISO 5 critical process zone.
2. Disinfection is typically manual in a standard RABS involving interaction with the process for cleaning.
3. Traditionally, RABS come in two types (Fig. 23-3): "passive" where there is no in-process open door access; and "active" where, under certain validated system configurations and control conditions, access may be included.
4. With RABS, if component entry is needed after disinfection, then aseptic transfer devices are used to prevent recontamination of the critical process zones.
5. RABS should include environmental monitoring of critical zones to ensure operators are alerted to deviations from performance levels of particulate and microbiological contamination, so that action can be taken.

ISOLATOR CONTAMINATION CONTROL ATTRIBUTES

1. The physical barrier is typically controlled at positive pressure with clean-air filtration providing air exchanges and particulate cleanup for an ISO 5 critical process zone.
2. The ability to bio-decontaminate to a high level with a combined cleaning and sporicidal process often validated to achieve 6 log reduction of Geobacillus stearothermophilus biological indicator challenges using a sporicidal gassing process, for example, hydrogen peroxide vapor.

(A)

(B)

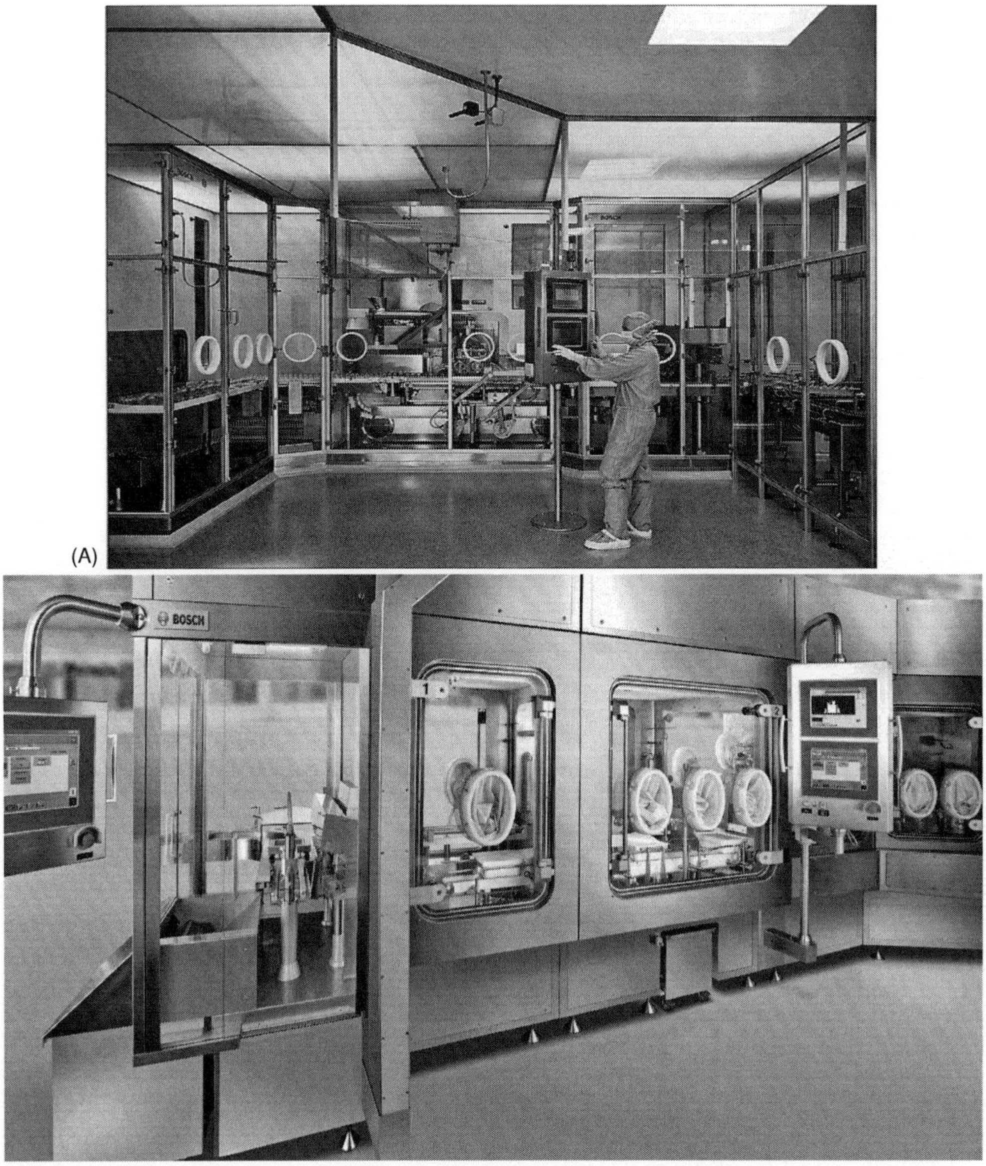

Figure 23-4 A and B Example of a large-scale filling line within an isolator. *Source*: From Ref. 8. Parenteral Drug Association and Davis Healthcare Int'l Pub, LLC. Courtesy of Robert Bosch, GmbH.

3. There is no operator-human access to the isolator critical zone after the sporicidal bio-decontamination process and during any subsequent processing of process transfer steps.
4. All product contact parts are cleaned and sterilized in place (CIP/SIP) or enter the sporicidally disinfected isolator system using aseptic transfer devices. Closed processing post gassing maintains sterility or product contact parts and prevents recontamination of the critical process zones during all process operations and transfers.
5. The ability to environmentally monitor the critical zones to assure any deviations from performance levels of particulate and microbiological contamination are alerted and action taken.

Figure 23-5 Scene inside an isolator filling sterile product. *Source*: Courtesy of Robert Bosch, GmbH.

IMPORTANT ASPECTS OF A HIGH-QUALITY ISOLATOR

Isolators are expensive, both initial costs and long-term operating costs. In order to make sure that fiscal responsibility is practiced, the following considerations are important in selecting an isolator and protecting its investment (9).

1. Chamber leak tightness
   a. To protect operators from excess exposure to sanitizing agents used with the isolator
   b. To minimize opportunities for particle ingress
   c. To protect operators from hazardous drug products being processed within the isolator
2. Materials of construction—must be cleanable and resistant to cleaning agents. Materials of choice are polyvinyl chloride or low carbon stainless steel, polished to an appropriate surface finish and passivated. For rigid windows, Lexan®, acrylic plastics, or tempered glass are used. Appropriate materials of construction also apply for air handling and filling equipment. Minimize use of elastomeric and plastic material and avoid silicone as these materials are well known to absorb sanitizing chemicals used for bio-decontamination of the isolator.
3. Product requirements—Certain products like those containing pharmaceutical proteins will be sensitive to bio-decontamination solutions. Therefore, the isolator must have optimal chamber aeration processes to reduce sanitizing agent levels that will not be harmful to the product.
4. Environmental monitoring—Isolator design must consider the methods, locations, and frequency of environmental monitoring, including both microbial and particle monitoring. Particle monitoring requires either chamber penetrations to sample the air or portable equipment that can be bio-decontaminated. Ideally, the isolator is designed so that automated monitoring equipment are integrated within the isolator. All parts of the isolator must be accessible to placement of microbial monitoring systems; thus, the location of the glove ports is critical.
5. Glove testing—Gloves are tested by visual inspection and automated testing using pressure pulses. Removal of gloves for inspection must be minimized as gloves can be damaged. The design of the isolator determines the ease of glove testing, for example, how easy it is to detect glove defects looking through the glass door, how easily gloves can be removed, and how easily automated testing can occur if/when other processes are occurring in the isolator.
6. Controlling particle generation—Particle generation is ubiquitous, but can be minimized through optimal design that does not allow particle from the surrounding environment to enter the chamber. Also items inside the chamber (filling and stoppering equipment, tubing,

other parts) should be designed to minimize particle generation. This includes opening of packaging containing sterile items as the opening process likely generates particles.

7. Material handling—The ease of transferring materials in and out of an isolator largely determines whether a process can be performed within an isolator. The role of the "mousehole" is very important for continuous transfer during filling operations.
8. Bio-decontamination method integration—Bio-decontamination of isolators is automated, either standalone or integrated. Standalone systems consist of a sanitizing agent injection system, a fan for moving the air, tubing to connect to the ports on the isolator, and a desiccant system for controlling humidity. Integrated systems are installed in the isolator structure. Sanitizing agent injection and humidity control are performed by the bio-decontamination system while airflow control is performed by the air handling unit of the isolator.
9. Equipment in the chamber—Equipment within the isolator must be designed to be cleaned and bio-decontaminated. Connections from the equipment to external sites (e.g., power supply) must be as seamless as possible. Equipment interventions must be minimal so equipment needs to be durable. Equipment must be designed for easy access during operation. Adjustments to equipment must not require a high degree of manipulation. Typically, equipment is specially designed to fit and work easily within isolators. Making connections, changing parts, correcting failures, and other manipulations are designed to be performed with one gloved hand.
10. Airflow—Isolators are either closed or open. Closed isolators are fully sealed while open isolators have mousehole openings for continuous processing. Unidirectional airflow is required for correct operation of open isolators. Close isolators can have no airflow or airflow that is either unidirectional or even turbulent. Airflow definitely is required for optimal bio-decontamination to assure proper distribution of sanitizing agent, humidity, temperature, and sufficient aeration. Large isolators usually require full HVAC systems to assure optimal airflow and distribution.

While isolators have been implemented in the industry, progress has been slower than initially anticipated. There are several reasons for this slow growth and acceptance:

1. General regulatory and industry caution because of the relative novelty of isolator technology.
2. Regulatory agencies have insisted so far that isolators be located in classified environments (usually at least Class 100,000/EC Grade D). This discouraged investment by some in isolator technology because it was originally thought that classified environments would not be necessary.
3. Initial promotion that isolator technology could create a truly sterile environment and, thus, allow a much greater claim for sterility assurance proved not to be true. Isolators tend to have small leaks, particularly at the glove ports and gloves or half suits. The industry has learned the hard way that for aseptic processing, sterility assurance levels for isolators are not much greater than conventional Class 100/EC Grade A filling operations.
4. Validation of isolators has been more difficult than expected. For example, it is difficult to convince reviewers that contamination will not occur despite constant movement of materials in and out of the isolator, the occasional need to manipulate equipment, and the problem of pinhole leaks. The significantly increased time and resources required to validate and maintain isolators have discouraged many companies from investing in these systems.

VALIDATION OF BARRIER ISOLATION AND ASSOCIATED STERILIZATION SYSTEMS

Like any other process in the pharmaceutical industry, barrier isolation must be shown to reproducibly deliver the desired result. Because of their complexity, there are several parameters to consider in the design and validation of isolation systems (Table 23-2) (10). United States Pharmacopeia (USP) Chapter <1208> provides guidance for the design and validation of isolator systems for use in sterility testing. The guidelines in USP <1208> are summarized below, as well as common practices in the validation of sterility testing isolators. For a complete description of <1208> consult the most current USP revision.

The steps and considerations that are essential to the isolator validation and design are outlined below.

Table 23-2 Functional Specifications to Be Validated for Isolators

Function	Examples
Air supply specifications	Air change rate
	Air velocity
	Particulate air specification
	Recirculation rate
	Temperature and humidity
	Aeration of the decontaminating agent
Leak testing	Pressure decay test
	Tracer gas detection test
Ergonomics	Eventual and uneventful situations
Rapid transfer ports	Seal integrity
Facility requirements	Classification of isolator room
	Temperature and humidity control
	Process utilities
User requirements	Sterility assurance—sterilization and decontamination methods
	Cleaning
	Containment
	Environment control and monitoring
	Microbiological monitoring
	Process simulation

Source: From Ref. 10.

DESIGN

Isolator design includes materials of construction (rigid vs. flexible; compatible with cleaning agents and sterilants), size (especially considering length of human arms and ability to reach all parts of the isolator interior), ease of cleaning and sanitization and/or sterilization, ergonomics (e.g., height and length of gloves), lighting, ability to connect to other systems (e.g., sterilization tunnels), and other considerations that isolator manufacturing companies are well aware of now after years of experience overcoming many problems. An isolator needs to be equipped with filters capable of microbial retention. HEPA filters are required, but ultra low particulate air filters may be substituted. While the isolator is at rest, it must meet the particulate requirement for an ISO 5/Grade A area. There is no particulate requirement while the unit is in operation during a sterility test, and there is no requirement for air velocity or air exchange rate. The isolator should be leakproof, but it may exchange air with the surrounding environment. While direct openings to the surrounding environment should be avoided, air overpressure can be employed to maintain sterile conditions within the isolator. Air overpressure should also be employed to help avoid ingress of non-sterile air in the event of an unexpected leak.

Location

The isolator does not need to be installed in a classified clean room, but the surrounding room should be limited to essential staff. Environmental monitoring of the surrounding room is not required.

The surrounding room should have sufficient temperature and humidity control to maintain operator safety and comfort, to allow for proper operation of the associated sterilizer (the air should exhaust to an outside source for safety reasons) unit, and to allow for proper operation of the isolator. The temperature within the room should be as uniform as possible to avoid the formation of condensation within the isolator.

Installation Qualification

The installation qualification (IQ) should include a detailed description of all of the mechanical aspects of the system such as dimensions, internal configuration, serial numbers of the equipment, blueprints, purchase orders, electrical supply, specifications, exhaust, vacuum supply, and equipment manuals. All documentation should be reviewed for accuracy. The documentation that is recommended is discussed next (10).

Equipment

The listed equipment is listed with critical design specifications. The IQ should verify that the appropriate design specification was received and that all equipment was installed per the manufacturer's requirements.

Construction Materials

The critical components of the system are checked for compliance with the design specification and for compatibility with the method of sterilization.

Instruments

System instruments are listed with their calibration records.

Utility Specifications

All utilities that are required for operation, as defined in the operating manuals and diagrams, are verified. Any connection between electrical and/or exhaust systems are inspected and verified to conform to specifications.

Filter Certification

HEPA filters are tested and certified, and copies of the certifications are included.

Computer Software

All computer software is listed with name, size, and version number. Any master copies should be properly labeled and stored should the need for a backup arise.

Operational Qualification

The operational qualification (OQ) step verifies that the isolator system operates within conformance to functional aspects.

Operational Performance Check

All alerts and alarms should be tripped and verified that they function properly.

Isolator Integrity Check

The integrity of the isolator should be verified to be free from leaks. The leak test is important to preclude contamination, and for operator safety. The overpressure set point should be established and shown to be maintained during operation.

Sterilization Cycle Verification

Verification of relevant temperature and/or humidity control during the sterilization cycle should be demonstrated. Humidity may be especially important, depending on the type of sterilizing gas that is used. The concentration and distribution of the sterilizing gas should be measured using chemical indicators. After sterilization, verification that the sterilizing gas has been removed (by aeration) to an acceptable level should be demonstrated by quantitative methods.

Sterilization Cycle Development

Upon completion of the OQ, cycle development parameters are established to achieve sterilization of the isolator unit. Sterilization should be demonstrated by use of the bioburden approach, through the use of biological indicators (BI) of a known concentration, or the half-cycle approach.

Note: *Bacillus stearothermophilus* is an appropriate choice as a BI in this application as it is more resistant, in the case of vapor phase hydrogen peroxide (VPHP), than most environmental isolates (11).

Performance Qualification

The performance qualification (PQ) verifies that the systems are functioning in compliance within its operational requirements. Upon completion of the PQ phase, sterilization efficacy should be established.

Sterilization Validation

The interior surfaces of the isolator, articles within the isolator, and equipment in the isolator, and sterility test articles should be rendered sterile after processing. BI kills of 10^3–10^6 are commonly used in the pharmaceutical industry, as associated with sterility testing isolators.

Since most pharmaceutical companies produce many different package presentations, the minimum load (an empty chamber) and maximum load (the maximum number of test articles) approach is helpful when performing the PQ of the isolator.

False Negative Evaluation

Because certain materials are adversely affected or absorbed by sterilizing agents, it is important to demonstrate that the method of sterilization would not destroy any microbial life since the point of a sterility test is to detect low levels of organism in the finished product.

Various containers, media, and rinsing fluids should be inoculated with low levels of organisms (<100 CFUs of either specific microorganisms listed in USP <71> or indigent microorganisms identified with the facility environmental monitoring systems) and subjected to the sterilization cycle. After sterilization the level of inoculum should be verified to prove that organisms are not destroyed during the cycle. In the event that organisms are adversely affected, a new container closure should be selected before use with the isolator system.

ADVANCES AND TRENDS IN MANUFACTURING PROCESSES AND EQUIPMENT

Injectable product manufacturing is booming because of the growth of new biopharmaceuticals and small molecule anticancer drugs. Companies with existing injectable product manufacturing capabilities either need to upgrade aging equipment and even remodel facilities or choose to outsource manufacturing to the rising presence of good manufacturing practice-compliant parenteral contract manufacturing companies. Biotechnology growth has given rise to many new companies, many of which are virtual, so they need to find contract manufacturers for the production of clinical supplies and eventual commercial product.

While growth seems to be more a matter of replacing aging equipment and upgrading current manufacturing facilities, some major pharmaceutical companies are building new facilities in offshore locations like Ireland and Puerto Rico. Smaller pharmaceutical and biotechnology companies use contract manufacturers. Contract manufacturers are major purchasers of new equipment.

Modular Construction

Some discussion of modular construction was presented in Chapter 14. Modular construction has become a design standard for a number of parenteral product companies worldwide. Standardized rooms are constructed to meet strict engineering guidelines while incorporating flexibility in size, classification, and utilization (Figs. 14-4 and 14-5). Modular construction has been adopted by Baxter BioPharma Solutions (BPS), named the 2006 Facility of the Year Award winner by the International Society of Pharmaceutical Engineers, Interphex and Pharmaceutical Processing magazine. The manufacturing portion of this facility was a design/build project provided by Pharmadule AB, a provider of modular facilities for the pharmaceutical and biopharmaceutical industry. The facility provides large-scale syringe filling, aseptic formulation, vial filling, lyophilization, terminal sterilization, and flexible formulation capacity for a variety of challenging products such as insoluble solutions and vaccines.

Many other companies, among them Merck and Cambrex, are moving to modular construction, installing prefabricated modular walls and ceilings rather than the classical studs and dry wall construction approach. One main reason for the growing popularity of modular construction is the claim that modules are cleaner than dry wall, and thus the potential decrease in the source of particles in clean rooms (12). The trend in modular construction is growing, not only because of higher quality, but also because of decreased time to complete construction. Costs for modular facilities are higher than permanent facilities, but the time savings can override the increased construction expense.

Processing

Improved filling technologies are helping to increase quality assurance. Fast and efficient filling technologies increase speed, but remain gentle on the product to minimize shear effects. The growth and demand for pre-filled syringes has caused a major growth in high-speed syringe filling equipment. The newest syringe filling equipment are capable of filling up to 600 syringes per minute.

High potency compound processing also is on the rise, requiring equipment and facilities that protect the operators. This has also accelerated the growth of isolation systems.

The biopharmaceutical industry has made a myriad of other improvements to streamline production and improve quality assurance. Freeze-drying technologies, including automated loading and unloading systems, help to minimize the potential for inadvertent contamination of the product.

Control of particulate matter can be assured through the combination of valid cleaning procedures, dedicated equipment, tight control of air handling systems, excellent training in aseptic techniques, and automated or semi-automated visual inspection machines. Weight checking, inspection technologies, and labeling and finishing operations also are being automated.

Disposable Technology

Disposable systems have been used in sterile processing for many years. Examples include plastic tubing (typically platinum-curing silicone) instead of steel pipes, expandable bags instead of stainless steel vessels, plastic pinch clamps, and, of course, filters, and even filter assemblies.

Today, the possibility of the entire manufacturing process being composed of disposable systems and materials is conceivable. A fully disposable system would include disposable plastic bags replacing stainless steel or glass tanks, disposable capsule filtration systems, and currently used plastic tubing. There are many important considerations with respect to disposable components meeting pharmaceutical specifications, bag construction and chemical inertness, barrier properties of disposable materials, costs, safety, quality and compliance, utilities, environmental impact, risk-benefit analysis, and validation (13).

The role of disposables in sterile product manufacturing has many advantages and a bright future (14). Disposable filling systems allow faster filling-line implementation, easier validation, and ensured sterility at the point of product introduction into the sterile vial (15). Since everything is pre-assembled and pre-sterilized (gamma irradiation), there is no need for aseptic assembly of pumps, reservoirs, tubing, and needles. Also, sterility assurance is enhanced because of elimination of operator intervention in aseptic connections and components. Filling capacities can range from 0.5 to 100 mL with filling rates up to 60 fills per minute per needle.

While still in its infancy, this technology likely will become state of the art, particularly for product mixing and holding. Like any relatively new technology, perceived and known advantages often outweigh disadvantages until significantly more experiences with the technology take place. Barrier technology is a good recent example where initially it was thought (and highly promoted) that this technology would enable aseptic filling to occur in non-classified rooms where the isolator was located and that validation would prove that such technology would match the same level of sterility assurance as terminal sterilization. These claims, of course, proved not to be exactly true.

A summary listing advantages and disadvantages of disposable technology is provided in Table 23-3 (16). The concerns about chemical compatibility (leachables, extractables, material qualification, product compatibility) have been discussed by Samavedam et al. (17).

Aseptic Connections and Sampling Methods

A single pharmaceutical company will make as many as 25,000 aseptic connections in a single year (18). The number of samples to be taken aseptically might be even more than this high number. Each connection or sample runs the risk of introducing contamination. In response, several vendors have made available new devices for making connections (Fig. 23-6) and taking samples more easily and quickly (Fig. 23-7). Major vendors such as Pall (Kleenpak™), Millipore (NovaSeptum®), BioQuate, Asepco, Stedim, and others provide these devices and their websites can be easily accessed for more information and technical literature. A review of sterile connectors for bioprocessing provided a comparison of five sterile connectors—Kleenpak™ by Pall, Lynx S2 S by Millipore, Opta SFT-1 by Sartorious Stedim, ReadyMate by GE, and Pure-Fit SC by Saint Gobain—to help biopharmaceutical manufacturers determine which connector is best for a particular application (19).

Table 23-3 Advantages and Disadvantages of Disposable Technologies

Advantages	Disadvantages
Cleaning—eliminates need	Cost per batch—requires new components for every batch
Sterilization—pre-sterilized components	Dependence on vendors—supply and timing issues
Engineering—design already included in unit operation	Material compatibility—potential formulation component incompatibilities and extractables
Equipment installation—reduces or eliminates need to fabricate equipment	Waste—how to destroy
Utility requirement—no requirement for CIP/SIP	Application—cannot replace all unit operations; unsuitable for some products
Space—more efficient use, much less storage requirements	
Labor—pre-cleaned and pre-sterilized components; less setup and turnaround time	
Quality—removes potential for cross-contamination, reduced bioburden	

Source: From Ref. 16.

Pall Kleenpak™ Connectors

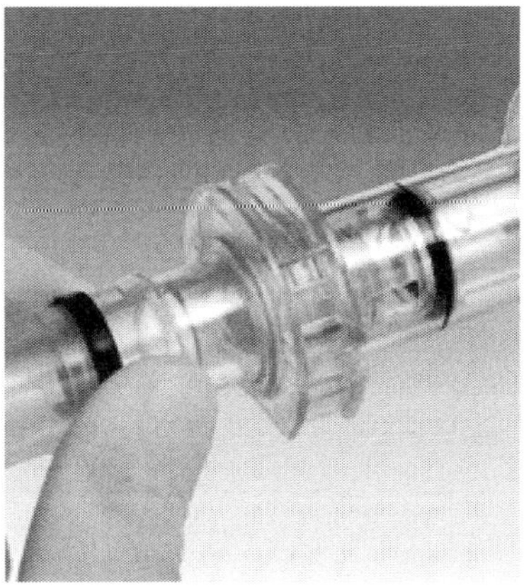

- Connects two separate presterilized pathways
- Male and female connector
- Vented peel away strip that protects the port and maintains sterility of the sterile fluid pathway
- Vented to allow steam penetration and to prevent tubing collapse after steam sterilization
- Aseptic connections can be made in seconds vs minutes
- Eliminates the need for cleaning and cleaning validation

Figure 23-6 Example of an aseptic connection device. *Source*: Courtesy of Pall Corporation.

NovaSeptum® System

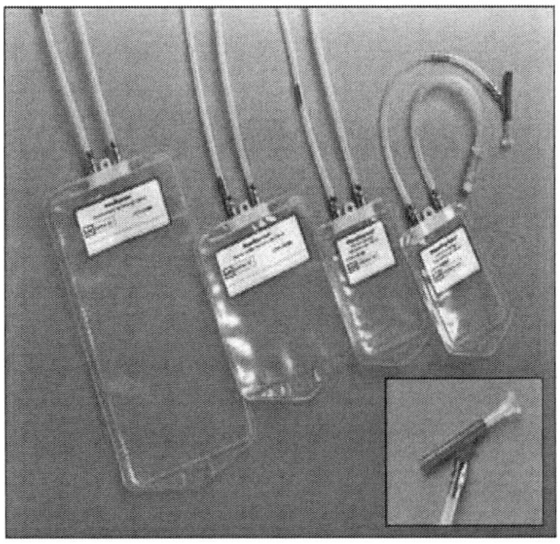

- Presterilized, disposable, totally enclosed system
- Eliminate CIP/SIP
- Multiple sampling units
- Multiple applications (sterility test, endotoxin test, chemical analysis, pH)
- Multiple sizes 50 to 1000 mL
- Conforms to USP Class VI requirements

Figure 23-7 Examples of aseptic sampling systems. *Source*: Courtesy of Millipore Corporation.

Rapid Microbiology Systems

Microbiological testing is ubiquitous throughout the parenteral industry—water system monitoring, environmental monitoring, preservative challenge testing, bioburden testing, and sterility testing, to name a few. However, we still rely on the same technology that Louis Pasteur and his colleagues used in the 19th century; that is, incubating samples with one or more sterile growth media for a period of time from 5 days up to perhaps 2 weeks, and observing the sample for evidence of microbial growth. These methods are limited by slow microbial growth rates, unintended selectivity of media, and the inherent variability of microorganisms in their response to culture conditions.

Rapid microbiological methods comprise several different technology platforms that offer significantly reduced turnaround time (hours instead of days), opening the possibility of process microbiological testing in real time. They offer increased sensitivity, accuracy, precision, and reproducibility. Some of these methods do not require microbial growth, thus eliminating the need for incubation. Some allow detection of single cells, and many detect stressed organisms that are not detected by conventional methods. Some of these technologies are automated, miniaturized, and offer high-throughput processing. These technology platforms are categorized in Table 23-4 (20).

The adenosine triphosphate (ATP) bioluminescence and fluorescent labeling methods have been studied extensively for application in the sterile products industry. The basic principle of the ATP bioluminescence method is summarized in Figure 23-8. Basically, if the test sample contains microorganisms, their growth will release ATP from cells and the generated ATP will react with the luciferin substrate/luciferase enzyme from the firefly *Photinus pyralis* to generate light that can be measured with a highly sensitive and accurate luminometer. One commercial example using this method is the PallChek™ Rapid Microbiology System with a sensitivity of 1 CFU. Less than 1 minute is required for immediate quantification of contaminants while product release testing requires less than 24 hours.

The fluorescent labeling method (ChemScan® RDI by Chemunex, Paris) uses a filter to trap any microorganisms in the sample on the membrane surface. A substrate is added that is enzymatically cleaved by any viable organism present. Only viable cells with intact membranes have ability to achieve this cleavage and retain the fluorescent label. This cleavage produces a fluorochrome that is retained within the cytoplasm of single cells, including spores. The membrane surface is then analyzed in the ChemScan RDI instrument where a laser scans the entire surface of the membrane in less than 3 minutes and detects all fluorescent events using sensitive photomultipliers.

Despite the impressive advantages offered by rapid microbiological methods, the progress in introducing these technologies into parenteral manufacturing and quality control operations has been slow. One reason could be the lack of clear guidance as to establishing the suitability of rapid methods as a replacement for current methodology. Another factor could be the enhanced sensitivity of rapid methods relative to old methodology—some have voiced concerns that, as a result, microbial limits would have to be raised, and there is a perception that regulatory agencies would not accept this. Another factor could be that pharmaceutical companies are focused on cost containment and headcount control, and this culture is not consistent with development and implementation of new technology. Another may be that industry, as a whole, is uncertain as to just how to go about gaining regulatory approval for improved microbial test methods. Regardless of the reasons for hesitancy, there is no denying that efficient pharmaceutical manufacturing is a critical part of an effective healthcare system, and the industry must make development and implementation of improved microbial test methods a priority. The changing regulatory environment presents the industry with an opportunity to aggressively pursue these technologies.

With the advent of process analytical technologies and the need for information faster without waiting for days, rapid microbiology systems have evolved and will continue to gain greater application in the parenteral industry. FDA guidelines for aseptic processing state that "other suitable microbiological test methods (e.g., rapid test methods) can be considered for environmental monitoring, in process control testing, and finished product release testing after it is demonstrated that the methods are equivalent or better than traditional methods

Table 23-4 Rapid Microbial Method Platforms (courtesy of Dr. Michael Miller (20))

Platform	Description and examples
Growth-based methods	Measurement of biochemical or physiological parameters associated with microbial growth • Measure changes in electrical impedance in solution due to changes in electrolyte composition—bioMérieux Bactometer® • Measure CO_2 production resulting from microbial growth—BacT/ALERT® • ATP bioluminescence, utilizing the luciferin–luciferase reaction—Milliflex® Rapid System, Celsis Advance™ Luminometer, PallChek™ Microbiology system • Monitors changes in kinetic reactions resulting in color change—Omnilog® or turbidity–VITEK® 2 • Digital imaging of growth on agar surfaces—Rapid Micro Growth Direct™
Viability-based methods	The use of viability stains or markers for detection and enumeration of microorganisms without the requirement for cell growth • Solid-phase cytometry—Esterases in viable cells cleave a substrate to form a fluorophore—Chemunex Scan RDI® • Flow cytometry—viability market, cells pass through flow chamber, detect fluorescence and light scattering signals—Advance Analytical (AATI) Rapid Bacterial Detector 3000, Chemunex D-Count®, and BactiFlow® systems
Artifact-based methods	Analysis of cellular components using highly selective and sensitive methods • Fatty acid analysis—MIDI Sherlock® • Mass spectrometric methods—Waters MicrobeLynx™, Ciphergen Biosystems • Endotoxin detection systems using Limulus Amebocyte Lysate—see chap. 28
Nucleic acid-based methods	Gene amplification and detection platforms that detect presence or absence of a specific microorganism according to strain differentiation • 16 S rRNA typing—DuPont Qualicon RiboPrinter® • Polymerase chain reaction to find specific DNA sequence—DuPont Qualicon BAX® • Polymerase chain reaction + mass spectrometry—Sequenom MassARRAY®, Ibis TIGER Universal Pathogen Sensor • Transcription-mediated gene amplication—MilliPROBE, bioMérieux Nucleic Acid Sequence Based Amplification
Micro-electro-mechanical systems (MEMS)	Integration of mechanical, electrical, fluidic, and optical elements, sensors, and actuators on common silicon substrate (Lab-on-a-Chip) technologies • Bacterial Barcodes DiversiLab Microbial Typing System • Affymetrix GeneChip® • CombiMatrix CustomArray™ • Ambri-ICS™ Chip • STMicroelectronics In-CheckTM Chip • BioForce NanoArray™

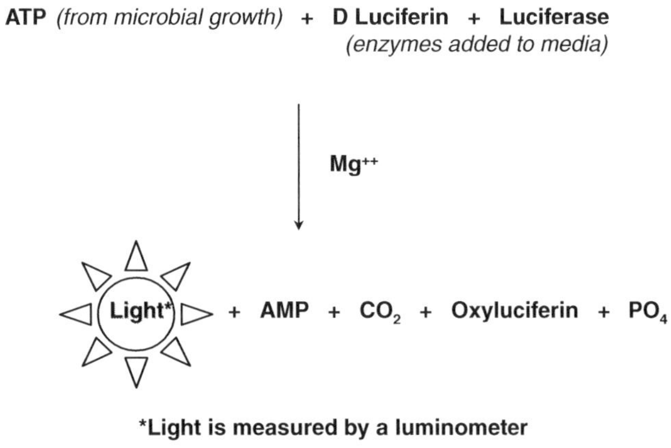

Figure 23-8 Basic principle of ATP bioluminescence rapid microbial detection system.

(e.g., USP)." Specifically, rapid microbiological test methods have been explored for the following applications:

1. Early release of product
2. Environmental monitoring
3. Raw material and process monitoring
   a. Microbial limit tests
   b. Testing of Water for Injection systems
   c. Bioburden testing for products to be terminally sterilized
4. Sterilization validation testing of biological indicators
5. Antimicrobial effectiveness testing.

Closed Vial Filling

As introduced on page 292 Aseptic Technologies has developed a closed-vial filling system, called "Crystal" technology (Fig. 23-9) (21). Gamma sterilized fully stoppered and empty vials are filled with sterile solution using a needle (typically 13G) passing through the stopper. After vial filling and needle withdrawal, a laser system reseals the hole in the rubber. The rubber is made of thermoplastic elastomer. Both vial and stopper are injection-molded and assembled by robots in an ISO 5 (Class 100/Grade A) environment. Following laser sealing, a top ring (seal) is added and the vial is gamma-irradiated.

The main advantage of this technology is the fact that the entire process from vial manufacturing to liquid filling to final sealing of the vial is entirely closed and, therefore, sterility assurance is optimized. There are many other positive claims made by the vendor. The main disadvantage(s) are the investments involved in conversion of conventional, even barrier-based, filling lines to this new type of filling line, added costs associated with materials and sterilization, added studies to determine the effects of gamma irradiation on the product, the use of plastic rather than glass, and the uncertainties of the use of a thermoplastic elastomer. This technology is certainly promising, but like any other uniquely new technology, it will take many years for the sterile product industry to be convinced and make the investments in a completely different approach to filling product into vials.

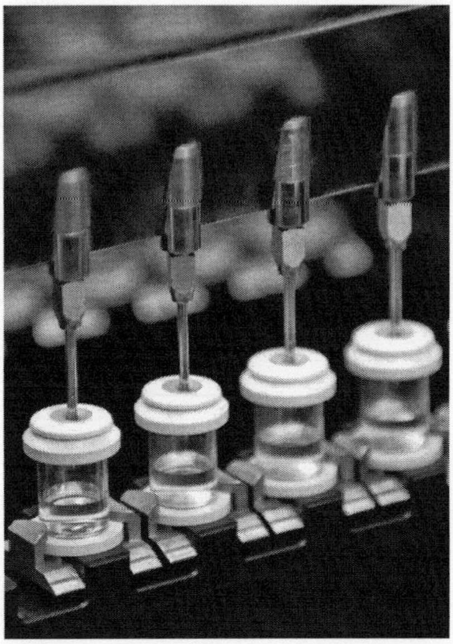

Figure 23-9 Closed vial filling. *Source*: From Ref. 22.

REFERENCES

1. Lysfjord J, ed. Practical Aseptic Processing: Fill and Finish. Vols 1 and 2. Bethesda, MD: Parenteral Drug Association/Davis Healthcare, 2009.
2. Farquharson GL. Isolators for pharmaceutical applications. In: Swarbrick J, ed. Encyclopedia of Pharmaceutical Technology. Vol 4. London, UK: Informa, 2007:2133–2141.
3. Food and Drug Administration. Guidance for Industry: Sterile Drug Products Produced by Aseptic Processing—Current Good Manufacturing Practice, September 2004.
4. Agalloco J, Akers J, Madsen R. Aseptic processing: A review of current industry practice. Pharm Tech 2004; 28:126–150.
5. Akers JE, Kokubo M, Oshima Y. The next generation of aseptic processing equipment. Pharm Tech 2006; 30:32–36.
6. Agalloco JP, Akers J. RABS and advanced aseptic processing. Pharm Tech 2006; 30:S24–S30.
7. Chiarello K. Pharma industry drives innovation in barrier/isolation design. Pharm Tech 2004; 28:44–54.
8. Lysfjord J. Restricted access barrier systems (RABS) and the application for aseptic processing. In: Lysfjord J, ed. Practical Aseptic Processing: Fill and Finish. Vol 2. Bethesda, MD: Parenteral Drug Association/Davis Healthcare, 2009:273–293.
9. Isberg E. Top 10 things to consider when purchasing an isolator: ten often overlooked mechanical and operational considerations that are important to investigate when selecting an isolator (Special Report: Keep It Clean). Pharm Proc. 2006; 23:8–14.
10. Design and validation of isolator systems for the manufacturing and testing of health care products. Technical Report No. 34. PDA J Pharm Sci Tech 2001; 55(5):1–23.
11. Kokubo M, Inoue T, Akers J. Resistance of common environmental spores of the genus bacillus to vapor hydrogen peroxide. PDA J Pharm Sci Tech 1998; 52:228–231.
12. Valigra L. Modular cleanroom. Pharm Formul Qual 2008; 11(6):15–18.
13. Aranha H. Disposable systems. Bio Proc Int 2004; 2:S6–S16.
14. Pora H. Simplifying validation of disposable technologies with presterilization by gamma irradiation, Guide to advances in bioprocessing: Disposables. Bio Pharm Int 2006; 19:36.
15. Blanck R. Disposable filling—A new trend for sterile pharmaceuticals. Pharm Manufacturing Packing Sourcer 2002; Autumn, 34–36.
16. Hodge G. Disposable components enable a new approach to biopharmaceutical manufacturing. Bio Pharm Int 2004; 17:38–49.
17. Samavedam R, Goldstein A, Schieche D. Implementation of disposables: Validation and other considerations. Am Pharm Rev 2006; 9:46–51.
18. Haughney H, Aranha H. Considerations for use in aseptic processing of pharmaceuticals. Pharm Tech 2003; 27(Supplement on Aseptic Processing):16–21.
19. Strahlendorf K, Harper K. A review of sterile connectors. Bio Pharm Intern Guide 2009; 22:28–35.
20. Miller MJ. Rapid microbiological methods in support of aseptic processing. In: Lysfjord J, ed. Practical Aseptic Processing: Fill and Finish. Vol 2. Bethesda, MD: Parenteral Drug Association/Davis Healthcare, 2009:169–219.
21. Verjans B, Thilly J, Vandecasserie C. A new concept in aseptic filling: Closed-vial technology. Pharm Tech 2005; (Supplement on Aseptic Processing), 29:S24–S29. http://www.aseptictech.com/aseptic/. Accessed June 13, 2010.
22. Verjans B. A new technology for aseptic filling of injectable drugs. Controlled Environments, October 2007. http://www.cemag.us/article/new-technology-aseptic-filling-injectable-drugs. Accessed June 13, 2010.

24 | Stability, storage, and distribution of sterile drug products

Consider the fact that the large majority of injectable and ophthalmic drugs are in the solution state, either "ready-to-use" or reconstituted. If not solutions, injectables or ophthalmics exist in a dispersed system state (e.g., suspensions, emulsions, gels) where the drug does not exist in a dry state environment. Therefore, sterile drugs are much more prone to chemical and physical degradation mechanisms than their oral solid dosage form counterparts. Major degradation mechanisms include hydrolysis, oxidation, and physical deterioration such as protein aggregation and visible particle formation. Other degradation mechanisms include photolysis and compatibility problems with packaging surfaces. Stability issues and stabilization have been covered in previous chapters (8–11). This chapter discusses good manufacturing practice (GMP) requirements for stability studies and submission of stability data for new drug applications or abbreviated new drug applications. Also, good practices for storage and distribution of sterile drug products will be covered.

GMP regulations (1) state two main requirements with respect to stability studies:

- (211.166) "There shall be a written testing program designed to assess the stability characteristics of drug products. The results of such stability testing shall be used in determining appropriate storage conditions and expiration dates."
- (211.167) "An adequate number of batches of each drug product shall be tested to determine an appropriate expiration date and a record of such data shall be maintained."

While these regulations provide the GMP basis for stability studies and establishment of expiration dates based on stability data, they do not provide details for specific requirements for submitting stability data for drug product approvals including stability requirements for bulk drugs (active pharmaceutical ingredients). Thus, the need for stability guidelines arose quickly and the first FDA stability guidelines were published in 1987 (2). Since then, the guidelines have been revised with respect to coverage within the guidelines and specific requirements for different dosage forms and different drug product submission categories (e.g., NDA, ANDA) (3–6). Requirements for bulk stability studies will not be covered.

Basic requirements for design and interpretation of stability studies for finished dosage forms are given in Table 24-1 and include the following (5):

- Full-term stability studies on at least three primary batches that represent the marketed formulation, package, and validated production process.
- Two of the three batches should be at least pilot scale, typically assumed to be at least 10% the size of the commercial batch. The third batch can be smaller if justified.
- Each batch should use a different lot of active pharmaceutical ingredient, where possible.
- Each active ingredient strength and container size must be stability tested unless the manufacturer can justify the bracket or matrix approach (discussed later).
- Photostability testing needs to be performed on one primary batch (6,7).

Stability testing should cover the following as a function of storage time and temperature:

- Physical, chemical, biological, and microbiological attributes
- Preservative (antimicrobial and/or antioxidant) content
- Functionality tests (breakloose and glide forces) for products packaged in syringe and cartridge containers.

The analytical procedures used to measure these attributes must be fully validated and stability indicating (8–12). Also differences between acceptance criteria for release of a batch and acceptance limits at the end of the expiration dating period must be justified. For example, if the potency at expiry be $\geq$95% of label claim and the acceptance limit for potency is 98% of

Table 24-1 Basic Requirements of ICH Stability Guidelines

- Three batches, two of which are pilot plant scale
- Data on lab-scale batches not acceptable as primary information
- Twelve-month stability data at the time of regulatory submission
- 25 ± 2°C, 60 ± 5% RH
- Accelerated testing is defined as six months at 40 ± 2°C, 75 ± 5% RH
- If significant change occurs at accelerated conditions, then can use an intermediate condition: 30 ± 2°C, 60 ± 5% RH
- Different batches of drug substance
- Three month testing frequency first year; six months second year

label claim, there must be data to support the fact that <3% loss in potency will occur during the product's shelf life.

If significant change occurs at any time during six-month accelerated testing, additional testing at the intermediate storage condition should be conducted (unless 30°C ± 2°C / 65% RH is the long-term condition, then there is no intermediate condition). Examples of significant changes include the following:

- Five percent change in assay from initial;
- Failure to meet potency criteria for biopharmaceuticals
- Degradation product exceeding acceptance criterion
- Failure to meet acceptance criteria for appearance, physical attributes, functionality
- Failure to meet acceptance criterion for pH
- Failure to meet acceptance criterion for dissolution for 12 dosage units.

Once determined by a manufacturer that the formulation, primary package, and process have been finalized (typically or ideally via a document signed off by all appropriate representatives from development, manufacturing, and quality), stability data must be generated on the first three production batches. Ideally, at time of regulatory submission, stability data cover the proposed shelf life from three production batches. Long-term stability studies will be conducted through the proposed product shelf life. Accelerated six-month stability data should be available, or if the drug is too unstable for six months at these stressful conditions, then shorter time periods might be acceptable with the appropriate label precautions regarding storage conditions for the relatively unstable product.

If, at the time of regulatory submission, there are no or incomplete data from at least production size batches of the product, then the commitment must be made to place the first three production batches on long-term (through proposed shelf life) stability testing as well as six-month accelerated stability testing.

The design and implementation of valid stability studies will include a justifiable sampling plan for assurance of unbiased selection of samples. At least two unit containers will be sampled at each sampling time. Assay of composites is allowable, although must be justified. Sampling intervals are usually standard (0, 1, 3, 6, 12, 18, and 24 months), although more frequent sampling pulls might be necessary. The number of sample replicates should be increased at the later sampling times.

Establishment of expiration dating period is typically computed by determining the best-fit linear regression analysis of real-time stability data and establishing the one-side lower 95% confidence time point (Figure 24-1). Expiration dating must use percent of label claim, not the percent of the average initial potency from the definite stability test lots. There are many sources of variability in stability testing, so the concern is to overestimate the expiration date. Equality of potency degradation slopes and intercepts from definitive stability test data must be compared to assure that the stability data are repeatable. Although rare, if there is little degradation and analytical variability (must be justified), then there is no need to apply formal statistical analyses.

For freeze-dried products after reconstitution, the standard stability period is 24 hours at room temperature and anywhere from 48 hours to two weeks at refrigeration, depending on chemical and physical stability. Long-term reconstituted stability may have more of a

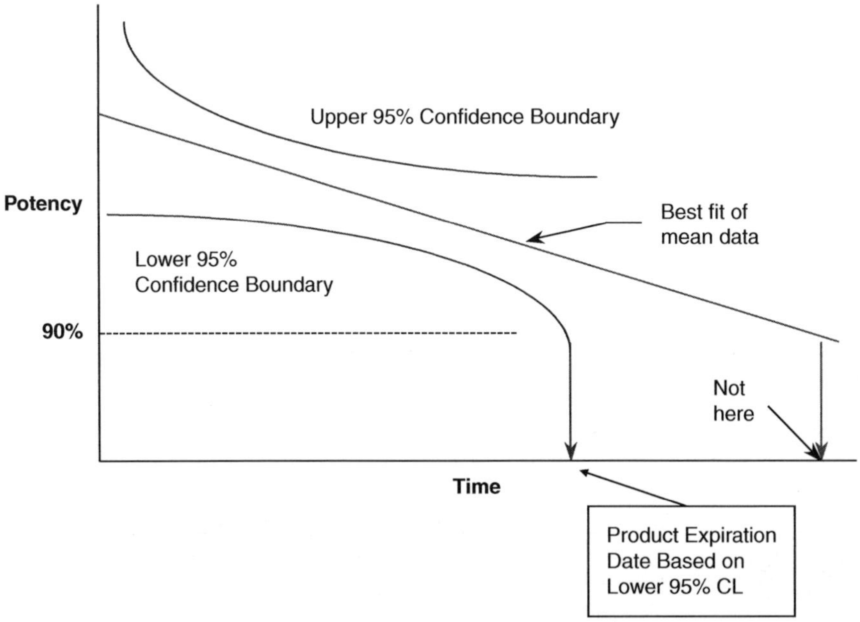

Figure 24-1 Regression analysis of stability data.

microbiological issue than chemical/physical stability issue. In fact, European guidelines (13) require immediate usage (not more than three hours after reconstitution) if the product is not preserved with antimicrobial preservatives or to use within 28 days if the product is preserved.

Many drugs are sensitive to light and are protected from excessive light exposure via packaging (e.g., amber glass, amber bags, always maintaining drug product in its secondary package) and storage in dark conditions. Data on light stability should be an integral part of *stress* testing and should be conducted on at least one primary batch of the drug product (5,6). The unprotected bulk drug, fully exposed drug product, & drug product in marketed package are all to be stress tested for light stability. It must be demonstrated that the product is adequately protected from exposure to light if it is unstable in the presence of light. The light source can be any source with output similar to D65 (outdoor daylight) or ID65 (indoor daylight) standards defined in ISO 10977 (1993). Typical light sources include the following:

- Cool, white fluorescent lamp
- Xenon or metal halide lamp
- Near-UV fluorescent lamp.

The exposure requirements are an overall illumination of >1.2 million lux hours and integrated and a near-UV energy of >200 watt hr/m2. Such exposure conditions are approximately equal to a five-month exposure period under normal room light (800 lux) for 10 hr/day.

After a product has been approved for commercial use, the first three commercial lots are stability tested. Afterwards, a minimum of one lot per year of each dosage form and each container-closure system of the product is placed on long-term stability testing. Special stability testing must be conducted whenever changes or deviations occur unless it is determined that stability testing not required (14).

Inherent in all aspects of design of appropriate stability studies, including sampling, sample times, number of replicates, defining validation of analytical methods, and evaluation of all data is the use of valid statistical procedures (15). Regression analysis is considered an appropriate approach by the FDA in evaluating stability data for any quantitative attribute and establishing a retest period or shelf life (16). Other examples of statistical methods employed in stability testing include poolability tests (17, 18) and statistical modeling (19).

Table 24-2 Examples of FDA 483 Observations on Stability Testing

- Stability failure not properly investigated, not conducted in a timely manner, conclusions questionable.
- Storage incubators not monitored; relative humidity not monitored
- Degradation products not identified
- Preservative efficacy test not done at end of stability period
- Reconstitution stability not evaluated at end of stability period
- No data on impurities
- No characterization, quantification, toxicology, clinical effects of degradation products
- Impurities found after assay was changed
- Unknown degradants discovered during stability testing, nothing done to study what degradant was and impact on product safety and quality.
- Change in upper-limit specification without QC unit appropriately justifying the change
- No data to correlate preservative efficacy test results to a HPLC assay for the preservative
- Very slow (>10 mo) investigation of unknown peaks in stability samples
- No data assuring sterility of product at end of shelf life
- Inadequate data supporting stability of reconstituted freeze-dried product, particularly at end of shelf life.

FDA INSPECTIONS AND 483 OBSERVATIONS RELATED TO STABILITY TESTING

Being a major quality system, stability testing and documentation are reviewed carefully by FDA and other government regulatory inspectors worldwide. Stability testing protocols can be complicated, especially for lyophilized products that must be tested both after reconstitution and after storage as solutions in potentially different diluents at different temperatures. Examples of 483 observations based on inspector concerns after reviewing stability testing protocols, data, and documentation supporting expiration dating are given in Table 24-2.

FDA STABILITY GUIDELINES APPLICABLE TO STERILE PRODUCTS

Storage requirements for drug products in semi-permeable and impermeable containers, and for drug products that require refrigeration or need to be frozen are as follows (2) (see also Table 24-3):

1. Storage requirements for drug products in semi-permeable containers
   (a) Accelerated: 40°C ± 2°C/NMT25% ± 5% relative humidity (RH)
   (b) Intermediate: 30°C ± 2°C/65% ± 5% RH
   (c) Long-term: 25°C ± 2°C/40% ± 5% RH
2. For drug products in impermeable containers, any controlled or ambient humidity condition
3. Storage requirements for refrigerated drug products
   (a) Accelerated: 25°C ± 2°C/60% ± 5% RH
   (b) Long-term: 5°C ± 3°C, monitor, not control, RH
4. Storage requirements for frozen drug products
   (a) Accelerated: 5°C ± 3°C/ambient humidity
   (b) Long-term: −20°C ± 5°C

BRACKETING AND MATRIXING (20)

Bracketing is allowed where stability data are obtained for the smallest and largest container and closure to be commercially marketed provided that intermediate packaging is of comparable composition and design. Storage must take place with the primary package both in the upright

Table 24-3 Storage Condition Requirements for ICH Stability Studies

Length of study	Storage conditions	Minimum time period covered by data at submission date
Long term	25°C ± 2°C/40% ± 5% RH or 30°C ± 2°C/65% ± 5% RH	12 months
Intermediate	30°C ± 2°C/65% ± 5% RH	6 months
Accelerated	40°C ± 2°C/NMT75% ± 5% RH	6 months

and inverted or on the side configurations unless there is clear validation that the container-closure system does not impact drug product quality. Typically, ongoing stability studies use inverted or on the side positions.

Matrixing reduces the number of stability samples and tests. The FDA must be consulted before implementing the study design. Matrixing applies typically for stable products with little variability in analytical methods. It is not required to have assay performed on all three lots at intermediate time points. Testing might only require two of three container-closure types within given strength of active ingredient. Matrixing does not apply to initial and final time points, test parameters, different formulations, and dosage forms.

EXTRACTABLES AND LEACHABLES AND STABILITY TESTING

Extensive extractable studies should be performed as part of the qualification of the container-closure components, including labels, adhesives, and ink. Use various solvents, elevated temps, and prolonged extraction times in conducting these studies. Adsorption or absorption of drug product components must be evaluated during stability studies.

Leachables have been covered in Chapter 7, but some reinforcement here. Leachables are potentially problematic with drugs stored in plastic syringes where components from the plastic or from the label migrate into the product. Leachables can occur from glass and rubber closures including tip caps of syringes and cartridges.

Stability studies must evaluate the sterility integrity of tip cap or needle, sterility integrity of the stopper, syringeability, and transportation fluctuations in temperature and pressure. For terminally sterilized products stability studies must evaluate and validate the terminal sterilization cycle with respect to minimization of stopper movement.

STABILITY STUDIES OF STERILE PRODUCTS CONTAINING ANTIMICROBIAL PRESERVATIVES

For products containing antimicrobial preservatives, acceptance criteria must be established for the chemical content of the preservative. Such criteria must assure that sufficient preservative activity will remain throughout the product shelf life as well as when the product is in use. Microbial challenge studies must be conducted using a preservative level less than the minimum amount specified as acceptable. The first three production batches should be tested with microbial challenge assay at start and end of stability period.

ASSURANCE OF STERILITY AND STABILITY TESTING

Stability studies of sterile products are unique from other dosage forms in that sterility must be monitored throughout the stability-testing period. A sterility test is performed at beginning of stability-test period and testing to assure integrity of the container-closure system must be conducted annually and at expiry. The container-closure test used must be a validated test, either microbiologically based or chemically based (see Chapter 30). There must be an established sensitivity to show the amount of leakage necessary to detect a failed barrier in the container-closure system.

Another unique characteristic of stability testing of sterile products is the need for pyrogens/bacterial endotoxins testing at the beginning and the end of the stability period. Sterile solid or ampoule products only need initial data.

STABILITY ASSAYS FOR DIFFERENT TYPES OF STERILE PRODUCTS

The FDA stability guidelines list various product characteristics that should be monitored during stability testing depending on the type of dosage form.

Solution Dosage Forms (Drug Injection Products)

Stability data should be generated for:

- The active ingredient
- If present, the antimicrobial preservative
- Appearance, especially any changes in color and clarity
- Degradation products

- Particulate matter (USP <788>)
- pH
- Sterility
- Pyrogenicity.

Sterile Dry Products That Will Be Reconstituted into Solutions (Drug for Injection Products)

Stability data must be obtained both from the solid state, e.g., immediately after reconstitution, and from the solution state, e.g., during specified time that the drug remains in solution before it must be administered.

Immediately Following Reconstitution

- Residual moisture right before reconstitution
- The active ingredient
- Appearance—clarity, color
- Particulate matter
- Reconstitution time
- pH
- Sterility—annual determination
- Pyrogenecity

At Various Times After Reconstitution

- The active ingredient
- Appearance—clarity and color
- pH
- Particulate matter

Dispersed system products such as Drug Injectable Suspension Products and Drug Injectable Emulsion Products

Most of the same assays as listed above for solution dosage forms are to be done with the following special tests specific to the type of dispersed system dosage form.

Suspensions

- Particle-size distribution
- Redispersibility
- Rheological properties

Emulsions

- Mean globule size and distribution
- Phase separation
- Viscosity

Products Contained in Prefilled Syringes

Syringe products will be evaluated during stability testing for the appropriate quality attributes, depending on the type of dosage form. In addition, certain functionality and integrity tests should be part of the stability-testing program:

- Applied extrusion (break) force—pressure required to move the rubber plunger
- Glide force (movement of the product through the barrel of the syringe)
- Syringeability
- Leakage

Combination Products

If drug product is to be combined with a diluent or other drug product, compatibility in admixture must be determined in upright and inverted/on-the-side orientations typically immediately after combination, then 6 to 8 hours later and after 24 hours or longer storage. Evaluations should include

- Appearance
- Color
- Clarity
- Active ingredient potency
- Degradation products
- pH
- Particulate matter
- Interaction or compatibility data in container/closure system
- Sterility

Stability data must be obtained with the lowest and highest concentrations of the drug in each diluent including using three different lots of the drug product.

All Sterile Dosage Forms

There are other basic data related to stability throughout the shelf life of the product that should/must be known:

- Continued assurance of sterility, with emphasis on container/closure integrity
- For terminally sterilized products, stability following exposure to at least the maximum specified process lethality (i.e. the maximum sterilization equivalent time (F_0) exposure—see Chapter 17)
- Inclusion of testing for extractables/leachables if other qualification tests have not provided sufficient information or assurance from plastics and rubber
- Interaction of administration sets and dispensing devices.

Stability Data Requirements As a Function of Development Phase

Phase 1

A brief description is required of the stability study and test methods used to monitor stability of drug product during Phase 1 clinical studies. Stability data are considered preliminary in that there are neither requirements for detailed stability data nor stability protocols. Stability data must show that the drug will remain stable during the course of the trial.

Phase 2

It is expected that Phase 1 and 2 stability data will provide sufficient information to develop the final formulation and select the most appropriate container and closure system by end of Phase 2. Studies started for Phase 1 will continue, if possible, or new studies must be started to support shelf of the product used throughout Phase 2 studies.

Phase 3

Emphasis should be on testing final formulations in their proposed market packaging and manufacturing site. A final stability protocol must be well defined prior to initiation of Phase 3 studies. Scientists also must establish the appropriate linkage between preclinical and clinical batches of the drug substance and drug product, and the primary stability batches in support of the proposed expiration–dating period.

Need for Extra Stability Studies Following NDA/ANDA Approval

There are three possibilities for reporting changes (filing requirements) to an approved NDA or ANDA that typically required stability data to be generated to support the change (14):

1. Data provided in the product Annual Report—change does not require prior FDA approval.
2. Changes-being-effected (CBE) supplement—data reported to FDA, FDA has 30 days to provide any comments; otherwise manufacturer can proceed to implement the change.
3. Prior approval supplement—manufacturer must obtain FDA approval before implementing the change.

There are many types of changes that require additional stability data:

- Manufacturing process of drug substance
- Manufacturing site(s)
- Formulation
- Addition of new strength for drug product
- Equipment
- Batch size
- Reprocessing step(s)
- Container closure
- Stability protocol.

It is beyond the scope of this chapter to discuss the types of changes that fall under each of the three filing requirements. However, in general, if any changed is viewed to affect the critical quality attributes of the drug product, stability data must be generated and reviewed by FDA before the change is approved.

Extra stability data also must be obtained if any manufacturing deviation occurs where the critical quality attributes of the product are in question. Examples of manufacturing deviations likely to necessitate the need for stability data generated on the affected product batch include:

- Time limits exceeded
- Administrative (release) limit not met despite data within regulatory limit
- Freeze dry cycle not met
- Finished product stored outside chill room for a period of time exceeding procedure
- Equipment breakdown, product remains in tank for extended period of time.

Distribution and Storage

Once unit dosage forms are filled into primary packaging and are stoppered, sealed, inspected, labeled, and placed in the appropriate secondary packaging, issues of distribution and storage come into play. Storage of finished dosage forms is of concern at the place of manufacture prior to distribution (shipping), the warehouse or distribution center as an intermediate storage place, and at the hospital or other final destination place (including the home). Distribution of finished products must be controlled with respect to temperature fluctuations and handling (i.e., stress, shear) during at least five transportation transfers (21,22):

- Preparing the products for transport
- Loading and unloading products into shipping equipment
- Loading and unloading products from one shipping equipment to another
- Receipt of products
- Handling of products between transportation (airport or harbor transit).

The need for good practices in storage and handling of parenteral drug products cannot be overemphasized. While all drug products must maintain strength and quality after manufacture until usage, parenteral drug products have the added requirement to remain sterile and maintain their high-purity characteristics. Difficulties encountered in the adequate storage and handling of parenteral drug products include temperature excursions, outdated shelf lives, and glass hairline cracks (not detectable) that may lead to microbial contamination.

The shipping process must be validated to assure product stability. Maps of the shipping process must be known to understand potential risk points to product stability. The product's susceptibilities to temperature excursions must be known and controlled. The packaging system must have the appropriate protective properties, e.g., light protection, insulation, cold storage compatibility. Actual temperature exposures during shipment must be monitored in order to assess any excursions and time exposure on product stability.

For temperature-sensitive products, as many biopharmaceutical products are temperature sensitive, cold chain distribution systems have become routine. Systems exist to assure that temperature-sensitive products are protected and remain at the proper temperatures throughout their distribution. Many guidelines exist for achieving these requirements, but at the publication of this book, official regulatory guidance in cold chain distribution standards is still lacking.

Risks always exist that distributed products experience freeze-thaw cycle(s). Thus the effects of these potential exposures on product stability must be studied. Risks also exist for distributed products being exposed to temperatures above 40°C, the ICH Q1A maximum temperature for accelerated stability studies. Again, distribution studies must be done to assure that such temperature excursions do not occur, or if they do, the effect on product stability is well documented. For any product, brief exposures to temperature extremes and the effect on long-term product stability and quality must be understood.

REFERENCES

1. United States Food and Drug Administration, Title 21 of the U.S. Code of Federal Regulations: 21 CFR 211—Current good manufacturing practice for finished pharmaceuticals.
2. Guidance for Industry: Submitting Documentation for the Stability of Human Drugs and Biologics. Rockville, MD: Food and Drug Administration, 1987.
3. Guidance for Industry (Draft): Stability Testing of Drug Substances and Drug Products. Rockville, MD: Food and Drug Administration, 1998.
4. Guidance for Industry: Q1A Stability Testing of New Drug Substances and Products. Rockville, MD: Food and Drug Administration, September, 1994 and first revised in August, 2001.
5. Guidance for Industry: Q1A(R2) Stability Testing of New Drug Substances and Products. Rockville, MD: Food and Drug Administration, 2003.
6. International Conference on Harmonization: ICH Q1B Guidelines for the Photostability Testing of New Drug Substances and Products. Federal Register 1997; 62:27115–27122.
7. Thatcher SR, Mansfield RK, Miller RB, et al. Pharmaceutical photostability: A technical guide and practical interpretation of the ICH guideline and its application to pharmaceutical stability, part I. Pharm Tech 2001; 25(3):98–110.
8. International Conference on Harmonization (ICH) of Technical Requirements for the Registration of Pharmaceuticals for Human Use, ICH Q2A, Validation of analytical procedures: Methodology, adopted in 1996, Geneva
9. International Conference on Harmonization (ICH) of Technical Requirements for the Registration of Pharmaceuticals for Human Use, ICH Q2B, Validation of analytical procedures: Definitions and terminology, Geneva, 1996.
10. Guidance for Industry, Bioanalytical Method Validation, FDA, May, 2001.
11. General Chapter <1225>, Validation of Compendial Procedures, United States Pharmacopeia, Rockville, MD: The United States Pharmacopeial Convention, Inc.
12. Diana, FJ. Method validation and transfer. In: Huynh-Ba K, ed. Handbook of Stability Testing in Pharmaceutical Development. New York, NY: Springer Science, 2009:163–188.
13. The European Agency for the Evaluation of Medicinal Products, Committee for Proprietary Medicinal Products, Annex to ICH/CPMP Note for Guidance on Stability Testing of New Active Substances and Medicinal Products, Maximum Shelf-Life for Sterile Products After First Opening or Following Reconstitution, June 1997, http://www.ema.europa.eu/pdfs/human/qwp/015996en.pdf. Accessed June 14, 2010.
14. Guidance for Industry: Changes to an Approved NDA or ANDA, FDA, November 1999.
15. Chow S-C. Statistical Design and Analysis of Stability Studies. Boca Raton, FL: Chapman and Hall/CRC, 2007.
16. Guidance for Industry: Evaluation of Stability Data, FDA, February, 2002.
17. Ruberg SJ, Stegeman JW. Pooling data for stability studies: Testing the equality of batch degradation slopes. Biometrics 1991; 47:1059–1069.
18. Ruberg SJ, Hsu JC. Multiple comparison procedures for pooling batches in stability studies. Technometrics 1992; 34:465–472.
19. Fairweather W, Lin TD, Kelly R. Regulatory, design, and analysis aspects of complex stability studies. J Pharm Sci 1995; 84:1322–1326.
20. Guidance for Industry. Q1D Bracketing and Matrixing Designs for Stability Testing of New Drug Substances and Products, FDA, January, 2003.
21. Seevers RH, Bishara RH, Harber PJ, et al. Designing stability studies for time/temperature exposure. Am Pharm Outsourcing 2005, 6(5), 18–23.
22. Technical Report No. 39, Guidance for Temperature-Controlled Medicinal Products: Maintaining the Quality of Temperature-Sensitive Medicinal Products through the Transportation Environment. PDA J Pharm Sci Tech 2007: 61(suppl).

BIBLIOGRAPHY

Ahern TJ, Manning MC. Eds, Stability of Protein Pharmaceuticals. New York: Plenum, 1992.

Baertschi SW, ed. Pharmaceutical Stress Testing—Predicting Drug Degradation. London: Informa Healthcare, 2005.

Cartensen JT. Drug Stability. 2nd ed. New York: Marcel Dekker, 1995.

Grimm W, Krummen K, eds. Stability Testing in the EC, Japan and the USA. Germany: Wiss. Verl. mbH Stutt, 1993.

Huynh-Ba K, ed. Handbook of Stability Testing in Pharmaceutical Development: Regulations, Methodologies, and Best Practices. New York, NY: Springer, 2008.

International Conference on Harmonization documents related to stability testing:

- ICH Q1B Photostability Testing
- ICH Q1C Stability Testing of New Dosage Forms
- ICH Q1D Bracketing and Matrixing Designs for Stability Testing
- ICH Q3A Impurities in New Drug Substances
- ICH Q3B Impurities in New Drug Products
- ICH Q5C Stability Testing of Biotech Products
- ICH Q6A Specifications for Chemical Substances
- ICH Q6B Specifications for Biological Substances.

Lachman L, DeLuca PP, Akers MJ. Kinetic principles and stability testing. In: Lachman L, Lieberman HA, Kanig J, eds. Theory and Practice of Industrial Pharmacy. Philadelphia, PA: Lea & Febiger, 1986:760–803.

Manning M, Chou D, Murphy B, et al. Stability of protein pharmaceuticals: An update. Pharm Res 2010; DOI: 10.1007/s11095–009-0045–6.

United States Pharmacoepiea, General Information <1049>, Quality of Biotechnological Products: Stability testing of Biotechnological/Biological Products.

United States Pharmacoepiea, General Information <1150>, Pharmaceutical Stability.

United States Pharmacoepiea, General Information <1079>, Good Storage and Shipping Practices.

25 | Good manufacturing practice

Good manufacturing practice regulations (also referred to as GMP, GMPs, current GMPs, cGMPs) were first proposed by the United States government in 1963, following congressional passage of the Kefauver–Harris amendment in 1962, in turn, following the thalidomide tragedy in the United Kingdom in 1960–1961. These regulations described the basic requirements for the manufacturing and packaging and distribution of finished pharmaceutical products.

In 1971, the "Rules and Guidance for Pharmaceutical Manufacturers and Distributors" or "Orange Guide" (cover of the book was orange) was published that provided GMP guidelines for pharmaceutical manufacturers in the United Kingdom. The Orange Guide was revised in 1997. One of the authors of the Orange Book was John Sharp. In his book (1), Sharp expressed the need for good manufacturing practice regulations arose from the interactions of

1. The severe limitations of end-product testing as a determinant of the quality of medicines.
2. The high potential hazard of even a very small proportion of defective units.
3. The low level of probability that the ultimate consumer (patient) would be able to detect a defective product before it would be too late.

In fact, his book contains a quote (p. 12) that is repeated, in part, here: When a maintenance technician was asked what he thought was the purpose of cGMPs, his answer was "You always have to remember the poor bloke that's going to take the bloody stuff." Mr. Sharp stated that he regarded this statement as one of the most profound statements on Pharmaceutical Quality Control that he ever heard. Indeed the purpose of GMP is to ensure that pharmaceutical products are safe and effective and it is the responsibility of product manufacturers overseen by government regulatory bodies such as the FDA of the United States and the European Medicines Agency (EMEA) of the European Union.

Compliance to cGMPs ensures that pharmaceutical products taken by or administered to humans and animals meet or exceed minimum requirements of the following five attributes:

SAFETY
IDENTITY
STRENGTH
PURITY
QUALITY

Those in the pharmaceutical industry refer to these attributes simply as SISPQ. Every regulation and every guidance statement are aimed at meeting one or more of these attributes. Meeting these attributes ensures that every unit of pharmaceutical product is safe and effective.

It took over 15 years for the Food and Drug Administration (FDA) and the U. S. pharmaceutical industry to agree on terminology and descriptions in cGMP regulations for Congress to finally legalize these regulations effective from September 29, 1978. These regulations are found in the Code of Federal Regulations (CFR) Part 210 and Part 211[1] (2). Interestingly, there also was a CFR Part 212—Good Manufacturing Practices for Large Volume Parenterals (GMP LVP)—but were never enacted as law as were Parts 210 and 211. However, many components of CFR 212 were adopted by the sterile products industry such as facility requirements and sterilization validation requirements.

The European Union GMPs are described in "EU Guidelines to Good Manufacturing Practice: Medicinal Products for Human and Veterinary Use" (3). The first edition of the Guide was published in 1989 that included an annex on the manufacture of sterile medicinal products.

[1] Other GMP CFR references are CFR 606 for blood and related components, CFR 600 and 610 for biologics, and CFR 820 for devices.

The second edition was published in 1992 and the latest edition was published in 2008. The 2008 edition was divided into two parts: Part I for medicinal products for human and veterinary use and Part II for active substances used as starting materials.

During the 15 years of the FDA and industry debating on the final wording of the GMP regulations, the industry had fought vigorously for the GMP regulations not to be substantive; rather allowing the industry to have freedom to interpret and apply the regulations. Once GMPs became official in 1978 and FDA inspections began to enforce their compliance, the industry found that more help was required to better understand how the FDA was interpreting the regulations. Thus, guidance documents began to be issued, giving the industry more specific information on FDA expectations, especially in areas of validation and documentation (Table 25-1 contains a partial list of FDA Guidance documents). Guidance documents are not legally binding, but the basic attitude of the FDA is that if the industry chooses not to follow the guidance documents, it needs to justify why not.

GMPs are enforced in the United States by the FDA, under Section 501(B) of the 1938 Food, Drug, and Cosmetic Act (21USC351). GMPs are legally considered industry standards such that

Table 25-1 Examples of Relevant FDA Guidance Documents for GMP Compliance

- Guidance for the submission of documentation for sterilization process validation in applications for human and veterinary drug products
- Guidance for industry: sterile drug products produced by aseptic processing—current good manufacturing practice
- Guideline for validation of Limulus Amebocyte Lysate Test as an end product endotoxin test for human and animal parenteral drugs, biological products, and medical devices
- Compliance program guidance manual 7356.002 A, sterile drug process inspections
- Guide to inspections of lyophilization of parenterals
- Guide to inspections of high purity water systems
- Guide to inspections of microbiological pharmaceutical quality control laboratories
- Guide to inspections of sterile drug substance manufacturers
- Draft guidance for industry on process validation: general principles and practices—11/18/2008
- Draft guidance for industry: submission of documentation in applications for parametric release of human and veterinary drug products terminally sterilized by moist heat processes—8/5/2008
- International Conference on Harmonisation (ICH); guidance for industry: Q3 A impurities in new drug substances—6/5/2008
- Guidance for industry: container and closure system integrity testing in lieu of sterility testing as a component of the stability protocol for sterile products—2/22/2008
- International Conference on Harmonisation (ICH); draft guidance: Q10 pharmaceutical quality system—7/12/2007
- Guidance for industry: quality systems approach to pharmaceutical CGMP regulations—9/29/2006
- International Conference on Harmonisation (ICH); guidance for industry: Q9 quality risk management—6/1/2006
- International Conference on Harmonisation: (ICH); guideline for industry: Q2 A text on validation of analytical procedures—3/1995
- International Conference on Harmonisation (ICH); guidance for industry: Q2B validation of analytical procedures: methodology—5/19/1997
- International Conference on Harmonisation (ICH); guidance for industry: Q5E comparability of biotechnological/biological products subject to changes in their manufacturing process—6/29/2005
- Guidance for industry: nonclinical studies for the safety evaluation of pharmaceutical excipients—5/18/2005
- International Conference on Harmonisation (ICH); guidance for industry: Q1E evaluation of stability data—6/7/2004
- International Conference on Harmonisation (ICH); guidance for industry: Q1 A(R2) stability testing of new drug substances and products—11/20/2003
- International Conference on Harmonisation (ICH); guidance for industry; Q1D bracketing and matrixing designs for stability testing of new drug substances and products—1/15/2003
- Guidance for industry: container closure systems for packaging human drugs and biologics; questions and answers—5/13/2002

Check http://www.fda.gov/cber/guidelines.htm for latest versions.

Table 25-2 Organization of United States and European Union Good Manufacturing Practice Regulations

United States (Food and Drug Administration)[a]	Europe (European Union)[b]
Resources	Resources
Organization and personnel	Personnel
Buildings and facilities	Premises and equipment
Equipment	
Methods and materials	Methods and materials
Control of components, containers and closures	Documentation
Production and process controls	Production
Packaging and labeling control	Quality control
Laboratory control	Contract manufacturing and analysis
Documentation and distribution	Quality management
Holding and distribution	Qualified persons
Records and reports	Complaints and product recall
Returned/salvaged drug products	Self-inspection

[a]United States: The Code of Federal Regulations, Title 21, Food and Drugs, Parts 210 and 211, cGMP in Manufacturing, Processing, Packing, or Holding of Drugs and Finished Pharmaceuticals.
[b]Europe: Volume 4 of "The rules governing medicinal products in the European Union" contains guidance for the interpretation of the principles and guidelines of good manufacturing practices for medicinal products for human and veterinary use laid down in Commission Directives 91/356/EEC, as amended by Directive 2003/94/EC, and 91/412/EEC respectively.

continued violations of these standards can be used to prosecute violators in courts of law. The World Health Organization (WHO) version of GMP is used by pharmaceutical regulators and the pharmaceutical industry in over 100 countries worldwide, primarily in the developing world. The European Union's GMP (EU-GMP) enforces more compliance requirements than the WHO GMP, as does FDA's version in the United States.

Both U.S. and E.U. GMP regulations are organized into three main sections (Table 25-2). Such an organization demonstrates to some extent what both regulatory groups especially emphasize. For example, the U.S. GMPs emphasize documentation a little more than the European GMPs while Europe emphasizes Quality Management. However, as time has passed, requirements for GMP compliance have merged to essentially the same emphasis regardless of manufacturing location.

Because GMP regulations were first proposed in 1963 and enacted as law in 1978 with relatively minor changes, it is important to consider the "c" in cGMP. GMPs are *current*, which means that the meaning and application of GMPs change as the industry changes. Four main areas of change in the industry include the following:

1. Scientific and technological advances—Examples include computer systems, computer process control, paperless manufacturing, electronic signatures, barrier isolation technologies, biotechnology medicine manufacturing, and many other examples.
2. Adverse events—Examples include product tampering incidents, product recalls caused by lack of understanding or lack of control of processes, contamination incidents, and needle safety precautions.
3. Inspection activities and findings—Examples include the generic drug scandal of the late 1980s, lack of GMP compliance in manufacturing of active pharmaceutical ingredients, dealing with out-of-specification data, poor documentation practices, lack of aseptic process validation and many other validation studies, and many of other examples.
4. Industry practice—Examples include introduction of laboratory management systems, improvements in all aspects of manufacturing (equipment advances, automation, inspection, etc.), new drug delivery systems (e.g., micro-and nano technologies), new analytical methods, and many other examples.

Interpretation and application of cGMPs have had to adjust and adapt to all these sources of change over the years. Often, this has led to confusion and controversy that has taken months to years to resolve. Undoubtedly, changes will always be the rule rather than the exception so cGMPs will always evolve over time.

The generic drug scandal of the late 1980s propelled the enhancement of cGMP compliance enforcement. Until the scandal, GMP compliance inspections occurred on a routine basis, sometimes two or more years after a new drug application (NDA) or abbreviated new drug application (ANDA) was submitted and approved by the FDA. During and after the scandal, starting in 1990, the FDA introduced "preapproval GMP inspections" where local and/or regional GMP compliance inspections occur soon after the regulatory dossier has been submitted. Prior to approving the application FDA headquarters would need to receive a recommendation from the GMP compliance inspectors. Therefore, even today cGMP compliance now impacts approvability of NDAs and ANDAs (shown schematically in Fig 5-2, p. 51).

Preapproval inspections required the manufacturer to accelerate final product/process validation studies. Prior to preapproval inspections, manufacturers would submit process information that had not yet been done. Indeed, often manufacturers would wait for FDA approval of a NDA before finalizing equipment selection/purchase or even building the facility to manufacture the commercial product. Preapproval inspections required manufacturers to finalize the manufacturing process with evidence of validation of the process at commercial scale prior to submitting the NDA.

During the first few years of preapproval GMP inspections, roughly 40% of NDAs and ANDAs were withheld approval because of GMP compliance problems uncovered at the manufacturing facility and/or testing laboratory. The major reasons for failing preapproval inspections in the early 1990s were in order of frequency:

1. Facility noncompliance
2. Laboratory noncompliance
3. Any discrepancy suggesting fraud or deception
4. Lack of data supporting process control
5. Clinical batch analytical and performance data do not correlate to data from production batches
6. Lack of acceptable validation controls
7. Excessive number of cGMP problems.

Generally, the most common problem found during GMP inspections is the failure to follow written standard operating procedures (SOPs). This has been true since cGMP regulations became enforced and backed by law. Another common GMP noncompliance problem is the failure to follow good documentation practices (see Chapter 26). Table 25-3 lists a wide variety of specific 483 observations found during GMP compliance inspections of sterile manufacturing drug plants.

The GMP inspection process can be relatively simple if the FDA inspection team finds no problems with GMP compliance during the inspection. However, if problems are found, depending on their severity and/or frequency, the following sequence of regulatory activity can occur:

1. Inspections are usually preapproval or annual or biannual visits.
2. Inspections can also be prelicensing inspections (for biologics), follow-up inspections from previous inspection or stimulated by some problem, e.g. complaint and recall.
3. At the conclusion of an FDA inspection, the inspection team communicates its observations to the inspected company via Form FDA-483 (or simply, "483").
4. Upon returning to the FDA District Office, the inspection team writes an Establishment Inspection Report (EIR) that elaborates and expands on the inspection observations and links the observations to the evidence collected to support them. The EIR is reviewed, and if the conditions it describes are serious enough in the minds of the reviewing officials, a Warning Letter may follow.
5. A Warning Letter may also be issued if the FDA is not satisfied with the timing or content of the firm's response(s) to the 483 observations.
6. A Warning Letter differs from a 483 in several important respects. A 483 represents the observations of the inspection team (or lone investigator, if such is the case). A Warning Letter indicates that higher level FDA officials have reviewed the inspection findings and have concluded that the findings warrant further formal notification to the inspected company that FDA believes serious violations may exist.

Table 25-3 Examples of Specific 483 Observations During cGMP Compliance Inspections of Sterile Drug Manufacturing Plants

- Lack of time limits between product manufacture and filling
- Lack of filter validation for microbial challenge and extractables
- Lack of SOP for how sterility test samples are obtained
- Failure to collect sufficient number of samples to evaluate for particulate matter based on valid statistical plan
- Failure to design and maintain WFI system in certain buildings
- Failure to establish sufficiently detailed and validated instructions for batch reprocessing
- Failure to validate the environment where filling machines located
- Levels of viable and nonviable particulates observed during production cannot be related to media fills or other qualification studies
- Lot released although active failed initial potency test with no investigation—several examples
- Investigations started, but not completed with lots failing certain specs, e.g., pH
- No investigation of a failed initial sterility test
- No validation of aseptic connection
- No sterility data of product at end of shelf life
- Failure to clean and sanitize various types of equipment
- No assurance that all lots released are free from critical glass defects. Inspection process not capable of detecting cracks in glass vials
- Inspections repeated until lots passed without reacting to excessive rejections
- No data on media used in media fill capable of supporting growth of environmental isolates
- Lyophilization media fill inadequate because did not simulate pulling and released vacuum
- Air samples taken only when people not in the room
- Lack of cleaning validation for certain equipment
- Equipment use logs not maintained
- Changes made without applying change control procedures
- Lack of management knowledge and experience in cGMP
- No corrective actions despite fact that 11 product sterility failures had occurred over past two years with same organism causing the failures
- Sterility test procedures do not show proof that low levels of organisms can be recovered
- Inadequate monitoring of clean rooms, air pressure differentials
- Sterilization processes not validated
- Many failures of QC unit to properly investigate deviations such as capping defects, particulates not identified, labeling errors, environmental excursions, WFI microbial levels too high, assay failures
- Lack of authority of QC unit
- QC unit failed to review production records for errors and deviations and failed to fully investigate deviations
- Sterility failures—at least nine batches
- Did not account for gram negative microbes detected after fumigation
- Media fills did not include known interventions
- Aseptic connections not validated
- Failure to establish a system for maintaining equipment to control aseptic conditions
- Failure to follow appropriate written procedures designed to prevent microbial contamination of drug products
- Failure to establish accuracy, sensitivity, specificity, and reproducibility of test methods
- Inadequate investigations of bioburden failures
- Deficiencies in aseptic practices by personnel
- Operators performing setup, sterile filtration, and aseptic dispensing did not apply proper aseptic techniques
- Failed to demonstrate effectiveness of cleaning and disinfection
- Inadequate controls to prevent cross-contamination of potent compounds
- Inadequate facility design, e.g., porous drywall-like material not easily cleaned
- Rust-like substances seen in several locations in different rooms
- Lack of smoke studies
- Many observations of poor aseptic practices by operators
- Failure to conduct bacterial filtration retention validation for all aseptically filled products
- Lack of adequate sanitization of items brought into Class 100 areas
- Failure to investigate thoroughly or have written records of batch exceptions and failures (e.g., sterility test failures, media fill failures)

Table 25-3 Examples of Specific 483 Observations During cGMP Compliance Inspections of Sterile Drug Manufacturing Plants (*Continued*)

- Failure to replace faulty HEPA filters
- No air flow pattern testing in aseptic areas
- Lack of classified environments after vial stoppering
- No justification of selected surfaces for EM
- Smoke studies do not fully demonstrate air flow movement away from work surfaces during personnel activities
- No attempt made to determine correlation between a product defect (visible precipitate) and consumer complaints
- Inadequate investigations/response to other product complaints; no root cause determined
- Failure to follow timeframes in SOPs for completing investigations
- Particle monitoring locations not very close to filling zones and exposed product
- No preventive maintenance schedules for stability chambers, freezers, etc.
- Personnel with factory scrubs and dedicated shoes allowed access to common personnel hallways where nongowned personnel are also located
- Inadequate investigations of OOS results

Abbreviations: WFI, water for injection; EM, environmental monitoring; OOS, out-of-specification.

7. A Warning Letter has a twofold purpose: (*i*) to stimulate voluntary corrective action and (*ii*) to establish a background of prior warning should further regulatory action by FDA be needed at a later date.
8. Warning letters should be issued only for violations of "regulatory significance." The threshold for determination of what constitutes "regulatory significance" is that failure to adequately and promptly achieve correction to the warning letter may be expected to result in enforcement action ... the warning letter would be appropriate to document prior warning if adequate corrections are not made and subsequent enforcement action is warranted, that is, injunction or prosecution.
9. Both the 483 and the Warning Letter are serious documents that warrant a prompt and thoughtful reply. Companies often tend to rush replies at the expense of careful consideration of the issues. Many companies believe that a rapid response to a 483 will prevent a Warning Letter. In certain cases this may be true, but a rapid 483 reply is no guarantee that a Warning Letter will not follow. A poorly written 483 response, on the other hand, may very well increase the likelihood of a subsequent Warning Letter.

Hundreds of Warning Letters were issued during the first few years (1990–1994) after initiating this documentation as a regulatory enforcement activity. These years also witnessed a huge number of product recalls, seizures, injunctions, and prosecutions as the FDA purged the industry of unscrupulous pharmaceutical companies and individuals. Issuance of 483s and Warning Letters are still common today, but not as bad as many years ago.

The FDA also began an enforcement activity called "Consent Decree" for companies that are in so much out of compliance with GMPs that it will take months to years to return to compliance. The FDA issues consent decrees to resolve long-term and significant GMP noncompliance problems with the intent to ensure production and distribution of safe and effective drug products. The consent decree outlines steps that the company must take to become compliant with penalties for failing to meet conditions and schedules. Consent decrees are issued because of recurrent failures:

- To ensure products meet quality standards prior to release
- To conduct adequate lab investigations
- To maintain lab equipment
- To keep adequate records
- To complete validation of products
- Equipment cleanliness
- Insufficient employee training
- Poor QC oversight and practices.

Many pharmaceutical companies, including some of the largest in the world, have received consent decrees.

Reasons for issuance of 483 observations during either preapprovable or general biannual GMP inspections usually fall under one of the following categories (4):

- 211.22(d) QC unit responsibilities and SOPs not in writing or not fully followed.
- 211.192 Lack of or incomplete investigation into batch discrepancies or failures.
- 211.113(b) Appropriate sterile manufacturing procedures not established or fully followed.
- 211.113(b) Inadequate validation of sterile manufacturing.
- 211.160(b) Lack of scientifically sound laboratory controls.
- 211.192 Investigation into batch discrepancies or failures did not extend to other products that may have been affected.
- 211.42(c)(10)(iv) Inadequate system for environmental monitoring.
- 211.100(b) Manufacturing SOPs not followed or documented at time of performance.
- 211.63 Inadequate equipment design, size, and/or location.
- 10.211.67(a) Inadequate equipment cleaning, sanitizing, and/or maintenance.

Table 25-3 provides some specific examples of 483 observations issued to manufacturers of sterile drug products. There are many publications that can be consulted to keep current with 483 observations and FDA GMP compliance activities, one of the best being the GMP Letter (5).

GMP APPLICATION FOR PHASE I AND II CLINICAL MANUFACTURING

Generally, GMP regulations have applied to the manufacturing of early phase investigation drug products. However, manufacturing processes are not fully validated during Phase I product manufacturing. European GMP guidelines (Annex 13) state that manufacturing practices should be "appropriate to the stage of development of the product." Quality systems must be in place to ensure that investigational drug products meet basic requirement of safety, identity, strength, purity and quality. In fact, because so little is known about drugs in early clinical development, quality control laboratories and systems used to ensure quality are even more important.

LATEST GMP REVISIONS

As stated earlier, GMPs are always evolving. Besides specific requirements in the regulations, the overall focus and philosophy of GMP compliance has changed over the years. In the early years of GMP compliance (1960s and 1970s) the main focus was on the product. In the 1980s and 1990s the emphasis focused on the process. In the 21st century the emphasis has moved to quality systems (Fig. 25-1). Quality systems are discussed in chapter 26.

The latest revisions to the GMP regulations were enacted in 2008 to harmonize U.S. GMPs with other international regulations as well as to simply update the regulations. The main changes were as follows:

1. In 211.48(a) the Environmental Protection Agency (EPA) water standard was eliminated since the FDA does not want to refer strictly to the "Primary Drinking Water Regulations." This standard was replaced with the requirement that all water be "safe for human consumption."
2. Make it clear in writing that equipment is to be sterilized [211.67(a)].
3. Omission of the second personnel check in case of automatic equipment (211.68).
4. Elimination of the reference to asbestos-containing filters from 211.72.
5. The requirement that depyrogenization of sterile containers should be validated [211.94(c)]. This is industry practice, but was not previously clearly specified in GMP law.
6. The addition of bioburden testing as an example of in-process testing that should be carried out (Section 211.110, Sampling and testing of in-process materials and drug products).
7. The requirement to validate procedures to prevent microbiological contamination of all aseptic processes (not just sterilization processes), in Section 211.113 (a) bringing the regulation into line with existing guidance and industry practice, such as the FDA Aseptic Processing Guidelines.

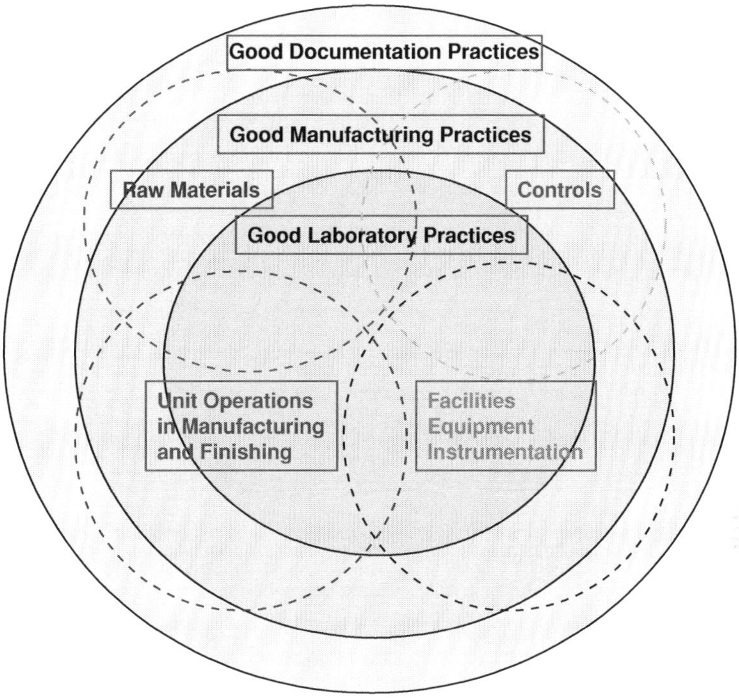

Figure 25-1 Basics of GMP quality systems.

8. The requirement that certain procedures be carried out by one person and checked by a second person may be satisfied by the use of an automated system, if one person verifies that the operations are performed accurately by such equipment. Such verification by a second Individual is being updated for the following sections—211.68, 211.101, 211.103, 211.182, and 211.188, but interestingly, not 211.194 that covers laboratory results.

EUROPEAN UNION GMP COMPLIANCE AND STERILE PRODUCT INSPECTIONS

The EU GMP Guide Annex 1 Sterile Medicinal Products is the main GMP document for GMP compliance that is used by EMEA inspectors. The guideline often is updated with the last update occurring in March of 2009. Among the most recent changes are

- Classification table for environmental cleanliness of clean rooms
- Media simulations
- Bioburden monitoring
- Capping of freeze-dried vials.

EU inspections are primarily done by the United Kingdom Medicines and Healthcare Products Regulatory Agency (MHRA). The top three deficiencies found by EU inspectors usually fall into one of these categories

1. Documentation problems with quality systems and procedures
2. Documentation problems in manufacturing
3. Problems with facility design and maintenance.

EMEA and MHRA inspectors are especially interested in the role of corporate management in GMP compliance. The increased emphasis on corporate management involved with quality systems and GMP compliance somewhat conflicts with the traditional role of the European "Qualified Person" and who is ultimately responsible for the assurance of safe, effective, and high-quality products being released for commercialization.

The main question being pursued during GMP inspections by European inspectors is "what system does the company have in place to assure that the company complies with GMP" (i.e., "to ensure a robust supply of medicines have been manufactured and controlled in such a way as to ensure that they are of a quality fit for the intended purpose") (6). MHRA declares its high interest in making sure that senior management of a company has significant responsibility for GMP compliance. Like the FDA, the inspector's concern is not so much whether or not problems occur during manufacturing and testing, but much more so how these problems were investigated, resolved, and Corrective Action and Preventive Action (CAPAs) applied.

ISO

The International Standard Organization (ISO) represents 162 countries in the world, is non-government although many of its member work for government organizations, and develops and publishes standards applicable worldwide. More than 17,500 standards have been developed covering all kinds of technologies and processes. ISO standards have no legal authority in any country unless a country chooses to adopt the ISO standard for a particular process rather than develop its own. An example might be the application of ISO clean room standards (7) (e.g., ISO 5–8) replacing the former Federal Standards 209 for air classification. Other ISO documents applicable to the sterile products industry are given in Table 25-4 with the table not intended to be complete.

GMP AND ASEPTIC PROCESSING

As was emphasized in chapter 21, the large majority of small-volume sterile products are produced by aseptic processing where there exists a much higher probability of adventitious contamination compared with products sterilized by terminal sterilization. Because of severe limitations with validation methods used to predict the probability of products becoming contaminated during aseptic processing, FDA GMP inspections of sterile product manufacturing facilities are frequent and comprehensive. Experience has shown that some manufacturers continue to have difficulty in satisfying regulatory requirements evidenced by significant 483 findings, regulatory letters, and potential product recalls and seizures (8). Johnson and Farquharson have provided a broad perspective on global regulatory compliance applied to aseptic processing (9).

Table 25-4 International Standards Organization (ISO)—Some Standards Related to Good Manufacturing Practices Relevant to the Sterile Products Industry

ISO 9000	Quality management system in production environments
ISO 9001	Quality management
ISO 10993	Biological evaluation of medical devices
ISO 11135	Sterilization of health care products—ethylene oxide
ISO 11137	Sterilization of health care products—radiation (3 parts)
ISO 11607	Packaging for terminally sterilized medical devices (parts 1 and 2)
ISO 11737	Sterilization of medical devices (parts 1 and 2)
ISO 13408	Aseptic processing of health care products
-1	General requirements
-2	Filtration
-3	Lyophilization
-4	Clean-in-place technologies
-5	Sterilization-in-place technologies
-6	Isolator systems
ISO 14161	Biological indicators
ISO 14644	Clean rooms and associated environments
ISO 14698	Biocontamination control
-1	General principles and measurement of biocontamination of air, surfaces, liquids, and textiles
-2	Evaluation and interpretation of biocontamination data
-3	Methodology for measuring efficiency of cleaning and/or disinfection processes of inert surfaces bearing biocontamination
ISO 14937	Sterilization of health care products—general requirements
ISO 17664	Sterilization of medical devices—reprocessing and resterilization
ISO 17665	Sterilization of health care products—moist heat

FINAL COMMENT

The author was involved a few years ago as an expert witness in a court case involving GMP compliance. In preparing for the deposition, the study by Barr et al. (10) was found to be the most valuable and informative resource on history, development, and application of current good manufacturing practices.

REFERENCES

1. Sharp J. Good Manufacturing Practice: Philosophy and Applications. Buffalo Grove, IL: Interpharm Press, 1991.
2. FDA. Code of Federal Regulations, Title 21, Part 210 and Part 211, "Current Good Manufacturing Practice for Finished Pharmaceuticals". Washington, DC: U.S. Government Printing Office, 2009.
3. European Commission, Enterprise and Industry Directorate-General. The Rules Governing Medicinal Products in the European Union, Volume 4 EU Guidelines to Good Manufacturing Practice: Medicinal Products for Human and Veterinary Use, Annex I, Manufacture of Sterile Medicinal Products, 2008 Revision.
4. PDA Newsletter, 2008 (January), Parenteral Drug Association, 23–24.
5. GMP Letter, FDA News, 300 N. Washington St., Suite 200. Falls Church, VA 22046-3431. http://www.fdanews.com/newsletter?newsletterId=17. Accessed June 14, 2010.
6. EU realigning inspections on risk basis, International Pharmaceutical Quality, July/August, 2008, Vol 2(4):9, 12.
7. Cleanrooms and associated controlled environments—Part 1: Classification of air cleanliness, International Standard Organization (ISO), ISO 14644-1: 1999, Geneva, Switzerland.
8. Tetzlaff RF. FDA Regulatory inspections of aseptic processing facilities. In: Olson WP, Groves MJ, eds. Aseptic Pharmaceutical Manufacturing. Buffalo Grove, IL: Interpharm, 1987:367–401.
9. Johnson R, Farquharson G. Regulatory background to aseptic processing. In: Lysfjord J, ed. Practical Aseptic Processing: Fill and Finish. Vol 1. River Grove, IL: Parenteral Drug Association/DHI Publishing, 2009:3–18.
10. Barr DB, Celeste CC, Fish RC, et al. Application of pharmaceutical CGMPs. Washington, DC: FDLI, 1997.

26 | Quality assurance and control

The importance of undertaking every possible means to ensure the quality of the finished product cannot be overemphasized. Every component and step of the manufacturing process must be subjected to intense scrutiny to be confident that quality is attained in the finished product. The responsibility for achieving this quality is divided appropriately in concept and practice into Quality Assurance (QA) and Quality Control (QC). QA relates to the studies made and the plans developed for ensuring quality of a product prospectively, with a final confirmation of achievement. QA covers all aspects of quality that includes QC, manufacturing, distribution, and inspections. QC embodies the carrying out of these plans during production and includes all of the tests and evaluations performed to be sure that quality exists in a specific lot of product.

The principles for achieving quality are basically the same for the manufacture of any pharmaceutical dosage forms. Quality systems pertinent to sterile drug manufacturing and control are listed in Table 26-1. Quality systems have evolved from principles of QC and the product and QA and the process (Fig. 26-1) (1). Table 26-2 list examples of quality systems that might be part of research and development activities.

DOCUMENTATION

An injectable products manufacturer is only as good as its documentation practices (Table 26-3). The most important documents to a manufacturer are the master file, the batch record, process logbooks, and material logbooks.

Master File

The master file is the perpetual record of production and control cycles on all batches of a particular product.

Batch Record

The batch record is the complete record of manufacture, finishing, control, and distribution of a single batch of product. The batch record contains the following:

- Formulation information
- Control numbers for each component with QC Unit approval
- Start and completion times for each operation
- Chemical weight checks
- Identity of all processing equipment
- Complete details of the entire process
- Labeling requirements
- In-process sampling procedures, test requirements
- Material accountability (reconciliation)

Reconciliation typically is a source of continual angst for batch release because it is rare that 100% of the starting material (weight basis) is recovered (accounted for) at the end of the batch process. Sources of loss or forgetting to account for loss include the following:

- Formulated bulk solution: Samples for in-process assay, residual solution left in the bottom of the tank, and solution lost or absorbed in transfer systems and the filtration apparatus.
- Filling process: Losses due to spills, filter assembly, residuals in tubing, and errors in accounting for vials filled, samples for testing, and vials rejected.

Weighing the filter assembly before and after filtration can be a good practice for measuring solution left in the assembly, but this is time-consuming and many companies do not implement this practice. More careful accounting for solution lost or samples removed can reduce the

Table 26-1 Examples of Quality Systems Relevant to Sterile Product Manufacturing and Quality

Quality system	Specific examples
Document control	All information management systems
	Electronic data management systems
	Logbooks
	Records and record retention
	Document reviews
Batch records	Batch record release
	Exception (deviation) reporting
	Rejected product
	Returned product
	QA hold
Change Control	Standard operating procedures
	Work orders
	Validation change control
	Process validation change control
	Regulatory commitment documents
	Vendor documents
	Facility shutdown procedures
	Bill of material changes
	Master batch record revisions
Validation	Sterility assurance validation
	Finishing validation
	Business and facility systems validation
	Manufacturing equipment IQ/OQ/PQ
Production systems	Scheduling
	Equipment coordination
	Preparation
	Formulation
	Filling
	Capping
	Lyophilization
	Sampling
	Inspection
	Packaging
	Label control
Laboratory control	Quality control chemistry—raw material, in-process finished product, stability, equipment release, LIMS
	Quality control metrology
	Method validation
	Method transfer
	Quality control microbiology
	Environmental and personnel monitoring
	Sterility
	Bioburden
	LIMS
	Microbial (genus/species) identifications
	Classified area performance qualifications
	Water for injection
	Nitrogen and other compressed gases
	Clean steam
Material management, warehousing, identification, and traceability	Receiving
	Raw material sampling/testing/release
	Damaged material
	Returned product
	Shipping
	Storage and labeling
	Material transfer
	Inventory control

(*Continued*)

Table 26-1 Examples of Quality Systems Relevant to Sterile Product Manufacturing and Quality (*Continued*)

Quality system	Specific examples
Purchasing controls and supplier management	Purchase orders
	Bill of material
	Global supplier quality
	Supplier corrective action reports
	Inventory control
Training	Curricula
	Course review
	Compliance metrics
	Record keeping
	Departmental training programs
	Training effectiveness
	Performance qualifications
New project implementation	Potential project evaluation
	New product introduction
	Process validation
	Cleaning validation
	Technology transfer
Periodic product quality evaluation	Release data
	All batches that failed specifications
	Production deviations and investigations
	Adequacy of all corrective actions
	Complaint data
	Stability data
	In-process control data
	Internal limits
	Changes in processes and/or methods
	Regulatory specification control
	Product recalls
	Returned and salvaged products
	Review of previous evaluations
Regulatory affairs	Change notifications
	Regulatory hold
	Pre and post-approval submissions
	Drug master files
	Site master files
	Product recalls
Quality management systems	Responsibilities of the quality unit
	Management review meetings
	Regulatory inspection commitments
	Internal audits
	Corrective and preventive actions

Abbreviations: IQ, installation qualification; OQ, operational qualification; PQ, performance qualification; LIMS, laboratory information management system.

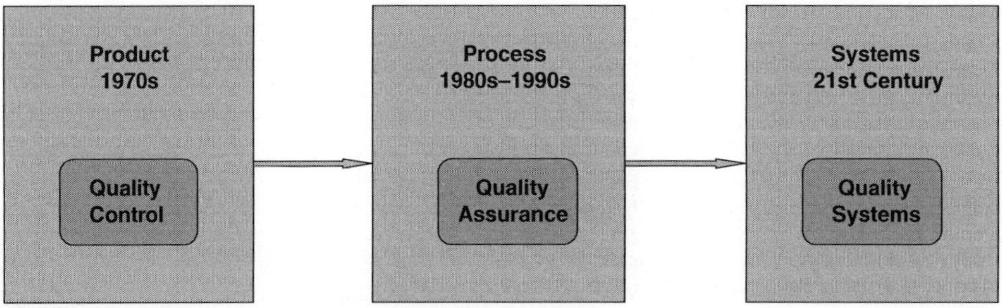

Figure 26-1 Models of quality. *Source*: From Ref. 1.

Table 26-2 Research and Development Quality Systems

- Change control
- Risk management
- Protocols and reports
- Reference standards
- Supplier quality
- Technology transfer
- Nonconformance, exceptions, and corrective and preventive actions
- Out-of-specification and out-of-trend data
- Training
- Compendial compliance
- Quality management reviews
- Harmonized testing/acceptance criteria (ICH/ISO)
- Method validation
- Stability
- Complaint investigations
- Chemical and reagent expiration dating
- Good documentation practices
- Transportation studies
- Plastic and elastomeric closure testing
- Lab equipment IQ/OQ/PQ
- Setting product limits

potential for unaccepted ranges of reconciliation. Generally, ranges of reconciliation are 97% to 103% of starting material by weight. Ranges tend to be broader the larger the batch size (again the larger the weight of starting material).

Process Logs

Process logs include the following:

- Equipment records for equipment cleaning, sterilization, calibration, and maintenance
- Component processing records
- Filling tickets
- Labeling control tickets
- Product release information.

Table 26-3 Some Examples of Good Documentation Practices (GDP) "Do's and Don'ts"

Do's	Don'ts
Take your time to record original data clearly and legibly in ink (if written) or per electronic requirements	Cover data with whiteout Scratch over data Write over data Erase data Destroy data
Record original data as soon as you observe it	Record data from memory Write in a result of a check that was not made in the first place Record results before a check is made
Record only your own results	Record anyone else's check
Record data accurately	Round data outside specification Change data to meet tolerance levels
Ask supervisor about any questions	Guess at the data
Report immediately any GDP violations to supervision	Turn a blind eye if another employee is violating GDP

Material Logs

Material logbooks contain a complete inventory of every lot of raw material used in a batch, how much was used, expiration or re-evaluation dating periods, and test results from raw material QC.

Documentation and GMP Inspections

The verification of the quality of a good documentation program (Table 26-3) results from a successful FDA inspection, usually conducted during preapproval new drug application (NDA) inspections (PAI). Table 26-4 provides an example of all the documentation that should be in place and readily accessible for a PAI.

QUALITY MANAGEMENT SYSTEM

Quality management system (QMS) is a complete set of interacting components that direct and control the sterile product facility toward the desired quality objectives and regulatory compliance. The following components or systems are periodically reviewed to ensure maintenance and continuously improve product quality.

1. External audits
2. Internal audits
3. Corrective and preventive actions (resulting from audits as well as from responses to batch exceptions)
4. Product complaints
5. Adverse events
6. Manufacturing batch exceptions (deviations)
7. Environmental monitoring trends
8. Batch record metrics (e.g., reworks and rejected lots)
9. New or revised regulatory requirements
10. Supplier audits
11. Hold data
12. FDA 483 and/or Warning Letter commitments
13. Service and repair data
14. Change control
15. Training
16. Customer satisfaction trends
17. Validation records and status
18. Follow-up actions from previous QMS reviews.

Grazal and Lee compared FDA, EU, and International Conference on Harmonization (ICH) requirements and guidance for product annual reviews and product quality reviews (2) (Table 26-5). These records and data are reviewed by senior management representatives from Manufacturing, Engineering, Technical Services, Regulatory Affairs, and the QC Unit. Typically the reporting departments present their analyses of records/data since the last review meeting. Data are presented, trends identified, key contributing factors identified, action plans are proposed, and discussions conducted with owners and action-owning departments to ensure that the right actions are identified and reviews are documented, reported, and tracked.

PHARMACEUTICAL QUALITY SYSTEM

The ICH Pharmaceutical Quality System (PQS) Guideline Q10 was finalized in June 2008. It emphasizes a pharmaceutical company's quality system throughout the lifecycle of a product that includes the following:

- Pharmaceutical development—Drug substance, novel excipients used in new formulations, formulation and the container closure system, delivery system, manufacturing process and scale-up, and analytical method.
- Technology transfer—Transfer of new products from development to manufacturing and transfers within and between manufacturing and testing sites for marketed products.

Table 26-4 Documentation Typically Required for Readiness of Preapproval GMP Inspections

General category	Specific topics	Documentation items needed
General	Introductions and purpose	Organization chart, facility floor plans
	Process flow/description	Facility floor plans
	Presentation on NDA product	Slide show
	Plant tour	Incoming warehouse through finished product distribution
Technology transfer	Development report	Discuss report
	Analytical method transfers	Analytical and microbiological methods
	FDA inspection review	Previous FDA 483s, EIRs
	CMC review	CMC document
Validation and process improvement	Validation master plan	Schedule, qualification/validation, maintenance, requalification
	Product process validation	Protocol, raw data, summary reports
	Equipment preparation	IQ/OQ/PQ protocols, data, summary reports
		Arrangements for autoclave controls, testing
	Bulk formulation preparation	Sanitization and transfer protocols, data, reports
	Filling line validation	Protocols, data, reports
		System process capabilities and reliability
	Packaging line validation	Protocols, data, reports
		Inspection equipment
		Coding equipment
		Vision systems
		Bar code readers
	Cleaning validation	Validation approach, protocols, data, reports
		Analytical methods
	QC laboratory validation	Laboratory autoclave
		Water systems
		Instrumentation IQ/OQ/PQ
	Packaging systems	Glass syringe and rubber plunger data, reports
Materials assurance and control	Shipping studies	Temperature mapping
		SOPs
		Inventory and distribution records
		Cold supply chain for storage and shipping conditions
	Warehouse	Receiving, quarantine, release SOPs
		Work in progress
		Finished product inventories
	Filling/packaging components	Incoming material inspection
		Specifications
		Records
		Approved supplier list
Manufacturing process assurance and control	Class 100 (ISO 5) material,	SOPs for facility design for filling area providing Class 100
	Personnel and product flows and transfers	Syringe transport and loading procedures
		Sterilization/sanitizing and transfer control of materials from lesser to higher classified areas to prevent contamination
	Manufacturing order and batch records	Master manufacturing order
		Batch records
		Executed batch records
	Environmental monitoring	Pressure differentials
		Temperature and relative humidity in classified areas

(*Continued*)

Table 26-4 Documentation Typically Required for Readiness of Preapproval GMP Inspections (*Continued*)

General category	Specific topics	Documentation items needed
		Alarm systems
		Daily checks
	In-process controls and testing	Pressure hold method for drum integrity
		Weight checks
		Turbidity
		Stopper placement
	Holding times	SOPs and MBRs describing time limits
		Filled product exposure while on filling line
		Storage of sterilized equipment
		Storage of containers and closures
		Qualification data supporting time limits, especially bioburden
	Autoclave management	Routine control of autoclaves—daily/weekly/monthly tests
		Testing applied after routine and breakdown maintenance
	Facility or area shutdown and contingency procedures	SOPs and systems for deviation/change control addressing atypical conditions posed by shutdown of air handling systems or other utilities
		SOPs for returning facility to operating conditions following shutdown and impact on facility
Labeling/packaging assurance and control	Line clearance	SOPs, log books, investigations
	Syringe inspection and operator qualification	Protocol, operator qualification, and training records
	Lot/expiry coding	Systems and equipment
		Monitoring of printing devices
	In-process line checks/	SOPs
	Inspections	Packaging job order records
		Electronic/visual verification systems
	Reconciliations/ accountability	Acceptance criteria, investigations
Supportive analytical data	Environmental microbial	SOPs, air and surface sampling and testing procedures
	Monitoring program	Alert and action limits for environment and personnel
		Investigation procedures
		Shift coverage, monitoring frequency, summaries
		QA data review
		Near term and long-term trend reports
	Environmental microbial monitoring trend reports	Data generated by location, shift, room, operator, etc.
		Special data reports (isolate search)
		Significant changes in microbial flora
		Definition of system for regular notification and update
	Critical area nonviable particle counting system	Qualification studies that determine locations of particle counting probe
	Non-viable particle monitoring	SOPs, layout of system, documentation
		Provisions for continuous monitoring
		Qualification of sampling sites selection
		Investigation requirements for excursions
		Corrective action plans
		Evaluation of trending data
Process simulations	Study design	SOPs, protocols, master batch record
		Operator media fill participation records
	Frequency/number of runs	SOPs, summary list

Table 26-4 Documentation Typically Required for Readiness of Preapproval GMP Inspections (*Continued*)

General category	Specific topics	Documentation items needed
	Size/duration of runs; speed	Protocols
	Interventions	SOPs describing type of interventions
		Assessment of these practices during media fills
	Media, incubation, examination	Incubation conditions and duration requirements
		SOP for media fill units inspection
		Qualification of media inspectors; records
		Interim and final inspection requirements
		Damaged units
		Media filled units from initial aseptic setup (before fill)
		Side-by-side comparison of master batch record versus sterile media fill master batch record
	Interpretation	SOPs supporting documentation and justification describing circumstances and requirements to abort and/or invalidate sterile media fill
		Acceptance criteria
	Failure investigations	SOP
Sterility testing	Sterility testing	Release testing SOPs and test result summaries
		Sampling and incubation conditions
	Investigation of sterility positives	SOPs for sterility test failure
Microbiological laboratory testing	Container/closure integrity	Protocol and summary reports for microbial bacterial challenge for container/closure system
	Validation	SOPs and reports for all microbiological tests and assays
	Sanitization system	SOPs for handling, holding times, use of sporicidal agents
		Disinfectant efficacy qualification and requalification
		Cleaning and sanitization SOPs
		Assessing adequacy of environmental monitoring program
	Microbial identification	SOPs for identifying isolates including automated and instrumentation systems
	Microbial media	Growth promotion procedures and records for EM, sterility testing, and media fill testing
	Components	Test methods for packaging components
	Biological indicators	SOPs describing storage and confirmation testing requirements (count, *D* value) for biological indicators by site and contract testing laboratories
		BI supplier qualification requirements
Analytical chemistry	Analytical method transfers	SOPs, analyst training records
Testing	Specifications	Incoming and release data (C of As, components, finished product release)
	Routine testing	Test methods and procedures performed for all WIP and finished products (e.g., ID, turbidity, appearance, volume of injection test methods)
Master batch records	MBRs	Filling records, packaging records
Packaging job orders	MIRs	Manufacturing deviation and investigation reports associated with all batches
	Product batch record review	Review of all executed product batch records
	Batch release procedures	SOPs for review of environmental and personnel monitoring data
		Data related to acceptability of output from support systems (e.g., HEPA filters, HVAC, WfI, steam generation)

(*Continued*)

Table 26-4 Documentation Typically Required for Readiness of Preapproval GMP Inspections (*Continued*)

General category	Specific topics	Documentation items needed
		Data from proper functioning of equipment (e.g., batch alarms report, integrity of various filters)
Change control	Product specific	All related manufacturing and laboratory instrumentation change controls and associated product related issues
	Change control program	SOPs, change control log
		Examples related to process/equipment/employees
Quality training	Training programs	SOP, job specific training files
		GMP training files, evaluations, assessments
	Personnel training, qualification and monitoring	SOPs and systems documenting appropriate training conducted before operators permitted to enter aseptic manufacturing areas
		Fundamental training—aseptic techniques, clean room behavior, microbiology, hygiene, gowning, patient safety hazards posed by nonsterile drug products
		Specific written procedures for aseptic manufacturing area operations
	Aseptic technique interventions	SOPs and systems documenting appropriate training conducted for aseptic techniques and maintaining proper gowning control
	Gowning certification	Qualification SOPs, training records
		Summary report for all sterile manufacturing employees
Quality audit and assessment	Internal audits	SOPs, schedules, audit reports
	External audits	SOPs, schedules, audit reports
	Customer complaint program	SOPs, database for recording/tracking investigations
	Supplier qualification/auditing	SOPs, schedules, qualification reports
		Audit reports
		Quality agreements
		BSE/TSE MSQs for all components
HVAC system	Drawings and schematics	Approved and dated "as built" drawings
	Validation	IQ/OQ validations
		Dynamic smoke studies—documentation and videotape recording including in situ air pattern analysis and evaluation on impact of aseptic manipulations, interventions, and equipment design
		Original and any evaluation of subsequent equipment configuration changes
	Specifications	Area differential pressures
		Temperature and relative humidity
		FDA requirements
	Air filtration HEPA	SOPs and systems defining equipment specifications, test methods, and acceptance criteria
	Room classifications	Air balancing report
		Semiannual HEPA filter certifications
Water/clean steam systems	Drawing and schematics	Approved and dated "as built" drawings
	WFI/clean steam validation	IQ/OQ/PQ validation data
	Specifications	Water/clean steam systems
Compressed gas systems	Nitrogen and compressed air	SOPs and systems describing purity, microbiological and particle quality after filtration
		Filter integrity testing procedures
		Change frequencies
		Failure investigation procedures

Table 26-4 Documentation Typically Required for Readiness of Preapproval GMP Inspections (*Continued*)

General category	Specific topics	Documentation items needed
Calibration program	Calibration program for production and laboratory equipment and instrumentation	Calibration SOP
		Methods and master schedule
		OOS investigations
		Calibration master standards
Preventive maintenance program	PM program for production and laboratory equipment and instrumentation	PM program SOP
		Routine assignments and master schedule
Exit discussion	Audit closeout meeting	Action items
		Corrective action plan
		Timelines

Abbreviations: EIR, establishment inspection report; SOP, standard operating procedure; EM, environmental monitoring; BI, biological indicator; WIP, work in process; ID, identity; MBR, master batch record; Wfl, water for injection; HVAC, heating, ventilation, air conditioning; BSE/TSE, Bovine spongiform encephalopathy/transmissible spongiform encephalopathy; OOS, out-of-specification; PM, preventive maintenance; C of A, Certificate of Analysis; CMC, chemistry, manufacturing, and control; MIR, manufacturing investigation report.

Table 26-5 Product Annual Reviews and Product Quality Reviews: Comparison of FDA, EC, and ICH Requirements

Objective	FDA: product annual review 21 CFR 211.180(e)	EC: product quality review (1.5)	ICH Q7A API; product quality review (2.5)
Determine appropriateness of, and/or need to change, product specifications	Required	Required	Not specified
Appropriateness of starting material specifications	Not specified	Required	Not specified
Determine the need to change manufacturing procedures	Required	Not specified	Not specified
Determine the need to change manufacturing control procedures	Required	Not specified	Not specified
Verify consistency of the existing processes	Not specified	Required	Required
Determine the need to revalidate the production process	Not specified	Required (also specified in EU GMP Annex 15)	Required
Highlight trends	Expected, but not specified	Required	Not specified
Identify product and process improvements	Not specified	Required	Not specified
Identify corrective actions	Expected, but not specified	Required	Required
Frequency and procedure for performance of review	FDA: product annual review 21 CFR 211.180(e)	EC: product quality review (1.5)	API Q7A API: product quality review (2.5)
Annual	Required	Required	Required
Account for previous reviews	Expected, not specified	Required	Not specified
Grouping by product type, e.g., solids, liquids, injections	Not specified (specifically not allowed)	Allowed	Not specified

(*Continued*)

Table 26-5 Product Annual Reviews and Product Quality Reviews: Comparison of FDA, EC, and ICH Requirements (*Continued*)

Objective	FDA: product annual review 21 CFR 211.180(e)	EC: product quality review (1.5)	ICH Q7A API; product quality review (2.5)
Items for PAR/PQR review	FDA: product annual review 21 CFR 211.180(e)	EC: product quality review (1.5)	ICH Q7A API; product quality review (2.5)
Representative number of batches, whether approved or rejected	Required	Not specified	Not specified
All batches that failed specifications	Not specified	Required	Required
Inclusion of export only products	Expected, not specified	Required	Not specified
Complaints	Required	Required	Required
Recalls	Required	Required	Required
Returned products	Required	Required—quality-related returns	Required—quality-related returns
Salvaged products	Required	Not specified	Not specified
Investigations	Required	Required	Required
Critical/significant deviations/nonconformances	Expected, not specified	Required	Required
Product stability program and adverse trends	Expected, not specified	Required	Required
Starting materials and packaging materials, especially new sources	Not specified	Required	Not specified
Critical in-process controls	Expected, not specified	Required	Required
Changes to processes	Expected, not specified	Required	Required
Changes to analytical methods	Not specified	Required	Required
Marketing authorization variations submitted, granted, or refused	Not specified	Required	Not specified
Postmarketing commitments	Not specified	Required	Not specified
Impurity profile comparison to historical data and regulatory submission	Not specified	Not specified	Expected, not specified
Qualification of status of relevant equipment and utilities (e.g., HVAC, water, compressed gases)	Not specified	Required	Not specified
Technical agreements are up to date	Not specified	Required	Not specified
Adequacy/effectiveness of corrective actions for significant deviations or nonconformities	Not specified	Required	Required
Adequacy/effectiveness of preventive actions for significant deviations or nonconformities	Not specified	Required	Not specified
Adequacy of any previous product process or equipment corrective actions (from previous product quality reviews)	Not specified	Required	Not specified

Source: From Ref. 2.

- Commercial manufacturing—Procurement of materials, design and use of facilities, utilities, and equipment, actual production of the product, final package and labeling, and QC and assurance (product release, storage, and distribution).
- Product discontinuation—How all documentation and samples are retained and continued product assessment.

The scope of this guideline encompasses pharmaceutical drug substances and drug products including biotechnology and biological products throughout the entire product lifecycle. It encompasses such things as outsourcing activities, specific management activities, and all the quality systems already described such as change control, periodic product quality evaluation, and all other quality data monitoring, trending, and responses based on trends.

Section 2 of the guideline emphasizes the importance of management responsibilities in all stages of the product lifecycle. Indeed, management responsibility is a key issue for the success of Q10. Management commitment, including resourcing, internal communication, review, and oversight of all activities is the main determinative of the level of quality of a company.

The major elements or pillars of the PQS applying at all stages include the following:

- Process performance and product quality monitoring system(s)
- Corrective action/preventative action (CAPA) system
- Change management system
- Management review.

A term called "enablers" is used to describe knowledge management and quality risk management, again applicable throughout the product lifecycle stages, that support PQS goals of achieving product realization, establishing and maintaining a state of control, and facilitating continual improvement. Table 26-6, adapted from ICH Q10, summarizes the application of these four specific quality elements for the four main components of the product lifecycle (development, technology transfer, manufacturing, and product discontinuation).

The effectiveness of the PQS can be confirmed during a GMP regulatory inspection at the manufacturing site. This was covered in some detail in chapter 25.

ICH Q10 is the companion document along with two other ICH quality guidelines—pharmaceutical development (Q8) and risk management (Q9). The main objectives of the PQS include the following:

1. To establish, implement, and maintain and set of processes that provides a product with the quality attributes to meet the needs of patients, health care professionals, regulatory authorities, and internal customers.
2. To establish and maintain a state of control using effective monitoring systems and test systems for process performance and product quality.
3. To facilitate continual improvement of the quality of the product and process that includes, for example, reduction of variability, allows for innovation, and enables appropriate enhancement of the marketed product.

Many companies prepare, follow, and upkeep Quality Manuals that are typically huge master documents describing such items as the company quality policy, the scope of the quality system, management responsibilities within the quality system, and identification of all processes within the PQS, especially detailing how all these processes link together and are interdependent.

It is outside the scope of this chapter to provide more information on Quality Manuals, quality policies, management responsibilities, and other components of a Total Quality System, but there are plenty of references available that do provide such detail. Table 26-7 provides an organized listing of ICH Quality Guidelines (3).

QUALITY RISK ASSESSMENTS

Risk management and assessments have been broadly and specifically applied in pharmaceutical manufacturing, be it any type of dosage form or device or the active pharmaceutical ingredient. Basically risk management enables the identification of critical areas or product/process vulnerabilities. It is a systematic process for the assessment, control, communication, and review

Table 26-6 Application of the Four Specific Pharmaceutical Quality System Elements to the Four Main Components of the Product Lifecycle

Quality system element	Development	Technology transfer	Manufacturing	Product discontinuation
Process performance and product quality monitoring	Conduct quality risk management and monitoring to establish a control strategy for manufacturing	Monitor carefully scale-up activities for successful transfer to manufacturing and further develop control strategy	Apply a well-designed system to ensure performance within a state of control and to identify improvement areas	Continue stability studies to completion. Take appropriate action on marketed product according to regional regulations
Correction action and preventative action	Product and process variability explored to help when such actions are required later	Effective system for feedback and continual improvement	Apply CAPA, then evaluate effectiveness of the actions taken	Continue with impact on product remaining on the market and similar products
Change management system	Change definitely part of development process so documentation needs to be excellent. Formality of change management process will increase as product moves forward in development	Provides documentation of adjustments made to the process during technology transfer operations	Must be in place for commercial manufacturing. Oversight by the Quality Unit to provide assurance of appropriate science and risk-based assessments	Any changes afterwards should go through appropriate change management
Management review	Management needs to review adequacy of product and process design	Perform reviews to ensure that the developed product and process can be manufactured at commercial scale	Should be a structured system and support continual improvement	Continue to review product stability data and product complaints

Source: Adapted from ICH Q10.

of risks to the quality of the drug product. Such factors identified can help the establishment of risk controls. Quality risk management is an important part of science-based decision making essential for quality management of pharmaceutical manufacturing.

One of the most important applications of risk management in the sterile product field is the identification and potential control of risk factors involved in aseptic product processing. There are many risks associated with all the variables associated with aseptic processing—personnel hygiene, techniques, and attitudes; design, qualification and validation of processes, equipment, and facilities; cleaning methods, air and water quality, handling and sterilization of raw materials, sterilization and depyrogenation process validation, plus many other variables related to manufacturing in general—mixing, aseptic additions, dose control, lyophilization, labeling, secondary packaging, and so on.

One specific example is provided here on how to assess the risk of an aseptic filling process. A common tool used is called Failure Mode Effect Analysis (FMEA). Fishbone analysis can also

Table 26-7 ICH Quality Guidelines

Stability	
Q1A (R2)	Stability testing of new drug substances and products
Q1B	Stability testing: photostability testing of new drug substances and products
Q1C	Stability testing of new dosage forms
Q1D	Bracketing and matrixing designs for stability testing of new drug substances and products
Q1E	Evaluation of stability data
Q1F	Stability data package for registration applications in climatic zones III and IV
Analytical validation	
Q2 (R1)	Validation of analytical procedures: test and methodology
Impurities	
Q3A (R2)	Impurities in new drug substances
Q3B (R2)	Impurities in new drug products
Q3C (R3)	Impurities: guidelines for residual solvents
Pharmacopoeias	
Q4	Pharmacopeias
Q4A	Pharmacopoeial harmonization
Q4B	Evaluation and recommendation of pharmacopoeial texts for use in the ICH regions
Q4B Annex 1	Evaluation and recommendation of pharmacopoeial texts for use in the ICH Regions on residue on ignition/sulfated ash general chapter
Q4B Annex 2	Evaluation and recommendation of pharmacopoeial texts for use in the ICH Regions on test for extractable volume of parenteral preparations general chapter
Q4B Annex 3	Evaluation and recommendation of pharmacopoeial texts for use in the ICH Regions on test for particulate contamination: subvisible particles general chapter
Q4B Annex 4A	Evaluation and recommendation of pharmacopoeial texts for use in the ICH Regions on microbiological examination of nonsterile products: microbial enumeration tests general chapter
Q4B Annex 4B	Evaluation and recommendation of pharmacopoeial texts for use in the ICH ICH regions on microbiological examination of nonsterile products: tests for specified micro-organisms general chapter
Q4B Annex 4C	Evaluation and recommendation of pharmacopoeial texts for use in the ICH ICH regions on microbiological examination of nonsterile products: acceptance criteria for pharmaceutical preparations and substances for pharmaceutical use general chapter
Q4B Annex 5	Evaluation and recommendation of pharmacopoeial texts for use in the ICH ICH regions on disintegration test general chapter
Q4B Annex 6	Evaluation and recommendation of pharmacopoeial texts for use in the ICH ICH regions on uniformity of dosage units general chapter
Q4B Annex 7	Evaluation and recommendation of pharmacopoeial texts for use in the ICH ICH regions on dissolution test general chapter
Q4B Annex 8	Evaluation and recommendation of pharmacopoeial texts for use in the ICH ICH regions on test for sterility general chapter
Quality of biotechnological products	
Q5A (R1)	Viral safety evaluation of biotechnology products derived from cell lines of human or animal origin
Q5B	Quality of biotechnological products: analysis of the expression construct in cells for production of r-DNA derived protein products
Q5C	Quality of biotechnological products: stability testing of biotechnological/biological products
Q5D	Derivation and characterization of cell substrates used for production of biotechnological/biological products
Q5E	Comparability of biotechnological/biological products subject to changes in their manufacturing process
Specifications	
Q6A	Specifications: test procedures and acceptance criteria for new drug substances and new drug products: chemical substances (including decision trees)

(*Continued*)

Table 26-7 ICH Quality Guidelines (*Continued*)

Q6B	Specifications: test procedures and acceptance criteria for biotechnological/biological products
Good manufacturing practice	
Q7	Good manufacture practice guide for active pharmaceutical ingredients
Pharmaceutical development	
Q8 (R1)	Pharmaceutical development
Quality risk management	
Q9	Quality risk management
Pharmaceutical quality system	
Q10	Pharmaceutical quality system

be applied. FMEA in assessing the risks involved in aseptic processing can be divided into the following questions or statements for each processing step or unit operation.

1. Identify the potential failure, for example, sterility or endotoxin failure
2. Assign a severity value. The severity of sterility or endotoxin failure is high
3. What are potential causes of the failure(s)? Each potential cause for failure should be evaluated individually
4. What is the probability that the failure will happen and if it does happen, will it result in the loss of sterility and/or unacceptable endotoxin levels? Occurrence values generally are high (often occurrence), medium (periodic occurrence), or low (occurrence is seldom). There is some subjectivity to determining these values, but historic process capability data will help to make an accurate judgment.
5. What are the existing procedural or design controls that can detect, reduce, or eliminate the cause of the failure from occurring? These controls will dictate a "detectability ranking."
6. Assign a "detectability value" by determining the likelihood that the cause will be detected by controls in place prior to product release. The detectability ranking is opposite of ratings for severity and occurrence. A high detectability has a ranking of low, that is, the cause will likely be detected. If there are no controls, the problem will not be detected resulting in a high-risk ranking.
7. Finally, there is a "risk-prioritization ranking (RPR)" that measures the overall risk of the process step or item by combining the individual risk values. Risk values may be quantitative (e.g., high risk = 7–10; medium risk = 4–6; low risk = 1–3) or qualitative (high, medium, and low). In either case, the RPR can be obtained by combining all the values.

For example, for a steam sterilization process, one potential cause for failure is excess air left in the autoclave. Existing controls are standard operating procedures, training of personnel, and validated sterilization cycles. The severity value is *high*. The occurrence value is *medium* in that this potential cause for failure can happen sometimes. The detection value is *high* because there is no way to know if pockets of air have not been removed. Therefore, the overall risk of this process failure is *high*. Any high-risk conclusion is not acceptable as is most medium-risk conclusions whereas low-risk conclusions are acceptable most of the time. In order to reduce this risk, added steps are added to utilize multiple presteam vacuum pulses to ensure air removal. The occurrence value now becomes *low* and because a faulty vacuum cycle can be detected the detectability value is *low* (likelihood that process failure cannot be detected). Therefore, the overall risk factor now becomes *low* and the risk is acceptable.

Quality By Design/Process Analytical Technologies

Quality by Design (QbD) represents a new regulatory philosophy based on predefined quality targets and deep understanding of formulations and processes based on prior knowledge, experimental data, and published literature, as opposed to empirical determination of

Table 26-8 Current Paradigm Compared to Quality by Design Paradigm

Current paradigm	Quality by design paradigm
• Quality is tested into the product • Product specifications are based on batch testing results • Validation "freezes" the process • Process improvements require preapproval	• Quality is designed into the product • Employs real-time quality control based on process analytical technology • Product specifications are based on "fitness for use" and process capability • Process changes within the established design space do not require preapproval • Process validation is redundant

performance criteria based on analysis of experimental data. Table 26-8 briefly contrasts the QbD paradigm with the current regulatory environment.

A key element of QbD is the concept of design space (Fig. 26-2), which is a multidimensional space encompassing combinations of product design and processing variables that provides assurance of suitable product performance. The QbD approach is intended to provide regulatory relief throughout the lifetime of a product by allowing product and process changes that fall within the design space to be implemented without prior approval. Design space is proposed by the applicant and is subject to regulatory review and approval.

While the principle of QbD is simple and appealing, the actual development, scale-up, and commercialization of pharmaceutical products present a significant challenge to pharmaceutical scientists and engineers. Establishing a meaningful design space requires aggressive experimentation on a small scale, particularly since the supply of active pharmaceutical ingredient (API) is generally limited. In order for these small-scale experiments to be meaningful, scale-up must be understood at a very sophisticated level.

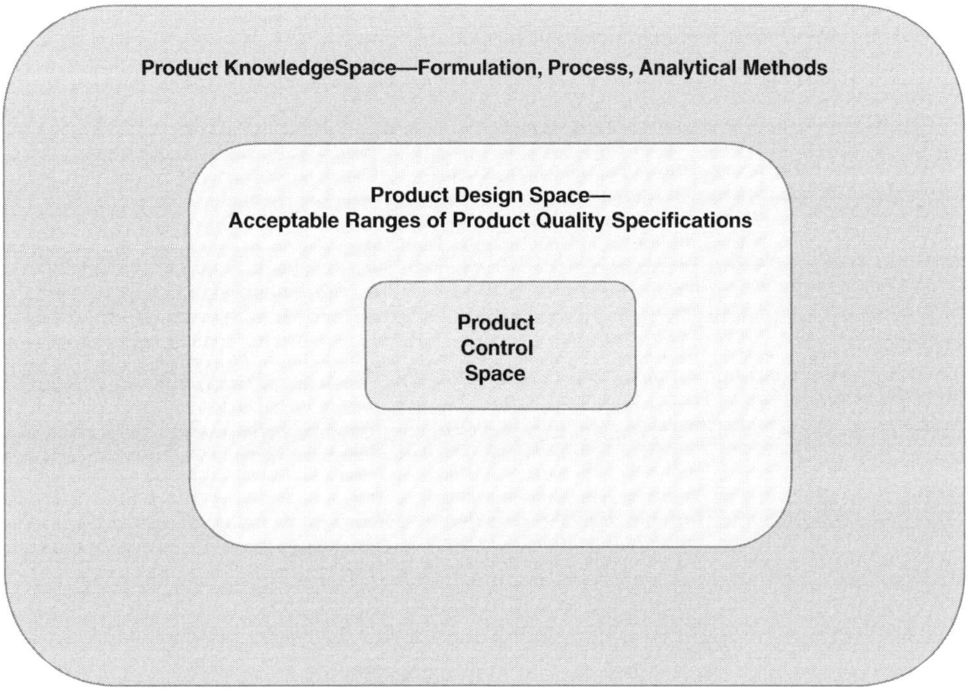

Figure 26-2 Illustration of quality by design.

While there have been many oral presentations, articles, and case studies in trade magazines by both regulators and industrial pharmaceutical scientists, most have dealt with solid oral dosage forms (4,5 and references therein). There are good reasons for this, including the greater number of unit operations involved in solid oral dosage forms and because solid oral dosage forms are generally preferred over injectable products. However, parenteral scientists must adapt to the QbD paradigm, develop design space approaches, and deal with the uncertainties of scale-up in order to make QbD part of the way we do business (6). In some ways, the job of parenteral scientists may be easier than that of their counterparts in pharmaceutical solids, particularly when dealing with simple, ready-to-use sterile aqueous solutions. There are relatively few process variables, and scale-up for such systems is relatively straightforward. However, for more complicated parenteral dosage forms, such as freeze-dried products or disperse systems, scale-up problems are very significant and perhaps every bit as demanding as scale-up of pharmaceutical solids.

In July 2008, FDA Deputy Director Barry Cherney, from the CDER Office of Biotechnology Products Division of Therapeutic Proteins, provided the basic principles of QbD (7). One of the overarching principles of QbD is that products are designed to maximize their efficacy while minimizing their adverse events.

1. Designing the appropriate product requires knowledge of the mechanism of action of the API, the biological characteristics of the product that affects its safety, understanding all the activities of biotechnology APIs, knowledge of the attributes of the API and how they affect therapeutic performance, knowledge of impurities and their impact on quality, safety, and efficacy, and knowledge of how the formulation impacts product performance. All this knowledge should be used in risk assessment to design the product formulation and process.
2. Processes should be designed to be robust and consistently deliver the desired product. This requires knowledge of the critical attributes of the raw materials, knowledge of the operating parameters, and knowledge of the output performance parameters, particularly in terms of the critical quality attributes of the product.
3. Where does all this knowledge come from?
   - Scientific literature
   - Previous experience
   - Previous platforms
   - Developmental studies using animal models to look at how attributes relate to safety and efficacy
   - Design of experiments to evaluate process parameters
   - Mining of clinical and nonclinical data.

QbD is not about regulatory relief; that is, there will still be the need to perform all the necessary studies and document all data. Basically, QbD is putting more emphasis on the manufacturer to perform all the necessary studies prior to initial submission in order to know everything possible about the product. FDA and other regulatory bodies are not going to be responsible for bringing up problems and providing guidance. This is becoming the manufacturer's responsibility. Regulatory relief will be experienced later because of the lack of need to submit supplements and obtaining approval from regulatory agencies. Also regulatory relief will have the benefit of manufacturers seeing the significant benefits of QbD. There will be (or should be) fewer batch rejections, investigations, recalls, and regulatory inspections. QbD will allow for expedited implementation of process changes and manufacturing processes being much more adaptable. Productivity will be increased and overall quality will be significantly improved.

Process analytical technology (PAT) is an integral part of QbD, since the paradigm relies on use of real-time process monitoring and control as a part of the overall control strategy. It is incumbent on parenteral scientists and engineers to explore novel process sensors and to pursue a much more sophisticated level of process monitoring. If PAT is used to monitor critical performance outputs, it would be expected that there would be less dependency on control of operating parameters during processing.

While the pharmaceutical industry has made modest efforts over the years to improve in-process testing, it was not until the FDA started to emphasize in-process testing (using the moniker PAT) via release of its PAT guidance document in 2004 that the industry seriously started to pursue new quality measurement systems. Such systems not only involved in-process measurement tools but also emphasized the need for process understanding, risk-based management, integrated systems thinking, and real-time product release.

Process monitoring tools are evolving toward nondestructive methods using spectroscopic techniques such as Fourier Transform infrared spectroscopy, near infrared spectroscopy, Raman spectroscopy, and fluorescence. The application of these techniques has primarily focused on blend uniformity, dryness, hardness, and other physical properties of solid oral dosage forms. These methods work well except for samples that do not possess a suitable chromophore, are opaque, or are dispersed systems. High Resolution Ultrasonic Spectroscopy is an up-and-coming technique that can analyze a variety of properties such as composition, aggregation, gelation, crystallization, dissolution, sedimentation, and particle size.

An advancement in PAT applied to freeze-drying is the development of a near IR–based mass flow meter that provides instantaneous real-time measurement of the mass flow rate of water vapor from the chamber to the condenser (8). This provides a scale-independent link between small-scale and large-scale equipment that should facilitate scale-up through quantitation of equipment capability. As a process development tool, this mass flow meter facilitates process optimization by quantifying, for example, the effect of chamber pressure on sublimation rate. The instrument also shows promise as a means of measuring product temperature without the need for temperature sensors in individual vials of product.

Another advancement in PAT in the past few years is the development of noncontact check weighing systems, which are capable of performing 100% check weighing based on magnetic resonance technology. Such systems are suitable for both liquids and solids.

REFERENCES

1. Edwards AG. Quality systems for drugs and biologics. Pharm Technol 2008; 32:60–74.
2. Lee JY, Grazal JG. Product annual/quality review: US–EU comparative analysis and interpretations. Pharm Technol 2008; 32:88–104.
3. International Conference on Harmonization, Geneva, Switzerland, http://www.ich.org/cache/compo/363-272-1.html. Accessed June 15, 2010.
4. Lionberger RA, Lee SL, Lee L, et al. Quality by design. Concepts for ANDAs. AAPS J 2008; 10:268–276.
5. Kourti T. Quality by design in the pharmaceutical industry. Am Pharm Rev 2009; 12.
6. Jameel F, Khan M. Quality by design as applied to the development and manufacturing of a lyophilized protein product. Am Pharm Rev 2009; 12.
7. Cherney B. International Pharmaceutical Quality. Bethesda, MD, 2008:6–13.
8. De Beer TRM, Vercruysse P, Burggraeve A, et al. In-line and real-time process monitoring of a freeze drying process using Raman and NIR spectroscopy as complementary process analytical technology (PAT) tools. J Pharm Sci 2009; 98:3430–3446.

BIBLIOGRAPHY

FDA. Guidance to Industry: Quality Systems Approach to Pharmaceutical CGMP Regulations. Rockville, MD: FDA, 2006.

FDA. Guidance on Process Analytical Technology (PAT)—A Framework for Innovative Pharmaceutical Development, Manufacturing, and Quality Assurance. http://www.fda.gov/downloads/Drugs/GuidanceComplianceRegulatoryInformation/Guidances/ucm070305.pdf. Accessed June 15, 2010.

ICH Q8. Guidance for Industry, Q8 Pharmaceutical Development, Food and Drug Administration, May 2006, ICH. http://www.fda.gov/downloads/RegulatoryInformation/Guidances/ucm128029.pdf. Accessed June 15, 2010.

ICH Q9. Quality Risk Management. http://www.ich.org/lob/media/media1957.pdf. Accessed June 15, 2010.

ICH Q10. Pharmaceutical Quality System. http://www.ich.org/lob/media/media3917.pdf. Accessed June 15, 2010.

Paulson B. Quality system principles. International Pharmaceutical Quality, 2 (4) July–August, 2008.

Pharmaceutical cGMPs for the 21st Century—A Risk Based Approach, FDA, Final Report, Fall, 2004. http://www.fda.gov/Drug/DevelopmentApprovalProcess/Manufacturing/QuestionsandAnswersonCurrentGoodManufacturingPracticescGMPforDrugs/ucm137175.htm. Accessed June 15, 2010.

Quality risk management for aseptic processes. Technical report No. 44. PDA J Pharm Sci Technol 2008; 62(1 Suppl):2–39.

Vinther A, Grace MB. Quality systems. In: Lysfjord J, ed. Practical Aseptic Processing: Fill and Finish. Vol 2. Bethesda, MD: Parenteral Drug Association, 2009:3–30.

27 Microorganisms and sterility testing[1]

All lots of injectables in their final containers must be tested for sterility, except for products that are allowed to apply parametric release. [2] The United States Pharmacopeia (USP) prescribes the requirements for this test for official injections. Portions of the USP sterility test chapter <71> have been harmonized with the corresponding texts of the European Pharmacopeia and/or the Japanese Pharmacopeia. The FDA uses these requirements as a guide for testing official sterile products. The primary official test is performed by means of filtration, but direct transfer is used if membrane filtration is unsuitable. To give greater assurance that viable microorganisms will grow, if present, the USP requires that all lots of culture media be tested for their growth-promotion capabilities. However it must be recognized that the reliability of both test methods has the inherent limitations typical of microbial recovery tests. Therefore, it should be noted that this test is not intended as a thoroughly evaluative test for a product subjected to a sterilization method of unknown effectiveness. It is intended primarily as a check test on the probability that a previously validated sterilization procedure has been repeated or to give assurance of its continued effectiveness.

In the event of a sterility-test failure, the immediate issue concerns whether the growth observed came from viable microorganisms in the product (true contamination) or from adventitious contamination during the testing (a false positive). The USP does not permit a retest, unless specific evidence is discovered to suggest contamination occurred during the test. Therefore, a thorough investigation must be launched to support the justification for performing the retest and assessing the validity of the retest results relative to release of the lot of product.

It should be noted that a *lot* with respect to sterility testing is that group of product containers that has been subjected to the same sterilization procedure. For containers of a product that have been sterilized by autoclaving, for example, a lot would constitute those processed in a particular sterilizer cycle. For an aseptic filling operation, a lot would constitute all of those product containers filled during a period when there was no change in the filling assembly or equipment and which is no longer than one working day or shift.

When the term "sterile" appears on the label of a parenteral product, it means that the batch or lot from which the sample originated passed the requirements of the USP Sterility Test <71> (or other national compendial sterility-test requirement). The USP sterility test provides an estimate of the probable, not actual, sterility of a lot of articles. The actual product itself administered to a patient has not been tested for sterility. The sterility test is a destructive test; thus it is impossible to test every item for sterility. This presents a major limitation of the sterility test. Sterility is based on the results of the testing of a small number of batch samples assuming that these samples are representative of every article from the batch not tested for sterility. The question of the sample being representative of the whole will always be an uncertainty. Furthermore, another limitation of the sterility test is the finite frequency of accidental (or inadvertent) contamination of one or more samples during the performance of the testing procedures. Regardless of the perfection attempted in the attitudes and techniques involved in sterility testing, accidental contamination will occur with a given percentage of tests conducted. The use of barrier isolation technology (compared to a conventional clean room) by the pharmaceutical industry has greatly reduced the chance of accidental contamination that can yield a positive sterility test.

[1] This chapter is an update of the Sterility Test chapter from Reference 1

[2] Parametric release means that a lot of product, if terminally sterilized by a well-defined, fully validated sterilization process, has a sterility assurance level sufficient to omit the sterility test for release. See United States Pharmacopeia General Chapter <1222> Terminally Sterilized Pharmaceutical Products—Parametric Release.

In light of these and other limitations of the USP sterility test, why is it still a requirement of and enforced by the Food and Drug Administration (FDA) and other regulatory agencies? The most important and obvious reason is to provide some means, albeit a small one, of end-product testing to protect the consumer from being administered with a contaminated injectable product.

An exception to end-product sterility testing involves terminally-sterilized large volume parenterals, which have been exposed to sterilization conditions experimentally validated to assure product sterility well beyond the capability of sterility testing to detect contamination, while products that are terminally sterilized usually have a sterility assurance level (SAL) of at least 10^{-6}. Release of products without end-product sterility testing but based on validation of the sterilization process is called *parametric release* (Chapter 17, p. 265).

While the sterility test does not assure sterility of every article, it does provide the FDA as well as the manufacturer and the user with some end-point check that a representative sample of the batch does not disclose the existence of a high proportion of contaminated units in a lot or batch. End-product sterility testing also presents a reliable means of checking the sterility of a product that has been sterilized by marginal sterilization processes such as an aseptic filtration.

The USP chapter <71> on injections states that preparations for injection meet the requirements under "Sterility Tests." After meeting these requirements, that is, all media vessels incubated with product sample reveal no evidence of microbial growth (turbidity), the tested product may be judged to meet the requirements of the test. If evidence for microbial growth is found, the material tested has *failed* to meet the requirements of the test for sterility. Retesting is only allowed if there is unequivocal proof that the failed result was due to operator or accidental contamination. The FDA has stringent requirements for sterility retesting (2).

Evidence for microbial growth is determined by visual evaluation of a vessel containing the product sample in the proper volume and composition of nutrient solution. Provided that the growth conditions are optimal—proper nutrients, pH, temperature, atmosphere, sufficient incubation time, etc.—a single microbial cell will grow by geometric progression until the number of microbial cells and their metabolic products exceeds the solubility capability of the culture medium. Manifestation of this "overgrowth" is visualized by the appearance of a cloudy or turbid solution of culture media. A noxious odor may also accompany the turbid appearance of the contaminated media. The sterility test is failed by a product that generates turbidity in a vessel of culture medium, while the same lot of medium without the product sample shows no appearance of turbidity.

Parenteral drug administration was a routine practice in the early 1900s. For example, insulin was discovered in 1921 and was, as it is today, administered by subcutaneous injection. Yet, the first official compendial requirement of sterility testing of drugs administered by the parenteral route did not appear until 1932 in the *British Pharmacopoeia.* Sterility tests were then introduced in the 11th edition of the USP and in the sixth edition of the *National Formulary* (NF) in 1936. Since 1936, significant changes and improvements have occurred in the official sterility-test requirements.

In 1978, the final approved regulations of the FDA-authored cGMPs were published. Sterility testing was briefly mentioned under section 211.167, "For each batch purporting to be sterile, there shall be appropriate tests to determine conformance to such requirements." To elaborate on this requirement and to address more specific issues confronted by both industry and the FDA in manufacturing and control of aseptically produced drug products, the FDA published its "Guidelines on Sterile Drug Products Produced by Aseptic Processing" in 1987 and revised in 2004 (2).

SAMPLING FOR STERILITY TESTING

The sterility of a parenteral product lot is checked by a statistically valid sampling procedure. After years of experience, most manufacturers of parenteral products will sterility test 10 to 20 units of product per lot. The number of units tested may be doubled where the deliverable volume is 1 mL or less. The number of units sampled depends on the number of units in the batch, the volume of liquid per container, the method of sterilization, the use of a biological indicator system, and the good manufacturing practice requirements of the regulatory agency for the particular product. For example, if the batch size is greater than 500 articles, a minimum of 20 units is sampled. If the final batch size is between 100 and 500 articles then not fewer than 10

Table 27-1 Minimum Number of Product Units to be Tested in Relation to the Batch Size of Finished Product Units

Finished product	Number of product units per batch[a]	Minimum number of units to be tested for each medium[b] (Unless otherwise justified and authorized)
Parenteral preparations	NMT 100 containers	10% or 4 containers, whichever is greater
	More than 100 but less than 500 containers	10 containers
	More than 500 containers	2% or 20 containers, whichever is less
Large volume parenterals	Any number	2% or 20 containers, whichever is less
Antibiotic solids	Pharmacy bulk packages (<5 g)	20 containers
	Pharmacy bulk packages (≥5 g)	6 containers
	Bulks and blends	See bulk solid products
Ophthalmic and other noninjectable preparations	NMT 200 containers	5% or 2 containers, whichever is the greater
	More than 200 containers	10 containers
	Single dose containers	Apply scheme shown above for parenteral preparations
Catgut and other surgical sutures for veterinary use		2% or 5 packages, whichever is greater, up to a maximum total of 20 packages
Bulk solid products	Up to 4 containers	Each container
	More than 4 but not more than 50	20% or 4 containers, whichever is greater
	More than 50 containers	2% or 10 containers, whichever is greater

[a] If batch size is unknown, use the maximum number of items prescribed.

[b] If the contents of one container are enough to inoculate the two media, this column gives the number of containers needed for both the media together.

of the articles are sterility tested although there are minimum requirements for sterility testing of biologics. For large-volume parenteral (LVP) products (volume > 100 ml per container), at least 2% of the batch or 10 containers whichever is less are sampled. Table 27-1 provides the number of units to be sterility tested on the basis of batch size. Table 27-2 provides requirements for the minimum quantity of product to be used in the sterility test based on quantity per container. Specifics of the conductance of the USP sterility test for different types of dosage forms and devices are not covered in this chapter, but can be found in Chapter or Section <71> of the most recent revision of the USP.

CULTURE MEDIA

The USP describes two primary types of culture media to be used in the sterility testing of parenteral products. One type is called Fluid Thioglycollate Medium (FTM) (Fig. 27-1 and Table 27-3). FTM provides both aerobic and anaerobic environments within the same medium with its primary intention being able to culture anaerobic bacteria. Thioglycollate and L-cysteine are antioxidants or reducing agents that maintain anaerobiosis in the lower levels of the culture tube. FTM solution has a two-color appearance. The pinkish color of the top part of the solution is indicative of the presence of resazurin sodium, an oxygen-sensitive indicator. The pink color should consume no more than one-third of the medium volume. Because of the need for two environments in the same test tube or container, the ratio of surface to medium depth is very important. To provide adequate depth for oxygen penetration, a 15-ml volume of FTM must be contained in a test tube of the dimensions 20 × 150 mm. A 40-ml volume of FTM is to be contained in 25 × 200 mm test tubes and 75–100 ml FTM in 38 × 200 mm test tubes.

Devices containing tubes with small lumina are sterility tested using an alternate thioglycollate medium in which the agar and resazurin sodium are deleted. The same medium is used for turbid or viscous parenterals. Without the agar the medium will not interfere with the viscosity of the product or be as resistant in filling small lumina. Since the medium will be turbid,

Table 27-2 Minimum Quantity to be Used for Each Sterility-Test Medium

Type of finished product	Quantity per container	Minimum quantity to be used (Unless otherwise justified and authorized)
Liquids	Less than 1 mL	Whole contents of each container
	1–40 mL	Half of the contents of each container, but not less than 1 mL
	Greater than 40 mL, but ≤ 100 mL	20 mL
	Greater than 100 mL	10% of the contents of the container, but not less than 20 mL
	Antibiotic liquids	1 mL
Insoluble preparations, creams, and ointments to be suspended or emulsified		Use the contents of each container to provide not less than 200 mg
Solids	Less than 50 mg	The whole contents of each container
	50 mg or more, but less than 300 mg	Half the contents of each container, but not less than 50 mg
	300 mg to 5 g	150 mg
	Greater than 5 g	500 mg
Others	Catgut and other surgical sutures for veterinary use	3 sections of a strand (each 30 cm long)
	Surgical dressing/cotton/gauze (in packages	100 mg per package
	Sutures and other individually packaged single use material	The whole device
	Other medical devices	The whole device, cut into pieces or dissembled

the presence of a color indicator would not be seen anyway. For oily products, FTM is slightly modified by the addition of 1 ml Polysorbate 80 to 1 liter of the media. Polysorbate 80 serves as an emulsifying agent to permit adequate dispersal of a lipophilic product in a hydrophilic growth medium.

The other primary USP/NF culture medium for the sterility testing of parenterals is called soybean-casein digest (SCD) or trypticase soy broth (TSB) medium (Fig. 27-1 shows a nonsterile

Figure 27-1 (*Left*) Sterile fluid thioglycollate medium (*Right*) non-sterile trypticase soy broth. *Source*: Courtesy of Ryan Cool, Baxter BioPharma Solutions.

Table 27-3 Formulations of Sterility-Test Media

Fluid thioglycollate medium	
L-Cystine	0.5 g
Sodium Chloride	2.5 g
Dextrose monohydrate/Anhydrous	5.5/5.0 g
Agar	0.75 g
Yeast extract (water soluble)	5.0 g
Pancreatic digest of casein	15.0 g
Sodium thioglycollate	0.5 g
Or Thioglycolic acid	0.3 mL
Resazurin sodium solution (1 in 1000)	1.0 mL
Freshly prepared	
Purified water	1000 mL
Soybean-casein digest medium (aka trypticase soy broth)	
Pancreatic digest of casein	17.0 g
Papaic digest of soybean meal	3.0 g
Sodium chloride	5.0 g
Dextrose monohydrate/Anhydrous	2.5/2.3 g
Dibasic potassium phosphate	2.5 g
Purified water	1000 mL

TSB container). TSB has a slightly higher pH (7.3 ± 0.2) than does FTM (7.1 ± 0.2), considered a better nutrient for fungal contaminants. TSB promotes growth of fungi and bacteria, and is also considered a better medium for slow-growing aerobic microorganisms than FTM.

Other media have been proposed to replace or be substituted for FTM and/or TSB and these can be found in the USP. For example, concentrated brain heart infusion broth has been suggested as an alternative to FTM and TSB when large-volume parenterals are directly inoculated with culture medium.

After preparation of culture media solutions, a validated steam sterilization process is applied. If media are to be stored, storage temperature should be between 2°C and 25°C in sterile, airtight containers. The length of storage time must be validated.

When membrane filtration is used for the sterility test, a diluting fluid must be used to rinse the filtration assembly in order to ensure that no microbial cells remain anywhere but on the filter surface. The diluting fluid may also be used to dissolve a sterile solid prior to filtration. Diluting fluid A, D, and K formulas are listed in the USP. Diluting fluids are intended to minimize the destruction of small populations of vegetative cells during the pooling, solubilizing, and filtering of sterile pharmaceutical products.

Both FTM and SCD media need to be modified for sterility testing by direct transfer of penicillin and cephalosporin antibiotics. To containers of each medium, transfer aseptically a quantity of β-lactamase sufficient to inactivate the amount of antibiotic in the specimen under test.

TIME AND TEMPERATURE OF INCUBATION

No ideal incubation time and temperature condition exists for the harvesting of all microorganisms. Most organisms grow more rapidly at 37°C than at lower temperatures. However, a temperature of about 23°C may reveal the presence of some organisms that might remain undetected if incubations were done at higher temperatures. The Division of Biologics Standards of the National Institutes of Health discovered that a pseudomonad contaminant in plasma grew in FTM at 25°C, but was killed at 35°C (3). As a result of this finding, the incubation temperature range of FTM was lowered from 32°C–35°C to 30°C–35°C as required by the USP.

The current time and temperature incubation requirements of the USP and EP sterility tests are found in Tables 27-4A and 27-4B. Incubation in TSB is accomplished at 20°C to –25°C because of favorable growth of fungal and slow-growing aerobic contaminants at this temperature range. The time of incubation for sterility testing by membrane filtration is 14 days

Table 27-4A Time and Temperature Incubation Requirements of the USP Sterility Test

Medium	Test procedure	Time (Days)[a]	Temperature (°C)	How product is sterilized
FTM	Direct transfer	14	30–35	Steam or aseptic process
	Membrane filtration	7	30–35	Terminal moist heat
	Membrane filtration	14	30–35	Aseptic process
TSB	Direct transfer	14	20–25	Steam or aseptic process
	Membrane filtration	7	20–25	Terminal moist heat
	Membrane filtration	14	20–25	Aseptic process

[a]Time is the minimum number of incubation days. Additional incubation time may be required if the nature of the product is conducive to produce a "slow-growing" contaminant.

Table 27-4B Time and Temperature Incubation Requirements of the EP Sterility Test

Medium	Test procedure	Time (Days)[a]	Temperature (°C)	How product is sterilized
FTM	Direct transfer	21 (14 + 7)	30–35	Steam or aseptic process
	Membrane filtration	7[b]	30–35	Terminal moist heat
	Membrane filtration	14	30–35	Aseptic process
TSB	Direct transfer	21 (14 + 7)	20–25	Steam or aseptic process
	Membrane filtration	7[b]	20–25	Terminal moist heat
	Membrane filtration	14	20–25	Aseptic process

[a]Time is the minimum number of incubation days. Additional incubation time may be required if the nature of the product is conducive to produce a "slow-growing" contaminant.
[b]A seven-day incubation period is only permissible where authorized or dictated in the European Medicines Evaluation Agency (EMEA) submission. In general, a 14-day incubation period is required for all products, which are required to meet the EP sterility test.

plus four more days to detect growth in media used as negative controls after adding a challenge organism.

Optimal detection conditions for 5 to 50 CFU of nine different microorganisms (aerobic and anaerobic bacteria and molds) were reported to be 22°C to 32°C over 14 days using soybean-casein digest and thioglycollate broths (4).

STERILITY-TEST METHODS

The USP and EP sterility tests specify two basic methods for performing sterility tests—the direct transfer or direct inoculation method and the membrane filtration method, with a statement that the latter, where feasible, is the method of choice. In fact, in some cases, membrane filtration may be the only possible choice.

Direct Transfer Method

The direct transfer (DT) method is the more traditional sterility-test method. Basically, the DT method involves three steps:

1. Aseptically opening each sample container from a recently sterilized batch of product.
2. Using a sterile syringe and needle to withdraw the required volume of sample for both media from the container.
3. Injecting one-half of the required volume sample into a test tube containing the required volume of FTM and the other half volume of sample into a second test tube containing the required volume of TSB.

The DT method is simple in theory, but difficult in practice. The technician performing the DT test must have excellent physical dexterity and the proper mental attitude about the concern for maintaining asepsis. The demand for repetition in opening containers, sampling, transferring, and mixing can potentially cause fatigue and boredom with a subsequent deterioration in operator technique and concern. As this occurs, the incidence of accidental product sterility-test contamination will increase.

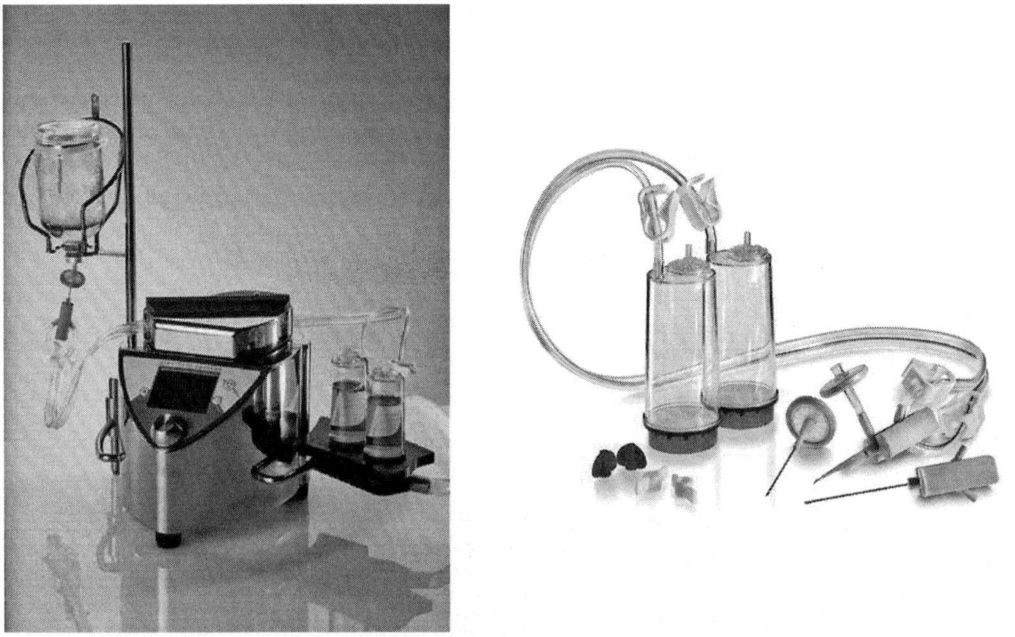

Figure 27-2 Membrane filtration sterility test device. *Source*: Courtesy of Millipore Corporation.

The USP and EP tests require a minimum volume of sample per container volume to be transferred to a minimum volume of each culture medium (Table 27-1). The sample volume must be a sufficient representation of the entire container volume and the volume of medium must be sufficient to promote and expedite microbial growth, if present. Adequate mixing between the sample inoculum and the culture medium must take place to maximize interaction and facilitate microbial growth.

Membrane Filtration Method

The membrane filtration (MF) sterility test became official in the 18th edition of the USP in 1970. It has since become the more popular and widely used method over the DT method and, when feasible for pharmacopeial articles, should be preferred.

The successful employment of this technique requires more skill and knowledge than that required for the DT method. Five basic steps are involved in the use of the MF sterility-test method:

1. The filter unit (Fig. 27-2) must be properly assembled and sterilized prior to use.
2. The contents of the prescribed number of units are transferred to the filter assembly under strict aseptic conditions.
3. The contents are filtered with the aid of a vacuum or pressure differential system.
4. The appropriate type and volume of culture media is added to the canister.
5. The canister is incubated according to the medium used.

A suitable membrane filter unit consists of an assembly that facilitates the aseptic handling of the test articles and allows the processed membrane to be removed aseptically for transfer to appropriate media or an assembly where sterile media can be added to the sealed filter and the membrane incubated in situ. A membrane suitable for sterility testing has a rating of 0.45 μm, and a diameter of approximately 47 mm. These membranes have hydrophobic edges or low product- binding characteristics that minimize inhibitory product residue, and it is this residue that interferes with requirements of the validation test for bacteriostasis and fungistasis. For products that do not contain inhibitory substances, membranes without hydrophobic edges can be used, but wet them prior to testing.

The MF method offers at least four advantages over the use of the DT method. They are as follows:

1. Greater sensitivity.
2. The antimicrobial agent and other antimicrobial solutes in the product sample can be eliminated by rinsing prior to transferring the filter into test tubes of media, thereby minimizing the incidence of false negative test results.
3. The entire contents of containers can be tested, providing a real advantage in the sterility testing of large-volume parenterals and increasing the ability to detect contamination of product lots containing very few contaminated units.
4. Low-level contamination can be concentrated on the membrane by filtering large volumes of product. This results in faster reporting of test results since MF requires only seven- day incubation (for most terminally sterilized products).
5. Organisms present in an oleaginous product can be separated from the product during filtration and cultured in a more desirable aqueous medium.

Interpretation of Results

No visible evidence of microbial growth in a culture medium test tube, after subjecting the sample and medium to the correct procedures and conditions of the USP and EP sterility test, may be interpreted that the sample representing the lot is absent of intrinsic contamination. Such interpretation must be made by those having appropriate formal training in microbiology and having knowledge of several basic areas involved in quality control sterility tests:

- Industrial sterilization methods and their limitations
- Aseptic processing
- Statistical concepts involved in sampling lots for representative articles
- Environmental control procedures used in the test facility.

If microbial growth is found, or if the sterility test is judged to be invalid because of inadequate environmental conditions, the sterility test may be repeated. However, this introduces a controversial and somewhat complicated subject.

STERILITY RETESTING

Sterility retests have been allowed by the USP since sterility testing became a USP requirement (XI edition, 1936). Specific definitions of first and second stage sterility re-testing were introduced in USP XX (1980). While sterility retesting is allowed per the CFR, retesting is no longer allowed per USP as of the eight supplement released in May 1998. The FDA has repeatedly reaffirmed that it supports the USP position provided that industry shows due diligence in their investigations of initial sterility-test failures. Sterility retesting and investigation of initial sterility-test failures should be done with the highest degree of diligence and responsibility on the part of high-level management of the parenteral industry.

FDA GUIDELINES ON STERILITY TESTING

The September, 2004 FDA Guideline on Sterile Drug Products Produced by Aseptic Processing contains a fair amount of direction regarding conductance, evaluation, limitations, interpretation, and retesting requirements of the USP sterility test (2). This was highlighted in Chapter 21, p. 322, but for the sake of completeness, some redundancy will be covered here. The testing laboratory environment should employ facilities and controls comparable to those used for the filling and closing operations (e.g., Class 100 air conditions for critical operations where a sterile product is exposed to the environment). The limitations of the USP Sterility Test cause the FDA considerable concern with respect to sampling plans and any positive test result that may occur. In investigation of sterility-test failures—positive test results, the guidelines state that

> When persuasive evidence showing laboratory error is absent, or when available evidence is inconclusive, firms should err on the side of safety and batches should be rejected as not conforming to sterility requirements.

Table 27-5 What Must be Checked During Investigation of Sterility-Test Failure to Determine Assignable Cause

Manufacturing Facility
Media fill validation records
Sterilization records
Environmental data
Bioburden data
HEPA filter certifications
Sanitization records
Filter integrity records
Equipment maintenance records
Manufacturing ticket review
Operator training records
Sterile certification of purchased sterile raw materials
Sterility-test facility
Sterilization records
Environmental data
HEPA filter certifications
Sanitization records
Operator training records
Sterility-test control data

This statement has caused some consternation among QC groups in the pharmaceutical industry because assurance of sterility is so difficult to prove with absolute certainty.

Investigations of sterility-test failures should consider every single factor related to the manufacture of the product and the testing of the product sample. Table 27-5 shows a representative list of factors to be investigated by QC both in the manufacturing areas and in the sterility-test laboratory to determine how a sterility-test failure could have occurred. Most of the time, there is no concrete conclusive evidence pinpointing where the contamination occurred and, thus, QC must make a decision based on philosophical positions and retrospective history of the manufacturing and sterility-test areas.

The FDA aseptic guidelines indicate that persuasive evidence of the origin of the contamination should be based on the following:

1. The identification of the organism in the sterility test (genetic typing may be useful or required.
2. The laboratory's record of tests over time
3. Monitoring of production area environments
4. Product presterilization bioburden
5. Production record review
6. Results of sterility retest.

Identification of the Organism in the Sterility Test

Not only the genus, but also the species of the isolated organism will provide invaluable information concerning the organism's habitat and its potential resistance to the product formulation and sterilization methods. If the organism is one normally found on people, then the investigation can focus on employee hygiene, washing and gowning techniques, and aseptic techniques. Identification of the organism can be compared to historic microbial databases for the manufacturing and testing areas to assess probabilities of where the organism originated. Obviously, if the organism identified had been isolated before in the production area, but never in the testing area, then the production area would be implicated as the source of the organism and the test would be judged as a true sterility-test failure. Identification of the organism allows the manufacturer to perform further testing to determine if the organism is sensitive to the product formulation, particularly if the product contains an antimicrobial preservative. If an organism that was isolated from a product that was terminally sterilized and whose resistance to terminal sterilization is *proven* to be below the microbial reduction produced by the sterilization cycle, then it can reasonably be deduced that the organism did not originate from the product. Knowledge of whether the organism identified is an aerobe or anaerobe would be important if the product were one that contained antioxidants or was overlaid with nitrogen. For example, if the organism were a strict anaerobe and the product was flushed with nitrogen prior to sealing, then it must be strongly suspected that the organism originated during the manufacturing process and was protected by the nitrogen overlay within the product container.

CONTROL IN STERILITY TESTING

Control of the quality of the environment under which the sterility test is performed is of extreme importance. The training and experience of personnel conducting the sterility test must also be controlled with regard to their understanding, use, and attitude toward strict aseptic technique. The types of control of sterility testing include the following:

1. Positive control of the culture media, that is, the testing of the growth-promoting quality of each lot of media
2. Negative control of the culture media, that is, testing the sterility of the media
3. Control of the product itself, that is, obtaining knowledge about the bacteriostatic and/or fungistatic activity of the product prior to its being subjected to a sterility test
4. Specific controls when using the MF technique.

The absence of growth in sterility-test samples at the completion of the test indicates that the product is sterile insofar as assumptions and limitations of the test are considered, that is, it meets the requirements of the test. However, this conclusion can be made only with the assurance that growth would have occurred during the sterility-test period, had microorganisms actually been present. The USP growth promotion test is designed to serve as a positive control for each lot of sterility-test media. Each lot is inoculated with 10 to 100 colony-forming units (CFU) of the microorganisms listed in Table 27-6. Growth of these microorganisms must occur in the appropriate medium within seven days' incubation. The evidence of growth in duplicate test containers compared with the same lot of medium containing no microbial inoculum qualifies the test medium to be used for sterility-test purposes. The USP allows for the growth promotion test to be the positives control run simultaneously with the actual sterility test with the understanding that the test becomes invalid if the medium does not support the growth of the inoculated microorganisms. However, if tested media are stored, additional tests are prescribed for particular storage conditions.

Negative Controls

Negative controls consist of containers of culture media without addition of product sample or microbial challenge. The purpose of negative control samples is to verify the sterility of the medium before, during, and after the incubation period of the sterility test. If microbial growth is detected with a negative control, the medium was not sterilized properly, contamination was introduced accidentally during the test procedure, or there exists an inefficiency in the

Table 27-6 Test Microorganisms Required by the USP for use in the Growth Promotion and Bacteriostasis/Fungistasis Test Used in Sterility Testing

Medium	Test microorganism	Incubation temperature (°C)	Condition
Fluid thioglycollate	*Staphylococcus aureus* (ATCC No. 6538)	32.5 ± 2.5	Aerobic
	Pseudomonas aeruginosa (ATCC No. 9027)[a]	32.5 ± 2.5	Aerobic
	Clostridium sporogenes (ATCC No.11437)[b]	32.5 ± 2.5	Aerobic
Alternative thioglycollate[c]	*Clostridium sporogenes* (ATCC No. 19404)	32.5 ± 2.5	Anaerobic
Soybean-casein digest	*Bacillus subtilis* (ATCC No. 6633)	22.5 ± 2.5	Aerobic
	Candida albicans (ATCC No.10231)	22.5 ± 2.5	Aerobic
	Aspergillus niger (ATCC No. 16404)	22.5 ± 2.5	Aerobic

[a]An alternative microorganism is *Micrococcus luteus* (ATCC 9341).
[b]An alternative to Clostridium sporogenes, when a nonspore-forming microorganism desired is *Bacteroides vulgatus* (ATCC 8482).
[c]Use for sterility test of devices that have tubes with small lumens.

container or packaging system. If microbial growth occurs in a negative control and there is an absence of evidence of accidental contamination, there becomes a clear indication for retesting the product. USP has added verbiage that the sterility test must be carried out under aseptic conditions. Aseptic conditions are achieved without affecting any microorganisms that should be revealed during the test. The working conditions in which the tests are performed are monitored regularly by appropriate sampling of the working area and by carrying out appropriate controls.

Bacteriostatic and Fungistatic Testing

If a sterility test is negative (no growth), there must be the assurance that growth was not inhibited by the antimicrobial properties of the product itself. The USP provides a procedure for determining the level of bacteriostatic and fungistatic activity of a product or material prior to its being tested for sterility by the direct transfer or membrane filtration test. Basically, the procedure calls for adding product to containers of culture media in volumes corresponding to those that would be used for testing the product containing 10 to 100 of the microorganisms listed in Table 27-6 and comparing with medium-inoculum controls without the product. If the material possesses bacteriostatic or fungistatic activity, then the product-media will show decreased or no microbial activity compared to control culture media. If this is the case, then procedures must take place for the proper inactivation of these bacteriostatic/fungistatic properties. Either a suitable sterile inactivating agent must be found or the material and medium must be adequately diluted to overcome the static effects. If at all possible, the membrane filtration test should be applied for those materials found to be bacteriostatic or fungistatic. Where membrane filtration is used, similar comparisons are made of incubated filters through which product and suitable diluting fluid have been passed, each containing the same added microorganisms.

Controls for Membrane Filtration Techniques

The MF test relies on the ability to produce sterile equipment and to have aseptic conditions under which to conduct the test. Three basic control procedures are recommended in separate experiments:

1. The membrane filters are challenged after their sterilization cycle for their ability to retain microorganisms.
2. The exposure times for agar settling plates used to monitor the environment are validated.
3. The cleaning procedures used to remove bacteriostatic and/or bactericidal residues from equipment following the MF test must be validated. This is especially important for the equipment involved in the sterility testing of antibiotics.

VALIDATION OF THE STERILITY TEST

For every product that is tested for sterility, the sterility-test method must be validated for that product. What this means, simply, is that prospective validation studies must be performed to collect data to prove that the sterility test can detect microbiological contamination in the product. Validation of the sterility test for a particular product involves adding small but known concentrations ($\leq$100 CFU) of various microorganisms to the final rinse and then demonstrating recovery of the organisms using the sterility-test methodology. Table 27-6 provides the test organisms required by USP and EP. The EP and USP chapters on sterility testing are now considered to be "harmonized", if the practitioner desires to test a product for sterility release, and that product is required to meet EP and USP requirements, there are several key points to consider.

Organisms from Table 27-6 will have to be used in bacteriostasis and fungistasis test, and *Bacillus subtilis* must be tested in both FTM and TSB. Upon incubation of the challenge containers, all bacteria must show visible growth within three days, and all fungi must show visible growth within five days of the test.

Even if the product is terminally sterilized, the final sterility test must incubate for 14 days (if the membrane filtration technique is used) to satisfy the EP requirement. ATCC 19404 must be used for the *C. sporogenes* challenge to satisfy the EP requirement, while ATCC 11437 must be used *for the C. sporogenes* to satisfy the USP requirement when performing the bacteriostatic and

Table 27-7 Probability of Accepting a Batch as Sterile Assuming the Contamination Rate to be Constant at 0.1%

Sterility-test sample size	Batch size		
	1000	**2000**	**5000**
10	0.99	0.99	0.99
20	0.98	0.98	0.98
50	0.95	0.95	0.95

fungistatic tests. Where the DT test is employed, the initial transfer test is required to incubate for 14 days for EP, but only 7 days for the USP test.

LIMITATIONS OF THE STERILITY TEST

The USP referee sterility test suffers from at least three limitations:

1. The invariant uncertainty that the small sample used in the test reliably represents the whole lot
2. The inability of the culture media and incubation conditions to promote the growth of any and all potential microbial contaminants
3. The unavoidable problem of occasional accidental contamination of the sterility-test samples.

The Problem of Sampling and Statistical Representation

The probability of accepting lots having a given percent contamination is related to the sterility-test sample size rather than to batch size (Ref. 1 and all references therein). For example, if a batch is 0.1% contaminated (one nonsterile unit in 1000 units) and 10 units are sampled for a sterility test, the probability of finding one of those 10 samples to be the one contaminated unit in 1000 is not significantly different if the batch size were 1000, 2000, or 5000. Increasing the sample size from 10 to 20 to 50 units per batch, however, affects the probability of accepting the batch as sterile to a more significant degree than does the increase in batch size, assuming that the increase in batch size does not increase the level of contamination. This phenomenon is depicted in Table 27-7. The probability rate does not change as the batch size is increased, but does change as the sample size is increased. Of course, a key factor is that the contamination rate remains at 0.1% as the batch size increases. This, in reality, may not be true, especially for aseptically filled products. Hence, if the contamination rate increases with batch size, the probability of acceptance decreases for the same sample size.

The relationship of probability of accepting loss of varying degrees of contamination to sample size is given in Table 27-8. Three details may be learned assuming the data in Table 27-8 to be real: (*i*) as the sample size is increased, the probability of accepting the lot as sterile is decreased; (*ii*) at low levels of contamination, for example, 0.1%, the odds of ever finding that one contaminated sample in 1000 units are so small that one must face the fact that lots are

Table 27-8 Relationship of Probabilities of Accepting Lots of Varying Assumed Degrees of Contamination to Sample Size

Number of samples tested (n)	Probability of accepting the lot as a function of assumed contamination rate from 0.1 to 20 percent					
	0.1	**1**	**5**	**10**	**15**	**20**
10	0.99	0.91	0.60	0.35	0.20	0.11
20	0.98	0.82	0.36	0.12	0.04	0.01
30	0.95	0.61	0.08	0.01		
100	0.91	0.37	0.01			
300	0.74	0.05				
500	0.61	0.01				

Table 27-9 Probability of Finding at Least One Nonsterile Unit in a Sample Size of 20 Subjected to a Sterility Test

Assumed percent nonsterile units in the lot	Probability of finding at least one nonsterile unit
0.10	0.01980
0.05	0.00995
0.02	0.00399
0.01	0.00199
0.005	0.00100
0.002	0.00040
0.001	0.00020

going to be passed as sterile but somewhere, at some time, some patient is going to receive that nonsterile sample (even at a contamination rate of 1% with 20 sterility-test samples, it must be realized that such a lot will be passed as sterile 82% of the time); and (*iii*) realistically, a batch must be grossly contaminated for the sterility test to detect it. This fact was concluded at a 1963 conference on sterility testing in London (5) in which experts in sterility testing recognized that the lowest contamination rates that can be detected with 95% confidence are 28% with a sample size of 10, 15% with a sample size of 20, and 7% with a sample size of 40 units.

A sample size of 20 units is shown in Table 27-9. As an example, if it is assumed that only one unit in a batch of 100,000 units is contaminated (0.001%), the probability that the one contaminated unit is among the 20 sterility-test samples taken at random is 0.0002, or two times in one million sterility tests. As the assumed level of sterility assurance is increased, i.e., going from 0.1% to 0.001% to 0.000001%, it is absurd to expect that the sterility test will ever fail as long as the number of test samples is in the 20 to 50 range. Even if the sterility-test sample size were 160 units, the odds of failing the test at the 0.001% sterility assurance level is only approximately 15%.

Problem of Supporting the Growth of Microbial Contaminants

No single medium will support the growth of all microbial forms, that is, bacteria, molds, fungi, and yeasts. FTM will not recover very low levels of some aerobic spore formers such as *Bacillus subtilis.* TSB gave more efficient recovery of small numbers of *B. subtilis* and *Clostridium sporogenes* spores than in FTM. TSB, being strictly an aerobic medium, will not support the growth of the genus *Clostridia.* On the other hand, while FTM effectively supports the growth of various strains of *Clostridia*, it has been reported that sodium thioglycollate is toxic to *Clostridia* and this antioxidant should be replaced by cysteine hydrochloride.

TSB is incubated at 20°C to 25°C to permit the adequate growth of facultative organisms such as enterobacteria (*Escherichia coli, Salmonella, Shigella, Proteus, Serratia marcescens,* and *Flavobacterium*) and many yeasts. FTM is incubated at 30°C to 35°C to detect mesophilic bacteria. These sterility media, therefore, are not incubated at temperatures conducive to the growth of psychophiles (predominantly pseudomads) and thermophiles (predominately bacilli). TSB and FTM do not contain the necessary nutritional ingredients to support the growth of obligate halophiles, osmophiles, or autotrophs.

Problem of Accidental Contamination

Growth that occurs in sterility-test media must be ascertained to have originated from the test sample and not from the culture media or from an external source during the execution of the test. Such a determination can be made only to a limited extent. The use of negative controls eliminates one source of contamination, that being a result of nonsterile culture media. Thus, a positive sterility-test result is concluded to be true (the test sample is contaminated) unless it can be shown to be false (contamination was accidently introduced during the test procedure). The problem of false positives is widespread and cannot be completely eliminated.

The percentage of false positive sterility test results has decreased significantly thanks to the use of barrier isolation technology (next section). The most common types of microbial contaminants found in false positive sterility-test samples are human borne. Indeed, the single largest contributor of accidental contamination in sterility-test samples results from lack of strict adherence to good aseptic techniques by the person or people conducting the test. False positive sterility tests result also from contaminants located in the environment (air and surfaces, especially if barrier isolators are not used), and/or equipment used in conducting the test (e.g., nonsterile membrane filter assemblies, scissors, forceps, other devices that somehow are contaminated).

ISOLATION STERILITY-TEST UNITS

As previously discussed, false positive sterility tests occur because of inadvertent contamination of the sample in the sterility-test laboratory. Such contaminations are of a finite probability as long as human manipulation is involved. Concerns over such unreliabilities of the sterility test have given rise to new technologies designed to remove as much as possible the human element involved in sterility testing.

The use of hard-walled isolators has become the most recent trend in sterility testing. Isolators are made of polyvinyl chloride supported externally by a framework of stainless-steel rods. The barriers are accessed by the operator through either glove sleeves or half suits. Materials are introduced into and removed from these barriers through a double door transfer port where both sides of the double door are sterilized. Room air enters and exits through a High Efficiency Particulate Air (HEPA) filter system. All sterility-test operations occur within the barrier system.

One of the major aspects of the isolation chamber is the sterilization and its validation of all surfaces within the chamber and product containers and other items brought into the chamber. The original method of surface sterilization was the use of peracetic acid as a spray. The most commonly used method of surface sterilization today is VPHP (vapor phase hydrogen peroxide). The VHP 1000® manufactured by the Steris Corporation (formerly AMSCO) (Fig. 27-3) is widely used in the pharmaceutical industry in conjunction with barrier isolation systems. VPHP is less corrosive to metals such as stainless steel than is peroxyacetic acid.

The advent of isolation chambers and robotic sterility-test systems has challenged the long-held level of acceptability of false positives. The historical generally acceptable level of false positive sterility tests was in the vicinity of 1.0%, although in the past couple of decades,

Figure 27-3 Example of sterility test isolator with vapor phase hydrogen peroxide generator. *Source*: Courtesy of Baxter Healthcare Corporation.

the expected rate of false positives using standard cleanroom sterility-test work stations has been closer to 0.1%. Cloué and Wagner reported that the average false positive rate of sterility tests using cleanroom technology was 0.5% to 1.0% while that for isolation technology was 0% (6)

REFERENCES

1. Akers MJ, Larrimore DS, Guazzo, DM. Sterility testing. In: Marcel Dekker, ed. Parenteral Quality Control: Sterility, Pyrogen, Particulate, and Package Integrity Testing. 3rd ed. London: Informa Healthcare, 2003.
2. Guidance for Industry: Sterile Drug Products Produced by Aseptic Processing—Current Good Manufacturing Practice. FDA, September, 2004, http://www.fda.gov/downloads/Drugs/GuidanceComplianceRegulatoryInformation/Guidances/ucm070342.pdf. Accessed June 15, 2010.
3. Bowman FW. Application of membrane filtration to antibiotic quality control sterility testing. J Pharm Sci 1966; 55:818–821; The sterility testing of pharmaceuticals. J Pharm Sci 1969; 58:1301–1308.
4. Bugno A, Pinto TDA. The influence of incubation conditions in sterility tests. PDA J Pharm Sci Tech 2003; 57:399–409.
5. Council of the Pharmaceutical Society of Great Britain. Round Table Conference on Sterility Testing, London, 1963:B31.
6. Cloué P, Wagner CM. Sterility test isolators. In: Lysjford J, ed. Practical Aseptic Processing: Fill and Finish. Vol 2. Bethesda, MD: Parenteral Drug Association, 2009:86.

BIBLIOGRAPHY

European Pharmacopeia, Sterility Testing, 2.6.1

The Pharmaceutical Inspection Convention and Pharmaceutical Inspection Co-operation Scheme (PIC-PIC/S), Document PI-012–3, Recommendation on Sterility Testing

United States Pharmacopeia, Sterility Tests Chapter <71>. Rockville, MD: United States Pharmacoeial Convention, Inc.

28 Pyrogens and pyrogen/endotoxin testing[1]

Pyrogens are products of metabolism of microorganisms. The most potent pyrogenic substances (endotoxins) are constituents (lipopolysaccharides, LPS) of the cell wall of gram-negative bacteria (e.g., *Pseudomonas sp, Salmonella sp, Escherichia coli*). Gram-positive bacteria and fungi also produce pyrogens but of lower potency and of different chemical nature. Gram-positive bacteria produce peptidoglycans while fungi product β-glucans, both of which can cause nonendotoxin pyrogenic responses. Endotoxins are LPS that typically exist in high-molecular-weight aggregate forms. However, the monomer unit of LPS is less than 10,000 Da, enabling endotoxin easily to pass through sterilizing 0.2-μm filters. Studies have shown that the lipid portion of the molecule is responsible for the biological activity. Since endotoxins are the most potent pyrogens and gram-negative bacteria are ubiquitous in the environment, especially water, this discussion focuses on endotoxins and the risk of their presence as contaminants in sterile products.

Pyrogens, if present in parenteral drug products and injected into patients, can cause fever, chills, pain in the back and legs, and malaise. While pyrogenic reactions are rarely fatal, they can cause serious discomfort and, in the seriously ill patient, shock-like symptoms that can be fatal. The intensity of the pyrogenic response and its degree of hazard will be affected by the medical condition of the patient, the potency of the pyrogen, the amount of the pyrogen, and the route of administration (intrathecal is most hazardous followed by intravenous, intramuscular, and subcutaneous). When bacterial (exogenous) pyrogens are introduced into the body, LPS targets circulating mononuclear cells (monocytes and macrophages) that, in turn, produce proinflammatory cytokines such as interleukin 2, interleukin 6, and tissue necrosis factor. Besides LPS, gram-negative bacteria also release many peptides (e.g., exotoxin A, peptidoglycan, and muramuyl peptides) that can mimic the activity of LPS and induce cytokine release. The Limulus Amebocyte Lysate (LAL) test, discussed later, can only detect the presence of LPS. It has been suggested that a new test, called Monocyte Activation Test (MAT), replaces LAL as the official pyrogen test because of its greater sensitivity to all agents that induce the release of cytokines that cause fever and a potential cascade of other adverse physiological effects (1). While the MAT may not be a complete replacement for the LAL test, it is expected to be an official test in the European Pharmacopeia (chap. 2.6.30).

The control and sources of pyrogens and the elimination of pyrogenic contamination are covered in chapter 13.

When injected into humans in sufficient amounts, pyrogens will cause a variety of adverse physiological responses (Table 28-1). The most common or recognizable response is an increase in body temperature, from which the name "pyrogen" is derived (Greek "pyro" = fire; "gen" = beginning). Pyrogenic responses rarely are fatal unless the patient is very sick and the dose is very large. Nevertheless, pyrogens are considered toxic substances and should never be injected knowingly. Pyrogen contamination of large-volume parenteral solutions is especially serious because of the large amounts of fluid administered to people whose illnesses must be of the severity to warrant the use of such large volumes.

Pyrogens come from microorganisms. All microbial forms produce pyrogen; however, the most potent pyrogen originates from gram-negative bacteria. The entity primarily involved in pyrogenic reactions in mammals is the LPS from the outer cell membranes of gram-negative bacteria. Another name for LPS is endotoxin. Although not entirely correct, the names pyrogen, LPS, and endotoxin are routinely used interchangeably. Figure 28-1 is a schematic representation of the three cell wall layers of a gram-negative microorganism (2). The outer membrane shown in the figure is not found in gram-positive bacteria. This structure contains the LPS moiety that interacts with the coagulable protein of the amebocytes of the horseshoe crab, a phenomenon from which evolved the LAL test.

[1] This chapter is an update from Ref. 1.

Table 28-1 Adverse Physiological Effects of Pyrogens in Humans

Primary
1. Increase in body temperature
2. Chilly sensation
3. Cutaneous vasoconstriction
4. Pupillary dilation
5. Piloerection
6. Decrease in respiration
7. Rise in arterial blood pressure
8. Nausea and malaise
9. Severe diarrhea
10. Pain in the back and legs
11. Headache

Secondary
1. Cutaneous vasodilation
2. Hyperglycemia
3. Sweating
4. Fall in arterial blood pressure
5. Involuntary urination and defecation
6. Decreased gastric secretion and motility
7. Penile erection
8. Leucocytopenia, leucocytosis
9. Hemorrhage and necrosis in tumors
10. Altered resistance to bacterial infections
11. Depletion of liver glycogen
12. Rise in blood ascorbic acid
13. Rise in blood nonprotein nitrogen and uric acid
14. Decrease in plasma amino acids

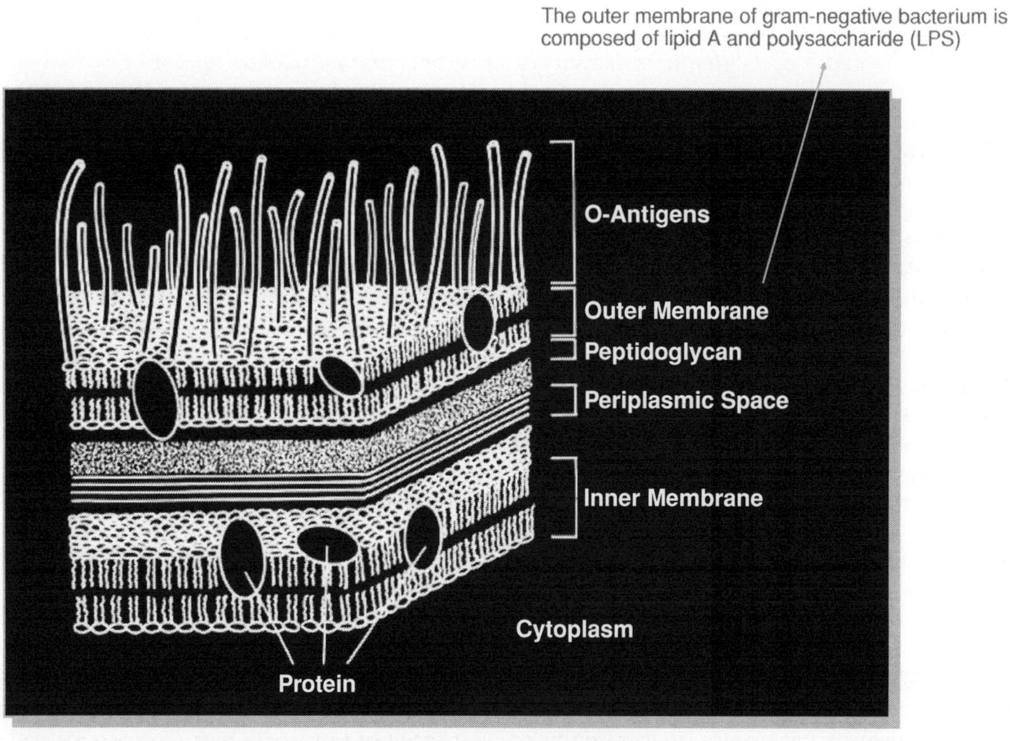

Figure 28-1 Gram-negative bacterial cell wall.

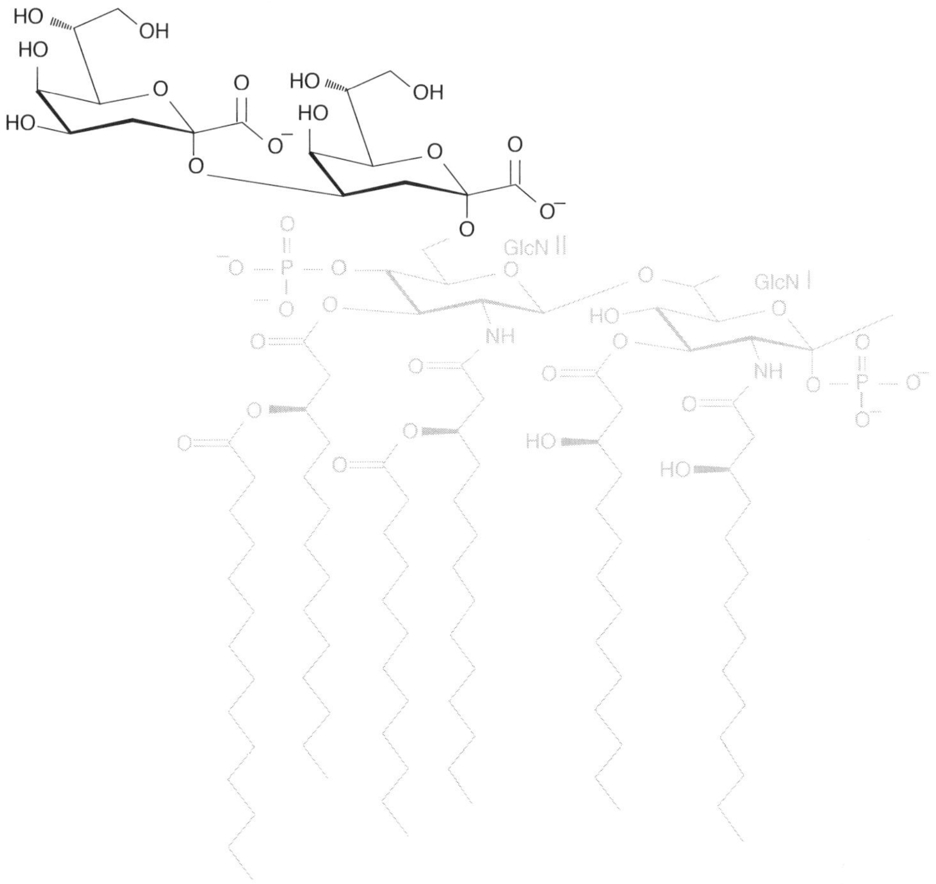

The basic lipopolysaccharide of *E. coli.* incorporating lipid A. (Gray line portion of the structure)

Figure 28-2 Lipid A. *Source*: Courtesy of Williams Christie. http://lipidlibrary.aocs.org/Lipids/lipidA/index.htm.

LPS, extracted and recovered as a colloidal suspension, may be split by mild acid hydrolysis into lipid A and degraded polysaccharides. Lipid A is composed of B-1, 6-glucosamine disaccharide units with α-hydroxymyristic acid replacing one of the amino hydrogens, and fatty acids replacing hydrogen in some of the –OH groups (Fig. 28.2). Each two glucosamine units are separated by two phosphate moieties forming a linear polymer. Lipid A alone lacks biologic activity, yet LPS is toxic, probably because polysaccharide increases the aqueous solubility of lipid A. When lipid A is separated from the polysaccharide component of endotoxin, it loses more than 99.9% of its pyrogenic activity in rabbits (3,4).

Freedom from pyrogenic contamination characterizes parenteral products in the same manner as sterility and freedom from particulate matter. Preventing the presence of pyrogens is much preferred over removing pyrogens in parenteral products. Preventing pyrogenic contamination primarily involves the use of ingredients, solvents, packaging materials, and processing equipment that have been depyrogenated initially, then employing correct and propel procedures during the entire manufacturing process to minimize the possibility of pyrogen development.

BRIEF HISTORY

The pyrogenic response has been known since 1865 when it was reported that an injection of distilled water produced hyperthermia in dogs. It was not until 1923 that Florence Seibert (5,6) recommended that all pharmaceuticals be tested for pyrogens. Seibert used the rabbit as the animal model for detecting the presence of pyrogens in injectables. Seibert also demonstrated conclusively that pyrogens originate from water-borne organisms are heat resistant, filterable, and can be eliminated from water by distillation. The pyrogen test became an official United States Pharmacopeia (USP) quality control test for parenterals in 1942. The rabbit pyrogen test methodology officially recognized in compendial standards has remained essentially unchanged. The LAL test for endotoxin became an official USP test in 1985. Today the LAL test, or more commonly called the Bacterial Endotoxin Test, has pre-empted the rabbit test as the USP method of choice for detection of endotoxin in parenteral products. Also, the more sensitive and accurate LAL assay for testing raw materials, in-process pyrogen control or pharmaceuticals and medical devices, and end-product evaluation of devices, is small and large-volume parenteral products (7). The LAL test also is widely used in the validation of depyrogenation of dry-heat sterilization processes.

GENERAL DESCRIPTION OF THE USP PYROGEN TEST

The pyrogen test is designed to limit to an acceptable level the risks of febrile reaction in the patient to the administration, by injection, of the product concerned. The test involves measuring the rise in temperature of rabbits following the intravenous injection of a test solution and is designed for products that can be tolerated by the test rabbit in a dose not to exceed 10 mL/kg injected intravenously within a period of not more than 10 minutes. For products that require preliminary preparation or are subject to special conditions of administration, follow the additional directions given in the individual monograph or, in the case of antibiotics or biologics, the additional directions given in the federal regulations.

All apparatuses—glassware, containers, syringes, needles, etc.—and all diluents used in performing the pyrogen test must themselves be free from pyrogenic contamination. Heat-durable items such as glass and stainless steel can be depyrogenated by exposure to dry-heat cycles at temperatures greater than 250°C for at least 30 minutes. Negative controls utilize the diluent rather than the product sample as the injection, with the diluent being exposed to the same procedure and materials as the product sample. The use of negative controls with each pyrogen test is not standard practice because of prior knowledge and assurance that materials used in the test are nonpyrogenic.

Rabbits are used as pyrogen test models because they physiologically respond similarly to pyrogens as do human beings. Rabbits and humans respond identically on a nanogram per kilogram basis to pyrogenic quantities of endotoxin. Rabbits for pyrogen testing are not used more frequently than once every 48 hours, nor prior to two weeks following a maximum rise of its temperature of 0.6°C or more while being subjected to the pyrogen test or following its having been given a test specimen that was adjudged pyrogenic.

The albino rabbit is the most widely used rabbit, particularly strains from New Zealand and Belgium. It is essential that the rabbit colony be treated with utmost care. The environment in which the rabbits are housed must be strictly controlled with respect to temperature, humidity, lighting, and potential contamination of air, surfaces, and feed. Any new shipment of rabbits should be quarantined and monitored for one to two weeks following receipt of the shipment for the presence of illness and/or disease.

Rabbits must become accustomed to being restrained in their cages and being handled both in the rectal insertion of the thermocouple and the injection of the test product. The normal basal body temperature of rabbits ranges between 38.9°C and 39.8°C (102.0–103.6°F). Rabbit baseline temperature is established by measuring rectal temperature during the conductance of several "sham" tests (following the entire pyrogen test procedure using pyrogen-free sodium chloride solution as the injection sample). Rabbits may become tolerant to pyrogenic activity after repeated injections of endotoxin. It is for this reason that a rabbit showing a rise of its body temperature of 0.6°C or more during a pyrogen test cannot be used again as a pyrogen test animal for at least two weeks.

USP PYROGEN TEST PROCEDURE

The USP procedure (8) requires the test to be performed in a separate area designated solely for pyrogen testing and under environmental conditions similar to those under which the animals are housed and free from disturbances likely to excite them. Withhold all food from the rabbits used during the period of the test. Access to water is allowed at all times, but may be restricted during the test. If rectal temperature-measuring probes remain inserted throughout the testing period, restrain the rabbits with light-fitting stocks that allow the rabbits to assume a natural resting posture. Not more than 30 minutes prior to the injection of the test dose determine the "control temperature" of each rabbit. This is the base for the determination of any temperature increase resulting from the injection of a test solution. In any one group of test rabbits, use only those rabbits whose control temperatures do not vary by more than 1 degree from each other, and do not use any rabbit having a temperature exceeding 39.8°C.

Unless otherwise specified in the individual monograph, inject into an ear vein of each of three rabbits 10 mL of the test solution per kilogram of body weight, completing each injection within 10 minutes after the start of administration. The test solution is either the product, constituted if necessary as directed in the labeling, or the material under test treated as directed in the individual monograph and injected in the dose specified therein. For pyrogen testing of devices or injection assemblies, use washings or rinsings of the surfaces that come in contact with the parenterally administered material or with the injection site or internal tissues of the patient. For example, 40 mL of sterile, pyrogen-free saline, test solution (TS) at a flow rate of approximately 10 mL/min is passed through the tubing of each of 10 infusion assemblies. Ensure that all test solutions are protected from contamination. Perform the injection after warming the test solution to a temperature of 37 ± 2°C. Record the temperature at 1 and 3 hours and 30-minute interval in between subsequent to the injection.

Noise represents a major problem in maintaining and using rabbits for pyrogen testing. The room in which the tests are conducted should be as free from noise and activity as possible. Anything that causes excitement in the rabbit potentially can produce a 0.2°C to 1.0°C rise in body temperature that may not return to normal for 60 to 90 minutes. During the pyrogen test, which could last four to six hours, the rabbits should be restrained with a minimum of discomfort. Restraint should be confined to the neck and head of the rabbit to facilitate the test dose injection into the ear vein and to permit the rabbit comfortable movement of its legs and back.

Dose administration is accomplished using a sterile syringe and 20 to 23 gauge needle. The size of syringe will depend on the dose volume. The USP requires a dose of 10 mL/kg body weight unless otherwise specified in the individual monograph. For example, Phytonadione Injection, USP, pyrogen test dose is 2 mL/kg while Protamine Sulfate Injection, USP, requires only 0.5 mL/kg containing 10 mg/mL. Some injectable monographs specify the pyrogen test dose on a weight–weight basis, for example, the dose of Diazepam Injection, USP is 0.25 mg/kg.

The test solution must be warmed to 37°C prior to injection. The ear vein is swabbed with alcohol (70%), which not only disinfects but also improves visibility of the vein. Employing correct technique in making the injection can preserve vein longevity, with the thumb at the site of injection to retard bleeding and scarring.

Rectal temperatures are recorded at one, two, and three hours subsequent to the injection. During the test period rabbits and equipment should be checked periodically.

Test Interpretation

The solution may be judged nonpyrogenic if no single rabbit shows a rise in temperature of 0.5°C or greater above its control temperature. If this condition is not met the test must proceed to a second stage. There is no longer a second condition involving the sum of individual temperatures. In the second stage, five additional rabbits are given a new preparation of the same test sample as the original three rabbits. The solution may be judged nonpyrogenic if not more than three of the eight rabbits showed individual temperature rise of 0.5°C or more.

The USP rabbit pyrogen suffers from several limitations that established the opportunity for the LAL test as a possible alternative for the rabbit test as an official pyrogen test procedure.

In Vivo Model

A test method that uses a living animal as its model certainly must submit to a number of problems offered by biological systems. Variability in biological systems poses a great problem.

Table 28-2 Dose Dependency of Rabbit Pyrogen Test

E. Coli endotoxin concentration (ng/mL)	Volume solution injected (mL/kg)	USP total temperature (°C)	Mean temperature increase (°C)[a]	Standard deviation (°C)[b]	Coefficient of variation (%)
3.125	1.0	7.80[c]	0.975	0.246	25.2
1.56	1.0	4.75[c]	0.594	0.218	36.7
1.00	1.0	3.70[c]	0.462	0.158	34.2
0.78	1.0	1.40	0.144	0.208	144.4
0.39	1.0	1.00	0.088	0.187	212.5
0.195	1.0	1.20	0.150	0.065	43.3

Eight rabbit pyrogen test results in saline with *E. coli* 055:BS using 3 to 5 kg rabbits.
[a]Negative rabbit temperature values were excluded from total temperature increase determinations according to USP.
[b]Negative rabbit temperature values were included in the determinations of means and standard deviations to properly reflect total variability.
[c]Failed USP test criteria of 3.7°C total increase.
Source: From Ref. 9.

No two rabbits will possess exactly the same body temperature or respond identically to the same pyrogenic sample. Rabbits are extremely sensitive and vulnerable to their environment. This translates into an expensive proposition in terms of facilities, control of the environment, and training of the animal. Pyrogen testing of rabbits is not only expensive but also laborious. Several hours are consumed in performing the pyrogen test including a great amount of preliminary effort in preparing the animals. Rabbits must be fed and watered properly, cages cleaned to prevent disease, and time spent in training the animals to adapt to the conditions of the pyrogen testing facility and the test itself.

Rabbit Sensitivity to Pyrogens

The pyrogenic response in rabbits is dose dependent. The greater the amount of pyrogen injected per kilogram body weight, the greater the temperature increases in rabbits (Table 28-2) (9).

A collaborative study initiated under the auspices of the Health Industry Manufacturers Association (HIMA) demonstrated that rabbits from 12 laboratories consistently failed (pyrogenic) the test at = 1.0 ng/mL doses (10 mL/kg of 10 ng/kg endotoxin) of *E. coli* 055:B5 endotoxin, and all colonies passed (no pyrogenicity) at the 0.156 ng/kg dose (or 0.156 ng/mL using a 10 mL/kg dose) (10). The same study reported that the "average" rabbit colony will attain a 50% pass/fail rate with 95% confidence at an endotoxin level above 0.098 ng/mL (10 mL/kg dose). The LAL test generally will detect endotoxin levels of 0.025 ng/mL or less. Thus, the rabbit test is less sensitive to endotoxin than the LAL test is.

Sensitivity of the rabbit bioassay for endotoxin appears to fall in the range of 1 to 10 ng/kg (11,12). Greisman and Hornick (11) found that the threshold pyrogenic dose of *E. coli* endotoxin for both rabbits and humans is 1.0 ng/kg of body weight. This holds true regardless of the volume of pyrogenic solution administered because of the dose (rather than concentration) dependency of the rabbit response to pyrogen.

Interferences of the Rabbit Pyrogen Test

Many products administered parentally cannot be tested for pyrogens with the rabbit test because of the interferences they create in the rabbit response to pyrogens, if they are present in the product. Any product having a pyretic side effect, such as cancer chemotherapeutic agents, will interfere with the rabbit response. Several products are inherently toxic to the rabbit (e.g., cytotoxic compounds) and must be diluted to concentrations far below the pharmacologically effective dose of the drug.

Despite these major limitations and the insurgence of the LAL test, it must not be forgotten that the USP rabbit pyrogen test for decades has nobly served as a sufficiently sensitive test for pyrogens and has helped to eliminate pyrogenic contamination from drugs reaching the marketplace, though most pharmaceutical and device manufacturers currently use the LAL test for the pyrogen test. The official "referee" test according to USP and EP is the LAL gel clot test.

The Limulus Amebocyte Lysate Test

Credit for discovering the interaction between endotoxin and the amebocyte lysate of the horseshoe crab, *Limulus polyphemus*, belongs to Levin and Bang (13). Basing their work on earlier research by Bang (14), these workers were involved in the study of clotting mechanisms of the blood of lobsters, fish, and crabs. Autopsies of dead horseshoe crabs revealed intravascular coagulation. The clotted blood was cultured and found to contain gram-negative bacteria such as *E. coli* and *Pseudomonas*. Further tests showed that amebocyte cells of the horseshoe crab's blood were extremely sensitive to the presence of endotoxin, the toxic substance liberated by the disintegration of bacterial cells. The substance in the amebocytes responsible for reacting with endotoxin is known to be a clottable protein. In lysing the amebocyte cells by osmotic effects, a most sensitive biochemical indicator of the presence of endotoxin was produced, hence the name LAL test.

L. polyphemus (Fig. 28-3) is found only at specific locations along the east coast of North America and the coasts along Southeast Asia. The hearts of mature crabs are punctured and bled to collect the circulating amebocyte blood cells (Fig. 28-4). Carefully performed, this procedure is not fatal to the crab, and upon proper restoration, the crab can be used again. Since amebocytes act as activators of the coagulation mechanism in the crab, an antiaggregating agent must be added to inhibit aggregation. *N*-ethylmaleimide is the most commonly used antiaggregant.

Amebocyte cells are collected and washed by centrifugation and lysed using distilled water. Lysing can also be done with ultrasound, freezing and thawing, and grinding in a glass tissue homogenizer (15). After lysing, the suspension is cleared of debris by centrifugation and the supernatant is lyophilized. Lyophilization is necessary for stability purposes. LAL reagent

Figure 28-3 *Limulus polyphemus* and collection of its amebocytes. *Source*: Courtesy of Associates of Cape Cod.

Figure 28-4 Blood from horseshoe crab after centrifugation. (Horseshoe crab blood color is blue due to copper being the metal in the oxygen carrying hemocyanin, as opposed to iron in hemoglobin. White pellets at bottom contain the amebocytes.) *Source*: Courtesy of Associates of Cape Cod, Inc.

is extremely sensitive to heat and even in the lyophilized state must be stored in the freezer (16). Upon reconstitution, LAL has a shelf life of one month's storage at freezing conditions.

LAL Reaction Mechanism

Endotoxin or a suitably prepared lipid-A derivative of endotoxin activates a proenzyme of LAL having a molecular weight of 150,000. Activation also depends on the presence of divalent metal cations such as calcium, manganese, or magnesium. It has been shown that the sensitivity of the LAL assay for endotoxin detection can be increased 10 to 30 times by using LAL reagent containing 50 mM magnesium (17).

The activated proenzyme, related to the serine protease class containing such enzymes as thrombin, trypsin, and factor Xa, subsequently reacts with a lower molecular weight protein fraction (MW = 19,000–25,000) contained also in the LAL substance. The lower molecular weight fraction, called coagulogen, is cleaved by the proenzyme into a soluble and insoluble subunit. The insoluble subunit appears as a solid clot, a precipitate, or a turbid solution, depending on the amount of insoluble coagulogen by-product formed.

Therefore, the coagulation reaction requires three factors in addition to endotoxin. These three factors—a clotting enzyme, clottable protein (coagulogen), and certain divalent cations—are found in the LAL reagent. A schematic representation of the LAL reaction mechanism is found in Figure 28-5 (18).

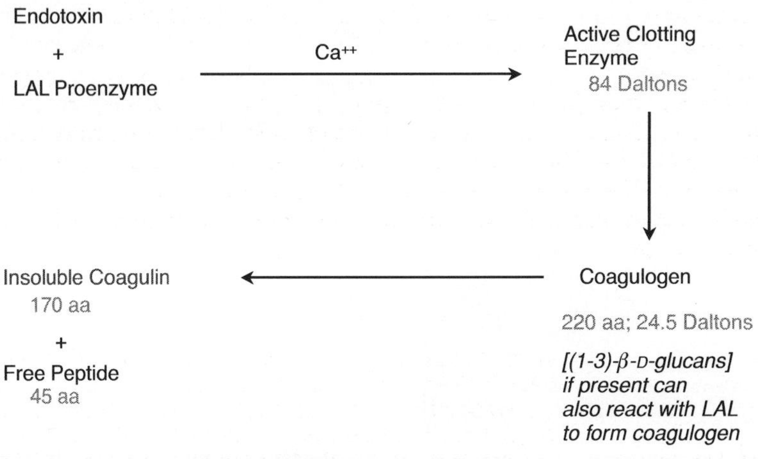

Figure 28-5 LAL reaction mechanism (simplified).

LAL Test Procedure

While the LAL test is a relatively simple procedure, especially when compared with the USP rabbit test, certain specific conditions must be met. These include the following:

- All materials that will come into contact with the LAL reagent or test sample must be thoroughly cleaned and depyrogenated.
- The reaction temperature cannot be outside the range of 36°C to 38°C.
- The pH of the reaction mixture must be within the range of pH 5–7.
- The reaction time should be no longer than one hour.
- Each test must be accompanied by positive and negative controls.

The basic procedure of the LAL test is the combination of 0.1 mL test sample with 0.1 mL LAL reagent. After one-hour incubation at 37°C, the mixture is analyzed for the presence of a gel clot. The LAL test is positive, indicating the presence of endotoxin, if the gel clot maintains its integrity after slow inversion of the test tube containing the mixture (Fig. 28-6).

Standards

For drugs, biological products, and medical devices, the endotoxin standard is called the United States Standard Endotoxin or the USP Reference Standard Endotoxin (RSE). The first RSE lot was designated as Lot EC-2 and had a defined activity of 1 Endotoxin Unit (EU)[2] in 0.2 nanograms (ng) of the standard (19). One vial of RSE contains 10,000 EUs.

When the USP selected the FDA endotoxin standard (purified LPS from *E. coli* 0113) as the new USP reference standard (with established potency in endotoxin units) this gave manufacturers the opportunity to standardize their own control endotoxin standard (CSE) against the USP RSE.

If a manufacturer chooses to use an endotoxin preparation (CSE) other than the U. S. RSE, the CSE will have to be standardized against the RSE. What this means is that the CSE reaction in the rabbit, its uniformity, its stability, and its interaction to a particular LAL lot all must be determined and related to these same characteristics of the RSE.

Validation of the LAL Test

To validate the use of the LAL test for any application requires two determinations: initial qualification of the laboratory and inhibition or enhancement properties of the product on the LAL–endotoxin interaction. Extensive details of LAL test validation requirements are found in the "Guideline on Validation of the Limulus Amebocyte Lysate Test as an End-Product Endotoxin Test for Human and Animal Parenteral Drugs, Biological Products, and Medical Devices" (20).

Qualification of the laboratory simply involves using the selected test method (gel clot end point, chromogenic, and end point turbidimetric, or kinetic turbidimetric techniques) to determine its variability, to test new lots of lysate before use, and to qualify the ability of the analyst(s) to conduct the test. The LAL reagent used must have a confirmed potency (sensitivity). This is achieved by combining the particular reagent with a series of concentrations of RSE or CSE endotoxin bracketing the stated sensitivity (EU/mL) of the LAL reagent. Use four replicates per concentration of endotoxin. The series of endotoxin concentrations are prepared by twofold dilutions of the RSE or CSE endotoxin using LAL-negative water for injection. Following incubation and end point determination (manual or instrumental), the sensitivity of the LAL reagent will be confirmed if the test results are positive to within one twofold dilution of the state label potency.

[2] It has become accepted practice to use Endotoxin Units (EU) as the more desirable expression of endotoxin strength than weight or concentration terms. The use of EU will allow any endotoxin type or lot to be used as a reference lot because its activity can always be related to the original U.S. Reference Standard lot. This chapter will use the EU term as much as possible, but most literature references cited will use the weight or concentration terms as reported in the published articles. It is noted in the USP and EP that 1 EU = 1 IU (international unit).

Table 28-3 Examples of Minimum Valid Concentration (MVC) and Minimum Valid Dilution (MVD) Calculations (20)

MVC determination

$$\text{MVC} = \frac{\lambda M}{K}$$

λ = Sensitivity of LAL reagent in EU/mL
M = Rabbit dose or maximum human dose/kg
K = 5.0 E/kg (0.2 EU/kg for intrathecal drugs)
If LAL sensitivity (ë) was 0.065 EU/mL, and the maximum human dose were 25 mg/kg, then the MVC would be:

$$\text{MVC} = \frac{0.065\,\text{EU/mL} \times 25\text{mg/kg}}{5.0\,\text{EU/kg}} = 0.325\,\text{mg/mL}$$

If this dose were to be given intrathecally, the denominator would be 0.2 EU/kg
MVD determination

$$\text{MVC} = \frac{\text{Potency of product}}{\text{MVC}} = 1:61.5$$

If the potency of a product were 20 mg/mL, the MVD would be:

$$\text{MVD} = \frac{20\,\text{mg/mL}}{0.325\,\text{mg/mL}} = 1:61.5$$

Therefore, this product can be diluted to 61.5 times its original volume and still be able to detect the lower endotoxin concentration limit by the LAL test

Inhibition/enhancement testing must be performed on undiluted drug products or diluted drug products not exceeding the maximum valid dilution value (Table 28-3) (20). At least three production batches of each finished product should be tested. The product is spiked with various known amounts of RSE (or CSE), bracketing the sensitivity of the lysate used, using four replicate reaction tubes per level of endotoxin. The same number of tubes is used for drug product containing no added endotoxin and for control water for injection samples also spiked with various known amounts of RSE or CSE. The LAL test procedure is carried out manually or instrumentally.[3] The end points (E in units per milliliter) are then observed and recorded for all replicate samples.

The end points are determined followed by computation of the geometric mean of these end points. Geometric mean is

$$\frac{\sum E(\text{endpoints})}{f(\text{number of replicates})}$$

calculated both for the control and test samples. An illustration is given in Table 28-4 (18). The geometric means of the product sample and the water control sample are compared. If the product sample mean is within twofold of the control mean sample, the drug product is judged not to inhibit or enhance the LAL–endotoxin reaction. For example, if the product sample showed a geometric mean of 0.4 EU/mL and the water control mean of 0.2 EU/mL, the LAL test is valid for that product.

If endotoxin is detectable in the untreated specimens under the conditions of the test, the product is unsuitable for the inhibition/enhancement test. Either endotoxin must be removed by ultrafiltration or further dilution can be made as long as the minimum valid dilution (MVD) is not exceeded and the inhibition/enhancement test repeated. If the drug product is found

[3] The FDA validation guideline contains specific directions for inhibition/enhancement testing depending on the technique used—gel clot, inorganic and endpoint turbidimetric, and kinetic turbidimetric.

Table 28-4 Example of Geometric Mean Determination for a Small-Volume Injectable Product Undergoing LAL Testing for Endotoxin[a] (18)

	Gel end point results for specimen dilutions				
Replicates (*f*)	**Unity**	**0.5**	**0.25**	**0.125**	**End point dilution factors (*E*)**
1	+	+	+	−	0.25
2	+	+	−	−	0.5
3	+	+	−	−	0.5
4	+	+	+	−	0.25
5	+	+	−	−	0.5
					$\Sigma E = 2.0$

[a]Geometric mean $= \frac{\sum E}{f} = \frac{2.0}{5} = 0.4$.

to cause inhibition or enhancement of the LAL test, the following courses of action can be taken (20):

- If the drug product is amenable to rabbit testing, then the rabbit test will still be the appropriate pyrogen test for that drug.
- If the interfering substances can be neutralized without affecting the sensitivity of the test or if the LAL test is more sensitive than the rabbit pyrogen test, then the LAL test can still be used.
- For those drugs not amenable to rabbit pyrogen testing, the manufacturer should demonstrate that the LAL test can detect the endotoxin limit established for the particular drug. If the limit cannot be met, the smallest quantity of endotoxin that can be detected must be determined.

There are various miscellaneous requirements in the procedures for validating the LAL test:

- Use positive and negative controls in all tests.
- Use the highest and lowest drug concentrations for drug products marketed in three or more concentrations.
- Use three lots of each drug concentration for the validation tests.
- If the lysate manufacturer is changed, the validation test must be repeated on at least one unit of product.
- The LAL reagent should have a sensitivity of at least 0.25 EU/mL.
- The endotoxin control must always be referenced to the RSE.
- Any change in the product formulation, manufacturing process, source of formulation ingredients, or lot of lysate necessitates a revalidation of the LAL test for the product.

The possibility of a device inhibiting or enhancing the LAL–endotoxin reaction is determined by extraction testing of each of three device production lots. The extract solution must be pyrogen-free water or saline to which known amounts of standard endotoxin, bracketing the sensitivity of the lysate, have been added. Depending on the type of device, extracts may be obtained by flushing, immersing, or disassembling, then immersing the device with the endotoxin-spiked solution. The LAL test results of the extract should not be different than the results of testing standard solutions containing endotoxin that have not been exposed to the device.

Endotoxin highly adsorbs to container surfaces. Recovery of endotoxin occurs with polystyrene containers while the worst for recovering endotoxin were polypropylene containers. In fact, regardless of extraction method, less than 1% endotoxin was ever recovered from polypropylene containers. Bonosilicate glass allowed higher recovery than flint glass (21).

Manual LAL Test Procedure

Four or more replicate samples at each level of the dilution series for the test samples are used in most cases. The pH of the reaction mixture must be between 6.0 and 7.5 unless

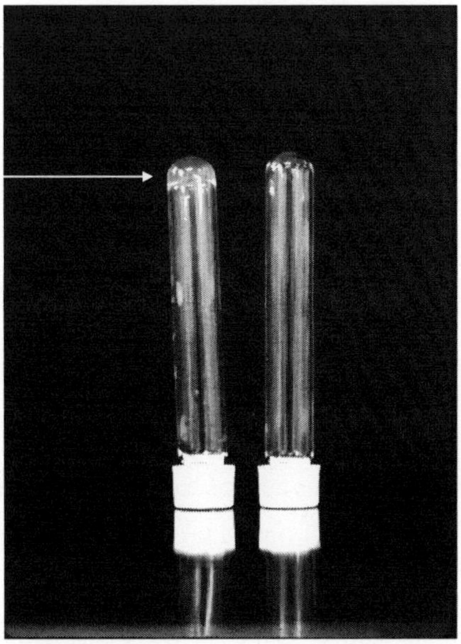

Figure 28-6 LAL positive and negative test result.

specified differently in the particular monograph. The pH may be adjusted by addition of sterile, endotoxin-free 0.1 N sodium hydroxide or 0.1 N hydrochloric acid or suitable buffers.

Test tubes, usually of the dimensions 10 by 75 mm, are filled with an aliquot, usually 0.1 mL, of reconstituted LAL reagent, and the same aliquot volume of the test sample. In other test tubes, equal volumes of LAL reagent and endotoxin standard are combined. Positive controls (LAL reagent sample containing a known concentration of endotoxin) and negative controls (LAL reagent + equal volume of sterile, pyrogen-free solvent) are run simultaneously with the test samples and endotoxin standards.

When the equal volumes are combined, the test tube is swirled gently. The tube is placed in a constant temperature water bath with temperature controlled at 37 $\pm$ 1°C. Incubation times ideally last 60 $\pm$ 2 minutes. While incubating, the test tubes must never be disturbed for fear of irreversibly disengaging the gel clot if it has formed. Careful removal of the incubated test tubes for gel clot analysis is extremely important.

The degree of gel formation can be determined by either direct visual observation or instrumental analysis. Visual observation starts by carefully removing the test tube from the incubator, then carefully inverting (by 180°) the test tube and visually checking for the appearance of a firm gel. A positive reaction is characterized by the formation of a firm gel that does not break or lose its integrity during and at the completion of the inversion process (Fig. 28-6). A negative result is characterized by the absence of a gel or by formation of a viscous gel that does not maintain its integrity during the inversion process.

Instrumental Tests

Direct visual observation of the gel end point relies on the subjective interpretation of the observer and, unless twofold serial dilutions are performed, provides only a qualitative (yes or no) measurement of the endotoxin present in the sample. Analysis of the gel end point by instrumental methods offers several advantages, including single-tube quantitation and objectivity. Additionally, instrumental methods can be automated, resulting in increased speed, efficiency, and adaptation to computer control.

Two basic instrumental methods are available for LAL testing. One method is based on turbidimetric measurement of gel formation while the other method is based on colorimetrically measuring a chromophobic substance produced during the LAL–endotoxin reaction.

An automated LAL test system is based on the measurement of color intensity of the LAL gel end point. This system is called the Chromogenic LAL assay system (Fig. 28-7). Test sample

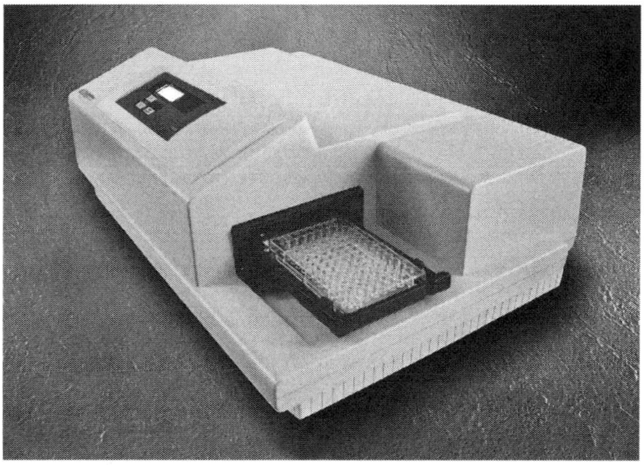

Figure 28-7 Chromogenic LAL test system. *Source*: Courtesy of Associates of Cape Cod, Inc.

is mixed with LAL reagent and incubated at 37°C for a period of time (usually 10 minutes). A substrate solution containing a color-producing substance is then mixed with the LAL test sample and incubated at 37°C for an additional three minutes. The reaction is stopped with 50% acetic acid. The color absorbency of the sample mixture is determined spectrophotometrically at 405 nm. The more intense the color, the greater the absorbance value measured. Endotoxin concentration can then be determined from a standard plot of absorbance versus endotoxin concentration in ng/mL or EU/mL

Miller reviewed the most recent methods for rapid detection of endotoxin (22). Endosafe™ PTS (Charles River Laboratories) is a hand-held LAL test system for rapid detection of endotoxin by quantitative kinetic chromogenic spectrophotometry. The PyroSense™ online system (Cambrex) monitors endotoxin detection in purified water and water for injection loops. This system employs a kinetic assay with a fluorescence reader that uses a recombinant Factor C that does not exhibit glucan interfering activity.

For laboratories responsible for conducting multiple LAL tests, automation practically becomes a necessity. Automation employs all the advantages of instrumental analyses, including greater precision and sensitivity. Technology has advanced to the point where the LAL test can be performed automatically using robotic systems. Such a system will automatically dilute a stock reference endotoxin standard for construction of a five-point standard curve, make sample dilutions to the proper testing concentration, and perform chromogenic substrate LAL assays in duplicate. In 48 minutes, the automated system assays three samples and a reference standard in duplicate along with a water blank. The method can be sensitive to a detection limit of 0.003 endotoxin units per milliliter with 30 minutes of incubation. Assay precision is approximately 6%. The major disadvantages of automated LAL testing systems are their cost and complexity.

The LAL test requirements for lack of pyrogenicity or critical endotoxin concentration will be met if there is no formation of a firm gel at the level of endotoxin specified in the individual monograph. For instances where instrumental analyses have been done, the sample will pass the LAL test if not more than the maximum permissible amount of endotoxin specified in the individual monograph present in the sample. Additionally, the confidence limits of the assay must not exceed the limits previously specified for the instrumental analysis.

Endotoxin Limits in Parenteral Articles

Endotoxin limits are necessary because bacterial endotoxin is ubiquitous and expected to be present in all articles at some level. The question is—what level is safe? This becomes the endotoxin limit.

The first FDA draft guideline for LAL testing of drugs proposed an endotoxin limit for all parenterals of 0.25 EU/mL (23). This limit was vehemently opposed by the parenteral drug industry because the limit was arbitrary, based on concentration rather than endotoxin quantity per dose, and did not permit sufficient dilution of small-volume parenterals known to inhibit the LAL test reaction.

The Parenteral Drug Association proposed an alternative endotoxin limit based on rabbit or human dose (24) that FDA accepted and became part of the new FDA draft guideline for end product testing published in December 1987 (25). The new endotoxin limit is:

$$\frac{K}{M} = \frac{\text{Threshold pyrogen dose (TPD)}}{\text{Maximum rabbit or human dose}}$$

where the TPD has been defined as 5 EU/kg, the lower 95% confidence limit of the average dose found to produce a pyrogenic response in rabbits and humans (26). For drugs administered intrathecally, where pyrogenic contamination can be much more dangerous, the TPD is 0.2 EU/kg.

The maximum rabbit or human dose is that dose administered per kilogram of body weight of rabbit or man[4] in a single hour period, whichever is larger. For example, if a drug of a concentration of 1 mg/mL has a maximum human loading of 25 mg/kg while the rabbit pyrogen test dose is 10 mg/kg, the maximum dose used in the denominator of the endotoxin limit equation would be the human dose of 25 mg. On the other hand, were the above human dose only 2.5 mg/kg, then the rabbit dose of 10 mg would be the larger of the two doses. The endotoxin limit for the two examples would be:

$$\text{EU} = \frac{5\,\text{EU/kg}}{25\,\text{mg/kg}} = 0.2\,\text{EU/mg}$$

$$\text{EU} = \frac{5\,\text{EU/kg}}{10\,\text{mg/kg}} = 0.5\,\text{EU/mg}$$

For devices, the endotoxin limit is 0.1 ng/mL of extract solution.

Four classes of drugs are exempted from the endotoxin limit defined by K/M:

- Compendial drugs for which other endotoxin limits have been established.
- Drugs covered by new drug applications, antibiotic Form 5 and Form 6 applications, new animal drug applications, and biological product license where different limits have been approved by the Agency.
- Investigational drugs or biologics for which an investigational new drug application (IND) or investigational new animal drug application (INAD) exemption has been filed and approved.
- Drugs or biologics that cannot be tested by the LAL method example.

Maximum doses per kilogram and the corresponding endotoxin limits for a large number of aqueous injectable drugs and biologics on the market are listed in Ref. 25.

LAL Sensitivity

LAL sensitivity is defined as the lowest concentration of a purified endotoxin that will produce a firm gel, which will remain intact when inverted carefully after one hour of incubation at 37°C. (LAL sensitivity is also expressed as how many times its sensitivity is greater than the rabbit test.) In general, it seems to be well established that the LAL test is sensitive to picogram quantities of endotoxin and that LAL is from 5 to 50 times more sensitive than the rabbit to the presence of endotoxin, depending on the type of comparative study conducted.

In earlier years, the LAL test is at least five times more sensitive to purified endotoxin than the rabbit test (28). Improvements in LAL production and formulation methodology increased the sensitivity of LAL 10 to 50 times greater than the rabbit test (29). These numbers were based on a gel time of one hour and a rabbit test dose of 1 mL/kg. The ability of LAL to detect *E. coli* endotoxin in pyrogen-free distilled water was found to be 100 times more sensitive than the rabbit test (30).

[4] Body weight of average human considered to be 70 kg. For pediatric indications, body weight needs to be adjusted for age of child, for example, average weight of 3-year-old child is 15 kg, 6-year-old child 20 kg, and 8-year-old child 25 kg). For preclinical animal studies, endotoxin limits for injections are based on the following body weights: mouse 0.03 kg, gerbal 0.09 kg, rat 0.45 kg, rabbit 4 kg, monkey 8 kg, baboon 12 kg (27).

LAL Test Specificity

Whereas sensitivity is the ability of a test to give positive reactions in the presence of the material tested, specificity is the ability of a test to give positive reactions with only the material tested (15). The sensitivity of LAL toward endotoxin is undisputed. However, its specificity in reacting solely with endotoxin is its most controversial characteristic.

In 1973, Elin and Wolff (31) first reported the possible lack of specificity of the LAL test for bacterial endotoxin. Substances found to cause lysate gelatin included thrombin, thromboplastin, ribonucleases, and polynucleotides such as polyriboadenylic acid and polyribouridylic acid. Positive reactions (false lysate gelation) have been reported between LAL and peptidoglycans, streptococcal exotoxins, synthetic dextrans, lipoteichoic acids, the dithiols, dithiothreitol and dithioerythritol, and immonoglobulins administered intravenously [32, and all references therein].

Substances called (1→3)-β-D-glucan-sensitive factors can activate LAL to produce false positive results for the presence of endotoxin (33). This particular glucan is found in cell walls of fungi and can be a contaminant in air, water, and processed materials. Cellulosic materials may also leach this glucan. Interestingly, very small amounts of beta-glucan (1–1000 ng/mL plasma) will trigger gelation while greater amounts of beta-glucan (1 mg/mL plasma) will not (34).

Lysate manufacturers have specific reagents available that neutralize the potential effects of (1→3)-β-D-glucans. Buffer diluent formulations used to reconstitute LAL will render the reagent insensitive to glucan interference by blocking the Factor G pathway of the endotoxin-clotting cascade. A specific (1→3)-β-D-glucan chromogenic reagent is available for quantitation of potential glucan leached from filters or from rDNA yeast protein production or sampled from the air. An endotoxin specific turbidimetric lysate that is not reactive with (1→3)-β-D-glucan also is available.

LAL Test Limitations

The LAL test offers many advantages compared with the USP rabbit test including greater sensitivity, reliability, specificity, and application. Unquestionably, the LAL test fills the need for a simple, sensitive, accurate, and inexpensive method for detecting bacterial endotoxin. It certainly offers itself as an excellent alternative or supplemental method of the official USP rabbit test for pyrogen. However, it is not without limitations or problems.

The greatest limitation of the LAL test is the problem of interference of the lysate–endotoxin interaction that is caused by a variety of drugs and other substances (35). Inhibition of the lysate–endotoxin interaction is the number one factor limiting the applicability of the LAL test (36). The LAL gelation reaction is mediated by a clotting enzyme that is heat labile, pH sensitive, and chemically related to trypsin (37). Inhibition is caused by any material known to denature protein or to inhibit enzyme action. Inhibition by many drug components can be overcome by dilution or pH adjustment. Of course, dilution reduces the concentration of the endotoxin and places greater demand on the sensitivity of the LAL reagent to detect diluted amounts of endotoxin.

Tests for inhibition or activation basically involve the use of positive controls. Product samples are "spiked" with known endotoxin levels, preferably the same levels used in standards prepared for sensitivity determinations. The end point of detection for the product sample should be no different from the end point for the standards series. In other words, if the lowest standard detectable level of endotoxin is 0.025 ng/mL, this level must also be detectable by the same lot of LAL reagent in the product sample. If inhibition is found to occur, serial dilutions of the product sample are made until the appropriate dilution is found that no longer modifies the gelation reaction.

Inhibition of LAL Test

Thirty percent of drug products do not inhibit the LAL test (producing an increase in the expected gelation onset time) (37). Of the majority of products that do inhibit the test, 97% of the problems can be resolved because the inhibition is concentration dependent. Simple dilution usually can overcome inhibitory properties of drug products against the LAL–endotoxin reaction.

LAL test inhibition is considered significant if the positive control varies by more than a twofold dilution from the standard in water. Inhibition acts on endotoxin, not the LAL reagent, that is, inhibition is often a failure to recover inadequately dispersed liposaccharide (aggregation of purified endotoxin).

Primary ways in which drug products inhibit the LAL test are as follows:

- Suboptimal pH
- Aggregation or adsorption of control endotoxin spikes
- Unsuitable cation concentrations
- Enzyme or protein modification
- Nonspecific LAL activation.

Other concerns or limitations of the LAL test are as follows:

- LAL is dependable only for the detection of pyrogen originating from gram-negative bacteria.
- Being an in vitro test, the LAL test cannot measure the fever-producing potential of endotoxin present in the sample.
- The sensitivity of LAL varies appreciably with endotoxins from various microbial sources.
- Difficulty comparing the sensitivity of the LAL test and the rabbit test because the rabbit assay is dose dependent while the LAL test is concentration dependent.
- Gel formation can be difficult to interpret and can be broken upon the slightest vibration.

LAL Test Variability

There are several sources of variability that can affect the accuracy and reliability of the LAL test. It is for these reasons that validation is so important and why the FDA produced its validation guideline for the LAL test (20).

Reagent Variability

There are significant differences in LAL reagent formulation from manufacturer to manufacturer (38). Although all LAL reagents are standardized to the USP RSE, both manufacturing processes and formulation differences account for variations seen in real-world endotoxin test situations. Major differences in reagent preparation include addition of the following: divalent cations, albumin, buffers, and surface-active agents. Some manufacturers allow the crude reagent to age, adjust coagulogen concentration, and perform chloroform extraction to remove inhibitors and increase sensitivity.

Method Variability

LAL reagents are designed specifically for optimal activity in each of the major LAL test systems. Thus, lysate-drug product compatibility may change when switching from one test method to another using the same lysate manufacturer.

Product Variability

The majority of parenteral products will interfere with the lysate–endotoxin reaction although most of these interferences can be overcome by dilution (39).

Laboratory Variability

Type of glass and/or plasticware used, equipment calibration procedures, recalibration procedures, purity of water used, dilution procedures, and other different laboratory procedures all contribute to LAL test variability (21). Differences in handling (degree of agitation) and storage of parenteral products prior to LAL test analysis can markedly affect test results.

As a reiteration, to control all these sources of variability, the FDA guideline says "The USP inhibition/enhancement tests must be repeated on one unit of the product if the lysate manufacturer is changed. When the lysate lot is changed, the two lambda positive control is used to re-verify the validity of the LAL test for the product" (20).

For an LAL reagent to be compatible with the FDA Guidelines for LAL evaluation of drugs, devices, and biologicals and with the USP Bacterial Endotoxin Test, the reagent should have a stabilized sensitivity of 0.12 EU/mL. This sensitivity should be referenced to an *E. coli*-delivered LPS such as the USP RSE from *E. coli*. An LAL reagent should be buffered to accommodate small changes in pH of the test solution and be stabilized for divalent cations. The reagent also should be specific for endotoxin and should exhibit a clear and accurate end point.

LAL Test Applications

From a modest beginning of detecting endotoxin in blood, LAL test application has expanded into a variety of laboratory and clinical situations. New or improved usage of the LAL test appears in the literature on a monthly basis. Methodology has become more standardized, reference standards more accepted, and automatic instrumental analysis has been developed. LAL testing for endotoxin in the parenteral field has become standard practice.

The LAL test has been used as an indicator of endotoxin contamination in pharmaceuticals, radiopharmaceuticals, biologics, devices, diseases caused by gram-negative bacteria (endotoxemia), food, and drinking water.

Pharmaceuticals administered by the intrathecal route represented a drug class that most urgently needed the LAL test for endotoxin detection (29) since endotoxin injected into intrathecal spaces can be at least 1000 times more potent in producing a febrile response than administered by the intravenous route (40). Such pharmaceuticals included

- Dyes such as methylene blue and fluorescein for detecting cerebrospinal fluid (CSF) leakage
- Contrast media for visualization of CSF pathways
- Cancer chemotherapeutic agents such as methotrexate for treatment of leukemic meningitis
- Antibiotics such as gentamicin for septic meningitis
- Radiopharmaceuticals for radionuclide cisternography, a procedure wherein a small volume of radiotracer is administered intrathecally to study CSF dynamics by means of nuclear imaging devices.

VALIDATION OF ENDOTOXIN REMOVAL

Validation of dry-heat sterilization and depyrogenation cycles based on the destruction of endotoxin can be accomplished through the employment of the LAL test (41–45). This could not be accomplished practically using the USP rabbit test. This has resulted in a FDA requirement for a 3-log reduction in endotoxin levels in materials being dry-heat sterilized (43).

Analogous to biological indicators used for validation and routine monitoring of sterilization processes, there are "endotoxin indicators" that can be used in the validation and routine control of endotoxin reduction processes (44).

Validation of deyprogenation of rubber closures by washing/rinsing procedures involves first proving that endotoxin challenges applied to rubber closures can be recovered. The use of sonification techniques has proven quite useful to extract applied endotoxin from rubber closures. Typically at least 10,000 EU is applied to each of 10 stoppers in a depyrogenated flask containing 100 to 200 mL water for injection (WFI). After rotary shaking and sonification for a period of time, the endotoxin recovered is assayed a kinetic LAL test method. Recovery of at least 20% of the initial inoculum must be obtained.[5]

REFERENCES

1. Das REG, Brugger P, Patel M, et al. Monocyte activation test for pro-inflammatory and pryogenic contaminants of parenteral drugs: Test design and data analysis. J Immunol Methods 2004; 288:165–177.
2. Good CM, Lane HE Jr. The biochemistry of pyrogens. Bull Parenter Drug Assoc 1977; 31:116–120.
3. Davis BD, Dulbecco R, Eisen HN, et al. Microbiology. 3rd ed. Hagerstown, MD: Harper and Row, 1981:85.

[5] If 10,000 EU is inoculated on each closure and 10 closures extracted in 100 mL WFI, the theoretical recovery would be 100,000 EU in 100 mL (1000 EU per mL). If the extraction recovered at least 200 EU/mL, there is assurance that any endotoxin contamination greater than 1 EU per closure on incoming rubber stoppers would be detected.

4. Kennedi E, Laburn H, Mitchell D, et al. On the pyrogenic action of intravenous lipid A in rabbits. J Physiol 1982; 328:361–370.
5. Siebert FB. Fever producing substance found in some distilled waters. Am J Physiol 1923; 67:90–104.
6. Siebert FB. The cause of many febrile reactions following intravenous injections. Am J Physiol 1924; 71:621.
7. Pearson FC. Limulus amebocyte lysate testing: Comparative methods and reagents. In: Groves MJ, Olson WP, Anisfeld MH, eds. Sterile Pharmaceutical Manufacturing: Applications for the 1990s. Vol. 2. Buffalo Grove, IL: Interpharm Press 1991:185–197.
8. United States Pharmacopeia, General Chapter <151>, United States Pharmacopeial Convention, Inc. Bethesda, MD.
9. Mascoli CC, Weary ME. Limulus amebocyte lysate test for detecting pyrogens in parenteral injectable products and medical devices: Advantages to manufacturers and regulatory officials. J Parenter Drug Assoc 1979; 33:81–95.
10. Dabbah R, Ferry E Jr, Gunther DA, et al. Pyrogenicity of E. coli 055:B5 endotoxin by the USP rabbit test: A HIMA collaborative study. J Parenter Drug Assoc 1980; 34:212–216.
11. Greisman SD, Hornick RB. Comparative pyrogenic reactivity of rabbit and man to bacterial endotoxin. Proc Soc Exp Biol Med 1969; 131:1154–1158.
12. Dare JG, Mogey GA. Rabbit responses to human threshold doses of a bacterial pyrogen. J Pharm Pharmacol 1954; 6:325–332.
13. Levin J, Bang FB. The role of endotoxin in the extracellular coagulation of Limulus blood. Bull Johns Hopkins Hosp 1964; 115:265–274.
14. Bang FB. A bacterial disease of Limulus polyphemus. Bull Johns Hopkins Hosp 1956; 98:325–351.
15. Marcus S, Nelson JR. Tests alternative to the rabbit bioassay for pyrogens. Dev Biol Stand 1977; 34:45–55.
16. Cooper JF, Hochstein HD, Seligman EG. The Limulus test for endotoxin (pyrogen) in radiopharmaceuticals and biologicals. Bull Parenter Drug Assoc 1972; 26:153–162.
17. Tsuji K, Steindler KA. Use of magnesium to increase sensitivity of Limulus amoebocyte lysate for detection of endotoxin. Appl Environ Microbiol 1983; 45:1342–1350.
18. Cooper JF, Neely ME. Validation of the LAL test for end product evaluation. Pharm Technol 1980; 4:72–79.
19. Hochstein HD. The LAL test versus the rabbit pyrogen test for endotoxin detection. Pharm Technol 1981; 5:37–42.
20. U. S. Food and Drug Administration. Guideline on validation of the LAL test as an end product endotoxin test for human and animal parenteral drugs, biological products, and medical devices. Fed Regist 1988; 53:5044.
21. Novitsky TJ, Schmidt-Gengenbach J, Remillard JF. Factors affecting recovery of endotoxin adsorbed to container surfaces. J Parenter Sci Technol 1986; 40:284–286.
22. Miller MJ. Rapid microbiological methods in support of aseptic processing. In: Lysfjord J, ed. Practical Aseptic Processing: Fill and Finish. Vol 2. Bethesda MD: Parenteral Drug Association/Davis Healthcare, 2009:181–182.
23. U. S. Food and Drug Administration. Licensing of Limulus amebocyte lysate: Use as an alternative for rabbit pyrogen test. Fed Regist 1977; 42:57749.
24. Parenteral Drug Association. Parenteral Drug Association Response to FDA Draft Guideline for the Use of Limulus Amebocyte Lysate. Information Bulletin #3, March 1980.
25. U. S. Food and Drug Administration. Guideline for Validation of Limulus Amebocyte Lysate Test as an End Product Endotoxin Test for Human and Animal Parenteral Drugs, Biological Products and Medical Devices. Rockville, MD: FDA, 1987.
26. Tsuji K, Steindler KA, Harrison SJ. Limulus amebocyte lysate assay for endotoxin for detection and quantitation of endotoxin in a small-volume parenteral product. Appl Environ Microbiol 1980; 40:533–538.
27. Malyala P, Singh M. Endotoxin limits in formulations for preclinical research. J Pharm Sci 2008; 97:2041–2043.
28. Cooper JF, Levin J, Wagner HN Jr. Quantitative comparison of in vitro and in vivo methods for the detection of endotoxin. J Lab Clin Med 1971; 78:138–148.
29. Cooper JF. Principles and application of the Limulus test for pyrogen in parenteral drugs. Bull Parenter Drug Assoc 1975; 29:122–130.
30. Nyerges G, Jaszovszky S. Reliability of the rabbit pyrogen test and of the Limulus test in predicting the pyrogenicity of vaccines in man. Acta Microbiol Acad Sci Hung 1981; 28:235–243.
31. Elin RJ, Wolff SM. Nonspecificity of the Limulus amebocyte lysate test: Positive reaction with polynucleotides and proteins. J Infect Dis 1973; 128:3.

32. Akers MJ, Larrimore DS, Guazzo DM. Pyrogen testing. In: Dekker M (now Informa Healthcare), ed. Parenteral Quality Control: Sterility, Pyrogen, Particulate, and Package Integrity Testing, 3rd ed. New York: Marcel Dekker, Inc., 2003:183–196.
33. Ohno N, Emori Y, Yadomae T, et al. Reactivity of Limulus amoebocyte lysate towards (1>3)-β-D-glucans. Carbohydr Res 1990; 207:311–318.
34. Kambayashi J, Yokota M, Sakon M, et al. A novel endotoxin-specific assay by turbidimetry with Limulus amoebocyte lysate containing beta-glucan. J Biochem Biophys Methods 1991; 22:93–100.
35. vanNoordwijk J, DeJong Y. Comparison of the LAL test with the rabbit test: False positives and false negatives. Dev Biol Stand 1977; 34:39–43.
36. Dawson ME. Interference with the LAL test and how to address it. LAL Update, Associates of Cape Cod, Inc. October 2005. http://www.acciusa.com/pdfs/newsletter/LAL%20Update%20Vol%2022_No3%20rev%20001.pdf. Accessed June 15, 2010.
37. Cooper JF. Resolving LAL test interferences. J Parenter Sci Technol 1990; 44:13–15.
38. Cooper JF. Ideal properties of a LAL reagent for pharmaceutical testing. In: Watson L, Novitsky, eds. Bacterial Endotoxins: Structure, Biomedical Significance, and Detection with the Limulus Amebocyte Lysate Test. New York: Alan R. Liss, 1985:241–249.
39. Twohy CW, Duran AP, Munson TE. Endotoxin contamination of parenteral drugs and radiopharmaceuticals as determined by the Limulus amebocyte lysate method. J Parenter Sci Technol 1984; 30:190–201.
40. Bennett IL Jr, Petersdorf RG, Keene WR. Pathogenesis of fever: Evidence for direct cerebral action of bacterial endotoxins. Trans Assoc Am Physicians 1957; 70:64.
41. Akers MJ, Ketron KM, Hompson BR. F value requirements for the destruction of endotoxin in the validation of dry heat sterilization/depyrogenation cycles. J Parenter Sci Technol 1982; 36:23–27.
42. Tsuji K, Harrison SJ. Dry heat destruction of lipopolysaccharide: Dry heat destruction kinetics. Appl Environ Microbiol 1978; 36:710.
43. Guidance for Industry, Sterile Drug Products Produced by Aseptic Processing—Current Good Manufacturing Practice, U.S. Department of Health and Human Services, Food and Drug Administration, September, 2004. http://www.fda.gov/downloads/Drugs/GuidanceComplianceRegulatoryInformation/Guidances/ucm070342.pdf. Accessed June 15, 2010.
44. LAL Users Group. Preparation and use of endotoxin indicators for depyrogenation process studies. J Parenter Sci Technol 1989; 43:109–112.
45. Berzofsky R, Schieble L, Williams K. Validation of endotoxin removal from parenteral vial closures. Biopharm 1994; 7:58–66.

BIBLIOGRAPHY

Current practices in endotoxin and pyrogen testing in biotechnology, The Quality Assurance/Quality Control Task Group, Parenteral Drug Association. J Parenter Sci Technol 1990; 44:39–45.
LAL Update. Associates of Cape Cod, Inc. monthly newsletter.
Williams KL, ed. Endotoxins: Pyrogens, LAL Testing, and Depyrogenation. 3rd ed. New York: Informa Healthcare, 2007.

29 Particles and particulate matter testing

The thrust of this chapter focuses on testing methods and requirements for particulate matter in sterile products. Some coverage on increasing concerns about the effect of particles (especially subvisible) on biopharmaceutical product safety and quality will supplement what was presented in Chapter 8.

No quality control test presents more difficulties for quality control specialists than inspection and analysis of injectable solutions for the presence of foreign particulate matter, just as no challenge is greater during manufacturing than producing products without the presence of particles. How you get rid of them and how you measure what remain are huge challenges to the sterile products industry. The oldest, yet most commonly used, test for particulate matter evaluation involves human visual examination. Such examination is subjective, time-consuming, and limited in the types of sterile products and containers that can be inspected. This has stimulated many studies regarding ways of not only improving efficiency of human inspection but also developing and improving methods of detecting particulate matter electronically. Additional specifications are required for subvisible particulate matter content and analysis in large-volume injections for single-dose infusion and in small-volume parenterals [United States Pharmacopeia (USP) Chapters <788> for Parenteral Products and <789> for Ophthalmic Solutions]. There are also subvisible particle requirements in the European Pharmacopeia (EP or PhEur), Japanese Pharmacopeia (JP), and other compendia organizations. Also in USP <788> "Particulate Matter In Injections," the following statement is made: "Particulate matter in injections and parenteral infusions consists of mobile undissolved particles, other than gas bubbles, unintentionally present in the solutions." As pointed out by Das and Nema (1), this is noteworthy for biological dosage forms, as it does not discriminate the source of particles—either intrinsic or extraneous.

Focus on inspection and detection of visible particulate matter was covered in Chapter 22. This chapter discusses subvisible particulate matter testing with added discussion of the history and significance of particulate matter in injectable products.

Why are injectable solution products to be free of visible evidence of particulate matter? Primarily, lack of particulate matter conveys a clean, quality product, indicative of the high quality standards employed by the product manufacturer. Moreover, particulate matter is known as a potential hazard to the safety of the patient undergoing parenteral therapy. While there still seems to be a lack of sufficient clinical data to incriminate particles as producers of significant clinical complications during parenteral therapy, it is a universal belief in the health care field that particulate matter does present a clinical hazard and must be absent from the injectable solution. Also, as further discussed later, there may be implications of the role of particulate matter in causing immunogenic reactions due to protein aggregation in biopharmaceutical solutions.

PARTICULATE MATTER CONCERNS IN INJECTABLE PRODUCTS

After the inclusion of the first injectable product in the USP (12th edition) in 1942, Godding expressed the need for standards in the visual inspection of particulate matter (2). The 13th edition of the USP gave a detailed method for inspecting an injectable solution against a white and black background using a light intensity between 100 and 350 foot-candles at a distance of 10 in. Interestingly, the method described in the 13th edition is still widely used in manual inspections for evidence of visible particulate matter.

The "rule-of-thumb" standard that a person with 20/20 vision under inspection conditions should be able to detect particles having sizes of approximately 30 μm came from a report by Brewer and Dunning (3). This detection limit has persevered since 1957, although later research suggested that inspectors should actually be able to see particles in the size range of 20 μm (4).

In the early 1950s, a number of reports began citing evidence of biological hazards produced by foreign injected materials. Among the materials found to cause pulmonary granulomata or emboli were cotton fibers and cellulose. Glass particles and their potential hazard were studied by Brewer and Dunning (3) with no evidence of foreign body reactions in animals. These and other reports led to the classic work done by Garvan and Gunner published in 1963 and 1964 (5,6). These Australian physicians showed that foreign body granulomas could be produced experimentally in the lungs of rabbits following the administration of 500 mL saline solution contaminated with visible particulate matter. Most commercial intravenous solutions inspected contained particle contamination and the source of most of the particles was attributed to the rubber closure. For every 500 mL of particle-contaminated intravenous solution injected into a rabbit, 5000 granulomas appeared in the lungs. Garvan and Gunner further found that similar granulomas appeared in the postmortem examinations of the lungs of patients receiving large volumes of intravenous fluids. Their comments included the possibility that postoperative pulmonary infarction was a result of particulate thrombosis. The repercussions of Garvan and Gunner's reports have stimulated numerous studies on the analysis and potential clinical hazards of particulate matter that continue to this day.

A collaborative study conducted by the Pharmaceutical Manufacturers Association (7) involved the intravenous injection of varying quantities and sizes of inert polystyrene spheres into hundreds of rats, then performing necropsies at various periods of time from 1 hour to 28 days following injection. The results were as follows:

- Thirteen of 18 rats injected with 8×10^6 particles per kg at a particle size of 40 μm died within 5 minutes.
- Rats showed normal blood studies, organ weights, and pathologic criteria after being injected with either 8×10^6 particle size 0.4 to 10 μm or 4×10^5 particles per kg of particle size 40 μm.
- Particles in the 4 μm size range were found in the lung, liver, and spleen.
- Particles in the 10 μm size range were found in the lung primarily, although particles were found in five other organs.
- Particles in the 40 μm size range were found in the lungs and myocardial tissue.

It was concluded that nonreactive particles administered intravenously over a broad size range and up to dosages that produced death were without clinical or tissue toxicity. Much disagreement resulted over this conclusion, especially because of the artificial nature of the type of particle studied. However, the same size-dependent localization of particles in different organs was found in the case of glass particles derived from breaking the necks of glass ampuls (8). Large particles (>20 μm) were retained mostly in the lungs of mice while smaller particles (5–10 μm) were found in the liver, spleen, and kidney. No glass particles were found in the brain.

In a study of 173 patients undergoing cardiac catheterization and/or surgery, 14 (8%) had fiber emboli in routine autopsy sections (9). The embolized fiber often resulted in narrowing or occlusion of the involved blood vessel. Three cases of myocardial infarction were associated with embolic fibers. Fibers were believed to have originated from various materials used in surgery and from drug solutions. It was concluded that particulate matter is a hazard and all steps must be taken to prevent its inadvertent administration.

In a repeated double-blind study of 146 patients, a significant reduction in the incidence of infusion phlebitis was seen when patients were administered intravenous fluids filtered through an in-line 0.45 μm filter (10).

Barber (11) cited the review article by Pesko (12) as the best focused review of the literature on the hazards of particulate matter. Some interesting facts from this article include

1. Size, number, rate of introduction, and type of particle entering the bloodstream will all contribute to what harm, if any, the particle(s) actually produce. Some particles might cause allergic reactions.
2. The health condition of the person receiving solutions containing particulate matter also greatly matters with respect to potential harm of these particles.
3. Some particles will cause an inflammatory response. The potential harm of this inflammatory response depends on where the particles end up in the body and, if they are located in a vital organ, what is that organ's capacity to compensate for the insult caused by the foreign matter.

Freedom from visible evidence of particulate matter is a basic, essential characteristic of injectable products. Such a characteristic imparts three significant qualities to the product:

1. Significance to the manufacturer—lack of particulate matter indicates good production technique and a high quality product.
2. Significance to the user—lack of particulate matter indicates a clean product that is safe to the patient and conveys high quality standards employed by the manufacturer of the product.
3. Clinical significance—lack of particulate matter results in minimal concerns of potential hazards resulting from particles entering the circulatory system

NATURE AND SOURCES OF PARTICULATE MATTER

Anything that directly or indirectly comes in contact with a parenteral solution, including the solvent and solutes composing the solution itself, represents a potential source of particulate contamination. Foreign matter in formulation components, packaging, and originating from the environment (air, equipment, personnel) exacerbated by shear stresses all are sources of particulate matter in final product solutions (refer to Table 22-8). The smallest capillary blood vessels are considered to have a diameter of approximately 7 μm. Thus, all particles having a size equal to or greater than 7 μm can conceivably become entrapped in and occlude a blood capillary. Most particulates potentially can be this size and, obviously, represent a hazard to the health of a patient administered parenteral medications containing these contaminants. Plus, as emphasized in Chapter 8 and at the end of this chapter, particles much smaller, even less than 1 μm, are implicated and in the future might be proven to be the cause of protein aggregation in vitro and in vivo resulting in immunogenic reactions.

NUMBER AND SIZE OF PARTICLES

While it is desirable to prepare and use parenteral products completely free from particulate matter, it must be admitted that this ideal state is not possible.[1] All parenteral products contain some level of particulate matter contamination. The question is "how many particles of what type and of what size?"

Thomas Barber (13) addresses this question throughout his book. Following are excerpts taken from his book with the page number referenced:

- Particulate matter present in parenteral solutions and medical devices has been an issue in the pharmaceutical industry since the introduction of injectable preparations and remains unavoidable, even with today's well controlled manufacturing processes. (p. 2)
- Correctly or incorrectly, the particulate matter burden of a product has been taken by some healthcare practitioners, academic investigators, and regulatory personnel as an indicator of overall product quality. This is unfortunate, since particulate matter is, realistically, only a single parameter by which product suitability or conformity may be judged. (p. 2)
- The particle content of IV fluids in plastic containers has been repeatedly found to be lower than that of glass containers. (p. 14)
- There are three very important concepts embodied in the definitions provided by the USP, JP, and BP. Particulate matter current exists at extremely low levels in injectable products so that there is no demonstrable evidence of adverse patient effects. The material cannot be monotypic, but rather results from a variety of sources inherent in a GMP-controlled production process. The material is not amendable to chemical analysis due to the small mass that it represents and its heterogeneous composition. Thus, the appropriate analytical enumeration of this material must be sensitive physical tests that detect size and quantitate the material based on its optical properties. (pp. 20–21)

[1] However, the sterile product industry must keep pursuing approaches and disciplines to keep minimizing the presence of particles through proper selection of equipment and materials, higher quality packaging, best practices in cleaning techniques, and training and awareness of personnel.

- The occurrence of low numbers of heterogeneously sourced particles is inevitable in the manufacture of injectable products and medical devices. (p. 22)
- The USP requires that injectables be "essentially free" of visible particulate matter. The allowable particle burden of units tested must ultimately be judged with respect to the acceptable small quantity of visible particulate matter that may be present in units produced by parenteral manufacture under cGMP conditions. The most important aspect of the visual inspection procedure is the detection of any particulate matter that is related to solution degradation or any particulate matter present in sufficient quantity or of sufficient size to constitute a non-GMP condition. (pp. 261–262)
- Despite rigorous cleaning procedures either at the vendor's plant on in-house, glass vials and stoppers occasionally will bear or contain a single visible particle following filling, stoppering or lyophilization, and these particles may become free in the solution. Such single visible particles of random isolated occurrence are analogous to the "allowable" particle burden under USP <788>. A unit bearing such an isolated single visible particle must be rejected. On the basis of the low level of occurrence of such visible particulates in the batch of manufactured materials from which the unit came, however, the batch may be still be considered essentially, substantially, and practically "free" of visible particulate material. (p. 262)
- It is important for the pharmaceutical industry and regulators to recognize that there are no particle-free parenteral solutions. There is, furthermore, no evidence of any patient issue related to infusion of a small number of inert particles with the current or previous USP limits. (p. 273)

PARTICULATE MATTER STANDARDS

The first reference to particulate matter in the USP occurred in the eighth edition in 1905 Diphtheria Antitoxin, a hypodermic injection product, was described as a "transparent or slightly turbid liquid." Not until 1936, in the National Formulary (NF), sixth edition, was the term "clearness" defined for parenteral products: "Aqueous Ampul Solutions are to be clear; that is, when observed over a bright light, they shall be substantially free from precipitate, cloudiness or turbidity, specks or fibers, or cotton hairs, or any undissolved material."

The words "substantially free" caused interpretative difficulties; thus, in 1942, the NF, seventh edition, provided a definition: "substantially free shall be construed to mean a preparation which is free from foreign bodies that would be readily discernible by the unaided eye when viewed through a light reflected from a 100-watt mazda lamp using as a medium, a ground glass, and a background of black and white." It was also in 1942 that the 12th edition of the USP contained its first particulate matter standard:

> Appearance of Solution or Suspension Injections which are solutions of soluble medicaments must be clear, and free (note the absence "substantially") of any turbidity or undissolved material which can be detected readily without magnification when the solution is examined against black and white backgrounds with a bright light reflected from a 100-watt mazda lamp or its equivalent.

The requirement that every injectable product in its final container be subjected individually to visual inspection appeared in the 13th edition of the USP.

What particle size is truly visible to the human eye? Visible has the connotation of particles being seen with the unaided eye. The unaided eye can discern, at best, particles at sizes of about 40 to 50 μm. Detection of smaller particles cannot be accomplished assuredly with the USP physical inspection test. Health care professionals became increasingly concerned about the aspect of intravenous solutions, especially large-volume parenterals, contaminated with particles too small to be seen with the unaided eye, yet still hazardous when introduced into the veins of a recumbent patient. In the mid-1970s, the USP and Food and Drug Administration (FDA) cosponsored the establishment of the National Coordinating Committee on Large-Volume Parenterals (NCCLVP). The NCCLVP then established a subcommittee on methods of testing for particulate matter in LVPs (now called large-volume injectables—LVIs). Ultimately, the efforts of this subcommittee resulted in the establishment of the USP microscopic assay procedure for the determination of particulate matter in LVIs for single-dose infusion and set upper limit acceptable particle standards at particle sizes of 10 and 25 μm. These two sizes were also subsequently used as size standards for particulate matter in small-volume injections (SVIs).

Japanese Method for Inspection and Analysis of Particulate Matter

Requirements for freedom of parenteral solutions from the presence of particulate matter are very strict in Japan. Inspection of individual containers for any visible evidence of particulate matter is done much more rigorously. For example, an inspector in a typical Japanese pharmaceutical company will take up to 10 seconds inspecting a single vial of a parenteral solution. Contrast this with inspectors in a typical fast-speed American parenteral manufacturer who will inspect 50 to 150 vials per minute for evidence of particulate matter.

The Japanese technique for preparation and testing of solutions for the presence of particulate matter by microscopic analysis deserves some attention. The meticulousness of their preparation techniques is impressive. For example, all materials (forceps, petri dishes, filtration funnels) used in filtering solutions are first sonicated for at least 5 minutes, then washed thoroughly with particle-free water three times. The membrane filters used for the blank controls are washed thoroughly using a very rigid procedure involving starting at the top of the nongridded side of the filter, sweeping a stream of particle-free water back and forth from top to bottom, then repeating this on the gridded side of the filter. After inserting the filter into the filter holder base and installing the funnel, the entire system is rinsed twice with particle-free water taking care not to allow the rinsings to pass through the filter. Further rinsings are completed with the water vacuumed through the filter. Interestingly, this water is introduced into the funnel using an injection syringe filtered with a 0.45-μm filter. The maximum allowable number of particles for the entire membrane filter pad used as the blank control is 3 $\geq$ 10 μm and 1 $\geq$ 50 μm using a suitable microscope with 40× and 100× magnification with incident light at an angle of 20 μm.

Sample test solutions are handled in the same way. Five vials are filtered through the same filter pad. Some Japanese companies even use a filter pad that is only 4 mm in diameter. Filter pads, after vial contents have been filtered through the pads, are photomicrographed, usually at the 40× magnification. Test results are judged by visual comparison of the test filter pads with reference photographs of previous test samples judged by the quality control department to represent the particulate quality desirable with the product sample.

USP <788> PARTICULATE MATTER TESTING IN INJECTABLE PRODUCTS

There are two possible procedures for determination of particulate matter—the light obscuration particle count test (preferred test) and the microscopic particle count test.

All parenteral solutions, regardless of route of administration (i.e., even if administered subcutaneously or intramuscularly) or container (e.g., syringes and cartridges that used to be accepted because of the high levels of silicone that were counted as particles), are required to pass the USP <788> test by one of the two methods with the light obscuration method being preferred. Also included are reconstituted solutions from sterile solid dosage forms, although the EP allows higher particle limits for reconstituted solutions. The USP <788> is harmonized with EP requirements, although one interesting special statement from the EP not contained in the USP is as follows: "In the case of preparations for subcutaneous or intramuscular injection, higher limits may be appropriate" (EP 2.9.19).

While particle counting is required for clear solution products, including reconstituted powders, what about other dosage forms such as suspensions, emulsions, or even solutions that are colored? No regulatory guidelines exist, but it is possible to apply certain techniques during the development stages of these dosage forms to ascertain the potential existence of extraneous foreign matter. Microscopy is most often used after filtration and sufficient rinsing where foreign particles can be differentiated from drug particles. Other techniques used to separate drug or other solid or nonaqueous components in the vehicle supernatant have included centrifugation, sedimentation, sieving, and direct solution probe microscopy. For batch release of injectable dispersed systems (i.e., suspensions), visual inspections by human inspectors are the only legitimate approach to detect foreign particulate matter. Inspectors are trained to allow dispersed systems to stabilize (particles settle), and then carefully inspect both the supernatant and the bottom of the container for evidence of any visible foreign matter. Perhaps, future innovations will enable routine testing of suspensions for subvisible foreign particulate matter.

Types of products that are exempt from the requirements of <788> include radiopharmaceutical preparations, parenteral products for which the labeling specifies the use of a final filter prior to administration provided that scientific data are available to justify this exemption, and parenterals packaged and labeled exclusively for use as irrigating solutions.

Parenteral preparations that have reduced clarity or are highly viscous, such as emulsions, colloids, and liposomal products, can be evaluated for particulate matter by microscopy. Extremely viscous products, for example, emulsions, colloids, liposomal products, and products containing high amounts of polymers and/or oily solutions, cannot be examined by either electronic or microscopic methods, but USP <788> allows for quantitative dilution with an appropriate diluent to reduce viscosity until one method can be used. Microscopy also can be used for products whose formulations generate air bubbles when drawn into the electronic particle counter sensor.

As is true with all end product quality testing, results obtained from a few samples cannot be extrapolated with certainty with all other product units from that batch. USP states that statistically sound sampling plans must be used to have the highest confidence possible that sample results represent the level of particulate matter in the remainder of the batch product units.

After several years of collaborative effort among laboratories from the FDA, universities, and pharmaceutical manufacturers, a method became official in the first supplement of the USP (19th edition) in 1975 for the particulate matter analysis and release specifications for single-dose large-volume parenterals (LVI). The original method involved the filtration of 25 mL of solution through an ultraclean membrane filtration assembly, then observing the membrane and counting entrapped particles on its surface under a microscope using 100× magnification.

Since the advent of the microscopic test for LVI solutions, the particle load in these solutions was substantially reduced. This was recognized in the early 1980s such that attention turned to establishing subvisible particle standards for SVIs. By the USP XXI (July, 1985), the electronic light obscuration test method was introduced for SVIs. Until around 2000, LVIs were evaluated for subvisible particulate matter using the microscopic method while SVIs were primarily evaluated by the light obscuration method. However, effective with the USP 25, both LVI and SVI solutions are now primarily evaluated for the presence of subvisible particulates by the light obscuration method. If the light obscuration method cannot be used, then the microscopic method is allowed. Many companies perform both methods in measuring the amounts and types of particulate matter in their parenteral solutions and reconstituted powders. Also, effective with the USP 25, both LVIs and SVIs followed the same general USP guidelines (test apparatus, calibration, test environment, test procedures, and calculations) for both the light obscuration particle count test and the microscopic particle count test.

There continues to be debate over the acceptance criteria for particles ≥10 and ≥25 μm as well as why are these two particle sizes the measuring or threshold sizes and not smaller sizes? Recalling discussion introduced in Chapter 8 about the possible role of extremely small particles, even as small as 0.1 μm, on protein aggregation and the potential relationship between aggregation and immunogenicity, it seems likely that sometime in the future, there will be changes in both particle size standards and particle number acceptance criteria.

Development of the SVI Subvisible Particle Test

In the early 1980s, there was a general consensus that these LVI limits were too strict for SVI solutions and, in fact, SVIs should not have particle limits because volumes administered are much smaller than those for LVIs, and health hazards from injected particulates were not unequivocally established. Nevertheless, the USP sponsored studies to establish particle limits for SVI solutions reasonable from both a safety standpoint and a quality control standpoint achievable by the sterile products industry.

Two SVI particle limit proposals were published in 1983 (USP Pharmacopeial Forum, November–December 1983, pp 3729–2735). One was based on particles per container and the other based on particles per milliliter.

	Number of particles ≥10 μm	Number of particles ≥25 μm
Proposal 1	10,000 per container	1000 per container
Proposal 2	250 per mL	70 per mL

The particles per container proposal was based on the following rationale: the addition of up to five containers of any SVI to a 1-L LVI solution should not increase the number of particles by more than double those allowed by the USP limit (at that time) for LVI solutions:

50 particles per mL ≥10 μm or 50,000 particles in 1 L ≥10 μm
5 particles per mL ≥25 μm or 5000 particles in 1 L ≥25 μm

If five additives, each containing no more than 1000 particles per container ≥25 μm, were admixed with the 1-L LVI containing 5000 particles ≥25 μm, the total particles ≥25 μm would be 10,000, which would be the maximum allowable particle number per admixed solution. At 10 μm, the total particle number with five additives in a 1-L LVI would be 100,000, which would be no more than double that of the LVI alone. Therefore, the particles per container proposal was based upon concern more for the cumulative particulate insult the patient might receive than for the number of particles per milliliter of solution and indeed the particles per container proposal prevailed.

While many laboratories preferred to employ the LVI microscopic method for counting particles in SVI products, the USP XXI introduced the use of an electronic liquid-borne particle counter system. Initial controversy over the test resulted in a postponement of the test becoming official until July 1985. The major complaint of the new USP method centered around the use of the HIAC-Royco electronic particle counter. Like any electronic counting device, particles cannot be seen or characterized, cannot accurately measure a particle's longest dimension (i.e., measures all particles as spheres), and will count silicon and air bubbles as particles and standardization/calibration of the instrument can be difficult. Also, many manufacturers objected to being forced to use an instrument that at that time was available from only one major U.S. supplier.

Other concerns over the proposed USP test for SVI particulate matter included lack of a sufficient data base from which limits were established, lack of validation of the USP proposed method, the basis for requiring particle limits for some products but not for others in individual monographs, problems in specific details in the calibration, preparation and determination sections of the test, and the lack of consistency between the LVI and SVI tests for particulate matter. The USP Chapter <788> requirement for particulate matter in SVI became effective in 1986. USP Section 789 for Ophthalmic Solutions was introduced in 2007. The test called for the use of an electronic liquid-borne particle counter system utilizing a light obscuration–based sensor with a suitable sample-feeding device.

The Pharmacopoeial Discussion Group of the European Directorate for the Quality of Medicines and Health Care (EDQM) created an overview chapter replacing <788> and applicable other compendial organizations (personal communication and subsequent documentation from Scott Aldrich, formerly of Pfizer and member of the USP Expert Committee on Parenterals, 2008). Much of the specific operational and system suitability methodology was removed from the harmonized <788> document. Significant in the harmonization of the particle counting methods are

1. The use of a two-stage acceptance scheme for subvisible particulate matter, using light obscuration (extinction) as the preferred method, with membrane microscopical counting utilized secondarily, for lots failing Stage I or in certain cases, as an alternate method. Visible particulate matter content is evaluated by visual inspection for sterile injectable products.
2. Adoption of common definition of SVI and LVI and utilization of either SVI or LVI subvisible particle limits for nominal 100 mL volume injections.

USP <788> is now harmonized with the corresponding texts of the EP and JP.

ELECTRONIC PARTICLE COUNTERS

The limitations of human inspection and microscopic analytical methods in the detection of particulate matter in injectable products have necessitated the use and advancement of electronic particle counting methods in the pharmaceutical industry. In 1986, the USP adopted for the first time an electronic particle counting method to be used in particulate matter testing of SVIs. Much controversy over the type, standardization, and limitations of electronic particle counting methods has continued over the years. The 1987 International Conference on Particle

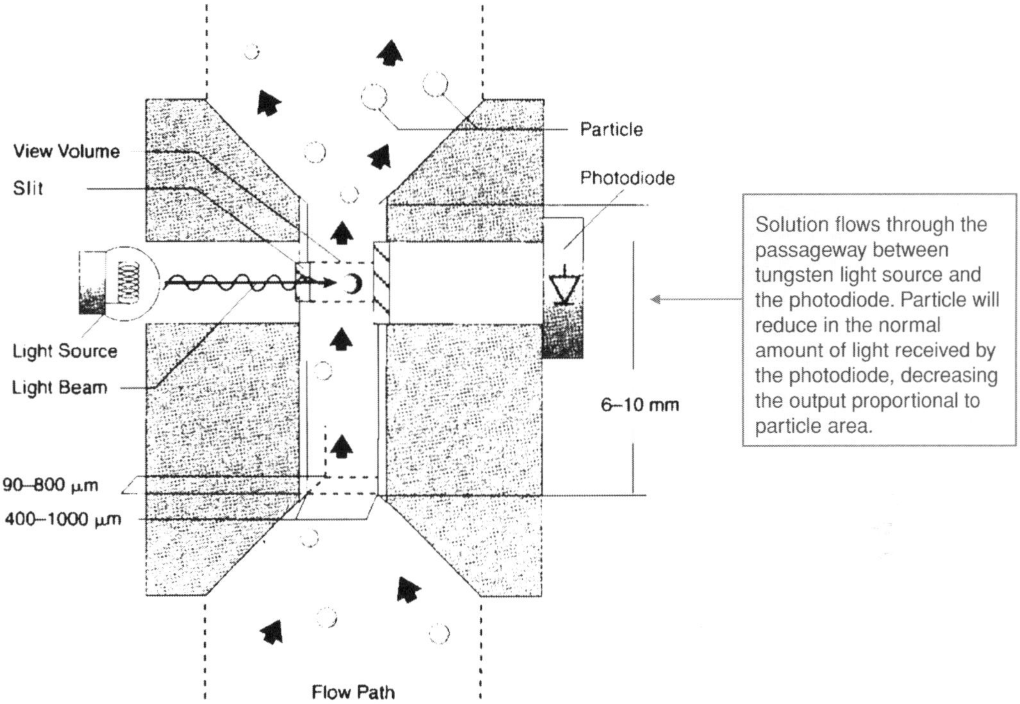

Figure 29-1 Light Obscuration Principle. *Source*: Reproduced with permission from Barber TA. Control of Particulate Matter Contamination in Healthcare Manufacturing. Denver, CO: Interpharm Press, 2000:278.

Detection, Metrology, and Control concluded with the general perception that there remained many measurement problems with electronic particle counters. However, the 1990 Particle Conference closed on a strongly optimistic note that the basic error mechanisms have been identified and that accurate, replicable particle data are within reach (14). These continuing advancements and problems will be reviewed in this section.

Two major advantages of electronic particle counters are their automated characteristics and the rapidity at which they accomplish particulate measurement. Two major disadvantages hinder electronic particle analysis from becoming a more acceptable means of measuring particulate contamination: they cannot differentiate among various types of particles and they measure particle size differently than microscopic methods.

Principle of Light Obscuration

A schematic representation of the light-obscuration principle is shown in Figure 29-1. A tungsten lamp produces a constant collimated beam of light that passes through a small rectangular passageway and impinges onto a photodiode. In a clear passageway, the light intensity received by the photodiode remains constant.

Liquids can flow through the passageway between the light source and the photodiode. If a single particle transverses the light beam, there results a reduction in the normal amount of light received by the photodiode. This reduction of light and the measurable decrease in the output from the photodiode is proportional to the area of the particle interrupting the light flow. Thus, the light-obscuration principle measures particle size based on the diameter of a circle having an equivalent area.

HIAC (High Accuracy Instruments Division) particle counters employ the light-blockage principle in the detection and quantifying of particulate matter in injectable solutions (Fig. 29-2). These instruments count approximately 4000 particles per second. HIAC counters use sensors having size measurement ratios of 1:60. In other words, a 1 through 60 μm sensor can measure particles from 1 to 60 μm, while a 2.5 through 150 μm sensor can measure particles ranging

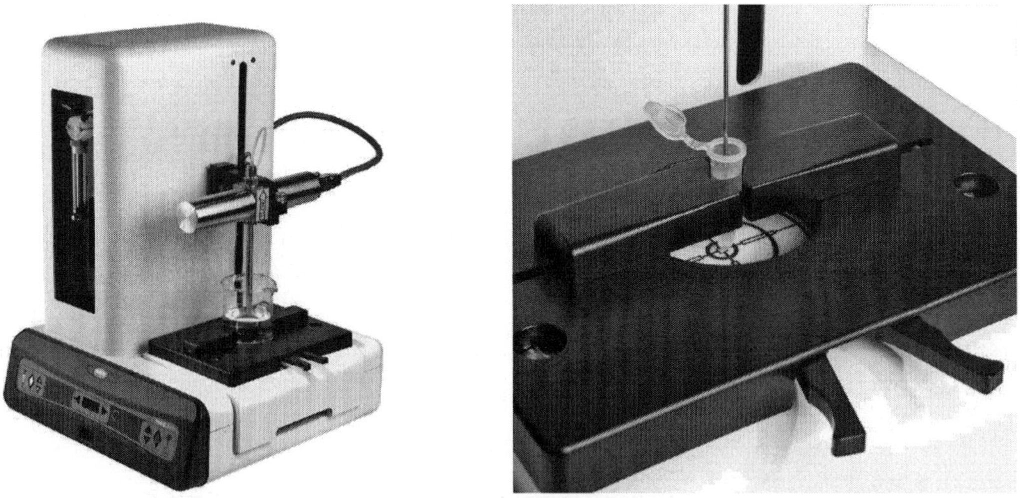

HIAC 9703+™ with close-up view of small container holder for small samples (< 1 mL)

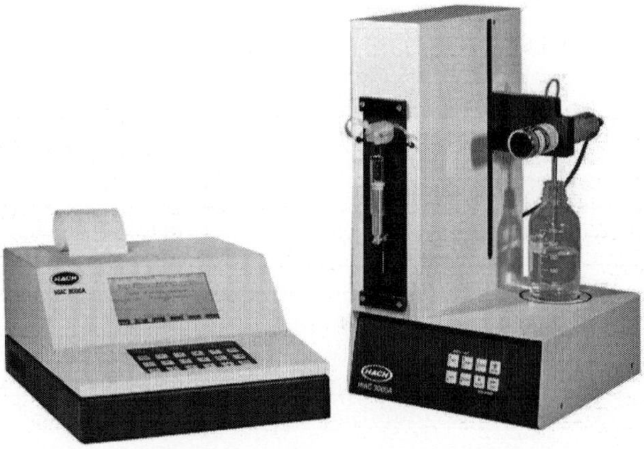

HIAC Model 8103™

Figure 29-2 HIAC light obscuration instruments. *Source:* Courtesy of Hach Company.

from 2.5 to 1.50 μm. Channel numbers on the counter are selected and calibrated according to the size range desired.

Increasingly, over the past several years, HIAC systems have progressed in technological advances and user application in the particle analysis field. Advantages for using HIAC particle counters have outweighed the disadvantages (Table 29-1).

LIGHT OBSCURATION PARTICLE COUNT TEST

The USP allows that any suitable apparatus based on the principle of light blockage can be used provided that the apparatus allows for an automatic determination of the size of particles and the number of particles according to size. The apparatus is calibrated using dispersions of spherical particles of known sizes between 10 and 25 μm. These standard particles are dispersed in particle-free water (defined in USP Reagent Specifications).

Table 29-1 Advantages and Disadvantages of Light Obscuration Particulate Matter Measurement Methods

Advantages
• Particles are counted automatically. • Parenteral solutions, either electrolytes or nonelectrolytes, could be counted. • The instrument was easy to calibrate and use. • Replication of counts was good. • Ability to vary the volume of samples as desired for counting. • Dilution method of counting permitted counting of both "clean" and "heavily contaminated" solution. • Direct method of counting permitted counting of crystallized soluble particles.
Disadvantages
• Instrument is relatively expensive as compared to equipment used for counting by optical microscope. • Particulate contaminants cannot be identified. • Large and/or fibrous particles may block the sensor opening. • Air bubbles are counted as particulate matter. • Dilution method of counting does not permit counting of crystallized soluble materials because dilution solubilizes crystals.

Prior to using the USP procedure, three preliminary tests are to be done:

1. Determination of sensor resolution—use dispersions spherical particles of known sizes between 10 and 25 μm. These certified standard particles are dispersed in particle-free water. Instrument vendors may provide the calibration and/or performed periodically in the QC laboratory. Calibration comprises three steps: (*i*) collect data from 5000+ counts each, for seven standards of known nominal median diameter, in aqueous suspensions; (*ii*) determine mean and standard deviation at each size; (*iii*) determine the relationship between particle size and detected signal.
2. Sensor flow rate—certify that the actual flow rate is within the manufacturer's specification for the particular sensor used.
3. Sample volume accuracy—since particle count varies directly with the volume of fluid sampled, sampling accuracy must be known and be within $\pm 5\%$.

Test Environment

At a minimum, the light obscuration apparatus must be located within the confines of a high-efficiency particulate air (HEPA)-filtered laminar flow workbench. However, most industrial particle test laboratories have separate and contained work areas where these instruments are located. Furthermore, strict standard operating procedures are followed for entering the work area, with personnel appropriately gowned and samples introduced following procedures to remove all extraneous particulate matter.

All glassware, rubber closures, and other samples entering the testing environment must be scrupulously cleaned as described in the USP. Final rinsing of all materials is performed using filtered distilled water with the filter porosity being 1.2 μm or finer.

Prior to determining particle counts of the actual samples, counts from particle-free water are determined. Particle-free water counts are determined to ensure that the environment is suitable for the actual product test and that the glassware used is properly cleaned. Five samples each of 5 mL of particle-free water are tested for particulate matter using the test method described in the next paragraph. From these five × 5 mL samples (25 mL total), if the number of particles of 10 μm or greater size exceeds 25, the precautions taken for the test are insufficient. All preparatory steps must be repeated until the environment, glassware, and water are suitable to pass this baseline test.

Test Procedure

Prior to placing test samples within the confines of a laminar flow work area containing the light obscuration particle counter, the exterior surfaces of the samples are rinsed with filtered distilled or deionized water. Since samples need to be protected from environmental contamination after cleaning until analyzed, having a specially built clean room or module offers significant advantages. Technique to withdraw samples is very important to minimize particle contamination and introduction of air bubbles. Rubber closed vials may be sampled either by needles penetrating the closure or by completely removing the closure. This also holds true for vials containing sterile powders in that diluent can be added either via needle penetration or first removing the rubber closure. If samples are pooled, then it is better to remove the rubber closure.

If the volume of the product in the container is less than 25 mL, then at least 10 units of product are used for the LO test. Each unit of product is inverted slowly at least 20 times. The outer surfaces of the container openings are cleaned using a jet of particle-free water and the closure is removed, obviously avoiding any contamination of the contents. Gas bubbles can be eliminated by waiting 2 minutes or applying sonification prior to sampling. The 10 units are opened and the contents pooled into a cleaned container. The total volume in the pooled container must be at least 25 mL. If necessary for container volumes less than 2.5 mL, the pooled sample can be diluted with particle-free water or with an appropriate particle-free solvent. The pooled container is then gently swirled, and four aliquots of at least 5 mL of solution are withdrawn and injected into the LO counter sensor. The first sample measured by the electronic counter is not counted, but the next remaining samples are counted.

If the volume of the product in the container is 25 mL or more, but less than 100 mL, individual containers may be tested. Fewer than 10 product units may be tested as long as an appropriate sampling plan is applied. The same rules apply as above in that four portions of at least 5 mL each are tested with the first portion discarded.

Freeze dried and other sterile powdered filled samples are reconstituted by first removing the rubber closure without contamination, using particle-free water or other appropriate filtered diluent at the required volume. The rubber closure is then replaced and the product manually agitated to ensure complete dissolution of the drug product. Reconstituted samples can be pooled as described earlier with the total volume of pooled sample being at least 25 mL, four samples of at least 5 mL tested with the first sample discarded.

For large-volume injections, single units are tested with the sampling and testing procedures being the same as those for small-volume injections having a volume of 25 mL or more.

Calculations

Calculations for pooled and individual samples from small-volume injections and calculations for individual unit samples from large-volume injections are determined using the following formula:

Small-volume pooled samples: $PV_t/V_a n$
Small-volume individual samples: PV/V_a
Large-volume individual samples: P/V

where P is the average particle count obtained from the portions analyzed, V_t is the volume (mL) of pooled sample, V_a is the volume of each portion analyzed, V is the volume of the tested unit, and n is the number of containers pooled.

Interpretation

The SVI injection meets the requirements of the USP subvisible particle test if the average number of particles present in the units tested does not exceed the limits shown in Table 29-2. Table 29-3 provides subvisible particulate matter limits for worldwide compendial requirements. With time, these are always subject to change. Limits for particles in ophthalmic solutions are also given and note that ophthalmic solutions, regardless of volume, have particle limits per milliliter, not per container.

Also, the number of particles detected at 25 μm need to be added to the number detected at 10 μm. For example, if a solution is measured to have 7 particles at 10 μm and 3 particles at 25 μm, the correct results are reported as 10 particles ≥10 μm and 3 particles ≥25 μm.

Table 29-2 Subvisible Particulate Matter Limits for Small-Volume Injectables Measured by Light Obscuration

Product	≥10 μm	≥25 μm
Small-volume injections in containers with nominal volume ≤100 mL	6000 particles per container	600 particles per container
Large-volume injections in containers with nominal volume >100 mL	25 particles per mL	3 particles per mL
Ophthalmic solutions	50 particles per mL	5 particles per mL

If these limits are exceeded, the USP requires that the product be tested by the Microscopic Particle Count Test.

It is possible that the above maximum particle limits for small-volume injections could be reduced as they already have in the past. It is extremely rare that sterile product solutions prepared in modern manufacturing facilities contain particles more than 1000 and 100 for 10 and 25 μm, respectively, although the limits at the time of this publication remained at 6000 for 10 μm and 600 at 25 μm.

MICROSCOPIC PARTICLE TESTING

The present USP method provides both qualitative and quantitative data on particulate content in LVI and SVI solutions. Particles not less than 10 μm can be counted, sized, and described in terms of their shape and, at times, their nature, for example, a cotton fiber, piece of glass, or metal sliver. Photographs of the filter membrane further provide a permanent record of the particulate test results.

Considerable care and skill are required for preparing the membrane, cleaning the glassware and equipment used in the procedure, and using the microscope (Fig. 29-3). This presents a major disadvantage and motivates pharmaceutical manufacturers to develop and validate alternative methods employing automation, electronic counting instrumentation, or both.

Table 29-3 Comparison of Compendia for Particulate Matter Standards

Compendia	Large-volume (LVI) or small-volume injectable (SVI)	Method of particle counting	Limits
USP	LVI	Light obscuration	25 particles/mL ≥10 μm 3 particles/mL ≥25 μm
	LVI	Microscopic	12 particles/mL ≥10 μm 2 particles/mL ≥25 μm
	SVI	Light obscuration	6000 particles/container ≥10 μm 600 particles/container ≥25 μm
	SVI	Microscopic	3000 particles/container ≥10 μm 300 particles/container ≥25 μm
EP	LVI	Light obscuration	25 particles/mL ≥10 μm 3 particles/mL ≥25 μm
	SVI solutions	Light obscuration	6000 particles/container ≥10 μm 600 particles/container ≥25 μm
	SVI solutions after reconstitution of sterile powders	Light obscuration	10,000 particles/container ≥10 μm 1000 particles/container ≥25 μm
BP	LVI	Coulter counter	1000 particles/mL ≥2 μm 100 particles/mL ≥5 μm
		Light obscuration	500 particles/mL ≥2 μm 80 particles/mL ≥5 μm
JP	LVI	Microscopic	20 particles/mL ≥10 μm 2 particles/mL ≥25 μm

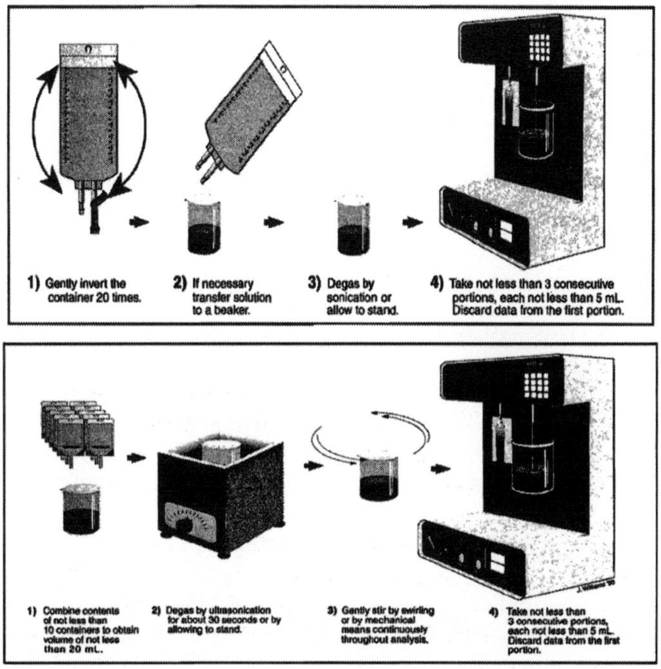

Figure 29-3 Microscopic Particle Measurement Method. *Source*: Reproduced with permission from Barber TA. Control of Particulate Matter Contamination in Healthcare Manufacturing. Denver, CO: Interpharm Press, 2000:113.

Laminar Airflow Hood

All operations and manipulations must be performed under a certified laminar flow hood equipped with HEPA filtered air in a class 100 environment. Working in a laminar airflow (LAF) environment can never replace the necessity for rigid clean technique in sample preparation and analysis. Prior to conducting a test, the hood must be cleaned with an appropriate solvent, preferably 70% ethanol or 70% isopropyl alcohol. The HEPA filter itself is not cleaned because of potential damage to the filter surface. The hood should have a built-in sink or some accommodation for collection and disposal of solvents used in the filtration process.

Introduction and Use of Equipment in the Laminar Airflow Hood

The USP demands the use of "scrupulously" clean glassware and equipment for the particle test. The word "scrupulous" means the following:

1. Rinse glassware and equipment successively with (*i*) warm detergent solution, (*ii*) hot water, (*iii*) water, and (*iv*) isopropyl alcohol. The first supplement of the 19th edition of the USP listed a fifth rinse with trichlorofluoroethane (Freon 113). Freon was eliminated in the 20th edition procedure because of concern about its toxicity in a closed environment, and harm to the ozone layer.
2. Rinsing technique is important. Glassware and equipment must be rinsed starting at the top of the vertically held object and working downward in a back-and-forth manner. Water rinsing may be done outside the LAF hood, but the final isopropyl alcohol rinse must be performed within the hood.
3. After rinsing, all objects must dry under the hood upstream of all other operations. This helps to ensure that few, if any, extraneous particles adhere to the drying object.

Gloves

The USP requires the use of suitable, nonpowdered gloves for the particle test. Gloves are important in protecting the hands from the dehydrating effects of isopropyl alcohol. However, gloves may create more problems than they solve. Using gloves of improper size will promote problems in careful handling of glassware and equipment. Gloves also produce a false sense of security resulting in less than ideally careful manipulations in the LAF hood. The greatest potential limitation of gloves is the contribution they can make to particulate contamination, even after adequate rinsing. Thus, this requirement continues to be controversial.

Membrane Filter and Assembly

Membranes

The USP specifies a gridded or nongridded black or dark gray or filter of suitable material compatible with the product. The filter must have a porosity of at least 1.0 μm. Explicit instructions are provided in the USP for rinsing the membrane filter. In the 19th edition of the USP, Freon was used as the rinsing agent. In the 20th edition, water replaced Freon. Rinsing of a vertically held filter (using forceps) is accomplished using filtered water sprayed from a pressurized container. Rinsing of the membrane with filtered water starts at the top of the nongridded side, sweeping a stream of water back and forth across the membrane surface from top to bottom. This process is repeated on the gridded side of the membrane. Pressures exceeding 2 psi may damage the delicate membrane.

The rinsing solvent is checked for particle counts, serving as the blank determination in the testing portion of the USP procedure. It must be assumed that no dispensing vessel will provide a particle-free solvent. While the membrane filter on the nozzle will effectively remove particles above the rated porosity of the filter (usually 1.2 μm), particles on the downstream side of the filter on the nozzle will shed into the dispensed solvent. Of course, there is always the possibility of a misplaced or torn membrane filter on the dispenser nozzle.

Filter Assembly

The appropriately rinsed membrane filter is placed with the grid side up on the filter holder base. Great care is taken when the filtering funnel is situated on the base so that the membrane is not rumpled or torn. Prior to placing this assembly on the filtering flask, the unit is rinsed thoroughly and carefully with filtered water from the pressurized solvent dispenser. After allowing time for the rinse fluid to drain the filter, the apparatus is then secured on top of the filter flask.

Test Preparation

Containers to be tested for particulate matter must be inverted 20 times before the contents are sampled. Agitation has been shown to affect particle size distribution so the 20-fold inversion procedure must be consistent. After rinsing the outer surface of the container with filtered water, the closure is removed. One can never be certain that removal of the closure will not introduce extraneous particles. Careful aseptic and clean technique must be adhered to as much as possible.

After the closure has been carefully removed, the contents are swirled before 25 mL of samples is transferred to the filtering funnel. After standing for 1 minute, a vacuum is applied to filter the 25 mL sample. An additional 25 mL sample of water is then applied to the sides of the funnel to rinse the walls of the funnel. The stream of filtered water should not hit the filter membrane for fear of tearing the membrane. The rinse fluid then is filtered via vacuum. Unfortunately, particles tend to adhere to the underside of the filter assembly top and to the O-rings used between the filter base and filter funnel.

The funnel section of the assembly is carefully removed. The membrane is lifted away from the base using forceps and placed on a plastic Petri slide containing a small amount of stopcock grease or double-sided adhesive tape. The cover of the Petri slide is placed slightly ajar atop the slide to facilitate the membrane drying process. The slide then is placed on the micrometer stage of the microscope for visual analysis.

Particle Count Determination

Examination of the entire membrane filter surface for particulates may be accomplished using a precisely aligned and calibrated microscope. The microscope should be binocular, fitted with a 10× objective, and have one ocular equipped with a micrometer able to measure accurately particles of 10 and 25 μm linear dimension.

Particles are counted under 100× magnification with the incident light at an angle of 10° to 20°. Obviously, this is a slow and tedious process requiring patience and dedication on the part of the microscopist. Use of higher magnification, up to 400×, may be necessary occasionally to discern discrete particles from agglomerates or amorphous masses. Sometimes, particles not visible with dark field reflected light are very easily observed by means of bright field illumination at 45° polarization.

Two sizes of particles are counted, those having effective linear dimensions ≥10 and ≥25 μm. The counts obtained from the sample membranes are compared with counts obtained from a membrane treated exactly like the sample membrane minus the filtration of the product sample. Blank membrane counts rarely are zero. However, if 5 or more particles ≥25 μm and/or more than 20 particles 10 μm are counted on the blank membrane, the test is invalidated and it signifies a serious problem in one or more of the following areas: poor technique, filter breakdown in the solvent dispenser, poorly cleaned membranes, poorly cleaned filter assemblies, and/or HEPA filter leaks. The problem must be resolved before particle testing can resume.

If the USP limit of not more than 12 particles per mL ≥10 μm and not more than 2 particles per mL ≥25 μm is exceeded, the large-volume injection product fails the USP test for particulate matter. For small-volume parenterals, the test fails if more than 3000 particles/container 10 μm and/or 300 particles per container 25 μm is exceed. For ophthalmic solutions, the limit for particles measured by microscopy is 50 per mL ≥10 μm, 5 per mL ≥25 μm, and 2 per mL ≥50 μm. These limits are also stated in the ICH Q4B document.

Analysis by microscopic techniques suffers from several disadvantages—it is very time-consuming, requires technical expertise, and, because of the manpower requirements, can be very expensive. The major method for determining subvisible particulate matter in parenteral solutions, including reconstituted sterile powders, is the light obscuration technique. However, if any dispute arises regarding fulfillment of USP particulate matter specifications, such disputes must be settled by applying the official USP microscopic method.

LVI particulate matter standards in other countries governed by other compendia will be reviewed in a section at the end of this chapter.

COMPARISON OF ELECTRONIC AND MICROSCOPIC PARTICLE COUNTING METHODS

Difficulties in comparing particle-counting methods result from differences in the way in which different methods determine particle size and distribution. For example, the microscopic method measures size as the longest linear dimension of the particle. The principle of light blockage, utilized by the HIAC particle counter, expresses size as the diameter of a circle of equivalent area as the actual area consumed by the particle. Particle counting by electrical resistance (Coulter Counter) treats the particles as a three-dimensional object and measures the volume consumed by the particle. Thus, the microscope, HIAC, and the Coulter Counter methods size particles in one, two, and three dimensions, respectively.

It is virtually impossible to correlate instrumental and microscopic particle counts directly for irregularly shaped particles (15). As long as the particle is a sphere, all methods will size the sphere equally (Table 29-4). However, as the particle shape deviates from sphericity, the

Table 29-4 Summary of Sphericity Correction Factors Based on Longest Linear Dimension (15)

Shape	D_O Longest dimension	D_H Horizontal projection	D_A Light blockage	D_V Electrolyte displacement
Sphere	1.00	1.00	1.00	1.00
Cube (1:1:1)	1.00	0.90	0.95	0.88
Equant (3:2:1)	1.00	0.88	0.81	0.62
Prolate ellipsoid (2:7:1)	1.00	0.87	0.61	0.52
Flake (4:4:1)	1.00	0.90	0.81	0.55
Rod (3:1 diameter)	1.00	0.81	0.62	0.52
Fiber (rigid, 10:1)	1.00	0.64	0.36	0.25

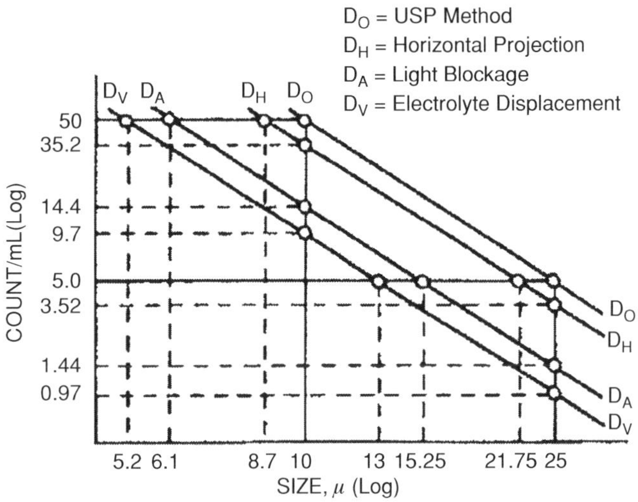

Figure 29-4 Log count versus log size corrections for sizing and counting prolate ellipsoids. *Source*: From Ref. 15.

size measurement by the three alternate approaches will differ, sometimes drastically, from the value obtained by the USP microscopic method. For example, if the solution sample contained 50 ellipsoid particles with their longest linear dimension equaling 10 μm, the HIAC will yield a count of $50 \times 0.61 = 30.5$ particles. In fact, this HIAC value may be an overestimate because the 0.61 correction factor considers only size (10 μm), not the actual number of particles. Assuming that the size–count relationship follows the conventional log–log relationship, the theoretical HIAC count of 50 ellipsoid particles of 10 μm size would be only 14.4 particles. Figure 29-4 provides the explanation. The USP microscopic method follows a log–log distribution, yielding a straight-line slope between 10 and 25 μm. Assuming the HIAC method to follow the same log–log distribution between 10 and 25 μm, its slope will be parallel to the USP slope. However, the HIAC correlation factor for ellipsoid particles theoretically is 0.61 that of the USP method. Thus, the starting point for the HIAC method is not 10 μm but 6.1 μm on the log–log graph. Therefore, following a parallel relationship with the slope of the USP method, the HIAC method yields a theoretical particle count value of 14.4 particles at the point intersecting the vertical line from the particle size of 10 μm.

Lim et al. (16) filtered various small-volume parenteral solutions and counted particles using the manual counting method under the microscope and the electronic Millipore MC method. In products with relatively few particles, both methods gave similar results. In products containing a high number of particles in the size range of 5 to 25 μm, the electronic method detected more particles.

CURRENT ISSUES WITH ELECTRONIC PARTICLE COUNTERS

The single greatest obstacle in using LO counters is their inaccurate measurement of both particle number and particle size (17). This is not because of design flaws or engineering defects with these counters, but rather because of the basic principle on which these instruments operate. Particle counts result from a series of interactions between a particle moving at high velocity and an intense light beam in the counter's sensor. Whenever a particle crosses the light beam, the intensity of light that reaches the photodiode is reduced and an amplified voltage pulse is produced. The amplitude of the pulse is approximately proportional to the area of the particle projected onto a plane normal to the light beam, and the particle size is recorded by the counter as the diameter of a sphere having an equivalent projected area. When particles are few, large (>5 μm), and spherical, good numerical accuracy is possible. However, when particles are many, small (>5 μm), and nonspherical, inaccuracies will result. A particle's residence time in the view volume usually is too short to allow the sensor to detect more than one aspect of the particle, and consequently, the LO measurement is based on the light that is obscured by the particle according to its orientation when it enters the counter's view volume. Other issues with electronic particle counters.

Solution Flow Rates

Flow rates will greatly affect count accuracy. Slower rates result in longer pulse durations, increased probability of electronic noise effects on count pulse, and possible increases in apparent particle size. Faster flow rate pulses may not rise to full height, resulting in undersizing (18).

Nonspherically shaped Particles

Such particles may produce significant errors in sizing accuracy of electronic particle counters. Because particles of irregular shape are viewed in random aspect as they pass through the sensor of a counter, the size recorded typically will be less than that defined by the maximum area of light obscuration (Table 29-4).

As differences between the refractive index of the particle and the refractive index of the solution containing the particle increases, the measured particle size will increase. A particle in water will have a greater refractive index between the two than the same particle in a concentrated solution of dextrose. Thus, these particles in water will be measured by the light obscuration sensor to have greater size and greater number than the same particles in the concentrated dextrose solution.

Calibration Errors

Errors during calibration may occur because calibration is done with monosized spherical latex particles that provide a very narrow range of known mono-shaped particle size. This introduces a calibration bias when measuring actual and largely unknown sizes and shapes of particles in parenteral solutions. The error introduced nearly always results in particle measurements being smaller than they should be. However, to attempt to calibrate counters with nonspherical particles adds greater difficulties because of their nonuniformity, dispersal difficulties, and differences in chemical composition and optical properties; the calibration "value" would be practically meaningless.

Coincidence Effects

This phenomenon occurs when two or more particles are counted as a single larger particle. This problem can be most easily detected by comparing dilutions of the same sample; if an increase in total counts occurs with the diluted sample, coincidence counts are probably the cause. Eradication of coincidence effects is difficult; the only reasonable method for obtaining valid data with such solutions is to do microscopic analyses.

Immiscible Fluids and Air Bubbles

These artifacts are counted as particles; thus, sources of error for light obscuration and other electronic particle counting methods. The primary source of immiscible fluid is silicone, usually very small (1 μm) microdroplets. Other immiscible liquids detected and counted as particles by electronic particle counters include leachables from packaging (e.g., plasticizers), oils from manufacturing equipment, and lower polarity impurities from the active pharmaceutical ingredient. Silica (inorganic) and silicone fragments and extracts from process tubing may also be present in the formulation, and may positively alter the particle count data. Only in significant numbers do silicone microdroplets produce significant errors in particle measurement. Air bubbles are also problematic, but the USP states gas bubbles can be eliminated by allowing the solution "to stand for 2 minutes or sonicating." Such degassing probably does not remove all microscopic bubbles nor reduce the dissolved air content in the solution.

Sampling variability, as with any quality-control test relying on sampling procedures, must also be recognized as a source of error with electronic particle counting. Sampling-associated factors that adversely affect particle counting are caused by particle stratification effects, by a small sample volume relative to the total sample volume, and by the low numbers of particles per milliliter that typically are counted in a parenteral solution. Adequate agitation of the product container prior to collecting samples must be properly done to minimize the effects of sampling variability.

Nearly every scientific paper featuring the use of a particle test method, whether it is visual, microscopic, electronic, manual, or automatic, will highlight major limitations to the method. For example, visual examination by human beings is limited by its tedium and subjectivity. Microscopes often are improperly calibrated. Electronic particle counters count air bubbles as particles. For LVIs, the USP relies on membrane filtration in which particles from the equipment, environment, or personnel involved in conducting the test inadvertently become deposited on the filter.

Other Potential Problems

Particle contents of injection containers vary considerably between the date of production and a later date when the same containers are tested again. Storage causes particle agglomeration. Mechanical agitation breaks up the agglomerates, resulting in counts that cannot reproduce the original count or replicate one another on the same date of testing. Freshly prepared solutions seemed to give more stable counts. It was suggested that only the manufacturer, who can reproduce the handling of its products, use particle counting as a meaningful control method.

Agitation or shaking will increase the number of particles in a parenteral solution. Blanchard et al. (19) found that the slope and number of particles per milliliter greater than 1 μm in a log–log plot of number against diameter depended on the degree of agitation. Agitation of LVI by 20 hand inversions, as required by the USP procedure, removed particulate matter from the surface of the container, thus increasing the total number of particles greater than 1 μm. Yet the relative size distribution of particles was not altered significantly. Agitation for 30 minutes disintegrated agglomerates, greatly increased the number of particles with diameters less than 1 μm, and brought about a corresponding decrease in the number of particles exceeding 1 μm in diameter. Particle-counting procedures must be carried out that do not impose a sheer force upon the particles and affect the reproducibility of the test results.

PARTICULATE MATTER AND BIOPHARMACEUTICAL SOLUTIONS

Accurate measurement of particles in therapeutic biopharmaceutical solutions is especially important because of the potential safety problems associated with the effects of small particles on protein aggregation. Protein aggregation is believed to be one of the causes of immune responses in patients administered these products.

Protein and other biopharmaceutical molecules form particles with a huge range of sizes (1 mm down to 1 nm, a range of 1-million size units) and shapes. A major challenge is to find particle counting and sizing methods that can comprehensively characterize this huge range in actual biopharmaceutical solution dosage forms.

USP and EP tests for subvisible particulate matter (light obscuration and light microscopy) are not sufficiently adequate to detect submicron particles that may/will form in biopharmaceutical solutions that could result in the eventual formation of large subvisible and eventual visible protein aggregates. Thus, especially during early development of biopharmaceutical solutions, other methods for detection of particulate matter formation might be more useful. Such methods include laser diffraction particle analyzers, polarization intensity differential scattering, dynamic image analysis, and Raman spectroscopy (1). And, of course, during formulation development of biologic products, methods such as size-exclusion chromatography and dynamic light scattering are very important to measure soluble protein aggregates as a function of formulation, processing effects, final packaging, and storage stability.

While compendial and regulatory standards exist for particulate matter in solutions released from manufacture for use commercially or in clinical studies, there are no specific standards for particulate matter in solutions just prior to injectable administration. A survey of commercial biopharmaceutical products' package inserts contained in the Physicians Desk Reference revealed a wide variety of statements for the acceptability of visible particulate matter and use of syringe or in-line filters just prior to product injection (Table 29-5). Note the following:

- Different product solutions require different filter porosity sizes, ranging from 0.2 to 15 μm.
- Some products are incompatible with in-line filters

Table 29-5 Sterile Product Package Insert Statements Regarding Particulate Matter and/or Filter Usage for Solutions Prior to be Injected (Specific product names purposely not identified)

Excerpts of statements regarding use of filters
• Withdraw solution from vial using an enclosed sterile filter needle • Withdraw solution into a syringe through a low protein binding 0.2 or 0.22 μm filter • Withdraw solution into a syringe and filter the injection using a sterile, nonpyrogenic, low protein binding 0.2 or 0.22 μm filter (filter vendor and type may be provided) • PVC infusion set equipped with an in-line, low protein binding 0.2 μm filter • Colorless solution that may contain translucent particles, use a 0.22 μm low protein-binding filter to be in-line between the syringe and the infusion port. • Reconstituted solution is not transparent, any undissolved particulate matter is difficult to see when inspected visually. Therefore, terminal filtration through a sterile 0.45 μm or smaller filter is recommended. • The reconstituted solution can be filtered through a 0.8 μm or larger pore size filter • A separate IV line equipped with a low protein binding 1.2-μm terminal filter must be used for administration of the drug. • Occasionally, a very small number of gelatinous fiber-like particles may develop on standing. Filtration through a 5.0 μm filter during administration will remove the particles with no resultant loss in potency. • Withdraw the necessary amount of product from the ampoule into a syringe, filter with a sterile, low protein binding, nonfiber releasing 5 μm filter prior to dilution. • Use a 15 μm filter • Administered through an intravenous line using an administration set that contains an in-line filter (pore size 15 μm). A smaller in-line filter (0.2 μm) is also acceptable. • Reconstituted product does not need to be filtered. If a filter is used, it should be a 15 μm filter or larger • Do not use an in-line filter • Do not filter the reconstituted solution. • Do not use filter needles during the preparation of the infusion.
Excerpts from statements regarding appearance of particles
• Solution should be clear to slightly opalescent and colorless to pale yellow. A few translucent particles may be present. Do not use if there is particulate matter in the solution. • Parenteral drug products should be inspected for visible particulate matter and discoloration prior to administration. If particulate matter is present or the solution is discolored, the vial should not be used. • Parenteral drug products should be inspected visually for particulate matter and discoloration prior to administration, whenever solution and container permit. If visibly opaque particles, discoloration or other foreign particulates are observed, the solution should not be used. • If visibly opaque particles, discoloration or other foreign particulates are observed, the solution should not be used. • There should be no visible gel-like particles in the solution. Do not use if foreign particles are present. • Solution should be clear immediately after reconstitution. Do not inject if the reconstituted product is cloudy immediately after reconstitution or after refrigeration (2–8°C/36–46°F) for up to 14 days. Occasionally, after refrigeration, small colorless particles may be present in the solution. This is not unusual for solutions containing proteins. • Because product is a protein, shaking can result in a cloudy solution. The solution should be clear immediately after reconstitution. Do not inject if the reconstituted product is cloudy immediately after reconstitution or refrigeration. Occasionally, after refrigeration, small colorless particles may be present in the solution. This is not unusual for solutions containing proteins. • The solution should be clear immediately after removal from the refrigerator. Occasionally, after refrigeration, you may notice that small colorless particles of protein are present in the solution. This is not unusual for solutions containing proteins. If the solution is cloudy, the contents must not be injected. • There should be no visible gel-like particles in the solution. Do not use if foreign are present. It is acceptable to have small bubbles or foam around the edge of the vial. Do not use if the contents of the vial do not dissolve completely by 40 minutes.

(*continued*)

Table 29-5 Sterile Product Package Insert Statements Regarding Particulate Matter and/or Filter Usage for Solutions Prior to be Injected (*Continued*)

Other examples of package insert statement related to particles or filters
• Some loss of potency has been observed with the use of a 0.2 μm filter
• After reconstitution, product should be inspected visually before use. Because this is a protein solution, slight flocculation (described as thin translucent fibers) occurs occasionally after dilution. The diluted solution may be filtered through an in-line low protein-binding 0.2 μm filter during administration. Any vials exhibiting opaque particles or discoloration should not be used.
• Thin translucent filaments may occasionally occur in reconstituted product vials, but do not indicate any decrease in potency of this product. To minimize formation of filaments, avoid shaking the vial during reconstitution. Roll and tilt the vial to enhance reconstitution. The solution may be terminally filtered, for example, through a 0.45 μm or smaller cellulose membrane filter.
• Since reconstituted product is not transparent, any undissolved particulate matter is difficult to see when inspected visually. Therefore, terminal filtration through a sterile 0.45 μm or smaller filter is recommended.
• Occasionally, a very small number of gelatinous fiber-like particles may develop on standing. Filtration through a 5.0 μm during administration will remove the particles with no resultant loss in potency. Some loss of potency has been observed with the use of a 0.2 μm filter.

Source: Physicians Desk Reference, 2004.
Note: Use of word "product" or "solution" replaces actual product stated in insert.

- Some products require rejection if visible particles are seen, but most do not require rejection, rather use the described filter
- There is no standard of language or grammar used, filter descriptions, or particle descriptions.

A 2007 special article in *Hospital Pharmacy* listed 53 commercial drug products that require a filter for their preparation in the hospital pharmacy and/or administration of the product to a patient (20). This publication likely will publish updates of this article in the future.

REFERENCES

1. Das T, Nema S. Protein particulate issues in biologics development. Am Pharm Rev 2008; 11:52–57.
2. Godding EW. Foreign matter in solutions for injection. Pharm J 1945; 154:124.
3. Brewer JH, Dunning JHF. An in vitro and in vivo study of glass particles in ampuls. J Am Pharm Assoc 1947; 36:289.
4. Leelarasamee N, Howard SA, Baldwin HJ. Visible particle limits in small-volume parenterals. J Parenter Drug Assoc 1980; 34:167–174.
5. Garvan JM, Gunner BW. Intravenous fluids: A solution containing such particles must not be used. Med J Aust 1963; 2:140.
6. Garvan JM, Gunner BW. The harmful effects of particles in intravenous fluids. Med J Aust 1964; 2:1.
7. Geisler RM. The biological effects of polystyrene latex particles administered intravenously to rats: A collaborative study. Bull Parenter Drug Assoc 1973; 27:101.
8. Hozumi K, Kitamura K, Kitade T, et al. Localization of glass particles in animal organs derived from cutting of glass ampoules before intravenous injections. Microchem J 1983; 28:215–226.
9. Dimmick JE. Fiber embolization: A hazard of cardiac surgery and catheterization. N Engl J Med 1975; 292:685.
10. DeLuca PP, Rapp RR, Bivins B, et al. Filtration and infusion phlebitis: A double-blind prospective clinical study. Am J Hosp Pharm 1975; 32:101.
11. Barber TA. Control of Particulate Matter Contamination in Healthcare Manufacturing. Denver, CO: Interpharm Press, 2000:9.
12. Pesko LJ. Physiological consequences of injected particles. In: Knapp JZ, Barber TA, Lieberman AW, eds. Liquid and Surface Borne Particle Measurement Handbook. New York, NY: Marcel Dekker, 1996; 661–685.
13. Barber TA. Control of Particulate Matter Contamination in Healthcare Manufacturing. Denver, CO: Interpharm Press, 2000.
14. Knapp JZ, Barber TA. Overview of the international conference on particle detection, metrology and control. J Parenter Sci Technol 1990; 44:257–263.

15. Schroeder HG, DeLuca PP. Theoretical aspects of particulate matter monitoring by microscopic and instrumental methods. J Parenter Drug Assoc 1980; 34:183.
16. Lim YS, Turco S, Davis NM. Particulate matter in SVIs as determined by two methods. Am J Hosp Pharm 1973; 30:518.
17. Barber TA. Limitations in light-obscuration particle counting as a compendial test for parenteral solutions. Pharm Technol October 1988; 12:34–52.
18. Rebagay T, Schroeder HG, DeLuca PP. Particulate matter monitoring II: Correlation of microscopic and automatic counting methods. Bull Parenter Drug Assoc 1977; 31:150.
19. Blanchard J, Schwartz JA, Byrne JM. Effects of agitation on size distribution of particulate matter in large-volume parenterals. J Pharm Sci 1977; 66:935.
20. Anonymous. Drugs to be used with a filter for preparation and/or administration. Hosp Pharm 2007; 42:378–382.

BIBLIOGRAPHY

Knapp JZ, Barber TA, Lieberman A, eds. Liquid and Surface-Borne Particle Measurement Handbook. New York, NY: Marcel Dekker, 1996.

Akers MJ, Larrimore DS, Guazzo DM. Pyrogen testing, In: Parenteral Quality Control: Sterility, Pyrogen, Particulate, and Package Integrity Testing, 3rd ed. London (and all references therein): Marcel Dekker (now Informa Healthcare), 2003.

Q4B Evaluation and Recommendation of Pharmacopoeial Texts for Use in the ICH Regions; Annex 3, Test for Particulate Contamination: Subvisible Particles General Chapter, Food and Drug Administration, January 2009.

30 Sterile product-package integrity testing

*Dana Morton Guazzo**

Package integrity, also called container–closure integrity, is the measure of a primary package's ability to keep the product in (including vacuum or inert gas headspace, if present) and to keep potential microbial, particulate, and chemical contaminants out. Package integrity is a requirement that must be met throughout the product's life cycle, beginning from early development phases. A variety of tests are available for use by the pharmaceutical industry to measure parenteral product-package integrity, although no one test can be recommended for all parenteral package integrity testing. Historically, microbial ingress tests were considered the definitive standard, although regulatory agencies increasingly prefer validated physical test methods less subject to variability. Container–closure integrity verification of all units in marketed product lots has become a reality for many dosage form packaging types.

U.S. AND EU REGULATIONS AND GUIDANCES

Prior to the mid-1990s only sterility of the packaged product was required by the U.S. Food and Drug Administration (FDA) as verification of package integrity. Since 1994, the U.S. FDA issued several Guidances for Industry addressing this topic. First, the 1994 U.S. FDA Guidance for Industry describing sterilization process validation submission documentation requires a demonstration of a container–closure system's ability to maintain the integrity of its microbial barrier, thus indirectly verifying a drug product-package's sterility through its shelf life. Sterility testing alone is insufficient for this purpose (1).

Then in 1999, the FDA issued a comprehensive guidance discussing container and closure systems for packaging human drugs and biologics (2). Pharmaceutical packaging should be shown suitable for its intended use, including protection—the ability of the container–closure system "to provide the dosage form with adequate protection from factors (e.g., temperature, light) that can cause degradation in the quality of that dosage form over its shelf life." Package integrity-related causes of degradation cited include loss of solvent, exposure to reactive gases (e.g., oxygen), absorption of water vapor, microbial contamination, and contamination by filth. Package suitability verification in any new product submission must therefore include package integrity study results, and specifically, data extended throughout the product's full shelf life.

A 2008 FDA Guidance for Industry addresses the issue of integrity testing as part of pre and postapproval stability protocols for sterile biological products, human and animal drugs, including investigational and bulk drugs (3). As noted, stability testing must include a method(s) that supports the continued capability of containers to maintain sterility. While sterility testing satisfies this requirement, the Guidance acknowledges practical and scientific limitations to this approach, allowing the substitution of other integrity tests in stability protocols. Good scientific principles are recommended in selecting integrity tests, taking into consideration the container–closure system, product formulations, and, where applicable, routes of administration. How the method relates to microbial integrity should be noted.

The 2008 revision to Annex 1 of the European Union Good Manufacturing Practices (GMPs) for sterile products states that "Containers closed by fusion, e.g., glass or plastic ampoules should be subject to 100% integrity testing. Samples of other containers should be checked for integrity according to appropriate procedures" (4). Additionally, "Containers sealed under vacuum should be tested for maintenance of that vacuum after an appropriate, pre-determined period." Concerning stoppered vials, "Vials with missing or displaced stoppers should be rejected prior to capping." Another reference to integrity testing in the EU GMPs states: "Filled containers of parenteral products should be inspected individually for extraneous contamination or other defects." Direction is given for human inspection, and "where other

* This chapter contributed by Dr. Dana Morton Guazzo, RxPax, LLC.

methods of inspection are used, the process should be validated and the performance of the equipment checked at intervals."

The 2004 U.S. FDA Sterile Drug Products Aseptic Processing GMPs delineate similar standards to those in the European GMPs (5). Referring to inspection of container–closure systems, "Any damaged or defective units should be detected, and removed, during inspection of the final sealed product. Safeguards should be implemented to strictly preclude shipment of product that may lack container-closure integrity and lead to nonsterility. Equipment suitability problems or incoming container or closure deficiencies can cause loss of container-closure system integrity. For example, failure to detect vials fractured by faulty machinery as well as by mishandling of bulk finished stock has led to drug recalls. If damage that is not readily detected leads to loss of container-closure integrity, improved procedures should be rapidly implemented to prevent and detect such defects." Appendix 2 entitled Blow-Fill-Seal Technology states the following: "Container closure defects can be a major problem in control of a BFS operation. It is critical that the operation be designed and set-up to uniformly manufacture integral units. As a final measure, the inspection of each unit of a batch should include a reliable, sensitive, final product examination that is capable of identifying defective units (e.g., leakers). Significant defects due to heat or mechanical problems, such as wall thickness, container or closure interface deficiencies, poorly formed closures, or other deviations should be investigated in accordance with §§ 211.100 and 211.192."

PDA Technical Report No. 27

The Parenteral Drug Association (PDA) published a technical resource to offer clarification about selection of appropriate container–closure integrity test methods for different types of packaging (6). This report summarizes package leakage concepts and critical leak specifications and discusses the need to consider package integrity for the life of the product beginning in early product development. Eighteen different integrity tests are described and referenced. These are linked to a decision tree to help the reader in selecting the most appropriate methods. While the PDA Technical Report No. 27 is not an official regulatory document, for years it provided a valuable resource when first selecting package integrity tests. However, given the rapid developments in leak testing technologies during the last decade, the reader is advised to also consult more current sources for newer developments in package integrity testing.

LEAKAGE UNITS OF MEASURE

Leakage is mathematically defined as the rate at which a unit of gas mass (or volume) flows into or out of a leak path under specific conditions of temperature and pressure. The units of measure commonly used in many literature references to specify leakage rate are standard cubic centimeters per second (std cm^3/sec or std cc/sec). According to the international metric system of units (SI nomenclature) leakage is measured in pascal cubic meters per second ($Pa \cdot m^3$/sec). In both expressions, units of gas mass (std cc and $Pa \cdot m^3$) indicate the quantity of gas (air) contained in a unit of volume at sea level atmospheric pressure (101 kPa). The std cc/sec is the more common unit of measure. To convert to std cc/sec from $Pa \cdot m^3$/sec, the SI units should be multiplied by a factor of 9.87, or approximately 10. When expressing leakage volumetrically, rather than in mass flow units, test pressure and temperature conditions should be specified.

CRITICAL LEAK RATE AND SIZE

All parenteral product packaging must maintain product sterility by preventing the ingress of microorganisms. Therefore, the "critical leak rate" is generally understood to mean that leak rate, corresponding to a leak path that will permit microbial ingress. Pharmaceutical, medical device, and food packaging scientists have worked for many years to define this "critical leak rate" and its corresponding "critical leak size." Research results and conclusions derived from these studies vary widely. Differences seem to be colored by the perceived microbial ingress risk to product quality. For example, medical device experts who rely on nonporous as well as porous barrier material packaging are generally concerned with air-borne, rather than liquid-borne microbial challenges. Food packaging scientists are concerned with liquid-borne microbial ingress; however, food products often have a relatively short shelf life, may include antimicrobial preservatives, and are always ingested rather than injected, making the tiniest integrity breaches of a few microns or less of minor concern. On the other hand, pharmaceutical parenteral product packaging scientists tend to define critical leak size as any defect that might theoretically allow the ingress of even one microorganism under the most severe challenge conditions. With this in mind, the following discussion of pharmaceutical sciences research addressing critical leaks is offered.

Direct Comparison of Microbial Ingress to Physicochemical Leak Tests

In the late 1980s, researchers led by Dana Morton designed a series of tests to determine the gaseous, liquid, and microbial barrier properties of the classic parenteral vial package, that is, vial/closure compression seal systems (7,8). Test packages consisted of a simulated vial, fashioned from stainless steel, stoppered with disc-shaped closures made of various elastomers, either uncoated or laminated with various fluorocarbon or polypropylene-based polymeric materials, and sealed at a range of capping forces. Test packages were mounted onto a manifold equipped with a differential pressure transducer. The manifold with test package was pressurized with filtered nitrogen to an initial target value. Package leakage rate was measured by monitoring the test system's pressure drop over time. With this device, various elastomeric closures applied to the test vial across a wide range of compression forces were tested for their ability to affect a seal. Measured gas flow rates ranged from 10^{-3} to 10^{-7} Pa · m^3 · sec^{-1} (or 10^{-2} to 10^{-6} cc/sec) at 3 psig differential pressure test conditions.

After gaseous leakage rate determination, test packages were transferred to a separate test manifold for evaluating the seal's microbial barrier properties. Test packages, filled with a saline lactose broth suspension of *Pseudomonas aeruginosa*, were inverted so that the finish area was immersed in sterile saline. The test packages were pressurized to 3 psig for 15 minutes, replicating the differential pressure decay test conditions. *P. aeruginosa* migration from the package into the immersion fluid was determined using a filter plate count method. *P. aeruginosa* was shown incapable of passing through vial/closure compression seals that exhibited gas leakage rates of less than 10^{-5} Pa · m^3/sec (or 10^{-4} std cc/sec). Interestingly, there were vials that failed to allow microbial ingress even at gaseous leakage rates significantly higher than the critical leakage cut-off rate.

In the same publication, Morton determined the likelihood of liquid leakage across the test package seal. Test vials mounted on a manifold were filled with aqueous copper sulfate solution, and immersed, closure-end down, into distilled water. After pressurization for 15 minutes at 3 psig, copper sulfate presence in the water was measured by atomic absorption. The test was calculated to be capable of detecting as little as 0.1 μL of copper solution in the immersion water. No packages of gas leak rates less than 10^{-5} Pa · m^3 · sec^{-1} demonstrated microbial or liquid tracer leakage. Interestingly, liquid passage occurred for every package exhibiting gas leakage at or above this rate limit, while microbial leakage only occurred sporadically, with the number of colony forming units moving across the seal bearing no relation to the gas flow rate.

Later, Kirsch et al. worked to correlate helium leakage rate to the probability of microbial ingress using a liquid challenge media (9,10). While Morton et al. investigated leakage across a vial/closure compression seal, Kirsch and team studied leakage through glass micropipettes of various sizes imbedded in the walls of glass vials. A population of vials containing leak paths ranging in nominal diameter from 0.1 to 10 μm was flooded with helium and subsequently tested for helium leak rate using a mass spectrometry leak rate detector. These same vials were then filled with sterile media, and immersion challenged for 24 hours at 35°C with a saline lactose suspension of 10^8 to 10^{10} colony forming units of *B. diminuta* (*Brevundimonas diminuta*) and *Escherichia coli*, additionally spiked with magnesium ion tracer. Prior to challenge, test vials were thermally treated to eliminate airlocks within the micropipette lumen and establish a liquid path between the microbial challenge media and the test units' contents. After immersion, test vials were incubated at 35°C for an additional 13 days.

Kirsch's results showed that microbial ingress probability decreased as hole size and helium leakage rate decreased. Yet even under such extreme challenge conditions, only 3 of 66 test vials with log leak rates less than −4.5 std cc/sec failed the microbial challenge, consistent with the vial/closure interface leakage results reported by Morton. The probability of microbial ingress dramatically dropped from over 60% to about 10% within the helium log leak rate range of log −3.8 to −4.5 std cc/sec, which roughly corresponds to a leak nominal diameter of 0.4 to 1.0 μm. No ingress occurred through holes of helium leakage rate between 10^{-5} and $10^{-5.8}$ std cc/sec.

Because the immersion challenge media also contained magnesium ion tracer, Kirsch was able to use this same body of research to explore the relationship between liquid leakage, verified by the presence of magnesium in the packages, and the likelihood of microbial ingress (11). He concluded that both liquid leakage and microbial ingress are probabilistic occurrences. For any given leak, liquid passage was more likely to occur than microbial ingress. However, even at relatively large gas leak rates greater than 10^{-4} std cc/sec liquid leakage at times failed to occur. Microbial ingress only occurred when liquid leakage was also present, but liquid leakage did not guarantee microbial ingress. Thus, it was concluded that microbial ingress through a leak rated at $<10^{-2}$ std cc/sec requires liquid penetration through the leak path. And liquid leakage likely depends on variables such as liquid surface tension, defect diameter, leak morphology, leak surface conditions, environmental contaminants blocking the leak, and procedural technique.

Thus, the work by Kirsch et al., backed by the results of Morton et al., is often cited to support a critical leak rate specification of anywhere from 10^{-5} to $10^{-5.8}$ std cc/sec helium leak rate, when measured at one atmosphere differential pressure and standard temperature conditions, for rigid, nonporous parenteral packages. In addition, both works support the supposition that where liquid passage is prevented, liquid-borne microbial ingress is also blocked.

Burrell et al. adapted the ISO 8362-2 Annex C dye ingress method (12) in order to compare this standard dye leak test with a liquid immersion microbial challenge test for vial package integrity determination (13). Positive controls were created by inserting polyimide-coated glass microtubes ranging in internal diameter from 2 to 75 μm through the elastomeric closures of 5 mL vial packages. The dye ingress method used a solution of 1% FD&C Red No. 40, analyzed in the test packages by spectrophotometry. The microbial challenge used included *E. coli* suspension ($\geq 10^8$ cfu/mL) in saline. Both dye ingress and microbial ingress tests included package immersion for 30 minutes at 22 inHg (75 kPa) vacuum, followed by rapid vacuum release, and 30 minutes of immersion at ambient pressure. There was no attempt to eliminate airlocks in the microtubes. Results showed the dye ingress test and the microbial challenge test were equally sensitive. Dye and microbial ingress occurred in at least half the units with microtubes 10 μm in diameter. No leakage of any kind was detected in packages with smaller defects (2 and 5 μm). All units of microtubes $\geq$ 20 μm demonstrated dye leakage and microbial ingress. Therefore, it was concluded that the ISO dye ingress method was equally sensitive to the specified microbial challenge test performed under identical challenge conditions, given a microtube leak path.

Keller and team published a study exploring the relationship between critical leak size and package sterility (14). In this case, aerosolized microbial challenge was used. Leaking package models were created using nickel microtubes, 7 mm long, with inner diameters of 2, 5, 7, 10, 20, and 50 μm, each placed through the elastomeric septa of a small glass cell encased in a glass water jacket. Negative controls utilized solid tubes. Sterilized test cells filled with nutrient broth were placed tube-end down in an aerosol chamber to ensure liquid broth contact with the microtube opening. Test cells were challenged with an aerosol of motile *Pseudomonas fragi* microorganisms (approximately 10^6 cfu/cm^3) during a 30 minute come-up period, followed by 5 minutes at static conditions. Special ports added to each test cell enabled simulated package exposure to various controlled pressure/vacuum/temperature conditions during the biochallenge. A randomized block design allowed independent measurement of each test variable's influence on test package sterility. Considering all test variables, results showed microbial ingress can occur through microtubes as small as 5 μm in diameter; 2 μm tubes and negative controls showed no growth in any case. Test conditions that promoted broth flow into or through the tubes correlated to higher risk of microbial ingress; the greater likelihood for liquid flow, the greater the sterility loss risk. For instance, static conditions in which no differential pressure was applied only triggered microbial ingress through 2 of 9 tubes, 50 μm wide. Factors that promote product liquid flow and therefore increase risk of packaged product sterility loss include defect size, liquid product surface tension, and the pressures imposed on the package during processing, distribution, and storage.

In conclusion, all studies described illustrate the probabilistic nature of microbial ingress through package defects. Microbial challenge tests require carefully designed and conducted procedures using relatively large test sample populations to support convincing conclusions.

Numerous studies have attempted to pinpoint the critical leak size that corresponds to risk of product sterility loss. Results vary, with some studies implicating leak paths as small as 0.2 μm, while others imply leaks 10 μm and larger. Regardless, and perhaps most importantly, all research shows that liquid presence in the smallest defects is a prerequisite for microbial entry. Therefore, research seems to encourage a shift away from direct correlation of a given leak test to microbial ingress, toward the comparison of a leak test method's ability to detect defects capable of liquid passage—a less probabilistic, more easily verified parameter.

Indirect Comparison of Microbial Ingress with Physicochemical Leak Tests

Two published works compared vacuum decay leak testers' ability to find leaks previously sized by helium mass spectrometry. As the same test package population had been previously tested for microbial ingress risk, these data were used to indirectly determine the ability of the vacuum decay testers of that day to detect such defects (15,16).

In the previous section, "Direct Comparison of Microbial Ingress with Physicochemical Leak Tests," those leak test methods capable of accurate and sensitive detection of liquid passage were shown to be more reliable and sensitive indicators of package sterility risk than microbial challenge tests performed under the same test conditions.

Another indirect comparison approach applicable for packages sealed under vacuum is based on the predicted flow of gas that would occur if a defect of a given size were present in an evacuated vial package. This concept is explained more fully under Test Methods Frequency Modulation Spectroscopy. Briefly, laminar gas flow theory can be used to predict the rise in pressure inside such an evacuated package, given leak paths of various widths. The change in pressure over time for such a package can be correlated to defect size, and therefore, indirectly correlated to microbial ingress risk.

LEAK TEST VALIDATION

Calibrated Leak Standards

Calibrated reference leak standards are an important validation protocol component when evaluating leak test methods that rely on tracer gas flow through leaks, for example, helium mass spectrometry. Calibrated physical leaks are designed to deliver tracer gas at a known flow rate. There are two main categories of such standard leaks: (*i*) reservoir leaks that contain their own tracer gas supply, and (*ii*) nonreservoir leaks that rely on tracer gas addition during testing. Calibrated gas leaks perform by one of two methods. Either the leakage rate depends on the permeation of specified materials by certain gases, or an orifice is present allowing specified gas flow rates under prescribed differential pressure conditions. Often tracer gas detection systems, such as helium mass spectrometry instruments, incorporate internal reference standards to verify test system functionality.

Other leak test instruments that rely on air movement for leak detection, for example, vacuum decay leak testers, may utilize either a calibrated variable rate flowmeter or a calibrated fixed size orifice to introduce air leakage into the test chamber during equipment qualification or start-up.

Whenever possible, leak test instrument performance should be challenged using such calibrated standards. The Nondestructive Testing Handbook, Volume 1 Leak Testing (17) is an excellent resource for precautions and limitations regarding calibrated leak usage. While calibrated leak standards provide valuable instrument functionality and sensitivity information, leak test method validation is not complete without studies verifying the method's ability to differentiate between known positive and negative control test packages.

Positive Control Test Packages

Proper leak test method validation requires a demonstration that the integrity test method can successfully detect leaks in positive control, with-leak packages. Often this seemingly simple and clear requirement has been misinterpreted. For example, a common misconception is that a media-filled package used for a growth promotion check in a microbial challenge test is equivalent to a positive control test sample. While a growth promotion test proves that the packaged media can support microbial growth, it does not prove that bacteria would or could enter the package through a leak path. Another false perception is that a gas leak calibration standard, for example, a fixed orifice or an airflow meter used to introduce air into a vacuum decay test chamber, satisfies the positive control test requirement. Certainly, such a tool is important during system qualification and test method development, as it correlates equipment response (pressure rise) to a known challenge (airflow rate). However, it does not prove that the method can detect leaks of various sizes or types at various locations on the package. Finally, a dye or liquid tracer ingress test's limit of detection is performed by challenging the inspector or the inspection system with test samples previously spiked with known amounts of tracer element or dye. However, like the other examples, this alone fails to guarantee that the liquid tracer challenge method will effectively drive detectable amounts of tracer liquid or dye into a test package through known leaks. A highly sensitive detection system able to find tracer element in parts per billion is useless if the immersion challenge used to drive the tracer into the package is ineffective.

Useful leak test methods must detect leaks present in the package itself. Therefore, leak test method development and validation should include a population of negative and positive control test samples. A positive control is a known with-leak test package. Conversely, a negative control package is one made using components meeting dimensional and quality specifications, optimally assembled according to standard procedures, believed to be void of leaks or defects that may cause leakage.

DEFECT TYPES

Simple ways commonly used to create positive control test samples involve inserting microtubes or needles through package walls, placing wires or film between sealing surfaces, or adhering thin metal plates with microholes over package surface openings. These types of defects are inexpensive, simple to create, and give a quick assessment of a leak test's capabilities. Microtubes, microholes, and needles have fixed diameters and lengths, therefore researchers expect test results to infer detectable leak path sizes. This assumption is based on ideal gas and mass flow equations. However, vapor condensation or airlocks in the smallest bore microtubes can block liquid leakage flow resulting in variable test results. Also, such positive controls do not truly represent defects most likely to occur in actual product packages. Liquid or microbial migration around or through an item foreign to the package (e.g., needle, film, microhole, or microtube) may be very different from leakage through an actual defect located in or between package components.

A study by Morrical et al. illustrated this very point, by comparing helium leakage and microbial ingress through two types of defects in glass vial packages (18). One defect type consisted of a laser-drilled microhole in a thin metal plate mounted on a holed stopper, capped on each test vial. Microholes ranged in diameter from 0.5 to 15 μm. The other leak type was a copper wire placed along the sealing surface between the elastomeric closure and the glass vial. Wire thicknesses ranged from 10 to 120 μm. Helium trace gas leakage was detected using mass spectrometry. The microbial challenge test included a suspension of *Serratia marcescens* ($\geq 10^8$ cfu/mL). Challenge conditions consisted of one hour at 0.4 bar vacuum followed by one hour at 0.4 bar overpressure. Both test methods showed different leakage behavior for the two positive control types. Helium leak rates through the microholes matched theoretical predictions for gas moving through an orifice, whereas helium flow rates through the wired samples displayed complex, less predictable, gas flow dynamics. Microbial ingress occurred in at least a portion of the samples with microholes $\geq$ 4 μm (helium leakage rate $\geq 6.1 \times 10^{-3}$ mb · L · sec^{-1}), while units with holes $\leq$ 2 μm ($\leq 1.4 \times 10^{-3}$ mb · L · sec^{-1}) saw no microbial leakage. Microbial challenge results for hand-capped vials with wire defects demonstrated microbial leakage for wire diameters $\geq$ 20 μm (helium leakage rate $\geq 2.2 \times 10^{-5}$ mb · L · sec^{-1}).

Foreign objects inserted into a package to create a leak path can provide useful and quick leak test method assessments. However, whenever possible, final test method validation should include positive control test samples with defects simulating actual leaks likely to occur. For example, typical vial package defects may include glass cracks or breaks, misaligned or misshapen closures, and poorly crimped seals. Therefore, positive controls may include glass vials with a laser-drilled hole to simulate vial breakage. Including defects positioned above and below the liquid fill level is important if the leak test method's performance is a function of liquid or gas presence in the leak path. Scoring the vial finish might represent another type of glass defect. Removing slices along a closure's sealing surface, or loosely capping parenteral vial packages can replicate closure and seal defects, respectively. Pouch or bag positive control samples might include pinholes, open seals, channeled or wrinkled seals, weak seals, "burned" seals, and seals with trapped product inclusions. Ophthalmic dropper bottle positive controls could include loose caps, missing or poorly inserted dropper tips, defective tips or caps, and pinholes in the bottle.

With the exception of laser-drilled hole defects, the positive controls described earlier will not necessarily provide information about the exact sizes of detectable leaks, but they will help define detectable leak locations and types. Risks inherent in this approach include the possibility that the leak test would not find all the nonhole positive controls, and that the irregularities in defects' shapes or sizes may not permit statistically sound method reliability and sensitivity assessments. Nevertheless, including such positive controls in leak test method feasibility and optimization studies can provide invaluable information on the method's capabilities. Knowing this may give insight into ways of limiting the occurrence of actual defects not readily found by the chosen leak test method.

DEFECT SIZES

Published studies using microtubes or other artificial means to create leaks have unfortunately resulted in an expectation that all leak test methods need to detect defects as small as 0.2 μm in diameter, otherwise, the test method cannot compare with microbial ingress.

The first problem with this premise is creating defects 0.2 μm in size. Experience says naturally occurring leaks in packages smaller than a few micrometers wide are extremely rare, if they occur at all. Also, defects are not hole shaped, but are complex tortuous paths. Microscopic imaging verifies that even laser-drilled holes through the walls of glass vials or syringes are really a convoluted matrix of capillaries and chambers. Companies that laser drill holes certify their size by comparing the rate of pressurized gas flow through each hole to flow rates through standard orifices in thin metal plates. Generally, the smallest possible laser-drilled holes through small volume glass or plastic containers are about 5 μm in nominal diameter; smaller holes are difficult to make and readily clog. The smallest feasible holes through flexible laminates or films vary anywhere from about 2 to 10 μm in diameter, depending on the material. Without a way of creating and sustaining holes sized below these practical limits, positive control test samples with smaller defects are not possible.

The other factor complicating this requirement is that even typical microbial ingress tests cannot find 0.2 μm defects. Microbial ingress tests by Kirsch et al. (10) only found submicron sized defects in a very small fraction of samples, under extreme challenge conditions, after meticulous measures to eliminate leak path plugs and airlocks. The risk of microbial ingress rose significantly for defects > 1 μm, exceeding 80% probability for defects about 5 μm, and approached 100% probability for 8 μm defects. All defects considered in this analysis were those already confirmed as allowing liquid passage. In the absence of liquid passage, no microbial ingress occurred with any size defect (10,11). Research by Burrell et al. linked microtube defects ≥ 10 μm to a significant chance of dye and microbial ingress (13), while Keller's work implicated microtube leaks ≥ 5 μm when challenged with aerosolized organisms (14). Morrical detected microbial ingress in a portion of vial packages topped with thin metal plates having microholes ≥ 4 μm (8).

Therefore, positive control leaks should be as small as reasonably possible, given the type of package, the package dimensions, and the materials of construction. Generally speaking, parenteral product-package positive control test units used for checking the lower limit of sensitivity of physicochemical leak test methods include defects ≥ 5 μm in diameter.

Often researchers incorrectly assume that leak test method sensitivity is defined by the smallest detectable defect sizes. In fact, leak test methods generally have a larger leak size detection limit as well. For example, helium leak detection methods able to find the smallest submicron leaks may fail to find gross leaks that permit helium escape prior to or even during initial vacuum pumpdown. Some helium mass spectrometry units are equipped with pressure sensors and timers to detect such gross leaks during this test cycle vacuum pumpdown phase. Positive control sample populations should therefore include larger defects as well as smallest defects to represent the full range of anticipated leak sizes. If the primary leak test method fails to find the full range of package types and sizes of concern, then a second method may prove valuable. For example, high voltage leak detection may prove successful for 100% leak detection of a liquid product contained in a flexible pouch. However, larger gaps in the seal causing package collapse and liquid loss may be missed by this leak test system, making it important to screen for such larger leaks via visual inspection, weight checks, or some other means. Ironically, while the pharmaceutical industry has focused on finding the smallest package defects with better technologies, product recalls triggered by gross package integrity failures continue to occur.

PACKAGE INTEGRITY TEST METHODS

Many leak test methods exist that are applicable to pharmaceutical package systems. The goal of this chapter is not to describe all such methods. Instead, a review is offered of a few of the most commonly used methods and those proven to be most valuable to the pharmaceutical industry. These methods are presented in alphabetical order.

Bubble Tests

A bubble test is performed by immersing the package in water, drawing a vacuum, and observing for bubbles. Alternatively, a pressure source can be inserted into the package, allowing package pressurization during immersion. Immersion fluid surfactants improve method sensitivity. Bubble tests are quick and useful for leak presence and location confirmation in a laboratory setting. Smallest leaks may be missed if leaking gas dissolution rate in the immersion fluid is faster than bubble formation rate. Trapped gas on package seal surfaces may be confused for leaks. A common mistake when testing flexible packages is to fail to restrict test package volume, allowing package ballooning or expansion during vacuum exposure. Expansion will cause a drop in internal package pressure, eliminating the differential pressure necessary for bubbling to occur. A bubble test is a very useful forensic testing tool, but because it is destructive and test results are variable, it should not be used to access finished product quality.

Dye or Liquid Tracer Tests

A liquid tracer leak test consists of immersing test packages in a solution of either dye or other chemical tracer, then allowing time for liquid to migrate through any leaks present while pressure and/or vacuum are applied. After the liquid challenge, test packages' contents are checked for liquid leakage either visually or by using an appropriate analytical method. Liquid tracer leak tests are relatively inexpensive, simple to perform and conceptually easy to understand. However, the test is destructive to the package, and results may vary considerably. Dye or liquid tracer tests are inappropriate for testing product that may enter the market or clinic due to the risk of product contamination incurred by the method.

Test method parameters that promote greater liquid tracer test sensitivity include longer immersion times, increased pressure and vacuum conditions, smaller volumes inside the test package, and lower surface tension challenge liquids. Debris, airlocks, and event clogs of proteinaceous product may easily hamper leak path liquid migration. Restraining package part movement (e.g., prefilled syringes), or package expansion (e.g., flexible pouches) during vacuum exposure helps keep package internal pressure constant, thus ensuring consistent leakage driving forces.

Method development requires verification of dye or tracer compatibility with the package and its contents. Methylene blue is commonly used, but other chemicals specifically chosen for product compatibility are acceptable (19). Dyes may quickly fade or adsorb onto package surfaces shortly after leak testing; therefore, time gaps between testing and inspection or analysis should be limited and specified. Any dye or tracer detection method also requires validation. Human inspection is considered less reliable than analytical detection techniques. For the best visual inspection results, use qualified inspectors trained to follow defined inspection procedures in well-lit, controlled inspection environments. Inspection procedures should dictate lighting intensity and color, inspection angle, background color(s), background luster, inspection pacing, and any comparator negative control package(s) used. Inspector qualification protocols should entail accurate segregation of packages containing trace amounts of dye from negative controls in a randomly mixed, blinded test sample population. A multisite study led by Wolf demonstrated how differences in inspector capabilities and inspection environments play a significant role in interpreting dye ingress test results (20).

Numerous published leak test studies incorporate dye or liquid tracer test methods (8,11,13). U.S. compendia (21), EU compendia (22), and ISO international standards (12) all specify methylene blue dye ingress tests for demonstrating punctured closure reseal properties. But before using such closure reseal methods for whole package integrity testing, test parameters should be optimized and the methods validated using known positive and negative control packages. The importance of this was demonstrated in the study by Wolf et al., in which 1-mL water-filled syringes with laser-drilled defects in the barrel wall ranging in nominal diameter from 5 to 15 μm were leak tested according to the closure resealability dye ingress tests described in the U.S. and EU compendia and in ISO standards. None of these standard test methods permitted accurate identification of all defective syringes (20).

Electrical Conductivity Tests

The electrical conductivity leak test, also termed high voltage leak detection (HVLD), attempts to pass a high-frequency high voltage electrical current from an electrode positioned near the test package to a ground wire positioned at the far end of the package opposite the probe. Test packages made of plastic, glass, or elastomer are relatively resistant to current (i.e., insulating or nonconductive), and so allow minimal current to pass from electrode to ground—approximately 1 to 4 volts. If, however, a package leak is present near the electrode, with liquid product relatively conductive at or near the leak, a spike in measured current passing through the package will occur.

Möll et al. described test method development and validation of an electrical conductivity test used for gel-filled low-density polyethylene ampoules (23). Positive controls consisted of ampoules with laser-drilled holes positioned at the most likely zones for leaks to occur: the sealing zone at the ampoule bottom, and the top tear-off area. The voltage setting and the sensitivity or "gain" setting were the two parameters optimized to establish a window of operation that finds all defective ampoules and rejects few, if any, good ampoules. Replicate testing of a randomized population of negative and positive control test samples took place over three days. On each day of operation, the HVLD test successfully "failed" all 210 positive control ampoules (150: 5–10 μm; 60: 10–20 μm) and "passed" 3830 negative controls. A dye ingress test confirmed the presence of defects in two of three so-called negative controls consistently rejected by HVLD. Therefore, the electrical conductivity test correctly identified all defective units and falsely rejected only one negative control sample.

Recent studies have shown that HVLD is able to detect loosely capped stoppered vial packages, despite the absence of package component defects (24). The same work showed the method's ability to defects clogged with proteinaceous active compound, defects not detected by the vacuum decay method. In addition, multiple exposures to HVLD tests had no deleterious effect on three proteinaceous active substances, although additional testing would be required to fully qualify the method's product compatibility.

For obvious reasons, electrical conductivity is not appropriate for testing flammable liquid products. In addition, only leak paths near detectors are identifiable; therefore, either package surfaces are checked using multiple detectors or only the areas of greatest risk for leakage are monitored. Package rotation during testing may be required in order to capture defects around a package's circumference. Test method validation for a given product-package requires demonstration of the test's ability to detect leaks at all likely package locations.

Given HVLD's ability to rapidly and cleanly test a wide variety of product-package systems for the smallest leaks, this method's use is expected to expand in the future.

Frequency Modulation Spectroscopy

Frequency modulated spectroscopy (FMS) is a rapid, nondestructive analytical method suitable for monitoring oxygen and water vapor concentrations as well as evacuated pressure levels in the headspace of sterile product containers. Over the last 10 years, the technology has found commercial application in the pharmaceutical industry for leak detection (25), moisture monitoring (26), and oxygen monitoring (27). Systems for rapid nondestructive headspace analysis were first introduced to the pharmaceutical industry in 2000 (28), and are now routinely used in product development, process development, and commercial manufacturing.

The key to these test systems are diode laser devices fabricated to emit wavelengths in the red and near infrared regions of the electromagnetic spectrum where molecules such as oxygen and moisture absorb light. Containers made of glass (amber or colorless) as well as translucent plastics allow the transmission of near infrared (IR) diode laser light and are compatible with FMS test methods.

The underlying principle of laser absorption spectroscopy is that the amount of light absorbed by a molecule at a particular wavelength is proportional to the gas concentration and the gas pressure. Therefore, FMS technology works by tuning the wavelength of light to match the internal absorption wavelength of a molecule and recovering a signal where the amplitude is linearly proportional to gas density (e.g., headspace oxygen and moisture) and the signal width is linearly proportional to gas pressure (e.g., vacuum level in the headspace of a sealed vial). Briefly, laser passes through the gas headspace region of a sealed package; light is absorbed as a function of gas concentration and pressure; the absorption information is processed using phase-sensitive detection techniques; a mixer demodulates the radio frequency signal; the output voltage, proportional to the absorption lineshape, is digitally converted and further analyzed by a microprocessor, yielding final test results. Demodulated absorption signals can be used to accurately measure package headspace oxygen content, moisture vapor content, and total pressure. In general, measurements of higher headspace pressure require higher levels of moisture in the vial headspace.

A variety of diode laser-based system configurations can accommodate process monitoring and control and/or inspection of individual containers for oxygen, moisture, or vacuum. Lighthouse Instruments, Inc., of Charlottesville, Virginia provides benchtop systems for laboratory use, as well as at-line, fully automated systems for 100% monitoring, control, and inspection. Typical measurement times can be varied from 0.1 to 1 second corresponding to line speed throughput of 60 to 600 vials per minute. Maximum machine speeds will depend on the details of a particular application. Key parameters that impact maximum speed are container diameter and reject specification. Both faster speeds and smaller diameter packages increase measurement standard deviation.

Test systems are calibrated using National Institutes of Standards and Technology (NIST) traceable standards of known gas concentration or pressure. Standards are constructed from the same containers used to package the pharmaceutical product, so that calibration represents containers identical to the test sample containers. For example, an oxygen-monitoring instrument would utilize standards of known oxygen concentration in containers of the same type and diameter as test sample containers. Datasets of standards measurements versus certified values enable calibration constant or calibration function generation. Subsequent measurements of unknown samples use this calibration information to convert measured absorption signals into meaningful values of headspace gas concentration and/or gas pressure. System measurement performance is demonstrated by repeatedly testing a set of gas or pressure standards.

FMS offers invaluable insight for monitoring and controlling aseptic manufacturing processes. Oxygen-sensitive products typically require an inert gas headspace, and lyophilized products often require either vacuum or inert gas headspace. Using FMS for nondestructive testing of all such packages immediately post assembly guarantees the presence of the inert gas or vacuum content, and permits efficient culling of product not meeting specifications.

By testing sealed product at a later time post packaging, FMS technology can also verify container–closure integrity, or absence of leakage. In the case of product sealed with an inert gas overlay, oxygen leakage into the container will be a function of diffusive flow, driven by the greater oxygen partial pressure outside the container. Following Fick's laws of diffusion, given the test package volume and the initial oxygen partial pressure at the time of package assembly, and an assumed leak path length equivalent to the package wall depth, oxygen ingress as a function of time can be predicted if leaks were present. Caution is advised, however, when attempting to predict package integrity for periods longer than a few days according to diffusion kinetics. Over time, packages are exposed to pressure differentials from changes in altitude or weather, or even by doors opening and closing, all of which drive faster, convective flux leakage, thus complicating such projections.

Another scenario in which FMS methods can predict leakage includes a vial containing lyophilized product stoppered under vacuum. In this case, the differential pressure between the evacuated container and the atmosphere will drive air into the package according to either molecular or laminar flow kinetics, depending on the leak path diameter, the mean free path length of the leaking gas, and the package initial internal pressure. Vacuum loss predicted for such a package given various theoretical leak path sizes can be compared with measured vacuum loss by FMS. If the actual FMS measured vacuum loss is less than predicted given leak presence, then package integrity can be assured. Therefore, FMS spectroscopy is a reliable and sensitive approach for rapidly and nondestructively verifying the integrity of every evacuated container unit both upon package sealing and as a function of stability.

The reader is advised to consult other more detailed references for more information on how to predict package headspace loss, and how to interpret these predictions given headspace content analysis results using FMS technology (29).

Helium Mass Spectrometry

Helium mass spectrometry is a type of tracer gas detection method, in which helium leakage is detected by mass spectrometry. Helium may be introduced into the test package by either flooding the package with helium prior to final assembly, injecting helium into the assembled package (requiring injection site sealing), or pressurizing the package with helium, driving helium into the package.

The helium-filled test package is then placed inside a hermetically sealed test chamber, a vacuum is drawn, and the rate of helium leakage is quantitatively measured. Alternatively, the test package may be scanned with a sniffer probe connected to the mass spectrometer. The sniffer probe method allows for leak location detection; however, leak rate determination is less precise. The sniffer probe technique is also useful for packages that cannot tolerate high vacuum test conditions.

Helium mass spectrometer leak detection is widely used, including automotive, electronic, refrigeration, and medical device industries. Numerous texts, American Society for Testing and Materials (ASTM) standards, and technical resources exist describing specific applications. A similar trace gas method based on hydrogen gas detection is also popular, especially in lieu of helium gas cost and availability. These tracer gas detection methods are the most sensitive of all leak detection methods, making them invaluable in package design, development, and assembly optimization work, as well as leakage forensics studies.

But in order to rely on the quantitative test results obtained, tracer gas content within the package at the time of testing must be known. For example, if 50% of the package headspace consists of tracer gas, the helium leak rate results will be half the true package leak rate. Loss of tracer gas can occur quickly if package assembly post flooding is not immediate or optimized, if injection sites post helium addition are insufficiently sealed, or if the package has leaks that allow gas loss prior to testing. Some leak testers have a programmable gross leak detection mode option, in which the rate of pressure drop during the initial test evacuation phase is monitored; longer evacuation times caused by gross leaks trigger the test to abort. Another approach includes test package headspace analysis using a nondestructive technique such as FMS, immediately prior to helium leak testing. True package leak rate is calculated by multiplying the mass spectrometer measured helium leak rate by the helium headspace content fraction determined by FMS.

Microbial Challenge Tests

A microbial challenge test procedure includes filling containers with either growth-supporting media or product, followed by closed container immersion in a bacterial suspension or exposure to aerosolized bacteria or bacterial spores. Test containers are incubated at conditions that promote microbial growth, and container contents are then inspected for evidence of microbial growth. Very simply, positive challenge organism growth is indicative of package leakage.

For decades, the pharmaceutical industry has relied upon microbial challenge methods for validating package integrity. It seems logical to use a microbial challenge method to prove a package's ability to preserve product sterility. Indeed, there are situations where it is valuable to determine the ability of a package design or seal to prevent actual microbial ingress. Microbial challenge tests are one of the few appropriate tests for integrity verification of porous barrier materials and tortuous path closure systems.

But for the most part, one is cautioned against relying solely on microbial challenge methods for package integrity verification. Leak paths several fold wider than a microorganism will not guarantee microbial ingress, as numerous studies have shown (8,10,13,14). On the other hand, the rare occurrence of microbial grow-through across a package's fitted seam during an exceptionally severe biochallenge may negate the use of an otherwise acceptable container–closure system, even though such a challenge does not realistically portray naturally occurring phenomena. Conversely, inappropriately designed microbial challenge tests can easily make bad packages look good. Short exposure times, minimal or no differential pressure application, small test sample populations, and positive control packages with very large leaks all help samples with questionable seals pass a microbial challenge test, thereby falsely implying package integrity. In some cases, reliance on such tests has kept leery companies from adopting more reliable, physicochemical leak test methods, despite suspected product-package integrity problems. Finally, because microbial challenge tests are destructive, they cannot give any indication of package integrity for actual marketed product packages.

Currently, no standard microbial challenge test method exists. The following discussion explores factors to consider when designing a microbial challenge test.

Challenge Mode

If a package is able to tolerate liquid immersion, then this approach is generally favored for parenteral package system testing, as it presents the greatest challenge to package seals. Aerosol challenge testing is most appropriate for packages that rely on tortuous paths, or seals not intended to prevent liquid leakage. Food and medical device industries often prefer aerosolized challenge testing. Static testing, where packages filled with media are simply stored in normal warehouse conditions or in stability storage chambers, affords no definitive bacterial challenge and no significant pressure differential to the seals. If such long-term storage of media-filled units is part of an integrity verification program, then some known bacterial challenge to the packages at the end of the storage period is appropriate.

Challenge Parameters

Liquid immersion challenge tests preferably include vacuum/pressure cycling simulating pressure variations anticipated during product life processing, distribution, and storage. These cycles will enhance flow of packaged media into any leak paths present, thus encouraging potential microbial ingress. For this reason, package position during the challenge test should ensure packaged media contact with seal areas. An aerosol challenge test chamber size and design should guarantee uniform distribution of viable aerosolized bacteria or spores around the test packages, considering factors such as chamber temperature and humidity, as well as airflow patterns and speed.

Challenge Microorganism

Liquid challenge organism size, mobility, and viability in the packaged media are important factors for consideration. Bacteria concentration in the challenge media at the initial time point should ensure a high concentration of viable organisms at the test's conclusion (e.g., $\geq 10^5$ cfu/mL at end of test). Bacteria used in published immersion challenge studies include, but are not limited to, *E. coli*, *S. marcescens*, *Clostridium sporogenes*, *P. aeruginosa*, *Staphylococcus epidermidis*, *Brevundimonas diminuta*. When performing aerosol challenge tests, aerosolized microorganism concentration and uniformity are important factors, as well as viability in the packaged media. Reportedly, aerosol challenge testing commonly uses *Bacillus atrophaeus* spores and *P. fragi* microbes.

Growth Promotion Media

All challenge tests require test containers filled either with growth-promoting media or product that supports microbial growth. The product formulation itself or a product placebo is preferred as it most closely simulates the product-package system. However, this may be impractical when validating a variety of products that use similar packaging. Verification of the media's growth-promotion capability at the completion of the package integrity test is important, especially if the test sample holding time is lengthy.

Test Package Preparation

Two approaches are possible for preparing sterile packages for testing. Either previously sterilized package components are aseptically filled with the growth-promoting vehicle, or media-filled packages are terminally sterilized. If feasible, the sterilization procedures and package assembly processes chosen should mirror those used for the actual product. Otherwise, the test package and seal may differ in some respect from the marketed product-package system. For example, vial package capped closures exhibit a certain amount of sealing force on the vial land seal surface. This residual seal force will noticeably decay upon terminal steam sterilization, thus potentially changing the seal quality (30,31). Similarly, plastic bag test samples exposed to gamma irradiation post heat sealing may not represent product bags normally sealed using ethylene oxide sterilized materials.

Microbial Growth Verification

Microbial growth as evidenced by cloudiness in the package may be detected visually or with instrumentation. In the case of product-filled packages, verification of nonsterility may require aseptic filtration and filter plating for microorganism identification. Any nonsterile package contaminants are generally identified to verify the challenge microorganism as the source of contamination.

Test Package Population Size

There is no guarantee of microbial ingress even in the presence of relatively large defects. Microbial ingress is a notoriously probabilistic phenomenon. For this reason, a valid test requires a relatively large population of test samples and positive controls.

Positive and Negative Controls

All leak test validation protocols, including microbial challenge tests, require positive control or known-leaking packaging in the test package population to demonstrate the test's leak detection ability. Because even significant leak pathways will not always demonstrate microbial leakage, a large database of samples is needed to minimize the risk of false–negative results. Despite the best efforts, microbial challenge tests may yield erratic results that do not reliably correlate to leak size or presence.

Residual Seal Force

Residual seal force (RSF) is not a leak test method, but it is included in this discussion since compendial and regulatory guidances reference RSF as a package integrity test method option, and because RSF is a valuable tool in parenteral vial package assembly optimization and verification.

RSF is defined as the compression force exerted by an elastomeric stopper or closure on the sealing surface of a container, typically a parenteral glass or plastic vial. This compressive force ensures package integrity at the stopper/vial interface. RSF is established when the stopper is crimped onto the vial finish, and is a function of elastomeric viscoelastic properties, capping machine head pressure, package component stack height dimensions, and aluminum seal skirt length. Because closures are viscoelastic in nature, the RSF will decrease somewhat as a function of time, processing procedures, and elastomer composition (30,31).

RSF values can be determined indirectly using a constant rate of strain stress tester, also called a universal tester. Genesis Machinery Products, Inc. markets automated residual seal force tester (ARSFT) that works according to the same principle. To perform a RSF test, a specially designed aluminum cap is placed on top of the sealed vial and placed on a compression load cell of the universal tester or ARSFT. The vial is then slowly compressed at constant rate of strain, and a stress-deformation response curve is generated. The RSF is the force where the slope of the curve demonstrates a noticeable decrease; in other words, the second derivative reached a maximum value.

Published research by Morton et al. first described RSF measurement using a universal tester and demonstrated the method's usefulness. RSF measurements were proven sensitive to differences in elastomeric formulation, elastomeric stopper manufacturing lots, time post capping, and exposure to terminal steam sterilization (30,31). Lower RSF values correlated to increased package leakage rate measured by a pressure decay method, a bubble test, a liquid tracer test, and a microbial challenge test (8).

Later, a team of researchers lead by Ludwig optimized the RSF method by modifying the metal cap anvil placed on top of a vial (32,33). Rounding the top of the metal cap anvil helped to make a more uniform compression of slightly imperfect vials and making the cap fit more tightly helped improve centering of the cap anvil onto the vial.

As previously mentioned, RSF itself is not a leak test method. RSF alone is insufficient for parenteral package integrity assurance. A package capped at optimum RSF may leak if the vial or stopper is defective. Conversely, research has shown that a vial package assembled using defect-free components may leak if poorly capped, as measured by RSF (24). RSF measurements are important when optimizing and verifying capping machine setup. No other method, including visual inspection or manually twisting the capped vial to determine "tightness" has proven reliable.

Vacuum Decay Leak Tests

Vacuum decay is a whole package, nondestructive leak test method that has grown in popularity over the past 20 years. Today's test systems range from small benchtop laboratory instruments to production scale 100% fully automated on-line machines. In a typical test cycle, the test package is placed in a test chamber, the chamber is closed, and vacuum is rapidly drawn to a target pressure level within an allotted time segment. After establishing vacuum, the test system is isolated from the vacuum source, and any subsequent pressure rise (vacuum decay) inside the test chamber is monitored. Pressure rise above baseline, or background noise level, signifies package headspace gas leakage, and/or vaporization of product liquid plugging leak path(s). Total test cycle time is normally less than 30 seconds, but may vary with the test system, the product-package tested, and the desired sensitivity level.

The test equipment, package test chamber, and testing cycle are unique to each product-package system, and are specified based on the package's contents (liquid or solid, with significant or little gas headspace), package morphology (flexible or rigid, porous or nonporous), and package size. Uniquely designed test chambers snugly enclose the test package, minimizing test chamber deadspace for maximum test sensitivity. Added features may be required to limit package movement or expansion during the test (e.g., prefilled syringes, flexible bags, or pouches), or to mask gas flow through porous barrier materials (e.g., paper or Tyvek®) (34). Test systems are configured either to detect leaks in packages filled with liquids, or to detect leak paths in packages containing gases and/or solids. Leak paths that risk liquid clogging, or "liquid leaks," require higher vacuum test conditions below the liquid's vaporization pressure, so that vaporized liquid yields a measurable rise in pressure. On the other hand, "gas leaks" are detectable at less severe vacuum settings.

Vacuum decay leak tester designs vary among instrument manufacturers. While most models rely on a single 1000 Torr gauge or absolute transducer, some instruments use a dual transducer system with either a 1000 Torr gauge or absolute transducer coupled with a more sensitive, higher resolution 10 Torr gauge transducer. One manufacturer that relies on the single-gauge transducer approach also incorporates special software that continually readjusts the no-leak baseline to account for atmospheric pressure changes and no-leak noise variations that can affect test sensitivity. Another vendor is able to eliminate atmospheric pressure variation concerns and the need for calculated baseline adjustments by utilizing an absolute pressure transducer as part of their dual transducer test system (35).

Test method development and instrument functionality checks often utilize either a calibrated fixed orifice leak or a calibrated airflow meter for artificially introducing leaks into the test chamber containing a negative control no-leak package. Airflow meters certified by the NIST or other recognized certification bodies are recommended for such purposes. The smallest airflow rate that triggers a rise in pressure above background noise level is the limit of detection for the leak test.

However, use of calibrated airflow standards alone is not sufficient for complete test method development and validation. Positive control, with-leak packages should be used as well, in order to best understand how test chamber pressure rise compares for various sized leak paths positioned in various package locations. For example, a gross leak in a package with minimal gas headspace volume may not be detected if the time allotted for initial vacuum is so long that all headspace gas is evacuated prior to the start of the pressure rise test phase. In another example, a plastic bottle with a pinhole-size leak in the induction seal, beneath the torqued screw thread cap may require additional time to draw out trapped air in the cap's threads, before leakage through the induction seal hole is observed. Further, leaks simulated using a calibrated flowmeter only represent gaseous leakage and not leakage from liquid-plugged leak paths. Generally, liquids clogging leaks quickly volatilize once test pressure falls below the liquid's vaporization pressure, triggering a rapid rise in test system pressure. Pressure rise quickly stops and perhaps fluctuates once the vaporized liquid's saturation partial pressure is reached. This difference in leak behavior between so-called liquid versus gas leaks often requires different testing parameters. Additionally, test system cleaning procedures should be in place in anticipation of test equipment contamination from liquid-filled leaking containers. Negative control, no-leak packages may be solid material, package-shaped models, but at some point, larger test populations of actual, filled, no-leak packages are important to ensure the validated baseline represents all possible package-to-package variations.

In the late 1990s, the functionality of Wilco AG vacuum decay leak test systems was explored by Kirsch, Morton, and a team of researchers from Wilco in two published research studies. For both studies, test samples consisted of glass vials with micropipettes affixed into the glass vials to simulate leaks. Test package leakage was quantified using helium mass spectrometry, a leak test method previously compared with liquid-borne microbial challenge tests. In the first study, air-filled vials were vacuum decay leak tested (15). The second study evaluated vials filled with various solvents that plugged the leak paths using an "liquid-filled container (LFC)" pressure rise or vacuum decay approach. This concept required the test pressure to be substantially lower than the vapor pressure of the packaged liquid (16). LFC method test results indicated potentially greater sensitivity when testing liquid-filled vials.

ASTM F2338-09 standard test method for nondestructive detection of leaks in packages by vacuum decay method (36) is a recognized consensus standard by the United States FDA, Center for Devices and Radiological Health (CDRH), effective March 31, 2006 (37). According to the FDA Consensus Standard Recognition Notice, devices that are affected include any devices that are sterilized and packaged. Packages that may be nondestructively tested by this method include: rigid and semirigid nonlidded trays; trays or cups sealed with porous barrier lidding materials; rigid, nonporous packages; and flexible, nonporous packages.

The ASTM method includes precision and bias (P&B) statements for various types of packages based on round robin studies performed at multiple test sites with multiple instruments. P&B studies have looked at porous lidded plastic trays, unlidded trays, and induction-sealed plastic bottles with screw caps. The most recent P&B studies used glass prefilled syringes; a publication fully describing this work appeared in 2009 (35). Test packages included empty syringes, simulating gas leaks; and water-filled syringes, simulating leaks plugged with liquid (liquid leaks). Laser-drilled holes in the syringes' glass barrel walls ranging from 5 to 15 μm in nominal diameter served as positive control leaks. The leak testers used incorporated a 1000 Torr absolute transducer coupled with a 10 Torr differential transducer, manufactured by Packaging Technologies & Inspection, LLC of Tuckahoe, New York. Two different test cycles were explored; one with a target vacuum of 250 mbar absolute for testing gas leaks only, and another with a target vacuum of about 1 mbar absolute for testing both gas and liquid leaks. P&B study results showed the leak tests reliably identified holes as small as 5 μm in both air-filled and water-filled syringes.

More recent research described in a public forum indicates that vacuum decay is at times limited in its ability to detect leaks in packages containing proteinaceous liquid products. Proteinaceous active in an aqueous formulation irreversibly clogged a large percentage of glass vial laser-drilled holes making their detection by vacuum decay impossible (24). This phenomena plus the challenge of cleaning test chambers contaminated from liquid leakage are disadvantages that should be considered prior to leak test method selection.

Weight Loss or Gain

Product-package weight change as a function of time and temperature is a practical integrity method that can be readily incorporated as part of product stability studies. This technique is especially useful for semipermeable packages containing volatile products or products prone to moisture sorption.

CONCLUSION

In recent years, regulatory bodies have encouraged the pharmaceutical industry to explore novel package integrity test methods that do not rely on traditional dye or microbial ingress tests. This movement has driven improvements in leak testing technologies and sparked exciting new developments. Today, rapid, sensitive, and nondestructive leak test methods exist for testing most pharmaceutical parenteral product-package systems. Vacuum decay is primarily useful for detecting leaks in packages having gas headspace, but unique applications exist for testing liquid-filled container–closures as well. Electrical conductivity or HVLD tests have demonstrated great potential for testing a large portion of liquid-filled packages. FMS is ideal for integrity testing clear or translucent containers sealed under low pressure or with an inert gas headspace. Other techniques exist that are important tools for laboratory use in package development and forensics testing, including helium mass spectrometry, dye or liquid tracer ingress, bubble tests, and weight change, checks for seal quality assurance, for example, RSF, are vital as well. These and other methods likely on the horizon provide a leak test method arsenal that can help ensure better quality products, with fewer recalls linked to container–closure integrity failures.

REFERENCES

1. Department of Health and Human Services. Submission Documentation for Sterilization Process Validation in Applications for Human and Veterinary Drug Products. Guidance for Industry. Rockville, MD: U.S. Food and Drug Administration, Center for Drug Evaluation and Research (CDER), Center for Veterinary Medicine CVM), 1994.
2. Department of Health and Human Services. Container Closure Systems for Packaging Human Drugs and Biologics, CMC Documentation, Guidance for Industry. Rockville, MD: U.S. Food and Drug Administration, Center for Drug Evaluation and Research (CDER) and Center for Biologics Evaluation and Research (CBER), 1999.
3. Department of Health and Human Services. Container and Closure System Integrity Testing in Lieu of Sterility Testing as a Component of the Stability Protocol for Sterile Products. Guidance for Industry. Rockville, MD: U.S. Food and Drug Administration, Center for Biologics Evaluation and Research (CBER), Center for Drug Evaluation and Research (CDER), Center for Devices and Radiological Health (CDRH), Center for Veterinary Medicine (CVM), 2008.
4. EudraLex. The Rules Governing Medicinal Products in the European Union, Volume 4. EU Guidelines to Good Manufacturing Practice, Medicinal Products for Human and Veterinary Use, Annex 1 Manufacture of Sterile Medicinal Products. Brussels, Belgium: European Union, European Commission, DG Enterprise and Industry—Pharmaceuticals Unit, 2008.
5. Department of Health and Human Services. Sterile Drug Products Produced by Aseptic Processing—Current Good Manufacturing Practice. Guidance for Industry. Rockville, MD: U.S. Food and Drug Administration, Center for Drug Evaluation and Research (CDER), Center for Biologics Evaluation and Research (CBER), Office of Regulatory Affairs (ORA), 2004.
6. PDA. Task Force on Container/Closure Integrity, Pharmaceutical Package Integrity. Technical Report 27. PDA J Pharm Sci Technol 1998; 52(suppl): 1–48.
7. Morton DK, Lordi NG, Ambrosio TJ. Quantitative and mechanistic measurements of parenteral vial container/closure integrity. Leakage quantitation. J Parenter Sci Technol 1989; 43:88–97.
8. Morton DK, Lordi NG, Troutman LH, et al. Quantitative and mechanistic measurements of container/closure integrity: Bubble, liquid, and microbial leakage tests. J Parenter Sci Technol 1989; 43:104–108.

9. Kirsch L, Nguyen L, Moeckley C. Pharmaceutical container/closure integrity I: Mass spectrometry-based helium leak rate detection for rubber-stoppered glass vials. PDA J Pharm Sci Technol 1997; 51:187–194.
10. Kirsch L, Nguyen L, Moeckley C, et al. Pharmaceutical container/closure integrity II: The relationship between microbial ingress and helium leak rates in rubber-stoppered glass vials. PDA J Pharm Sci Technol 1997; 151:195–201.
11. Kirsch L. Pharmaceutical container/closure integrity VI: A report on the utility of liquid tracer methods for evaluating the microbial barrier properties of pharmaceutical packaging. PDA J Pharm Sci Technol 2000; 54:305–313.
12. ISO 8362–2. Injection Containers for Injectables and Accessories—Part 2: Closures for Injection Vials, Annex C—Test method for closure/container integrity and self-sealing. Switzerland: International Organization for Standardization, 1988.
13. Burrell L, Carver MW, DeMuth GE, et al. Development of a dye ingress method to assess container-closure integrity: Correlation to microbial ingress. PDA J Pharm Sci Technol 2000; 54:449–455.
14. Keller S, Marcy J, Blakistone B, et al. Application of fluid and statistical modeling to establish the leak size critical to package sterility. J Appl Pack Res 2006; 1:11–21.
15. Nguyen LT, Muangsiri W, Schiere R, et al. Pharmaceutical container/closure integrity IV: Development of an indirect correlation between vacuum decay leak measurement and microbial ingress. PDA J Pharm Sci Technol 1999; 53:211–216.
16. Kirsch LE, Nguyen LT, Kirsch AM, et al. Pharmaceutical container/closure integrity V: An evaluation of the WILCO "LFC" method for leak testing pharmaceutical glass-stoppered vials. PDA J Pharm Sci Technol 1999; 53:235–239.
17. Boeckmann MD, Sherlock CH, Tison SA. Calibrated reference leaks. In: Moore PO, Jackson CN Jr, Sherlock CN, eds. In Nondestructive Testing Handbook. 3rd ed. Vol 1. Columbus, OH: Leak Testing, American Society for Nondestructive Testing, Inc., 1998:71–100.
18. Morrical BD, Goverde M, Grausse J, et al. Leak testing in parenteral packaging: Establishment of direct correlation between helium leak rate measurements and microbial ingress for two different leak types. PDA J Pharm Sci Technol 2007; 61:226–236.
19. Jacobus R, Torralba P, Moldenhauer J, et al. Development and validation of a spectrophotometric dye immersion test method used to measure container-closure integrity of an oil-based product. PDA J Pharm Sci Technol 1998; 52:110–112.
20. Wolf H, Stauffer T, Chen SCY, et al. Vacuum decay container/closure integrity testing technology. Part 2. Comparison to dye ingress tests. PDA J Pharm Sci Technol 2009; 63:489–498.
21. General Chapter <381> Elastomeric Closures for Injection. United States Pharmacopeia 32. Rockville, MD: United States Pharmacopeial Convention, Inc., 2009.
22. General Notice 3.2.9 Rubber Closures for Containers for Aqueous Parenteral Preparations, for Powders and for Freeze-dried Powders. European Pharmacopoeia 6.0. France: European Directorate for the Quality of Medicines and Healthcare, 2008.
23. Möll F, Doyle DL, Haerer M, et al. Validation of a high voltage leak detector to use with pharmaceutical blow-fill-seal containers—A practical approach. PDA J Pharm Sci Technol 1998; 52:215–227.
24. Orosz S, Guazzo D. Presentation. Annual Meeting of the Parenteral Drug Association, 2010, Packaging Science Interest Group, Glass Vial Finish Defects—Leak Detection and Product Risk Assessment, March 16, Orlando, FL.
25. Lin TP, Hsu CC, Kabakoff BD, et al. Application of frequency-modulated spectroscopy in vacuum seal integrity testing of lyophilized biological products. PDA J Pharm Sci Technol 2004; 58: 106–115.
26. Mahajan R, Templeton AC, Reed RA, et al. Frequency modulation spectroscopy—A novel nondestructive approach for measuring moisture activity in pharmaceutical samples. Pharm Technol 2005; 10:44–61.
27. Templeton AC, Han YR, Mahajan R, et al. Rapid headspace oxygen analysis for pharmaceutical packaging applications. Pharm Technol 2002; 7:41–46.
28. Veale J. Presentation. Annual Meeting of the Parenteral Drug Association, 2001, Validation of a Tunable Diode Laser Spectrometer for Non-Destructive Monitoring of Oxygen in the Headspace of Parenteral Containers, March 16, Orlando, FL.
29. Veale J. New inspection developments. In: Lysfjord J, ed. Practical Aseptic Processing: Fill and Finish. Vol 1. Bethesda, MD: DHI Publishing, LLC, 2009:305–327.
30. Morton DK, Lordi NG. Residual seal force measurement of parenteral vials. I. Methodology. J Parenter Sci Technol 1988; 42:23–29.
31. Morton DK, Lordi NG. Residual seal force measurement of parenteral vials. II. Elastomer evaluation. J Parenter Sci Technol 1988; 42:57–61.

32. Ludwig J, Nolan P, Davis C. Automated method for determining Instron residual seal force of glass vial/rubber closure systems. J Parenteral Sci Technol 1993; 47:211–218.
33. Ludwig J, Davis C. Automated method for determining Instron residual seal force of glass vial/rubber closure systems Part II. 13-mm vials. J Parenteral Sci Technol 1995; 49:253–256.
34. PTI Inspection Systems, http://ptiusa.com. Accessed June 16, 2010.
35. Wolf H, Stauffer T, Chen SCY, et al. Vacuum decay container/closure integrity testing technology. Part 1. ASTM F2338–09 precision and bias studies. PDA J Pharm Sci Technol 2009; 63:472–488.
36. ASTM F2338–09 Standard Test Method for Nondestructive Detection of Leaks in Packages by Vacuum Decay Method. West Conshohocken, PA: ASTM International, 2009.
37. United States Federal Register Notice, FR Notice (list #014), Docket No. 2004 N-0226, March 31, 2006, 71 (62).

31 Administration of injectable drug products

This brief chapter will highlight the advantages and disadvantages of the injectable route of administration and provide some detail about the most common routes of injectable drug administration (1)

ADVANTAGES OF THE INJECTABLE ROUTE OF ADMINISTRATION

There is no question that if a person could choose between taking a drug product by mouth versus by injection, the oral route would always be chosen. Of course, this choice is not possible for many drug products and even if it were a choice, there are good reasons why the injectable route would be preferable.

Rapid Onset of Action

Administration of drugs by the intravenous or intra-arterial routes provides 100% immediate bioavailability. Administration by the intra-arterial route is rare, but drugs are 100% bioavailable by this route. Intravenous administration, while the best guarantee of 100% bioavailability, is not the quickest way to administer drugs in life-threatening emergency situations because of difficulties in finding veins in such situations. Subcutaneous or intramuscular routes are the easiest administration routes and, therefore, the quickest approach. However, both routes require a drug-absorption step, and so the drug is not immediately available nor 100% bioavailable compared to the intravenous route.

Drug Cannot Be Administered Orally Due to Inactivation and/or Low Bioavailability

Most products of biotechnology (peptides and proteins) and other molecules (e.g. many chemotherapeutic drugs) are destroyed or simply not absorbed if administered by mouth. Destruction of biomolecules and many chemotherapeutic agents will occur either in the gastrointestinal tract (low pH, gastric enzymes) or after absorption where drugs pass through the liver where they are metabolized to inactive forms (first-pass metabolism).

Direct Injection to Site of Action

The injectable route ensures delivery of adequate concentrations of a drug to the diseased tissues or target areas of the body. Examples include

- direct intraventricular injection of an aminoglycoside for patients suffering from bacterial or fungal meningitis and/or ventriculitis;
- intra-arterial injection of an oncolytic drug immediately upstream from a solid tumor where the drug can be directly delivered; and
- intra-articular injection of steroid suspensions for immediate and prolonged treatment of inflammation.

Patient Unable to Take Medication by Mouth

Patients who are aspirating or who have had the upper gastrointestinal tract stream diverted or removed (carcinoma) cannot take oral medications. This also applies to patients who are unconscious (e.g., narcotic abuse, under anesthesia), uncooperative (e.g., psychotic patients), or having uncontrollable seizures (e.g., epileptic patients).

Administration Controlled by Physician

The injectable route of administration enables the physician or other health-care professional to exert control over certain pharmacologic parameters such as time of drug onset, blood levels, tissue concentrations, and rate of elimination of the drug from the body. For example, intravenous or direct intracardiac injections are used in emergency situations such as life-threatening hypotension, hypertension, or arrhythmias.

Better Control of Outpatient Compliance

Some patients cannot be depended upon to take their medicine when required.

To Provide a Local Effect

Especially for cancer drugs, local administration of the drug will eliminate toxic reactions were the drug to be administered systemically. Intrathecal injections and other injections directly to an organ will avoid the systemic toxic effects that would occur were cancer drug administered intravenously.

To Permit Rapid Correction of Fluid and Electrolyte Imbalances and to Supply Short- or Long-Term Nutritional Needs to the Patient

Patients suffering from severe dehydration or a heat stroke or patients whose intestinal tracts have been resected are examples of situations where the injectable route of administration is the only way to provide drug therapy.

DISADVANTAGES OF THE INJECTABLE ROUTE OF ADMINISTRATION

Fear of Needles

This is by far the greatest disadvantage of parenteral administration. Few human beings, if given the choice between taking a medication by mouth (or for that matter any other route) or by injection, the choice is obvious. Fear of needles will prevent some people from complying to the proper dosage regimen for their medication(s). Fear of needles is a significant marketing strategy for vendors of needle-free devices (see Chapter 4).

Inevitable Discomfort

Depending on needle size and injection technique, injections may cause discomfort, even pain. Also, pain may follow the injection if the drug product is irritating to the tissues. Lack of isotonicity may cause discomfort, especially for injections other than intravenous. Injections having pH extremes (i.e., pH <4 or >9) may cause pain depending on rate of injection. If the formulation injected contains an organic co-solvent for drug solubilization purposes, some irritation or discomfort will take place, again depending on rate of injection and also the amount of co-solvent injected.

Cannot Recover the Drug Once Injected

If the wrong drug or wrong concentration or volume of medication is injected, there is little to no option for correcting the error. Medication errors, therefore, for injectable medications are especially dangerous because of this inability to recover whatever was injected. Medication errors include not only administering the wrong drug product, but also, and more commonly, administering the wrong dose of a particular drug product.

Requirements for Aseptic Techniques

Manufacturers are required by good manufacturing practices, strictly enforced by governmental regulatory authorities, to produce and maintain sterile dosage forms in the marketed package. Sterility must be maintained when the sterile product is withdrawn from its package and either injected as is or, more often, combined with another delivery system, such as an intravenous (IV) bag or bottle and infused. If good aseptic techniques are not followed, potential adverse consequences could result due to accidental contamination of the medication. Practicing aseptic techniques, while not difficult, may not be followed strictly or personnel carelessness may be the problem. Administering sterile drug products certainly are not nearly as easy to accomplish as administering nonsterile dosage forms.

Table 31-1 Alphabetical Listing of Injectable Routes of Administration

Injection route	Comments
Hypodermoclysis	
Intra-arterial	Typically for vasodilator drugs in treatment of vasospasm, thrombolytic drugs for treatment of embolism, localized cancer drugs, and diagnostics
Intra-articular	Injection into joints for treatment of arthritis and other inflammatory problems
Intracardiac	Directly into the heart in cases of cardiac arrest, but rarely performed
Intracavernosal	Injection into penis
Intracerebral	Direct injection into cerebrum of the brain, typically for treatment of malignancies
Intracisternal	Injection primarily of diagnostic into cisternal space surrounding base of brain. Considered too dangerous by most neurosurgeons due to potential for brain damage
Intradermal	Vaccines, skin testing for allergenic reactions, treatment of burns and scars
Intralesional	Injection directly into a solid tumor
Intramuscular	Most convenient route, vaccines, antibiotics, long-acting formulations, suspensions, and oily injectables
Intraocular	Injection directly into the eye
Intraosseous	Infusions into bone marrow for indirect access to venous system. Alternative to IV when IV access difficult
Intraperitoneal	Injection or infusion into peritoneum for dialysis purposes
Intrapleural	Injection into the lung
Intrathecal	Injection into spinal column, usually for spinal anesthesia and chemotherapy
Intrauterine	Injection into the uterus
Intravenous	Most common route considering not only drugs, but also fluids including parenteral nutrition
Intraventricular	Injection or infusion into lateral ventricles of brain, treatment of meningitis and malignancies, and last resort for pain therapy for terminal cancer patients
Intravesical	Infusion into urinary bladder
Subcutaneous	Major route for chronically-administered injections like insulin, anti-arthritic injectable drugs, growth hormone, vitamins

Sources: Refs. 2 and 3.

Greater Expense of Injectable Drugs Compared to Same Drug Available Orally

It is much more costly to produce sterile drug products than any other type of pharmaceutical dosage form. Special requirements for facilities, air control, contamination control, cleaning and sanitization, aseptic manipulations, process validation, testing, and many other aspects of sterile processing and control all contribute to increased cost of manufacture, packaging, and release of sterile dosage forms. Regulatory requirements for process control and achieving the main attributes of GMP—safety, identity, strength, quality, and purity—also are more costly for the manufacturer of sterile products than for the manufacturer of nonsterile products.

ROUTES OF INJECTABLE ADMINISTRATION

Injections can be administered anywhere in the body. An alphabetical listing of parenteral routes of administration is given in Table 31-1. The most important or frequent routes of parenteral administration are discussed in this chapter.

Intravenous Route

The IV route involves direct administration of a drug solution, oil-in-water emulsion or nanosuspension. The largest globule size of an emulsion or the largest particle size of a nanosuspension cannot be larger than five microns, the approximate diameter of blood capillaries. A simple pharmacokinetic schematic is shown in Figure 31-1.

The IV route is the most common of all parenteral routes and is especially convenient for large-volume infusions. Advantages of the IV route of administration, compared to other routes, include the following:

1. Instantaneous response, especially essential in emergencies such as arrhythmias or seizures. Instantaneous response is also required in the restoration of electrolyte and fluid balances and in the treatment of life-threatening infections.

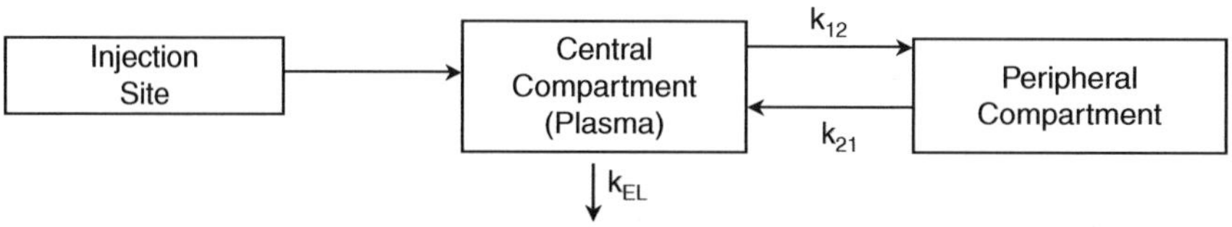

Figure 31-1 Intravenous route of parenteral administration.

2. Control (predictability) of dosage, for example, in the treatment of hypotension or shock situations.
3. Greater intensity of response since the entire dose if systemically available.
4. Less irritation since the IV route can tolerate ranges of tonicity and/or pH better than other parenteral routes.
5. Greater stability.
6. Only way to administer large-volume injections, especially essential for continuous nutrition where patients cannot be fed by mouth.
7. Avoids complications that might result with other routes, e.g., local inflammatory responses or hematomas with intramuscular (IM) injections.

Disadvantages specific to the IV route include potential for hemolysis, thrombosis, and precipitation at the injection site, all depending on the product formulation, injection procedure, and patient sensitivity. These and other hazards are covered in Chapter 32.

IV administration results in complete drug bioavailability. The shape of the drug plasma concentration versus time profile is determined by the rate of injection. Intermittent bolus and infusion injections will give curves shown in Figure 31-2(A) and 31-2(B). IV infusion at a constant rate over a period of time (minutes to hours), usually accomplished by an infusion pump/controller, will result in a drug concentration versus time curve shown in Figure 31-3. Controlled infusions are particularly important for drugs with narrow therapeutic index (small difference between therapeutic and toxic blood levels), and when effective blood levels are well defined (e.g., infusing aminophylline for treating asthma).

Infusions can be continuous or intermittent. Continuous infusions are better for drugs with narrow margins of safety where it is easier to adjust the rate depending on patient response to the medication. Drug classes administered by continuous infusions include anesthetics, antiarrhythmics, antihypertensives, neuromuscular blocking agents, vasopressin, and oxytocin. Intermittent infusions are commonly used in chemotherapy and may produce higher tissue levels than that seen with continuous infusions.

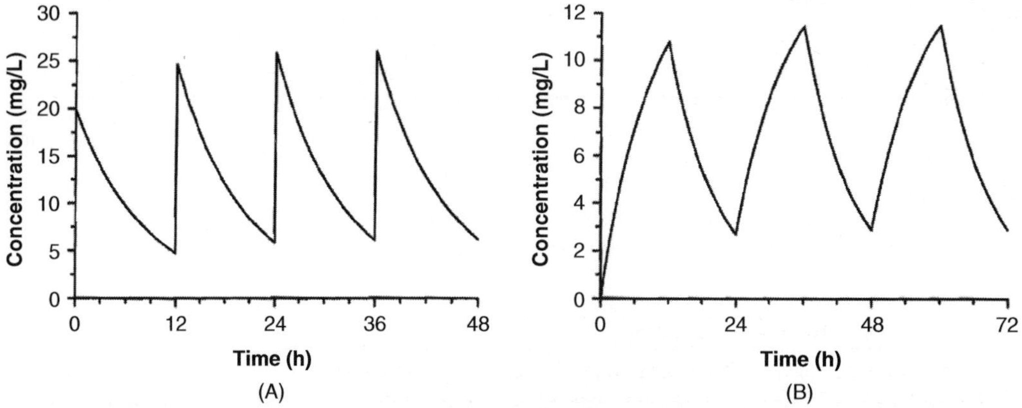

Figure 31-2 Plasma concentration vs time profile for a intermittent IV bolus dose (A) and intermittent IV infusion dose (B). *Source*: From Ref. 4.

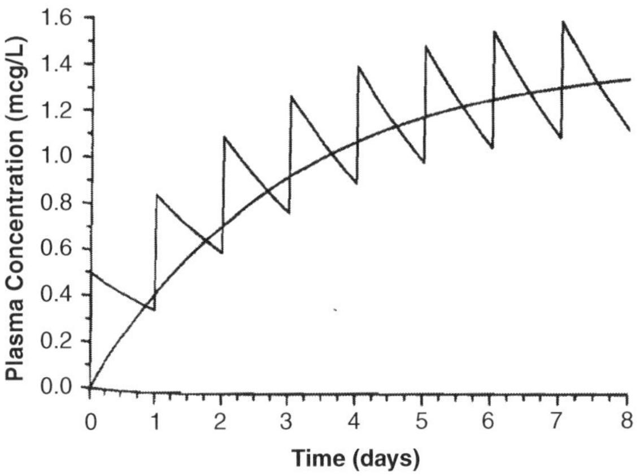

Figure 31-3 Plasma concentration vs time profile for a continuous infusion. *Source*: From Ref. 5.

Intramuscular Route

The IM route can use almost any muscle of the body. It is the most convenient parenteral route available. It is much easier to inject drugs into muscles than probably any other route. However, because of the need to use larger needles, the IM route can be the most painful of the common routes. The IM route provides a means of injecting drug formulations (aqueous or oily solutions or suspensions) designed for sustained (prolonged) or controlled release. The IM route is preferred over the subcutaneous route when a rapid rate of absorption is desired and over the IV route in cases where the drug should not be administered directly into the vascular system.

Many important drugs are administered primarily by IM injection—corticosteroids, lidocaine, aminoglycosides, diazepam, leuprolide depot formulations, and most, if not all, other sustained release formulations (see Chapter 3), because sustained release formulations are typically dispersed systems that cannot be administered into the blood stream directly. Also, the nature of a sustained release formulation is to reside at the site of injection for a period of time and drug slowly released from the injected formulation (depot) for diffusion into the circulatory system. Figure 31-4 depicts the effect of formulation on the blood level curves as an IM-administered drug product.

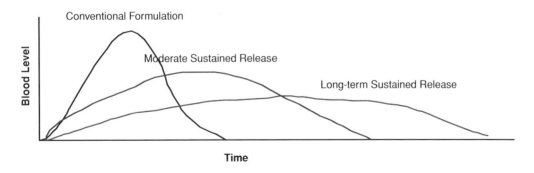

IM administration
--Solutions, suspensions, and emulsions
--Easier administration than IV
--Must be iso-osmotic
--Slower onset and prolonged action
--Absorption influenced by drug release and blood flow
--Formulation effects: solutions absorbed faster than aqueous suspensions, followed by oily solutions, oily suspensions and viscous oily suspension
--Particle size, salt form affects rate/extent of absorption of suspension

Figure 31-4 Blood level versus time plots for a conventional, moderate, and long-acting intramuscularly administered drug.

The most frequent muscle sites are the deltoid, gluteus maximus, and lateral thigh muscle. The larger the muscle, the larger can be the volume injected. Therefore, the gluteus maximus or lateral thigh muscles can be injected with volumes up to 5 mL while 2 mL is the volume limit for the deltoid muscle.

Although the IM route is easier to inject drug products, there are potential hazards in using this route. A primary potential problem is accidentally puncturing a blood vessel, especially an artery that might lead to an injection of a toxic drug or vehicle directly into the bloodstream and ultimately to an organ. Those taught to give IM injections are shown how to pull back on the plunger of a syringe and observe for the appearance of blood. If blood does appear, then the needle has penetrated a blood vessel and should be repositioned to a place where it is not within a blood vessel before injecting the product. Another potential problem with IM injections is accidental striking of a peripheral nerve resulting in nerve palsy. IM injections should not be given to patients with significant heart failure or shock because of poor uptake potential into the bloodstream.

Pharmacokinetically, the IM route is similar to the oral route. Drugs must dissolve if not already in solution, and be absorbed through mucosal tissue into the blood system. Thus, factors that affect drug absorption from the oral route also hold true for the IM route. Aqueous solubility, partition coefficient, particle size, dissolution rate, and concentration all affect drug absorption. In addition, blood flow is an important factor.

Drug products to be administered IM need to be isotonic and close to physiological pH. Since there is little opportunity for dilution of the injected formulation, extremes in these properties will cause significant pain and tissue damage.

Subcutaneous (SC) Route

The subcutaneous (SC) route involves injecting small volumes (usually less than 1 mL), just below the outer surface of the skin into the adipose tissue. This route also is relatively simple to administer drugs. The SC injection will enter the open space just beneath the skin and, like IM injections, slowly diffuse through the capillary bed and tissue into the circulatory system. Like IM injections, SC injections should be isotonic and close to physiological pH and the same factors that control rate of absorption of drugs from IM injection also are involved in SC absorption. Absorption from the SC route is slower and less predictable than from the IM route.

Insulin products are the most commonly used SC injections. The SC route is preferable to the IV route for insulin in order to avoid too intense and potentially dangerous response. Other drugs commonly administered by the SC route include vaccines, narcotics, epinephrine, and vitamin B12. The subcutaneous route is by far the simplest route of injectable administration and preferred for injectable products that can be self-administered (home health-care injections).

Medications that are highly acidic, alkaline, or irritating should not be administered by this route because of the potential inflammation and necrosis of tissue with obvious accompanying pain.

A special form of SC administration is called hypodermoclysis. This involves the infusion of large amounts of fluid into the subcutaneous tissues when IV sites are not available. Hypodermoclysis is not used much today, but used to be a common mode of replenishment of fluid and electrolytes in the very young and very old.

HYLENEX (recombinant hyaluronidase human injection) received FDA approval to facilitate the subcutaneous absorption and dispersion of other SC-injected drugs or fluids; for subcutaneous fluid administration; and as an adjunct in subcutaneous urography for improving resorption of radiopaque agents (6). When injected under the skin or in the muscle, hyaluronidase can degrade the hyaluronan gel, allowing for temporarily enhanced penetration and dispersion of other injected drugs or fluids.

OTHER ROUTES OF PARENTERAL ADMINISTRATION (Table 30-1)

Intradermal Injections

Intradermal injections primarily involve diagnostic agents, antigens, and vaccines. These products are injected (maximum volume 0.1 mL) into the dermis just beneath and adjacent to the epidermis. Drug absorption from intradermal injection is slower than that for SC injection.

Intra-Arterial (IA) Injections

The IA route can be very dangerous because the product administered into the artery goes directly to major organs without first going through the heart and lungs and without being adequately diluted. The heart and lungs serve as important screens to filter out any particles in the injection, so the body's natural defense against injected particles is obviated with IA injections. In IA injections, it is more difficult to insert the needle compared to that in IV administration. IA injections carry the risk of thrombosis, embolization, and gangrene. Accidental air infusion resulting in air embolism with consequent ischemia and/or infarction of the tissue supplied is more serious for IA than IV injections. Diagnostic agents (e.g., radiopaque substances for roentgenographic study of the vascular supply of various organs or tissues) and organ-specific cancer drug therapy often use the IA route. All the other injectable routes are direct injections into the specific organ or tissue to treat very serious diseases that cannot be treated efficiently by the other safer routes.

Intra-Articular Injections or Infusions

This route of administration is used for antibiotics, lidocaine, and corticosteroids needed in the synovial area of bodily joints to treat infection, pain, and inflammation.

Intrathecal (or Intraspinal)

This route of administration involves injections or infusions directly placed into the lumbar sac (intrathecal) location of the spinal cord. This route is used more frequently for diagnostic applications than for therapeutic purposes. Infusion of radiopaque products by this route assesses the potential obstruction of the subarachnoid space around the spinal cord and base of the brain as well as the possibilities of abscesses and tumors. Therapy usually involves the injection of a chemotherapeutic agent, often in conjunction with IV therapy.

Such injections into the cerebrospinal fluid must be essentially endotoxin free because of the special danger endotoxins pose to this kind of tissue. Calculations for endotoxin limits for all routes except intrathecal and intraspinal use a numerator value of 5 EU/kg (see Chapter 28, p. 428). For intrathecal and intraspinal routes, the numerator for that calculation becomes 0.2, a 25-fold greater safety factor for what is an acceptable limit of endotoxin in the product.

Intraocular Injections

Injectable (not topical) administration of drugs to the eye involves four different possible routes:

1. Anterior chamber—injection or irrigation directly into the anterior chamber of the eye.
2. Intravitreal—injection directly into the vitreous cavity of the eye.
3. Retrobulbar—injection around (not into) the posterior area of the eye globe that subsequently diffuses into the eye.
4. Subconjunctival—injection into the conjunctiva that subsequently diffuses into the eye.

These routes of administration are used to treat infections and inflammatory diseases of the eye that cannot be effectively treated by topical ophthalmic drug administration or by more conventional systemic administration. Other purposes for these special routes into the eye include anesthesia of the globe and papillary dilation with cycloplegic and mydriatic drugs. Obviously, great skill and care are required in administering drugs by these routes to avoid damage and infection to the eye.

Other routes of parenteral administration directly to specific organs or tissues include intra-abdominal (or intraperitoneal), intracardiac, intracisternal (space surrounding the base of the brain), intralesional, intrapleural, intrauterine, and intraventricular injections and/or infusions. These routes are relatively minor in frequency of use and are not covered further.

Table 31-2 provides basic information about specific requirements for injections given by several injectable routes of administration.

Table 31-2 Specific Requirements for Injections Given by Several Parenteral Routes of Administration

Route	Volume (mLs)	Comment
Intravenous	Any	Small volume injections not absolutely required to be isotonic like all other injectable routes need to be. Large volume IV injections must be isotonic.
Intramuscular	2	Any type of dosage form
Subcutaneous	2	Any type of dosage form, primary route for chronically-administered injectables and best route for self-administration
Intradermal	1–10	Best for diagnostics, vaccines, wound healing agents
Intra-arterial	2–20	Dangerous route with respect to particulate matter as injection leads straight to organs, no natural screening by lungs and heart
Intrathecal Intraspinal	1–4	Sensitive to endotoxins, much lower threshold pyrogen dose compared to IV route
Intraepidural	6–30	Solutions only
Intracisternal	1–5	Must be isotonic
Intra-articular	2–20	Must be isotonic
Intracardiac	0.2–1.0	Must be isotonic
Intrapleural	2–30	Must be isotonic
Intraocular Intravitreous	0.05–0.1	Must be isotonic

REFERENCES

1. Duma RJ, Akers MJ, Turco S. Parenteral drug administration: routes, precautions, problems and complications. In: Avis KE, Lieberman HA, Lachman L, eds. Pharmaceutical Dosage Forms: Parenteral Medications. Vol 1, 2nd ed. New York: Marcel Dekker, 1992:17–58.
2. FDA. Routes of Administration. http://www.fda.gov/Drugs/DevelopmentApprovalProcess/FormsSubmissionRequirements/ElectronicSubmissions/DataStandardsManualmonographs/ucm071667.htm. Accessed December 30, 2009. Accessed June 17, 2010.
3. Wikipedia. Routes of Administration. http://en.wikipedia.org/wiki/Route_of_administration. Accessed June 17, 2010.
4. Beringer PM, Winter ME. Clinical pharmacokinetics and pharmacodynamics. In: Beringer P, ed. Remington: The Science and Practice of Pharmacy. 21st ed. Philadelphia, PA: Lippincott, Williams and Wilkins, 2005:1200.
5. Beringer PM, Winter ME. Clinical pharmacokinetics and pharmacodynamics. In: Beringer P, ed. Remington: The Science and Practice of Pharmacy. 21st ed. Philadelphia, PA: Lippincott, Williams and Wilkins, 2005:1199.

32 Clinical hazards of injectable drug administration

Any route of drug administration has certain risks or hazards. Injecting drug products directly into blood or tissue offers greater risks and hazards compared with any other route of administration. All injectable routes of administration have risks and hazards, but the intravenous (IV) and, especially, the intraarterial (IA) routes are most hazardous. The most common injectable hazards will be presented in alphabetical order with all hazards listed in Table 32-1 (1).[1]

Although not strictly classified as a clinical hazard, pain upon injection and the fear of receiving an injection are the most common concerns of injectable drug administration. Science continues to investigate ways to reduce both physiological and psychological adverse reactions to pain upon injection as well as to develop better methods for predicting pain and/or irritation upon injection (2).

AIR EMBOLI

Air emboli result principally from IV infusions, particularly if infusions administration uses pumps that do not have active air alarms. If air enters the bloodstream it can occlude cerebral or coronary arteries, resulting in major strokes and potential death. Air can cause blood vessels of the lung to constrict that, in turn, causes pressure in the right side of the heart to rise. Air can then move on to the brain or coronary arteries.

Small amounts of air are not harmful, but 10 mL or more air injected into the bloodstream could be fatal. To minimize or eliminate air from entering the bloodstream, great care should be exerted to purge all air bubbles from the syringe or IV line prior to starting an injection or infusion and ensure that the system used remains airtight throughout their use. Perhaps the greatest advantage of using plastic bags composed of polyvinylchloride (PVC) for IV infusions is the great characteristic of PVC to collapse upon itself as the internal fluid is administered so that when all the fluid is gone, the collapsed bag will not have any air to release into the IV line. Other precautions employed to minimize the risk of air emboli include the following:

1. Discontinuing the infusion before the fluid drains from the IV tubing
2. Ensuring that all attachments fit tightly
3. Being careful to clamp off the first bottle or bag of fluid that empties in a Y-type administration line
4. The part of the body receiving the infusion should not be elevated above the heart that would create negative venous pressure
5. Permitting the infusion tubing to drop below the level of the body part if emptying occurs unobserved.

BLEEDING

Bleeding usually occurs in patients given injections who either have platelet deficiency or hemophilia. If bleeding tendency in a patient is known, the IV route may be safer than the intramuscular (IM) route because bleeding may be better controlled. Those patients with hemophilia or with vitamin K or platelet deficiencies should be given antihemophilic globulin, Factor VIII, vitamin K, and/or platelet transfusions prior to administration of parenteral products.

[1] Visual examples of clinical hazards of injectable administration of drugs may be found at the following websites: http://www.sciencephoto.com/, http://www.photoresearchers.com/main.html, http://catalog.nucleusinc.com, or simply use key words on a search engine site.

Table 32-1 Alphabetical Listing of Potential Hazards of Injectable Drug Administration

- Air emboli
- Bleeding
- Fever
- Hypersensitivity
- Incompatibilities
- Infiltration and extravasation
- Overdosage
- Pain and irritation
- Particulate matter
- Phlebitis
- Sepsis
- Thrombosis
- Thrombophlebitis
- Toxemia

HYPERSENSITIVITY

Hypersensitivity reactions occur under a variety of conditions, most commonly when the patient has been previously exposed to the drug and may be immediate or delayed. Immediate hypersensitivity reactions are associated with a "wheal-and-flare" skin lesion (urticaria), anaphylaxis, and/or an Arthus reaction (e.g., serum sickness). Delayed hypersensitivity reactions show a tuberculin-type reaction where the patient appears to be suffering from sepsis. Some protein and large polypeptide drugs (e.g., insulin, therapeutic antibodies) can directly stimulate antibody production. However, most drugs act as haptens, binding covalently to serum or cell-bound proteins, including peptides embedded in major histocompatibility complex (MHC) molecules. The binding makes the protein immunogenic, stimulating anti-drug antibody production, T-cell responses against the drug, or both (3).

Hypersensitivity reactions must be treated promptly, typically by discontinuing the use of the particular drug, sometimes needing to administer antihistamines, corticosteroids, or epinephrine.

INCOMPATIBILITIES

Incompatibilities are unfortunately a frequent problem occurring in parenteral therapy. The concern about incompatibilities resulted in texts published by Baxter, Abbott, and Trissel (4) that inform the pharmacist and other health care professionals what combinations of drugs potentially are incompatible. Incompatibilities cause drug precipitation in the container or in the administration set and, worse, could cause adverse side effects such as platelet aggregation, anaphylatoid reactions resulting in shock, and/or pulmonary infarctions.

INFILTRATION AND EXTRAVASATION

These hazards are caused by faulty needle injection technique. Infiltration is caused by the needle puncturing through a vein (or artery) or when an infusion cannula becomes dislodged from the vein and injecting or infusing the administered product into the surrounding tissues. Extravasation occurs when fluids seep out from the lumen of a vessel into the surrounding tissue. Damage to the posterior wall of the vein or occlusion of the vein proximal to the injection site are common causes of extravasation. Either of these events will result in increasing edema at the site of the infusion. Extravasation is especially traumatic in children.

This can be very painful, for example, in cases where potassium chloride is improperly injected. Hypertonic dextrose and solutions with pH differences from bodily pH also will cause pain if extravasation occurs. Obviously, extravasation with a cancer drug can be very dangerous. Infiltration can cause infection, phlebitis, thrombosis, or necrosis of the infiltrated tissue.

Hematoma, a type of extravasation incident, caused by faulty injection technique where the needle punctures the vein, is an ugly looking bruise that is usually not harmful, but should be treated with cold compresses and elevating the part of the body where the injection occurred.

OVERDOSAGE

Overdosage can be drug or excessive fluid overload. Drug overdosage results in some kind of toxic reaction. Fluid overdosage results in pulmonary edema. Because parenteral drug administration suffers a disadvantage that once the drug is administered, it cannot be retrieved or neutralized. Overdosage can be avoided with careful management of drug and fluid administration. If patients require prolonged fluid administration, pulmonary capillary wedge or central venous pressures are monitored to ensure against overload occurring.

PARTICULATE MATTER

This subject is covered thoroughly in chapter 29. The issue of particulate matter is primarily a reflection of product quality with respect to manufacturing, quality control, and inspection. Large numbers of particles of large sizes can be a serious problem for drug solutions administered by IV or, certainly, by IA routes. Particles are known to cause foreign body reactions. Particles greater than 5 μm in diameter can clog the smallest capillaries, especially in the lungs. Particles as low as 2 μm can form microthrombi (5). Regulatory concerns about the potential hazards of particulate matter, both large (visible) and small (subvisible), have resulted in standards for particle numbers in parenteral solutions. Any product with visible foreign particles must be rejected and subvisible particles at 10 and 25 μm have specifications for acceptable numbers that must be passed for each batch of product released.

PHLEBITIS

Phlebitis is a local inflammatory reaction usually associated with IV injection or infusion. Symptoms of phlebitis include redness (erythema) of the skin where the injection/infusion occurred, pain along the vein, edema, and hardness of the vein. Phlebitis caused by infusion can usually be reduced simply by slowing down the infusion rate.

This condition is exacerbated by long-term use of a device in contact with the blood vessel, for example, long-term usage of a needle and catheter. Phlebitis sometimes can be associated with infection and can result in thrombosis, thus referred to as thrombophlebitis. Thrombophlebitis clots can be fatal if they travel to the lungs causing pulmonary embolisms.

Phlebitis can result from extremes in solution pH, inherent irritating properties of the drug being injected or infused, drug insolubility at the injection site, particulate matter, extremes in osmolality, injecting too much volume of drug product for the blood vessel chosen, reaction with the catheter residing within the vein or artery, and general trauma. It has been documented that solutions of low pH, hyperosmotic infusion fluids, and not using an in-line particulate retaining filter appear to increase the risk and incidence of phlebitis. Another factor, perhaps the most common cause of phlebitis, is the lack of technical skill of the person administering the injection or infusion where clumsiness, hastiness, or simple careless technique in inserting the needle produces the irritation leading to phlebitis.

SEPSIS AND TOXEMIA

Sepsis results from

- microorganisms contaminating the product or delivery system that are subsequently injected; or
- microorganisms from the skin surface that are carried into the body when the product is injected, or
- microorganisms that migrate from the skin into the vein along a sleeve of an IV line (catheter) if present.

In fact, any indwelling device like a catheter or needle may serve as an attractive residence for circulating microorganisms where they will eventually depart from the foreign device and reseed the bloodstream. Sepsis may be localized forming an abscess and/or may be systemic producing septicemia and metastatic infections. Sepsis, not surprisingly, is the most dangerous of all potential hazards possible when administering drug products by the parenteral route.

The microorganisms that most commonly cause sepsis are those indigenous to the locale through which the infusion or injection passes. Skin bacteria such as *Staphylococci sp*,

Candida sp, Streptococci, Acinetobacter calcoaceticus, Pseudomonas, Serratia, Escherichia coli, Enterobacter, Klebsiella, and Proteus are examples of microorganisms found to cause sepsis.

If sepsis occurs in a hospital setting (nosocomial), it is often very difficult to destroy these organisms with conventional antibiotic therapy because of the resistance developed in these settings.

The best way to prevent sepsis from occurring is through careful aseptic procedures employed to prepare the skin for an injection and to manipulate and inject the sterile device. In-line bacterial retentive filters also help to prevent sepsis although there are other limitations in using these filters such as costs, potential for clogging, and they themselves can be a source of contamination, usually through improper insertional procedures. Also, as discussed in chapter 28, endotoxins resulting from gram-negative bacterial growth will not be removed by sterilizing filters.

With respect to endotoxins, the condition called toxemia results from an inadvertent infusion or injection of a biological toxin such as endotoxin. Endotoxins cause fever, leucopenia, circulatory collapse, capillary hemorrhages, necrosis of tumors, and other cascades of problems that can lead to death if the amount of endotoxin is high. The LD_{50} dose of endotoxin in mice is approximately 150 μg (6). A fatal dose of endotoxin in humans is unknown or not found in the literature, but a threshold pyrogenic dose in humans is 350 endotoxin units (5 EU/kg × 70 kg with 5 EU/kg being the threshold pyrogen dose—see page 428—and 70 kg being the average weight of an adult person). The level of endotoxin is controlled in parenteral drug products and sterile devices through the use of endotoxin limits as discussed in chapter 28. The Limulus Amebocyte Lysate test has proven to be a very sensitive and specific indicator for the presence of endotoxin in amounts much lower than known to cause pyrogenic responses in humans.

THROMBOSIS

Thrombosis is a blood-clotting problem that occurs at the site of injection with either an IV infusion or IV or intra-arterial injection. The thrombus formed may propagate proximally for a distance from the injection site. Complications arising from thrombus formation include emboli formation that may cause pulmonary infarction and secondary infection resulting in septicemia, endocarditis, and/or pneumonia.

Thrombosis occurring in an artery creates a more much serious complication than venous thrombosis. Gangrene of the tissues supplied by the artery could result, especially if collateral circulation around the thrombotic artery is inadequate.

Thrombosis can result from extremes in solution pH, inherent irritating properties of the drug being injected or infused, drug insolubility at the injection site, particulate matter, extremes in osmolality, injecting too much volume of drug product for the blood vessel chosen, reaction with the catheter residing within the vein or artery, and general trauma. Certain people and certain disease states, for example, systemic lupus erythematosis, are prone to react adversely to injections or infusions and the slightest irritation could cause a thrombotic reaction.

SUMMARY

Besides all of these hazards discussed in some detail, parenteral drug administration always is subject to potential serious and specific hazards or complications every time an injection is given. Every parenteral drug injected itself has specific potential side effects associated with its injection. Of course, these are required to be specified in the package insert. Indeed, studying package inserts of drug products administered by injection is the best way to learn and be aware of the potential hazards of injectable drug administration.

REFERENCES

1. Duma RJ, Akers MJ, Turco SJ. Parenteral drug administration: Routes, precautions, problems, complications, and drug delivery systems. In: Avis KE, Lieberman HA, Lachman L, eds. Pharmaceutical Dosage Forms: Parenteral Medications. Vol 1. 2nd ed. New York: Marcel Dekker, 1992:17–58.
2. Gupta PK, Brazeau GA, eds. Injectable drug development: Techniques to reduce pain and irritation. Englewood, CO: Interpharm Press, 1999.

3. Drug hypersensitivity, The Merck Manuals Online Medical Library. Merck Manual. http://www.merck.com/mmpe/sec13/ch165/ch165e.html. Last updated September, 2008. Accessed June 17, 2010.
4. Trissel LA. Handbook on Injectable Drugs. 15th ed. Bethesda, MD: American Society of Health-System Pharmacists, 2009.
5. Lehr HA, Brunner J, Rangoonwala R, et al. Particulate matter contamination of intravenous antibiotics aggravates loss of functional capillary density in postischemic striated muscle. Am J Respir Crit Care Med 2002; 165:514–520.
6. DeFranco AL, Locksley RM, Robertson M. The immune response to bacterial infection. In: DeFranco AL, Locksley RM, Robertson M, eds. Immunity: The Immune Response in Infections and Inflammatory Disease. London: New Science Press, 2007. Chapter 9.

33 Biopharmaceutical considerations with injectable drug delivery

Biopharmaceutics involves the study of the relationship between the formulation and properties of a drug product and its in vivo performance, primarily rate and extent of drug absorption from the injection site. Pharmacokinetics involves the study of the rates of absorption, distribution, metabolism, and excretion and the relationship of these rates to therapeutic results (Fig. 33-1). Drug dissolution and ultimate absorption does not apply to the intravascular (intravenous or intra-arterial) routes, but definitely applies for parenteral products administered by intramuscular (IM) and subcutaneous (SC) injection although other routes involving drug dissolution and absorption from the injection site can be studied. Rates of distribution, metabolism, and excretion apply to all parenteral routes of administration, but the rate of absorption does not apply to the intravenous route since the entire dose of drug injected already exists in the circulatory system.

DRUG ABSORPTION

For drug absorption to occur for all parenteral routes except for the intravascular route, the drug must be released from the dosage form. Drug release depends on physical and chemical properties of the drug, the dosage form, and on the physiologic environment where drug injection occurs. After the drug is administered as a solution or is dissolved from a suspension or solid injectable form, drug molecules passively diffuse through the capillary beds of biologic tissue or membranes and eventually enter the circulatory system. Drug molecules may also be transported by the lymphatic system that ultimately leads to the bloodstream. Discussion of the circulatory system later points out some fundamental aspects of drugs moving from the site of injection to the main systemic circulation of the body. The structure of biologic membranes is a discontinuous bilayer of phospholipids, oriented so that their polar heads are in contact with the external aqueous environment (Fig. 33-2).

Drugs are primarily absorbed by passive diffusion. Active transport systems can also help drugs to be absorbed, but passive diffusion is the mechanism of absorption for most drugs. Drug molecules in aqueous solution or after dissolution into aqueous solution will then dissolve in the lipid material of the membrane and diffuse through the membrane to reach the aqueous bloodstream. This points out that drug molecules need both lipophilic and hydrophilic properties. This is true for small molecules that exist as weak acid or base electrolytes, to be discussed later.

The rate of passive diffusion is based on Fick's first law:

$$\text{Diffusion flow } (dq/dt) = -DA(C_1 - C_2)/\ell$$

where A is the surface area of the membrane, D is a diffusion coefficient, $C_1 - C_2$ is the concentration gradient (concentration on either side of the membrane) and ℓ is the membrane thickness. Once the drug is absorbed, it disperses quickly into the systemic circulation so the concentration on the other side of the membrane is considered a "sink" and C_2 is essentially zero. Therefore, diffusion flow is dependent on drug concentration on the upstream side of the membrane and the diffusion rate equation is considered first order.

DRUG DISTRIBUTION

Distribution of drug molecules from site of injection into systemic circulation to organ, tissue, and cellular reception sites for therapeutic activity depends on several factors—drug lipophilicity, permeability of tissue membranes, the affinity of the drug to particular tissues and membranes, and blood flow supply to the tissues. Since most biologic tissues are well perfused by the circulatory system, measuring plasma concentrations of drugs will monitor the course of drug therapy. Drug distribution in preclinical and early clinical studies can

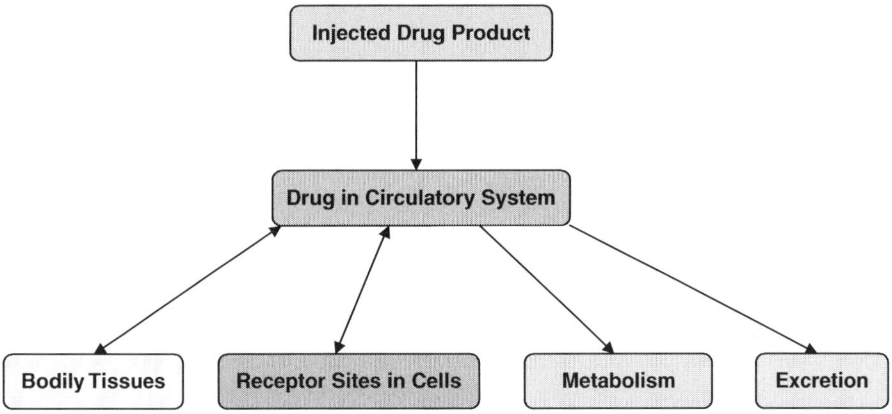

Figure 33-1 Pharmacokinetic Interplay.

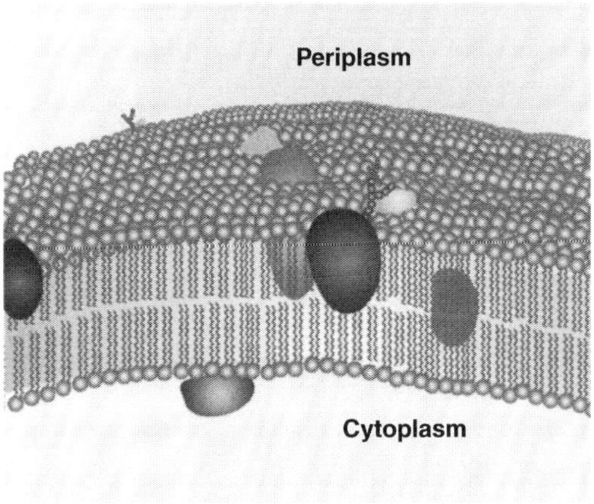

Figure 33-2 Biological membrane. *Source*: Courtesy of Open Learning Initiative Modern Biology Course, Carnegie Mellon University. William E. Brown and Gordon S. Rule, professors, Biological Sciences; Ornella Pagliano and Diana Bajzek, Office of Technology for Education.

be studied using radioisotope forms of the drug substance where radioactive labeling techniques can provide imaging results of drug distribution and concentration throughout all bodily tissues.

Drug distribution is highly dependent on it binding to plasma proteins, primarily albumin, but also the globulins, lipoprotein, and glycoprotein. Some drugs bind significantly to plasma proteins while other drugs do not (see Table 33-1) (1). A drug bound to a plasma protein is too big to pass across capillary walls and will not readily distribute into bodily tissues. Drugs bound to proteins cannot cross the blood–brain barrier.[1] Drug–protein binding is reversible so desorption of drug from the protein will eventually occur. Protein binding will significantly affect passive diffusion if 90% or more of the drug is bound to the protein. Passive diffusion will eventually desorb the drug and it will be available, but the rate of availability will be retarded. Plasma protein binding also will affect rates of metabolism since bound drugs will not interact with metabolic sites and rates of excretion since only unbound drugs can be filtered through the kidneys.

[1] Brain endothelial cells are "packed" closely together with the passageway or gap between cells very small or tight. These "tight junctions" form what is commonly called the "blood-brain" barrier that presents a resistant barrier to the movement of and delivery of drugs from the blood stream into the brain.

Table 33-1 Drug–Plasma Protein Binding

Drug	% Bound to plasma proteins (1)
Dicumarol	99.9
Warfarin	99.5
Phenylbutazone	99.5
Digitoxin	97
Diazepam	96
Phenytoin	87
Carbamazepine	75
Gentamicin	50
Penicillin G	50
Digoxin	23
Caffeine	10

DRUG METABOLISM AND EXCRETION

Drugs are eliminated from the body either by metabolism or excretion. Drug metabolism primarily occurs in the liver through processes of hydrolysis, enzymatic reduction, or oxidation. The liver, unlike the brain, does not have "tight junctions" between endothelial cells that allow all dissolved substances in the plasma, even plasma proteins, to pass from plasma into liver cells. Drug metabolism also occurs, to a lesser extent, in other organs including the skin, lungs, kidneys, and gastrointestinal mucosal cells. Drug excretion occurs primarily through the kidneys. Excretion can also occur, to a much lesser extent, through the bile, lungs, sweat, saliva, and breast milk. In general lipid-soluble drugs are metabolized in the liver and water-soluble drugs are readily excreted through the kidneys. The reason many drugs cannot be administered orally is that they are completely metabolized and rendered inactive in the liver prior to reaching the systemic circulation.

Drug elimination determines the biological half-life of the drug, that is, the time required for reduction of one-half of the drug concentration in the bloodstream. Since most drugs are eliminated by first order kinetics, where the rate of elimination (k_e) is dependent on drug concentration and as drug concentration decreases, the rate of elimination decreases, the biological half-life of a drug undergoing first-order elimination rate kinetics is expressed as

$$t_{1/2} = 0.693/k_e$$

Biological half-lives for a few injectable drugs are given in Table 33-2, just to illustrate how wide-ranging biological half-lives are and how important it is to know these values. Biological half-lives are published in the package inserts of commercial drug products, typically under the clinical pharmacology section of the insert.

PHYSICOCHEMICAL FACTORS AFFECTING DRUGS ADMINISTERED BY INJECTION

The bioavailability of drugs administered by IM or SC injection (or any other injectable route except for intravenous or intra-arterial) depends on several physicochemical and physiological factors. Physicochemical factors include solubility, partition coefficient, particle size, viscosity, and solid-state morphology, all of which affect the ability of the drug to diffuse passively from the injection site to the blood circulation. The primary physiological factor is blood flow with blood flow depending on the capillary bed density at the injection site.

Solubility

Drugs administered into muscle or subcutaneous tissue primarily rely on passive diffusion to be absorbed into the blood stream. To diffuse through tissue and be available to the bloodstream, the drug must be in solution. The solubility of drugs in aqueous solution varies from being completely soluble to partially or sparingly soluble to being insoluble. Drug solubility is dictated by its chemical structure, the orientation of its structure in water, and the propensity of its functional groups to interact with water molecules. Besides chemical structure, factors that

Table 33-2 Biological Half-Lives of Some Injectable Drugs

Drug	Half life ($t_{1/2}$) (Hours)
Adalimumab (Humira®)	240–480
Ampicillin	0.8
Cisplatin	30–100
Cephazolin	2.0
Clindamycin	2.4
Digitoxin	40
Epogen	4–13
Etanercept (Enbrel®)	70
Gentamicin	2
Infliximab (Remicade®)	184–228
Insulin	Intravenous injection–10 mins. Subcutaneous injection–4 hrs. Intramuscular injection–2 hrs.
Minocycline	10–12
Propofol	2–24
Tobramycin	2
Vancomycin	6

Source: From Ref. 2 and from the Physicians' Desk Reference, 64th ed., 2010, www.PDR.net.

affect drug solubility in aqueous solution include solution pH and the extent of ionization of the drug in solution.

If the drug is ionic (an electrolyte), it will form a charged species as a function of solution pH and, of course, can be synthesized as a soluble salt. The solubility of an electrolyte will depend on pH and the solubility of the ionized form at a given pH. Solubility versus pH profiles of drugs that are weak acid electrolytes (e.g., can form salts with cationic elements like sodium and potassium) and drugs that are weak base electrolytes (e.g., can form salts with anionic elements like HCl, sulfate, or phosphate species) are seen in Figure 33-3. These profiles are "Z-shaped" or "S-shaped" curves in that once the pH reaches a point where the salt dissociates, that is, the neutral free acid or base is formed, the solubility of the drug plummets. The dissociation constant (pK_a for a weak acid or pK_b for a weak base) indicates the pH at which

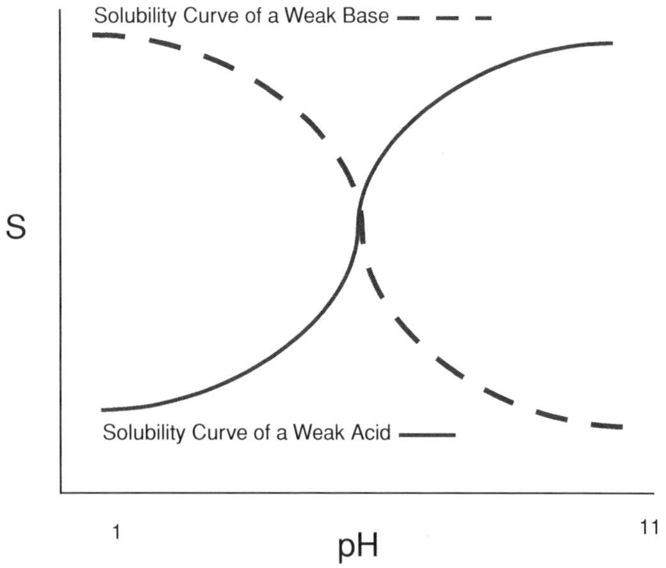

Figure 33-3 Solubility (S) versus pH profiles for salts of weak acids and weak bases.

the drug exists in equal parts of a dissociated (ionized) and undissociated (nonionized) species. Generally, one pH unit below the pK_a or above the pK_b and the drug becomes insoluble because the poorly soluble undissociated species predominates.

The relationships among pH, dissociation constant, pK_a or pK_b, and concentration of ionized (salt) and nonionized (free acid or base) are summarized by the following two equations:

Weak acid: $\text{pH} = \text{p}K_a + \log\,[(\text{salt concentration}/\text{weak acid concentration})]$
Weak base: $\text{pH} = \text{p}K_b + \log\,[(\text{weak base concentration}/\text{salt concentration})]$

These two equations show important facts related to drug solubility and pH:

1. When solution pH is the same as the dissociation constant of the drug (or buffer), then equal concentrations of ionized and nonionized species exist. This is a main principle of the effect of buffers on maintaining solution pH; that is, the buffer component is most effective when its dissociation constant is the same as the solution pH.
2. The degree of ionization of the electrolytic drug will change significantly as the difference between its pK_a or pK_b and solution pH becomes greater. After a three-unit difference between pH and pK_a/pK_b, there is little change in ionization since the relationship is on a logarithmic scale.
3. In the case of partition coefficient, discussed in the next section, the nonionized species of drug will easily partition into the lipid biologic membrane. Because the relationship between pH and the ratio of ionized to nonionized drug species is at equilibrium, when nonionized drug is partitioned into the lipid membrane, some of the ionized species will become nonionized to maintain equilibrium, and more nonionized drug will be absorbed. This dynamic process will continue until the entire injected drug has been absorbed.

If the injection site pH (pH 7.4 or slightly less) is greatly different than the pH of the injected drug solution, the drug will precipitate at the injection site. This might be desirable for a sustained release effect for drugs administered IM or SC, but not for intravenous injections where a physiologic effect is needed immediately and, of course, drug particles in the bloodstream are dangerous. Phenytoin is a classic example of a drug product that can easily precipitate at the site of injection. This concern exists for any drug product that is formulated with co-solvents used to solubilize the drug and any drug product that in poorly soluble at pH 7.4, but is rendered soluble via salt formation at low or high pH. The primary approach used to prevent precipitation at the injection site is to inject the drug product very slowly and allow adequate dilution by the circulatory system. For example, phenytoin should be injected at a rate no faster than 1 to 3 mg/kg/min. Vancomycin, an example of a salt of a weak base that is insoluble at physiological pH, should be injected at a rate no faster than 10 mg/min.

Partition Coefficient

Passive diffusion of drugs depends on the drug's partition coefficient. This is a property that compares the distribution of the drug between an aqueous phase and an oily or lipid phase. The partition coefficient is expressed as the concentration of a solute in the oil or lipid phase divided by the concentration of the solute in the aqueous phase at a defined pH. The higher the partition coefficient, the greater will be the solubility of the solute/drug in the oily phase. Drugs with high partition coefficients, when injected in biologic tissue, will readily pass from the aqueous drug formulation into the biologic tissue and have a higher rate of diffusion.

Ideally, a drug would have adequate solubility in water so that it can be easily formulated as an aqueous solution, yet it would have a high partition coefficient so that after IM or SC injection it readily perfuses into the tissues and is absorbed quickly.

Particle Size

Particle size will affect the rate of passive diffusion and extent of absorption. The equation for diffusion (Fick's Law):

$$(dq/dt) = DA(C_1 - C_2)/\ell$$

contains the term "A" that is the surface area available for diffusion. Particle size and surface area are inversely related. Thus, the larger the particle size, the smaller the surface area, and the lower is the rate of diffusion. Particle size not only affects rate of diffusion and absorption of drugs injected IM or SC but also affects the degree of re-suspension prior to injection and the syringeability/injectability of the suspended product. It is important not only to control the size of particles but also the size range. The smaller the particle size and range, the better will be the ability to resuspend, withdraw and inject the dose, and the rate of absorption at the injection site.

Viscosity

Viscosity is the resistance to free flow and is used to affect rate of absorption of IM and SC injections and topically applied ophthalmic medications. Depot formulations, using single or combination of polymers, in part, rely on viscosity effects of these large molecules to retard to the diffusion flow of the drug from the depot to the circulation. Topically applied ophthalmic medications usually contain viscous vehicles; for example, hydroxypropyl methylcellulose or polyvinyl alcohol, to adhere to the corneal epithelium for a longer time than solutions would otherwise without these polymers present, with the increased contact time allowing for more drug to be absorbed.

Solid State Morphology

In injectable suspensions, drugs exist as crystalline or amorphous solid state entities or mixtures of these states. The solid-state morphology will affect the rate of solution of these solid molecules. Crystalline drugs are known to have dissolution rates slower than the same drug in the amorphous state. Insulin is a great example of an important injectable drug that can be formulated as either completely amorphous (Semi-Lente formulations), completely crystalline (Ultra-Lente formulations), or combinations of crystalline and amorphous states (Lente). Each of these formulations have different rates of insulin release and availability with the amorphous Semi-Lente suspensions having the faster release rates and the crystalline Ultra-Lente having the slowest release rates.

Osmolality

Injectable products ideally should be iso-osmotic with biological fluid, and most commercial products are iso-osmotic in order to minimize pain and tissue irritation upon injection. Large differences in osmolality can effect passive diffusion. Hypotonic (hypo-osmotic) solutions will cause movement of the product solvent away from the site of injection since by the law of osmosis, product solution will move from a region of lower concentration to a region of higher concentration to equalize pressure on both sides of the biologic membrane. This movement of product solvent will cause drug concentration to increase at the injection site and the rate of passive diffusion will increase. The opposite phenomenon will occur with drug injections that are hypertonic (hyperosmotic) where the fluid from biological cells will flow to the drug solution to equalize pressure and drug concentration, and passive diffusion rate will decrease. Obviously isotonic formulated injectable products will have no effect on fluid movement and no effect on drug diffusion.

Injection Volume

Injection volume matters for all routes of administration except for intravenous. Fick's Law shows that diffusion rate is inversely proportional to injection volume [$dq/dt = K\ (A/V)$]; therefore, more rapid absorption is generally obtained when drugs are administered in smaller injection volumes. The larger the injection volume for injections within confined areas, e.g., IM/SC, capillary beds in the region will be compressed and the tissue surface area to volume ratio will be lowered. Since passive diffusion is directly proportional to surface area (A), larger injection volumes will decrease the rate of diffusion. Excessively large injection volumes in confined areas of the body will also increase local pressure and induce unnecessary pain. Table 31-2 provided the usual volume of injection based on site of injection. For IM injections in different muscle groups, the maximum volume given in the gluteus maximus region is 5 mL while only 2 mL is given in the deltoid muscle of the shoulder.

Molecular Size

Perhaps intuitively similar to the effect of injection volume, absorption rates of drugs injected IM/SC are inversely proportional to molecular weight. Small molecules (less than 1,000 Daltons) are readily absorbed through capillaries while larger molecules (several thousand Daltons and higher) must enter the circulatory system via the lymphatic system. Lymph flow is much slower than plasma flow, so drugs relying on the lymphatic system for absorption will be affected by the slower lymph flow.

TYPE OF DOSAGE FORM

Solubilized aqueous drug solutions will release the drug from the dosage form faster than any other type of dosage form. The order of drug release (fastest to slowest) as a function of dosage form is the following:

Aqueous solution *Fastest*
Aqueous suspension
Oily solution
Oil-in-water emulsion
Water-in-oil emulsion
Oily suspension
Sustained release *Slowest*.

Aqueous Solutions

For aqueous solutions, no dissolution step is required, so the drug is immediately available for absorption, unlike the other types of dosage forms. Aqueous solutions are typically simple formulations that are injected by a variety of parenteral routes. However, some aqueous solutions can be composed of soluble complexation systems where the drug is reversibly bound to a soluble macromolecule such as cellulosic polymer or polyvinylpyrrolidone. Drugs bound to these macromolecules must first be released as unbound drug before it can be absorbed.

Aqueous Suspensions

For drugs administered as aqueous suspensions, the drug is in the solid form and must first dissolve and then be released from the dosage form, prior to being available for diffusion and absorption. Drug dissolution rate is described by

$$\text{Rate of solution} = KS(C_s - C_t)$$

where C_s is the concentration of drug in a saturated solution (equilibrium solubility), C_t is the concentration of the drug in solution at time t, S is the surface area of the solid drug and K is a constant that reflects the diffusion constant for the drug in solution and a rate constant for the transfer of drug from the solid–liquid interface of the dosage form to the solution around the injection site. For all practical purposes, the rate of solution (dissolution) is dependent primarily on surface area and drug concentration in solution. Surface area is primarily dependent on the particle of the suspended drug in the aqueous suspension.

The fact that aqueous suspensions require drug dissolution and drug dissolution is dependent on drug particle size or, in the case of suspensions containing complexing agents, dependent on drug release from the complexing agent, is taken advantage of in formulating sustained or controlled released products (e.g., insulin suspensions and microsphere drug formulations).

Aqueous microsphere suspensions are ready-to-use or reconstituted microspheres where the drug is suspended in a bioerodible or biodegradable polymer. The drug typically will diffuse out of the microsphere at some defined rate (usually zero order) with drug diffusion taking place over a period of days to months.

The compactness (e.g., microsphere shape either truly spherical or needle-like) of the depot in the muscle will affect drug release and absorption rate.

Oily Solutions and Suspensions

Partition coefficient of the drug in solution or suspension is the most important factor affecting release of the drug from the oil into the tissue. For solutions, the drug is soluble in oil, so it will have a relatively slow release due to its naturally high partition coefficient. For suspensions, the drug is not soluble in oil, but has to dissolve in the oil phase, then partition from the oil phase to the aqueous tissue. Oily injectable suspensions are formulated as sustained release dosage forms, e.g., amoxicillin, pencillin procaine G, haloperidol decanoate, fluphenazine decanoate, testosterone enanthate, others (Chapter 6). Oily injection vehicles tend to remain in tissue as oily cysts for a long time. Olive oil is thought to reside the least amount of time in tissue while castor oil resides the longest.

Oil-in-Water and Water-in-Oil Emulsions

The first word of these two emulsions is the internal phase where the drug is dissolved. For oil-in-water emulsions the drug is dissolved in the oil either for solubilization purposes or for sustained release purposes. The drug must partition from the oil phase to the aqueous phase, then diffuse into the surrounding tissue, then diffuse and partition into the bloodstream. For water-in-oil emulsions, the drug is dissolved in water with these dosage forms formulated for sustained release purposes. It takes a longer period of time for drug to partition from the aqueous phase to the external oil phase and then to the surrounding tissue fluid.

Sustained or Controlled Release Dosage Forms

Included in this category of dosage form are complexes, polymeric systems, liposomes, and other microparticulate delivery systems. Many biopharmaceuticals have very short biological half-lives that are especially applicable to being formulated in sustained release dosage forms that will reduce the requirement for daily injections. Controlled release formulations are important for injectable drugs whose therapeutic doses are very small yet have a relatively high therapeutic index such that inadvertent dose dumping would not cause life-threatening reactions. Leuprolide acetate perhaps is the best example of an injectable drug that meets these requirements. The rate and extent of absorption of sustained and controlled released drugs from injection sites and from depot locations are heavily dependent upon formulation factors such as the type of polymer, drug solubility in the polymer matrix, partition coefficient, particle size, and other properties of the formulation (3).

PHYSIOLOGICAL FACTORS AFFECTING DRUGS ADMINISTERED BY INJECTION

Blood and Lymphatic Circulation

Table 33-3 gives a comparison of volume, rate, and pressure for various parts of the circulatory or arteriovenous system (4). What is especially interesting and relevant to injections is the fact that the capillary system is only 250 mL or 7% of the body's blood volume and has a flow rate much slower than other blood vessels, especially the arteries and vena cava vein. Injections other than intravascular depend on the capillary vessel system to transport drugs to the major blood vessels, so this further explains why IM/SC injections take time for the drug to reach the site of action.

Table 33-3 Comparison of Volume, Rate, and Pressure for Various Parts of the Circulatory or Arteriovenous System (4)

System	Volume (cm^3)	Rate (cm/sec)	Pressure (mm Hg)
Aorta	100	40	100
Arteries	325	40–10	100–40
Arterioles	50	10–0.1	40–25
Capillaries	250	0.1	25–12
Venules	300	0.3	12–8
Veins	2200	0.3–5	10–5
Vena Cava	300	30–5	2

The human circulatory system consists of approximately 3.6 billion capillaries with a total cross-section area of approximately 4500 cm^2. Dividing the number of capillaries by their cross-sectional area gives a value of 1.25×10^{-6} cm^2, an exceedingly small area for blood flow to occur. Transfer of nutrients and injected medications depends on this slow blood flow. During micro-circulation, fluid containing protein, lipids, various nutrients, and drugs, if injected, diffuses slowly but surely. Often, protein will leak out of the capillary system and will accumulate in tissue interstitial space. However, this is why the lymphatic system is so important because this system removes accumulated and residual protein from the interstitial. If the drug injected is a high molecular weight protein or additive, it too will be collected and moved by the lymphatic system.

Blood Flow (Vascular Perfusion)

In muscles, blood flow rates range from 0.02 to 0.07 mL/min per gram of muscle tissue (5). The higher the blood flow the higher the clearance rate of the drug from the injection site and the greater the rate of absorption. Blood flow is greater in smaller muscles than larger ones, so IM injections in the deltoid will absorb faster than IM injections in the gluteus maximus. Any factor that increases or decreases blood flow around the injection site will affect the rate of drug absorption. Absorption rates will increase if the skin around the injection site is massaged because massaging increases blood flow plus helps to increase the surface area available for drug diffusion and absorption. This effect is also true for any kind of exercising done after IM injection. The opposite effect will occur if blood flow is decreased. For example, local inflammation, often caused by reaction to the injection site trauma, can decrease blood flow and delay the absorption rate of IM injections. Co-administration of a vasoconstrictor, e.g., epinephrine, will decrease blood flow at the injection site and the rate of diffusion and absorption of drugs will be prolonged.

REFERENCES

1. Rowland M, Tozer TN. Clinical Pharmacokinetics: Concepts and Applications, 2nd ed. Philadelphia, PA: Lea & Febiger, 1989:141.
2. Rowland M, Tozer TN. Clinical Pharmacokinetics: Concepts and Applications, 2nd ed. Philadelphia, PA: Lea & Febiger, 1989:240.
3. Radomsky M. Product development principles of sustained-release injectable formulations. In: Senior J, Radomsky M, eds. Sustained-Release Injectable Products. Englewood, CO: Interpharm Press, 2000:13–23.
4. Motola S. Biopharmaceutics of injectable medication. In: Avis KE, Lachman L, Lieberman HA, eds. Pharmaceutical Dosage Forms: Parenteral Medications, 2nd ed. Vol. 1. New York: Marcel Dekker, 1992:73.
5. Tse FLS, Welling PG. Bioavailability of parenteral doses. I. Intravenous and intramuscular doses. J. Parenter Drug Assoc 1980; 34:409–421.

Index